Vol. 29. **The Analytical Chemistry of Sulfur and Its Compounds** (*in three parts*). By J. H. Karchmer

Vol. 30. **Ultramicro Elemental Analysis.** By Günther Tölg

Vol. 31. **Photometric Organic Analysis** (*in two parts*). By Eugene Sawicki

Vol. 32. **Determination of Organic Compounds: Methods and Procedures.** By Frederick T. Weiss

Vol. 33. **Masking and Demasking of Chemical Reactions.** By D. D. Perrin

Vol. 34. **Neutron Activation Analysis.** By D. De Soete, R. Gijbels, and J. Hoste

Vol. 35. **Laser Raman Spectroscopy.** By Marvin C. Tobin

Vol. 36. **Emission Spectrochemical Analysis.** By Morris Slavin

Vol. 37. **Analytical Chemistry of Phosphorus Compounds.** Edited by M. Halmann

Vol. 38. **Luminescence Spectrometry in Analytical Chemistry.** By J. D. Winefordner, S. G. Schulman and T. C. O'Haver

Vol. 39. **Activation Analysis with Neutron Generators.** By Sam S. Nargolwalla and Edwin P. Przybylowicz

Vol. 40. **Determination of Gaseous Elements in Metals.** Edited by Lynn L. Lewis, Laben M. Melnick, and Ben D. Holt

Vol. 41. **Analysis of Silicones.** Edited by A. Lee Smith

Vol. 42. **Foundations of Ultracentrifugal Analysis.** By H. Fujita

Vol. 43. **Chemical Infrared Fourier Transform Spectroscopy.** By Peter R. Griffiths

Vol. 44. **Microscale Manipulations in Chemistry.** By T. S. Ma and V. Horak

Vol. 45. **Thermometric Titrations.** By J. Barthel

Vol. 46. **Trace Analysis: Spectroscopic Methods for Elements.** Edited by J. D. Winefordner

Vol. 47. **Contamination Control in Trace Element Analysis.** By Morris Zief and James W. Mitchell

Vol. 48. **Analytical Applications of NMR.** By D. E. Leyden and R. H. Cox

Vol. 49. **Measurement of Dissolved Oxygen.** By Michael L. Hitchman

Vol. 50. **Analytical Laser Spectroscopy.** Edited by Nicolo Omenetto

Vol. 51. **Trace Element Analysis of Geological Materials.** By Roger D. Reeves and Robert R. Brooks

Vol. 52. **Chemical Analysis by Microwave Rotational Spectroscopy.** By Ravi Varma and Lawrence W. Hrubesh

Vol. 53. **Information Theory As Applied to Chemical Analysis.** By Karel Eckschlager and Vladimir Štěpánek

Vol. 54. **Applied Infrared Spectroscopy: Fundamentals, Techniques, and Analytical Problem-solving.** By A. Lee Smith

Vol. 55. **Archaeological Chemistry.** By Zvi Goffer

Vol. 56. **Immobilized Enzymes in Analytical and Clinical Chemistry.** By P. W. Carr and L. D. Bowers

Vol. 57. **Photoacoustics and Photoacoustic Spectroscopy.** By Allan Rosencwaig

Vol. 58. **Analysis of Pesticide Residues.** Edited by H. Anson Moye

Vol. 59. **Affinity Chromatography.** By William H. Scouten

Vol. 60. **Quality Control in Analytical Chemistry.** By G. Kateman and F. W. Pijpers

Vol. 61. **Direct Characterization of Fineparticles.** By Brian H. Kaye

Vol. 62. **Flow Injection Analysis.** By J. Ruzicka and E. H. Hansen

(*continued on back*)

Multielement Detection
Systems for
Spectrochemical Analysis

CHEMICAL ANALYSIS

A SERIES OF MONOGRAPHS ON
ANALYTICAL CHEMISTRY AND ITS APPLICATIONS

VOLUME 107

A WILEY-INTERSCIENCE PUBLICATION

JOHN WILEY & SONS

New York / Chichester / Brisbane / Toronto / Singapore

Multielement Detection Systems for Spectrochemical Analysis

KENNETH W. BUSCH
MARIANNA A. BUSCH

Department of Chemistry
Baylor University
Waco, Texas

A WILEY-INTERSCIENCE PUBLICATION

JOHN WILEY & SONS

New York / Chichester / Brisbane / Toronto / Singapore

Library of Congress Cataloging in Publication Data:

Busch, Kenneth W.
Multielement detection systems for spectrochemical analysis/
Kenneth W. Busch, Marianna A. Busch.
p. cm.— (Chemical analysis, ISSN 0069-2883; v. 107)
"A Wiley-Interscience publication."
Includes bibliographies and index.
1. Optical spectrometers. I. Busch, Marianna A. II. Title.
III. Series.
QD95.B87 1990
543'.0858—dc19 89-30004
ISBN 0-471-81974-3 CIP

Printed in the United States of America

10 9 8 7 6 5 4 3 2 1

"When any great design thou doest intend,
Think on the means, the manner, and the end."

Sir John Denham
English Poet
(1615–1669)

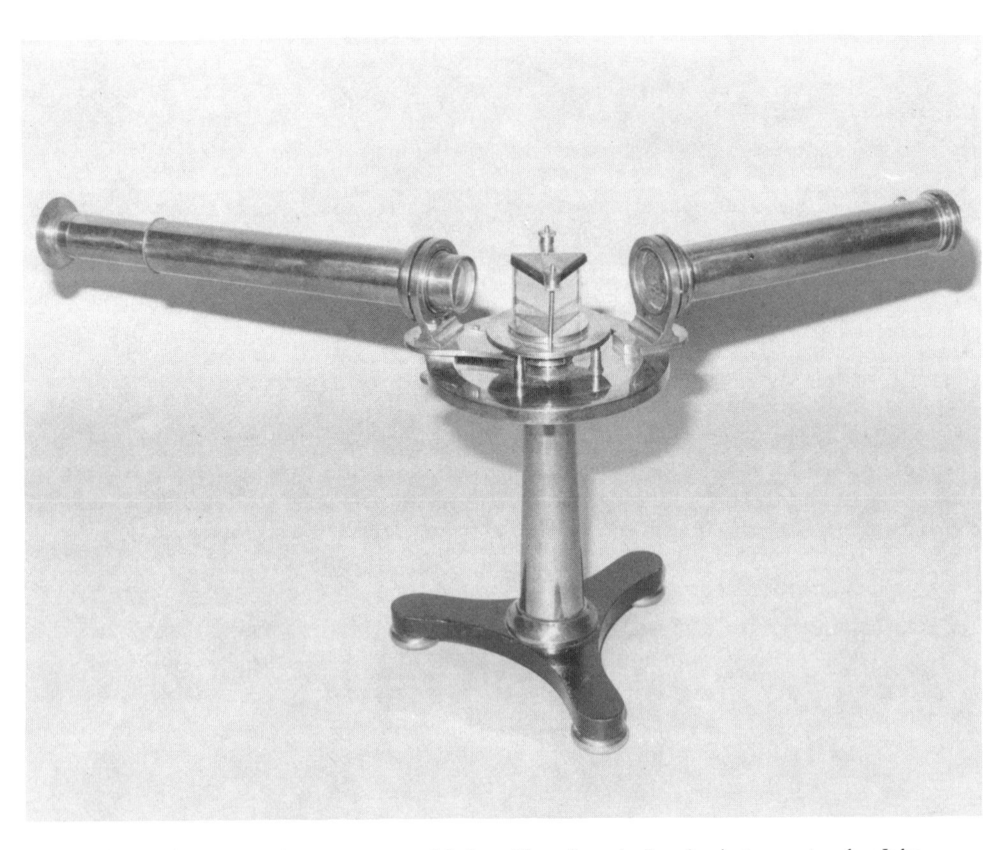

Nineteenth-century prism spectroscope fabricated from brass by London instrument maker John Browning (in authors' collection).

PREFACE

This book is intended to serve as an introduction to the general area of multichannel detection systems as they relate to multielement spectrochemical analysis. The term *spectrochemical analysis*, as used here, refers to use of the principles of optical spectroscopy for the purpose of performing elemental analyses. Therefore, the scope of this book has been limited to multichannel detection systems that operate in the ultraviolet–visible region of the spectrum, and the primary emphasis has been on an explanation of the fundamental principles of multichannel detection rather than on a comprehensive survey of the literature.

Although a number of books on experimental optical spectroscopy are available today, most do not deal with spectroscopic instrumentation as well as with modern detectors and detection systems. Therefore, a major goal in writing this monograph has been to produce a *unified* treatment of the subject that includes not only the fundamental principles necessary for understanding spectroscopic instrumentation such as diffraction and interference, but also a comprehensive discussion of modern image detectors. The book is intended for graduate students in analytical spectroscopy as well as for researchers in other disciplines who use optical spectroscopy for spectrochemical analysis. Rather than presenting a maze of bewildering spectroscopic relationships, every effort has been made to derive important optical and spectroscopic relations at a mathematical level that should be comprehensible to readers with an undergraduate background in chemistry or related sciences.

The original impetus for the book is an outgrowth of some research begun 15 years ago by one of us (KWB) during a two-year postdoctoral appointment at Cornell University with Professor George Morrison. At that time, the work centered on the potential role of the vidicon television camera tube as a multichannel detector for performing elemental analyses by atomic spectroscopy. During the course of this two-year period, a series of four papers was published dealing with the advantages and problems encountered in the spectroscopic application of silicon vidicons. It was clear then, and it is still true today, that the majority of array detectors available commercially have not been developed with spectroscopic applications in mind. As a result, the simple installation of an array detector in the focal plane of a dispersive spectrometer does not automatically guarantee the creation of a suitable analytical system for multielement analysis. The successful application of array detectors and other forms of multichannel detection to problems in multielement analysis can only be achieved with a fundamental understanding of all the various aspects of detector technology involved. As emphasized in Chapter 12, the development of viable multichannel detection systems for multielement analysis is actually a problem in systems engineering!

The book was written primarily for the research analytical scientist with an interest in instrumental development. As a result, it does not include a survey of multielement analyses for different sample types. The term *analytical science* is used instead of analytical chemistry to express the notion that the discipline extends beyond the traditional boundaries of chemistry. While the phenomenology associated with chemical measurements remains inherently chemical in nature, the means of achieving particular measurement objectives are essentially engineering in nature. The successful research analytical scientist today is, therefore, neither a pure chemist in the traditional sense nor a professional engineer but possesses manifestations or attributes of both. Engineering development of analytical systems without a thorough understanding of the underlying chemical phenomenology is pointless, while simple knowledge of phenomenology alone without a knowledge of the means of achieving particular measurement goals cannot lead to viable analytical systems.

For the research analytical scientist today whose training has been primarily in chemistry, the acquisition of the necessary background in optics, transform techniques, solid-state physics, television technology, electron optics, and array detector technology can generally be obtained only through reading the original literature or specialized monographs, which often assume a physics or engineering background. It is hoped that one of the unique features of this monograph will be its ability to bridge the gap between the physics and engineering aspects of multichannel detection and the traditional chemical background of the average analytical scientist today. To make the transition more comprehensible to the reader whose background is not in engineering or physics, the level of mathematical complexity has been kept to a minimum. Some basic understanding of electronics and the basic principles of atomic absorption and emission spectrochemical analysis is presumed, however, at the level of a good undergraduate instrumental analysis course.

In organizing the material covered in this monograph, we have attempted to provide a unified, structured, logical sequence of the topics that are essential to an understanding of multichannel spectroscopic detection. The book is organized into three parts. Chapters 1–7 deal with the optical principles upon which modern experimental spectroscopy is founded. Chapters 8–11 treat topics needed to understand the basic operation of detectors for optical spectroscopy. Chapters 12–15 discuss topics related to combining detectors with optical spectrometers to produce detection systems for multielement analysis.

Following a brief historical treatment of the subject and a discussion of the goals of spectroscopic research in Chapter 1, the reader is introduced to geometric optics and simple optical systems in Chapter 2. The important topic of diffraction is treated in Chapter 3, prior to discussing the basic principles behind dispersive spectrographs and spectrometers in Chapter 4. Chapter 5 lays the foundations of transform spectroscopy with a discussion of Hadamard transform because, from a pedagogical point of view, Hadamard spectrometers are most similar to the dispersive spectrometers discussed in Chapter 4. Then, following a discussion of interference in Chapter 6, the fundamental principles of Fourier transform spectroscopy are treated in Chapter 7.

Chapter 8, which describes the use of photographic detection in spectroscopy, begins a series of chapters on detectors and radiation detection. Chapter 9 introduces the principles of solid-state physics needed to understand the basic terminology of array detectors. Chapter 10 treats the basic principles related to vacuum image detector systems, while Chapter 11 is devoted to solid-state array detectors such as the photodiode array and the charge-coupled device.

The third part of the book begins with Chapter 12, which includes a discussion of the philosophy of instrumental development from the standpoint of introductory systems engineering. This is followed by a discussion in Chapter 13 of the aspects of analytical atomic spectroscopy that relate directly to the design of detection systems for multielement analysis. The last two chapters deal with the various transform and nontransform detection systems that have been developed and tested for use in multielement spectrochemical analysis.

The authors would like to thank Dr. James D. Winefordner, the series editor, for the invitation to write this monograph, and Baylor University for providing the summer support needed to complete the task. Special thanks are extended to Dr. George Morrison for arranging our summer appointments at Cornell University so that we could use the chemistry, physics, and engineering libraries.

KENNETH W. BUSCH
MARIANNA A. BUSCH

Ithaca, New York
May 1989

CONTENTS

Multielement Detection Systems for Spectrochemical Analysis

CHAPTER

1

HISTORICAL INTRODUCTION

Much of what we know about the fundamental nature of matter and the universe can be related, either directly or indirectly, to some form of spectroscopic observation. The discovery and development of the quantum theory of matter, for example, is a direct consequence of spectroscopic observations. Our knowledge of the composition and temperature of stars as well as whether they are traveling toward us or away from us are also determined from spectroscopic measurements. In the area of chemical analysis, spectroscopic measurements have been used for qualitative and quantitative analyses at both trace and ultratrace levels. As a result of the wide application of spectroscopic data, the techniques and principles of experimental spectroscopy are of interest to a wide variety of research workers, including biologists and medical researchers, theoretical chemists and physicists, analytical and environmental chemists, astrophysicists, geologists, materials scientists, and metallurgists, to name a few. The purpose of this work is to provide a unified treatment of the material needed to understand detectors and detection systems in modern optical spectroscopy that will be comprehensible to the diverse groups interested in the application of spectroscopic instruments. In order to understand and appreciate the developments occurring in spectroscopy today, it is worthwhile to review briefly some of the more important milestones in the development of spectroscopy.

1.1. THE EYE AS A DETECTOR

Philosophies on the nature of light and optics can be traced to the Greeks, who were aware of the law of reflection and knew of rudimentary lenses such as the "burning glass." It was not until 1621, however, that the law of refraction was discovered by Willebrod Snell, who died without ever publishing it. The discovery was included in a manuscript entitled, "Dioptrique," published anonymously in 1637 by Rene Descartes without mentioning Snell, although it is thought that Descartes had access to Snell's manuscript (Born and Wolf, 1987).

The next milestone in the development of spectroscopy was made by the young English philosopher who was soon to become Lucasian Professor of Mathematics at Cambridge University—Sir Isaac Newton. Newton (1672) was the first to use the Latin word *spectrum*, which means "image," to describe the series of colored images produced when sunlight from a small hole in a window shutter was allowed to pass through a lens and a prism and finally focus on a screen. In fact, the word *spectroscopy* is derived from the Latin word *spectrum*, which means "image," and the Greek word *skopos*, which means "to observe." Newton showed with his experiments that once light had been dispersed by a prism, it could not be further dispersed by a second prism. Newton's major contribution to spectroscopy was the demonstration that "white" sunlight was, in fact, a heterogeneous mixture of rays that could be sorted from one another by means of a prism.

For a century after Newton, little progress was made in the development of spectroscopy. Invariably, progress in experimental spectroscopy is linked to improvements in the optical performance of the instruments or to the availability of new detectors. The early 1800s saw developments in both of these areas.

More than anyone, Joseph Fraunhofer is probably responsible for the next series of improvements in Newton's spectroscope. Although he had little formal education and was apprenticed to a Munich glassmaker as a youth, Fraunhofer quickly became the greatest glassmaker and telescope constructor of his time. In 1814, Fraunhofer began a series of investigations whose goal was the precise measurement of the refractive indices of a series of different glasses. To accomplish this task, Fraunhofer added a theodolite telescope after the prism of the Newton spectroscope, as shown in Figure 1.1, to allow him to make precise measurements of angles. Using this improved spectroscope, Fraunhofer reported the observation that the solar spectrum was riddled with discontinuities that appeared as either strong or weak dark lines (Fraunhofer, 1817). Using his eye as a detector, Fraunhofer compiled a detailed description of the position of 700 of the dark lines and assigned letters of the alphabet to the more prominent ones. The so-called D-line of sodium (written Na_D), which is actually a doublet, is still used today in chemistry handbooks to specify the wavelength used in refractive index measurements.

The other great contribution made by Fraunhofer was in the development of the first diffraction grating. In an effort to improve the performance of optical instruments, Fraunhofer was led to a study of the phenomenon of diffraction by fine slits. These studies began with two slits and culminated in the discovery of the transmission diffraction grating. The first transmission grating produced by Fraunhofer was made by winding fine wire around two parallel screws with a very fine pitch. Using a wire transmission grating with 192 lines per centimeter, Fraunhofer succeeded in the first actual measure-

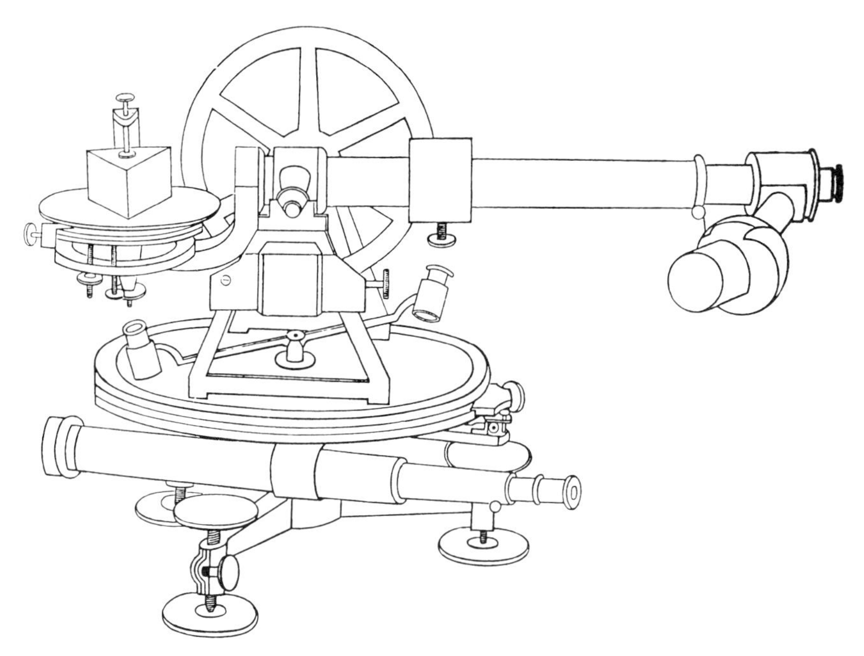

Figure 1.1. Fraunhofer spectroscope. [Reprinted with permission from H. Kayser (1900), *Handbuch der Spectroscopie*, S. Hirzel Verlag GmbH & Co., Stuttgart, Germany.]

ment of the wavelength of the sodium doublet. Prior to his untimely death at age 39, Fraunhofer successfully ruled the first glass transmission grating with 3000 lines per centimeter (Fraunhofer, 1823). His measurement of the wavelength of the unresolved sodium D line as 0.0005886 mm is a testimony to the accuracy of his measurements (Sawyer, 1963).

Although we take for granted today the idea that each atom and molecule has its own characteristic spectrum, this was not apparent to the spectroscopists of the early nineteenth century. Although the idea of characteristic spectra for different substances appears to have been hinted at by a variety of previous workers, the credit for the explicit and concise statement of the concept is usually given to a professor of physics at Heidelberg University during the 1850s, Gustav Robert Kirchhoff. In a series of papers beginning in 1859, Kirchhoff stated that a body that emits radiation over a certain frequency interval must also absorb radiation over the same frequency interval (Kirchhoff, 1859, 1860 a–c). In a classic experiment, he showed that the lines emitted when common salt was introduced into a gas flame were identical to the Fraunhofer D lines observed in the solar spectrum, and he explained that the dark Fraunhofer lines were, in fact, due to the absorption of the continuous solar spectrum by sodium atoms in the cooler outer atmosphere of the sun.

In collaboration with a professor of chemistry at Heidelberg University, Robert Wilhelm Bunsen, Kirchhoff applied the spectroscope as a means for performing qualitative chemical analyses (Kirchhoff and Bunsen, 1860 a, b, 1861 a, b). The Bunsen–Kirchhoff spectroscope used for these investigations, and shown in Figure 1.2, contained a number of significant improvements. Previous workers had illuminated the prism directly with light from the entrance slit. Light rays emerging from the entrance slit, however, do so as a diverging bundle of rays. In order to obtain high resolution from a prism, it is necessary that all the rays striking the prism do so at the same angle of incidence. To achieve this requirement, previous workers such as Fraunhofer placed the prism at a rather large distance from the entrance slit in order to obtain the required degree of collimation. In fact, Fraunhofer is reported to have placed his prism 24 feet from the entrance slit (Kayser, 1900)! Kirchhoff and Bunsen achieved the desired collimation effect by use of a lens placed between the entrance slit and the prism so that the lens was one focal length from the entrance slit. This simple innovation turned the spectroscope into a compact instrument that could be set on a bench.

Like Fraunhofer, Bunsen and Kirchhoff employed a telescope to observe the spectrum. Unlike Fraunhofer, however, who moved the telescope across the spectrum, Bunsen and Kirchhoff kept their telescope fixed and rotated the

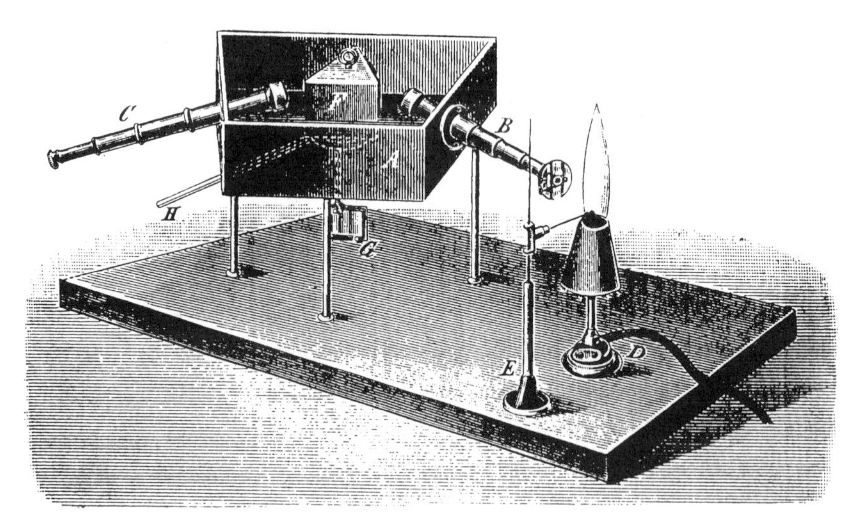

Figure 1.2. Kirchhoff–Bunsen spectroscope. [Reprinted with permission from H. Kayser (1900), *Handbuch der Spectroscopie*, S. Hirzel Verlag GmbH & Co., Stuttgart, Germany.]

prism instead. Prism angle was determined with the aid of a mirror attached to the base of the prism, which rotated as the prism was turned. A scale reflected by the mirror was viewed by another telescope to determine the prism angle. A fine wire in the spectroscope telescope could be made to coincide with any spectral line within the range of the instrument by rotating the prism. The angle of rotation of the prism was then determined by reading the scale position reflected into the reading telescope. This arrangement led to a compact instrument that could be used to determine the position of spectral lines reproducibly in terms of scale position. One disadvantage with the device, however, was that it did not locate the lines in terms of a universal standard such as the actual wavelength but merely reported their location in terms of relative scale position.

Another final innovation that should be mentioned regarding the Bunsen–Kirchhoff spectroscope that contributed to its utility was the use of the Bunsen burner coal gas flame as an excitation source. This source was especially convenient for visible observations with the eye because it was a steady excitation source that was relatively transparent. A platinum wire mounted on a stand was used to introduce analytical samples into the flame for spectroscopic observation.

Using their spectroscope, Bunsen and Kirchhoff were able to discover two new elements, which they named rubidium and cesium. Both names are derived from Greek roots and refer to the colors observed. Rubidium was named for the characteristic red doublet observed with the spectroscope while cesium was named for the sky-blue spectral line observed with the instrument. These discoveries were quickly followed by the discovery of thallium by Sir W. Crookes (1861) and the discovery of indium by F. Reich and T. Richter (1863). Both discoveries were made by spectroscopic observations.

Although the advent of the Bunsen–Kirchhoff spectroscope was an important step in the development of the emerging science of spectroscopy and laid the foundations for the future development of analytical spectroscopy, its use was limited by the rather narrow spectral response of the detector, that is, the eye. Further improvements in experimental spectroscopy would require better means of detection.

1.2. THERMAL DETECTORS

Actually the first use of a nonvisual means of detection of light radiation was made some 60 years prior to the work of Bunsen and Kirchhoff by the Royal Astronomer to King George III of England, Sir William Herschel (1800). Ironically, the discovery of the infrared region of the spectrum with which Herschel is credited was the result of attempting to develop appropriate glass filters to permit visual astronomical observations of the sun without damaging the eye. In his observations, Herschel had observed that some filters passed more heat than others, and he therefore resolved to study the matter systematically. Using an ordinary glass prism to form a solar spectrum and a mercury thermometer as a detector, Herschel discovered that the maximum heating effect of solar radiation lay well beyond the visible red end of the

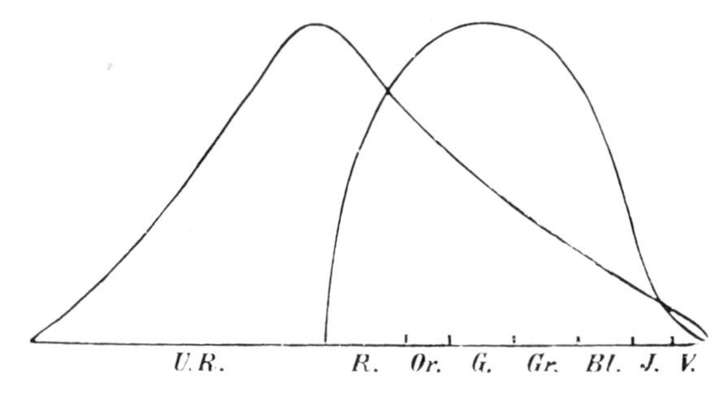

Figure 1.3. Results obtained by Herschel. Left-hand curve shows heating effect observed as function of wavelength. Right-hand curve shows intensity of visible region of spectrum. UR (ultraroth), infrared; G (gelbe), yellow; J (Jodinfarbig), purple [Reprinted with permission from H. Kayser (1900), *Handbuch der Spectroscopie*, S. Hirzel Verlag GmbH & Co., Stuttgart, Germany.]

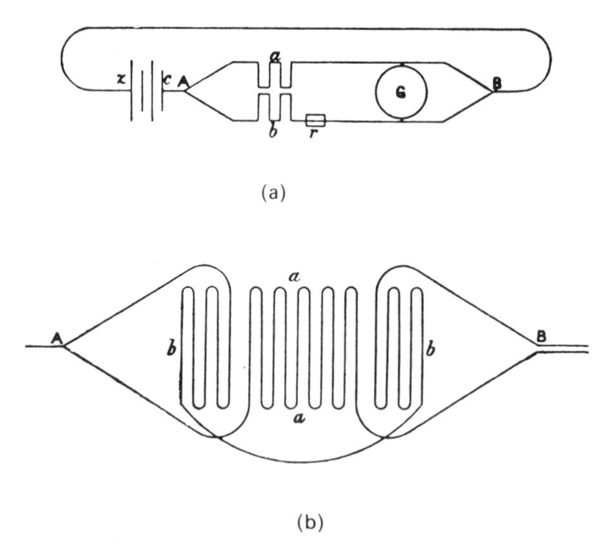

Figure 1.4. Langley's bolometer: (*a*) circuit; (*b*) resistance strips, *a*—detector, *b*—reference; (*c*) frame to hold resistance strips so that detector is exposed to radiation while reference (*b*) is shielded; (*d*) mounting tube to minimize effect of air currents. [Taken from E. C. C. Baly (1924), *Spectroscopy*, 3rd ed., Longmans, Green and Co., London, England, Vol. 1, Chapter 8.]

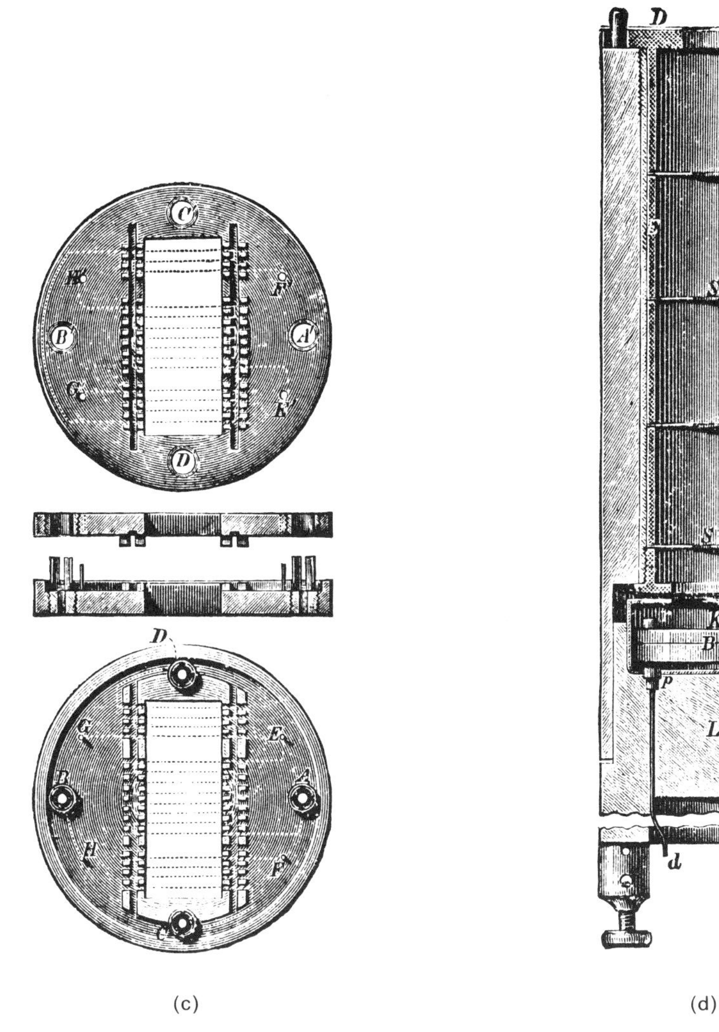

Figure 1.4 (*continued*)

spectrum in the region we now call the infrared (Hudson, 1969). Figure 1.3 shows the results Herschel obtained with his method.

Although Herschel did not actually use the term *infrared*, referring to the new spectral region by such terms as "invisible rays" and "dark heat," nor was he aware of the electromagnetic nature of the radiation, he did show that an ordinary thermometer could be used as a radiation detector. It was not until the 1830s that the thermometer was replaced by a new detector, the thermocouple developed by L. Nobili in 1830 (Nobili and Melloni, 1831). This device was a direct consequence of Seebeck's discovery of the thermoelectric effect in 1826 (Smith et al., 1957). The single thermocouple was followed quickly by the thermopile of Macedonia Melloni in 1833 (Melloni, 1833, 1835 a,b). The thermopile simply consisted of a number of thermocouples joined in series to produce a greater effect. By the end of the nineteenth century, Samuel Pierpont Langley had developed a bolometer detector, shown in Figure 1.4, that was 30 times more sensitive than Melloni's thermopile (Langley, 1881a,b). The bolometer is based on the change in resistance of a fine wire as it is heated. By incorporating the wire into a Wheatstone bridge, high sensitivities were possible.

These improvements in detectors resulted in fundamental discoveries about the infrared region of the spectrum. In the early years of the nineteenth century, for example, infrared research was hampered by a lack of sensitive detectors. On the basis of the results obtained with the detectors available at the time, it was believed that the long-wavelength limit of the infrared region of the spectrum was about 2 μm. As more sensitive detectors became available, this limit was extended. By 1880, the long-wavelength limit had been extended by P. Desains and P. Curie (1880) to 7 μm. By the close of the nineteenth century, the infrared region of the spectrum had been extended to nearly 20 μm.

1.3. PHOTOGRAPHIC DETECTION

While work on extending the long-wavelength limit of spectroscopic research was occurring, other workers were involved with extending our knowledge of short-wavelength radiation. It is actually quite difficult to determine who first thought of applying photographic detection to spectroscopic observations. As early as 1777, the German chemist, Scheele had discovered the existence of an invisible portion of the spectrum beyond the violet (Baly, 1912). Scheele was aware of the interesting property of silver chloride of changing from white to purple on exposure to sunlight. In a series of investigations on this property with different colors of the spectrum, Scheele discovered that the greatest change occurred when the salt was exposed to light at the extreme violet end of the spectrum.

Following Herschel's discovery of radiation beyond the red end of the spectrum, it was only natural to wonder whether similar invisible radiation existed beyond the violet. Only one year after Herschel's discovery, J. W. Ritter (1801) reported that the blackening of silver chloride, which had been previously studied by Scheele, was not limited by the violet end of the spectrum but extended into what we now refer to as the ultraviolet.

By 1842, E. Becquerel had photographed the solar spectrum by projecting it onto a strip of paper that had been coated with silver chloride (Baly, 1912). By this means, he was able to show the existence of Fraunhofer absorption lines in the long-wavelength region of the ultraviolet. By 1856, W. Crookes published a paper in the *Journal of the Photographic Society of London* entitled, "Photographic Researches on the Spectrum. The Spectrum Camera and Some of its Applications."

It was not long after this, that spectrography was applied to quantitative chemical analyses by W. N. Hartley (1884a–c). It took until the early portion of the twentieth century for the technique of quantitative analytical spectrography to finally mature, however. The key development in the use of photography as a means of detection for quantitative chemical analysis was the concept of using an internal standard described by W. Gerlach and E. Schweitzer in their monograph entitled, "Foundations and Methods of Chemical Analysis by the Emission Spectrum," published in 1931.

1.4. PHOTOELECTRIC DETECTION

The next step in the development of detection systems for spectroscopy was the development of photoelectric detection. Although photoelectric detection represents a developmental step in the evolution of detection systems, its origin occurred approximately during the same time period as the development of thermal detectors and photographic detection.

The father of photoelectric detection is Heinrich Hertz (1887a,b, 1888, 1889), who performed a series of experiments on the effect of electrical discharges in one oscillator circuit on a similar circuit that was not connected directly to the first. In these experiments, Hertz observed that a spark produced by the primary circuit was somehow transmitted through space so as to induce a spark in the secondary circuit even though the two circuits were not connected directly. It is a testimony to Hertz's powers of observation that he noticed a rather small effect that might have been overlooked had the work been conducted by someone else. Hertz noticed that the length of the spark induced in the secondary circuit was somewhat longer when the radiation from the primary spark gap was allowed to strike the secondary spark gap. By placing a slit and a quartz prism between the primary spark and the induced spark gap, Hertz was able to show that the effect occurred only when the prism was arranged to transmit light in the ultraviolet region of the spectrum (Zworykin and Wilson, 1934). Hertz then demonstrated that ultraviolet light from other sources was just as effective as radiation from the primary spark in causing the length of the secondary spark to increase. He also observed that the effect was more pronounced when (1) the ultraviolet light was incident on the negative terminal of the spark (i.e., the cathode), (2) the cathode was a large sphere or plate rather than a point, and (3) the cathode surface was freshly polished.

In the following year, E. Wiedemann and H. Ebert (1888) confirmed Hertz's results and showed conclusively that the effect took place entirely at the cathode of the gap being irradiated.

1.4.1. The Hallwachs Effect

In 1888 following Hertz's initial investigations with oscillatory circuits, Wilhelm Hallwachs performed a series of experiments to determine whether any fundamental effect could be observed when matter was irradiated with ultraviolet light. To avoid any association with spark gaps and electrical oscillations, Hallwachs designed the apparatus shown in Figure 1.5, which consisted of an insulated, polished zinc sphere connected to a gold-leaf electroscope. Using an arc as a source of light, Hallwachs was able to study the effect of ultraviolet irradiation of the sphere on the sphere charge as indicated by the electroscope. Whenever the insulated sphere was charged, the leaves of the electroscope would diverge. When the sphere was charged negatively and subsequently illuminated by the arc, the electroscope leaves would collapse, indicating a loss of negative charge by the sphere. If the sphere was charged positively, no loss of charge was detected by the electroscope when the sphere was irradiated by ultraviolet radiation. An uncharged sphere, however, was observed to become positive under ultraviolet irradiation. On

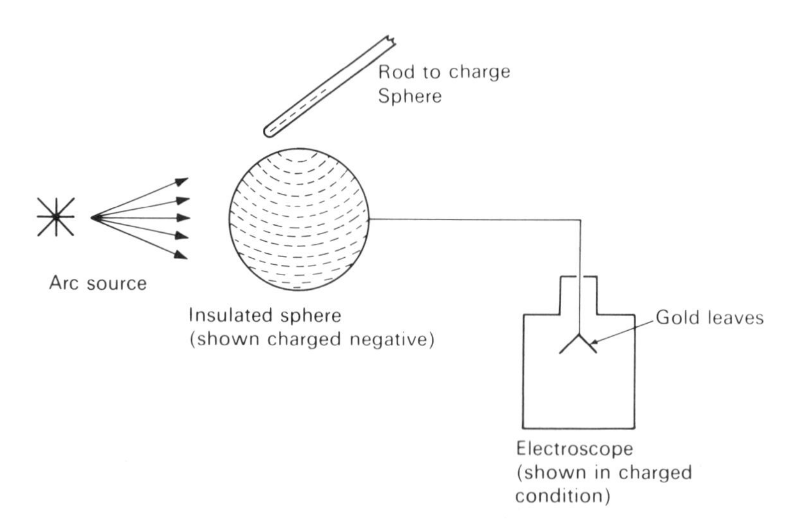

Figure 1.5. Hallwach's apparatus.

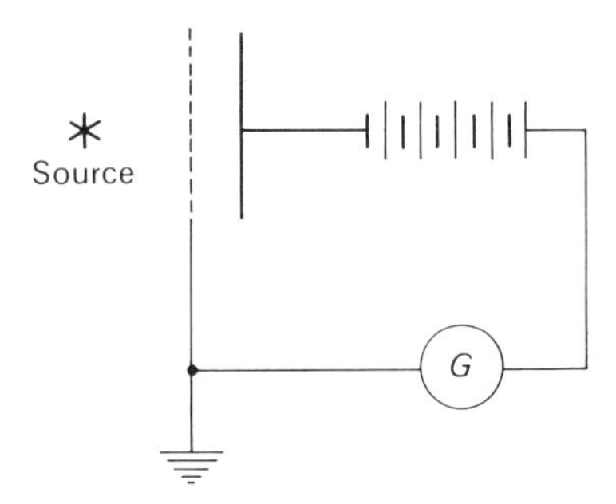

Figure 1.6. Circuit employed by Stoletow to measure photoelectric currents. [Reprinted from V. K. Zworykin and E. D. Wilson, *Photocells and their Applications*, 2nd ed., copyright 1934. By permission of John Wiley & Sons, Inc.]

the basis of these experiments, Hallwachs concluded that matter loses negative electricity upon exposure to ultraviolet light, and whatever the nature of the emission, it follows electrostatic lines of force. This phenomenon is known today as the *Hallwachs effect*.

Shortly after Hallwachs's study of the influence of ultraviolet radiation on a polished sphere, two other investigators, A. Righi (1888) and Stoletow (1890), independently demonstrated that a small current could be made to flow between a mesh grid and a polished metal plate when the plate was irradiated by light from a source, as shown in Figure 1.6. Righi called cells of this type that were capable of producing a current upon irradiation *photoelectric cells*.

1.4.2. The Work of Elster and Geitel

Subsequent studies of photoelectric phenomena revealed that magnesium, aluminum, and zinc produced the greatest effects when irradiated. Since these metals were relatively electropositive, it occurred to Julius Elster and Hans Geitel (1889) that more electropositive elements such as sodium and potassium might be even better. Although sodium and potassium are indeed more electropositive, they are also highly reactive and form oxides and hydroxides almost instantaneously upon exposure to air and water. Nevertheless, Elster and Geitel were able to overcome this difficulty in a rather ingenious manner. From previous studies with zinc, they learned that a *zinc amalgam* gave better results than zinc by itself. On this basis, they investigated the use of dilute *sodium* and *potassium amalgams* and found them to be considerably more sensitive than amalgamated zinc. In addition, they discovered in the course of their studies that sodium and potassium amalgams were photoelectrically sensitive to visible radiation! Figure 1.7 shows a schematic diagram of the amalgam photoelectric cell developed by Elster and Geitel (1890).

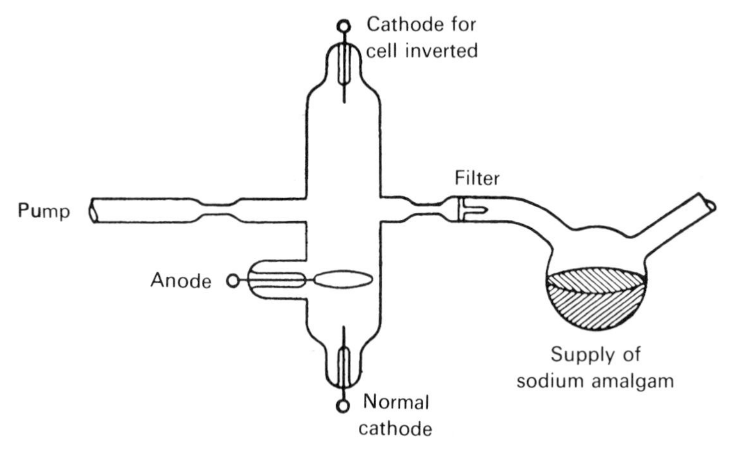

Figure 1.7. Amalgam photocell of Elster and Geitel. [Reprinted from V. K. Zworykin and E. D. Wilson, *Photocells and their Applications*, 2nd ed., copyright 1934. By permission of John Wiley & Sons, Inc.]

By the early 1900s, the first practical photocells began to appear. In 1904, Hallwachs developed a cell that produced a constant response over a period of several months. The cell consisted of an evacuated envelope with a cathode formed from a plate of copper that had been coated by copper oxide. Although the response was constant, the detector was not sensitive to visible radiation.

Simultaneously, Elster and Geitel continued to develop their alkali metal detector, and by 1913 they succeeded in improving the sensitivity of the device by two orders of magnitude. This was accomplished by "sensitizing" the alkali surfaces by exposing them to a glow discharge through hydrogen gas. It was observed that when the hydrogen glow discharge formed, the exposed surface of a potassium photocathode immediately became colored, reaching a peacock blue after several seconds.

The first modern photoemissive surface was described by L. R. Koller (1929, 1930) and consisted of a composite layer of cesium on cesium oxide on silver. The use of composite films marked a new era in the development of photoemissive technology and laid the foundations for modern multiplier phototubes. For a comprehensive discussion of the development of photoemissive materials, the reader should consult the monograph by Sommer (1968).

1.4.3. Photocurrent Amplification

In spite of the strides made in increasing the sensitivity of photocathode surfaces, even the most sensitive surface can pass only a minute amount of current without destroying the photoemissive layer. For this reason, the next challenge in phototube development was to find ways of amplifying the tiny photocurrent generated by vacuum photocells. This was accomplished in two basic ways.

One way of amplifying the photocurrent was to introduce a small amount of an inert gas into the cell envelope. If the electric field maintained between the cathode and anode of the cell and the gas pressure were adjusted so that the product of the field strength in volts per centimeter times the mean free path of the electron was equal to the ionization potential of the gas, secondary ionization of the gas would occur and could be used as a means of *gas amplification*. This idea led to the development of the gas-filled phototube, and while the gas cell was not as linear in its response as the vacuum phototube, it did provide a factor of 10 in amplification. The limit to the amount of amplification that could be obtained by this approach was set by the onset of the formation of a glow discharge within the cell.

The second way of amplifying the primary photocurrent from a phototube was to make use of the phenomenon of *secondary emission*. In 1918, A. W. Hull reported on a new type of thermionic tube with a plate between the emitting filament and the anode that resulted in an increase in the anode current. With this arrangement, electrons emitted from the filament could be accelerated by an appropriate voltage so that they struck the plate with sufficient energy to release *secondary electrons*. Since the number of secondary electrons produced was observed to be several times the number of primary electrons striking the plate, the use of intermediate electrodes could be used as a means of signal amplification.

It was not until 1935, however, that use of this form of signal amplification was applied to photoemissive detection. In 1935, H. Iams and B. Salzberg reported the development of a triode phototube having a single stage of amplification with a gain of 6. By 1940, J. A. Rajchman and R. L. Snyder had developed a nine-stage electrostatically focused multiplier phototube that has served as a model for photomultiplier tubes today.

1.5. PHOTOCONDUCTIVE DETECTORS

Still another class of devices for the detection of light radiation is based on the change in electrical conductance exhibited by certain materials when exposed to light. One of the first materials to be discovered to exhibit this property was selenium.

In 1817, while investigating a process for manufacturing sulfuric acid, the Swedish chemist J. J. Berzelius succeeded in isolating the element selenium from some residues that had accumulated (Berzelius, 1817, 1826). Of the various allotropes of selenium shown to exist, it was the gray lustrous metallic form that was discovered to possess the photoconductive property.

Since selenium occupies a position in the periodic table of the elements intermediate between the metallic elements and the nonmetallic elements, it is not surprising that it is actually a rather poor conductor of electricity. This fact led Willoughby Smith, in 1873, to employ tiny selenium rods as high-resistance elements in an experimental circuit he was investigating at a transatlantic cable station. Smith found that the resistance of the rods decreased whenever they were exposed to sunlight or artificial light. In a series of investigations on the fundamental nature of photoconduction, D'Albe showed in 1913 that current in a photoconductive circuit varies with the square root of the intensity of light. These experiments resulted ultimately in the wide variety of solid-state photoconductor detectors available today.

One disadvantage with the use of selenium as a photoconductor is that it is quite opaque to light. As a result, a primary goal in designing a selenium photoconductive cell is to expose as much area as possible to the radiation without making the resistance too large. This requirement was achieved in most early selenium photoconductive cells by employing a thin layer of selenium. Figure 1.8 shows a typical zigzag geometry for an early selenium

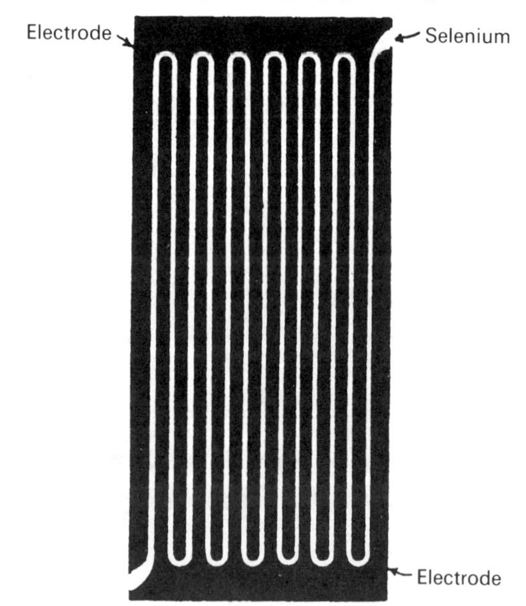

Figure 1.8. Selenium photoconductive detector. Selenium ribbon separates electrodes attached to diagonal corners. Black area represents conducting film. [Reprinted from V. K. Zworykin and E. D. Wilson, *Photocells and their Applications*, 2nd ed., copyright 1934. By permission of John Wiley & Sons, Inc.]

bridge (Zworykin and Wilson, 1934). The bridge was formed by first depositing a continuous conducting layer on a flat insulating substrate and then scratching a fine zigzag path diagonally across the plate so that electrodes attached to the diagonal corners were not directly connected electrically by the conducting layer. A thin film of vitreous selenium was then formed over the substrate and heated to convert the layer into a gray crystalline selenium. The bridge was then coated with paraffin or sealed in an evacuated tube to protect it from oxidation by the air and attack by water vapor. By designing the bridge in this manner, the only conducting path for electricity between the two diagonal corners of the cell was across the thin zigzag film of selenium that divided the two halves of the cell.

1.6. PHOTOVOLTAIC DETECTION

A final class of detectors whose origins need to be discussed in this historical introduction on the development of detectors and detection systems is the photovoltaic detector. Photovoltaic cells are detectors that produce a voltage or electromotive force when exposed to light. Although the exact details of the cell used in his experiment are not specified (Zworykin and Wilson, 1934), in 1839 E. Becquerel is reported to have observed the production of an electromotive force when one of two electrodes immersed in an electrolyte was irradiated with light. Becquerel, who published his observations in 1865, observed that the voltage was produced only when the electrode was irradiated and quickly disappeared when the light was removed.

Following the discovery of the Becquerel effect, little was done in the area of photovoltaics until the early 1900s when T. W. Case (1917) undertook a systematic study that led to the development of a cell consisting of two copper electrodes coated with a thin layer of cuprous oxide. When the electrodes were placed in a copper formate electrolyte and one electrode was exposed to sunlight, a voltage of 0.11 V was observed. The voltage obtained with this cell was not constant, however, and rapidly decreased as the cuprous oxide was oxidized to black cupric oxide. Because the electrochemical cell was reversible, it was possible to devise a method that would produce an alternating output by alternately exposing the two electrodes to light.

The next important advance in photovoltaic detection was the development of the *Sperrschicht* or barrier plane cell by Lange (1930a,b) at the Kaiser Wilhelm Institute in Berlin. This dry copper oxide cell consisted of a thin layer of crystalline red copper oxide produced on the surface of a metallic copper plate by heating with oxygen. A gold electrode was attached to the oxide surface by sputtering, and the voltage developed by the cell when exposed to light was measured between the gold electrode and the copper plate.

Although a complete explanation for the action of the barrier layer cell was not immediately available (see Chapter 9 for a discussion of photovoltaic devices), the presence of the rectifying interface between the metallic copper and the copper oxide layer was known. The importance of the barrier plane between the two materials was also appreciated, and it was shown by W. Schottky (1930) that the release of current carriers was confined to the junction between the metal and the oxide.

These studies eventually led to the development of the selenium barrier layer cell and ultimately to the development of the modern solid-state silicon photodiode discussed in Chapter 9.

1.7. THE DEVELOPMENT OF WAVELENGTH STANDARDS

Although Bunsen and Kirchhoff were able to show the existence of many terrestrial elements in the sun by comparing the spectral lines emitted when various salts were introduced into the flame with the Fraunhofer absorption lines in the solar spectrum, their determinations were not made in terms of the actual wavelength of the radiation but rather in terms of relative scale position obtained with their particular instrument. It was not until 1868, when A. J. Ångström published a map of the solar spectrum, which he entitled the Normal Solar Map, that an actual listing of the wavelengths of the Fraunhofer lines was made.

The wavelength measurements by Ångström were made using three carefully ruled gratings on glass prepared by F. A. Norbert and were expressed in units of a ten-millionth of a millimeter (Baly, 1912). This unit of length is referred to today as an Ångström unit. Although Ångström's measurements were reported to two decimal places, his measurements were expressed in terms of the Upsala meter and had to be corrected when it was determined that an error had been made when the Upsala meter was compared with the standard meter in Paris.

The importance of Ångström's measurements to the development of spectroscopy, however, cannot be overestimated. For the first time, it was possible to determine the wavelength of unknown spectral lines by comparison with the spectrum of the sun. All that was necessary was simple interpolation between the known lines of Ångström's map to determine the wavelength of any desired spectral line. As a result of Ångström's work, the measurements of different investigators could be compared and collected because they were all referenced to the same standard. Although all of the early wavelength determinations were made visually, the application of the dry gelatin photographic plate developed in 1870 led to improved accuracy in wavelength determinations.

The next advance in the accurate determination of wavelength was made by Henry A. Rowland of Johns Hopkins University, who made significant contributions to the technology of ruling the diffraction grating. Rowland realized that the key to the precise ruling of a diffraction grating by a dividing engine was the availability of a perfectly machined screw to advance the stylus. Rowland set about producing such a screw by starting with a screw of the desired pitch that was somewhat longer than needed. At the same time, a nut several inches long was made in four sections that accurately fit the male screw. By clamping the nut onto the male screw, it could be worked backward and forward along the length of the screw for a long period of time, continually tightening the nut as the work proceeded, until the errors in the pitch of the screw were averaged out along its entire length. When a sufficiently accurate screw had been prepared, it was installed in the carriage of the ruling engine and used to advance the position of the diamond ruling stylus between each stroke. Using a ruling engine maintained at constant temperature and driven automatically by clockwork, Rowland was able to rule a typical grating during the course of six days and nights.

In 1882, Rowland made another fundamental advance in experimental spectroscopy with his conception and subsequent development of the concave diffraction grating. By ruling a grating on a spherical mirror of speculum and mounting it in a specially designed spectrograph (see Chapter 4), Rowland was able to assemble an instrument with a constant dispersion where the various orders of spectra from the grating were superimposed. These two properties of Rowland's spectrograph enabled him to publish a series of wavelength tables of the solar spectrum that were accepted as standards for the next 20 years (Rowland, 1887a,b).

Not long after Rowland's pioneering work, A. A. Michelson of the University of Chicago succeeded in determining the absolute wavelength of three cadmium lines by means of an entirely new spectroscopic instrument, the interferometer (see Chapter 6). In a series of articles beginning in 1887, the same year Rowland published his wavelength lists, Michelson was able to establish the wavelengths of the three cadmium lines in terms of the standard meter at Paris and was able to show that the red cadmium line was one of the most monochromatic of all the lines he examined (1887, 1891, 1892).

The importance of Michelson's investigations with the interferometer extend beyond the bounds of spectroscopy. The extreme accuracy of the interferometric technique as a means of determining wavelength led the Eleventh General Conference of Weights and Measures, meeting in Paris in October of 1960, to adopt the red line of krypton 86 as the standard of length. It is especially ironic that Michelson's work has had such widespread impact. When Michelson was a young instructor of chemistry and physics at the U.S. Naval Academy, from which he graduated, he was asked by the Superintendent of the Academy why he wasted his time on his useless experiments! He received the Nobel Prize in 1907 for his work on light.

1.8. SELECTION OF THE APPROPRIATE SPECTROSCOPIC APPROACH

It should be apparent, even from the previous brief survey of the history of spectroscopy, that 400 years of development have produced a diversity of spectroscopic apparatus and approaches. In spite of its lengthy history, however, improvements and new developments continue to occur in experimental spectroscopy as research workers in various disciplines are forced to turn to less routine instruments for solutions to challenging new problems. Selection of the appropriate spectroscopic approach for a given study is by no means a trivial problem and depends ultimately on a thorough understanding of the principles of experimental spectroscopy and a knowledge of the advantages and limitations of various spectroscopic approaches. Failure to consider the critical spectroscopic parameters for a particular application may ruin an otherwise successful experiment. For example, a spectroscopic system designed to be used with a bright, stable source under conditions of moderate resolution may be totally useless with a relatively weak source or a transient source under conditions of high resolution.

The considerations involved in selecting an appropriate spectroscopic system for a particular problem are the subject of this monograph and will be developed in detail in subsequent chapters. At this time, however, it is worthwhile to examine some of the important considerations in a general way to gain an appreciation for the scope of the problem.

Figure 1.9 shows a schematic diagram of a generalized spectroscopic system consisting of a radiation source, a radiation analyzer, and a photosensitive detector. In order for this system to provide useful spectroscopic information,

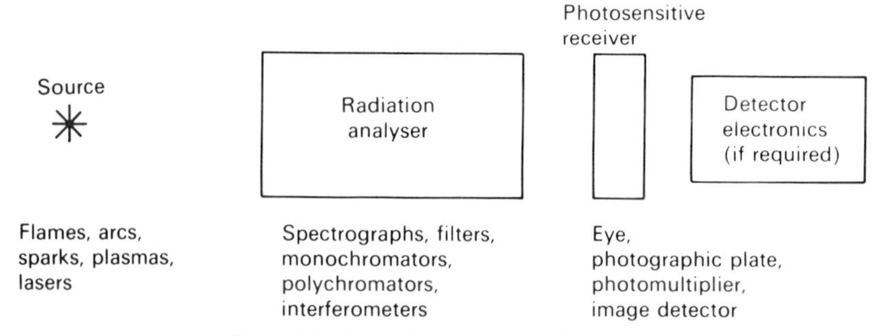

Figure 1.9. Generalized spectroscopic system.

the signal produced must be greater than the limiting system noise. The signal produced by a spectroscopic system is the product of three important factors[†]:

$$\text{Signal} = \text{source radiance} \times \text{luminosity of radiation analyzer} \times \text{responsivity of detector} \tag{1.1}$$

The system noise observed for a given experiment will be a function of the frequency response bandwidth of the detector electronics (see Chapter 9), and the bandwidth selected will ultimately determine the rate of information transmission by the system. Thus,

$$\text{Source radiance} \times \text{radiation analyzer luminosity} \times \text{detector responsivity} > \text{system noise } (\Delta f) \tag{1.2}$$

where the notation indicates that system noise is a function of frequency response bandwidth. If the system is detector noise limited, the noise equivalent power (NEP) of the detector can be used as a measure of the system noise (Chapter 9). Equation 1.2 is the fundamental equation that governs the performance of all spectroscopic systems and can be used to discuss the compromises that must be made in choosing an appropriate spectroscopic system for a particular study.

As in any design problem, there is usually some given parameter around which one must work. In spectroscopy, the source radiance is very often the factor that is fixed by the nature of the experiment under consideration. This means that the radiation analyzer luminosity, the detector responsivity, and the frequency response bandwidth must all be selected by the investigator so as to satisfy Eq. 1.2. The choice of detector responsivity (a measure of the signal produced per unit of radiant power) will be limited by the available detectors and, to a large extent, by the region of the spectrum under consideration. The frequency response bandwidth of the electronics will be determined by the rate of information transfer needed.

The final parameter to be determined is the radiation analyzer luminosity. The *luminosity*, or light-gathering ability, of an instrument is given by the product of the area of the entrance aperture of the system and the solid angle (see Chapter 2) of radiation collected by the system. In the language of optics,

$$\text{Luminosity} = \begin{array}{l}\text{Area of field stop}\\ \text{(usually area of}\\ \text{entrance slit)}\end{array} \times \begin{array}{l}\text{Solid angle subtended}\\ \text{by aperture stop at}\\ \text{field stop (solid}\\ \text{angle collected}\\ \text{by collimator)}\end{array} \tag{1.3}$$

[†] Each of these terms will be discussed in later chapters; see Chapter 2 for a discussion of radiance, Chapter 4 for a discussion of luminosity, and Chapter 9 for a discussion of responsivity.

The luminosity obtained with a given spectroscopic system depends on the resolving power needed to separate the spectral features under investigation. Even without a detailed understanding of the meaning of the term *resolving power*, it should be clear that the ability of a conventional spectroscopic system to separate two closely spaced spectral features will be inversely related to the entrance slit width. Therefore, within the limits set by diffraction (see Chapter 3), higher resolving power should be synonymous with a narrow entrance slit width. Luminosity, on the other hand, depends on the entrance slit area, which is the product of slit height and the slit width. Therefore, for a given spectroscopic system that collects a fixed solid angle of radiation, the luminosity will increase directly with entrance slit width. For a spectral source emitting a line spectrum†, the product of the luminosity of the radiation analyzer times the resolving power of the instrument will be independent of entrance slit width and will depend only on such fundamental design characteristics of the spectrometer as collimator size, focal length, and slit height and not on how it is used (i.e., the particular entrance slit width employed). The instrument parameters mentioned in the preceding discussion ultimately limit spectrometer performance, and for this reason the product of the resolving power times the luminosity of the system is referred to as the efficiency of the spectroscopic system (James and Sternberg, 1969),

$$\text{Efficiency} = \text{resolving power} \times \text{luminosity} \tag{1.4}$$

For a given spectrometer design, the efficiency of the system is approximately constant, and for this reason, the luminosity must be inversely related to the resolving power. Thus, it is not possible to increase simultaneously both the resolving power and luminosity without going to a spectroscopic system with a higher efficiency. The concept of efficiency is a useful figure of merit when comparing the performance of different spectroscopic systems. The concept of efficiency is not limited to conventional spectroscopic systems (i.e., those with an entrance slit and collimator) but can be extended to apply to any spectroscopic system including interferometers. Figure 1.10 shows the luminosity versus resolving power for various spectroscopic systems.

The preceding discussion serves to illustrate the trade-offs that must be considered in selecting a particular spectroscopic approach. If a high rate of information transmission is needed (say, to scan a given region of the spectrum in a certain time period), a large electronic bandwidth will be required. This means that the system noise will be relatively high. If a high resolving power is also needed (say, to resolve closely spaced atomic lines), a spectroscopic system with a high efficiency will be required to give an

† See Appendices A and B for a more detailed discussion of spectrometer efficiency.

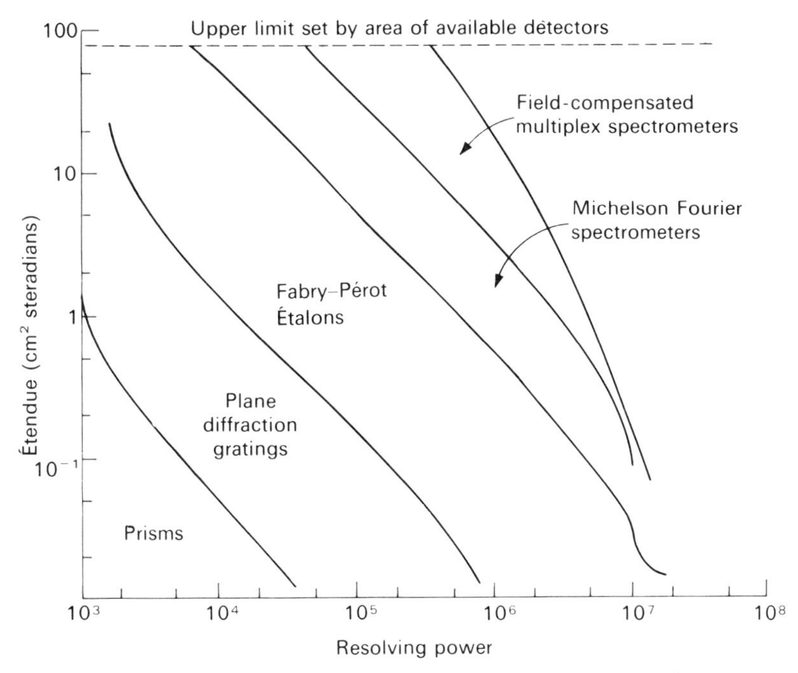

Figure 1.10. Luminosity versus resolving power. [Reprinted with permission from J. F. James and R. S. Sternberg (1969), *The Design of Optical Spectrometers*, Chapman & Hall, London, England].

adequate luminosity. Careful consideration will also need to be given to the selection of a detector with the highest possible responsivity over the spectral region of interest in order to satisfy Eq. 1.2. Finally, if it is not possible to satisfy Eq. 1.2 with the available radiation analyzers and detectors, efforts may need to be made to develop a spectral source (like an induction-coupled plasma) with a high spectral radiance. It should be clear that there is no single spectroscopic system suitable for all possible applications, and the ideal spectrometer for one application may be ill suited for another.

1.9. THE GOALS OF SPECTROSCOPIC RESEARCH

Table 1.1 gives the four primary goals of research in experimental spectroscopy. Progress in experimental spectroscopy depends on improvements in the optical design of spectrometers, spectral sources, and detectors. Of the four goals listed in Table 1.1, it was the first, improvement in resolution, that was pursued in the early development of spectroscopy. Within six years after the development of the Bunsen–Kirchhoff spectroscope, numerous multiple-

Table 1.1. Primary Goals of Spectroscopic Research

1. Improve resolution
2. Improve sensitivity
   a. Increase spectrometer efficiency
   b. Employ more sensitive detectors
3. Increase rate of information transmission
4. Develop improved spectral sources

prism spectroscopes were developed in an effort to improve resolution. Figure 1.11 shows a diagram of an early multiple-prism spectroscope developed in 1866. Using the new instruments, numerous exciting discoveries were made simply because of the availability of greater resolving power. The discovery of "fluted" or band spectra and the Zeeman effect are direct consequences of the availability of high-resolution instruments.

As with most instrumental developments, however, it was not long before most of the interesting and important research that could be conducted with

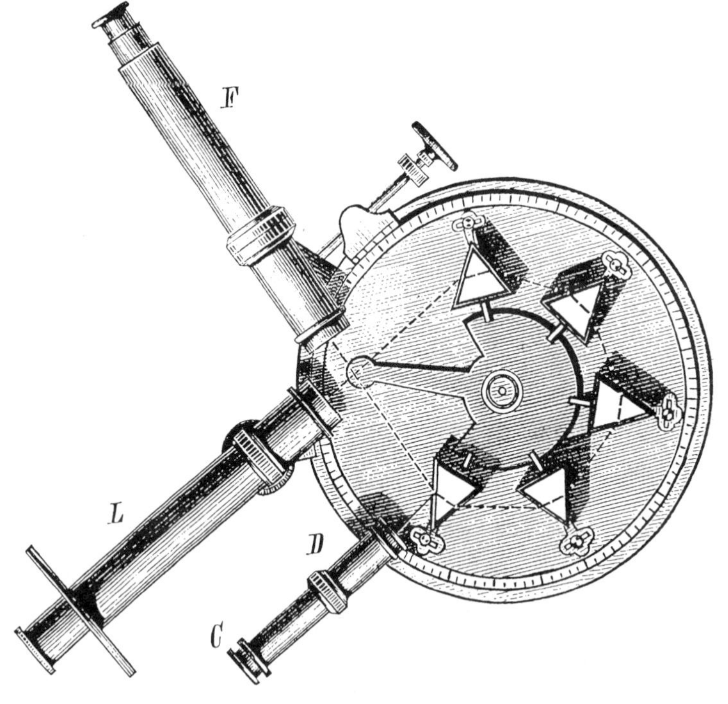

Figure 1.11. Multiple prism spectroscope. [Reprinted with permission from H. Kayser (1900), *Handbuch der Spectroscopie*, S. Hirzel Verlag GmbH & Co., Stuttgart, Germany.]

the available prism instruments had already been done. In a search for still higher resolving power, experimental spectroscopists were led to the development and refinement of the diffraction grating. As diffraction gratings became available, prism instruments were gradually replaced by higher resolving power diffraction grating instruments. As "blazed" gratings were developed (see Chapters 3 and 4), the efficiency (as expressed in terms of the resolving power–luminosity product discussed in Section 1.8) of grating instruments surpassed that of prism instruments.

As a result, the new grating instruments were more sensitive for a given resolving power than their prism counterparts. Having satisfied the resolution requirements of most investigators, at least for the time being, experimental spectroscopists turned their attention to improvements in sensitivity. With the advent of sensitive photoelectric detectors such as the photomultiplier and the availability of reliable amplifiers and power supplies, the photographic plate was gradually replaced as a detector. Not only was the photomultiplier more sensitive than the photographic plate, it was a real-time detector that provided a quantitative output directly without the need for photographic processing and densitometry. Although it was more convenient and sensitive and permitted time-resolved studies of spectral sources, the photomultiplier had to be used in conjunction with a monochromator having an exit slit. In contrast to the photographic plate, which is essentially a multichannel detector (ideal for complex spectra), the photomultiplier is a single-channel device that can monitor only one wavelength at a time. If information over a range of wavelengths is sought, some form of scanning instrument will be needed. Just as with the resolving power–luminosity trade-off, it soon became apparent that a similar trade–off existed between scan speed and signal-to-noise ratio. In addition, use of photoelectric detection, because it is a real-time process and not an integrating process like that obtained with the photographic plate, required the use of stable spectral sources if the spectrum was to be scanned with a monochromator.

These properties of photomultipliers led to two basic lines of investigation. One line of investigation was aimed at improving the rate of information transmission in photoelectric spectrometers. This line of research led to the development of polychromators having multiple photomultiplier tubes, each with its own exit slit. Although these instruments dramatically increased the rate of information transmission compared with a scanning monochromator, they were somewhat bulky, sensitive to temperature variations, and inconvenient to realign for different wavelengths.

Another approach to increasing the rate of information transmission was the development of various transform spectrometers (see Chapters 5 and 7). In addition to increased information transmission, these transform spectrometers also offered the promise of higher luminosities. The increase in luminosity of a transform spectrometer compared with a conventional scanning monochromator is known as *Jacquinot's advantage*. The increase in the rate of information transmission obtained with a transform spectrometer compared with a conventional scanning monochromator is known as *Fellgett's advantage*. Both of these topics will be discussed in detail in Chapters 5 and 7. Regardless of the nature of the transform technique used, however, all require the availability of a computer to transform the raw measurements into a recognizable spectrum. As a result, it was not until the ready availability of small computers that transform techniques became practical as a routine spectroscopic approach. In the case of *Fourier transform spectroscopy* (FTS), mathematical developments such as the *fast Fourier transform*, which greatly reduced the number of arithmetic operations required to perform the transform, allowed the complex mathematical algorithms to be carried out by small computers.

The other basic line of research alluded to earlier has been the development of stable spectral sources for use with photoelectric detection. The first modern attempt at producing a stable spectroscopic source was the revival of the combustion flame as an excitation source by H. Lundergårdh in the 1930s (1929, 1934). This work led ultimately to the development of the first commercial flame photometer. Although the flame was a convenient, stable source, it did not possess adequate energy to excite more than a handful of elements. In 1955, this led Sir Alan Walsh and independently C. T. J. Alkemade and J. M. W. Milatz to propose a method of analysis based on the flame as an atom reservoir rather than an excitation source as in flame photometry. The method was based on the atomic absorption properties of atoms that had been observed by Fraunhofer over 100 years earlier and is known today as *atomic absorption spectrophotometry* (AAS). The key to the success of the method was the use of spectral sources known as *hollow-cathode discharge tubes* (HCDTs), which emit very narrow lines. These sources had been developed earlier by H. Schüler in 1926 as spectral sources for hyperfine structure studies.

The development of atomic absorption spectrophotometry led to a renaissance of interest in experimental spectroscopy as a means of chemical analysis. For the first time, specific elemental analyses could be performed easily by any investigator without the tedious separations that were frequently necessary with solution absorption spectrophotometry.

In spite of the success of flame atomic absorption, it was soon discovered that the ordinary air–acetylene flame used as an atom reservoir did not, in many cases, supply enough energy to completely volatilize and atomize all samples. This led various investigators to study alternative hotter flames such as the nitrous oxide–acetylene flame. At the same time, other workers were studying the use of various electrical discharges and plasmas as an alternative to the use of flames.

This line of research ultimately resulted in the development of the induction-coupled plasma (ICP) as an emission source by Wendt and Fassel (1965) in the United States and Greenfield (1965) and Greenfield and co-workers (1964) in the United Kingdom. Because the temperature of the ICP was much higher than flame sources, the interferences caused by incomplete sample vaporization were virtually eliminated. In their place, however, was the increased likelihood of spectral interferences. The increase in the incidence of spectral interferences was a direct consequence of the much higher temperature of the ICP. It soon became apparent that the short-focal-length monochromators that had been used successfully in atomic absorption spectrometers with the hollow-cathode source did not provide adequate resolving power for use with the ICP. This led to a renewed search for detection systems with high resolution.

To take advantage of the ability of the ICP to excite a large number of elements simultaneously, various investigators began to study detection systems that would make possible simultaneous multielement analyses like those performed earlier with arc or spark excitation and photographic detection. What was sought, however, was not a return to photographic detection but an electronic detection system coupled with a computer to permit rapid determinations of as many as 20–40 elements simultaneously. One solution to this problem that has been employed in commercial spectrometers is the polychromator, or "direct reader", which had formerly been used with arc and spark excitation.

Another approach that has been taken to develop suitable multiwavelength detection systems for high-resolution simultaneous multielement determinations is the development of an electronic analog to the photographic plate. The availability of solid-state *image detectors* or *array detectors* (see Chapter 11) fabricated from silicon has led experimental spectroscopists to speculate about the possibility that these detectors could be used like an electronic "photographic" plate. Although it is still too early to determine exactly what effect the development of image detectors will have on spectroscopy, preliminary studies on the feasibility of using them in spectroscopy have already been conducted. Numerous problems must be solved before a satisfactory electronic analog to the photographic plate is developed, not the least of which is how to obtain high resolution with small detectors and still cover a useful wavelength interval. It is challenges such as these that provide the research problems in modern experimental spectroscopy.

It should be evident from the foregoing discussion that the science of experimental spectroscopy is still continuing to develop even after 400 years of study. Its vitality comes from the synergistic effect that technological innovations have on the discipline and the progressively more stringent requirements placed upon it by those who use it to solve ever more challenging research problems.

APPENDIX 1A

The concept of spectrometer efficiency is of sufficient importance to make it worthwhile to discuss the matter in more detail. We begin by examining Eq. 1.1 from a dimensional standpoint. The signal referred to in the equation will be expressed in terms of current in amperes. Detector responsivity is the signal produced per unit radiant power and will have units of amperes per watt. The unit of luminosity will depend on the nature of the source radiance. If the source is a line source and the detector is monitoring an isolated monochromatic line, the source radiance will be expressed in units of watts per centimeter squared per steradian. If the source emits a continuum, its spectral radiance will have units of watts per centimeter squared per steradian per nanometer. Therefore, in order to satisfy Eq. 1.1 dimensionally, the luminosity must have the following units:

Isolated Line Source

$$\mathrm{A} = \mathrm{W\,cm^{-2}\,sr^{-1}} \times \mathrm{cm^2\,sr} \times \mathrm{A\,W^{-1}}$$

Luminosity $\mathscr{L}_{\text{line}}$ has units of centimeter squared × steradians.

Continuum Source

$$\mathrm{A} = \mathrm{W\,cm^{-2}\,sr^{-1}\,nm^{-1}} \times \mathrm{cm^2\,sr\,nm} \times \mathrm{A\,W^{-1}}$$

Luminosity $\mathscr{L}_{\text{cont}}$ has units of centimeters squared × steradians × nanometers.

Thus the luminosity of a spectral system depends on whether an isolated line source is employed or whether the source emits a continuum. This distinction alters the way in which the efficiency of a spectroscopic system is expressed.

The efficiency of a spectroscopic system must be defined in such a way that it depends only on the fundamental properties of the instrument and not on the way in which the instrument is used. If it is assumed that the instrument is illuminated properly so that the optical system is filled with light, the amount of light entering the system will be governed by the area of the entrance aperture and the solid angle collected by the entrance optics. For a conventional spectrometer with a single entrance slit, the area of the entrance aperture will be simply the slit width w times the slit height h.

In comparing the performance of different spectroscopic systems, however, the resolving power of the instrument also must be considered. It would not be fair to compare the performance of one spectroscopic system employing a wide slit and therefore having a relatively poor resolving power with another instrument employing a narrow slit and therefore having a much higher resolving power. Since the resolving power obtained with a conventional spectroscopic system is inversely proportional to the entrance slit width and the light-gathering power or luminosity is directly proportional to the entrance slit width, the efficiency $\mathscr{E}$ must be defined in such a way that it is independent of slit width. In this way, it will depend only on the fixed parameters of the system.

If the efficiency of a spectroscopic system is defined in terms of the resolving power–luminosity product, $\mathscr{E} = R \times \mathscr{L}$, a figure of merit with the desired properties is obtained. The resolving power–luminosity product is independent of slit width and takes into account both the light-gathering power of the instrument as well as the resolving ability. Since the units of luminosity depend on the type of spectral source employed, the definition of efficiency must be modified to accommodate this fact.

For the isolated line case, the simple product of resolving power times luminosity gives the desired result, that is, functional independence from w. For a continuum source, however, the luminosity will be a function of the slit width squared. This dependence arises because $\mathscr{L}_{\text{cont}}$ depends on both the area of the entrance slit *and* the wavelength range transmitted by the instrument. Since the wavelength interval transmitted by the instrument is directly proportional to the slit width, $\mathscr{L}_{\text{cont}}$ is proportional to the slit width squared. To cancel this dependence, the efficiency for the continuum case must be taken as the product of the square of the resolving power times the luminosity, $\mathscr{E} = R^2 \mathscr{L}$, where $\mathscr{E}$ is the efficiency, R is the resolving power, and $\mathscr{L}$ is the luminosity.

APPENDIX 1B

The reader already familiar with the basic principles of spectroscopy may desire a more rigorous presentation of the concept of spectrometer efficiency. The rigorous derivation of the expression for the efficiency of a grating instrument may be performed easily using the equations developed in subsequent chapters. The derivation that follows is presented to give the reader an appreciation for the way in which spectroscopists employ fundamental principles to answer important questions. The definition and meaning of the various symbols used this appendix can be found by referring to the section in the book where each equation is developed.

The resolving power of a spectroscopic system is defined by Eq. 4.9 as

$$R = \lambda / \Delta\lambda \tag{1A.1}$$

For a grating instrument, λ is given by Eq. 3.74,

$$\lambda = (d/m)(\sin i + \sin \theta) \tag{1A.2}$$

while $\Delta\lambda$ can be expressed in terms of the spectral bandwidth given by Eqs. 4.12, 4.5, and 3.77 (see Table 4.1),

$$\Delta\lambda = wR_{\mathrm{D}} = wd \cos\theta / fm \tag{1A.3}$$

Substituting Eqs. 1A.2 and 1A.3 into 1A.1 gives

$$R = [\sin i + \sin\theta] f/w \cos\theta \tag{1A.4}$$

Equation 1A.4 shows that the resolving power is indeed inversely proportional to slit width w.

The luminosity of the instrument will be given by the area of the entrance slit times the solid angle of light collected by the instrument. The solid angle of radiation collected can be determined from Eqs. 2.2 and 2.6, where the focal length of the instrument is used as the distance r and A_g is the area of the grating:

$$\mathscr{L}_{\text{line}} = whA_g \cos\theta / f^2 \tag{1A.5}$$

Equation 1A.5 shows that $\mathscr{L}_{\text{line}}$ is directly proportional to the slit width w.

Multiplying Eq. 1A.4 by Eq. 1A.5 gives the expression for the efficiency of the instrument,

$$\mathscr{E} = A_g h [\sin i + \sin\theta]/f \tag{1A.6}$$

It can be seen from Eq. 1A.6 that the efficiency of a grating spectrometer with photoelectric detection is directly proportional to slit height h and grating area A_g and inversely proportional to focal length f. If the maximum useful slit height is taken as $f/50$, Eq. 1A.6 reduces to

$$\mathscr{E}_{\text{line}} = A_g (\sin i + \sin\theta)/50 \tag{1A.7}$$

For the case of the continuum source, $\mathscr{L}_{\text{cont}}$ will be given by

$$\mathscr{L}_{\text{cont}} = \mathscr{L}_{\text{line}}\,\Delta\lambda \tag{1A.8}$$

Substituting Eqs. 1A.5 and 1A.3 into Eq. 1A.8 gives

$$\mathscr{L}_{\text{cont}} = w^2 hdA_{\text{g}} \cos^2\theta/f^3 m \tag{1A.9}$$

The efficiency is obtained with the aid of Eq. 1A.4,

$$\mathscr{E} = R^2 \mathscr{L}_{\text{cont}} = hd\,A_{\text{g}}(\sin i + \sin\theta)/fm \tag{1A.10}$$

Equation 1A.10 shows that the efficiency $\mathscr{E}$ defined in this manner is independent of slit width. Substituting for slit height as discussed in the preceding gives

$$\mathscr{E}_{\text{cont}} = dA_{\text{g}}(\sin i + \sin\theta)/50m = \mathscr{E}_{\text{line}}(d/m) \tag{1A.11}$$

Equation 1A.11 shows that the efficiency of a grating spectrometer used with a continuum source depends on the grating constant d and the diffraction order m in addition to the other terms in Eq. 1A.7.

REFERENCES

Alkemade, C. T. J. and Milatz, J. M. W. (1955), *J. Opt. Soc. Am.*, *45*, 583.

Ångström, A. J. (1868), *Recherches sur le spectre normal du soleil*, W. Schultz, Upsala, Sweden.

Baly, E. C. C. (1912), *Spectroscopy*, Longmans, Green, London.

Becquerel, E. (1842), "Mémoire sur la constitution du spectre solaire," presented to the Academy of Sciences at the June 13th meeting.

Becquerel, E. (1865), *La Lumière Électrique*, *2*, 121.

Berzelius, J. J. (1817), *J. für Chemie und Physik*, *21*, 342.

Berzelius, J. J. (1826), *Poggendorff's Ann. Phys. Chem.*, *7*, 242.

Born, M. and Wolf, E. (1987), *Principles of Optics*, Pergamon, Oxford.

Case, T. W. (1917), *Trans. Am. Electrochem. Soc.*, *31*, 351.

Crookes, W. (1856), *J. Phot. Soc. Lond.*, *2*, 292.

Crookes, W. (1861), *Chem. News*, *3*, 193.

D'Albe (1913), *Proc. Roy. Soc.*, *89*, 75.

Desaines, P. and Curie, P. (1880), *Comptes Rendus*, *90*, 1506.

Descartes, R. (1637), *Dioptrique, Météores*, Leyden.

Elster, J. and Geitel, H. (1889), *Ann. Phys., 38*, 497.
Elster, J. and Geitel, H. (1890), *Ann. Phys., 41*, 161.
Elster, J. and Geitel, H. (1913), *Phys. Zeits., 14*, 741.
Fraunhofer, J. (1817), *Gilbert's Ann., 56*, 264.
Fraunhofer, J. (1823), *Ann. Phys., 74*, 337.
Gerlach, W. and Schweitzer, E. (1931), *Foundations and Methods of Chemical Analysis by the Emission Spectrum,* Adam Hilger, London, original German edition published in 1930.
Greenfield, S. (1965), *Proc. Soc. Anal. Chem., 2*, 111.
Greenfield, S., Jones, I. L., and Berry, C. T. (1964), *Analyst, 89*, 713.
Hallwachs, W. (1888), *Ann. Phys., 33*, 301.
Hallwachs, W. (1904), *Phys. Zeits., 5*, 489.
Hartley, W. N. (1884a), *Phil. Trans., 175*, 49.
Hartley, W. N. (1884b), *Phil. Trans., 175*, 325.
Hartley, W. N. (1884c), *Proc. Roy. Soc., 36*, 421.
Herschel, W. (1800), *Phil. Trans. Roy. Soc., 90*, 284.
Hertz, H. (1887a), *Ann. Phys., 31*, 421.
Hertz, H. (1887b), *Ann. Phys., 31*, 983.
Hertz, H. (1888), *Sitzungsber. Berlin Akad. Wis.*, presented at the meeting December 13.
Hertz, H. (1889), *Ann. Phys., 36*, 769.
Hudson, R. D. (1969), *Infrared System Engineering,* Wiley, New York.
Hull, A. W. (1918), *Proc. Inst. Radio Eng., 6*, 5.
Iams, H. and Salzberg, B. (1935), *Proc. Inst. Radio Eng., 23*, 55.
James, J. F. and Sternberg, R. S. (1969), *The Design of Optical Spectrometers,* Chapman & Hall, London.
Kayser, H. (1900), *Handbuch der Spectroscopie,* Hirzel, Leipzig.
Kirchhoff, G. R. (1859), *Monatber. Berlin Akad. Wis.,* 783.
Kirchhoff, G. R. (1860a), *Ann. Phys., 109*, 148, 275.
Kirchhoff, G. R. (1860b), *Phil. Mag., 20*, 1.
Kirchhoff, G. R. (1860c), *Ann. Chim. Phys., 58*, 254.
Kirchhoff, G. R. and Bunsen, R. (1860a), *Ann. Phys., 110*, 160.
Kirchhoff, G. R. and Bunsen, R. (1860b), *Phil. Mag., 20*, 89.
Kirchhoff, G. R. and Bunsen, R. (1861a), *Phil. Mag., 22*, 329.
Kirchhoff, G. R. and Bunsen, R. (1861b), *Ann. Chim. Phys., 62*, 452.
Koller, L. R. (1929), *J. Opt. Soc. Am., 19*, 135.
Koller, L. R. (1930), *Phys. Rev., 36*, 1639.
Lange (1930a), *Phys. Zeits., 31*, 139.
Lange (1930b), *Phys. Zeits., 31*, 964.
Langley, S. P. (1881a), *Nature, 25*, 14.

Langley, S. P. (1881b), *Proc. Am. Acad. Arts Sci., 16*, 342.

Lundergårdh, H. (1929), *The Quantitative Spectral Analysis of the Elements,* Vol. I, Fischer, Jena, Germany.

Lundergårdh, H. (1934), *The Quantitative Spectral Analysis of the Elements,* Vol. II, Fischer, Jena, Germany.

Melloni, M. (1833), *Poggendorff's Ann. Phys. Chem., 28*, 371.

Melloni, M. (1835a), *Poggendorff's Ann. Phys. Chem., 35*, 112, 277, 385.

Melloni, M. (1835b), *Ann. Chim. Phys., 60*(2), 418.

Michelson, A. A. (1887), *Phil. Mag., 24*, 463.

Michelson, A. A. (1891), *Phil. Mag., 31*, 338.

Michelson, A. A. (1892), *Phil. Mag., 34*, 280.

Newton, I. (1672), *Phil. Trans.,* No. 80 (Feb.), 3075.

Nobili, L. and Melloni, M. (1831), *Ann. Chim. Phys., 48*(2), 187.

Rajchman, J. A. and Synder, R. L. (1940), *Electronics, 13*, 20.

Reich, F. and Richter, T. (1863), *J. Prakt. Chem., 90*, 172.

Righi, A. (1888), *Phil. Mag., 25*, 314.

Ritter, J. W. (1801), *Gilbert's Ann., 7*, 527.

Rowland, H. A. (1882), *Phil. Mag., 13*, 469.

Rowland, H. A. (1887a), *Phil. Mag., 23*, 257.

Rowland, H. A. (1887b), *Am. J. Sci., 33*, 182.

Sawyer, R. A. (1963), *Experimental Spectroscopy,* Dover, New York.

Schottky, W. (1930), *Phys. Zeits., 31*, 913.

Schüler, H. (1926), *Zeits. Phys., 35*, 323.

Smith, R. A., Jones, F. E., and Chasmar, R. P. (1957), *The Detection and Measurement of Infrared Radiation*, Oxford Univ. Press, London.

Smith, W. (1873), *J. Soc. Tel. Eng., 2*, 31.

Sommer, A. H. (1968), *Photoemissive Materials—Preparation, Properties, and Uses,* Wiley, New York.

Stoletow (1890), *J. Phys., 9*, 486.

Walsh, A. (1955), *Spectrochim. Acta, 7*, 108.

Wiedeman, E. and Ebert, H. (1888), *Ann. Phys., 33*, 421.

Wendt, R. H. and Fassel, V. A. (1965), *Anal. Chem., 38*, 337.

Zworykin, V. K. and Wilson, E. D. (1934), *Photocells and Their Applications,* 2nd ed., Wiley, New York.

CHAPTER

2

PRINCIPLES OF OPTICS AND RADIOMETRY

Optics and spectroscopy are intimately connected so that it is virtually impossible to conduct a meaningful discussion of spectroscopic principles without a foundation in optics. Unfortunately, many analytical chemists who use spectroscopic instrumentation for analysis have had little or no formal training in optics, and it is often difficult to find the time to study the standard optical treatises (Born, 1933; Levi, 1968, 1980; Wood, 1934; Jenkins and White, 1957; O'Shea, 1985) to gain an appreciation of the relevant principles. The purpose of this chapter is to provide an introduction to the principles of optics and radiometry that are relevant to an understanding of spectroscopy in general and detection systems for multielement analysis in particular. The presentation of the subject will be somewhat abbreviated but will be written at a level that should be understandable to anyone with a scientific background but no previous experience in optics. The development will follow the standard presentation found in most optics texts.

We begin the discussion with a brief introduction to radiometry. Radiometry is the science concerned with the quantitative measurement of the energy content of a radiation field. One aspect of radiometry that is of interest to analytical spectroscopists is the determination of the manner in which the energy associated with a given radiation field flows through an optical system such as a spectrometer or spectrograph.

In order to discuss the performance characteristics of spectrometric systems on a scientific basis, it will first be necessary to review some basic aspects of radiometry and optics that are directly related to an understanding of spectrometric systems. The reader who wishes a more detailed treatment of these topics should consult the standard treatises on these subjects (Grum and Becherer, 1979; Boyd, 1983).

2.1. LIGHT AND ITS MEASUREMENT

We begin our discussion of light and its measurement by introducing a number of radiometric quantities that are used to specify the energy content of a radiation field. Figure 2.1 shows a point source of light at the center of a

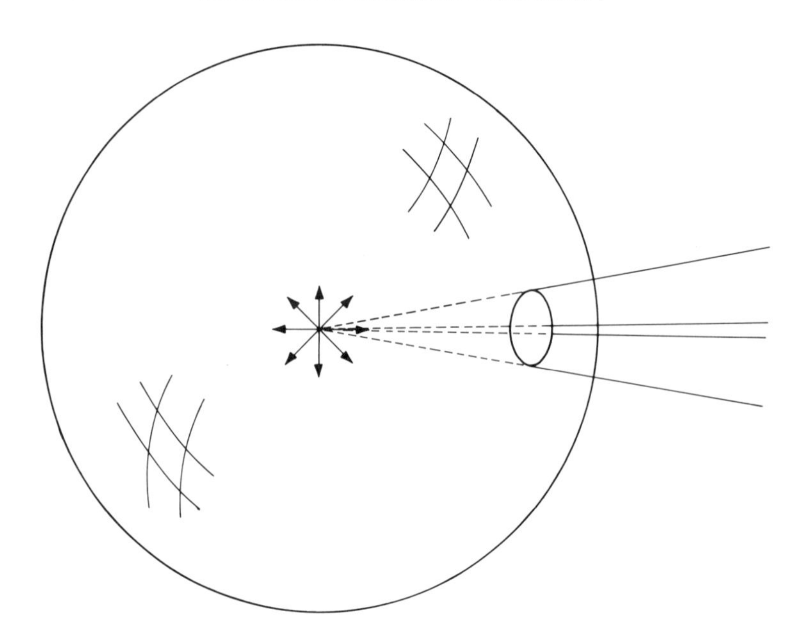

Figure 2.1. Point source of light at center of sphere.

sphere. If the light source is an incandescent filament of a tungsten light bulb, the radiation will be emitted isotropically to a first approximation. A source that emits radiation equally regardless of the angle of observation is known as a *Lambertian source*. Radiation emitted by a light source can be classified into two major categories depending on the amplitude and phase relationships that exist within the radiation field. For a thermal source such as a tungsten filament, each element of surface of the filament can radiate independently of the others. This results in the production of wave trains whose amplitudes and phases are randomly related. Such a radiation field is described as being *incoherent* and results from the stochastic combination of the independently emitted wave trains. At the other extreme, if a light source such as a laser emits radiation whose wave trains all have a constant phase relationship, this radiation is termed *coherent*. Between these two extremes, the radiation field can be described as *partially coherent*.

In specifying the various ways in which the energy emitted by the tungsten filament can be described, we assume that the source is incoherent and that the energy is propagated through the given optical system according to the laws of geometric optics.

The rate at which the tungsten filament radiates energy can be expressed in terms of its *radiant flux* F. The radiant flux is equivalent to the rate at which photons of a given energy pass through a specified region of space such as a hole in the sphere. Since the units of radiant flux are joules per second or watts, the radiant flux is sometimes referred to as the *radiant power*.

Naturally, the amount of energy received by a surface located some distance from the source will depend on a number of factors. The radiant flux per unit area that strikes a surface is known as the *irradiance* and can be defined as

$$E = dF/dA \tag{2.1}$$

where dF is the radiant flux that strikes an element of area, dA. The irradiance has units of power per unit area and is typically expressed in watts per square centimeter or some similar units. The power per unit area emitted by a source is defined in an analogous fashion and is known as the *emittance*.

It is clear from Figure 2.1 that the amount of radiation that can escape through a hole in the sphere will depend not only on the size of the hole but also on the radius of the sphere. The quantity that relates these two parameters in solid geometry is the concept of the solid angle. The solid angle is defined as the ratio of the spherical surface area of the opening to the square of the radius of the sphere and can be expressed as

$$\Omega = a/r^2 \tag{2.2}$$

where Ω is the solid angle in steradians, a is the surface area of the opening, and r is the radius of the sphere. Since the entire surface area of a sphere is given by $4\pi r^2$, there are 4π steradians in a sphere. For a right-circular cone whose half-vertex angle is α as shown in Figure 2.2, the solid angle subtended is given by

$$\Omega = 4\pi \sin^2(\alpha/2) \tag{2.3}$$

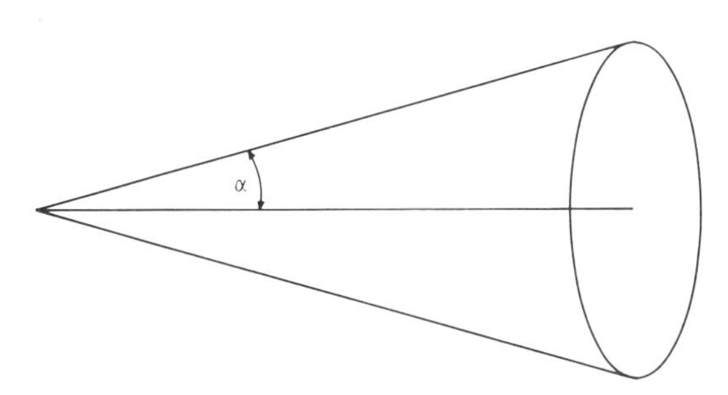

Figure 2.2. Right-circular cone with half-vertex angle α.

The amount of light a source emits in a particular direction can be expressed in terms of the *radiant intensity* I, given by

$$I = dF/d\Omega \tag{2.4}$$

where I has units of power per solid angle (watts per steradian).

The rate at which energy is emitted from the source in a particular direction from a given area of the source can be found by differentiating the radiant intensity with respect to area,

$$dI/dA = d^2F/dA\, d\Omega = L = dE/d\Omega \tag{2.5}$$

where L is the radiance or brightness. As shown in Figure 2.3, since the direction of motion of a wavefront is always perpendicular to the wavefront, an element of area can radiate or receive energy only in proportion to its projected area. The projected area is related to the actual area according to Lambert's cosine law, which is given by

$$dA_{\text{proj}} = dA \cos\theta \tag{2.6}$$

where θ is the angle between the surface normal and the direction of the rays.

Equations 2.1, 2.4, and 2.5 have been defined without regard for the spectral composition of the radiation. The effect of spectral composition can be accounted for by specifying the corresponding *spectral radiometric quantity*. Thus, the term *spectral radiance* refers to the radiance in the wavelength interval from λ to $\lambda + d\lambda$ and is defined as $L_\lambda = dL/d\lambda$ and has units of

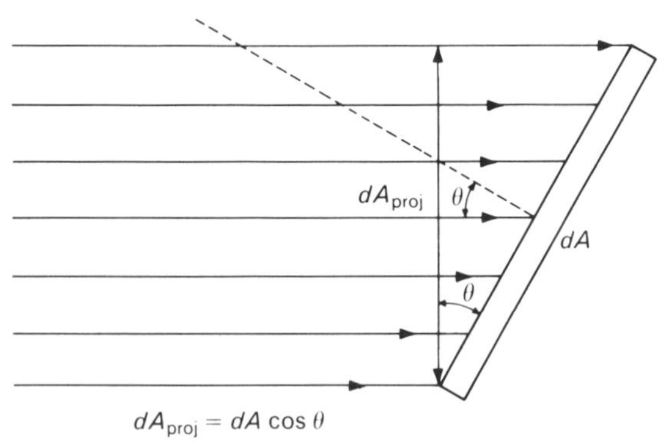

Figure 2.3. An optical element can radiate or receive energy only in proportion to its projected area.

watts $cm^{-2} sr^{-1} nm^{-1}$. The other radiometric quantities (Eqs. 2.1 and 2.4) can be defined as spectral radiometric quantities in a similar manner.

Equations 2.1–2.6 define the basic concepts needed to study the flow of energy through an optical system that obeys the laws of geometric optics. These equations are not valid for optical systems where interference or diffraction effects are significant.

Using these basic equations, it is possible to demonstrate some of the important laws of optics. For example, it is possible to show for a point source that the irradiance decreases as the square of the distance from the source. This phenomenon is the well-known *inverse-square law* of Lambert and is strictly valid only for a point source. From the definition for radiant intensity (Eq. 2.4) and the definition for a solid angle (Eqs. 2.2 and 2.6), one obtains

$$dF = I\,dA\cos\theta/r^2 \tag{2.7}$$

which upon rearranging gives

$$E = dF/dA = I\cos\theta/r^2 \tag{2.8}$$

which in turn shows that the irradiance varies inversely with the square of the distance. Intuitively, this must be the case since for a point source at the center of a sphere, the same amount of energy must pass through a sphere of any radius. Since the surface area of a sphere varies with the square of the radius, it follows that the energy per unit area must vary inversely with the square of the radius.

2.2. INTRODUCTORY GEOMETRIC OPTICS

Optical systems that obey the laws of reflection and refraction are said to follow geometric optics. Figure 2.4 shows a ray that strikes a reflecting surface at an angle θ with respect to the normal to the surface. The *law of reflection* states that the reflected ray lies in the plane established by the incident ray and the normal and the angle of incidence (measured from the surface normal) equals the angle of reflection (also measured from the surface normal). The plane defined by the incident ray and the normal to the surface is called the *incidence plane*.

Figure 2.5 shows what happens when a light ray strikes the boundary between two transparent substances that propagate light at different speeds. As the ray strikes the boundary, some of the radiation is reflected according to the law of reflection while some of the radiation penetrates the boundary and enters the second medium. The ray that penetrates into the second medium is

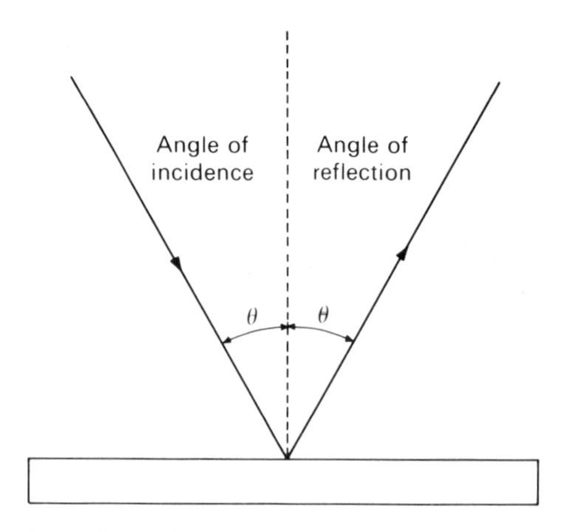

Figure 2.4. Law of reflection. Plane of incidence is the plane of the page.

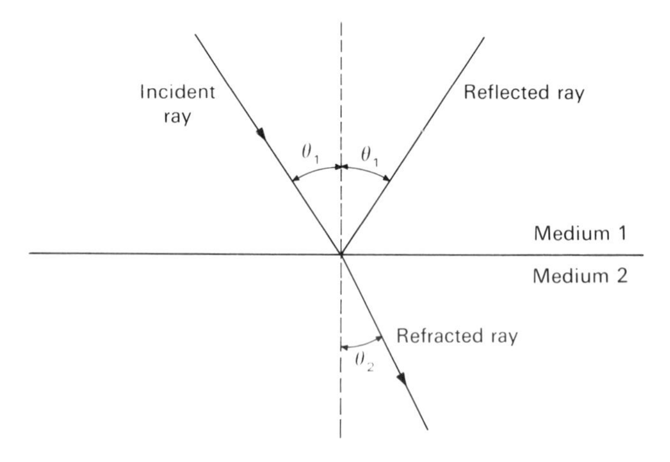

Figure 2.5. Snell's law of refraction.

known as the *refracted ray*. *Snell's law of refraction* states that the refracted ray lies in the plane of incidence and that

$$n_1 \sin \theta_1 = n_2 \sin \theta_2 \tag{2.9}$$

where n_1 and n_2 are the refractive indices of media 1 and 2, respectively, θ_1 is the angle of incidence, and θ_2 is the angle of refraction (measured from the surface normal).

The *index of refraction* is defined as the ratio of the speed of light in a vacuum to that in the given medium. As a consequence of this definition, values of the refractive index are always greater than 1. In addition, the refractive index also varies with the wavelength of the radiation. A plot of refractive index versus wavelength is characteristic of the material, and the slope of the plot ($dn/d\lambda$) is known as the *dispersion* of the material at a given wavelength. Materials whose refractive index increases with decreasing wavelength and whose rate of increase becomes larger at shorter wavelengths are said to follow *normal dispersion*. Figure 2.6 shows some representative normal dispersion curves.

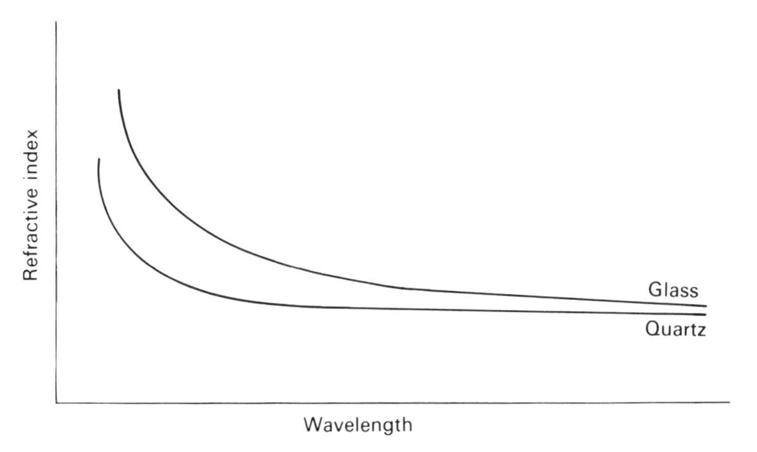

Figure 2.6. Generalized dispersion curves for glass and quartz.

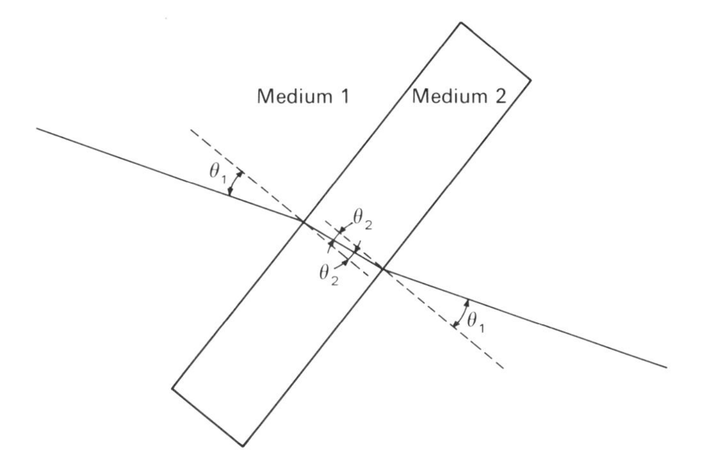

Figure 2.7. Effect of refraction with parallel boundary medium.

In general, the denser the medium, the higher the refractive index. As a consequence of Snell's law, light entering a more dense medium from a less dense one is refracted toward the normal.

Figure 2.7 shows what happens when light strikes a more dense medium with parallel boundaries and passes through it to reemerge once again into the original medium. A fundamental principle of optics is the *principle of reversibility*, which states that even for the most complicated optical paths if a ray is reversed, it will retrace its original path. As a result of this principle, light emerging from the opposite surface of the plane-parallel plate must emerge with the same angle to the normal as the original angle of incidence at the first surface. As a result of this encounter, the ray is undeviated from its original direction and simply suffers a lateral displacement.

2.2.1. Prisms

The situation is quite different if the two surfaces are inclined at some angle as in a prism. In contrast with the plane-parallel plate refractor, the geometry of the prism results in a net deviation of the incident beam from its original direction. Since the deviation is a result of refraction and since the refractive index is a function of the wavelength, it follows that the extent of deviation of the beam will be a function of the wavelength of the light.

Before discussing the refraction produced by a prism, it is well to recall a simple postulate from geometry shown in Figure 2.8. As can be seen from the figure, an exterior angle of a triangle is always equal to the sum of the two opposite angles.

Figure 2.9 shows an isosceles prism with an apex angle α. If the prism is irradiated by a parallel beam of monochromatic radiation that strikes the first surface with an angle of incidence i_1, it will be refracted by the prism according to Snell's law so that

$$\sin r_1 = (n/n')\sin i_1 \tag{2.10}$$

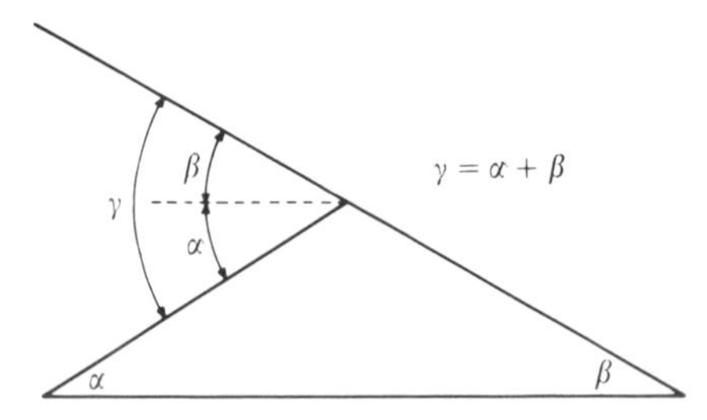

Figure 2.8. Geometric relation between exterior angle of triangle and two opposite angles.

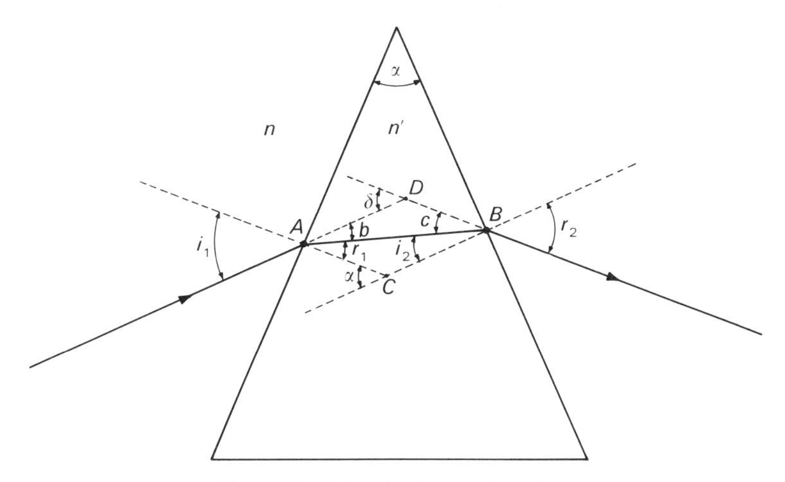

Figure 2.9. Refraction by isoceles prism.

$$r_1 = \sin^{-1}[(n/n')\sin i_1] \tag{2.11}$$

From the geometry of the prism, it is evident that the angle between the two normals to the surfaces is the apex angle α. Since α is an exterior angle of the triangle *ABC*,

$$\alpha = r_1 + i_2 \tag{2.12}$$

and

$$i_2 = \alpha - r_1 \tag{2.13}$$

From Snell's law, it follows that

$$\sin r_2 = (n'/n)\sin i_2 \tag{2.14}$$

and

$$r_2 = \sin^{-1}[(n'/n)\sin(\alpha - r_1)] \tag{2.15}$$

The total deviation of the beam is the angle δ between the incident and emerging beams. Since δ is an exterior angle of the triangle *ABD*,

$$\delta = b + c \tag{2.16}$$

where

$$b = i_1 - r_1 \tag{2.17}$$

and

$$c = r_2 - i_2 \tag{2.18}$$

Substituting Eqs. 2.17 and 2.18 into Eq. 2.16 and realizing that $\alpha = r_1 + i_2$ gives

$$\delta = i_1 + r_2 - \alpha \tag{2.19}$$

It is evident from Figure 2.9 that the minimum path for a given ray through the prism will occur when the refracted ray travels parallel to the prism base. This condition also results in the smallest angle of deviation δ and is known as the *angle of minimum deviation* δ_m. When a prism is illuminated under conditions of minimum deviation,

$$i_1 = r_2 \qquad r_1 = i_2 \qquad b = c \tag{2.20}$$

Prisms used for spectroscopic applications are nearly always used under conditions of minimum deviation for two reasons. Since the minimum-deviation condition results in the shortest path length through the prism, it results in the least light loss from absorption by the prism material. In addition, the minimum-deviation condition reduces astigmatism in the image that occurs with other illumination conditions if the incident beam is not perfectly parallel.

The angular dispersion for a prism may be obtained by taking the derivative of Eq. 2.19 with respect to wavelength, assuming a constant value for the angle of incidence. Since the apex angle for a given prism is also constant, the derivative is

$$d\delta/d\lambda = dr_2/d\lambda \tag{2.21}$$

$$\frac{dr_2}{d\lambda} = \frac{1}{n}\left(\frac{\cos i_2 \tan r_1}{\cos r_2} + \frac{\sin i_2}{\cos r_2}\right)\frac{dn'}{d\lambda} \tag{2.22}$$

where n is the refractive index of air, which can be taken as 1, and n' is the refractive index of the prism material. If the prism is used under conditions of minimum deviation so that $i_1 = r_2$ and $r_1 = i_2$, Eq. 2.22 reduces to

$$dr_2/d\lambda = d\delta/d\lambda = (2 \tan i_1/n')(dn'/d\lambda) \tag{2.23}$$

which is the usual equation given in spectroscopy texts. It should be remembered that Eq. 2.23 is strictly valid only for a prism under conditions of minimum deviation. Assuming this to be the case, Eq. 2.23 shows that the angular dispersion for a prism is inversely proportional to the refractive index of the prism material for the given wavelength and directly proportional to the dispersion of the prism material for the given wavelength. Angle i_1 is fixed by the condition of minimum deviation.

Equation 2.23 can be expressed in terms of the geometry of the prism itself rather than the angle of incidence. Figure 2.10 shows a prism illuminated under the conditions of minimum deviation, where s is the length of a face, B is the width of the prism base, and b is the width of the incident beam on the first face of the prism (i.e., the linear aperture). It is assumed that the entire face of the prism is filled with light. From Eq. 2.23, one can write

$$dr_2/d\lambda = [2 \sin i_1/(n' \cos i_1)](dn'/d\lambda) \tag{2.24}$$

Using Snell's law, Eq. 2.13, and the fact that $i_2 = r_1$ at minimum deviation, it can be shown that

$$\frac{dr_2}{d\lambda} = \frac{2 \sin r_1}{\cos i_1} \frac{dn'}{d\lambda} = \frac{2 \sin(\alpha/2)}{\cos i_1} \frac{dn'}{d\lambda} \tag{2.25}$$

Dividing the numerator and denominator by s gives

$$\frac{dr_2}{d\lambda} = \frac{2s \sin(\alpha/2)}{s \cos i_1} \frac{dn'}{d\lambda} \tag{2.26}$$

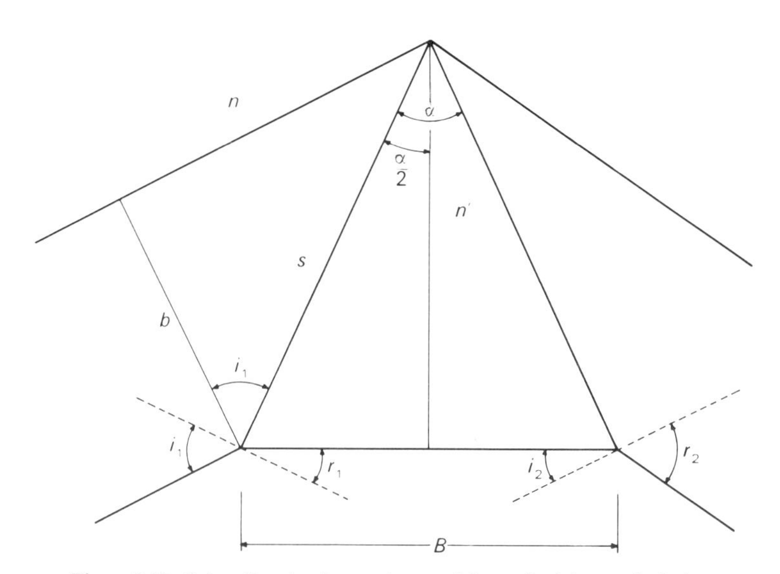

Figure 2.10. Prism illumination under conditions of minimum deviation.

From Figure 2.10, it can be seen that $2s \sin(\alpha/2) = B$ and that $s \cos i_1 = b$. Therefore, Eq. 2.23 becomes

$$dr_2/d\lambda = (B/b)(dn'/d\lambda) \tag{2.27}$$

where the first term is a factor related to prism geometry and the second term is the dispersion of the prism material.

On the basis of the preceding result, it would seem appropriate to select a prism material with the largest possible dispersion so as to achieve the largest possible angular dispersion according to Eqs. 2.23 and 2.27. Although this is certainly true from the standpoint of angular dispersion, Eqs. 2.23 and 2.27 do not take into account factors involving the transmission of light through the prism. Unfortunately, the slope of the refractive-index-versus-wavelength curve tends to increase most in the vicinity of an absorption band. As a result, a compromise between angular dispersion and optical transmission considerations must be made in selecting an appropriate prism material.

Similar considerations also affect the choice of an apex angle for the prism. We shall see later on that the resolving power of a prism (which is a measurement of the prism's ability to separate closely spaced lines) is directly proportional to the width of the base of the prism. Increasing the base width can be achieved by increasing the apex angle. Again light transmission losses mitigate against apex angles greater than 60° for two reasons. The most obvious reason is that increasing the apex angle increases the optical path length through the prism, which increases the losses due to absorption within

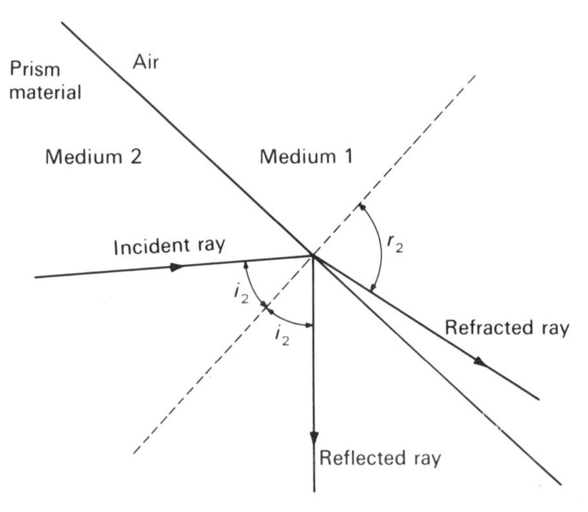

Figure 2.11. Internal reflection at second surface of prism.

the prism. The second factor involves internal reflection as the light strikes the second surface of the prism. This is shown in Figure 2.11.

According to Snell's law, as angle i_2 increases, angle r_2 approaches 90°. In the limit,

$$i_2 = \sin^{-1}(n/n') \tag{2.28}$$

when $r_2 = 90°$. When i_2 reaches this limit, the ray is totally internally reflected downward as indicated in the figure rather than emerging from the prism in the desired direction. The angle of incidence where total reflection occurs is called the *critical angle*. As i_2 approaches the critical angle, more light is internally reflected rather than being transmitted through the prism.

Other factors relating to the spectroscopic performance of prisms will be deferred until Chapter 4.

2.2.2. Lenses

So far we have discussed refraction at plane surfaces. We turn our attention now to the phenomenon of refraction at spherical surfaces such as occur with lenses. A thorough understanding of the image-forming properties of lenses is important in spectroscopy not only for understanding the principles behind spectroscopic instrumentation but also for understanding the factors involved in illuminating the spectrometer or spectrograph. Our discussion of lenses will be limited to *converging* or *positive* lenses since these are the type generally employed in spectroscopic work.

Figure 2.12 shows a typical equiconvex lens. Each ray that strikes the lens is refracted according to Snell's law. The *principal axis* of the lens is a line passing through the center of the lens and normal to both faces. The point where the principal axis intersects the lens surface is the *vertex* of the lens.

Every lens has two characteristic points located on the principal axis known as *focal points*. The focal point to the left of the lens is known as the *primary focal point*, while the focal point to the right of the lens is known as the *secondary focal point*. The primary focal point, being located prior to the refracting system, is said to be located in *object space*. The secondary focal point, being located after the refracting system, is said to lie in *image space*. The location of the primary focal point on the principal axis is such that rays emanating from it are rendered parallel to the optic axis after refraction by the lens. The secondary focal point is that point in image space on the principal axis through which rays that were originally parallel to the axis in object space are refracted after passing through the lens. For each focal point there is a plane perpendicular to the principal axis that passes through the focal point; this plane is known as the *focal plane*.

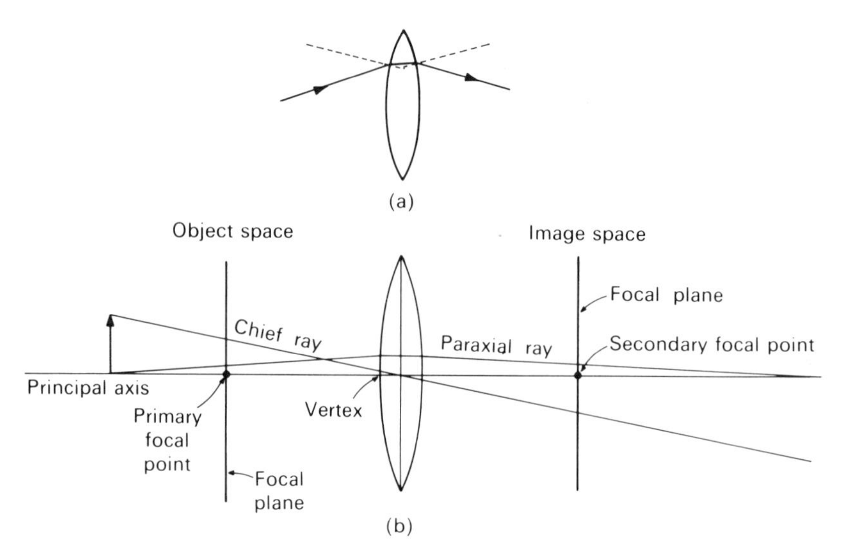

Figure 2.12. Ray diagram for equiconvex lens: (*a*) each ray that strikes lens is refracted according to Snell's law; (*b*) diagram illustrates terminology associated with ray diagrams for lenses.

In a ray diagram, a *chief ray* is any ray that passes through the center of the lens. The chief ray always passes through an equiconvex lens undeviated from its original path. This is true for an equiconvex lens because the chief ray always strikes the lens normal to its surface and also because both surfaces of the lens are parallel for any ray passing through the lenses' center.

Rays that lie close to the principal axis or make small angles to it are known as *paraxial rays.* Mathematically, paraxial rays are rays whose angles with respect to the optical axis are such that the cosines of the angles are approximately equal to 1 and the sines of the angles are approximately equal to the angles themselves.

Having introduced the necessary terminology, we are now in a position to discuss image formation by a thin lens. A *thin lens* is a lens whose thickness is small in comparison to the radii of curvature of both surfaces. If an object is placed to the left of the primary focal point of a convex lens, a *real image* of the object will form to the right of the secondary focal point. A real image is one that can be observed on an appropriate screen placed in the image plane. Because of the geometry associated with image formation by a single simple lens, the image that forms will be upside down, or *inverted.*

Figure 2.13 shows how an image of an object may be located on a ray diagram using the parallel-ray method. Using this method, three rays can be drawn (only two are actually needed) from any point on the object. The

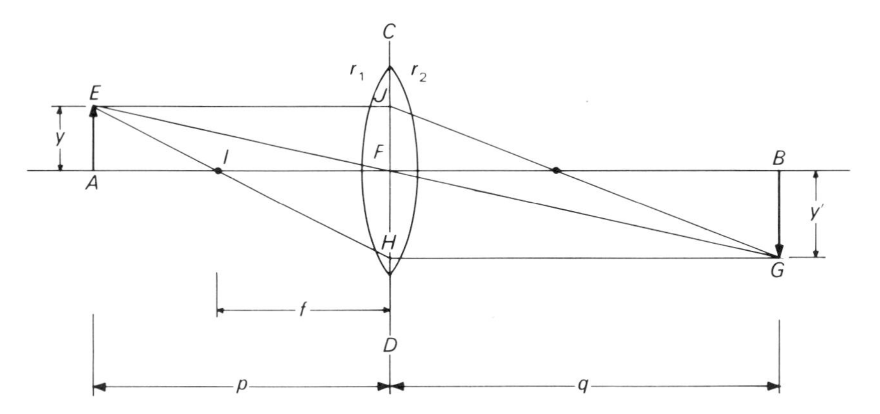

Figure 2.13. Parallel-ray method of image location with equiconvex lens.

intersection point of these rays in image space locates the position of that point in the image.

In constructing the ray diagram, the rays are drawn as if the refraction occurred at the vertical plane through the center of the lens rather than at the lens surface itself. Once the position of the lens has been established, the first line drawn is the principal axis (AB), and the primary and secondary focal points are then located on the axis. Line CD is then drawn through the center of the lens. According to the previous discussion, any ray parallel to the principal axis in object space (EJ) will be refracted through the secondary focal point in image space (JG). Similarly, any ray in object space that passes through the primary focal point (EI) will be refracted parallel to the principal axis in image space (HG). The intersection of these two lines is all that is required to locate the tip of the arrow in Figure 2.13. The chief ray from the tip of the arrow (EFG) simply passes undeviated through the lens and intersects with the other two lines at point G. For other graphical methods of determining the position of an image (such as the oblique-ray method), the reader should consult a standard optics text (see Jenkins and White, 1957).

The position of the image can also be calculated using the *lens formula* of Gauss,

$$1/p + 1/q = 1/f \tag{2.29}$$

where p is the lens-to-object distance, q is the lens-to-image distance, and f is the focal length. The focal length of a lens may be calculated with the aid of the *lens maker's formula*,

$$1/f = (n - 1)(1/r_1 - 1/r_2) \tag{2.30}$$

where n is the refractive index of the lens material for a given wavelength and r_1 and r_2 are the radii of curvature for the first and second surfaces.

In using these formulas several sign conventions should be kept in mind. Ray diagrams are typically constructed with the object on the left and the image on the right, although by the principle of reverse optics any pair of object and image points such as E and G are *conjugate points*, which means they are interchangeable. All object distances that are to the left of the vertex are positive and all image distances to the right of the vertex are positive. For a converging lens both focal lengths are positive. All convex surfaces encountered from left to right are taken as having positive radii of curvature while any concave surfaces encountered (from left to right) are considered to have negative radii of curvature. Image and object distances that lie above the principal axis are positive while those measured to points below the axis are negative.

Applying these conventions to the equiconvex lens shown in Figure 2.13, one finds that distances p, q, y, and f are all positive, while length y' is negative. The radius of curvature r_1 for the first surface is positive while the radius of curvature for the second surface, r_2, is negative since the surface appears concave from the left to the right direction. By observing the preceding sign conventions, no difficulty should be encountered in using Eqs. 2.29 and 2.30.

The *magnification* of a lens is defined as the ratio of the image dimensions to the object dimensions. Referring to Figure 2.13, it is evident that triangles AEF and BGF are similar triangles from which it can be seen that

$$y/p = -y'/q \qquad \text{or} \qquad -y'/y = q/p = m \tag{2.31}$$

where m is the magnification and the minus sign indicates an inverted image (i.e., y' is negative). Thus the magnification produced by the lens is equal to the ratio of the image distance to the object distance.

Equations 2.29 and 2.31 can be combined to give equations for the object distance, the image distance, and the total object-to-image distance in terms of the focal length of the lens and the amount of magnification desired:

$$p = f(1 + m)/m \tag{2.32}$$

$$q = f(1 + m) \tag{2.33}$$

$$p + q = f(1 + m)^2/m \tag{2.34}$$

Example 2.1. Suppose that a lens with a 10-cm focal length and a diameter of 2.5 cm is to be used to form a 1:1 image of the source on the entrance slit of a spectrograph. How far should the lens be placed from the slit? How far should the source be placed from the slit? What solid angle does the lens subtend?

From Eq. 2.33, the lens-to-image distance is calculated to be $10 \times 2 = 20$ cm for unit magnification. From Eq. 2.34, the source should be placed $10 \times 2^2 = 40$ cm from the entrance slit. It is clear from the example that for unit magnification $p = q = 2f$.

The solid angle subtended by the lens may be calculated using Eq. 2.3. From the geometry of the problem, the half-vertex angle is equal to $\tan^{-1}(1.25/20.0) = 3.58°$. Substituting this value in Eq. 2.3 gives 0.0122 sr. For small solid angles such as this, Eq. 2.2 can be used to calculate the solid angle by using the plane area of the lens as an approximation of the spherical area. Using this approach gives $= 4.90/(20.0)^2 = 0.0122$ sr.

Equation 2.30 implies another feature about image formation with a simple lens. Since n is a function of wavelength, it is clear that the focal length of a simple lens is also a function of wavelength, a fact responsible for *chromatic aberration*. In the preceding discussion, it was implicitly assumed that the refractive index of the lens material was constant. This is strictly true only for monochromatic light. When light of various wavelengths is passed through the lens, each separate wavelength will come to a focus at a different point with short wavelengths coming to a focus nearest the lens.

This phenomenon should be considered when focusing an image of a source on the slit of a spectrograph. An image that appears sharp to the eye may be completely out of focus as far as short-wavelength ultraviolet light (to which the eye does not respond) is concerned. To correct for this effect, it is better to calculate the required distances using Eq. 2.29 with the appropriate focal length for the given wavelengths of interest. The ratio of the focal lengths at two different wavelengths 1 and 2 can be obtained by taking the ratio of Eq. 2.30 at two different wavelengths:

$$f_1/f_2 = (n_2 - 1)/(n_1 - 1) \tag{2.35}$$

where n_1 and n_2 are the refractive indices at wavelengths 1 and 2. Eq. 2.35 allows calculation of the focal length of the lens at any wavelength as long as the focal length at one wavelength is known.

Another implicit assumption that was made in the previous discussion of image formation in lenses was that all the rays considered were paraxial rays. When paraxial rays are considered, a *stigmatic* image is produced. A stigmatic image occurs when a given point of the object is imaged as its conjugate point in the image. If rays that lie some distance from the optic axis are being

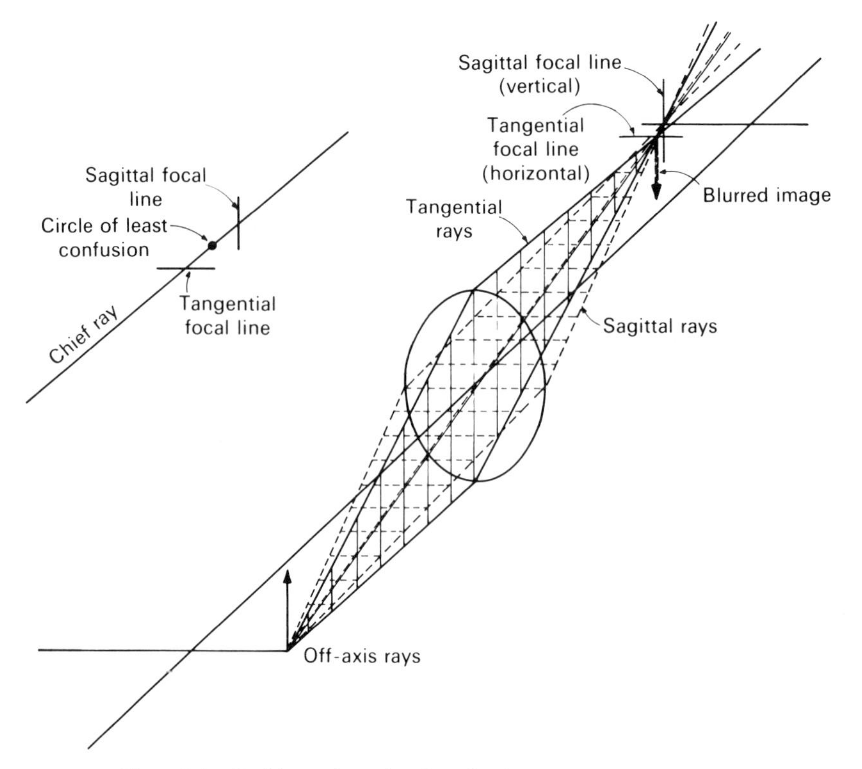

Figure 2.14. Positive astigmatism for off-axis rays striking single lens.

considered, one finds that it is not possible to bring them to a focus at a single point. Instead, one observes the formation of two perpendicular *focal lines* with a circle of least confusion between them. This condition is known as *astigmatism.* The formation of an astigmatic image is shown in Figure 2.14.

As shown in the figure, rays that strike the lens in the vertical, or *tangential*, plane come to a focus as a horizontal focal line indicated by T in the figure. Rays that strike the lens in the horizontal, or *sagittal*, plane come to a focus as a vertical focal line indicated by S. The distance between T and S measured along the chief ray is the *astigmatic difference.* As the rays get closer to the principal axis (i.e., become more paraxial), the astigmatic difference becomes smaller, eventually becoming zero for paraxial rays. If the focal line T occurs before focal line S as shown in Figure 2.14, the astigmatism is said to be positive.

2.2.3. Conservation of Radiance

It might be supposed from the foregoing discussion of image formation that it could be possible to increase the radiance of a source by using a lens to somehow concentrate the light. In fact, this is not the case at all. It is a fundamental law of optics that for an ideal lossless optical system where the image formed is in the same medium as the object, the brightness of the image equals the brightness of the source. For a real system where losses due to absorption and reflection occur, the brightness of the image will always be less than the brightness of the source.

This theorem can be illustrated by considering the image formation by a single lens as shown in Figure 2.15. In the figure, S is a luminous disk source of area A_1 whose distance from the lens is p. An image of the source of area A_2 is formed by the lens at a distance of q. It is assumed that the disk source, the lens, and the image are all normal to the optical axis and that the area of the lens is equal to R.

At first glance, it might appear from Figure 2.15 that the brightness of the image would be increased since the source is focused over a smaller area. In fact, the decrease in area is just compensated by an increase in solid angle.

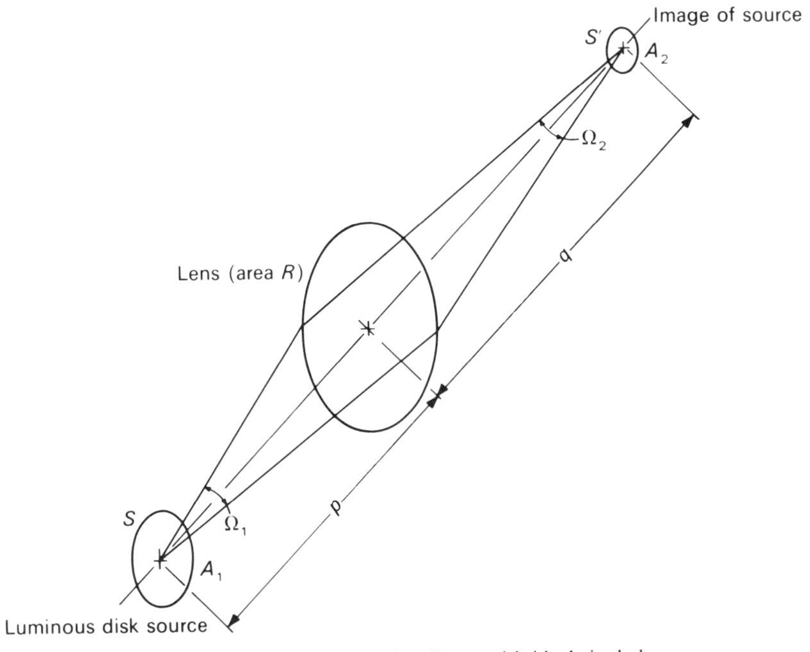

Figure 2.15. Conservation of radiance with ideal single lens.

This can be shown by considering the total flux from the source. From Eq. 2.5, the total flux emitted by a source will be given by

$$F = L_1 A_1 \Omega_1 \tag{2.36}$$

where F is the flux, L_1 is the radiance of the source, and Ω_1 is the solid angle the source makes with the lens. If the optical system is assumed to be an ideal lossless system, the flux from the lens to the image will be given by

$$F = L_2 A_2 \Omega_2 \tag{2.37}$$

where L_2 is the radiance of the image and Ω_2 is the solid angle subtended by the lens at the image. Equating Eqs. 2.36 and 2.37 and substituting for the individual quantities gives

$$L_1 A_1 \Omega_1 = L_2 A_2 \Omega_2 \tag{2.38}$$

$$L_1(\pi y^2)R/p^2 = L_2(\pi y'^2)R/q^2 \tag{2.39}$$

From Eq. 2.31, it is seen that $y' = y(q/p)$. Therefore,

$$\frac{L_1 \pi y^2 R}{p^2} = \frac{L_2 \pi (yq/p)^2 R}{q^2} = \frac{L_2 \pi y^2 R}{p^2} \tag{2.40}$$

From Eq. 2.40, it can be seen that $L_1 = L_2$. This result holds for any optical system as long as the source and the image are in the same medium. This theorem will be of importance in discussing the use of a condensing lens to illuminate the slit of a spectrograph.

2.2.4. Mirrors

Mirrors are often used in spectroscopic instrumentation because they do not suffer from chromatic aberration and also because it is easier to fabricate a large mirror compared with a lens of similar size. Compared with a large lens, the light transmission properties of a mirror system are generally superior because of the absence of absorption losses that occur in lenses.

Mirrors used for spectroscopic applications are different from the typical household mirror. In contrast to the ordinary household mirror where the reflecting surface is behind the glass, a mirror used for spectroscopic applications has the reflecting surface deposited on the front surface of the glass blank. Such a mirror is usually produced by vacuum deposition of a thin layer of a metal like silver or aluminum on a properly shaped glass blank and is known as a *front-surface mirror*. Since the reflecting layer of a front-surface mirror is exposed, it is subject to both physical and chemical attack. As a result, front-surface mirrors should not be exposed to chemical vapors or subject to rough handling.

In our discussion of mirrors, we limit ourselves to concave spherical mirrors since these are the type normally employed in spectroscopic instrumentation. Figure 2.16 shows a spherical concave mirror whose center of curvature is indicated by point C. Since C is the center of curvature, it follows that any line drawn from C to the mirror must be normal to the mirror surface at that point. Consider a parallel bundle of paraxial rays incident on the mirror as shown in Figure 2.16. Each ray that strikes the surface is reflected according to the law of reflection. For any ray, the angle of incidence with the mirror surface can be determined from the angle between the incident ray and a line drawn to the point of incidence from the center of curvature (angle ABC). According to the law of reflection, the angle of reflection will equal the angle of incidence, and the reflected ray will cross the principal axis at point F. By analogy with our discussion of lenses, point F is the focal point of the concave mirror. A plane normal to the principal axis and passing through point F is the focal plane. The distance from F to the vertex of the mirror is the focal length f and is taken as positive by analogy with a converging lens. In contrast to a lens that has both a primary and a secondary focal point, these points are coincident for a spherical mirror. As a result, the spherical mirror has but a single focal point. By the principle of reverse optics, a ray such as DB passing through the focal point and striking the mirror should be reflected parallel to the optic axis.

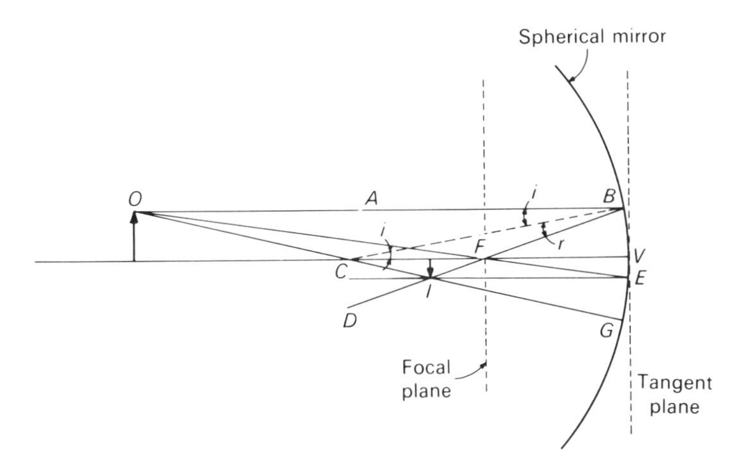

Figure 2.16. Ray diagram illustrating image formation with concave spherical mirror.

From Figure 2.16, it can be seen that angle ABC equals angle CBF. Since the angle of incidence equals the angle of reflection, $CF = BF$. For paraxial rays $BF \sim FV = f$. Since $CV = CF + FV$, $CV = BF + FV \sim 2FV = 2f$. Therefore, for paraxial rays,

$$f \sim \tfrac{1}{2}CV = -\tfrac{1}{2}r \tag{2.41}$$

where r is the radius of curvature of the mirror and is negative according to the sign conventions discussed in Section 2.2.2. The minus sign is introduced to keep f positive by analogy with the sign convention for a converging lens.

Image formation by a concave spherical mirror can be visualized by reference to Figure 2.17. Location of the image may be accomplished graphically by the parallel-ray method discussed in Section 2.2.2. Reflections are drawn from the tangent plane (see Figure 2.17) rather than from the mirror surface itself. By analogy with a lens, rays parallel to the optic axis are drawn through the focal point after striking the tangent plane. Rays that pass through the focal point and strike the tangent plane are drawn parallel to the principal axis. Principal rays that pass through the center of curvature strike the mirror at normal incidence and are simply reflected back upon themselves. The object and image distances are both measured with respect to the tangent plane.

The position of the image may also be calculated using the *mirror formula*,

$$1/p + 1/q = -2/r = 1/f \tag{2.42}$$

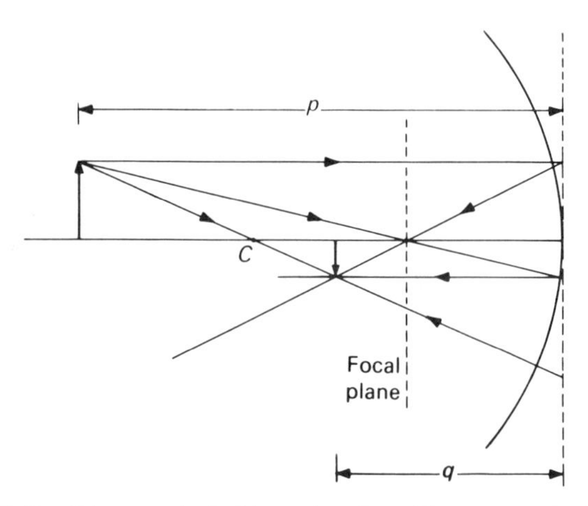

Figure 2.17. Parallel-ray method of image location with concave spherical mirror.

where p is the object distance, q is the image distance, r is the radius of curvature, and f is the focal length. Since the object and the image are both on the same side of the mirror, it is clear that object and image spaces coincide.

As with lenses, similar sign conventions prevail in using the equations for mirrors. By convention, distances measured from left to right are taken as positive. Since the object distance is measured from the object to the vertex and since the image distance is measured from the image to the vertex, both are positive for a spherical concave mirror because both object and image lie to the left of the mirror. Since the focal length is measured from the focal point to the vertex, it is also positive for a spherical concave mirror. By contrast, the radius of curvature is measured from the vertex to the center of curvature. This makes the radius of curvature negative for a concave mirror. Finally incident rays are drawn from left to right while reflected rays are drawn from right to left.

Figure 2.18 shows that the magnification can be determined from two similar triangles, ABV and DEV. From the figure, it can be seen that

$$AV/AB = DV/-DE$$

The minus sign indicates that DE is below the optic axis (i.e., the image is inverted). Since AV is the object distance p, DV is the image distance q, and AB and DE are the object and image heights, respectively,

$$m = q/p = -y'/y \tag{2.31}$$

which is identical with the result obtained for the converging lens.

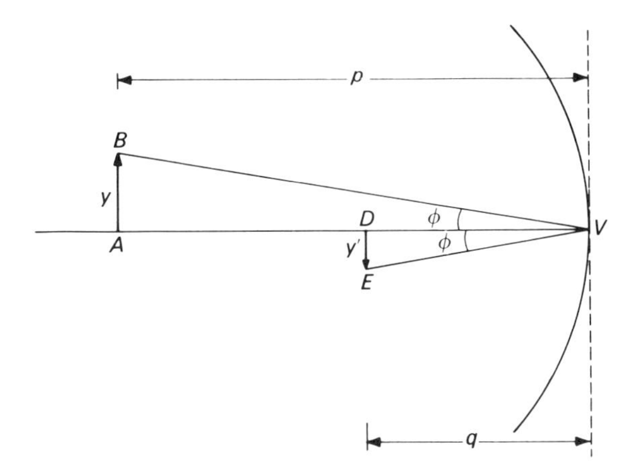

Figure 2.18. Magnification with concave spherical mirror.

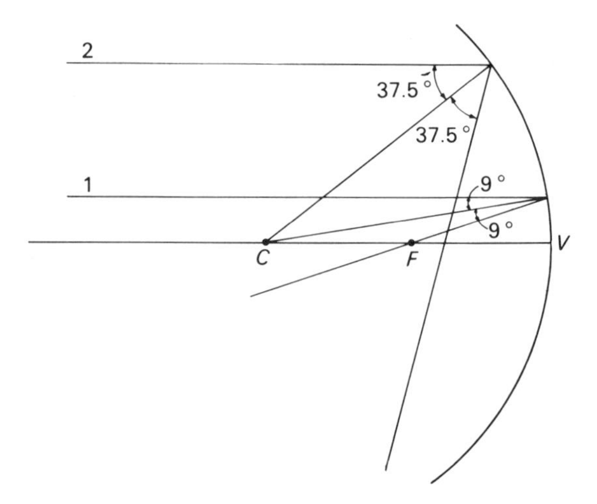

Figure 2.19. Spherical aberration with concave spherical mirror.

The preceding discussion of the image-forming characteristics of spherical mirrors is strictly true only for paraxial rays. When rays that lie at a distance from the optic axis are considered, one finds that they tend to cross the optic axis closer to the mirror, as shown in Figure 2.19. In the figure, ray 1 lies close to the optic axis (a paraxial ray) and strikes the mirror with an angle of 9°. Upon reflection at an angle of 9°, it passes through the paraxial focal point F located at $\frac{1}{2}CV$. By contrast, ray 2 is not a paraxial ray and strikes the mirror at an angle of 37.5°. Upon reflection at 37.5°, ray 2 does not pass through the paraxial focal point. This effect is known as *spherical aberration* and is common with spherical concave mirrors. As a result of spherical aberration, a circle of least confusion occurs between the paraxial focal point and the point where the nonparaxial rays cross the optic axis. Spherical aberration can be reduced by replacing the spherical mirror with a paraboloidal mirror, but this only increases the astigmatism encountered with off-axis rays.

As with lenses (see Section 2.2.2), spherical concave mirrors suffer from astigmatism when off-axis rays are considered. Astigmatism for off-axis rays with a paraboloidal mirror is particularly large.

2.3. SIMPLE OPTICAL SYSTEMS

So far in our discussion we have covered the performance characteristics of a variety of optical components. In this section we examine the effect of combining several optical components to form an optical system.

2.3.1. Thick Lenses

We begin our discussion of simple optical systems by considering the characteristics of a *thick lens*. In Section 2.2.2, a thin lens was defined as one whose thickness was small in comparison to the radii of curvature of both surfaces. By analogy, a thick lens is one that does not satisfy this condition. To understand the behavior of the thick lens, we must consider the image-forming properties of both surfaces independently. Figure 2.20*a* shows the image produced when light from a medium with refractive index n strikes a rod of a medium with a refractive index n' whose front surface is spherical with a radius r_1. The object distance (s_1) and image distance (s'_1) are both measured with respect to the transverse plane through the vertex A_1. The image formed by this system may be located graphically using the parallel-ray method (Section 2.2.2) or it may be determined mathematically with the following equation:

$$\frac{n}{s_1} + \frac{n'}{s'_1} = \frac{n' - n}{r_1} = \frac{n}{f_1} = \frac{n'}{f'_1} \tag{2.43}$$

where the subscript 1 refers to the first surface. Suppose now that the rod in Figure 2.20*a* is shortened so that the rays emanating from M encounter a second surface before they form an image at M'. This situation is shown in Figure 2.20*b*, where rays pass from a medium of refractive index n' to one with a refractive index n''. The second surface has a radius of curvature equal to r_2 and will satisfy the following equation:

$$\frac{n'}{s'_2} + \frac{n''}{s''_2} = \frac{n'' - n'}{r_2} = \frac{n'}{f'_2} = \frac{n''}{f''_2} \tag{2.44}$$

where the subscript 2 refers to the second surface. It should be noted that s'_2 equals s'_1 minus the thickness of the lens, t. In addition, s'_2 and r_2 are both negative according to the sign conventions discussed in Section 2.2.2. Rays 1 and 2 in Figure 2.20*b* correspond to rays 1 and 2 in Figure 2.20*a*. Point A_1 indicates the position of the first vertex, and the dotted line at M indicates the location of the original object. The arrow at M' indicates the location of the image in Figure 2.20*a*. The position of the image formed by the thick lens is determined by the intersection of lines BC and DF''_2. Ray DF''_2 must pass through F''_2 because ray 2 is traveling parallel to the optic axis when it encounters the second surface. Ray 1 is not traveling parallel to the optic axis, and its location after refraction by the second surface must be determined somewhat differently. From our discussion of lenses, we know that a ray passing through the center of curvature C_2 will strike the second surface at

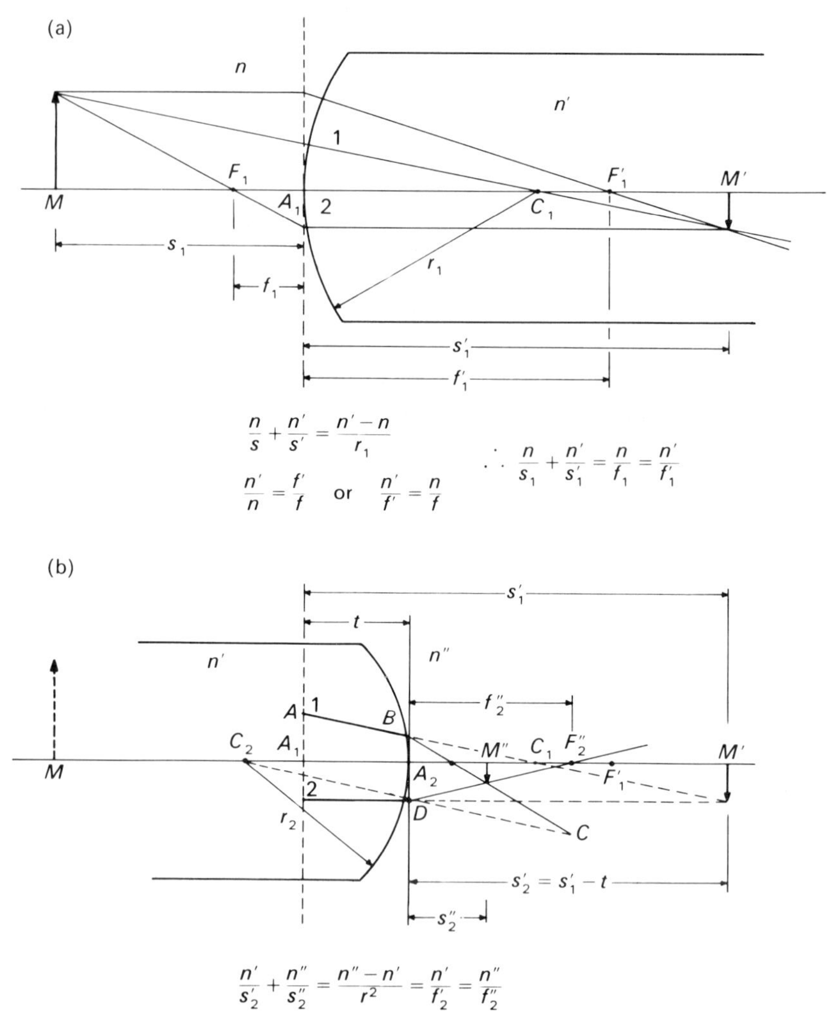

Figure 2.20. (*a*) Ray diagram for light striking rod of medium with refractive index n' whose front surface is spherical. (*b*) Rays emerging from short rod with spherical second surface. Both s_2' and r_2 are negative.

normal incidence and will therefore continue undeviated according to Snell's law. Furthermore, we know that a parallel bundle of rays that strikes a lens will come to a focus in the focal plane of the lens (F_2''). Therefore, if ray C_2C is drawn parallel to ray 1 and extended through the focal plane F_2'', ray BC must interesect it at the focal plane.

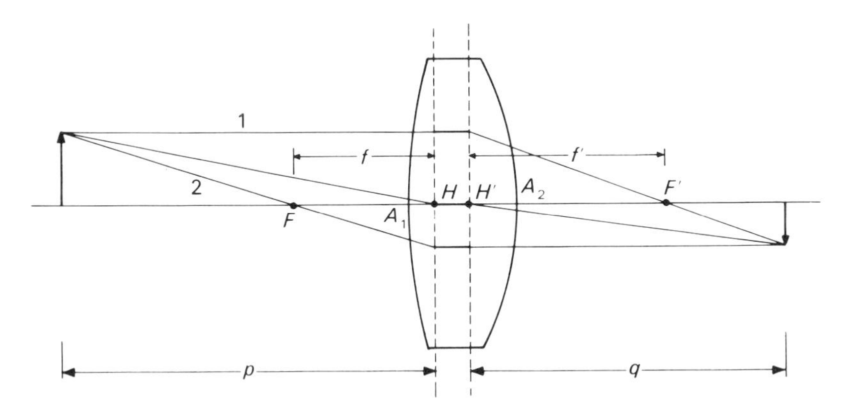

Figure 2.21. Image formation by thick lens.

Image formation by a thick lens is shown in Figure 2.21, where F and F' are the primary and secondary focal points of the lens, respectively. The focal lengths f and f' are measured from two planes passing through points H and H'. These points are known as the *primary* and *secondary principal points*, respectively, and the transverse planes through them are the *primary* and *secondary principal planes*. Since an object at the first principal plane is imaged at the second principal plane at unit magnification, these planes are sometimes referred to as *unit planes*. As indicated in Figure 2.21, the object distance p is measured from the object to the primary principal plane while the image distance q is measured from the secondary principal plane to the image. In constructing a ray diagram as shown in Figure 2.21, it should be noted that all rays in the region between the two principal planes are taken as traveling parallel to the optic axis. If the focal points F and F' are known along with the location of the principal planes, the ray diagram for a thick lens may be constructed by analogy with the parallel-ray method described in Section 2.2.2 for a thin lens with the object and image distances measured as described. Thus, ray 1, traveling parallel to the optic axis, will continue in this direction until it strikes the secondary principal plane whereupon it will be redirected through F'. Likewise, ray 2, which passes through F, will strike the primary principal plane and be redirected parallel to the optic axis.

The location of F and F' and the primary and secondary principal planes is illustrated in Figure 2.22. First the lens, located by vertices A_1 and A_2, is located on the optic axis MM'. Then the focal points for the first and second surfaces are located (F_1 and F'_1 for the first surface and F'_2 and F''_2 for the second surface). These points may be determined from Eqs. 2.43 and 2.44. Then ray AB is drawn from the object parallel to the optic axis. Upon striking the plane through A_1, the ray BC is directed toward the secondary focal point

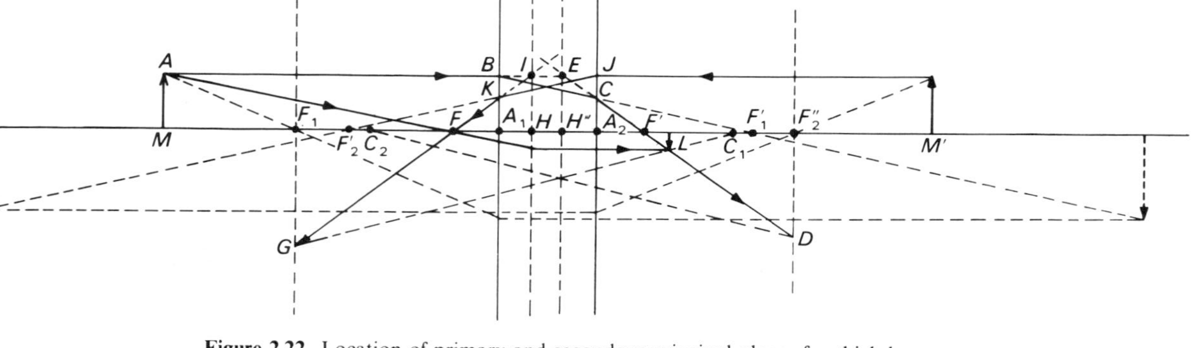

Figure 2.22. Location of primary and secondary principal planes for thick lens.

of the first surface, F'_1. Ray *CD* is located as described earlier by drawing a ray parallel to *BC* that passes through the center of curvature C_2 and intersects *BC* at the focal plane through F''_2. The location of the secondary principal plane is at point *E*, where *BJ* and *CE* intersect. The focal point *F'* of the lens is determined by the point where ray *CD* crosses the optical axis. The primary principal plane and the primary focal point *F* are located by reversing the process and drawing analogous rays from the object at *M'*.

2.3.2. Two Thin Lenses

With this background in mind, we can begin our discussion of simple optical systems by considering a system composed of two thin lenses, as shown in Figure 2.23. To locate the image of the object produced by this combination of lenses, the first step is to determine the position of the image produced by the first lens alone. This can be done easily using the parallel-ray method discussed in Section 2.2.2. Once this image has been located, the effect of the second lens on the rays can be determined. As can be seen from the figure, ray 1 is refracted by lens 1 so that it is parallel to the optic axis. Any ray traveling parallel to the optic axis will be refracted by lens 2 through its secondary focal point (ray 5). To determine how ray 4 is refracted by lens 2, ray 6 is drawn. Ray 6 is parallel to ray 4 and passes through the center of the lens undeviated, striking the focal plane of lens 2 at point *P*. Since a parallel bundle of rays must focus at the same point in the focal plane, ray 7 can be drawn to point *P*. The intersection of rays 5 and 7 represents the location of the image produced by the combination of lenses.

This same result can be achieved with the lens formula given by Eq. 2.29. From Figure 2.23, it can be seen that $p = 12$ units and $f = 6$ units for the first lens. According to Eq. 2.29, the image formed by lens 1 should be located 12 units to the right of lens 1, as shown in the figure. The image produced by lens

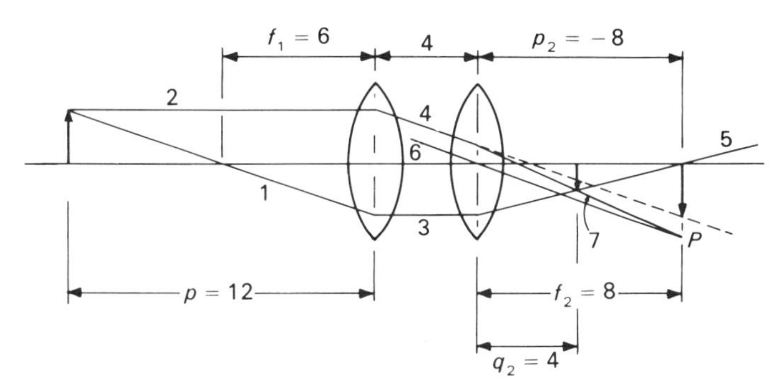

Figure 2.23. Ray diagram for two thin lenses.

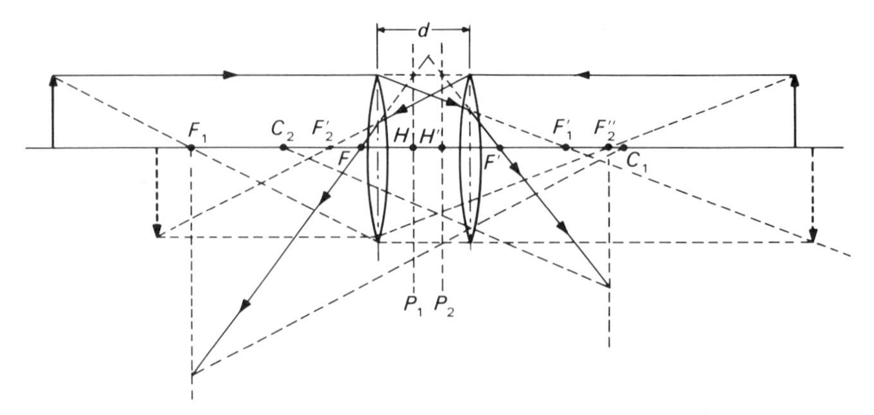

Figure 2.24. Effective focal lengths and location of principal planes for combination of two thin lenses.

1 now becomes the object of lens 2. Since the image is located 8 units to the right of lens 2, the object distance for lens 2 is $p = -8$. Substituting this distance in Eq. 2.29 along with the focal length for lens 2 gives an image distance q of 4 units. Thus a real, inverted image results from the combination of the two lenses with the image forming 4 units to the right of lens 2.

Figure 2.24 shows the image-forming properties of a combination of two thin lenses along with the effective focal lengths of the system and the location of the principal planes P_1 and P_2. These parameters have been located graphically according to the methods described in the preceding discussion. Table 2.1 summarizes some important formulas for thin lenses, thick lenses, and two thin lenses.

2.3.3. Limiting Apertures

The presence of limiting apertures or stops is of considerable importance in determining the light transmission properties of an optical system. These limiting apertures may be the rims of the lenses themselves or they may be deliberately introduced in the form of *diaphragms*. A diaphragm is a limited opening that controls the size of the bundle of rays that can pass through the plane where the diaphragm is located.

Another name for a limiting aperture is a *stop*. There are two general classes of stops known as *aperture stops* and *field stops*. An aperture stop limits the cone angle of rays from an axial object point that can pass through the system and therefore determines the amount of light that can be transmitted by the optical system. A field stop, on the other hand, determines the limit of the field of view (i.e., the extent of the object) that can be transmitted by the system.

Table 2.1. Some Gaussian Optics Formulas[a]

	Thin Lens	Thick Lens	Combination, Two Lenses
$\frac{1}{f}$	$(n-1)\left(\frac{1}{r_1}-\frac{1}{r_2}\right)$	$(n-1)\left[\frac{1}{r_1}-\frac{1}{r_2}+\frac{(n-1)d}{nr_1r_2}\right]$	$\frac{1}{f_1}+\frac{1}{f_2}-\frac{d}{f_1f_2}$
d_{p1}	0	$\frac{n-1}{n}\frac{tf}{r_2}$	$\frac{-fd}{f_2}$
d_{p2}	0	$\frac{n-1}{n}\frac{tf}{r_1}$	$\frac{fd}{f_1}$

[a] Symbols: n, refractive index; r, radius of curvature; d_{p1}, distance from first vertex to first principal plane; d_{p2}, distance from second principal plane to second vertex; d, t, lens thickness or lens spacing.

[Reprinted from Leo Levi, *Applied Optics, Vol. 1*, copyright 1968. By permission of John Wiley & Sons, Inc.]

Figure 2.25 shows a simple optical system composed of an aperture stop, a lens, and a field stop. It can be seen from the figure that the aperture stop limits the amount of light that can enter the system while the field stop limits the field of view.

The image of the aperture stop as seen from object space is the *entrance pupil*, while the image of the aperture stop as seen from image space is known as the *exit pupil*. These considerations are illustrated in Figure 2.26. Figure 2.26*a* shows the formation of two virtual images of the stop between the lenses. The images are located graphically by drawing a ray parallel to the

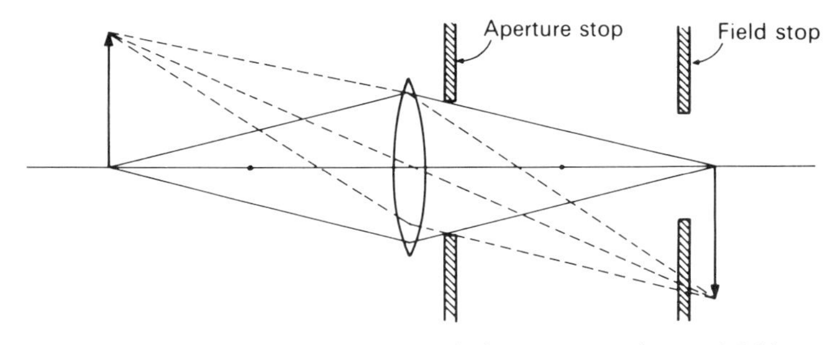

Figure 2.25. Simple optical system composed of aperture stop, lens, and field stop.

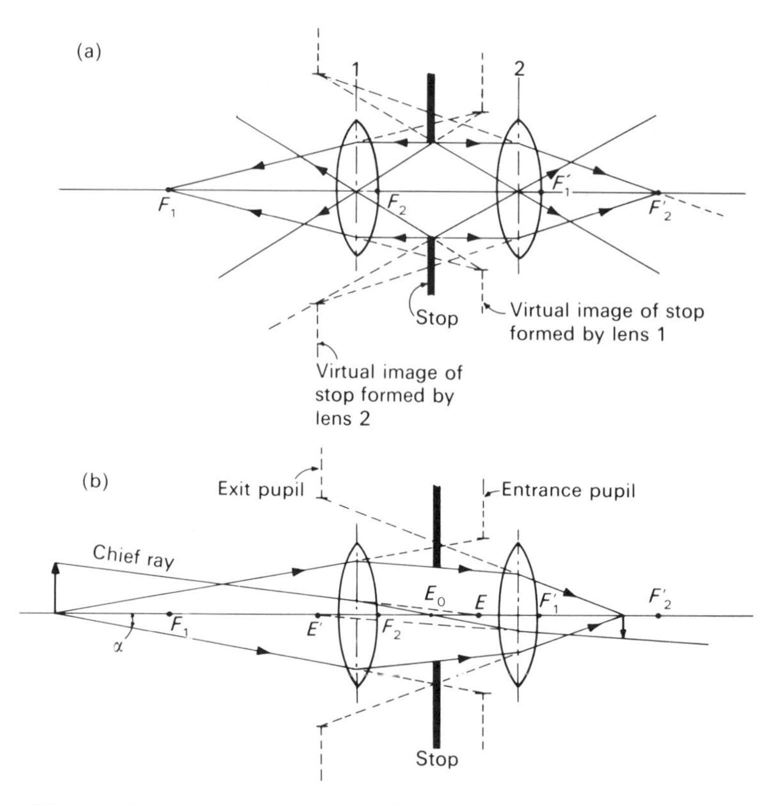

Figure 2.26. Location of entrance and exit pupils for simple optical system composed of two lenses with intermediate stop.

optic axis and then through the respective focal point for the given lens. A second ray from the object through the center of the lens locates the position of the virtual image at the point where the extensions of the two rays cross. The virtual image of the stop formed by lens 1 as seen from object space (looking through lens 1 from left to right) is the entrance pupil of the system. Similarly, it can be seen from the figure that the virtual image of the stop formed by lens 2 as seen from image space (i.e., looking through lens 2 from right to left) is the exit pupil.

Figure 2.26*b* shows the same system with the entrance and exit pupils drawn in as determined in Figure 2.26*a*. Notice that the maximum bundle of rays that can be accepted by the system is determined by the stop between the two lenses and not by the lenses themselves. The angle α set by the largest cone of rays that fall within the entrance pupil determines the *numerical aperture* of the system as given by

$$\mathrm{NA} = n \sin \alpha \tag{2.45}$$

where n is the refractive index of object space. Rays outside of this cone fall outside the entrance pupil and are blocked by the stop.

Any ray that passes through the center of the aperture stop (point E_0 in Figure 2.26*b*) is known as a *chief ray*. Looking from object space, the chief ray appears to have originated from the center of the entrance pupil (point E in Figure 2.26*b*). Looking from image space, the chief ray appears to have originated from the center of the exit pupil (point E' in Figure 2.26*b*). Thus points E_0, E, and E' are corresponding points.

In any given optical system, some aperture will limit the cone of chief rays able to pass through the system. This aperture will be the field stop because it will be the one that ultimately limits the field of view. The image of the field stop in object space is the *entrance window*. The image of the field stop in image space is the *exit window*. The angle subtended by the exit window in the plane of the exit pupil is the *field of view*. These considerations are shown in Figure 2.27. The entrance window has been located by determining the position of the real image of the field stop formed by lens 1. The figure shows that axial rays originating from the object and passing through the entrance window all ultimately pass through the optical system to produce the final image. It can be seen that this is not the case for all rays originating from the tip of the arrow. In this case, rays 1 and 2 do not pass through the entrance

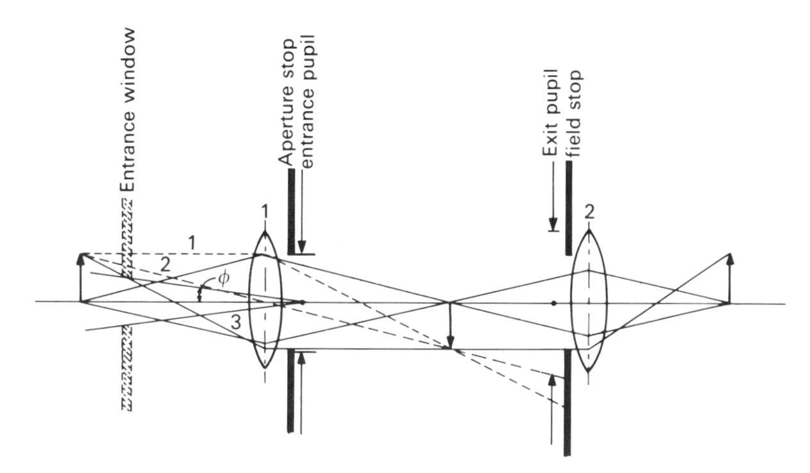

Figure 2.27. Location of entrance window for simple optical system composed of two lenses, aperture stop, and field stop.

window. Although they are transmitted by lens 1 and the aperture stop, they are blocked by the field stop and do not result in image formation. Only ray 3, which passes through the entrance window, ultimately passes through the system and results in image formation. As a result of the fact that not all the rays from all points on the object can pass through the optical system, there will be a gradual fading of the image for off-axial points. This reduction in image irradiance for the off-axis points as a result of the limited field of view is known as *vignetting*. In the figure, ϕ is the angle subtended by the entrance window from the entrance pupil. The angle 2ϕ is called the *field angle* or *angular field of view*.

2.3.4. Light-Gathering Power

From the previous discussion, it is evident that the amount of light that reaches the image from the object is dependent not only on the brightness of the object but also on the size of the optical components and the presence of any limiting stops. There are various quantitative ways of expressing the light-gathering power of an optical system. For example, the angle 2α (see Figure 2.26) is called the *angular aperture on the object side* and is a measure of the amount of light collected from the object. Another measure of the light-gathering power of an optical system is the numerical aperture defined by Eq. 2.45. For spectroscopic work, the most convenient measure of the light-gathering power of the spectrograph is the *f-number*, or *nominal focal ratio*.

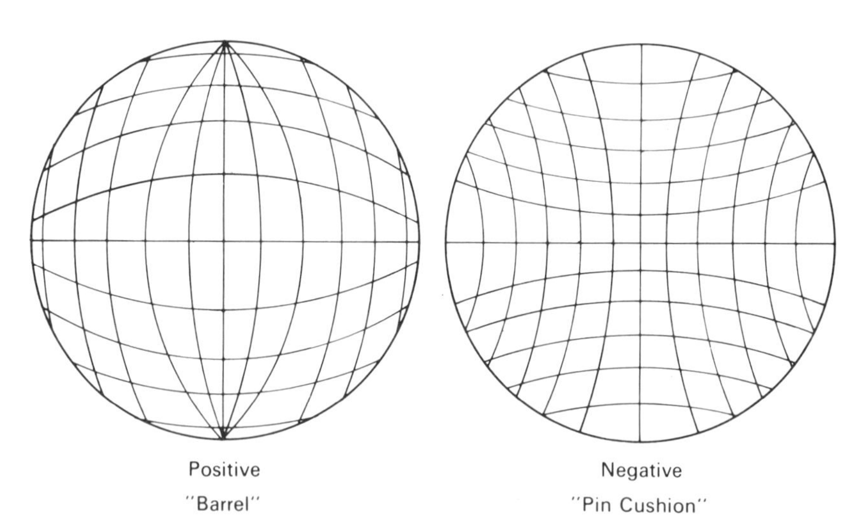

Figure 2.28. Effect of positive and negative distortion on image of square grid: (*a*) positive "barrel"; (*b*) negative "pincushion."

The f-number of an optical system is defined as

$$f/ = f/d \tag{2.46}$$

where f is the focal length of the objective and d is the diameter of the entrance pupil. From Eq. 2.3, it is apparent that all of the various ways of expressing the light-gathering power of the system are related to the solid angle collected by the entrance pupil.

2.3.5. Distortion

The phenomenon of distortion is often encountered in various image detector systems used for spectroscopic work and deserves brief mention. *Distortion* is the dislocation of the chief ray from the ideal image point to some other location. If the ideal image point lies farther from the axis than the actual image point, the distortion is considered *positive*. Conversely, if the ideal image point lies closer to the axis than the actual image point, the distortion is *negative*. Figure 2.28 shows the image of a square grid produced by positive and negative distortion. It can be seen from the figure that for positive distortion the off-axis points lie too close to the center, while for negative distortion they lie too far from the center. Because of the shapes produced by positive and negative distortion, they are often referred to as *barrel* and *pincushion* distortion, respectively.

REFERENCES

Born, M. (1933), *Optiks*, Springer, Berlin.

Boyd, R. W. (1983), *Radiometry and the Detection of Optical Radiation*, Wiley, New York.

Grum, F. and Becherer, R. J. (1979), *Radiometry*, Academic, New York.

Jenkins, F. A. and White, H. E. (1957), *Fundamentals of Optics*, McGraw-Hill, New York.

Levi, L. (1968), *Applied Optics—A Guide to System Design*, Vol. 1, Wiley, New York.

Levi, L. (1980), *Applied Optics—A Guide to System Design*, Vol. 2, Wiley, New York.

O'Shea, D. C. (1985), *Elements of Modern Optical Design*, Wiley-Interscience, New York.

Wood, R. W. (1934), *Physical Optics*, Dover, New York.

CHAPTER

3

PRINCIPLES OF DIFFRACTION

So far our discussion has been centered on the geometric relations between light rays as governed by the laws of reflection and refraction. These topics comprise the area of geometric optics and are sufficient to explain the action of prisms, lenses, and mirrors on a macroscopic level. In this chapter we turn our attention to questions related to the ability of an optical system to distinguish two closely spaced images and explain the principles behind diffraction and the diffraction grating. To understand these topics requires an understanding of the nature of light itself, which is the subject of *physical optics*.

3.1. PROPERTIES OF WAVES

A large number of the observed optical phenomena associated with the behavior of light can be explained on the assumption that light consists of waves. To begin the discussion of this topic, it is worthwhile to review some of the properties of waves. Figure 3.1 shows how a sine wave may be generated by the counterclockwise rotation of a vector of length A_0 around a circle at an angular frequency of ω. The angular frequency ω has units of radians per second and is equal to $2\pi f$, where f is the frequency in hertz. The quantity ϕ is known as the *phase angle*, which specifies the position of the vector at time $t = 0$. The length of the vector is the amplitude of the sine wave. The wave generated by the rotation of the vector can be completely described by

$$y = A_0 \sin(\omega t - \phi) \tag{3.1}$$

The phase angle can be expressed in either angular measure as discussed previously or in terms of the linear distance along the x axis by which the sine wave must be shifted. Since the distance traveled by the wave in a complete revolution of the vector is the wavelength λ and since a complete revolution is 2π radians, it follows that

$$(\phi/2\pi)\lambda = x \tag{3.2}$$

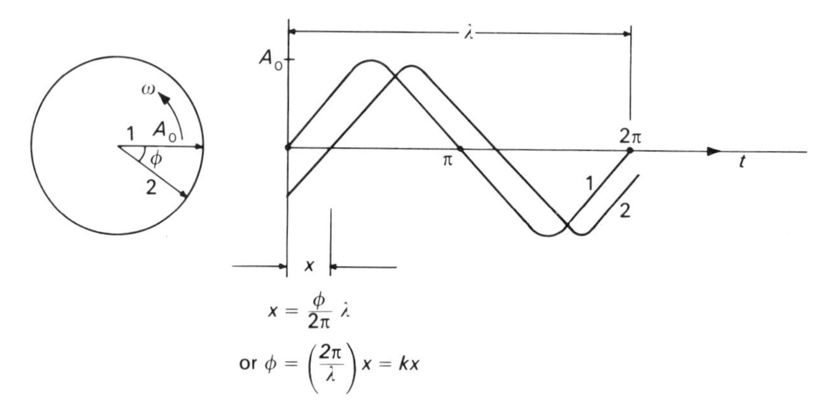

Figure 3.1. Generation of sine wave by counterclockwise rotation of vector.

$$\phi = (2\pi/\lambda)x = kx \tag{3.3}$$

where λ is the wavelength and k is known as the *propagation number*. Using Eq. 3.3, Eq. 3.1 becomes

$$y = A_0 \sin(\omega t - kx) \tag{3.4}$$

For two waves of the same frequency, the phase difference between them will be

$$\begin{aligned} \phi_2 - \phi_1 &= k(x_2 - x_1) = (2\pi/\lambda)(x_2 - x_1) \\ \delta &= (2\pi/\lambda)\Delta = k\Delta \end{aligned} \tag{3.5}$$

where $\delta = \phi_2 - \phi_1$ and $\Delta = x_2 - x_1$. The quantity δ refers to the *phase difference*, or the difference between the phase angles for the two waves. The quantity Δ refers to the *path difference* between the two waves.

The path difference referred to in Eq. 3.5 is actually the difference between the *optical paths* for the waves rather than the difference between the actual geometric paths. The optical path is defined as the geometric path d times the refractive index of the medium. The optical path refers to the distance that the light wave would have traveled in the given time had it been propagating through a vacuum. If a light wave is propagating through media of different refractive indices (and hence traveling at different speeds), the total optical path will be the sum of the optical paths in the various segments. Thus, if x is the total optical path,

$$x_j = \sum_i n_{ji} d_{ji} \tag{3.6}$$

Substituting Eq. 3.6 into Eq. 3.5 gives

$$\delta = k\left(\sum_i n_{2i} d_{2i} - \sum_i n_{1i} d_{1i}\right) \tag{3.7}$$

where the subscripts 1 and 2 refer to the two waves.

We must now relate the energy resulting from a wave such as that described by Eq. 3.1 to the amplitude of the wave. According to classical physics, the energy of a wave is proportional to the square of the amplitude. If a particle is displaced from its rest position by a continuous wave, it will acquire a certain total energy as a result of its motion. The total energy will be the sum of its potential energy and its kinetic energy. The kinetic energy will be a maximum when the velocity of the particle is greatest, and the potential energy of the particle will be a maximum when the instantaneous velocity is zero. The velocity of the particle may be determined by taking the derivative of the vertical displacement as given by Eq. 3.1:

$$dy/dt = \omega A_0 \cos(\omega t - \phi) \tag{3.8}$$

Since the cosine of an angle varies between 0 and 1, the maximum velocity will therefore occur when $\cos(\omega t - \phi) = 1$. The maximum kinetic energy will occur when the potential energy is zero, and as a result, the maximum kinetic energy will equal the total energy of the particle. The total energy of the particle will therefore be given by

$$\tfrac{1}{2} m v_{\max}^2 = \tfrac{1}{2} m \omega^2 A_0^2 \tag{3.9}$$

The total energy of the wave is therefore proportional to the square of its amplitude.

The total energy that flows per second per unit area perpendicular to the direction of travel has been defined as the irradiance E given by Eq. 2.1. Since E is proportional to the total energy of the radiation field, it follows that E is proportional to the square of the amplitude of the wave, as shown by Eq. 3.9. As was pointed out in Section 2.1, the irradiance of a spherical wave obeys the inverse-square law. To apply Eq. 3.1 to a spherical wave, it must be modified as follows:

$$y = (A_0/x)\sin(\omega t - kx) \tag{3.10}$$

where x is the distance from the source. In this way, E is proportional to A_0^2/x^2 and is consistent with Eq. 2.8, which shows that the irradiance varies inversely with the square of the distance. Here, A_0 is the amplitude for a sphere of unit distance.

3.2. DIFFRACTION

Up to now in our discussion, we have implicitly assumed that light rays always travel in straight lines. Although on a macroscopic level this is certainly true and is the basis for geometric optics, it can be readily demonstrated that whenever light waves pass through a small opening or travel past an obstacle, they tend to spread into regions not illuminated directly by the geometric ray. This phenomenon is known as *diffraction* and can be explained on the basis of *Huygen's principle*, which states that each point on a wavefront passing through a small aperture may be regarded as a point source of spherical waves. The application of Huygen's principle is illustrated in Figure 3.2 for light that strikes an opening when the source is at infinity. The diffraction observed when the source and the screen on which the pattern is observed are both located at infinity falls in the category of *Fraunhofer diffraction* and is the simplest to treat mathematically. In this section, we discuss Fraunhofer diffraction from a single slit.

Figure 3.3 shows the experimental arrangement for observing Fraunhofer diffraction from a single slit. Light from a monochromatic source is focused by a lens on a slit that acts as a secondary source. Light from the slit is *collimated* (i.e., made parallel) by lens L_1 before it strikes the diffracting slit S. Light from the diffracting slit is focused by lens L_2 onto the screen, which is placed at the secondary focal point of L_2. By using this optical arrangement, the condition that the source and screen both be located at infinity is satisfied. The pattern observed on the screen when the slit is illuminated consists of a series of light and dark lines that are parallel to the length of the slit. The brightest line is the called the *central maximum* and is twice as wide as the *secondary maxima*, which lie symmetrically on either side.

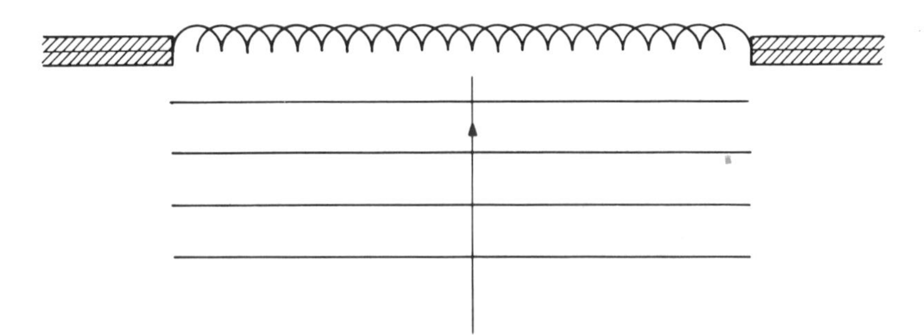

Figure 3.2. Application of Huygen's principle to wavefront passing through a small aperture.

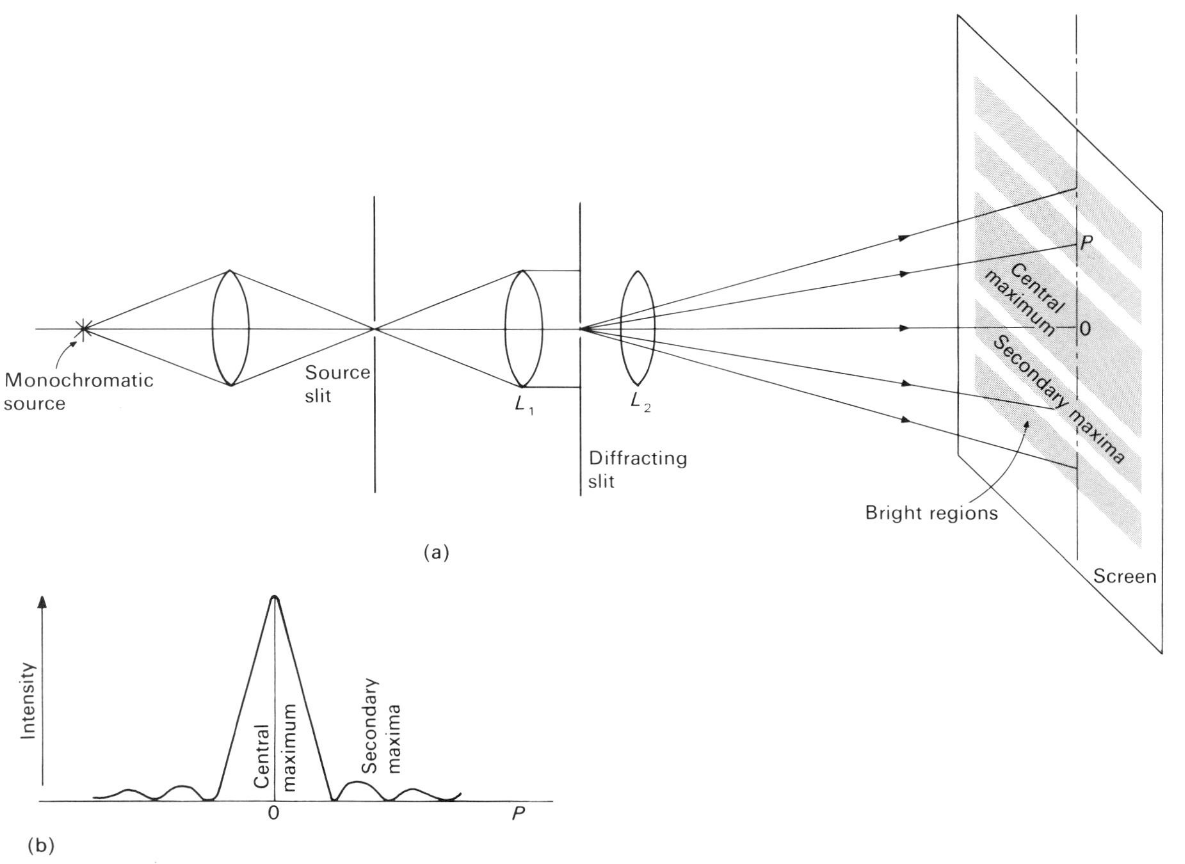

Figure 3.3. Experimental arrangement for observing Fraunhofer diffraction from single slit.

Since diffraction effects represent the limit of the performance of any optical system, they are important in discussing questions related to the ability of optical systems to distinguish closely spaced images. To obtain an understanding of the phenomenon, the single-slit diffraction pattern may be studied quantitatively with reference to Figure 3.3.

Figure 3.4 represents an enlargement of a portion of Figure 3.3, where b is the width of the slit that is illuminated by parallel monochromatic radiation from L_1. If we imagine, according Huygen's principle, that when a wavefront occupies the plane of the slit it forms a series of secondary spherical waves, then those parts of the secondary wavefronts traveling normal to the plane of the slit will be imaged at P_0 by lens L_2. Those traveling at some other angle θ will be imaged at P.

Consider a secondary wavelet emitted from an element of width ds located a distance s from the center of the slit C, as shown in Figure 3.4. If the wavelet is a spherical wave, its amplitude will be proportional to ds and inversely proportional to the distance to the screen, x. According to Eq. 3.10, a spherical wave of this type should be given by

$$dy = (A_0\, ds/x) \sin(\omega t - kx) \tag{3.11}$$

In Eq. 3.11, the factor kx (which is related to the phase angle of the wave) will depend on the position from which the wave originates compared to the origin C. If the wave originates from a point s below the origin, Eq. 3.11 may be written

$$dy = (A_0\, ds/x) \sin[\omega t - k(x + \Delta)] \tag{3.12}$$

where Δ is the path difference compared with a similar ray at the origin. Substituting for Δ gives

$$dy_- = (A_0\, ds/x) \sin[\omega t - kx - ks(\sin\theta)] \tag{3.13}$$

If the wave originates a distance of s above the origin, the corresponding equation will be

$$dy_+ = (A_0\, ds/x) \sin[\omega t - k(x - \Delta)] \tag{3.14}$$

$$dy_+ = (A_0\, ds/x) \sin[\omega t - kx + ks(\sin\theta)] \tag{3.15}$$

To obtain the total effect from all portions of the slit, we must integrate Eqs. 3.13 and 3.15 with respect to s. Because of the symmetry of the problem, it is

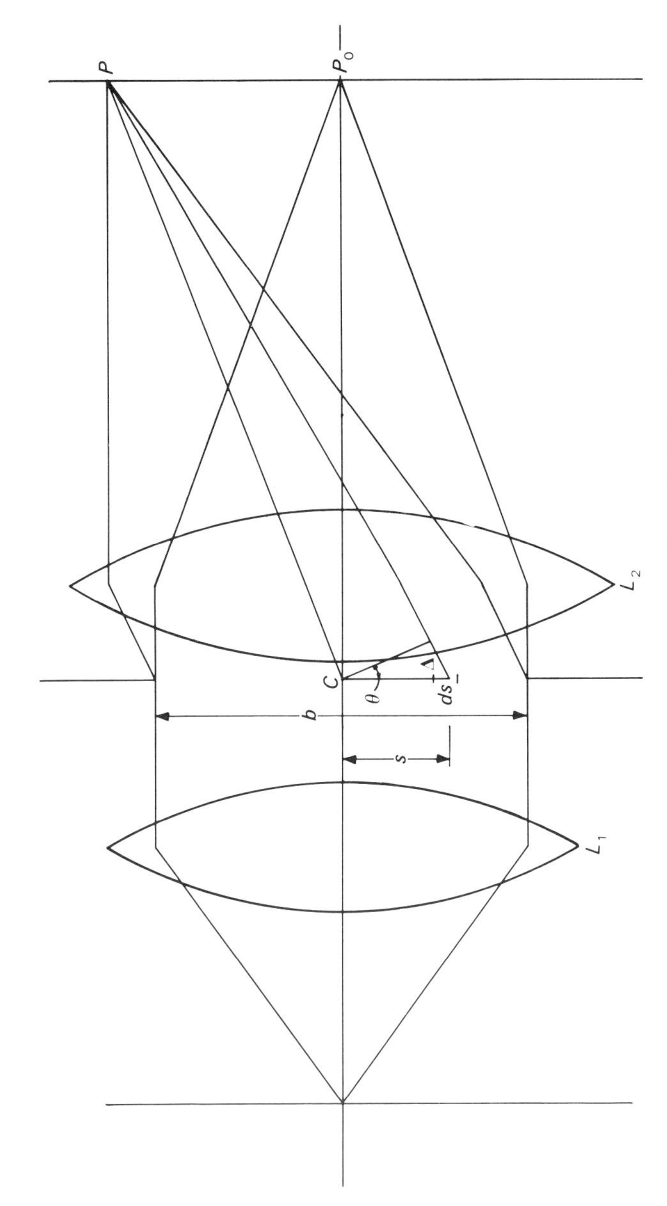

Figure 3.4. Geometry associated with Fraunhofer diffraction from a single slit. (After Jenkins and White, 1976.)

most convenient to integrate across the slit in a pairwise manner. Thus,

$$dy = dy_- + dy_+ \tag{3.16}$$

$$= (A_0 ds/x)\sin[\omega t - kx - ks(\sin\theta)] + (A_0 ds/x)\sin[\omega t - kx + ks(\sin\theta)] \tag{3.17}$$

Since $\sin a + \sin b = 2\cos\frac{1}{2}(a-b)\sin\frac{1}{2}(a+b)$, Eq. 3.17 can be written as

$$dy = (A_0\, ds/x)[2\cos(ks\sin\theta)\sin(\omega t - kx)] \tag{3.18}$$

Integrating Eq. 3.18 with respect to s from $s = 0$ to $s = \frac{1}{2}b$ gives

$$y = (2A_0/x)\sin(\omega t - kx)\int_0^{b/2}\cos(ks\sin\theta)\,ds \tag{3.19}$$

where x and θ can be regarded as constants. Thus,

$$y = \frac{(2A_0/x)\sin(\omega t - kx)}{k\sin\theta}\int_0^{b/2}\cos(ks\sin\theta)\,k\sin\theta\,ds \tag{3.20}$$

$$= \frac{(2A_0/x)\sin(\omega t - kx)}{k\sin\theta}[\sin(ks\sin\theta)]_0^{b/2} \tag{3.21}$$

$$= \frac{(2A_0/x)\sin(\omega t - kx)}{k\sin\theta}\left[\sin\frac{kb\sin\theta}{2}\right] \tag{3.22}$$

$$= \left(\frac{A_0 b}{x}\right)\frac{\sin[\frac{1}{2}(kb\sin\theta)]}{\frac{1}{2}(kb\sin\theta)}\sin(\omega t - kx) \tag{3.23}$$

Equation 3.23 is the equation for a spherical wave whose amplitude is a function of θ and is given by

$$\left(\frac{A_0 b}{x}\right)\frac{\sin[\frac{1}{2}(kb\sin\theta)]}{\frac{1}{2}(kb\sin\theta)} = \frac{L}{\beta}\sin\beta \tag{3.24}$$

where $L = A_0 b/x$ and $\beta = \frac{1}{2}kb\sin\theta = (\pi b\sin\theta)/\lambda$.

From Eq. 3.24, the irradiance produced on the screen will be proportional to

$$(L^2\sin^2\beta)/\beta^2 \tag{3.25}$$

Equations 3.24 and 3.25 may be used to explain the intensities observed in the diffraction pattern shown in Figure 3.5. As β approaches zero, the amplitude according to Eq. 3.24 will be given by

$$\lim_{\beta \to 0} \frac{L}{\beta} \sin \beta = \lim_{\beta \to 0} L \frac{\beta}{\beta} = L \tag{3.26}$$

since $\sin \beta$ approaches β as β becomes small. Therefore, the irradiance at the central or *principal maximum* will be proportional to L^2. Minima in the irradiance will occur when $\sin^2 \beta = 0$, while secondary maxima will occur when the derivative of Eq. 3.24 equals zero. Thus minima will occur at $\beta = \pm m\pi$, where m is an integer greater than zero. Taking the derivative of Eq. 3.24 and setting it equal to zero reveals that secondary maxima will occur when

$$\tan \beta = \beta \tag{3.27}$$

Although Eq. 3.27 reveals that the secondary maxima do not occur precisely halfway between successive minima as can be seen in Figure 3.5, not much error is introduced by assuming that secondary maxima occur when $\beta = \pm\frac{3}{2}\pi, \pm\frac{5}{2}\pi, \pm\frac{7}{2}\pi, \cdots$. If this assumption is combined with Eq. 3.25, the relative irradiances of the secondary maxima can be estimated as $(4/9\pi^2)L^2$,

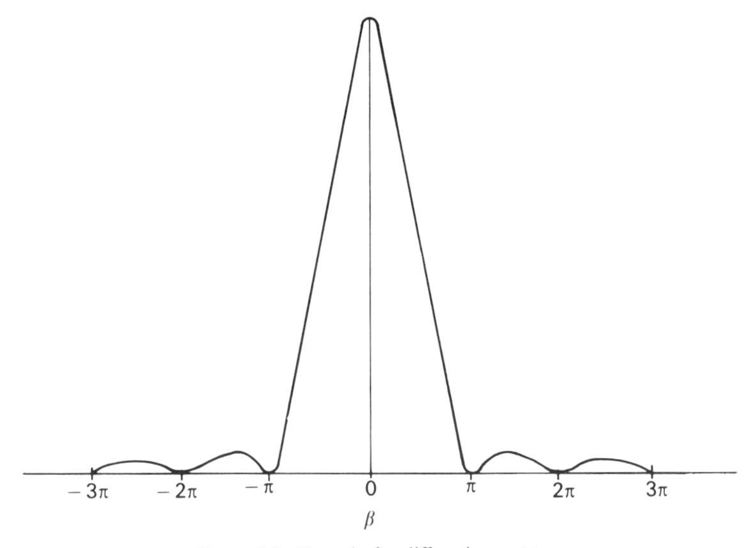

Figure 3.5. Fraunhofer diffraction pattern.

$(4/25\pi^2)L^2$, $(4/49\pi^2)L^2$ for the first, second, and third secondary maxima, respectively. Thus, the first secondary maximum would be predicted to have a relative irradiance of approximately 1/22.2 of the principal maximum, which corresponds to 4.51%. Using Eq. 3.27, the actual value can be shown to be 4.72% of the principal maximum. For our purposes, assuming the positions of the secondary maxima to be at $\pm 3/2\pi$, $\pm 5/2\pi$, $\pm 7/2\pi$, $\cdots$ is certainly adequate.

According to Eq. 3.24, the first minimum will occur when $\beta = \pi$. From the definition of β, this will occur when θ satisfies the following equations:

$$\beta = \pi = \tfrac{1}{2} kb \sin\theta = (\pi b \sin\theta)/\lambda \tag{3.28}$$

$$\lambda = b \sin\theta \sim b\theta \tag{3.29}$$

$$\theta = \lambda/b \tag{3.30}$$

Equation 3.30 gives the angular separation from P_0 on the screen to the first minimum. The linear distance d measured from P_0 on the screen to the first minimum will be given by

$$d = f \sin\theta \sim f\theta = f\lambda/b \tag{3.31}$$

where f is the focal length of the lens. This distance d is also the distance between successive minima since $\Delta\beta = \pi$ for any two successive minima.

From Eq. 3.31, it follows that the baseline width of the principal maximum will be $2f\lambda/b$. Thus the width of the central maximum will be directly proportional to the wavelength of the light and inversely proportional to the width of the opening responsible for the diffraction.

3.3. RESOLVING POWER

The diffraction pattern shown in Figure 3.5 reveals something about the ability of an optical system to form images of closely spaced objects. As seen from the foregoing discussion, the performance of any optical system is limited theoretically by diffraction effects at apertures within the system. As we shall see in Chapter 4, a conventional spectroscopic optical system forms images of the entrance slit in the focal plane of the instrument. As previously discussed, each image of the entrance slit formed by the optical system will be a diffraction pattern similar to Figure 3.5.

Figure 3.6 shows the effect of displacing two diffraction patterns whose central maxima have the same amplitude and base width. When the central

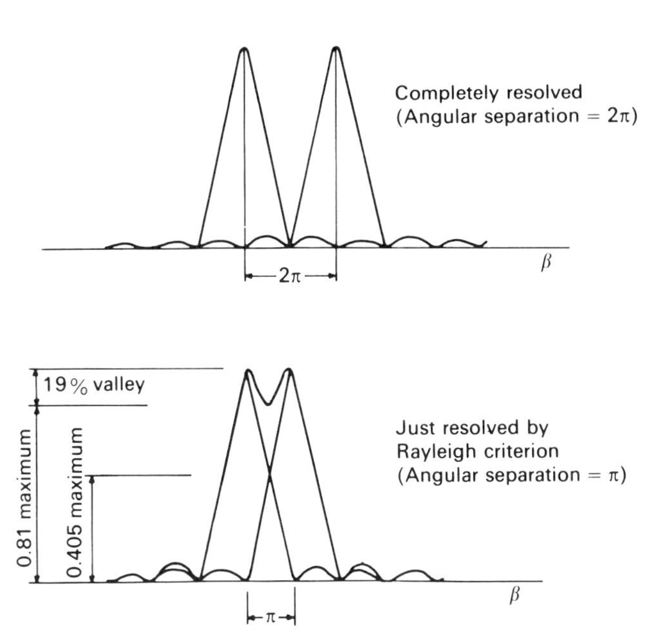

Figure 3.6. Effect of displacing two diffraction patterns whose central maxima have same amplitude and base width: (*a*) completely resolved, angular separation 2π; (*b*) just resolved by Raleigh criterion, angular separation π.

maxima are separated by $\beta = 2\pi$, the maximum of one pattern falls on the second minimum of the other, resulting in a zero irradiance between the two central maxima. This is the smallest angle ($\beta = 2\pi$) that will give a zero irradiance between the two maxima. Diffraction patterns separated by an angular separation of at least 2π are *completely resolved.*

For angles less than 2π, the central maxima will overlap and the total irradiance between them will be the sum of the irradiances from the individual diffraction patterns. Thus for separations much less than $\beta = 2\pi$, the central maxima will overlap to a significant extent and eventually coalesce into a single broad maximum.

In establishing his criterion for the minimum separation between two diffraction patterns in order to be considered *just resolved*, Rayleigh arbitrarily set the angular separation between the two diffraction patterns at $\beta = \pi$. In this way, the central maximum of one pattern falls on the first minimum of the second pattern. Since the curves cross at a point midway between their maxima ($\beta = \frac{1}{2}\pi$), the irradiance at that point will be given by

$$L^2(\sin^2\beta)/\beta^2 = L^2(4/\pi^2) = 0.405L^2$$

or 0.405 of the intensity of the central maximum. The total irradiance at the crossing point will be the sum of the contribution from each pattern, or 0.81 of either maximum. Thus when $\beta = \pi$, there will be a 19% valley between the two maxima. Although it may be possible theoretically to resolve diffraction patterns whose angular separation is less than $\beta = \pi$, Rayleigh's criterion is a convenient, albeit arbitrary, limit. The *resolving power*, or the ability of an optical system to produce images of objects that are close together, is generally based on *Rayleigh's criterion.* Images that satisfy Rayleigh's criterion are said to be *just resolved.*

Optical systems whose performance is limited by the properties of the light itself are *diffraction limited.* Once the diffraction limit of a given optical system has been reached, no further improvement in performance is possible for the given optical components.

3.4. DIFFRACTION GRATINGS

A *diffraction grating* consists of a regular array of parallel slits of the same width with equal spacing between successive slits. Fraunhofer was the first to study the diffraction of light with a grating made from an array of fine wires. Before discussing a typical diffraction grating that consists of an array of thousands of equidistant slits, it is worthwhile to consider the situation that exists for the smallest grating, which consists of an array of only two slits.

Figure 3.7 shows a schematic of the apparatus for observing Fraunhofer diffraction from a double slit. In this experiment, two slits of width b are

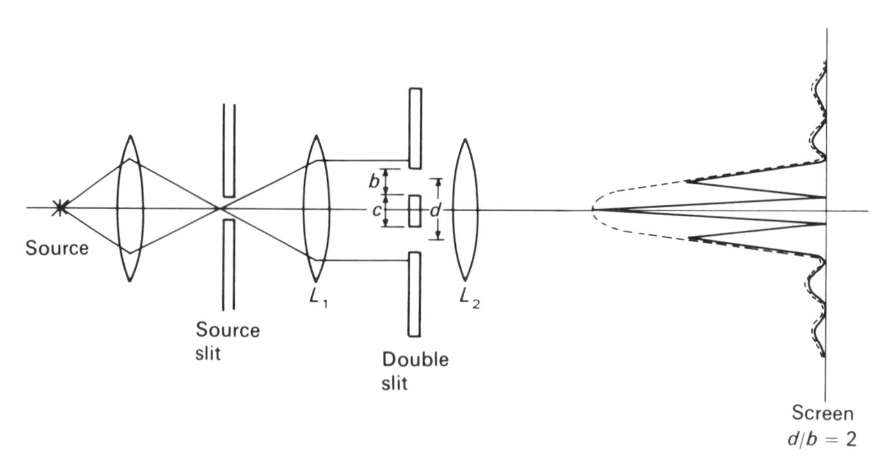

Figure 3.7. Experimental arrangement for observing Fraunhofer diffraction from double slit. (After Jenkins and White, 1976.)

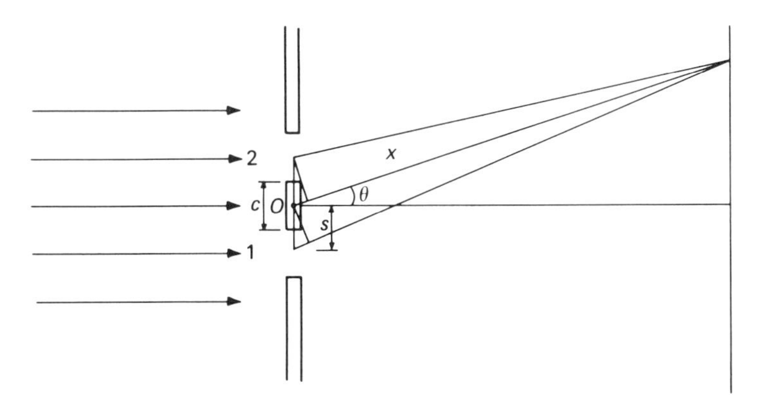

Figure 3.8. Geometric construction for investigating Fraunhofer double-slit diffraction pattern.

separated by an opaque section of width c, so that the distance between slit centers is $d = c + b$.

Figure 3.8 shows the geometric construction for investigating the double-slit diffraction pattern. In the figure, point O is located in the center of length c and will be taken as the origin. By analogy with the single-slit situation, a ray originating a distance s below the origin will travel $\Delta = s \sin \theta$ further than x. Similarly, a ray originating a distance s above the origin will travel $\Delta = s \sin \theta$ less than x. In the double-slit situation, rays originating from below the origin are emanating from slit 1, while those originating from above the origin are emanating from slit 2. To obtain the total effect from all portions of both slits, we must integrate the appropriate expression. Because of the symmetry of the situation, it is convenient to consider both slits simultaneously. Thus,

$$dy = dy_- + dy_+ \tag{3.32}$$

where dy_- and dy_+ represent rays emanating from homologous portions of slits 1 and 2, and the minus and plus signs are used to indicate elements below and above the origin, respectively. By analogy with the single-slit case,

$$\begin{aligned} dy = dy_- + dy_+ &= (A_0\, ds/x) \sin [\omega t - k(x + \Delta)] \\ &\quad + (A_0\, ds/x) \sin [\omega t - k(x - \Delta)] \end{aligned} \tag{3.33}$$

$$\begin{aligned} &= (A_0\, ds/x) \sin [\omega t - kx - ks \sin \theta] \\ &\quad + (A_0\, ds/x) \sin [\omega t - kx + ks \sin \theta] \end{aligned} \tag{3.34}$$

$$= (A_0\, ds/x) [2 \cos (ks \sin \theta) \sin (\omega t - kx)] \tag{3.35}$$

Since we are considering both slits simultaneously, we must integrate Eq. 3.35 with respect to s over one slit width, or from $s=\frac{1}{2}c$ to $s=\frac{1}{2}c+b$. Since $\frac{1}{2}c=\frac{1}{2}(d-b)$ and $\frac{1}{2}c+b=\frac{1}{2}(d+b)$, the integration can be carried out over these limits to give

$$y=\left(\frac{2A_0}{xk\sin\theta}\right)\{\sin[\tfrac{1}{2}k(d+b)\sin\theta]-\sin[\tfrac{1}{2}k(d-b)\sin\theta]\}\sin(\omega t-kx) \tag{3.36}$$

Since $\sin(a+b)-\sin(a-b)=2\sin b\cos a$, Eq. 3.36 can be written

$$y=(2A_0b/x)(\sin\beta/\beta)\cos\gamma\sin(\omega t-kx) \tag{3.37}$$

where $\beta=1/2kb\sin\theta=(\pi b/\lambda)\sin\theta$ and $\gamma=1/2kd\sin\theta=(\pi d/\lambda)\sin\theta$. Since $\beta=(\pi b/\lambda)\sin\theta$, it follows from Eq. 3.5 that the phase difference between the two edges of a given slit is 2β. Similarly, the phase difference between any two homologous points on different slits will be 2γ since $\gamma=(\pi d/\lambda)\sin\theta$.

The irradiance at any point P on the screen will be proportional to

$$4L^2(\sin^2\beta/\beta^2)\cos^2\gamma=4L^2(\sin^2\beta/\beta^2)\cos^2(\tfrac{1}{2}\delta) \tag{3.38}$$

where $L=A_0b/x$ and $\gamma=\frac{1}{2}\delta$. Equation 3.38 reveals that the irradiance is the product of two factors and will be zero when either factor is zero. The first factor corresponds to the diffraction pattern produced by a slit of width b. The second factor corresponds to the interference pattern produced between two beams of equal intensity with a phase difference of δ (see Eqs. 6.12 and 6.13). Thus, the pattern produced on the screen by the double slit can be thought to arise from a combination of diffraction and interference. The intensity of the fringes that arise from interference effects occurring with light from different slits is modulated by the diffraction pattern from a single slit. The result is a series of interference fringes whose intensities follow the single-slit diffraction envelope as shown in Figure 3.9.

The location of the various maxima and minima can be found from Eq. 3.38. The factor $\sin^2\beta/\beta^2$ will be zero when $\beta=\pm\pi, \pm2\pi, \pm3\pi, \cdots$ or when $b\sin\theta=\lambda, 2\lambda, 3\lambda, \cdots =p$, where p is any nonzero integer. If it is assumed that the slits are narrow, the approximate location of the maxima in the interference pattern will occur when $\gamma=0, \pm\pi, \pm2\pi, \pm3\pi, \cdots$ or when $d\sin\theta=0, \lambda, 2\lambda, 3\lambda, \cdots =m$, where m is the *order* of the interference and can take on values $0, \pm1, \pm2, \cdots$. The minima in the interference pattern will occur when the factor $\cos^2\gamma$ is zero. This occurs when $\gamma=\pm\frac{1}{2}\pi, \pm\frac{3}{2}\pi, \pm\frac{5}{2}\pi, \cdots$ or when $d\sin\theta=\frac{1}{2}\lambda, \frac{3}{2}\lambda, \frac{5}{2}\lambda, \cdots =(m+\frac{1}{2})\lambda$, where m is the order.

It is apparent from Eq. 3.38 that missing orders will occur whenever a minimum in the diffraction envelope coincides with a maximum in the

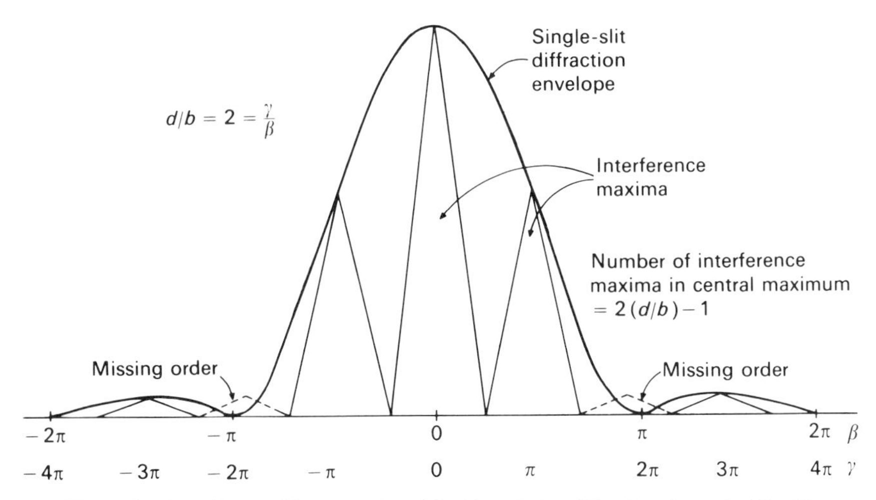

Figure 3.9. Interference fringes produced by Fraunhofer diffraction from double slit.

interference pattern. This will occur when

$$d \sin \theta = m \lambda \text{ maximum in the interference pattern}$$

and

$$b \sin \theta = p\lambda \text{ minimum in diffraction pattern}$$

are both simultaneously satisfied by the same value of θ. At this point, $d/b = m/p$. From the definitions of γ and β, it is apparent that d/b is also equal to γ/β.

Figure 3.10 shows a diffraction experiment conducted with a grating having four slits. As a result of the superposition of waves originating from each of the slits, the resultant amplitude and phase angle of the combined motion at P will be the vector sum of all the amplitudes originating from each of the slits. Figure 3.11 shows the vector sum of four amplitudes of equal magnitude where each amplitude is shifted from the previous one by a phase difference δ. This is precisely the situation that occurs with the diffraction grating where equidistant apertures of equal width produce equal amplitudes shifted by a constant phase difference. Thus, the problem of predicting the intensity pattern formed on a screen when a diffraction grating is illuminated by monochromatic light as shown in Figure 3.10 depends on determining the vector sum of the amplitudes originating from each slit. The mathematical basis for determining the vector sum is based on using *complex amplitudes*.

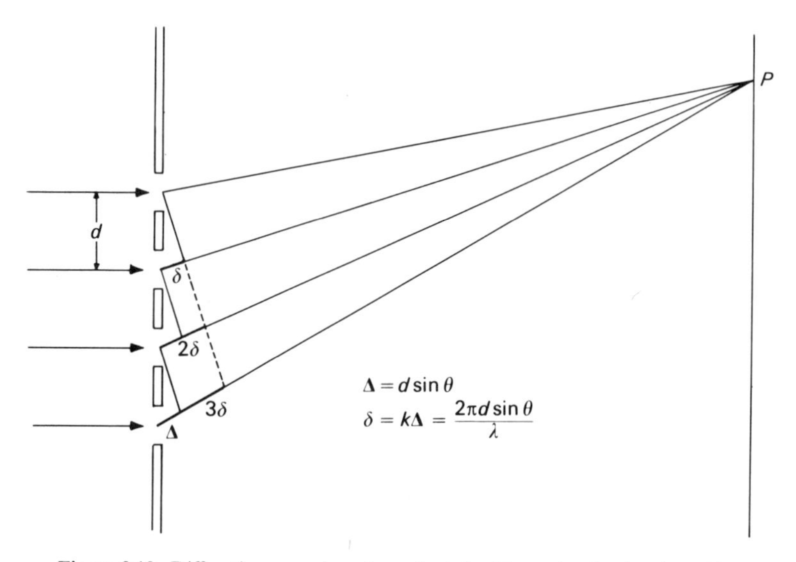

Figure 3.10. Diffraction experiment conducted with grating having four slits.

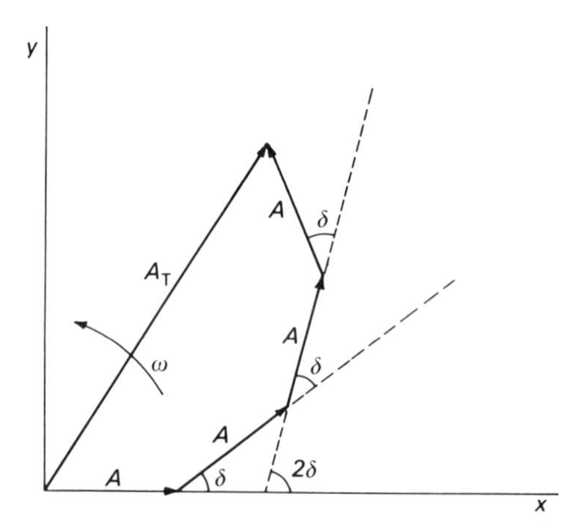

Figure 3.11. Vector sum of four amplitudes of equal magnitude each shifted from previous one by phase difference δ.

To unterstand the use of complex amplitudes, we need to recall that on the basis of Euler's theorem, Eq. 3.4 can be written in exponential form as

$$y = A_0 e^{i(\omega t - kx)} = A_0 e^{i\omega t} e^{-i\delta} \tag{3.39}$$

where $A_0 e^{-i\delta}$ is the *complex amplitude*. As can be seen from the vector sum in

Figure 3.11, the resultant amplitude from the combined motion of several waves is determined solely on the basis of the amplitudes and phase differences of the individual waves and is not influenced by the frequency of the wave. As a result, the factor $e^{i\omega t}$ that specifies the frequency of each wave can be neglected in our discussion since it will be the same for all the waves.

To see how complex amplitudes can be used to perform vector addition, the vector sum of two vectors, shown in Figure 3.12, will be considered on a geometric basis first. It can be seen from the figure that the resultant amplitude from the vector addition will be given by

$$A_T = [(y_1 + y_2)^2 + (x_1 + x_2)^2]^{1/2} = [(\Sigma y)^2 + (\Sigma x)^2]^{1/2} \tag{3.40}$$

and the phase angle will be

$$\tan \delta_T = \frac{y_1 + y_2}{x_1 + x_2} = \frac{\Sigma y}{\Sigma x} \tag{3.41}$$

Consider now a vector of magnitude A and phase angle δ as shown in Figure 3.13, where the abscissa is the real axis and the ordinate is the imaginary axis. Such a vector may be represented by

$$Ae^{i\delta} = x + iy \tag{3.42}$$

where $x = A\cos\delta$ and $y = A\sin\delta$. From Eq. 3.42, it follows that

$$A_T e^{i\delta_T} = A_1 e^{i\delta_1} + A_2 e^{i\delta_2} = x_1 + iy_1 + x_2 + iy_2 = (x_1 + x_2) + i(y_1 + y_2) \tag{3.43}$$

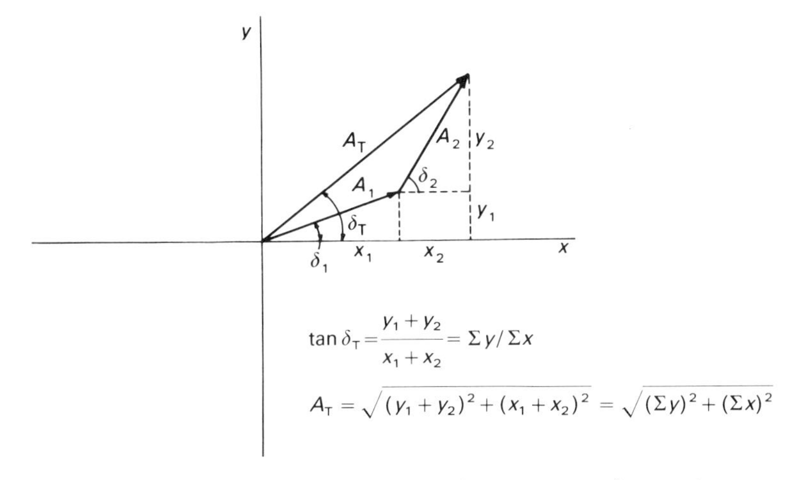

Figure 3.12. Geometric construction showing vector sum of two vectors.

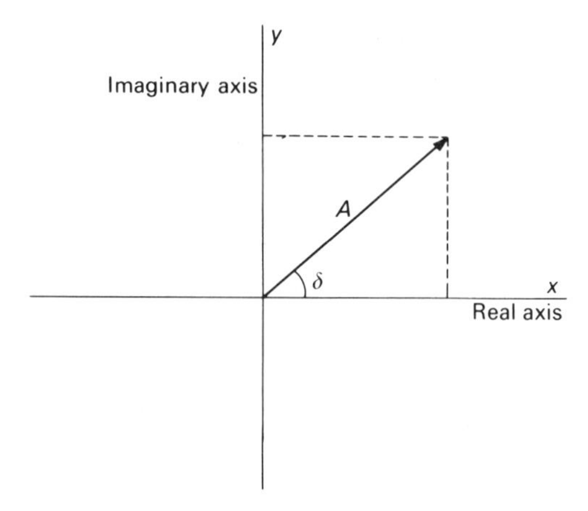

Figure 3.13. Vector representation in complex plane.

From the properties of complex numbers, the square of the total amplitude will be given by

$$A_T e^{i\delta_T} A_T e^{-i\delta_T} = A_T^2 = [(x_1 + x_2) + i(y_1 + y_2)][(x_1 + x_2) - i(y_1 + y_2)] \tag{3.44}$$

$$A_T^2 = (x_1 + x_2)^2 + (y_1 + y_2)^2 \tag{3.45}$$

Equation 3.45 is identical to that obtained from geometric considerations and shows that the mathematics associated with vector addition can be achieved conveniently with complex numbers. Thus the algebraic addition of complex amplitudes is equivalent to the vector addition of real amplitudes.

With this background in mind, we are now ready to discuss the intensity distribution from an ideal diffraction grating. Consider a grating of N slits of equal magnitude where each slit contributes an amplitude A and the phase angle changes by δ from one slit to the next. The total amplitude at any point P on a screen as shown in Figure 3.10 will be the vector sum of all the individual amplitudes from each slit. This vector sum can be obtained by the algebraic addition of the corresponding complex amplitudes. Thus,

$$A_T e^{i\delta_T} = A + Ae^{i\delta} + Ae^{i2\delta} + Ae^{i3\delta} + \cdots + Ae^{i(N-1)\delta} \tag{3.46}$$

Multiplying both sides of Eq. 3.46 by $e^{i\delta}$ gives

$$A_T e^{i\delta_T} e^{i\delta} = Ae^{i\delta} + Ae^{i2\delta} + Ae^{i3\delta} + \cdots + Ae^{iN\delta} \tag{3.47}$$

Subtracting Eq. 3.47 from Eq. 3.46 gives

$$A_T e^{i\delta_T} - A_T e^{i\delta_T} e^{i\delta} = A - Ae^{iN\delta} \tag{3.48}$$

from which it follows immediately that

$$A_{\mathrm{T}}e^{i\delta_{\mathrm{T}}}=A(1-e^{iN\delta})/(1-e^{i\delta}) \tag{3.49}$$

The irradiance at any point on the screen will be proportional to A_{T}^2, which can be determined by multiplying Eq. 3.49 by its complex conjugate. Thus,

$$A_{\mathrm{T}}e^{i\delta_{\mathrm{T}}}A_{\mathrm{T}}e^{-i\delta_{\mathrm{T}}}=A_{\mathrm{T}}^2=A^2(1-e^{iN\delta})(1-e^{-iN\delta})/(1-e^{i\delta})(1-e^{-i\delta}) \tag{3.50}$$

where

$$(1-e^{i\delta})(1-e^{-i\delta})=2-e^{i\delta}-e^{-i\delta} \tag{3.51}$$

$$=2-(\cos\delta-i\sin\delta)-(\cos\delta+i\sin\delta) \tag{3.52}$$

$$=2-2\cos\delta=2(1-\cos\delta) \tag{3.53}$$

By analogy, $(1-e^{iN\delta})(1-e^{-iN\delta})=2(1-\cos N\delta)$, and

$$A_{\mathrm{T}}^2=A^2(1-\cos N\delta)/(1-\cos\delta) \tag{3.54}$$

Since $1-\cos a=2\sin^2(\frac{1}{2}a)$,

$$A_{\mathrm{T}}^2=A^2[\sin^2(\tfrac{1}{2}N\delta)]/\sin^2(\tfrac{1}{2}\delta) \tag{3.55}$$

From our previous discussion, it was shown that $\gamma=\frac{1}{2}\delta$ so we can write

$$A_{\mathrm{T}}^2=A^2(\sin^2 N\gamma)/\sin^2\gamma \tag{3.56}$$

The complete expression for the intensity distribution from an ideal grating can now be obtained by substituting the Fraunhofer single-slit intensity distribution for A, giving

$$A_{\mathrm{T}}^2=(A_0b/x)^2\,(\sin^2\beta/\beta^2)(\sin^2 N\gamma/\sin^2\gamma) \tag{3.57}$$

Equation 3.57 may be written in terms of the grating width W, where $W=Nd$:

$$A_{\mathrm{T}}^2=(A_0/x)^2N^2d^2(b/d)^2\,(\sin^2\beta/\beta^2)(\sin N\gamma/N\sin\gamma)^2 \tag{3.58}$$

$$=(A_0/x)^2\,W^2(b/d)^2\,(\sin\beta/\beta)^2\,(\sin N\gamma/N\sin\gamma)^2 \tag{3.59}$$

Equation 3.59 reveals several important features about the intensity distribution expected with the ideal diffraction grating. The factor

($\sin N\gamma/N\sin\gamma$) is simply the N-slit interference pattern that arises from the interference effects produced by light from different slits. The factor $\sin\beta/\beta$ represents the single-slit diffraction pattern produced by a single aperture of width b. By analogy with our discussion of the double-slit experiment, the factor $\sin\beta/\beta$ represents the single-slit diffraction envelope that modulates the intensity of the N-slit interference pattern. In comparison with the double-slit "diffraction grating," the interference maxima produced with an N-slit grating are observed to become much sharper as the number of slits in the grating is increased. Finally, we see that the maximum intensity is proportional to the square of the grating width W.

Let us consider the factor $\sin N\gamma/N\sin\gamma$ of Eq. 3.59 in more detail. Although the numerator and the denominator of this factor become zero periodically, the numerator becomes zero more often than the denominator. For values of γ where both the numerator and the denominator simultaneously approach zero, the factor $\sin N\gamma/N\sin\gamma$ approaches a maximum equal to 1. This condition will occur when $\gamma = m\pi$, where m is the order and can take on values of 0, ± 1, ± 2, ± 3.

Thus, according to l'Hôpital's rule

$$\lim_{\gamma\to m\pi} \frac{\sin N\gamma}{N\sin\gamma} = \lim_{\gamma\to m\pi} \frac{N\cos N\gamma}{N\cos\gamma} = 1 \qquad (3.60)$$

Since γ is defined as $(\pi/\lambda)d\sin\theta$, the location of the *principal maxima* will occur when

$$\gamma = m\pi = (\pi/\lambda)d\sin\theta \qquad (3.61)$$

or when

$$m\lambda = d\sin\theta \qquad (3.62)$$

Equation 3.62 gives the position of the principal maxima and represents a simplified form of the *grating equation* for the case of light striking the grating at normal incidence. The quantity $d\sin\theta$ represents the path difference for light between homologous portions of successive slits, and where this path difference is equal to an integral number of wavelengths, constructive interference occurs, resulting in a principal maximum.

According to Eq. 3.60, principal maxima occur only when both the numerator and the denominator approach zero for the same value of γ. Since the numerator becomes zero more frequently than the denominator, there exist values of γ for which the numerator approaches zero but the denominator does not. These values of γ will result in zero intensity and will occur when $\gamma = p\pi/N$ excluding values of p equal to mN since these represent conditions for maxima. Thus, minima are expected when

$$\gamma=(\pi/\lambda)d\sin\theta=p\pi/N \tag{3.63}$$

or when

$$d\sin\theta=p\lambda/N=\lambda/N,\ 2\lambda/N,\ 3\lambda/N,\ \cdots,\ (N-1)\lambda/N,\ (N+1)\lambda/N \tag{3.64}$$

Thus for any order m, the principal maxima for a given wavelength will occur for $\gamma=m\pi$. As can be seen from Eq. 3.64, the adjacent minima will occur when $\gamma\pm\Delta\gamma=(mN\pm1)\pi/N$ or $m\pi\pm\pi/N$. For any order, the change in γ from the maximum to the adjacent minimum will be

$$\Delta\gamma=\pi/N \tag{3.65}$$

From Eq. 3.61, it follows that

$$\Delta\gamma=(\pi d/\lambda)\cos\theta\,\Delta\theta \tag{3.66}$$

Substituting Eq. 3.66 in Eq. 3.65 gives the angular half-width of a principal maximum as follows:

$$\Delta\theta=\lambda/Nd\cos\theta=\lambda/W\cos\theta=\lambda/B \tag{3.67}$$

Equation 3.67 shows that as the number of slits is increased, the interference maxima become narrower. The quantity $W\cos\theta$ represents the width of the diffracted beam from the grating, as shown in Figure 3.14.

As discussed in Section 3.3, the resolving power is based on the Rayleigh criterion for two diffraction patterns. Applying this criterion to the diffraction grating, a condition is sought where $\sin N\gamma_1/N\sin\gamma_1$ is a maximum while $\sin N\gamma_2/N\sin\gamma_2$ is a minimum for the same angle θ. From the previous discussion, $\gamma_1=m\pi$ and $\gamma_2=m\pi+\pi/N$. Thus,

$$m\pi=(\pi/\lambda_1)d\sin\theta \qquad \text{and} \qquad m\pi+\pi/N=(\pi/\lambda_2)d\sin\theta \tag{3.68}$$

from which it can be seen that

$$m\pi\lambda_1=(m\pi+\pi/N)\lambda_2 \tag{3.69}$$

Rearranging Eq. 3.69 gives

$$\lambda_2/(\lambda_1-\lambda_2)=\lambda/\Delta\lambda=R=mN \tag{3.70}$$

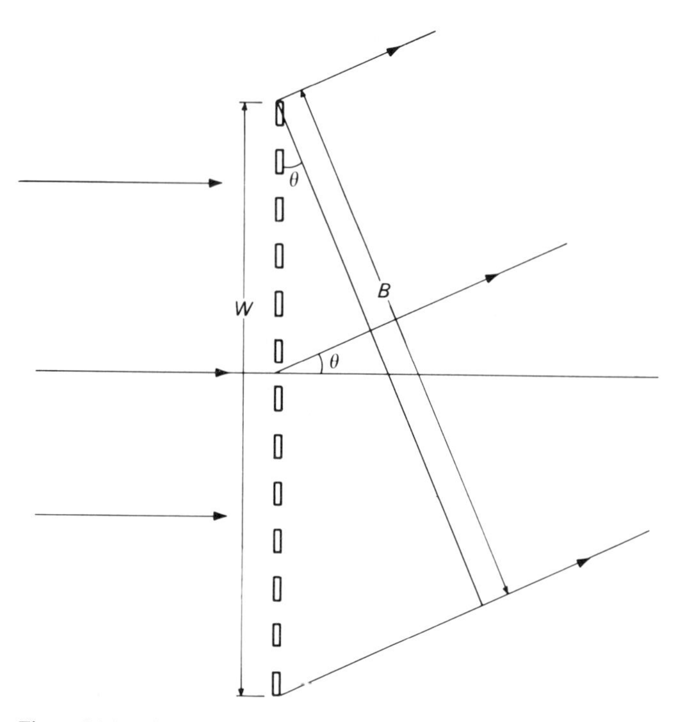

Figure 3.14. Width of the diffracted beam B from a grating of width W.

which is the usual equation given in spectroscopy texts for the resolving power of a grating. Substituting for m and N in Eq. 3.70 gives the following result:

$$R = (d \sin \theta/\lambda)(W/d) = (W \sin \theta)/\lambda \tag{3.71}$$

where the right-hand side of Eq. 3.71 represents the total number of wavelengths in the path difference between the extremes of the grating. Thus the resolving power of a grating depends on the number of wavelengths in the path difference between the extremes of the grating.

From Eq. 3.67, it is easy to show that

$$\Delta\theta/\Delta\lambda = \lambda/\Delta\lambda\ B \tag{3.72}$$

which upon rearrangement gives

$$R = B(\Delta\theta/\Delta\lambda) = W \cos \theta(\Delta\theta/\Delta\lambda) \tag{3.73}$$

Equation 3.73 shows that the resolving power of a diffraction grating is given by the product of the angular dispersion ($\Delta\theta/\Delta\lambda$) times the width of the diffracted beam. Equation 3.73 is a general one and can be derived directly from Eq. 3.29, where b becomes the width of the diffracting aperture (in this case, the projected width of the grating). It applies equally well to a prism where the quantity B in Eq. 3.73 refers to the width of the emerging beam (which equals $s \cos i$, where s is the length of the side of the prism) (see Section 2.2.1).

The grating equation presented by Eq. 3.62 is a simplified form of the complete equation and only applies to the case of normal incidence. The primary requirement for constructive interference is that the path difference from homologous portions of successive slits be some integral multiple of the wavelength. It can be seen from Figure 3.15 that when light is incident on a grating at an angle of incidence i and diffracted at an angle of diffraction θ, the path difference will be $d(\sin i + \sin \theta)$. Therefore, the *complete grating equation* that applies to any angle of incidence will be

$$m\lambda = d(\sin i + \sin \theta) \tag{3.74}$$

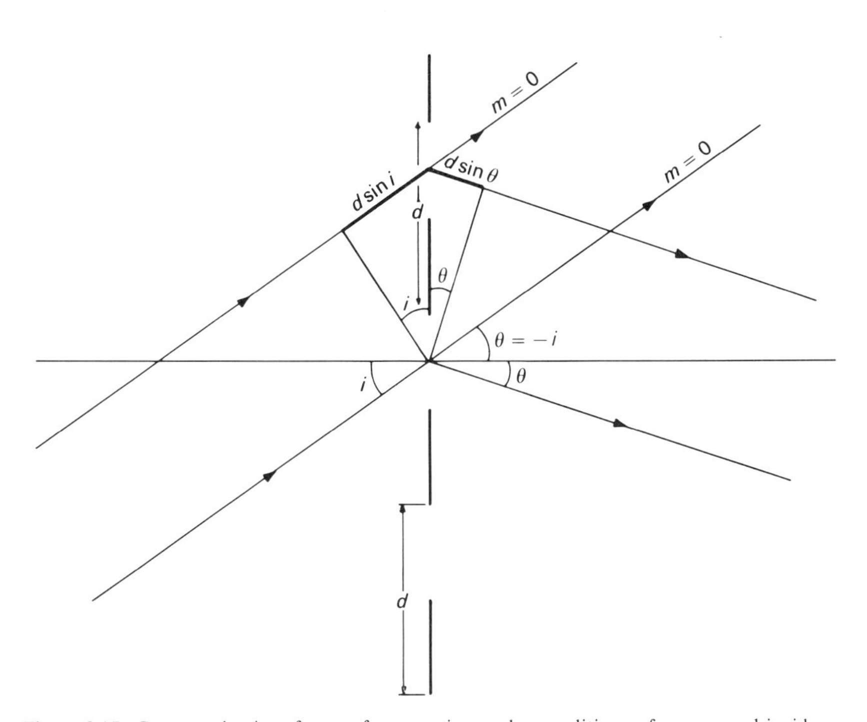

Figure 3.15. Constructive interference from grating under conditions of nonnormal incidence.

where m is the order, λ is the wavelength, d is the slit separation (or grating spacing) in units of wavelength, i is the angle of incidence measured from the normal, and θ is the angle of diffraction measured from the normal.

In using Eq. 3.74 several sign conventions should be kept in mind. By definition, the quantities λ, d, and i are always positive. The order m, on the other hand, can be zero as well as positive or negative. If m is negative, the right side of Eq. 3.74 must also be negative, and this can only occur when θ is negative and $|\theta|>|i|$. Since $\sin(-\theta)=-\sin\theta$, the quantity $d(\sin i-\sin\theta)$ will also be negative if $|\theta|>|i|$.

Equation 3.74 reveals that the positive and negative orders will be symmetrically displaced about the maximum, which occurs for $m=0$. When $m=0$, the wavelength dependence of Eq. 3.74 disappears and

$$\sin i=-\sin\theta=\sin(-\theta) \tag{3.75}$$

The maximum that occurs when $m=0$ has no wavelength dependence and simply represents the path of an undeviated ray. This maximum is called the *central image*. From Figure 3.15, it can be seen that to be consistent with Eq. 3.74, θ must be taken as negative whenever θ is on the opposite side of the normal from i.

Equation 3.74 also reveals another important aspect of diffraction gratings, namely, that θ is not unique for a single wavelength even when i and d are held constant. Thus when i, d, and θ are all fixed,

$$m_i\lambda_i=m_1\lambda_1=m_2\lambda_2=m_3\lambda_3=\text{const} \tag{3.76}$$

Equation 3.76 shows that for a given i and d, light of wavelengths λ_1, λ_2, and λ_3 will all be diffracted by the same amount, θ, and will overlap at the same position on a screen used to observe the diffraction pattern. This overlap is possible because λ_1, λ_2, and λ_3 all occur in different orders. The overlapping of spectra in different orders must be considered when diffraction gratings are used in spectroscopic applications. This topic will be discussed further in Chapter 4.

The angular dispersion of a diffraction grating can be determined easily from Eq. 3.74 by differentiating to give

$$m(d\lambda/d\theta)=d\cos\theta \tag{3.77}$$

which upon rearrangement gives the desired quantity

$$d\theta/d\lambda=m/(d\cos\theta) \tag{3.78}$$

Equation 3.78 reveals that for small diffraction angles, the angular dispersion of a diffraction grating is constant and equal to m/d. This is only an approximation, however, and the factor $\cos\theta$ should not be neglected for large θ. If the $\cos\theta$ dependence is considered, the angular dispersion will increase slowly as θ increases.

In Eq. 3.70, it was shown that the resolving power of a grating is given by $R = mN$. This equation can be rewritten in terms of the angles of incidence and diffraction by solving Eq. 3.74 for m and substituting in Eq. 3.70:

$$R = Nd(\sin i + \sin\theta)/\lambda = W(\sin i + \sin\theta)/\lambda \tag{3.79}$$

Equation 3.79 shows that the resolving power is not a function of the number of lines in the ruled width W for a given i and θ. If i and θ are kept constant and the number of lines in the ruled width W is decreased, this simply corresponds to using a coarser grating in a higher order. According to Eq. 3.79, the maximum theoretical resolving power will equal $2W/\lambda$ and will occur when $i = \theta = 90°$. This corresponds to grazing incidence and does not represent a practical situation. At best, a resolving power of two-thirds the theoretical is attainable (Jenkins and White, 1976). This corresponds to

$$R_{\max} \sim 4W/3\lambda \tag{3.80}$$

where W is expressed in wavelength units. According to Eq. 3.80, the maximum resolving power attainable with a grating 52 mm wide for light of 670 nm will be on the order of 100,000.

So far in our discussion, we have considered only transmission gratings where light passes through a regular array of slits. Most gratings used for spectroscopic applications are reflection gratings produced by cutting

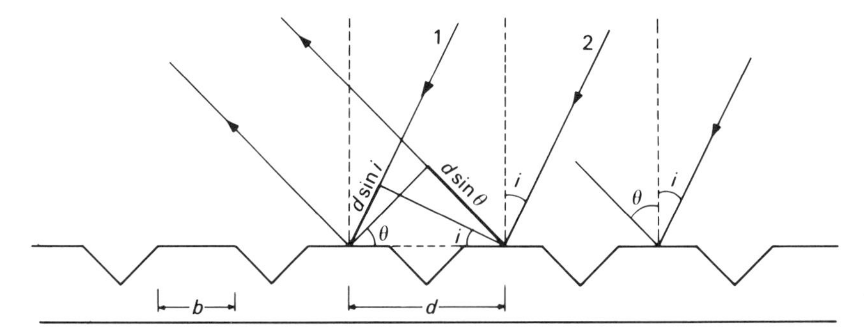

Figure 3.16. Ray diagram for an unblazed plane reflection grating: path difference between rays 1 and 2 $= d\sin i - d\sin\theta$; $m\lambda = d(\sin i - \sin\theta)$.

grooves in a suitable reflecting surface. In this case, the grooves correspond to the opaque sections of the transmission grating, while the mirrored sections correspond to the openings in the transmission grating. Figure 3.16 shows the ray diagram for a reflection grating. From the figure, it can be seen that the path difference between rays 1 and 2 equals $d(\sin i - \sin\theta)$ and the reflection grating obeys the same equation (Eq. 3.74) as the transmission grating with the same sign conventions discussed earlier.

Although a finely ruled diffraction grating such as that shown in Figure 3.16 is capable of high resolving power, the amount of the light that strikes the grating is divided among the various orders (on either side of the central image) and the central image itself. Figure 3.17 shows the intensity distribution expected from an ideal grating like that shown in Figure 3.16. In accordance with Eq. 3.59, the intensities of the various interference maxima are determined by the single-slit diffraction envelope corresponding to the given wavelength and the width b of the reflecting surface of each individual tiny facet. On the basis of Eq. 3.59, Figure 3.17 shows that the majority of the

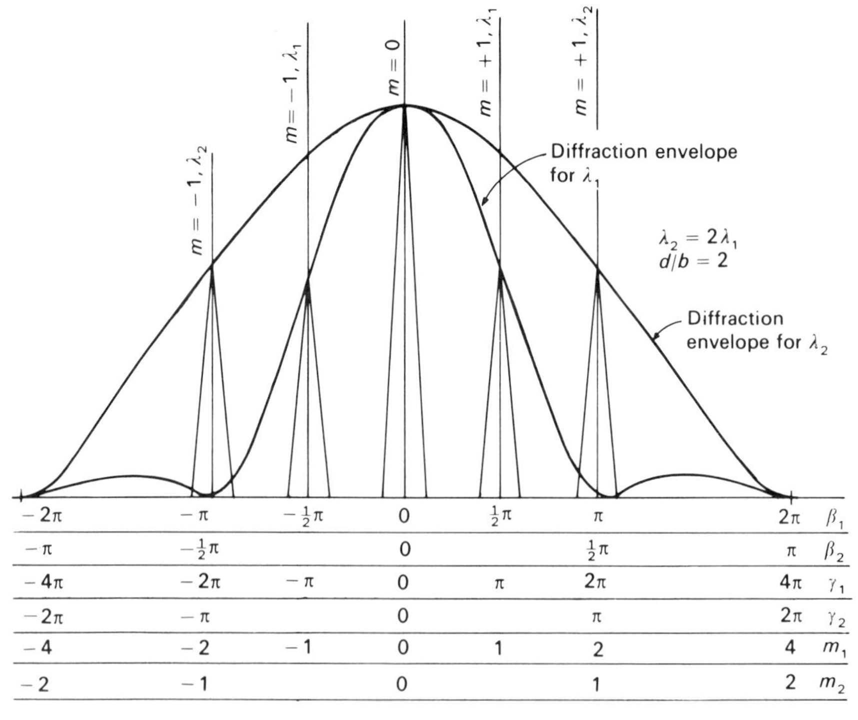

Figure 3.17. Relative intensities of various interference maxima for unblazed plane reflection grating.

light that strikes the grating ultimately goes to form the central image. This means that a major portion of light that strikes a grating such as that shown in Figure 3.16 does not contribute to the formation of a useful spectrum but instead is concentrated in the zeroth order. This intensity distribution would represent a severe shortcoming for the use of diffraction gratings in spectroscopy (particularly with sources of low intensity) if nothing could be done to alter it.

Fortunately, the intensity distribution from a grating may be altered by a process called *blazing*. Before discussing how a grating is blazed in physical terms, it is worthwhile to consider the process in terms of Eq. 3.59. Equation 3.59 reveals that the irradiance at any point in the focal plane is the product of the single-slit diffraction envelope and the N-slit interference pattern. The former is a function of β, while the latter is a function of γ. With an unblazed grating such as that shown in Figure 3.16, the zeroth-order interference maximum occurs when $\gamma = \beta = 0$. In other words, the zeroth-order interference maximum occurs at the point where the single-slit diffraction envelope is a maximum. This undesirable intensity distribution might be altered if a phase shift could be introduced between γ and β such that the first-order interference maximum would coincide with the maximum in the single-slit diffraction envelope.

Figure 3.18 shows the effect of introducing a phase shift of π between γ and β so that now when $\beta = 0$, $\gamma = \pi$ (for a given wavelength), which is the condition for the first-order interference maximum. If this condition can be realized, it can be seen from the figure that the intensity of the first-order interference maximum for the given wavelength will be greater than the zeroth-order maximum, which now has been shifted to one side of the single-slit diffraction envelope. In addition, introduction of such a phase shift causes the higher orders on one side of the central image ($m = 0$) to be more intense than the corresponding orders on the other side of the central image.

Figure 3.19 shows the theoretical intensity distribution from an ideal grating calculated from the single-slit diffraction envelope for a phase shift of π between γ and β. Figure 3.19 is plotted in terms of multiples of the wavelength for which $\gamma = \pi$. This wavelength, which coincides with the maximum in the single-slit diffraction envelope, is known as the *blaze wavelength*. As can be seen in Figure 3.19, wavelengths greater than the blaze wavelength fall off in relative intensity gradually, while wavelengths less than the blaze wavelength drop off in relative intensity rather sharply. For a more rigorous treatment of the intensity distribution expected from a diffraction grating, the reader should consult the comprehensive discussion by Stroke (1967).

For a reflection grating, the position of the maximum in the single-slit diffraction envelope will occur along the direct reflection from the individual

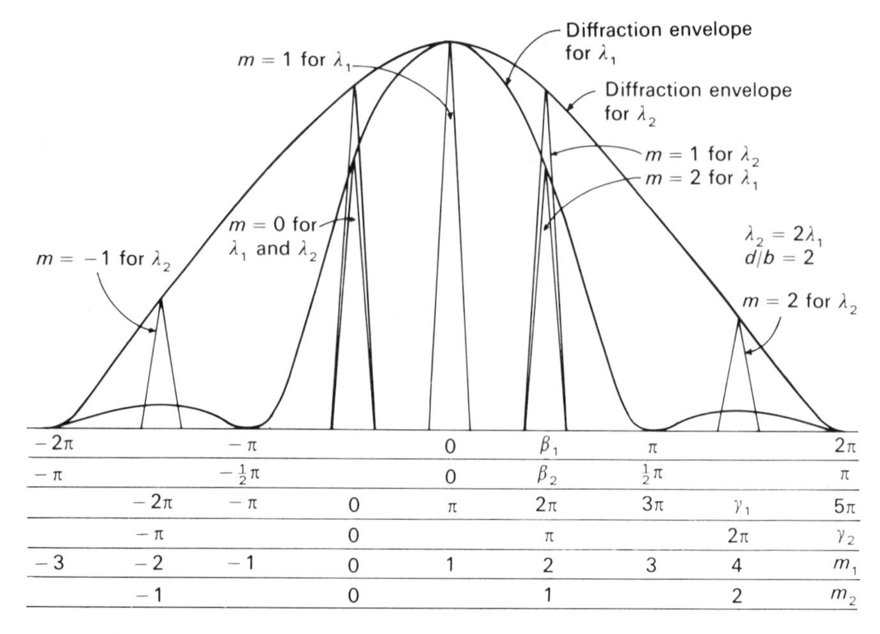

Figure 3.18. Effect of introducing phase shift on relative intensities of various interference maxima for plane reflection grating.

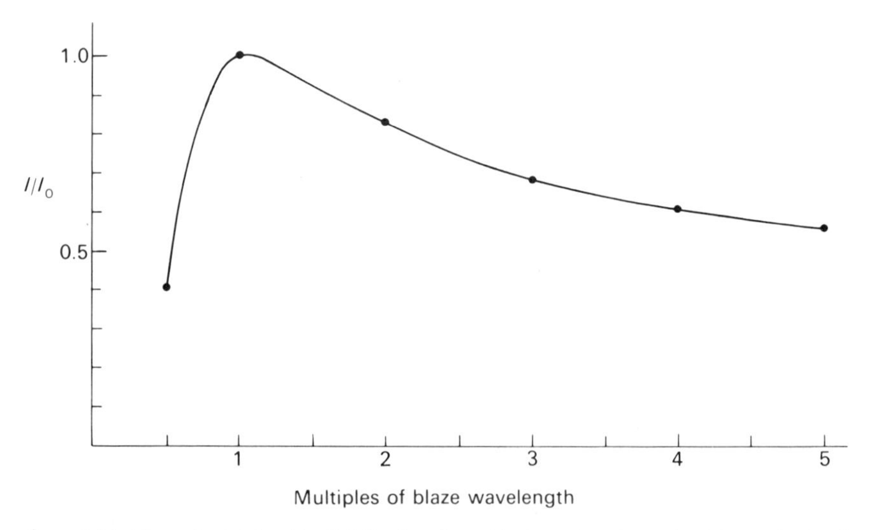

Figure 3.19. Theoretical intensity distribution from ideal grating determined from single-slit diffraction envelope for phase shift of π between β and γ.

facets of the grating. If this direct reflection can be made to coincide with a path difference of λ_B between successive slits, the phase shift conditions between γ and β discussed previously can be achieved. This situation can be made to occur if the reflecting facets of the grating can be inclined at an angle to the grating surface. This is shown in Figure 3.20. From the figure, it can be seen that when light is incident on the grating at an angle ϕ, the direct reflection from each facet will also occur at an angle of reflection ϕ to the grating. In other words, light striking the grating at this angle of incidence is reflected back on itself. If the grooves are shaped as shown in Figure 3.20 so that $\sin\phi = \lambda_B/2d$, the path difference between successive slits will be exactly λ_B. Thus the condition for first-order interference can be made to coincide with the maximum in the single-slit diffraction envelope. The angle is known as the *blaze angle* of the grating. If the grating is observed at this angle, all the individual facets will be normal to the observer and the grating will appear to "blaze" with reflected light.

If the grating is used so that $i \neq \theta \neq \phi$ as described, the blaze wavelength is shifted slightly from $\lambda_B = 2d \sin\phi$. The maximum in the single-slit diffraction envelope will still be located according to the law of reflection from each facet. The new blaze wavelength will correspond to the wavelength that, for a given angle of incidence, is diffracted according to the grating equation at an angle of diffraction that coincides with the angle of reflection from the facet. This situation is illustrated in Figure 3.21. From the figure, it can be seen that the law of reflection from each facet will be satisfied when

$$\theta - \phi = \phi - i \tag{3.81}$$

or when

$$\phi = \tfrac{1}{2}(\theta + i) \tag{3.82}$$

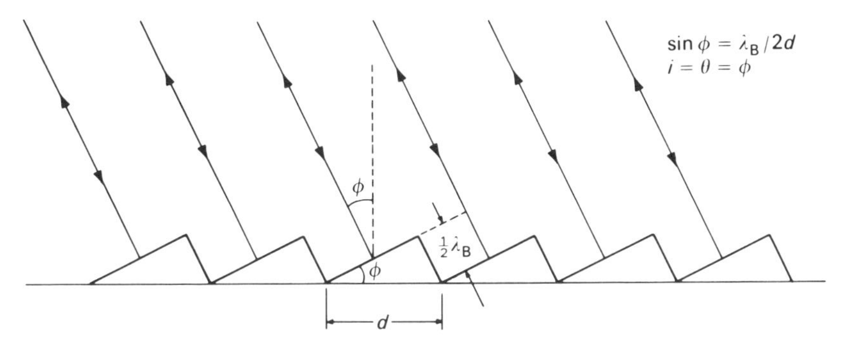

Figure 3.20. Microgeometry of groove facets for blazed plane reflection grating.

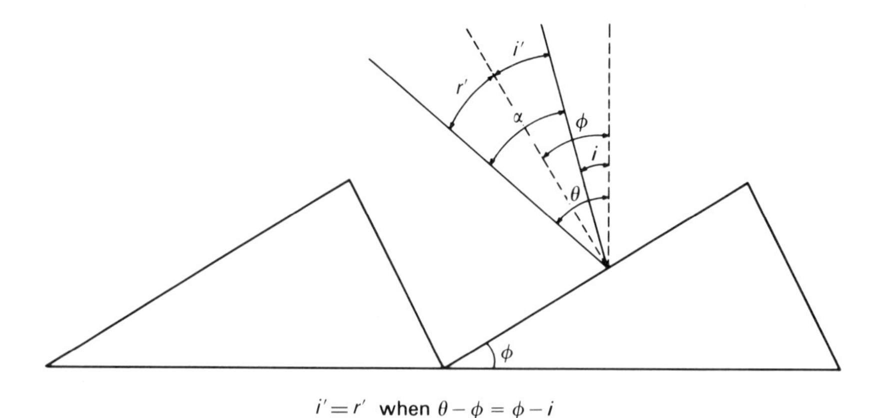

Figure 3.21. Angular relationship between incident ray and diffracted ray from facet of blazed plane reflection grating.

Using the trigonometric identity,

$$\sin i + \sin \theta = 2 \sin[\tfrac{1}{2}(\theta + i)]\cos[\tfrac{1}{2}(\theta - i)] \tag{3.83}$$

the grating equation can be written as

$$m\lambda'_{B} = 2d \sin \phi \cos(\tfrac{1}{2}\alpha) \tag{3.84}$$

where $\alpha = \theta - i$ and λ'_{B} is the new blaze wavelength. Since $\lambda_{B} = 2d \sin \phi$, the new blaze wavelength will be given by

$$m\lambda'_{B} = \lambda_{B} \cos(\tfrac{1}{2}\alpha) \tag{3.85}$$

Since according to Eq. 3.85 it takes an angle α equal to 52° to produce a 10% change in the blaze wavelength from that given by $\lambda_{B} = 2d \sin \phi$, the uncorrected value of λ_{B} is a reasonably good estimate of the actual blaze wavelength even if $i \neq \theta \neq \phi$.

Finally, it should be mentioned that reflection gratings can be ruled on concave surfaces to give a diffraction grating with collimating and focusing properties (see Stroke, 1967). The use of concave gratings in spectroscopic instrumentation will be discussed in Chapter 4.

REFERENCES

Jenkins, F. A. and White, H. E. (1976), *Fundamentals of Optics*, McGraw-Hill, New York.

Stroke, G. W. (1967), "Diffraction Gratings," in *Optische Instrumente*, Vol. 29, *Handbuch der Physik*, Springer-Verlag, Berlin.

CHAPTER

4

SPECTROGRAPHS AND SPECTROMETERS

In this chapter, the general optical principles developed in Chapters 2 and 3 will be applied to conventional spectrographic systems. The term "conventional" as used here refers to those systems that employ dispersive components such as gratings and prisms as opposed to transform-based forms of spectroscopy such as Fourier transform spectroscopy and Hadamard transform spectroscopy. These latter forms of spectroscopy will be discussed in Chapters 5 and 7.

A conventional spectroscopic system is an optical system that performs two basic functions simultaneously (Davis, 1970). The first function is the formation of images of the entrance aperture in the focal plane of the instrument, while the second function is the dispersion or alteration of the path of the radiation passing through the instrument as a function of wavelength. Taken together, the ideal spectroscopic system forms an image of the entrance aperture in the focal plane of the instrument for every wavelength of radiation incident on the entrance aperture.

Figure 4.1 is a schematic diagram of a typical spectrograph where the dispersing element may be either a prism or a diffraction grating. The system consists of four or five components depending on whether it is intended to monitor a range of wavelengths or to isolate a small wavelength interval. The first component is the entrance aperture, which is usually a narrow slit that forms the light source for the rest of the instrument. This is followed by a collimator, which may be either a lens or a mirror whose function is to render the light rays from the slit parallel. The collimator performs an important function since it is essential that all the light rays strike the dispersing element with the same angle of incidence. The dispersing element receives the parallel bundle of radiation from the collimator and alters the path of each wavelength slightly so as to produce an array of diverging parallel bundles with one parallel bundle for each wavelength. Each parallel bundle in the array of diverging parallel bundles that emerge from the dispersing device then strikes the focusing lens (or mirror) at a distinctive and slightly different angle and each is brought to a focus in the focal plane at distinctive and slightly displaced positions, resulting in a series of slit images, which we call a spectrum. Finally, if the instrument is a monochromator rather than a

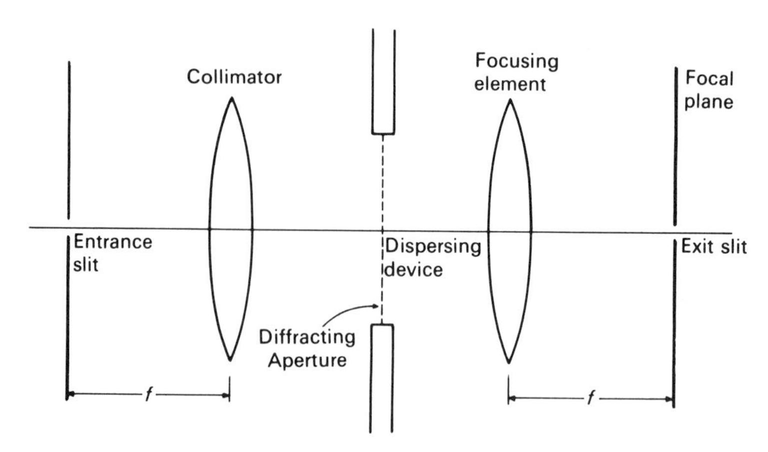

Figure 4.1. Schematic diagram of typical dispersion system for optical spectroscopy.

spectrograph, an exit slit will be located in the focal plane to transmit only a small wavelength interval to the detector.

If the entrance aperture is an infinitely narrow slit, the optical system described is analogous to that shown in Figure 3.3 for observing Fraunhofer diffraction from a single aperture. In this case, the diffracting aperture is the aperture of the dispersing device. According to elementary diffraction theory (see Section 3.2), the image that forms in the focal plane of the spectrograph when an infinitely narrow slit is employed will be a diffraction pattern characterized by a central maximum with secondary maxima located symmetrically on either side. The baseline width of the principal maximum under these conditions will be determined by Eq. 3.31 and will equal $2f\lambda/b$, where f is the focal length of the focusing lens or mirror, λ is the wavelength of the light, and b is the width of the dispersing device. Thus, regardless of whether the dispersing device is a prism or a diffraction grating, the larger the width of the dispersing device, the narrower the width of the central maximum. Since the resolving power of the system is determined by how closely two such central maxima can approach and still be recognized as two lines, it follows that the resolving power of a specroscopic system will improve as the width of the central maxima become narrower. As a result, the resolving power of a dispersive spectrograph such as that shown in Figure 4.1 is expected to increase as the width of the dispersing device is increased. This important concept will be discussed in detail in the next section.

The collimating, dispersing, and focusing functions can all be combined in one optical component by using a concave grating. A concave grating is one that has been ruled on a concave spherical mirror surface so that the grating not only disperses light according to the grating equation (Eq. 3.74) but also

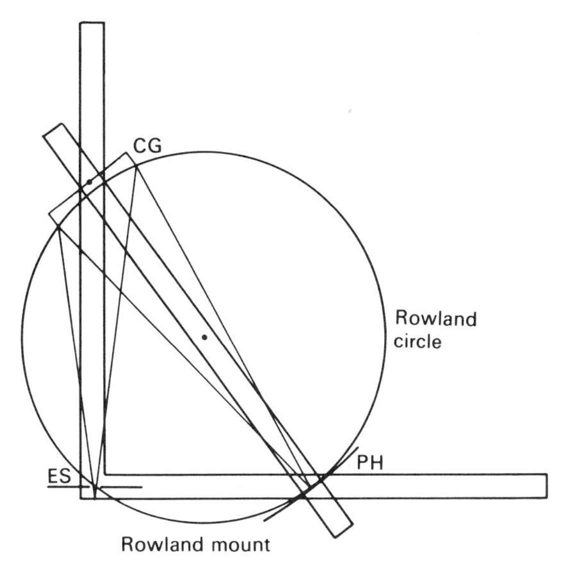

Figure 4.2. Rowland mounting for concave diffraction grating. CG, concave grating; ES, entrance slit; PH, plate holder.

serves as a focusing element. If the entrance slit, grating, and detector are all located at points tangent to a circle whose diameter is equal to the radius of curvature of the spherical mirror surface, the concave grating will form a 1 : 1 image of the entrance slit at the detector. The circle that represents the locus of corresponding object and image points is known as the *Rowland circle*. Figure 4.2 shows one possible configuration for a spectroscopic system employing a concave diffraction grating. Various alternative arrangements will be discussed later.

The advantage of the concave grating is that it reduces the spectroscopic system from five components, as described above, to three. By eliminating the need for lenses, it avoids absorption losses and chromatic aberration. Furthermore, since there is only one reflecting surface in the system, reflection losses are reduced to a minimum. For this reason, spectrometers employing concave gratings are especially useful for work in the vacuum ultraviolet (VUV) region of the spectrum where reflection losses are particularly important. As discussed in Section 2.2.4, however, spherical concave mirrors are particularly prone to astigmatism for off-axis rays.

If a plane grating or a prism is employed as the dispersing device, additional components must be used to collimate and focus the light within the spectrograph. Because of difficulties associated with producing large-diameter lenses that are free of chromatic and other aberrations and can transmit UV light, front-surface concave spherical mirrors are generally

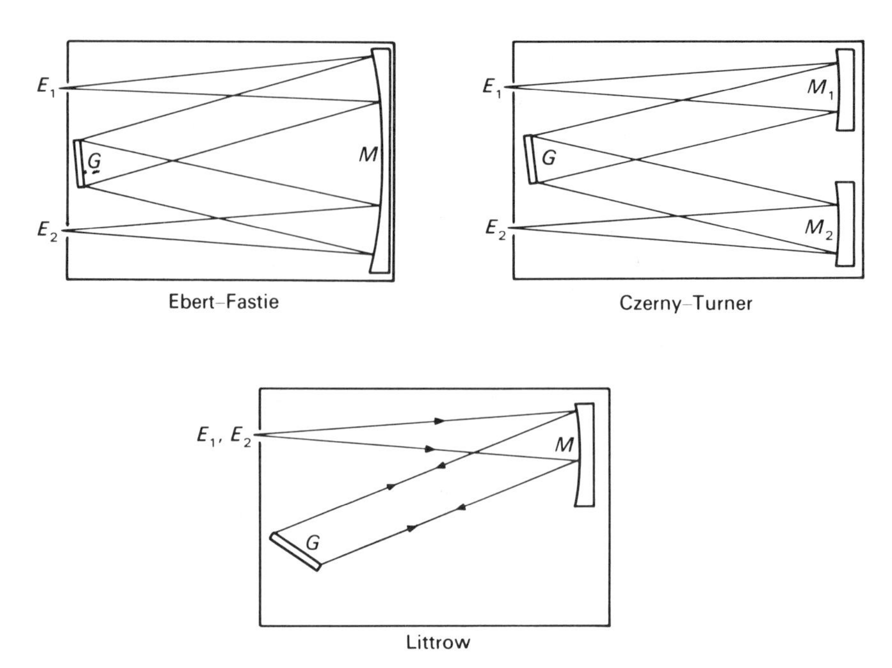

Figure 4.3. Common optical arrangements for spectroscopic systems employing plane gratings.

employed as collimating and focusing elements. As discussed in Section 2.2.4, the collimating and focusing mirrors must be located a distance of f from the entrance aperture or focal plane, respectively. Figure 4.3 shows a variety of common spectroscopic systems employing plane gratings.

4.1. FUNDAMENTAL PRINCIPLES

In considering the factors characterizing the performance of spectroscopic systems, four figures of merit are of importance: (1) the wavelength range of the instrument, (2) the angular (or alternatively, the reciprocal linear) dispersion, (3) the resolving power, and (4) the light throughput of the system.

The useful wavelength range of the instrument will be determined by the transmission and/or reflection characteristics of the optical components in the system as well as the response characteristics of the detector. Ultimately, the transmission characteristics of the atmosphere itself may limit the short-wavelength response of the system to about 180 nm. Since the short-wavelength vacuum UV region of the spectrum is not generally used for analytical work, spectroscopic systems for vacuum UV work will not be discussed. The typical range of most grating instruments available today is from 180 to 1000 nm.

4.1.1. Dispersion

Equations for the angular dispersion of prisms and gratings were derived in Chapters 2 and 3. Table 4.1 summarizes the relationships obtained. In analyzing the ability of a prism to separate radiation into its component wavelengths, Eq. 2.27 shows that the angular dispersion of prism instruments depends on two terms,

$$d\theta/d\lambda = (B/b)(dn/d\lambda) \tag{2.27}$$

where the first term is a function of the prism geometry and the second term is the dispersion of the prism material. Since this second term is a function of wavelength, the angular dispersion of the prism instrument is not constant and is a pronounced function of wavelength.

By contrast, the angular dispersion of a diffraction grating is given by Eq. 3.78,

$$d\theta/d\lambda = m/(d\cos\theta) \tag{3.78}$$

Table 4.1. Spectral Characteristics of Conventional Dispersive Systems

Property	Equation
	Prism
Angular dispersion	$d\theta/d\lambda = (2\tan i/n)dn/d\lambda = (B/b)\ dn/d\lambda$
Linear dispersion	$dl/d\lambda = f(2\tan i/n)\,dn/d\lambda = f(B/b)\ dn/d\lambda$
Reciprocal linear dispersion	$R_D = d\lambda/dl = n[2f\tan i(dn/d\lambda)]^{-1}$ $= b/[fB(dn/d\lambda)]$
Resolving power	$R = \lambda/\Delta\lambda = B(dn/d\lambda)$
Spectral bandwidth	$s = w_s n[2f\tan i(dn/d\lambda)]^{-1} = w_s b/[fB(dn/d\lambda)]$
	Grating
Angular dispersion	$d\theta/d\lambda = m/(d\cos\theta) = (\sin i + \sin\theta)/[\lambda\cos\theta]$
Linear dispersion	$dl/d\lambda = fm/(d\cos\theta) = f(\sin i + \sin\theta)/[\lambda\cos\theta]$
Reciprocal linear dispersion	$R_D = (d\cos\theta)/fm = (\lambda\cos\theta)/[f(\sin i + \sin\theta)]$
Resolving power	$R = mN = (W/\lambda)[\sin i + \sin\theta]$ $R_{max} \sim 4W/3\lambda$
Spectral bandwidth	$s = (w_s d\cos\theta)/fm = (w_s\lambda\cos\theta)/[f(\sin i + \sin\theta)]$

which is almost independent of wavelength. Although Eq. 3.78 is the one commonly given for the angular dispersion of a diffraction grating, it should be realized that for a given spectrometer geometry m and d are not both independent variables (Loewen, 1970). If the grating equation (Eq. 3.74) is solved for m/d and the result substituted in Eq. 3.78, the angular dispersion expression that results is

$$d\theta/d\lambda = (\sin i + \sin\theta)/(\lambda\cos\theta) \tag{4.1}$$

Equation 4.1 indicates that for a given wavelength, the angular dispersion is a function solely of the angles of incidence and diffraction. If the grating is used in a Littrow mount (see Figure 4.3) where the angle of incidence and the angle of diffraction are equal, Eq. 4.1 can be reduced to

$$d\theta/d\lambda = (2\tan\theta)/\lambda \tag{4.2}$$

It is evident from Eq. 4.2 that high angular dispersion requires a large angle of diffraction. Once θ has been selected, the choice between a finely ruled grating used in a low order and a coarsely ruled grating used in a high order must be made.

Although the angular dispersion of a spectroscopic system is a fundamental property, the *linear dispersion* and *reciprocal linear dispersion* are often of more practical use to spectroscopists. The *linear dispersion* refers to the linear separation of two wavelengths in the focal plane of the instrument and has the units of length per wavelength. The linear dispersion can be obtained from the angular dispersion, as shown in Figure 4.4. This figure shows two light rays emerging from the dispersing device with an angular separation of $d\theta$. These diverging rays are eventually brought to a focus in the focal plane by the focusing element at a distance of f from the mirror. The linear separation of the two images can be approximated by

$$dl = f\sin(d\theta) \sim f d\theta \tag{4.3}$$

The linear dispersion follows directly from Eq. 4.3 and is given by

$$dl/d\lambda = f(d\theta/d\lambda) \tag{4.4}$$

where $dl/d\lambda$ is the linear dispersion, f is the focal length of the focusing element, and $d\theta/d\lambda$ is the angular dispersion. The reciprocal of the quantity given by Eq. 4.4 is known as the *reciprocal linear dispersion* and has units of wavelength per unit length in the focal plane,

$$R_D = d\lambda/dl = (1/f)(d\lambda/d\theta) \tag{4.5}$$

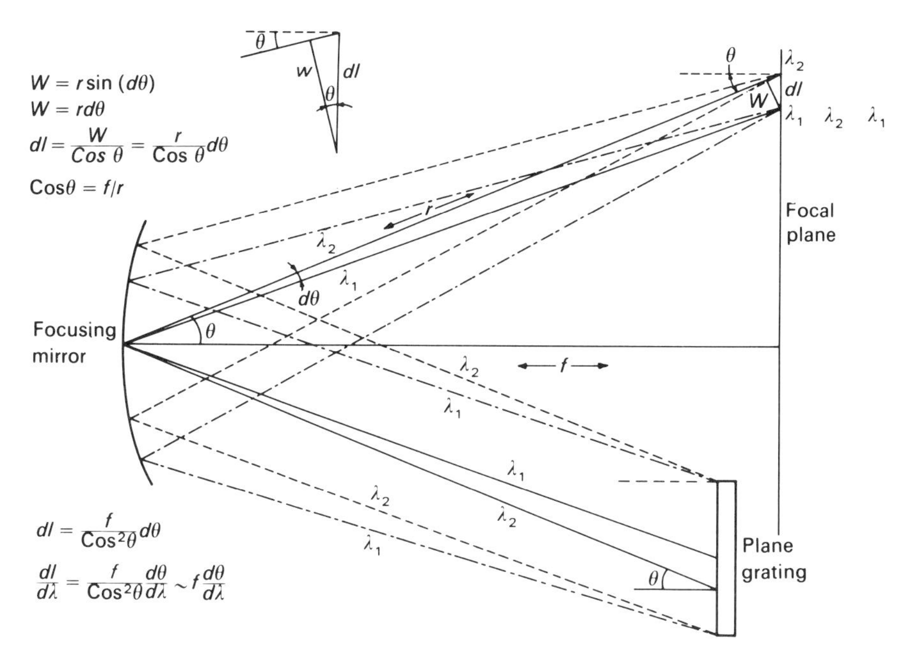

Figure 4.4. Geometric construction showing relationship between angular dispersion and linear dispersion for plane grating spectrometer.

Table 4.2. Approximate Reciprocal Linear Dispersion Values for Different Grating Systems

Focal Length (m)	Grating Ruling (lines/mm)	R_D (nm/mm)
0.25	1200	3.2
0.50	1200	1.6
1.0	1200	0.8
1.5	1200	0.5
3.0	1200	0.25
0.25	600	6.4
0.50	600	3.2
1.0	600	1.6
1.5	600	1.1
3.0	600	0.6

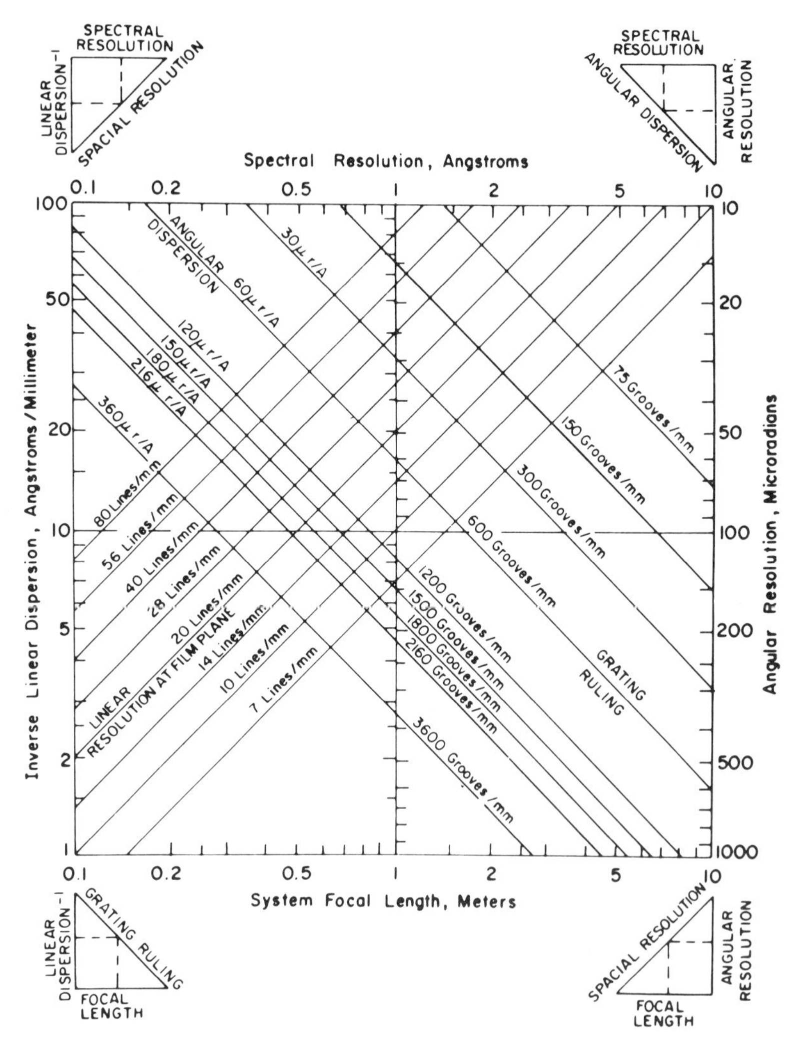

Figure 4.5. Nomograph of grating relationships in first order. Lower left-hand quadrant relates focal length to inverse linear dispersion as function of number of grooves per millimeter as outlined in diagram in that corner. Diagram in upper left-hand corner shows how to derive relationship between inverse linear dispersion and spectral resolution as function of detector resolution. Upper right-hand quadrant shows how to relate spectral resolution to angular resolution as function of angular dispersion. Lower right-hand quadrant is designed to relate angular resolution to focal length for various linear resolutions. [Reprinted with permission from *Diffraction Grating Handbook*, 1970, Bausch & Lomb, Inc.]

Table 4.2 gives some typical values for the reciprocal linear dispersion with different focal lengths and grating characteristics. Figure 4.5 gives a nomograph of grating relationships in the first order.

4.1.2. Resolving Power

The concepts of resolving power and spectral resolution are of considerable importance in spectroscopy and consequently deserve to be discussed in detail. The Rayleigh criterion for resolving power was discussed in Section 3.3 and briefly mentioned at the beginning of this chapter. As mentioned earlier, the image produced by a spectroscopic system when a narrow slit is used will be a diffraction pattern where the aperture causing the diffraction will be the aperture presented by the dispersing device (compare Figure 3.4 with Figure 4.1).

In Section 3.3, it was shown that for a diffraction-limited system two lines are just resolved according to the Rayleigh criterion when $\Delta\beta = \pi$. Since β is given by $\beta = (\pi b \sin\theta)/\lambda$,

$$d\beta = (\pi b \cos\theta \, d\theta)/\lambda \tag{4.6}$$

When $d\beta = \pi$,

$$\Delta\theta = \lambda/(b\cos\theta) \sim \lambda/b \tag{4.7}$$

where b is the width of the diffracting aperture and the approximation holds for small values of θ. Rearranging Eq. 4.7 gives

$$\lambda/\Delta\theta = b \tag{4.8}$$

Although we have discussed the criterion for the resolving power in terms of the diffraction images produced by the system, we have not, up to this point, formally defined the term. The *resolving power* of a spectroscopic system is a dimensionless number given by

$$R = \lambda/\Delta\lambda \tag{4.9}$$

where λ is the wavelength observed and $\Delta\lambda$ is the smallest wavelength difference that can just be distinguished according to the Rayleigh criterion. According to this definition and Eq. 4.8,

$$R = \lambda/\Delta\lambda = (\lambda/\Delta\theta)(\Delta\theta/\Delta\lambda) = b(d\theta/d\lambda) \tag{4.10}$$

For a prism under conditions of minimum deviation, it was shown in Section 2.2.1 that the angular dispersion is given by Eq. 2.27,

$$d\theta/d\lambda = (B/b)(dn/d\lambda) \tag{2.27}$$

Substituting this result into Eq. 4.10 gives the desired expression for the resolving power of a prism,

$$R = B(dn/d\lambda) \tag{4.11}$$

where B is the width of the base of the prism and $dn/d\lambda$ is the dispersion of the prism material. This equation applies rigorously only for the case of minimum deviation and if the entire face of the prism is illuminated.

For the case of a diffraction grating, it was shown in Section 3.4 that the resolving power is given by a number of equivalent expressions:

$$R = mN \tag{3.70}$$

$$R = W(\sin i + \sin\theta)/\lambda \tag{3.79}$$

$$R_{\max} \sim 4W/3\lambda \tag{3.80}$$

where the terms have been defined previously. It should be noted that the terms in Eq. 3.70, which is the usual equation given in most texts, are not independent (Loewen, 1970). Equation 3.79, which is a more useful expression, is obtained by solving the grating equation (Eq. 3.74) for m and substituting the result in Eq. 3.70. Equation 3.79 provides the basis for the use of echelle gratings (to be discussed later) because it implies that high resolution can be obtained even with a coarsely ruled grating if the grating is used in high orders. The maximum practical resolving power is obtained from Eq. 3.79 by substituting $i = \theta = 90°$ in Eq. 3.79 and multiplying by 2/3.

Before discussing the next topic, it is worthwhile to point out the relationship between resolving power and angular dispersion. For any dispersive system, the angular dispersion is equal to the resolving power divided by the linear aperture. This is clearly true for a prism used under conditions of minimum deviation, as can be seen from an inspection of Eqs. 4.11 and 2.27. In these equations, b is the linear aperture or cross-sectional width of the beam transmitted by the system. For a grating, the cross-sectional width of the beam is given by $b = W\cos\theta$, where W is the grating width and θ is the angle of diffraction. Dividing Eq. 3.70 by b gives the expression for angular dispersion (Eq. 3.78) where d is given by W/N.

4.1.3. Spectral Bandwidth

A spectroscopic system such as a monochromator that uses an exit slit to isolate the wavelength of interest actually transmits a range of wavelengths given by

$$s = wR_{\mathrm{D}} \tag{4.12}$$

where w is the exit slit width and R_{D} is the reciprocal linear dispersion. The quantity s given by Eq. 4.12 is known as the *spectral bandwidth* and represents the smallest wavelength interval the monochromator can isolate. The spectral bandwidth is often referred to by a number of names such as *spectral*

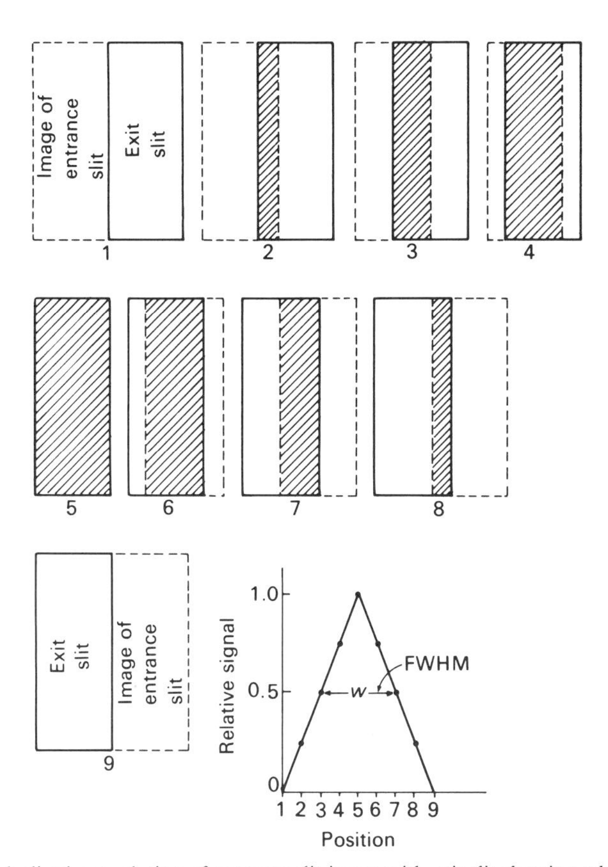

Figure 4.6. Idealized convolution of entrance slit image with exit slit showing relative detector response as grating is rotated.

resolution or simply *resolution*. These terms should not be confused with the resolving power discussed previously.

Figure 4.6 shows schematically how the signal from a detector placed after the exit slit varies as an isolated image of the entrance slit is rotated past the exit slit by slowly rotating the grating (Harrison, Lord, and Loofbourow, 1948). This situation can be realized by imaging a source emitting narrow atomic lines (such as a low-pressure mercury pen lamp) on the entrance slit of the monochromator. The monochromator is now set so as to almost pass one of the isolated spectral lines emitted by the source. As long as the image of the entrance slit corresponding to the given spectral line does not overlap the exit slit opening, no signal is registered by the detector. As the grating is rotated, the image of the entrance slit begins to overlap the exit slit opening and the detector registers a signal proportional to the extent of overlap. When the image of the entrance slit is coincident with the exit slit, the detector registers the maximum signal. Continued rotation of the grating now causes the image of the entrance slit to move past the exit slit, and the signal from the detector once again decreases in proportion to the extent of the overlap. When the image no longer overlaps the exit slit, the signal decreases to zero. In this example, the image of the entrance slit formed in the focal plane was assumed to be a geometric image of the slit, that is, the image was assumed to be rectangular rather than a diffraction pattern. This will be approximately true as long as the entrance slit is not too narrow.

It can be seen from Figure 4.6 that the signal generated by this procedure has a triangular shape that is often referred to as a *triangular slit profile*. From the geometry of the figure, it can be seen that the distance between the two half-intensity points on different sides of the triangle is equal to the slit width (the triangular slit profile will occur only when the entrance and exit slits are of

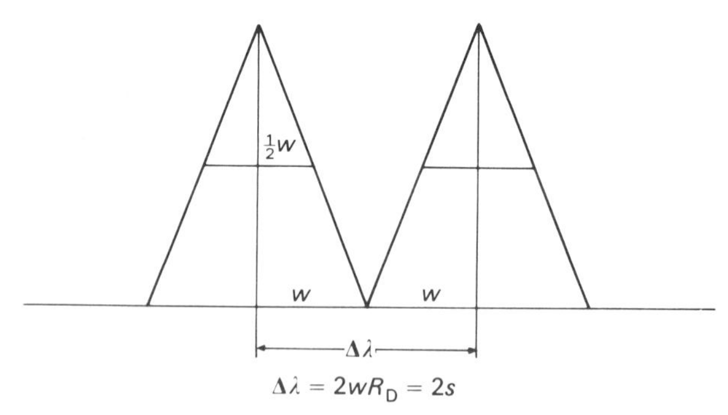

Figure 4.7. Wavelength separation required for complete resolution of two triangular slit profiles.

equal width). This width can be expressed in wavelength units by multiplying by the reciprocal linear dispersion as in Eq. 4.12. Thus the spectral bandwidth given by Eq. 4.12 is also known as the *half-intensity bandwidth* or the *full width at half maximum*, often abbreviated as FWHM by analogy with Figure 4.6.

Neglecting diffraction effects and optical aberrations, the half-intensity bandwidth can be used to estimate how closely two spectral lines can approach and still be completely resolved. Figure 4.7 shows that two triangular profiles must be at least $2s$ apart to avoid overlap. Thus the minimum wavelength separation between the two lines must be at least $2wR_D$.

4.1.4. Speed

The speed, or throughput, of an optical system can be expressed in a number of ways, although all are ultimately related to the solid angle of light collected by the instrument. In a typical instrument, the dispersing device is generally the limiting aperture, and the spectrometer is designed so that the dispersing device will be completely filled with light when the collimator is filled with light. Under these conditions, the f-number of the instrument will be determined by the area of the dispersing device and the focal length of the collimator. The f-number was defined in Chapter 2 by Eq. 2.46,

$$f/ \quad = f/d \tag{2.46}$$

where, in this case, f is the focal length of the collimator and d the equivalent diameter of the dispersing device. The equivalent diameter of the dispersing device can be obtained by finding the diameter of a circle whose area is equal to the area of the dispersing device. Thus,

$$d = 2(WH/\pi)^{1/2} \tag{4.13}$$

where W and H are the width and height of the dispersing device. If the dispersing device is oriented so that it makes an appreciable angle i with the incident beam from the collimator, the projected area of the dispersing device rather than the actual area should be used. The projected area is related to the actual area as shown in Figure 2.3 and is given by Eq. 2.6,

$$A_{\text{proj}} = A_{\text{actual}} \cos i \tag{2.6}$$

where i is the angle of incidence. This correction is not significant unless the grating is used with large angles of incidence. Neglecting the correction for the angle of incidence, a spectrometer with a 1-m focal length equipped with a grating with a ruled area of 102×102 mm has an aperture of $f/8.7$.

4.1.5. Factors Governing the Transmission of Radiant Energy through a Spectroscopic System

The determination of the manner in which the energy associated with a given radiation field flows through a spectroscopic system is related to the concept of speed. Although the nominal focal ratio is a good figure of merit for comparing the light-collecting ability of two spectrometers, the important factor from a spectroscopic point of view is the energy present in the slit images at the focal plane of the system. The effect caused by the presence of this energy in the focal plane of the spectrometer will depend on the mode of detection employed.

In comparing the flux-gathering ability of two spectroscopic systems (i.e., their *luminosities*), Jacquinot (1954) has shown that valid comparisons can only be obtained under conditions of equal resolving power for the two instruments.

If photoelectric detection by a photomultiplier is employed, the total flux passing through the exit slit onto the detector is the important factor. This photoelectric speed will depend on the radiance or brightness of the slit image, which has units of energy per unit area per unit solid angle. There are four factors that determine the radiance of the slit image (Harrison et al., 1948): (1) the radiance of the source itself, (2) the area of the source effective in illuminating the spectrum, (3) the solid angle of radiation from the source collected by the spectroscopic system, and (4) the transmission factor of the optical system.

In Section 2.2.3, it was shown that for an ideal lossless optical system the radiance of the image equals the radiance of the source. For this reason, the radiance of the slit image in the focal plane cannot exceed the radiance of the source itself at the given wavelength. Assuming that the spectroscopic system is properly illuminated so that the collimator is filled with light by the source and the optical system transmits all the energy present at the entrance slit to the focal plane of the instrument, the total energy striking the detector will depend on the area of the slit and the solid angle subtended at the exit slit by the focusing element. Since the signal from the detector will be proportional to the total energy striking the detector, the signal will be proportional to

$$dS = k(L\, dA\, d\Omega) \tag{4.14}$$

where S is the signal, k is a proportionality constant, L is the radiance of the slit image in the focal plane, A is the area of the exit slit, and Ω is the solid angle subtended by the focusing element at the exit slit. It is assumed, of course, that the detector is capable of receiving all the radiation transmitted through the exit slit. Thus for an isolated monochromatic spectral line, the signal will be proportional to the product of exit slit width and slit height (assuming the size of the entrance slit is equal to the size of the exit slit). If the spectrum monitored is a continuum, the signal will increase as the square of the exit slit width since the spectral radiance of the image is the quantity of importance. Thus,

$$dS = k(L\,dw\,dH\,d\Omega) \qquad \text{(monochromatic line)} \tag{4.15}$$

$$dS = k(L_\lambda\,dw\,dH\,d\Omega\,d\lambda) \qquad \text{(continuum source)} \tag{4.16}$$

$$dS = k(L\,dw^2\,dH\,d\Omega) \qquad \text{(continuum source)} \tag{4.17}$$

where L is the spectral radiance of the image. The factor w^2 arises because the wavelength range passed by the exit slit is equal to $d\lambda = wR_D$.

To increase the photoelectric speed of the monochromator, it is desirable to increase both the slit width and slit height as much as possible. It should be kept in mind that increasing slit width and slit height decreases the resolving power of the instrument, so that a compromise exists between optical throughput and resolving power. If both throughput and resolving power are important, the photoelectric speed of the instrument may be increased by increasing slit height if curved slits are employed (Fastie, Crosswhite, and Gloersen, 1958). For work requiring good resolving power, the slit height should be restricted to about 2 mm if straight slits are employed. The effect of slit length on resolution for curved and straight slits is shown in Figure 4.8.

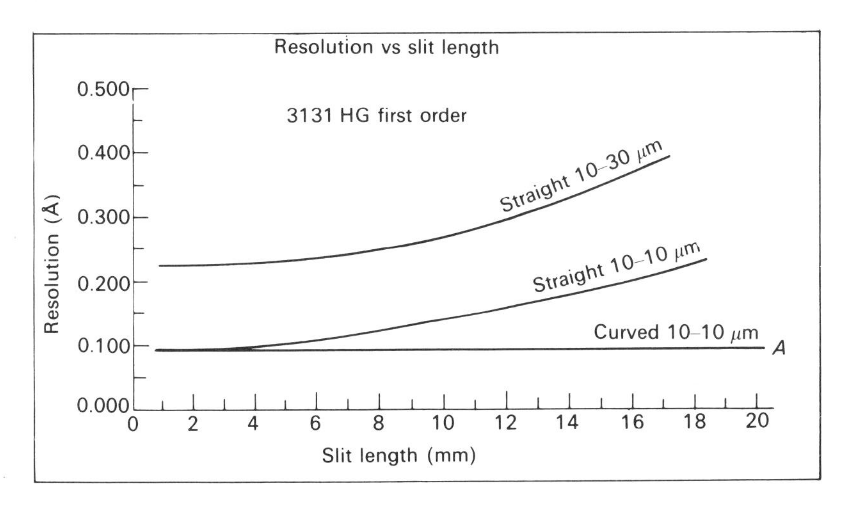

Figure 4.8. Effect of slit length on resolution for curved and straight slits. [Courtesy of Thermo Jarrell Ash Corporation.]

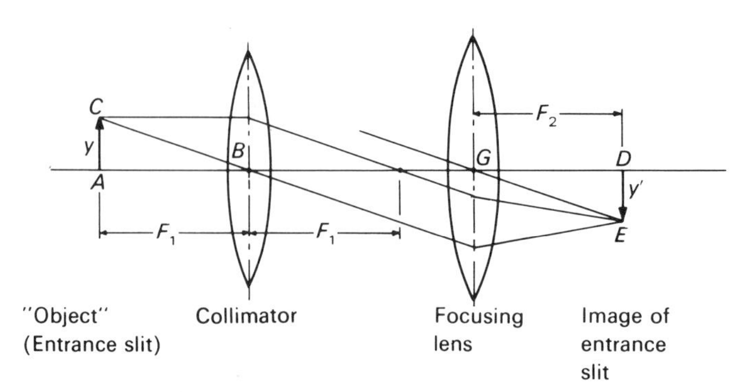

Figure 4.9. Image formation by typical spectroscopic system.

If phototographic detection or if an image detector is employed, the quantity of importance is not the total flux as was the case for photoelectric detection but the flux per unit area in the slit image, or the irradiance of the slit image. In this case, increasing the slit width or the slit height has no effect on the irradiance of the slit image and merely increases the size of the image. The irradiance of the slit image can be increased to a certain extent by forming a demagnified image of the entrance slit in the focal plane, thereby concentrating the energy over a smaller area. This may be accomplished by making the focal length of the focusing element shorter than the focal length of the collimator, as shown in Figure 4.9. Figure 4.9 shows the image-forming properties of a typical spectroscopic system. From the figure, it can be seen that triangles ABC and DEG are similar. As a result, the magnification of the slit image will be given by $m = F_2/F_1$, where $m < 1$ when $F_2 < F_1$. Since this will be true for magnification in both the horizontal and vertical planes, the area of the image in the focal plane will be related to the area of the entrance slit by

$$A_{\mathrm{im}} = A_{\mathrm{es}}\,(F_2/F_1)^2 \tag{4.18}$$

Decreasing the focal length of the collimator while maintaining the same diameter is equivalent to increasing the relative aperture of this component (decreasing the f-number). The extent to which this procedure will be effective in increasing the photographic speed of the instrument will be limited by the image size reduction that can be tolerated in the focal plane and by the maximum relative aperture of the focusing element available.

4.1.6. Grating Efficiency

One of the factors governing the transmission of radiant energy through a spectroscopic system that was mentioned in the previous section is the transmission factor. The transmission factor is a function of the absorption and reflection losses that occur in the system. For a grating instrument, the transmission factor is also related to the grating efficiency. For a reflection grating, the grating efficiency is defined as the ratio of the amount of monochromatic light diffracted into the given order to the specular reflection from a mirror coated with the same reflecting surface as the grating. In practice, the efficiency is measured with a double-monochromator system (Loewen, 1970), where the first monochromator supplies monochromatic light from various sources to a second monochromator containing the grating being tested. The signal for a given wavelength obtained with the grating is then compared with that obtained from a mirror with the same reflecting surface as the grating.

Figure 4.10 shows the results of this type of efficiency measurement for a reflection grating with 1180 grooves per millimeter blazed for 500 nm. This particular grating has an aluminum reflective surface on a Pyrex blank. Overcoating the surface of the grating with MgF_2 is useful in improving the short-wavelength performance. It should be mentioned that aluminized optics are subject to the formation of an oxide layer and can be adversely affected by fingerprints. Overcoating with MgF_2 provides a chemical protection from

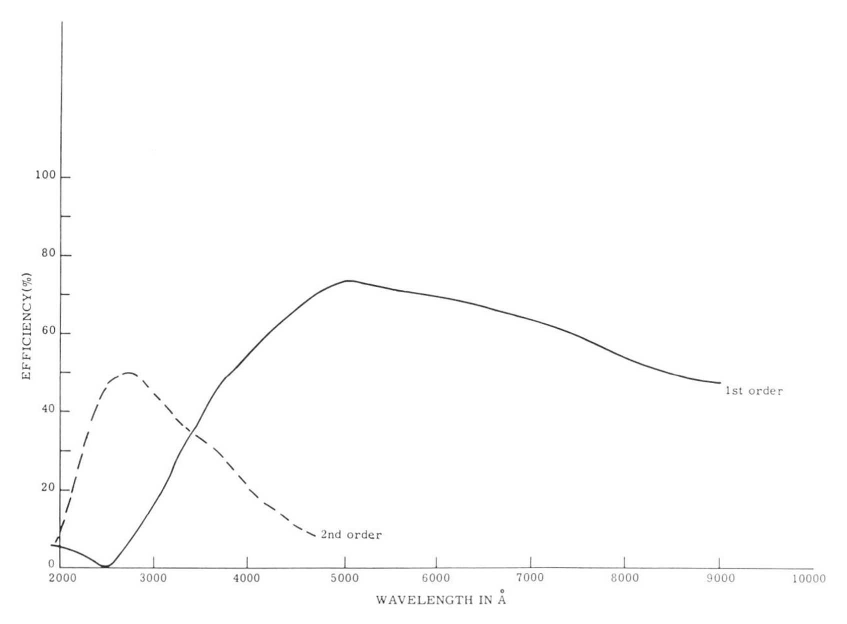

Figure 4.10. Experimentally determined grating efficiency for reflection grating with 1180 grooves per millimeter blazed for 500 nm. [Courtesy of Thermo Jarrell Ash Corporation.]

acids present in fingerprints, but gratings that have been coated in this fashion should not be washed in water because the coating has not been baked onto the surface as is commonly done with glass optics. As a result, such gratings should only be washed with spectral-grade organic solvents (Loewen, 1970). Before such drastic steps are taken, however, it would be wise to consult the manufacturer.

As discussed in Section 3.4, blazing the grating is a means of concentrating the diffracted light into a particular order by means of controlling the microgeometry of the grating grooves. When the grooves are cut at an angle so that the specular reflection from the individual groove facets for a given angle of incidence coincides with the angle of diffraction, a maximum in the grating efficiency is observed for that wavelength. As a rule of thumb for coarse gratings ($d > 2\lambda$), the efficiency of a grating in the first order drops to half its maximum value at $2/3\lambda_B$ and $3/2\lambda_B$, where λ_B is the blaze wavelength (Loewen, 1970). The same angle of incidence and diffraction that resulted in the maximum efficiency in the first order will also give a maximum efficiency in higher orders, as shown in Figure 4.10. Thus, for the same grating orientation, the maximum efficiency in the second order can be seen to occur at half the first-order blaze wavelength (i.e., 250 nm). As can be seen from Figure 4.10, the maximum efficiency in the second order is not as great as that in the first order, and the efficiency is a more pronounced function of wavelength.

4.1.7. Holographic Gratings

With the ready availability of a wide variety of diffraction gratings today, it is easy to take for granted the technological developments that were necessary to make modern diffraction gratings possible. Although the first diffraction grating is reported to have been constructed by the American astronomer David Rittenhouse in 1785 (Loewen, 1970), real development did not occur until the ninteenth century with the work of Professor Henry Rowland of Johns Hopkins University. Rowland's pioneering work on the development of sophisticated ruling engines needed to inscribe the grooves in a suitable substrate established the diffraction grating as a realistic tool for spectroscopy. The second major advance in this technology was the introduction of the interferometrically controlled ruling engine in 1955 by Stroke and Harrison of the Massachusetts Institute of Technology (Hayat and Pieuchard, 1973).

Improvements in the ruling of master gratings have led to the availability of finer ruled gratings with fewer imperfections. In fact, classically ruled gratings as fine as 3600 grooves per millimeter are available. Most gratings available today are so-called replica gratings, which are made from a master grating by making an impression of the master grating in a thin layer of clear resin. The thin layer is then stripped from the master and mounted on a rigid substrate material. The surface of the resin layer is then coated with aluminum, gold, or platinum depending on the wavelength region for which the grating is intended.

Within the past 10 years, advances in lasers and holography have made possible the production of diffraction gratings by an entirely new method. These so-called holographically recorded diffraction gratings have a number of improved characteristics such as lower stray light levels, higher signal-to-noise ratios, total absence of ghosts (to be discussed later), and the possibility of aberration correction for concave gratings (Hayat and Pieuchard, 1973). To produce a holographic grating, two beams of monochromatic light are made to produce an interference pattern on a photosensitive material that has been deposited on a glass substrate that is optically flat. The interference fringes produced on the photosensitive material are then processed chemically to produce the grooves. Finally, the reflective coating of aluminum is deposited by vacuum deposition. Using this procedure, gratings as large as 600 mm with up to 6000 grooves per millimeter have been produced.

In contrast to the classically ruled grating where the smooth facets of the groove faces are inclined to the surface by some angle (the blaze angle) giving rise to a triangular groove profile, early holographic gratings had a groove profile intermediate between a sinusoidal cross section and a saw-toothed cross section. As a consequence, the efficiency of these early holographic gratings was less than that obtained with classically ruled gratings, although the variation of efficiency with wavelength was less. Figure 4.11 shows a comparison between the efficiencies obtained from a classically ruled grating and an early holographic grating.

Within the past few years, however, great strides have been made in the production of holographic gratings (Lerner et al., 1980, 1983; Lerner and Laude, 1983), and it is now possible to produce blazed holographic gratings with efficiencies that approach classically ruled gratings. Such gratings are referred to as *ion-etched holographic gratings* and are produced by etching the glass substrate supporting the sinusoidal holographic photoresist mask with an argon ion beam. Using this procedure, triangular groove profiles have been obtained for blaze angles from 4° to 38° (Lerner and Laude, 1983).

As mentioned earlier, holographic gratings give superior stray light performance compared with classically ruled gratings. In addition, they show improved signal-to-noise performance and total freedom from ghosts. All of these improvements are due to the elimination of errors in the straightness, parallelism, and equidistance of the grooves found in classically ruled gratings. For example, *ghosts* are unwanted spectral signals that appear as false lines (Harrison et al., 1948). *Rowland ghosts* are due to periodic errors in

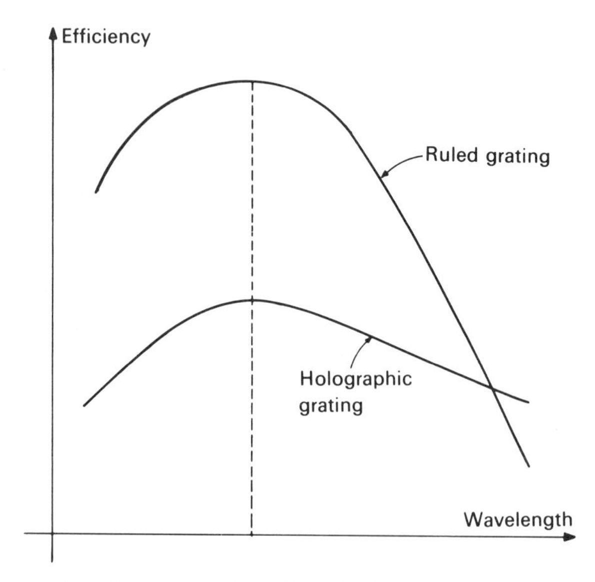

Figure 4.11. Comparison between grating efficiencies for classically ruled grating and early holographic grating. [Reprinted from G. S. Hayat and G. Pieuchard (1973), *Industrial Research, 40*, August. By permission of Cahners Publishing Co., New York.]

the screw that moves the diamond in the ruling engine. These periodic errors modulate the intensity of a given spectral line, producing side bands. Rowland ghosts are therefore found symmetrically about any strong line. *Lyman ghosts*, by contrast, are due to periodic vibrations originating outside the ruling engine and are often found widely separated from the parent line. Stray light from the grating arises from nonperiodic or random errors in the groove spacing and errors in the coplanarity of the facets. Since the holographic grating is a recording of a perfect optical interference phenomenon, all of the problems associated with errors in ruling are reduced or eliminated.

4.1.8. Free Spectral Range

Another parameter that characterizes the performance of grating instruments is the *free spectral range*, defined as the range of wavelengths that is free of overlap from adjacent orders. In Section 3.4, it was shown that according to the grating equation (Eq. 3.74), θ is not unique for a single wavelength even when i and d are held constant. Thus, even when θ, i, and d are all fixed, thereby holding the right-hand side of the grating equation constant, various wavelengths will be observed to overlap at the same position in the focal plane as indicated by Eq. 3.76. The effect of this overlap is shown in Figure 4.12.

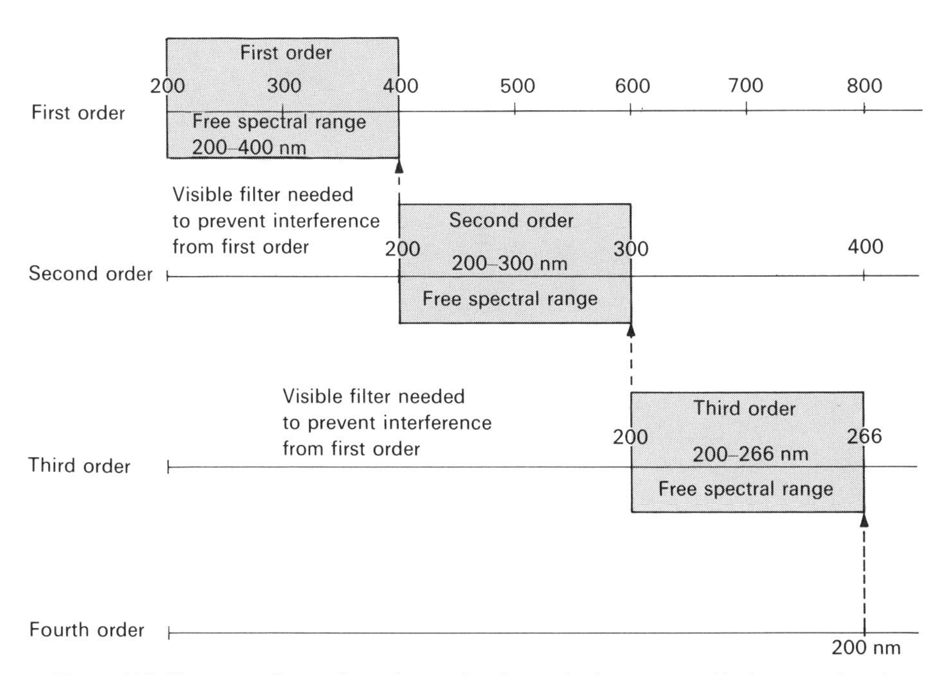

Figure 4.12. Free spectral range for various orders for grating instrument with short-wavelength limit of 200 nm.

The free spectral range of a grating instrument can be obtained directly from Eq. 3.76 by letting $m_1 = m_2 + 1$ and $\lambda_2 = \lambda_1 + \Delta\lambda$:

$$m_2(\lambda_1 + \Delta\lambda) = (m_2 + 1)\lambda_1 \tag{4.19}$$

Solving Eq. 4.19 for $\Delta\lambda$ gives the desired expression for the free spectral range,

$$\Delta\lambda = \lambda_2 - \lambda_1 = \lambda_1 / m_2 \tag{4.20}$$

Equation 4.20 shows that the free spectral range is directly proportional to the wavelength and inversely proportional to the orders under consideration. Suppose, for example, that the short-wavelength limit of an instrument (λ_1) is 200 nm in the first order. Equation 4.20 predicts that the free spectral range for this instrument will be 200 nm or from 200 to 400 nm. Beyond 400 nm, the second-order spectrum will overlap the first-order spectrum.

4.2. PLANE GRATING INSTRUMENTS

Because of the widespread application of the Czerny–Turner mounting, it is worthwhile to examine the geometry of this system in some detail. Figure 4.13

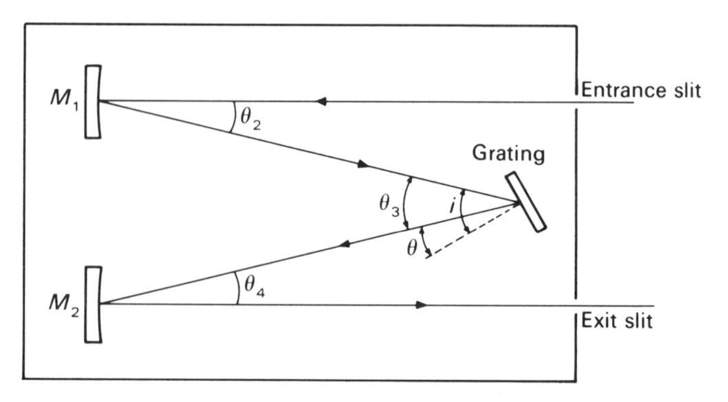

Figure 4.13. Optical arrangement for symmetrical Czerny–Turner grating spectrometer.

shows the optical arrangement for a so-called *symmetrical* system. With a symmetrical arrangement, angles θ_2 and θ_4 are equal. As discussed in Chapter 2, optical aberrations become more significant when components are used at large off-axis angles. One such aberration that is observed in this system is *coma*. Coma is a form of distortion that appears as a hump on the descending side of recorded peaks. In a symmetrical system, coma can be reduced by making angles θ_2 and θ_4 as small as possible. Using this approach, coma can be reduced to negligible proportions when the entrance and exit beams just barely clear the grating.

Although reducing the size of angles θ_2 and θ_4 is a perfectly satisfactory means of reducing coma for a Czerny–Turner monochromator, it will not work for a spectrographic system employing photographic detection or with a system having a large aperture. In both cases, the angle between the collimating mirror and the focusing mirror (θ_3) must be increased, thereby increasing angles θ_2 and θ_4. The solution to this problem was proposed by Fastie (1961), who showed that coma could be reduced to negligible proportions with an asymmetric arrangement even when large off-axis angles are employed. With the asymmetric configuration, angles θ_2 and θ_4 are no longer equal, and their ratio is given by

$$\theta_4/\theta_2 = \cos^3 i/\cos^3 \theta \tag{4.21}$$

where i is the angle of incidence on the grating and θ is the angle of diffraction.

Figure 4.14 shows the optical layout of an asymmetric Czerny–Turner mounting designed for photographic detection. By employing a large camera mirror that can accept a wide range of diffraction angles from the grating without attenuation, a wide unvignetted spectrum can be obtained.

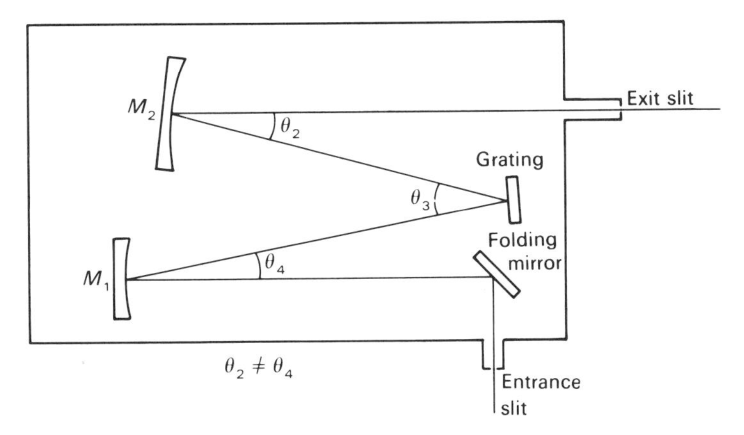

Figure 4.14. Optical arrangement for asymmetrical Czerny–Turner grating spectrometer.

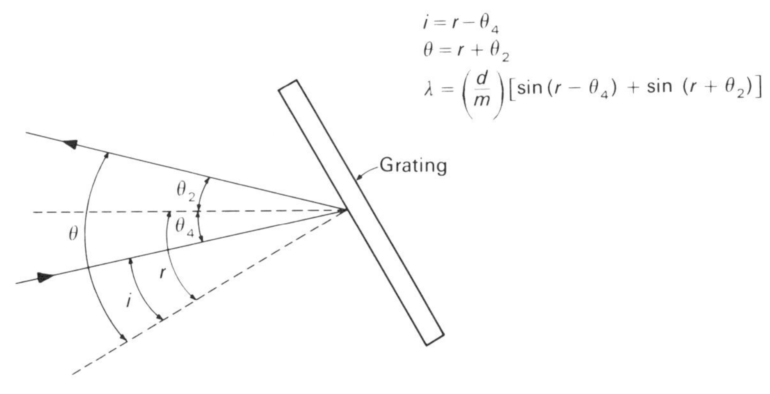

Figure 4.15. Geometric construction showing relationship between angles of incidence and diffraction and angle of rotation of grating. If $\theta_4 = 11.5$ and $\theta_2 = 13.75$, $m = 0$ when $r = -1.12°$.

Figure 4.15 shows how the angles of incidence and diffraction are related to the angle of rotation of the grating. From the figure, it can be seen that

$$i = r - \theta_4 \tag{4.22}$$

$$\theta = r + \theta_2 \tag{4.23}$$

where i is the angle of incidence to the grating, θ is the angle of diffraction, r is the angle of rotation of the grating as defined in the figure, and θ_2 and θ_4 have been defined previously in Figure 4.14. If Eqs. 4.22 and 4.23 are substituted into the grating equation (Eq. 3.74), the result is

$$m\lambda = d[\sin(r-\theta_4)+\sin(r+\theta_2)] \tag{4.24}$$

Since θ_2 and θ_4 are constants for the given optical layout, it can be seen that the first-order wavelength observed at a given position in the focal plane depends only on the grating constant d and the angle of rotation r. For a given grating, the wavelength observed at a given position in the focal plane depends only on the order m and the angle of rotation r.

Figure 4.16 shows a plot of $\sin(r-\theta_4)+\sin(r+\theta_2)$ versus rotation angle for an optical system with $\theta_4 = 11.5°$ and $\theta_2 = 13.75°$. For this system, the zeroth order from the grating will occur at $r = -1.12°$. From the figure, it can be seen that the function $\sin(r-\theta_4)+\sin(r+\theta_2)$ is a linear function of r out to about

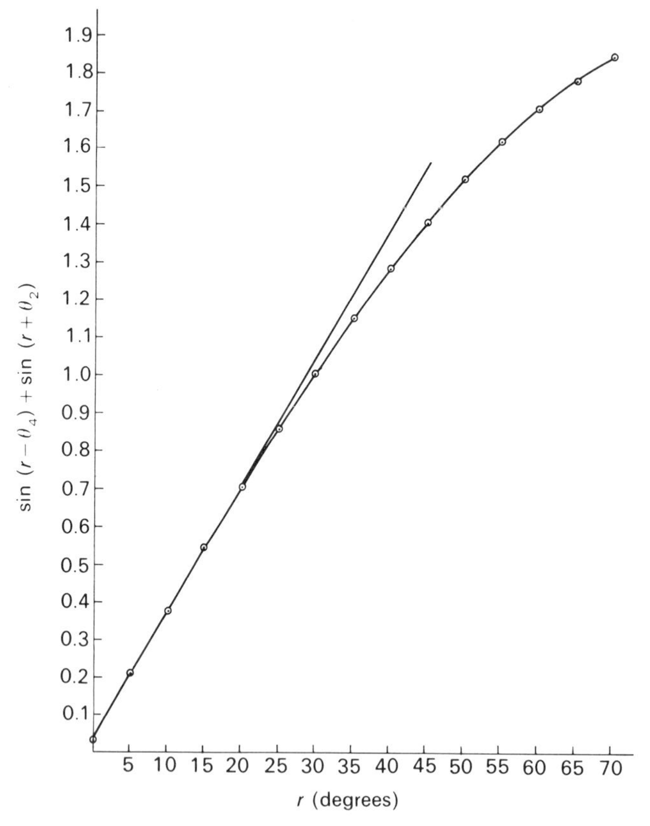

Figure 4.16. Plot of $\sin(r-\theta_4)+\sin(r+\theta_2)$ versus grating rotation angle for optical system with $\theta_4 = 11.5°$ and $\theta_2 = 13.75°$.

$r=30°$. In addition, it can be seen that the maximum possible angle of rotation for this particular design (i.e., for $\theta_2=13.75°$ and $\theta_4=11.5°$) will be 76°. Under this condition, $r+\theta_2=90°$. Although it is theoretically possible to rotate the grating through a large angle, the maximum practical angle of rotation will be determined by intensity considerations. For the system shown in Figure 4.14, the useful range of rotation angles varies from about zero to 40°. When a 1200-groove-per-millimeter grating is employed, this range of angles corresponds to a range from less than 200 to 1100 nm in the first order.

Figure 4.17 shows the effect of grating constant on the range of wavelengths covered by rotating the grating from zero to 30° for the spectrometer shown in Figure 4.14. It can be seen from the figure that the coarser the grating the longer the wavelengths that can be accessed by the system. If a 150-groove-per-millimeter grating is employed, it can be seen that the entire visible spectrum is compressed to a range of rotation angles from zero to 2°. For this grating, the major portion of the rotation angles corresponds to wavelengths in the infrared region of the spectrum.

From the geometry of the optical layout shown in Figure 4.14, it is possible to calculate the blaze angle needed to produce a given blaze wavelength. In Chapter 3, it was shown (Section 3.4) that for the blaze condition,

$$\phi-i=\theta-\phi \tag{3.81}$$

where ϕ is the blaze angle of the microfacets. Substituting for i and θ in terms of Eqs. 4.22 and 4.23 gives

$$\phi=\tfrac{1}{2}(2r+\theta_2-\theta_4) \tag{4.25}$$

Figure 4.18 illustrates the geometry of the situation. To calculate the required blaze angle corresponding to a particular wavelength, the value of r corresponding to the desired wavelength is determined from a plot of wavelength versus rotation angle (Figure 4.17) for the system under consideration and substituted in Eq. 4.25. Angles θ_2 and θ_4 are constants for the given spectrometer.

Equation 4.24 can be used to determine the change in wavelength for a given change in rotation angle. Substituting the appropriate trigonometric identities for the sum and difference of two angles in Eq. 4.24 gives

$$\lambda=(d/m)[(\cos\theta_4+\cos\theta_2)\sin r+(\sin\theta_2-\sin\theta_4)\cos r] \tag{4.26}$$

Equation 4.26 can be approximated by

$$\lambda=(d/m)(\cos\theta_4+\cos\theta_2)\sin r \tag{4.27}$$

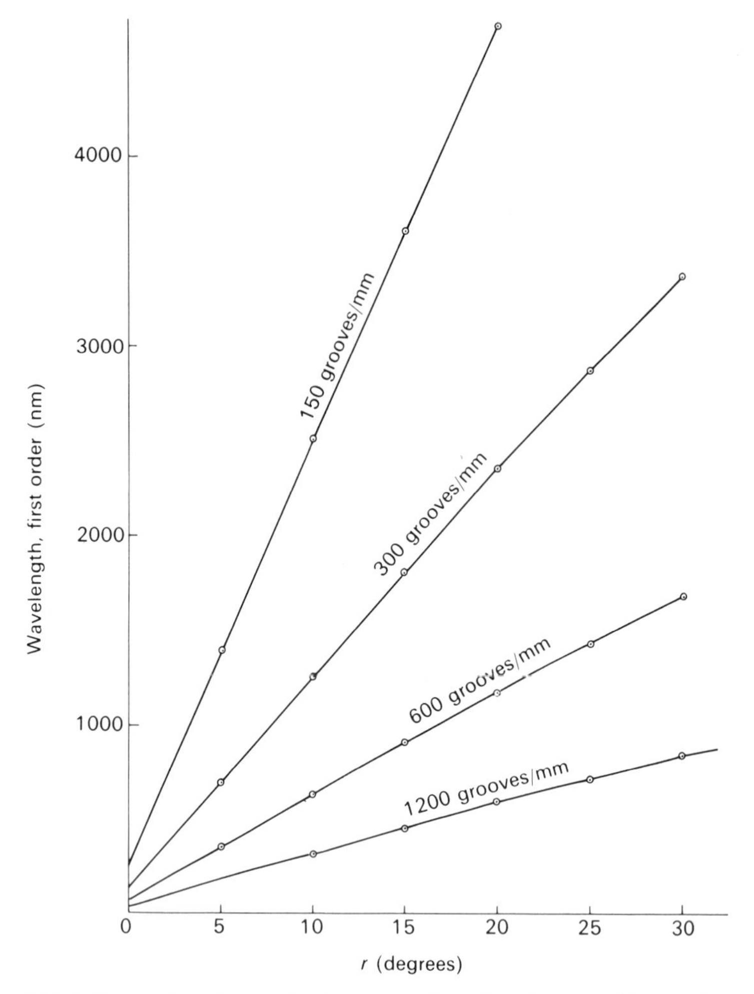

Figure 4.17. Influence of grating constant on range of wavelengths covered by rotating grating from 0° to 30° for spectrometer shown in Figure 4.14.

since the second term in Eq. 4.26 will be negligible because θ_2 and θ_4 are approximately equal. Equation 4.27 shows that a sine bar drive mechanism will provide a satisfactory means for producing a linear means of varying wavelength for scanning purposes.

To determine the tolerance needed in setting the rotation angle in order to achieve a particular tolerance in wavelength, Eq. 4.26 can be differentiated to give

$$d\lambda/dr = (d/m)[(\cos\theta_4 + \cos\theta_2)\cos r - (\sin\theta_2 - \sin\theta_4)\sin r] \qquad (4.28)$$

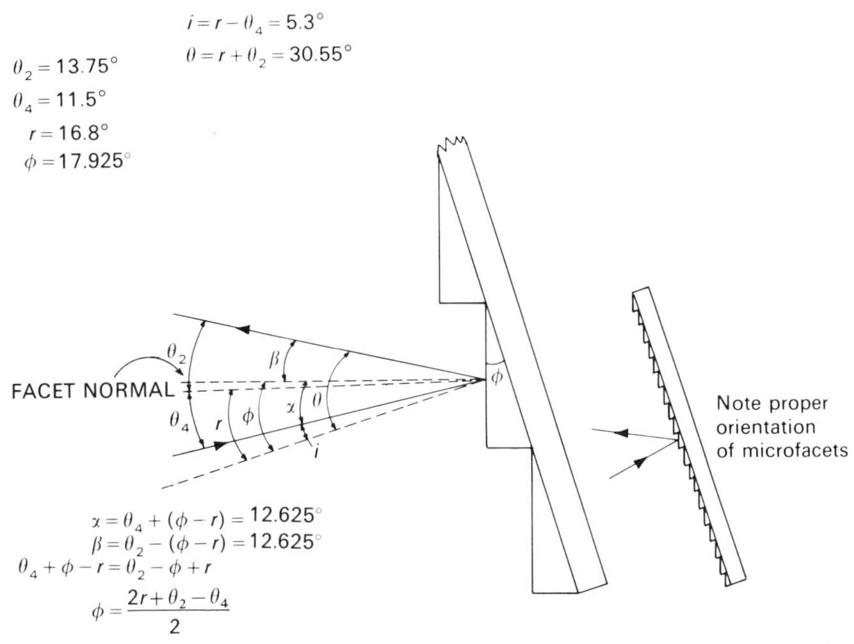

Figure 4.18. Geometric construction relating blaze angle of grating microfacets to grating rotation angle for spectrometer optical arrangement shown in Figure 4.14: α = angle of incidence to facet; β = angle of reflection from facet; i = angle of incidence measured from grating normal; θ = angle of diffraction measured from grating normal; r = angle of rotation of grating; ϕ = angle of microfacet with respect to grating surface.

Again the second term in Eq. 4.28 is negligible because θ_2 and θ_4 are approximately equal. Equation 4.28 can be used to get some idea of the tolerance required in accessing an atomic line reproducibly using a mechanical system that rotates the dispersion optics. If the width of an atomic line is taken as being on the order of 0.001 nm, Eq. 4.28 predicts that for a 1200-groove-per-millimeter grating with $\theta_2 = 13.75°$ and $\theta_4 = 11.5°$ the change in rotation angle, dr, would be approximately 6×10^{-7} rad, or 3×10^{-5} degrees!

4.3. ECHELLE SYSTEMS

The term *echelle grating* was introduced by Harrison (1949) to describe a specially ruled coarse grating capable of extremely high resolving power. The word *echelle*, which is French for "ladder," "scale," or "pair of steps," was chosen to indicate characteristics intermediate between an ordinary echelette grating and a *reflection echelon grating.*[†] At first glance, the use of a coarse grating to obtain high resolving power might seem unlikely, particularly in view of Eq. 3.70, which seems to imply that high resolving power is synonymous with finely ruled gratings (actually this is true only for a given order).

As mentioned under the discussion of resolving power (Section 4.1.2), Eq. 3.70 must be used with some caution since the terms m and N are not independent. The paradox can be resolved by Eq. 3.79, which was obtained from Eq. 3.70 by substitution of the grating equation (Eq. 3.74),

$$R = (W/\lambda)(\sin i + \sin \theta) \tag{3.79}$$

Equation 3.79 shows that the resolving power of a grating depends only on the grating width W and the angles of incidence and diffraction and is not dependent on the number of lines in the grating as long as the entire grating is filled with light. According to Eq. 3.79, high resolving power can be obtained by employing large angles of incidence and diffraction. If the grating is used in a Littrow mount where the angle of incidence equals the angle of diffraction, Eq. 3.79 reduces to

$$R = (2W \sin \theta)/\lambda = \Delta/t \tag{4.29}$$

where Δ is the optical path difference for two rays at the extremes of the grating and t is the width of a grating facet (See Fig. 4.19). Since many echelle spectrometers employ an angle of incidence of 63° compared with about 15° for an ordinary grating, Eq. 4.29 predicts that for the same wavelength the resolving power of the echelle system ought to be about 3.4 times that of the conventional grating spectrometer for gratings of similar width W.

Since the angular dispersion is equal to the resolving power divided by the linear aperture (Section 4.1.2), the angular dispersion under Littrow conditions will be

$$d\theta/d\lambda = (2 \tan \theta)/\lambda \tag{4.30}$$

since $b/W = \cos \theta$. These equations show that the resolving power depends on the grating width W and the angle of diffraction, while the angular dispersion

[†] A reflection echelon grating, first proposed by Michelson (Jenkins and White, 1957), consists of 20–30 plane parallel plates stacked together with an offset of about 1 mm. Since the plate thickness used was about 1–2 cm, which corresponds to about 20,000 wavelengths, a resolving power of about 600,000 should be possible with this grating. The reflection echelon has not been widely employed owing to the extreme difficulty in achieving the 1/80 of a wavelength uniformity between plates. Although commonly referred to as a "grating," the echelon is really a multiple-reflection interferometer.

depends only on the angle of diffraction. Thus, an echelle spectrometer with $\theta = 63°$ should have a dispersion 7.3 times that of a comparable conventional grating operated under Littrow conditions with $\theta = 15°$.

Figure 4.19 shows a diagram of a typical echelle grating employed in a Littrow configuration. From the figure, it can be seen that the grating consists of a series of steps with a wide face t and a narrow face s. Under normal conditions of illumination only the narrow face s is used, with illumination at or near normal to this face. Under normal incidence to s, the order m between successive grooves will be given by $m = 2t/\lambda$. Since $\tan\theta = t/s$, the angular dispersion will be given by

$$d\theta/d\lambda = 2t/s \tag{4.31}$$

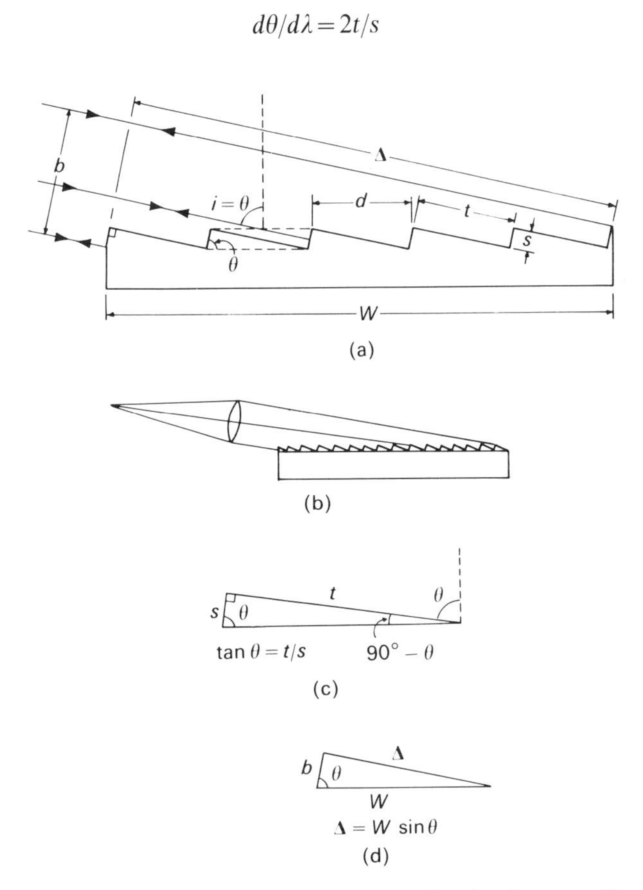

Figure 4.19. Echelle grating: (*a*) cross-section of echelle grating showing steps; (*b*) use of echelle grating in Littrow configuration; (*c*) geometric arrangement of grating microfacet; (*d*) geometric relationship between width of collimated beam b, grating width W, and optical path difference between two rays at extremes of grating (Δ).

Echelle gratings are generally characterized by an *r-number*, where r is the ratio of the groove width t to the groove height s. Thus, an $r/2$ echelle grating refers to one where $t/s = 2$. Used in the autocollimation mode (Littrow), such a grating would be illuminated at an angle of incidence of $\tan^{-1} t/s$, or 63°26′, as can be seen from Figure 4.19*c*. Since this angle of incidence is normal to the facet s and since the diffraction angle equals the angle of incidence in the Littrow mount, the blaze angle for the grating is 63°26′ since this also corresponds to the angle for the direct reflection from the facet. Table 4.3 gives the characteristics of some commercially available echelle gratings.

In most applications, echelle gratings are always used with some form of cross dispersion. To understand the need for this orthogonal dispersion, we must consider the free spectral range obtained with the echelle. In Section

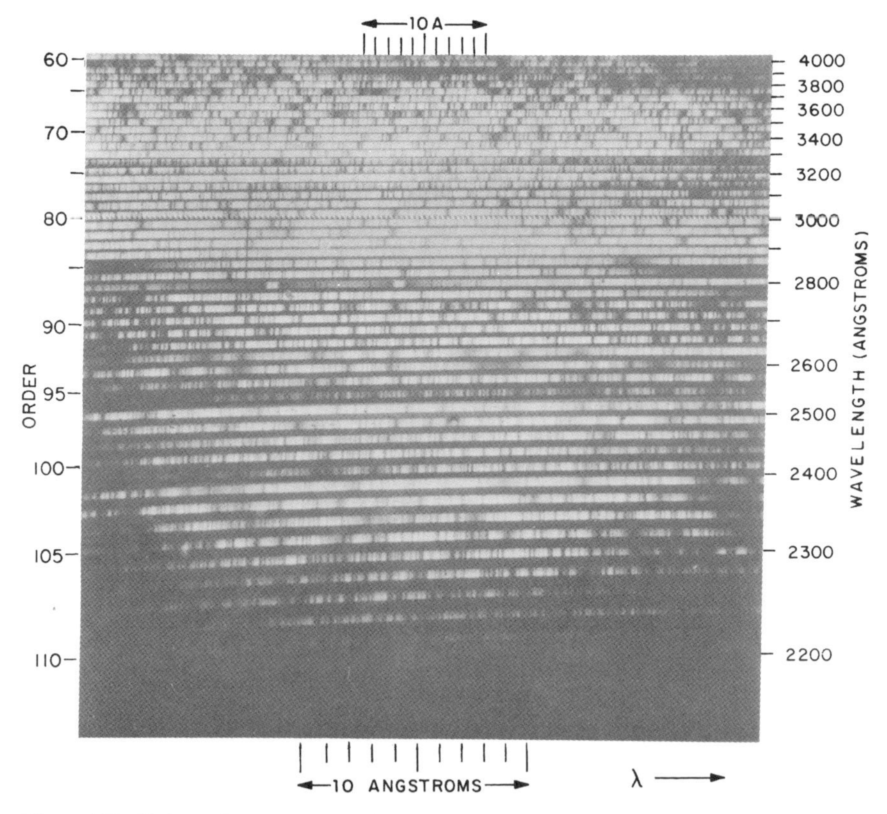

Figure 4.20. High-resolution spectrum from 220 to 400 nm obtained with echelle system and photographed onto film 25 mm square. [Reprinted from D. L. Garrett, S. D. Purcell, and R. Tousey (1962), *Appl. Opt.*, *1*, 726. By permission of the Optical Society of America.]

Table 4.3. Typical Commercial Echelle Gratings

Size (mm)	r-Number	Blaze Angle	Groove Spacing (groves/mm)
128×254	$r/2$	63°26′	31.6
128×254	$r/2$	63°26′	79
128×254	$r/2$	63°26′	316
128×254	$r/4$	63°26′	79

Source: Bausch & Lomb technical literature.

4.1.8, the free spectral range for a grating instrument was shown to be equal to λ/m. It is clear from this equation (Eq. 4.20) that since the echelle grating is used in very high orders, the free spectral range will be correspondingly small. For this reason, some form of cross dispersion is necessary to separate the overlapping orders from one another. The actual free spectral range for a given echelle grating can be determined by substituting $m = 2t/\lambda$ into Eq. 4.20,

$$F = \lambda^2/2t \tag{4.32}$$

Although the need for cross dispersion might, at first glance, appear as a severe disadvantage for analytical applications, the two-dimensional spectral format that results enables an enormous amount of spectral information to be concentrated over a relatively small area. Figure 4.20 shows a high-resolution spectrum from 220 to 400 nm obtained with an echelle system and photographed onto a 25-mm^2 film.

4.4. CONCAVE GRATING MOUNTINGS

As mentioned earlier, use of a concave grating eliminates the need for the collimating and focusing elements associated with plane grating systems. In this section, the important characteristics of five different optical configurations employing concave gratings will be described. Of the five mountings to be discussed, four are variations of the Rowland mounting, with the optical components located in various positions along the Rowland circle. All of the mountings except the Wadsworth suffer from varying degrees of astigmatism from the concave reflecting surface. In comparing the characteristics of the different mountings, it should be realized that the "ideal" or perfect optical configuration does not exist, and the best configuration will depend to a large degree on the nature of the experiment under consideration. Table 4.4

Table 4.4. Comparison of Concave Grating Mountings

Advantages	Disadvantages
	Rowland Mounting
Uniform dispersion	Limited spectral range available at given setting; high degree of astigmatism; higher order spectra not accessible; cumbersome, both plate and grating must be moved to change wavelength region.
	Abney Mounting
Plate and grating stationary; slit moved to different locations	Difficulty in moving source, condensing lenses, and other external equipment whenever slit is moved; similar to Rowland mounting.
	Paschen–Runge Mounting
Slit, grating, and plateholder all fixed; entire spectrum in focus, may be focused with single exposure on many plates; flexible, orders on both sides of grating accessible	Fairly bulky if whole circle is employed; astigmatic
	Eagle Mounting
Occupies minimum of space; astigmatism low; higher spectral orders may be accessed compared with Rowland mounting	Three adjustments required to change wavelength region monitored
	Wadsworth Mounting
Stigmatic; stigmatic range approximately one-sixth plate-to-grating distance	Dispersion for given grating cut by one-half compared with Rowland circle mounting; focal plane parabolic, curvature of plate must be changed from one wavelength setting to another; limited to coarser ruled gratings, 30,000 lines per inch maximum

summarizes the characteristics and advantages and disadvantages of the various arrangements.

4.4.1. Rowland Mounting

Historically, the Rowland mounting is the oldest of the common concave grating mountings. In the Rowland mounting (see Figure 4.2), the grating and plateholder are mounted at opposite ends of a rigid bar so that the grating is parallel to the plateholder and the grating-to-plateholder distance is fixed. The rigid bar, in turn, is arranged to move on rails placed at right angles to each other with the entrance slit located at the junction between the rails. As can be seen from Figure 4.2, this arrangement results in the three optical components being located on the Rowland circle. To change the wavelength range monitored by the plateholder, the rigid bar holding the grating and plateholder is moved to various locations as allowed by the orthogonal rails on which the ends of the bar are mounted. In this way, the angle of incidence to the grating is varied, while the angle of diffraction is maintained at zero.

The primary advantage of the Rowland mounting is the uniform dispersion obtained because the angle of diffraction is approximately zero (see Eq. 3.78). As a result of the uniform dispersion, this configuration is particularly useful for making wavelength measurements based on a limited number of standard lines.

On the negative side, the Rowland mounting is somewhat cumbersome to use as a result of the fact that both the plate and the grating must be moved to change the wavelength region observed. In addition, only a limited region of the spectrum can be observed at a given setting, and higher order spectra cannot be observed. As with most of the mountings employing classically ruled concave gratings, the Rowland mounting exhibits a high degree of astigmatism.

4.4.2. Abney Mounting

The Abney mounting, shown in Figure 4.21, is actually a variation of the Rowland mounting with the grating and plateholder at opposite ends of a rigid bar. In contrast to the Rowland mounting, the bar is maintained stationary (thereby overcoming one of the disadvantages of the Rowland mounting) and the slit is moved to various locations around the Rowland circle. To access different spectral regions, several slit positions are often provided in commercial embodiments of this configuration.

Like the Rowland mounting, the Abney mounting suffers from a high degree of astigmatism and is relatively bulky. Although the grating and plateholder are maintained stationary, this convenience is achieved at the

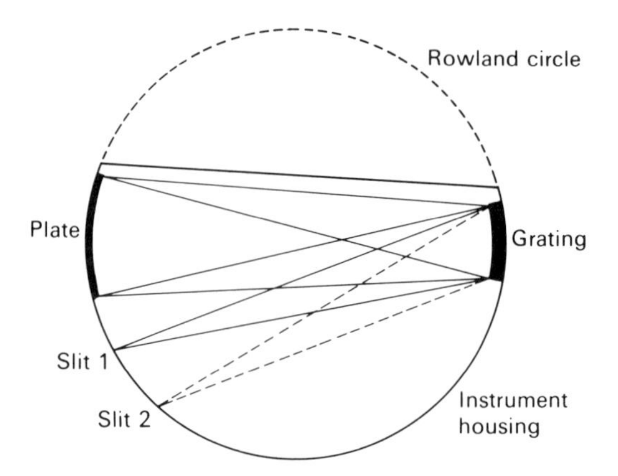

Figure 4.21. Concave grating used in Abney mounting.

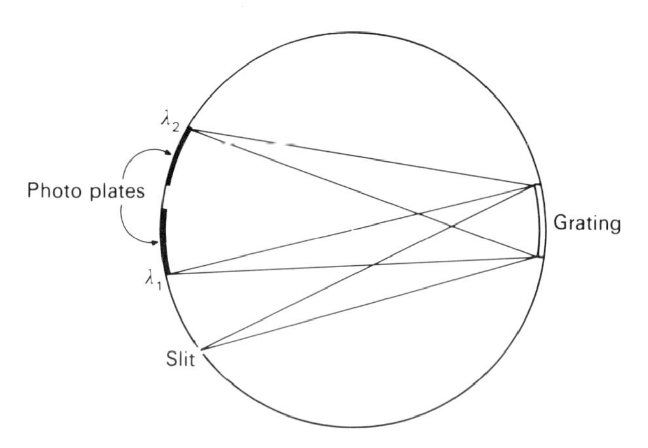

Figure 4.22. Concave grating used in Paschen–Runge mounting.

expense of having to move the source, condensing lenses, and any other external equipment whenever a different entrance slit is used. If this equipment is complex, the Abney mounting may actually be less convenient than the Rowland mounting.

4.4.3. Paschen–Runge Mounting

For high-resolution research applications in spectroscopy where space is not limited, the Paschen–Runge mount (Figure 4.22) has many advantages. A primary characteristic of this configuration is that the slit, grating, and plateholder are all fixed around the Rowland circle. In a typical research installation, the components are located in a light-tight room that serves as the enclosure for the spectrograph. A fixed, circular tract following the Rowland circle serves as a holder for the desired number of photographic plates. Since the plates, grating, and slit are all located on the Rowland circle, all parts of the spectrum are simultaneously in focus, and the entire spectrum may be observed with a single exposure by using many photographic plates. The arrangement is flexible in that it allows the operator to use only orders on one side of the normal, if desired, or the entire Rowland circle if orders on both sides of the grating normal are of interest. The major disadvantages are astigmatism and the amount of space required.

4.4.4. Eagle Mounting

The Eagle mounting, shown in Figure 4.23, is similar in many respects to the geometry associated with the Littrow mount (Figure 4.3). In this arrangement, the grating and plateholder are mounted at opposite ends of a rigid bar. To access different wavelength regions with the system, a series of three operations must be performed. First, the distance between the grating and plateholder must be selected. Then, the grating and plateholder must both be rotated to conform to the Rowland circle corresponding to the particular grating–plateholder separation. In commercial instruments these adjustments are usually done simultaneously by means of electric motors. The advantages of

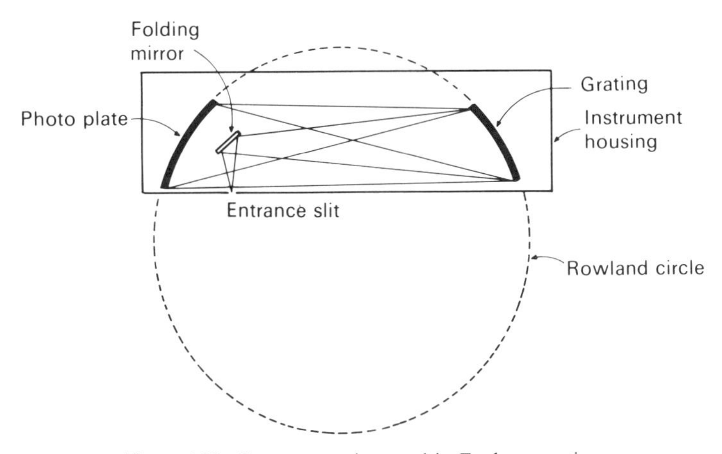

Figure 4.23. Concave grating used in Eagle mounting.

the Eagle mounting are that it is usually long and narrow and therefore occupies a minimum of space and, compared with the Rowland mounting, suffers less from astigmatism and can access higher order spectra.

4.4.5. Wadsworth Mounting

The final mounting we consider is the Wadsworth mounting shown in Figure 4.24. The major advantage of the Wadsworth configuration compared with the other arrangements that have been discussed is that the Wadsworth mounting is the only one that does not suffer from a high degree of astigmatism. This is accomplished by illuminating the grating with parallel light and observing the spectrum normal to the grating. With a properly designed system, a stigmatic spectrum with a range of about one-sixth the grating-to-plate distance is possible. In modern systems, light from the entrance slit is collimated by means of a concave mirror before it strikes the grating. In a typical system, the grating and plateholder are mounted at opposite ends of a rigid bar so that the plate is always parallel with the grating (i.e., the spectrum is always observed on the normal to the grating). Since the slit and concave mirror are fixed, rotation of the bar about point O in Figure 4.24 changes the angle of incidence to the grating, thereby changing the wave length region observed by the plate.

On the negative side, the Wadsworth mounting produces only half the dispersion obtainable with a given concave grating in a Rowland circle-based mounting. This is due to the fact that the focal plane obtained when the grating is illuminated with collimated light occurs at the focal length of the grating blank, which is half its radius of curvature. Thus, instead of coming to a focus on the Rowland circle (whose diameter is the radius of curvature of the grating blank), the Wadsworth system focuses at a distance approximately equal to the radius of the Rowland circle. Since the radius of the Rowland circle is half its diameter, the dispersion is reduced by a factor of

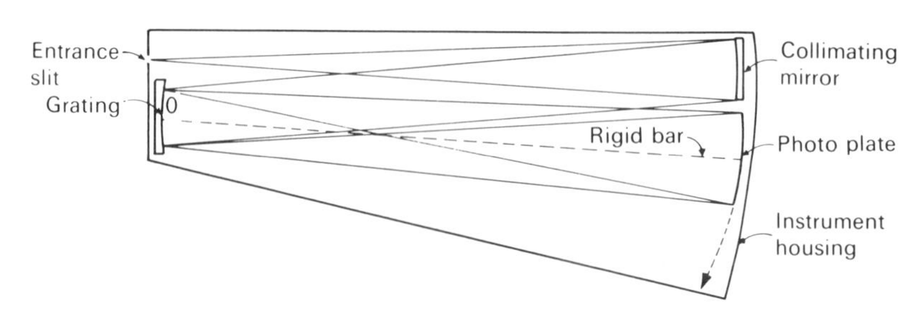

Figure 4.24. Concave grating used in Wadsworth mounting.

2. In addition, the focal plane obtained with the Wadsworth arrangement is parabolic. As a result of the parabolic focal plane, the plate-to-grating distance must be altered slightly as the bar holding the grating and plate-holder is rotated, and the curvature of the plate must also be changed. Finally, the Wadsworth mounting is limited to angles of incidence not greater than about 45°. This limits the Wadsworth mounting to gratings coarser than about 30,000 lines per inch if wavelengths out to about 800 nm are to be observed.

4.4.6. Selecting a Mounting

Probably the single most important factor to consider when selecting a concave grating mounting is the space available to house the instrument. If space is not limited, then either the Wadsworth or the Paschen–Runge mounting offer important advantages compared with the other alternatives. The choice between the Wadsworth and the Paschen–Runge arrangement will depend on whether astigmatism is an important factor in the particular experiment to be performed. If space is limited, the long, narrow configuration of the Eagle mounting may provide an attractive alternative.

4.5. SPECTROMETER ILLUMINATION

The performance obtained with a given spectrometric optical system is highly dependent on the manner in which the instrument is used. Indeed, the ultimate performance obtained with a given system may depend more on how the system is used than on the quality of the optical components themselves. To a large extent, however, the particular form of spectrometer illumination and slit width that is selected will depend on the nature of the experiment. For work requiring high resolution where line shape is an important factor, careful attention to the mode of spectrometer illumination and slit width will be necessary. For quantitative analytical spectroscopy, uniformity of spectral lines may be more important for accurate densitometry of photographic plates, and wide slits may be employed to give flat-topped line profiles that facilitate densitometry. Often in quantitative work, energy throughput may be an important factor, particularly in arc and spark work where the exposure time is limited. In some experiments, it may be desirable to sample radiation from selected regions of the spectral source. Because of the diverse nature of these various spectroscopic experiments, no one form of spectrometer illumination will be adequate for all spectroscopic work.

To begin our discussion of this topic, let us consider the case where high resolution is an important consideration. Theory and experiment have shown that there exists an optimum slit width that provides the best compromise between resolution and energy throughput. The compromise is such that using slit widths larger than the optimum merely degrades resolution without substantially increasing the intensity of the spectral line. Conversely, using slit widths more narrow than the optimum drastically reduces intensity without significant improvement in the resolution. The actual value for the optimum slit width in units of wavelength is known to depend on the mode of illumination of the slit (Stockbarger and Burns, 1933).

As discussed previously, maximum resolving power can be achieved only if the main aperture of the spectrometer is filled with light. Since many spectral sources of interest do not subtend a sufficient horizontal angle to fill the collimator, some form of diffraction broadening by the slit itself can be used to achieve the desired result. The width of the central maximum of a diffraction pattern produced by a diffracting aperture under conditions of Fraunhofer diffraction is given by

$$d = 2f\lambda / w \tag{4.33}$$

where d is the width of the central maximum (i.e., the distance between lateral minima), f is the focal length of the collimator, λ is the wavelength, and w is the slit width. As seen from Eq. 4.33, decreasing the slit width increases the width of the diffraction pattern produced at the collimator. In principle, the diffraction pattern can be made to fill any collimator by making the slit width sufficiently small; beyond a certain point, however, further reduction merely reduces intensity without increasing resolving power.

Schuster (1905) found that the best compromise between resolution and intensity occurred when the central maximum of the diffraction pattern produced on the collimator was twice as wide as the collimator itself. Thus,

$$d = 2D = 2f\lambda / w \tag{4.34}$$

or solving for w,

$$w_s = f\lambda / D \tag{4.35}$$

where D is the diameter of the collimator, and w_s is the so-called *Schuster slit width*. It can be seen by comparing Eqs. 3.31 and 4.35 that the Schuster slit width corresponds to a 1:1 geometric image of the slit in the focal plane, which would be one-half the width of the central maximum of the diffraction pattern produced by the main aperture (i.e., the collimator or dispersing element).

According to Eq. 4.35, the Schuster slit width for a 0.5-m spectrometer with a 62-mm grating should be 3 μm for 372 nm radiation. Experience has shown (Harrison et al., 1948; Sawyer, 1963; Slavin, 1971) that the optimum slit width for a given experiment will be between 1 and 2 times the Schuster slit width depending on the mode of illumination of the slit. When noncoherent radiation is used, the optimum occurs at approximately the Schuster slit width. When coherent radiation is used, the optimum occurs at approximately twice the Schuster slit width.

Noncoherent illumination of the slit is achieved when each point of the slit behaves as an independent radiator. In this way, rays from various parts of the slit reach the collimator in a random mixture of phases. This form of illumination can be achieved with a self-luminous slit or a spectral source of slit dimensions. It is approximated whenever an image of the source is focused on the entrance slit or when the source itself is positioned directly in front of the slit.

Coherent illumination of the slit occurs when the rays from the source all arrive at the slit in phase. This form of illumination can be approximated by a point spectral source placed a distance of 1–2 m from the slit.

In practice, the slit illumination is likely to be a mixture of the two limiting forms. As a result, it is common practice to calculate the critical slit width from Eq. 4.35. The effect of further slit widening is then determined experimentally by observation of the intensity as the width is varied from slightly less than the critical value to greater values. At the critical slit width, the brightness of the image in the focal plane will be observed to increase dramatically.

4.6. TYPES OF SLIT ILLUMINATION

As the slit is opened beyond the Schuster width (to, say, 2–4 times w_s or greater) the amount of diffraction that occurs at the slit is no longer adequate to fill the solid angle of the collimator. As a result, this task must be fulfilled by the illuminating system external to the spectrometer. Since diffraction effects at the slit are small, rays passing through the slit can be safely assumed to follow the principles of geometric optics. To obtain maximum resolution and energy throughput, radiation from the source must be made to fill the entire aperture of the slit and the collimator. Implicit in this statement is the proviso that when the collimator is filled with light, the dispersing element is also correspondingly filled with light. This condition will certainly be true in any properly designed spectroscopic system.

It was shown in Section 2.2.3 that for an ideal lossless optical system where the image formed is in the same medium as the object, the brightness of the image equals the brightness of the source. This law of conservation of radiance means that it is not possible to design some clever optical system for illuminating the slit of the spectrometer that will somehow produce a greater radiance than the source itself. As a result, if the source is large enough to fill the collimator of the spectrometer by itself, the simplest form of slit illumination is to place the source in front of the slit. The dimensions of the source needed to fill the collimator may be determined from geometric considerations, as shown in Figures 4.25 and 4.26. The condition that must be met to ensure complete filling of the aperture is that any line drawn from any portion of the collimator and passing through any portion of the slit must intersect an emitting region of the source. If the spectral source in question is not large enough to fulfill this condition, some form of condensing lens (or mirror) system will be necessary.

Before discussing the use of lens systems for spectrometer illumination, it is worthwhile to examine the geometries involved in Figures 4.25 and 4.26. Figure 4.25 shows a ray diagram for a plane perpendicular to the entrance slit of the spectrometer (i.e., a horizontal plane). As shown in the figure, the source

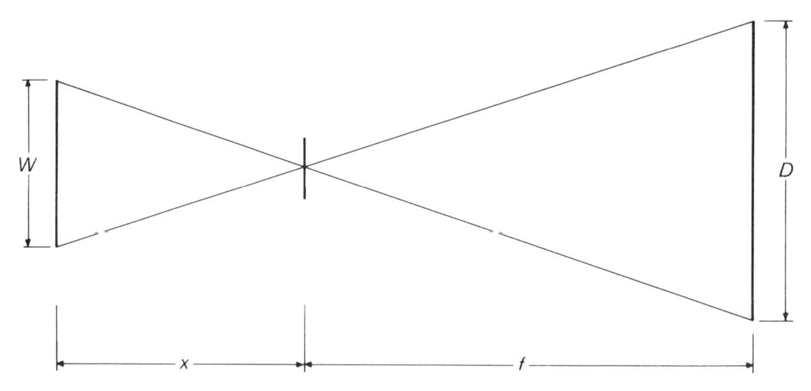

Figure 4.25. Marginal rays from source needed to fill spectrometer optics in horizontal plane.

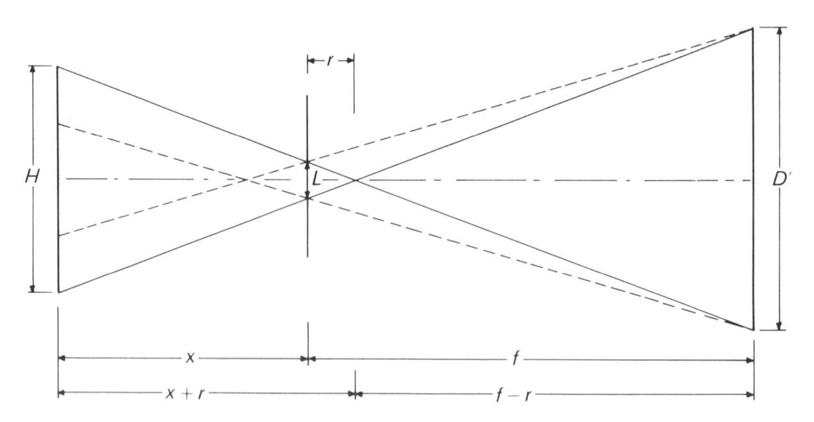

Figure 4.26. Ray diagram for vertical plane of spectrometer.

must have a width W in order to fill the collimator of width D and focal length f when placed a distance of x from the slit. From the geometry of the figure, it is apparent that

$$W/x = D/f \tag{4.36a}$$

or

$$W = Dx/f \tag{4.36b}$$

It is assumed in this calculation that the width of the slit itself is negligible.

Figure 4.26 shows a ray diagram for a plane parallel to the slit (i.e., a vertical plane). In this figure, D' is the height of the collimator, f is the focal length of the collimator, L is the length of the slit, and H is the height of the source needed to fill the collimator when placed at a distance x from the slit. From the geometry of the figure, it can be shown by similar triangles that

$$D'/(f-r) = H/(x+r) \tag{4.37}$$

and

$$L/r = D'/(f-r) \tag{4.38}$$

Solving Eq. 4.38 for r and substituting the result in Eq. 4.37 gives

$$H = [L(f+x)]/f + D'x/f = L[1 + (x/f)] + D'x/f \tag{4.39}$$

It is clear from Eqs. 4.36 and 4.39 that to completely fill the aperture of the spectrometer, the source must be larger in the vertical direction (H) than in the horizontal direction (W). This asymmetry in dimensions can lead to some difficulties in achieving the goal of filling the collimator. It should be kept in mind, however, that filling the collimator in the horizontal direction is essential for high resolution, while filling the collimator in the vertical direction is important for maximum energy throughput. These considerations should be kept in mind when compromises are necessary.

4.6.1. Use of a Single Condensing Lens or Mirror

It is often necessary in spectroscopic work to use some form of external optics to illuminate the entrace slit whenever the source itself is too small to satisfy the dimensions required by Eqs. 4.36b and 4.39. The external optics may consist of nothing more than a single lens or concave mirror or may consist of several components. The exact manner in which the external optics are arranged will depend on the desired result.

Let us begin our discussion by considering the situation where it is desired to focus an image of the source onto the entrance slit. This method of slit illumination is equivalent to placing the source itself at the slit position and is often used when spectral information from different regions of the source is to be examined. Depending on the nature of the experiment, it may be desirable to form an enlarged or a reduced image of the source at the entrance slit with the lens or concave mirror. The source-to-lens and lens-to-slit distances may be calculated from Eq. 2.29, while the desired degree of magnification will be determined by Eq. 2.31. Once the lens-to-slit distance is known, Eqs. 4.36b and 4.39 can be used to determine the dimensions of the lens required.

Suppose, for example, that it is desired to form a 1:1 image of a source on the entrance slit of a 0.5-m, $f/8.7$ spectrometer with an entrance slit that is 4 mm long using a single lens with a focal length of 10 cm. In achieving the desired goal, two questions must be answered. First, the diameter of the lens must be specified. In addition, the location of the source and lens on the optical rail must be determined. From the f-number and Eq. 2.46, we see that the diameter of the collimator is 57 mm. From Eq. 2.31 and the fact that unit magnification has been specified, we see that the lens-to-slit distance will be equal to the lens-to-source distance. Substituting this information into Eq. 2.29 reveals that the lens should be placed a distance of twice its focal length from the slit, or 20 cm. The source should then be located 20 cm from the lens. Substituting the lens-to-slit distance in Eq. 4.39 reveals that the lens should be 28 mm in diameter. If the lens-to-slit distance is substituted in Eq. 4.36b, a lens diameter of 23 mm is obtained. Thus, a lens with a minimum f-number of 4.4 or approximately half the collimator f-number will be necessary to fill the collimator in the horizontal direction. If maximum energy throughput is important, a lens with an f-number of 3.5 or approximately 0.4 times the collimator f-number will be needed.

In performing calculations of this type, it should be remembered that the focal length of a simple lens is a function of wavelength. As a result, what appears to be a focused image in the visible region may not be in terms of UV-light. If the refractive index of the lens material is known as a function of wavelength, the variation of focal length with wavelength can be calculated using Eq. 2.35, and the correct focal length can be used in the preceding calculations.

4.6.2. Uniform Slit Illumination

In the previous section, we discussed how a simple condensing lens or mirror could be used to form an image of the source on the entrance slit while simultaneously filling the entrance aperture of the spectrometer with light. Although this form of spectrometer illumination is useful for studying the radiation emitted from particular regions of a spectral source, it is not ideal for certain applications where it is desirable to have the entrance slit illuminated uniformly. Uniform slit illumination is particularly important in quantitative work where photographic densitometry is to be performed to obtain intensity data. In order to appreciate the requirements imposed by photographic work, it is worthwhile to briefly describe the process.

Quantitative spectrography is based on the principle that the amount of blackening of a photographic emulsion is proportional to the exposure film has received. The exposure, in turn, is given by the reciprocity relationship as the product of intensity times the length of the exposure. In order to achieve the goal of converting film blackening into relative intensity, an emulsion calibration must be prepared. This is generally accomplished by placing some form of intensity modifier in front of the entrance slit so that the exposure produced on the film along the length of the spectral line is varied in some known way that can be correlated with film blackening.

In order for this procedure to work, two conditions must be satisfied: (1) the spectrograph must be stigmatic and (2) the intensity modifier placed directly in front of the entrance slit must be uniformly illuminated. In this way, the only source of intensity variation along the slit will be due to the intensity modifier itself (i.e., not to variation of intensity from different regions of the source), and this intensity variation will be faithfully transmitted to the image in the focal plane by the stigmatic nature of the spectrograph. In addition, it is common practice to use an entrance slit width that will produce flat-topped spectral lines to facilitate densitometry.

Since most spectral sources do not emit uniformly from various regions within the source, uniform slit illumination is not likely to be obtained by focusing an image of the source on the entrance slit of the spectrometer. More nearly uniform slit illumination may be obtained by placing the condensing element directly in front of the slit and placing the source at such a distance from the condenser as to form a magnified image of the source on the collimator of the spectrometer. This form of spectrometer illumination is shown in Figure 4.27. The rationale behind this procedure is based on the condition that each portion of the source illuminates each portion of the

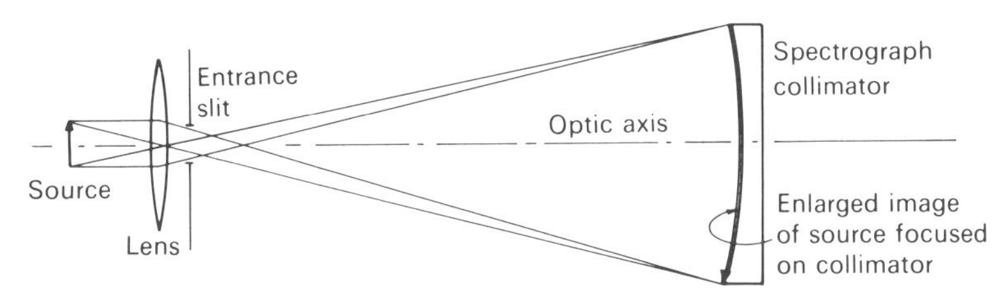

Figure 4.27. Optical arrangement for uniform slit illumination.

condenser and hence the slit. Although this condition is not strictly true because of small differences in path length from various regions of the source to various regions of the condenser and also as a result of small angular differences, it is true to a sufficiently good approximation to be useful for practical purposes. In order to provide the maximum energy transmission through the system, the magnified image of the source should be arranged to fill the collimator.

4.6.3. Use of Cylindrical Lenses

Since the geometry of the grating is often rectangular (i.e., wider than it is high), it is sometimes necessary to magnify the image formed on the collimator beyond that necessary to fill the vertical dimension of the collimator in order that the horizontal dimension of the grating is filled. This overmagnification means that only the central region of the source is sampled by the spectrometer. To avoid some of the problems encountered with the use of spherical condensers, cylindrical lenses are often used to provide uniform illumination of the entrance slit.

A cylindrical lens is a lens formed from an axial section of a cylinder so that it can refract only in a single plane. Thus, a cylindrical lens whose axis is vertical will focus rays only in the horizontal plane, as shown in Figure 4.28 (i.e., focusing will occur only in the plane perpendicular to the axis of the lens). There are two basic ways in which cylindrical lenses may be used to provide uniform slit illumination.

The first, and the simplest, procedure is to place the cylindrical lens directly in front of the entrance slit with the axis of the lens parallel to the slit. The

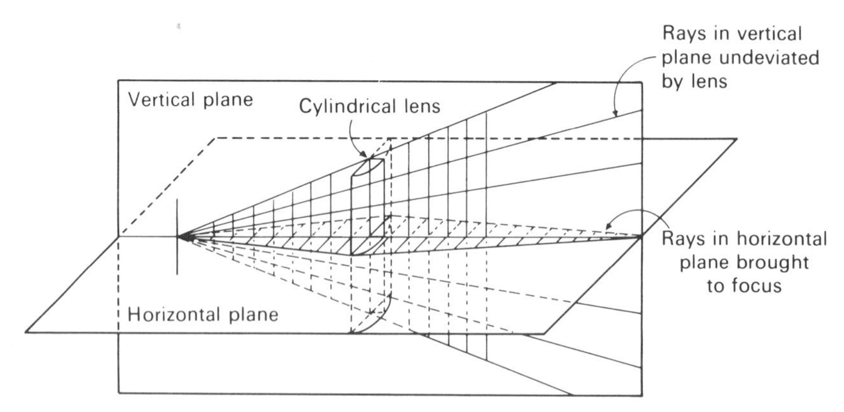

Figure 4.28. Ray diagram for cylindrical lens.

source is then positioned at such a distance from the lens as to form an enlarged image of the source on the collimator in the horizontal plane. Since the cylindrical lens arranged in the manner described has no focusing properties in the vertical plane, light from any portion of the source can strike any portion of the slit and hence any portion of the collimator. The only condition that must be met with this arrangement is that the vertical dimension of the source must be able to fill the vertical aperture of the spectrometer by itself without any help from the condensing lens.

If the source is unable to fill the vertical aperture of the spectrometer (as determined by Eq. 4.39) from its location in front of the entrance slit, a pair of cylindrical lenses whose axes are crossed may be employed. In this scheme, the lens closest to the slit is oriented with its axis horizontal, as shown in Figure 4.29. This lens focuses only in the vertical direction and is arranged to form an image of the source on the collimator with the magnification selected so as to fill the collimator in the vertical direction. The second lens is placed at some distance from the slit with its axis in the vertical plane (i.e., parallel to the entrance slit). This lens is selected to focus an image of the source as a vertical line on the entrance slit.

The optics of this arrangement can be determined by considering each lens separately. Starting with the first lens (L_1 in Figure 4.29), the degree of magnification required will be

$$m_1 = D/H = q/p = F/p \tag{4.40}$$

where the subscript refers to the first lens, m is the magnification, D is the vertical dimension of the collimator, H is the height of the source, F is the focal length of the collimator, and p is the distance from the source to the first lens. Substituting this information in the lens formula (Eq. 2.29) and solving for the focal length of L_1 gives

$$f_1 = F/(m_1 + 1) \tag{4.41}$$

If the source is an arc as commonly used in emission spectroscopy, it may be desirable to form a magnified image of the source on the entrance slit with L_2 in order to avoid fluctuations due to arc instability (i.e., arc "wandering"). This degree of magnification from L_2 will be given by

$$m_2 = S/K = q'/p' \tag{4.42}$$

This information, combined with the fact that $p' + q' = p = F/m_1$, gives

$$p' = F/[m_1(1 + m_2)] \tag{4.43}$$

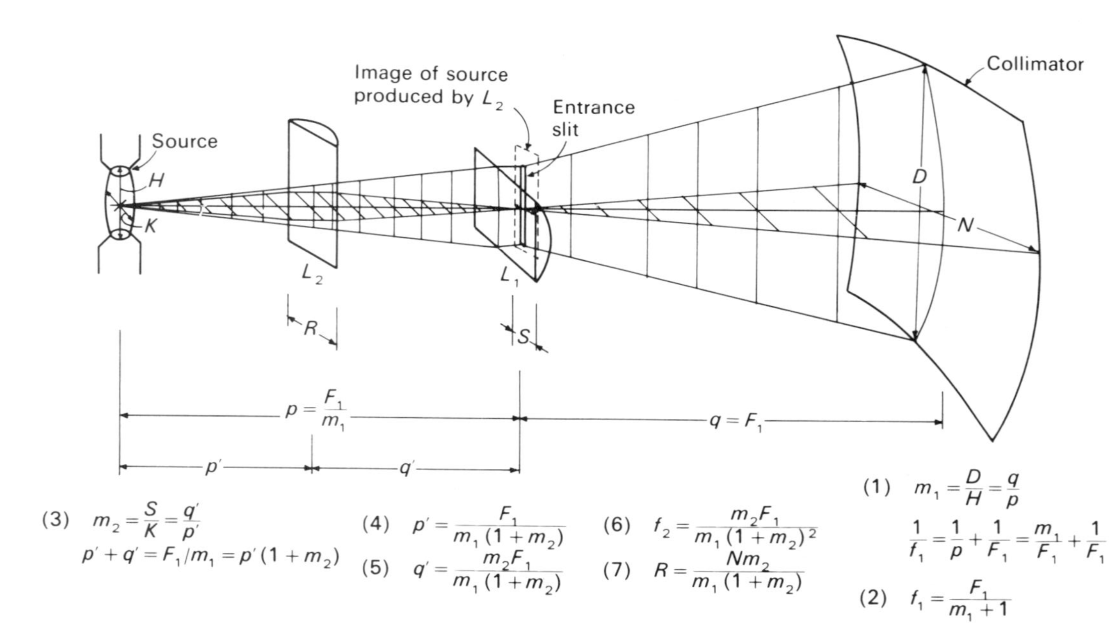

Figure 4.29. Use of crossed cylindrical lenses to fill spectrometer aperture.

and

$$q' = m_2F/[m_1(1+m_2)] \tag{4.44}$$

Substituting Eqs. 4.43 and 4.44 into the lens formula gives the required focal length for L_2,

$$f_2 = m_2F/[m_1(1+m_2)^2] \tag{4.45}$$

Finally, the width of L_2 required to fill the horizontal aperture of the collimator will be given by

$$R = Nm_2/[m_1(1+m_2)] \tag{4.46}$$

Equations 4.40–4.46 are all that are needed to completely determine the optical parameters of L_1 and L_2 and their location on the optical rail. Suppose, as an example, that we wish to fill the collimator of a 1-m spectrograph with light from a source that is 10 mm high and 4 mm in diameter. Let the dimensions of the collimator be 60 mm high and 100 mm wide. Starting with Eq. 4.40, we see that a magnification of 6 is required in the vertical direction. From Eq. 4.41, we find that the focal length of L_1 will be 143 mm for the wavelength of interest. If we choose a magnification of 2 for L_2 as a reasonable value to avoid problems with arc wandering, we see from Eqs. 4.43 and 4.44 that p' and q' will be 56 and 112 mm, respectively, and the source-to-slit distance required will be 168 mm. From Eqs. 4.45 and 4.46, we see that the focal length for L_2 will be 37 mm for the wavelength of interest and the lens should be at least 11 mm wide to fill the horizontal aperture of the spectrograph.

Thus, L_2 will form an image of the source in the horizontal plane (i.e., as a vertical line on the entrance slit). Since L_2 does not focus in the vertical plane, uniform slit illumination should be obtained. At the same time, L_1 will independently form an image of the source on the collimator in the vertical plane. As long as lens L_2 is at least as wide as predicted by Eq. 4.46, the horizontal aperture of the collimator should be filled. Use of two crossed cylindrical lenses provides a means of obtaining different degrees of magnification in the vertical and horizontal directions.

REFERENCES

Davis, S. P. (1970), *Diffraction Grating Spectrographs*, Holt, Rinehart, and Winston, New York.

Fastie, W. G. (1961), U.S. Patent 3,011,391, Dec. 5.

Fastie, W. G., Crosswhite, H. M., and Gloersen, P. (1958), *J. Opt. Soc. Am. 48*, 106.

Harrison, G. R. (1949), *J. Opt. Soc. Am., 39*, 522.

Harrison, G. R., Lord, R. C., and Loofbourow, J. R. (1948), *Practical Spectroscopy*, Prentice-Hall, Englewood Cliffs, NJ.

Hayat, G. S. and Pieuchard, G. (1973), *Industrial Research*, August, 40.

Jacquinot, P. (1954), *J. Opt. Soc. Am., 44*, 761.

Jenkins, F. A. and White, H. E. (1957), *Fundamentals of Optics*, McGraw-Hill, New York.

Lerner, J. M., Flamand, J., Laude, J. P., Passereau, G. and Thenenon, A. (1980), *Proc. Soc. Photo-Opt. Instrum. Eng. 240*, 82.

Lerner, J. M., Flamand, J., Laude, J. P. and Thevenon, A. (1983), *Proc. Int. Soc. Opt. Eng. 411*, 18.

Lerner, J. M. and Laude, J. P. (1983), *Electro-Optics*, May.

Loewen, E. G. (1970), *Diffraction Grating Handbook*, Bausch & Lomb, Rochester, NY.

Sawyer, R. A. (1963), *Experimental Spectroscopy*, Dover, New York.

Schuster, A. (1905), *Astrophys. J., 21*, 197.

Slavin, M. (1971), *Emission Spectrochemical Analysis*, Wiley-Interscience, New York.

Stockbarger, D. C. and Burns, L. (1933), *J. Opt. Soc. Am. 23*, 379.

CHAPTER

5

DISPERSIVE SPECTROMETERS THAT USE MASKS

A spectrometer is a device that produces a spectrum from a source of electromagnetic radiation. In chemical analysis, the source is a sample of material, and the spectrum is a presentation of the intensity of the emitted radiation as a function of one of its characteristic parameters, that is, energy, frequency, or wavelength. The typical arrangement of components in a conventional emission spectrometer is shown schematically in Figure 5.1.

Chapters 2–4 have discussed how prisms and gratings disperse the emitted electromagnetic radiation into its component wavelengths and how the optical arrangement of the mirrors and lenses direct this radiation within the spectrometer and onto a detector. Chapter 5 will begin a consideration of how this radiation should be detected, especially for the simultaneous monitoring of more than one wavelength in the spectrum. We begin by describing a number of the limitations of the conventional spectrometer and discuss how changes in the design of both the spectrometer itself and the procedure used to gather and process the spectroscopic data can often be helpful in overcoming certain of these limitations.

5.1. LIMITATIONS OF CONVENTIONAL SPECTROMETERS

As shown in Figure 5.1, however complex the source, a conventional spectrometer has only one pair of corresponding entrance and exit slits and therefore must detect the emitted radiation on a channel-by-channel basis. (A channel, or resolution element, is a specified narrow range of wavelengths making up part of the spectrum.) Conventional spectrometers are called scanning spectrometers because the entire spectrum can be obtained only by rotating the disperser, imaging each resolution element on the exit slit in a sequential manner, and monitoring the detector signal as a function of time.

The conventional spectrometer is highly inefficient in terms of both energy and time. Most of the radiation emitted by the source never reaches the detector, and if each channel must be examined individually, considerable time may be required to observe the entire spectrum. These limitations cannot

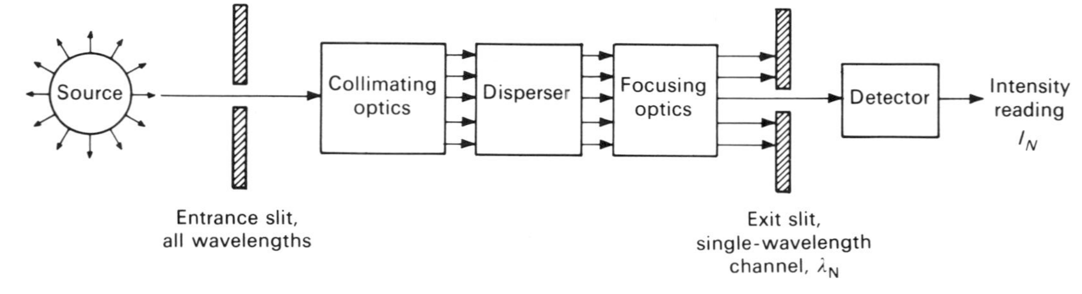

Figure 5.1. Schematic diagram of single-channel, scanning-type spectrometer. (After Harwit and Decker, 1974.)

be overcome without compromising instrument resolution and signal-to-noise ratio (SNR).

5.1.1. Optical Throughput and Resolution

The intensity of the signal obtained for a particular wavelength channel will depend on how much of the emitted radiation associated with that channel actually enters and passes through the spectrometer. Optical throughput (*luminosity* in the French literature) is a measure of the amount of radiation reaching the detector for a given amount leaving the source, that is, radiation transmitted by an instrument of a given aperture accepting a given solid angle,

$$\Theta = TA\Omega \tag{5.1}$$

where T is the transmission, A is the area of the pupil of the system, and Ω is the maximum solid angle accepted by the spectrometer. (See Eq. 2.3 and corresponding text.)

The effective throughput of the entire spectrometer is limited by that portion of the instrument with the smallest throughput. Thus, no optical arrangement can increase the amount of radiant energy reaching the detector if throughput is severely restricted at either the entrance or exit aperture of the spectrometer. Mertz (1965) has pointed out that this fact is equivalent to the second law of thermodynamics. If it were possible to increase throughput after radiation had passed through a small aperture, then it would be possible to concentrate the light energy emerging from the small opening of a blackbody radiator into an image of greater intensity and thereby transfer energy from a cooler region to a hotter one.

As discussed in Section 4.1.3, the resolution s of a spectrometer is a measure of the range of wavelengths transmitted in one wavelength channel. Resolution is calculated from Eq. 4.12,

$$s = wR_{\mathrm{D}} \tag{4.12}$$

where w is the width of the entrance or exit slit (assumed to be equal) and R_{D} is the reciprocal linear dispersion. To prevent spectral interference, it is desirable to keep the wavelength channels as small as possible. Since resolution improves with narrower slit widths, while the intensity associated with any given wavelength channel increases with wider slit widths, the spectroscopist is constantly faced with a choice of compromise conditions involving resolution and optical throughput.

5.1.2. Signal-to-Noise Ratio and Scanning Time

All spectroscopic intensity measurements are characterized by some degree of random error called noise. The precision associated with a particular intensity measurement is given by the SNR for that wavelength channel. Favorable SNRs are required for reliable analyses, and the lower limit for detection of a channel is often taken as SNR = 2.

Depending on the type of noise present in the spectrum (Section 5.6), the SNR of a signal can usually be improved by signal averaging, that is, scanning each channel for a longer period of time to allow the background noise to average out. Improvement in SNR using this technique is achieved only at the expense of a much slower scanning speed. Again, the spectroscopist is faced with a choice of compromise conditions, now involving the reliability of the intensity reading (SNR) and the time needed to take the measurement.

5.2. MONOCHROMATORS WITH MULTIPLE SLITS

In many cases, the need for compromise between spectral resolution and optical throughput can be eliminated by using multiple entrance and exit slits (Stewart, 1970). The resolution associated with each wavelength channel is determined by the width of each individual slit, but the optical throughput is determined by the total area associated with all slits present at either the entrance or exit aperture of the instrument (assumed equal at both apertures).

5.2.1. Monochromatic Incident Radiation

Figure 5.2 shows a multislit arrangement with two entrance slits (1 and 2) and two exit slits (1′ and 2′). The separation *d* or wavelength interval between the two entrance slits 1 and 2 is the same as the separation of the two exit slits 1′ and 2′. All four slits have the same spectral bandpass. As shown in Figure 5.2, each entrance slit produces its own image of the incident spectrum at the exit focal plane, but the two images are displaced from one another by an amount proportional to *d*.

Suppose the light striking the entrance slits is monochromatic. As the disperser is rotated, the two images of the single wavelength channel, λ_1, will be moved across the exit slits. At time T_1 (Figure 5.2*a*), the image of entrance slit 2 crosses exit slit 1′, generating a triangular slit function. At time T_2 (not shown), the image of entrance slit 2 moves to a position in between the two exit slits, and no radiation reaches the detector. At time T_3 (Figure 5.2*b*), the image of entrance slit 1 crosses exit slit 1′ while, simultaneously, the image of entrance slit 2 crosses exit slit 2′. Now the triangular slit function has the same

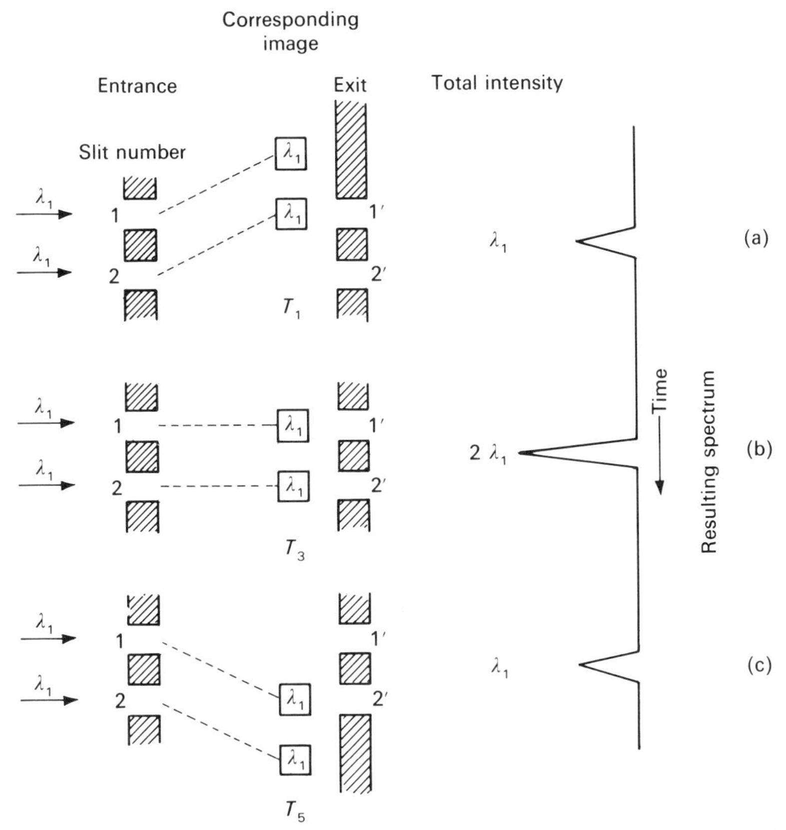

Figure 5.2. Slit functions for double entrance and exit slits shown at three different scanning times using monochromatic light. Images of two entrance slits form at exit focal plane but are shown somewhat ahead of focal plane for clarity.

resolution as that generated at time T_1 but is twice as intense. At time T_4 (not shown), the images of the entrance slits move to positions in between the two exit slits, and no radiation reaches the detector. Finally, at time T_5 (Figure 5.2*c*), the image of entrance slit 1 crosses exit slit 2′, generating a third triangular slit function that has the same intensity and resolution as that formed at T_1.

Analogous results are obtained for multislit spectrometers containing any number of corresponding entrance and exit slits: 1 and 1′, 2 and 2′, and so on. In general, if the incident radiation is monochromatic, then for n pairs of corresponding entrance and exit slits, a total of $2n-1$ triangular slit functions will be generated with relative intensities of $1:2:\cdots:n-1:n:n-1:\cdots:2:1$.

The most intense triangle always occurs in the middle of the sequence of slit functions, and the increase in spectral intensity for this central triangle is always proportional to n. The resolution obtained for any of the $2n-1$ triangles is identical to that obtained by a conventional spectrometer employing only one pair of corresponding entrance and exit slits that have the same spectral bandpass as those in the multislit arrangement.

5.2.2. Jacquinot's Advantage

The ability to increase optical throughput for a constant source intensity without loss in spectral resolution is called Jacquinot's advantage, after the French scientist Jacquinot (1959, 1960), who first recognized its implications for spectroscopy. For a multislit instrument, the value of Jacquinot's advantage (alternatively, the luminosity gain) at any wavelength is given by the ratio of the throughput of the multislit instrument to that of a single-slit instrument of equivalent resolving power. Assuming all slits are the same in both instruments, Jacquinot's advantage will be proportional to the number of pairs of corresponding entrance and exit slits that are open during the spectroscopic measurement.

5.2.3. Polychromatic Incident Radiation

The effect of a multislit arrangement on polychromatic incident radiation is shown in Figure 5.3. Again, two images of the incident radiation are produced at the exit focal plane, and these are displaced from one another by an amount proportional to d. At time T_1 (Figure 5.3*a*), radiation centered at λ_7 is imaged on both exit slits 1′ and 2′, and the signal received by the detector for this wavelength channel is twice as intense as that resulting from a comparable single-slit spectrometer. However, in addition to λ_7, radiation λ_5 also passes through exit slit 1′, and radiation λ_9 also passes through exit slit 2′.

As can be seen from Figures 5.3*a*–*c*, the effect of multiple slits on polychromatic radiation is different from the monochromatic case in that more than one wavelength channel is passed to the detector but similar in that one wavelength in the spectrum is selected for preferential treatment. All other wavelengths are either not passed to the detector at all or else are passed to a lesser extent. The identities of these additional wavelengths depend on the slit separation d and the reciprocal dispersion of the monochromator, R_D. To generalize, if radiation of wavelength λ_0 is centered at both exit slits (λ_7 in Figure 5.3*a*) and has a Jacquinot's advantage of 2, radiation at $\lambda_0 + R_D d$ and $\lambda_0 - R_D d$ will also be passed to the detector but without any improvement in optical throughput.

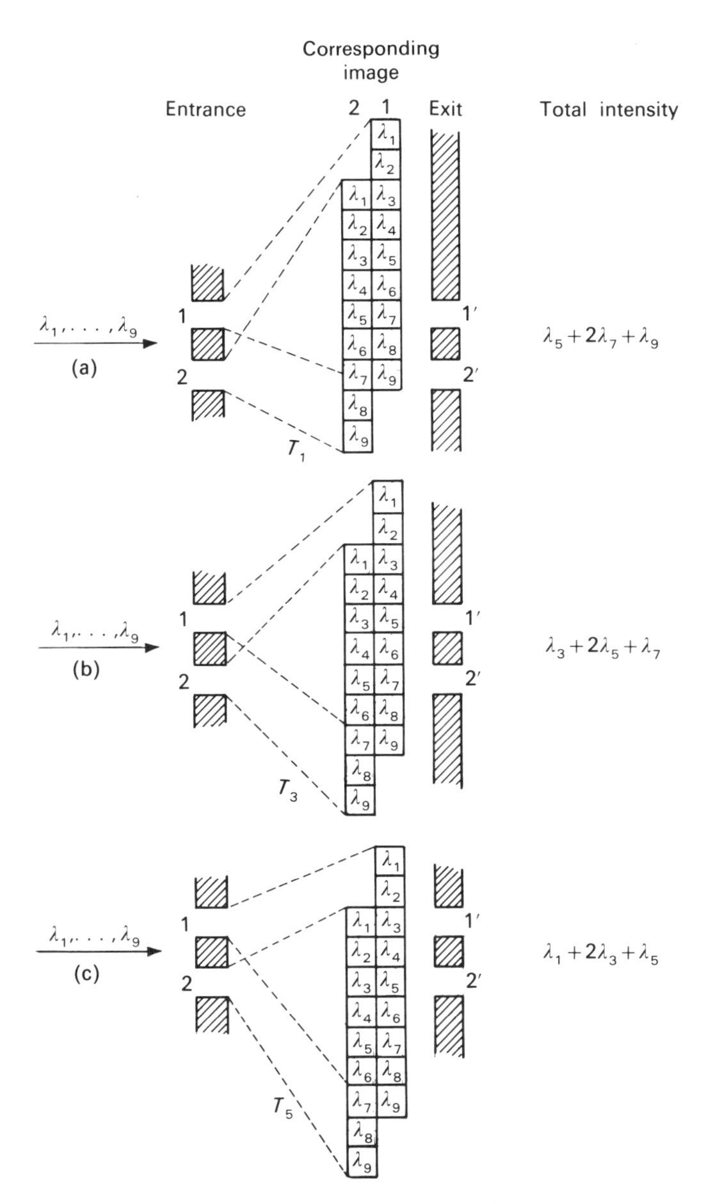

Figure 5.3. Channel intensities for two entrance and two exit slits shown at three different scanning times using polychromatic light. Images of two entrance slits form at exit focal plane but are shown somewhat ahead of focal plane and displaced from each other for clarity.

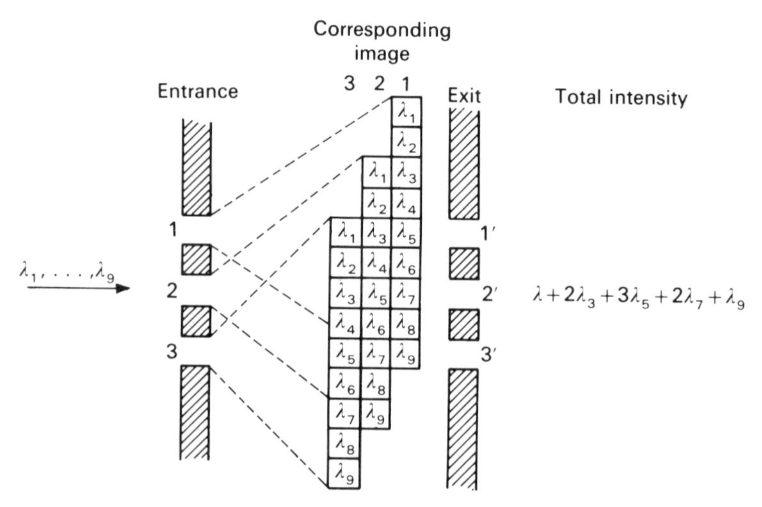

Figure 5.4. Channel intensities for three entrance and three exit slits shown at one scanning time for polychromatic light.

Figure 5.4 shows the effect of extending the number of pairs of corresponding entrance and exit slits to 3. Again, more than one wavelength is passed to the detector, but one wavelength, λ_5, is selected for preferential treatment. To generalize, for n pairs of corresponding entrance and exit slits, a total of $2n-1$ wavelength channels will be passed to the detector. If λ_0 is the wavelength channel with the highest luminosity gain, the additional wavelengths appearing in the spectrum will be $\lambda_0 + kR_D d$ and $\lambda_0 - kR_D d$, where $k = 1, 2, \cdots, n-1$. Jacquinot's advantage for all $2n-1$ channels is $1, 2, \cdots, n-1, n, n-1, \cdots, 2, 1$ in order of increasing or decreasing wavelength.

Figure 5.3 illustrates another important feature of multislit systems. By rotating the disperser, it is possible to scan through the entire spectrum of incident radiation, enhancing the signal for each of the component wavelength channels in turn, that is, λ_7 at T_1, λ_6 at T_2 (not shown), λ_5 at T_3, λ_4 at T_4 (not shown), λ_3 at T_5, and so on. Thus, the multislit spectrometer can be used to obtain a Jacquinot's advantage for all portions of the incident spectrum, but only by allowing more than the desired wavelength to appear at the exit slits and degrading the spectral resolution.

The appearance of extra wavelength channels creates a new problem—how to take advantage of the gain in luminosity for λ_0 using a multislit arrangement without loss of resolution due to overlapping images. One possibility for eliminating unwanted parts of the spectrum is to select a bandpass filter with a bandwidth that is narrower than the interval $2R_D d$. This filter would retain λ_0 but reject all radiation of other wavelengths.

A second possibility is to chop the radiation at both the entrance and exit slits and use a tuned amplifier to reject all radiation but that originating from the desired wavelength channel. To illustrate, suppose that the radiation entering slits 1 and 2 (Figure 5.3) and exiting slits 1′ and 2′ is chopped at the following frequencies.

ENTRANCE SLIT CHOPPING FREQUENCIES

$$f_1 = f_e \tag{5.2}$$

$$f_2 = f_e + f_d \tag{5.3}$$

EXIT SLIT CHOPPING FREQUENCIES

$$f_{1'} = f_x \tag{5.4}$$

$$f_{2'} = f_x + f_d \tag{5.5}$$

where the frequency f_d is the same in Eqs. 5.3 and 5.5 but f_e is different from f_x. As shown from the trigonometric identity given in Eq. 5.6,

$$\cos(f_a t)\cos(f_b t) = \tfrac{1}{2}[\cos(f_a + f_b)t + \cos(f_a - f_b)t] \tag{5.6}$$

simultaneously chopping a signal at two different frequencies, f_a and f_b, gives rise to two new signals that appear to be chopped at frequencies $f_a + f_b$ and $f_a - f_b$. Since the desired radiation, λ_0, that enters slit 1 will exit from slit 1′ while that which enters slit 2 will exit from slit 2′, the effective chopping frequency in both cases will be

$$\begin{aligned} f_{11'} &= f_e - f_x \\ f_{22'} &= (f_e + f_d) - (f_x + f_d) = f_e - f_x \end{aligned} \tag{5.7}$$

Undesired radiation will have chopping frequencies of

$$\begin{aligned} f_{12'} &= f_e - (f_x + f_d) = f_e - f_x - f_d \\ f_{21'} &= (f_e + f_d) - f_x = f_e - f_x + f_d \end{aligned} \tag{5.8}$$

If f_x is properly chosen so that none of the individual slit chopping frequencies and none of the sum frequencies coincide with $f_e - f_x$, an amplifier tuned to $f_e - f_x$ will detect only the desired radiation and reject all the rest. By rotating the disperser, it will be possible to image each wavelength channel on both exit slits in turn, thereby generating the entire spectrum at greater intensity but without any loss in spectral resolution.

One final point should also be noted. Multislit spectrometers tend to generate a great deal of extra, unused radiation at the detector. This will not affect the spectroscopic measurement provided the background or noise level in the detector is independent of signal intensity. The problem of noise and its effect on the precision associated with determining the intensity of a given wavelength channel will be briefly discussed in Section 5.6 and treated more fully in Chapters 9 and 13.

5.2.4. Golay's Dynamic Multislit Spectrometer

A somewhat more involved scheme for removing unwanted radiation through the use of binary codes was first introduced by Golay (1949). Although Golay initially described a system containing six entrance and six exit slits, a less complex example using only three pairs of corresponding entrance and exit slits will be used to illustrate the principle.

Suppose each entrance and each exit slit is either completely blocked (0) or completely open (1) at any given time and that the arrangement of open and blocked slits is varied in a synchronous fashion according to Figure 5.5. In this figure, four different arrangements at four different times are shown. Times T_1, T_2, T_3, and T_4 represent the start of sequential time periods of equal duration, and these periods are repeated on a regular basis (T_1, T_2, T_3, T_4, T_1, T_2, T_3, T_4, etc.).

As shown in Figure 5.5, radiation at λ_3 is passed to the detector during periods starting at T_3 and T_4, that is, for half of the regularly repeating cycle, $T_1, \cdots, T_4$. During T_3 and T_4, λ_3 has a luminosity gain that is proportional to the number of slits that are open during this part of the cycle. The effective chopping pattern of λ_3, is shown in Figure 5.6.

In comparison to the desired radiation, λ_3, all other wavelengths either do not reach the detector (i.e., λ_1 and λ_5) or else pass to the detector with a different chopping pattern (i.e., λ_2 and λ_4). (See Figure 5.6 for a comparison of chopping patterns.) If none of the chopping frequencies for unwanted radiation are harmonics of the chopping frequency produced for the desired radiation, then a tuned amplifier can be used to detect only the signal generated by λ_3.

The design of the on–off or (1,0) patterns in the entrance and exit slits can be generated as follows. Beginning with the 2×2 pattern,

$$a = \begin{bmatrix} 0 & 0 \\ 0 & 1 \end{bmatrix} \tag{5.9}$$

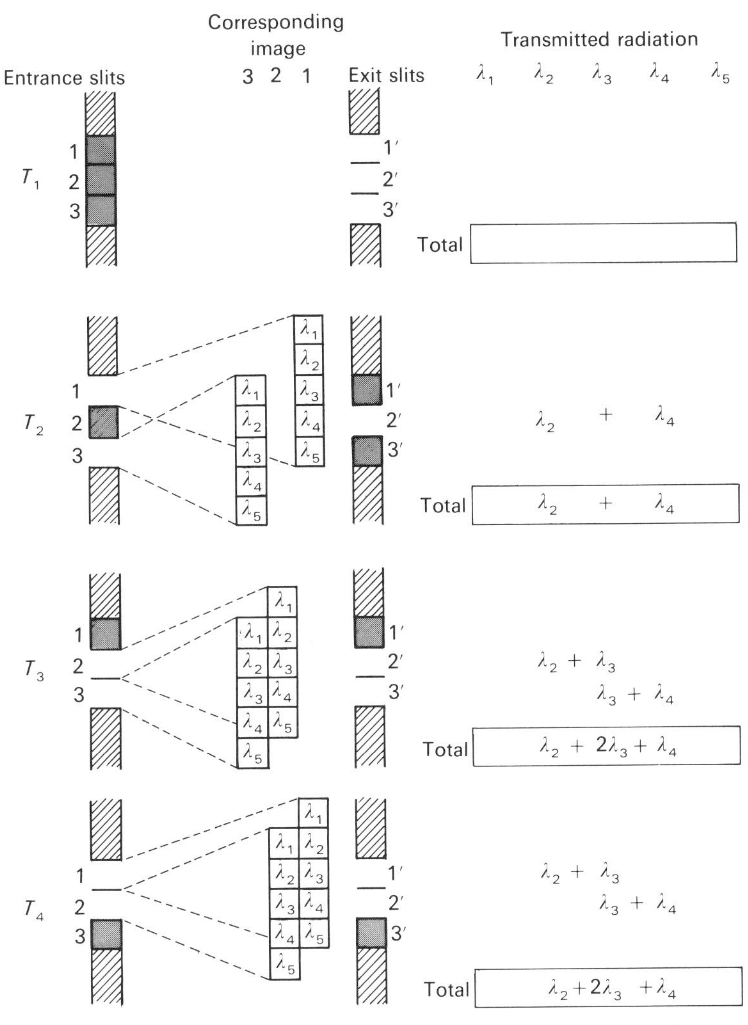

Figure 5.5. Channel intensities for Golay's dynamic multislit spectrometer using three entrance and three exit slits.

another identical pattern is placed to the right of it and below it. A square is completed by replacing the 0's by 1's and the 1's by 0's in the original pattern (5.9):

$$a' = \begin{bmatrix} 1 & 1 \\ 1 & 0 \end{bmatrix} \tag{5.10}$$

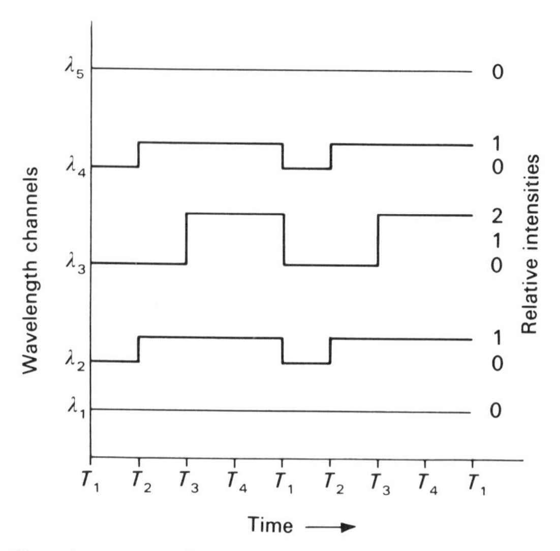

Figure 5.6. Chopping patterns for experimental arrangement shown in Figure 5.5.

and placing this opposite pattern in the lower right corner. The procedure yields,

$$\begin{bmatrix} a & a \\ a & a' \end{bmatrix} = \begin{bmatrix} 0 & 0 & 0 & 0 \\ 0 & 1 & 0 & 1 \\ 0 & 0 & 1 & 1 \\ 0 & 1 & 1 & 0 \end{bmatrix} \tag{5.11}$$

where a and a' are defined by Eqs. 5.9 and 5.10, respectively. If the first row (all zeros) is eliminated, the remaining three rows represent the on–off (1, 0) sequence of each of the three entrance slits, and the four columns give the

Table 5.1. Entrance and Exit Slit Patterns for Three-Slit Arrangement[a]

	Entrance				Exit			
	T_1	T_2	T_3	T_4	T_1	T_2	T_3	T_4
Slit 1	0	1	0	1	1	0	0	1
Slit 2	0	0	1	1	1	1	1	1
Slit 3	0	1	1	0	1	0	1	0

[a]After Golay (1949).

on–off (1, 0) pattern associated with the four time periods, $T_1, \cdots, T_4$, as shown in Table 5.1.

The pattern for the exit slits is obtained by exactly duplicating the right half of the entrance array (columns 3 and 4) but changing all 0's to 1's and all 1's to 0's in the left half of the array (columns 1 and 2.). The resulting exit slit patterns are shown in Table 5.1.

Larger patterns for more slits can be produced by repeating the process outlined in the preceding. If the process represented by Eq. 5.11 is repeated once more using the whole 4×4 array (Eq. 5.11) as a, we obtain

$$\begin{bmatrix} a & a \\ a & a' \end{bmatrix} = \begin{bmatrix} 0 & 0 & 0 & 0 & 0 & 0 & 0 & 0 \\ 0 & 1 & 0 & 1 & 0 & 1 & 0 & 1 \\ 0 & 0 & 1 & 1 & 0 & 0 & 1 & 1 \\ 0 & 1 & 1 & 0 & 0 & 1 & 1 & 0 \\ 0 & 0 & 0 & 0 & 1 & 1 & 1 & 1 \\ 0 & 1 & 0 & 1 & 1 & 0 & 1 & 0 \\ 0 & 0 & 1 & 1 & 1 & 1 & 0 & 0 \\ 0 & 1 & 1 & 0 & 1 & 0 & 0 & 1 \end{bmatrix} \tag{5.12}$$

If the first and fifth rows are removed, the resulting array represents the six entrance slit patterns given for eight sequential time periods in Golay's original paper (1949). The corresponding exit slit arrangements are generated as described previously, and the results are given in Table 5.2. In this six-slit arrangement, the desired wavelength channel will be passed to the detector during time periods $T_5, \cdots, T_8$. All other radiation either does not reach the detector at all or else reaches the detector for only one-fourth of the cycle and is chopped at a different frequency.

The results obtained from these schemes can be explained in the following general way. All of Golay's slit arrangements are pseudorandom sequences of 0's and 1's constructed in such a way that the desired radiation passing through two corresponding entrance and exit slits (1 and 1′, 2 and 2′, etc.) is always blocked either at the entrance or at the exit during the first half of the cycle ($T_1, \cdots, T_4$ in Table 5.2) but passes completely to the detector during the second half of the cycle ($T_5, \cdots, T_8$ in Table 5.2). As the time periods in each cycle are repeated, a regular chopping frequency equal to the cycle time is obtained for the desired radiation.

The pseudorandom patterns in each scheme were convenient to produce by Golay's computer and reduce the probability that any other, undesired radiation will pass to the detector. If such radiation does reach the detector during any part of the cycle, it will have a different chopping frequency than the desired radiation and can be filtered out. To obtain the entire spectrum, the disperser is rotated, thereby imaging a different wavelength channel on

Table 5.2. Entrance and Exit Slit Patterns for Six-Slit Arrangement[a]

	Entrance								Exit							
	T_1	T_2	T_3	T_4	T_5	T_6	T_7	T_8	T_1	T_2	T_3	T_4	T_5	T_6	T_7	T_8
Slit 1	0	1	0	1	0	1	0	1	1	0	1	0	0	1	0	1
Slit 2	0	0	1	1	0	0	1	1	1	1	0	0	0	0	1	1
Slit 3	0	1	1	0	0	1	1	0	1	0	0	1	0	1	1	0
Slit 4	0	1	0	1	1	0	1	0	1	0	1	0	1	0	1	0
Slit 5	0	0	1	1	1	1	0	0	1	1	0	0	1	1	0	0
Slit 6	0	1	1	0	1	0	0	1	1	0	0	1	1	0	0	1

[a]After Golay (1949).

corresponding entrance and exit slits and changing its chopping frequency to that of the tuned amplifier.

Using two separate, synchronously rotating masks with a 64×64 pattern in the place of the conventional spectrometer slits, Golay (1949) was able to experimentally demonstrate a luminosity gain of 10 without any loss in spectral resolution. While 10 was considerably less than the theoretically obtainable gain of 32, these experiments demonstrated that pseudorandom, binary codes could be used to eliminate unwanted radiation and that it was feasible to process more than one wavelength at a time by imaging large portions of a spectrum onto the exit focal plane of a spectrometer.

5.2.5. Golay's Static Multislit Spectrometer

Shortly after publication of his dynamic multislit spectrometer, Golay (1951) realized that tuned amplifiers were not needed to eliminate undesired radiation in his multislit spectrometer. Suitably designed entrance and exit masks could be used in conjunction with two ordinary detectors to produce a final signal that measured only the desired wavelength channel. The mask designs were based on pairs of complementary sequences of 0's and 1's as described in what follows.

Two sequences of 0's and 1's of the same total length are said to be complementary if the number of pairs of like elements in either sequence is equal to the number of pairs of unlike elements with the same spacing in the other sequence. For example,

$$a = 11 \tag{5.13}$$

$$b = 10 \tag{5.14}$$

represent two complementary sequences. There is one pair of like elements in a and one pair of unlike elements in b. Also, there are no pairs of unlike elements in a and no pairs of like elements in b. Other pairs of complementary sequences can be built up by recognizing that if a and b are complementary, then so are ab and ab', where the symbols a' and b' indicate that 0's and 1's have been interchanged, that is,

$$a' = 00 \tag{5.15}$$

$$b' = 01 \tag{5.16}$$

(Compare Eqs. 5.9 and 5.10.) The sequences a' and b' also from complementary pairs.

Table 5.3. Entrance and Exit Slit Patterns for Static Four-Slit Arrangement[a]

	General Design		Design for Figure 5.7	
Detector	Entrance	Exit	Entrance	Exit
A	a	a	1 1 1 0	1 1 1 0
	a'	a'	0 0 0 1	0 0 0 1
B	b	b'	1 1 0 1	0 0 1 0
	b'	b	0 0 1 0	1 1 0 1

[a] After Golay (1951).

Using the procedure described, we can build up larger complementary sequences such as $a=1110$ and $b=1101$, $a=11101101$ and $b=11100010$, and so on. (In the eight-membered a sequence there are three pairs of like elements and four pairs of unlike elements. In the b sequence there are four pairs of like elements and three pairs of unlike elements.) Complementary sequences may also be based on other starting values for a and b, such as $a=00$ and $b=01$, $a=111$ and $b=101$, and so on.

Given the pair of complementary sequences a and b and their opposites a' and b', four entrance and four exit masks are constructed and matched according to the arrangement given in Table 5.3. All four mask combinations are used simultaneously. The radiation passing through mask sets 1 and 2 is directed onto detector A, while the radiation from sets 3 and 4 is directed onto detector B. The final signal is the difference between the reading at detector A and detector B. Figure 5.7 shows the results obtained from this arrangement for the specific case $a=1110$ and $b=1101$.

As can be seen in Figure 5.7, mask sets 1 and 2 are constructed so that any radiation, λ_3 in this case, that is able to enter and exit from corresponding slits (1 and 1′, 2 and 2′, etc.) does so freely. However, mask sets 3 and 4 are constructed so that this same wavelength channel is always completely blocked. Radiation that is shifted across the mask to noncorresponding exit slits (1 to 2′, 3 to 2′, etc.) never reaches either detector or, due to the complementary nature of the codes making up the masks, reaches both detectors with equal intensity. Therefore, the difference between the two detector readings only measures the intensity of one wavelength channel, λ_3. The theoretical gain in luminosity is equal to the total number of slits open to detector A during the measurement (4 in this case). To scan the entire spectrum, the disperser is rotated and a different wavelength channel is directed through corresponding entrance and exit slits.

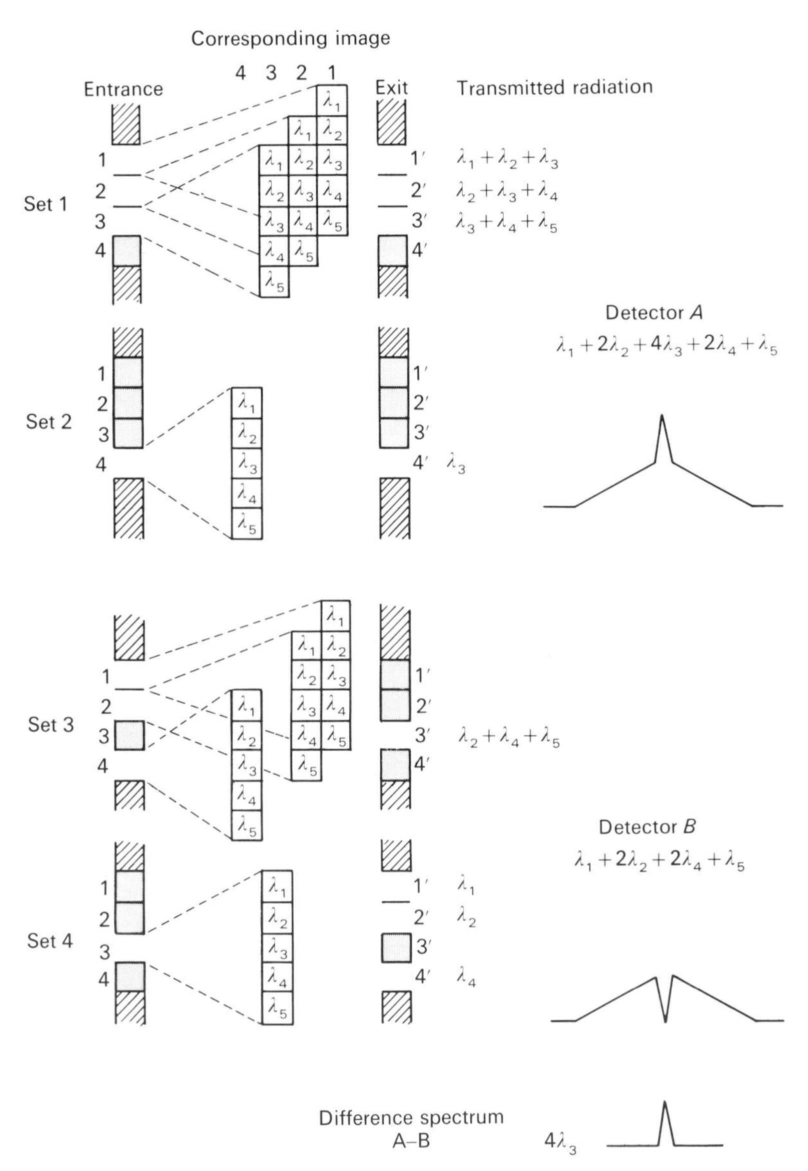

Figure 5.7. Channel intensities and slit functions for Golay's multislit spectrometer using four entrance and four exit slits. Sets 1 and 2 are measured at detector A while sets 3 and 4 are measured at detector B.

5.2.6. Girard's Grill Spectrometer

In spite of the potential advantages associated with increased optical throughput, multislit spectrometers received little attention until Girard (1960, 1963a, b) reported the development of another instrument that resembled Golay's static multislit spectrometer. Instead of using masks based on complementary series, Girard employed masks based on Fresnel zone plates.

A Fresnel zone plate can be easily constructed by drawing concentric circles on white paper with radii proportional to the square roots of whole numbers and making every other zone opaque (Figure 5.8*a*,*b*). If this image is photographed, the negative, when held in front of light from a distant point source, produces a large intensity at a point on its axis at a distance corresponding to the size of the zones and the wavelength of the light used. This bright spot is so intense due to the focusing properties of the plate that the plate is known as a Fresnel lens.

Girard placed a one-dimensional Fresnel zone plate or Girard grill (Figure 5.9*a*) in the entrance aperture of his instrument and another identical mask in the exit focal plane. An image of the entrance mask was produced at the exit mask for only one preferred wavelength, λ_0. Radiation of λ_0 was passed completely to the detector, while light at any other wavelength was severely attenuated. The exit mask was alternated with its complementary image, and as with Golay's spectrometer, the difference between the two signals gave the intensity of the desired radiation, λ_0 (Figure 5.10).

Girard later redesigned his mask and constructed an array containing several series of openings. These openings were equally spaced within a series, but each series had a different spacing (Figure 5.9*b*). For a very large series of openings, the mask becomes the hyperbolic array shown in Figure 5.9*c*, also known as the two-dimensional Girard grill. Two-dimensional grills of the type shown in Figure 5.9*c* were found to make much more effective masks than the one-dimensional grill.

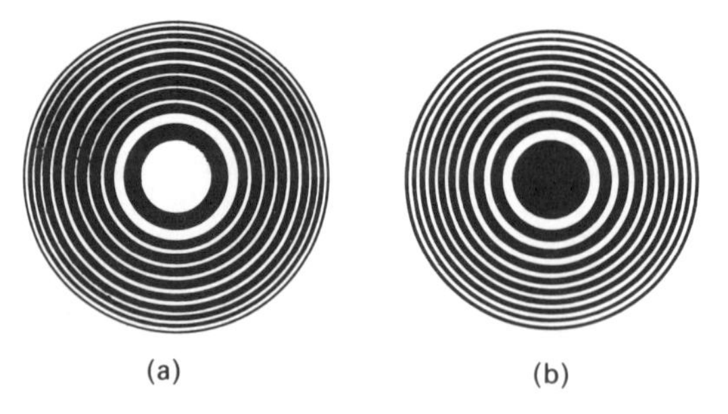

Figure 5.8. Fresnel zone plates. [Reprinted with permission from F. A. Jenkins and H. E. White (1976), *Fundamentals of Optics*, 4th ed., McGraw-Hill, New York.]

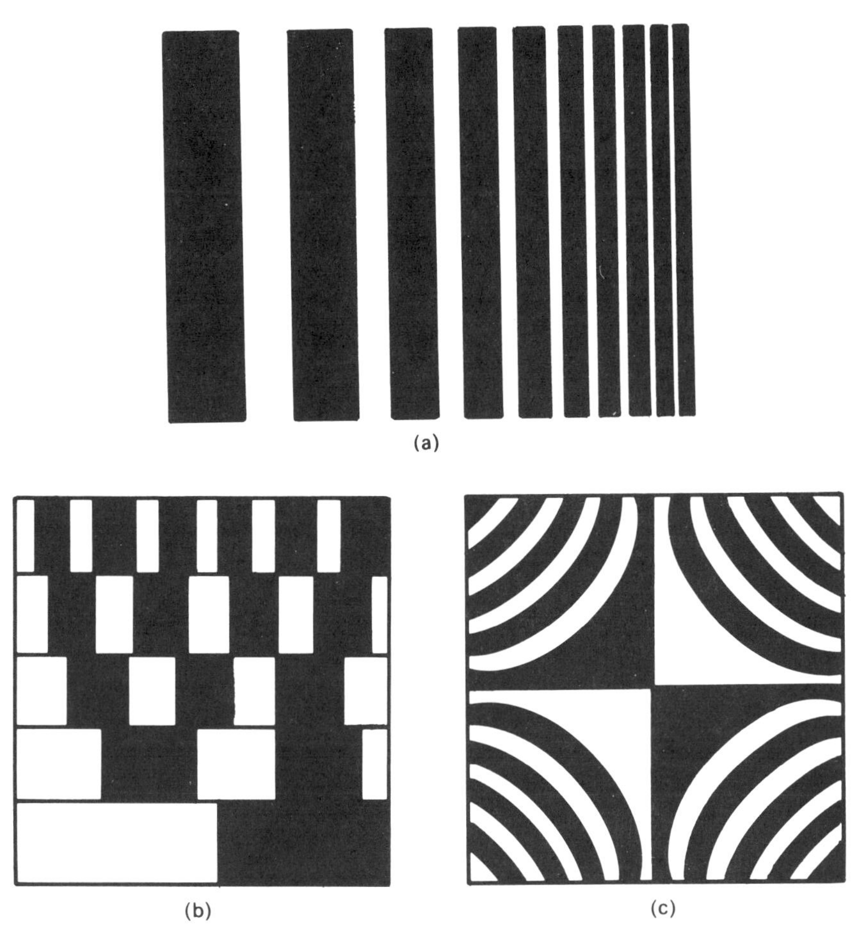

Figure 5.9. (*a*) One-dimensional Girard grill. (Reprinted from Stewart, 1970, p. 274, by courtesy of Marcel Dekker, Inc.) (*b*) Series of Girard grills. (*c*) Two-dimensional, hyperbolic Girard grill. [(*b*) and (*c*) reprinted from Stewart, 1970, p. 275, by courtesy of Marcel Dekker, Inc.]

The resolution of a grill spectrometer is determined by the smallest opening of the grill pattern and obeys the same equation (4.12) as the conventional single-slit spectrometer. The increase in throughput is theoretically the same as Golay's static multislit spectrometer. If the entire grill is of width W, then compared to a conventional spectrometer of the same resolving power with an entrance slit of width w,

$$\text{Luminosity gain} = \tfrac{1}{2}(W/w) \tag{5.17}$$

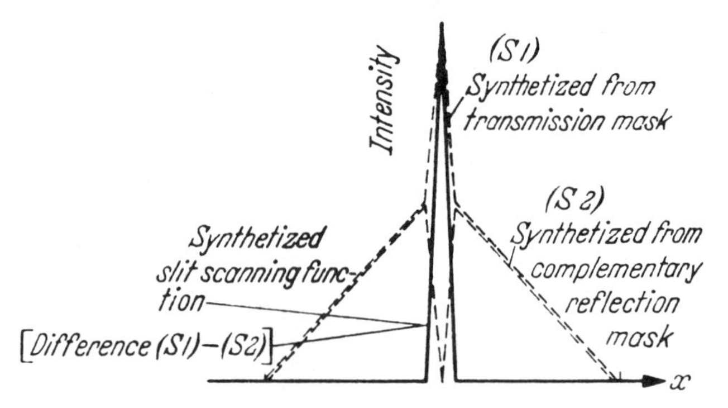

Figure 5.10. Generation of slit function using Girard's spectrometer. [Reprinted from G. W. Stroke (1967), "Diffraction Gratings," in *Handbuch der Physik; Optische Instrumente*, S. Flügge (Ed.), Vol. 29, p. 500, by permission of Springer-Verlag.]

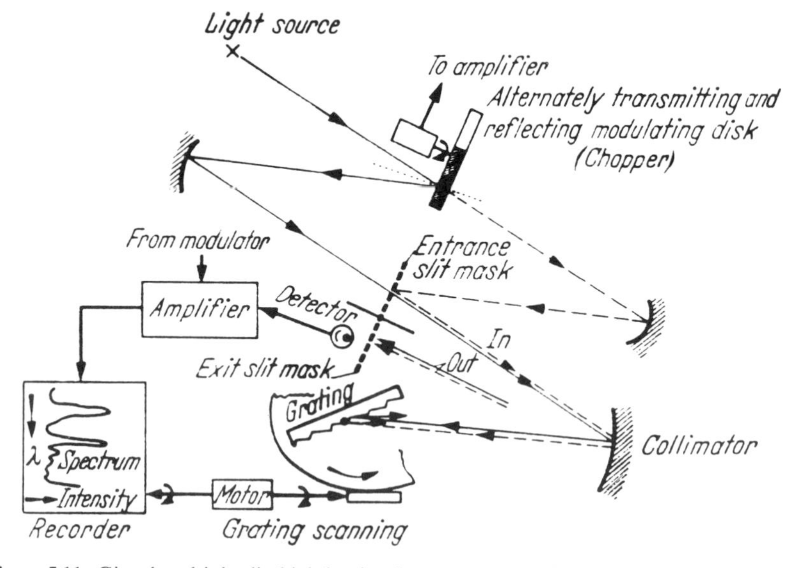

Figure 5.11. Girard multiple-slit, high-luminosity spectrometer. [Reprinted from G. W. Stroke (1967), "Diffraction Gratings," in *Handbuch der Physik; Optische Instrumente*, S. Flügge (Ed.), Vol. 29, p. 502, by permission of Springer-Verlag.]

The factor of $\frac{1}{2}$ occurs because only one-half of the total aperture W is actually open during each measurement.

Gains in luminosity in excess of 100 with no loss in resolution have been obtained in the infrared using the Girard instrument shown in Figure 5.11. A reproduction of a spectrum of methane obtained using this spectrometer is

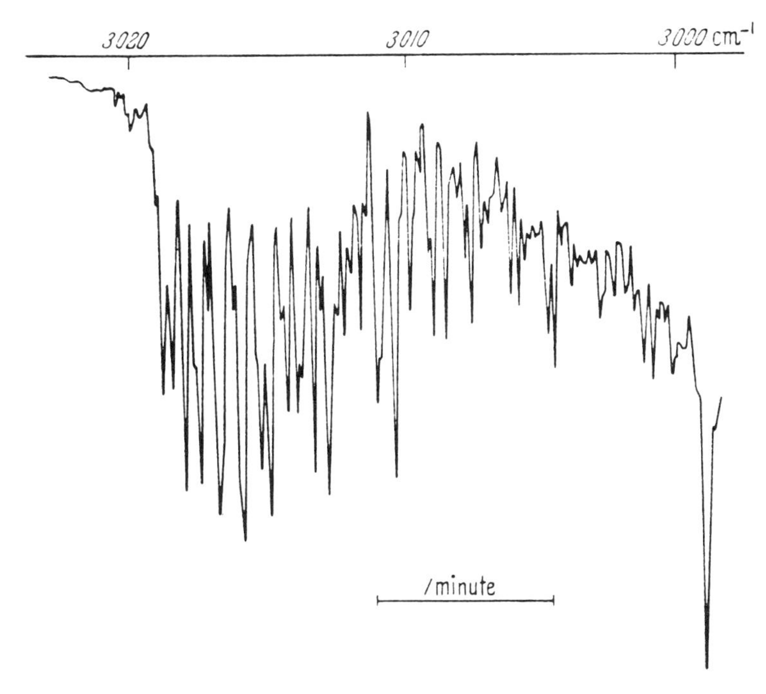

Figure 5.12. Emission spectrum of methane near 300°C obtained in Girard multiple-slit spectrometer (gas cell dimensions: length = 15 cm, pressure = 50 mm Hg at 20°C). [Reprinted from G. W. Stroke (1967), "Diffraction Gratings," in *Handbuch der Physik; Optische Instrumente*, S. Flügge (Ed.), Vol. 29, p. 502, by permission of Springer-Verlag.]

shown in Figure 5.12, and a comparison of recordings obtained using a single slit (0.05 × 30 mm) and using 30 × 30-mm masks is shown in Figure 5.13 (Stroke, 1967).

5.2.7. Other Dispersive Spectrometers with Jacquinot's Advantage

Girard's multislit spectrometer allowed significant gains in luminosity to be achieved with minimal data reduction—a significant advantage in the days prior to the advent of the modern computer. The same is also true for several other dispersive instruments developed and tested in the late 1950s and 1960s.

The SISAM (*spectromètre interférentiel à sélection par amplitude de modulation*) spectrometer described by Connes (1958, 1959, 1960) used two identical gratings (Figure 5.14) in place of the two mirrors in a Michelson interferometer (Section 7.1.1). For any particular wavelength the gratings can be adjusted to be in autocollimation so that oscillation of the compensator plate

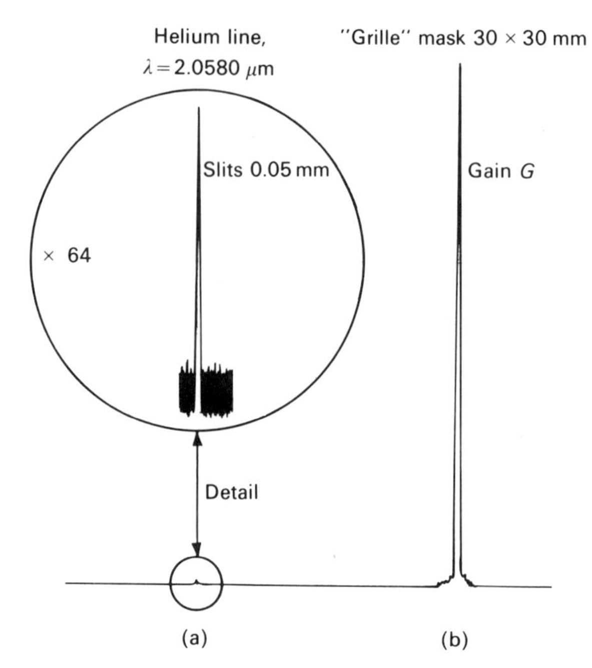

Figure 5.13. Comparison of luminosity in (*a*) single-slit grating spectrometer and (*b*) 30 × 30-mm grill, multiple-slit Girard spectrometer. Ratio of amplitudes is 30. [Reprinted from G. W. Stroke (1967), "Diffraction Gratings," in *Handbuch der Physik; Optische Instrumente*, S. Flügge (Ed.), Vol. 29, p. 503, by permission of Springer-Verlag.]

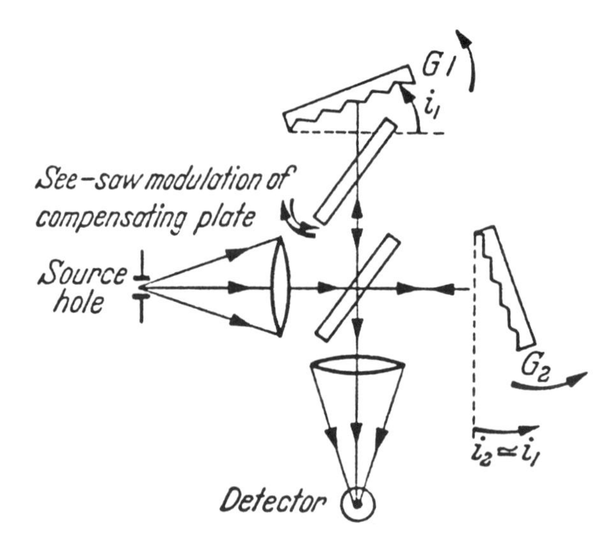

Figure 5.14. One form of SISAM spectrometer. [Reprinted from G. W. Stroke (1967), "Diffraction Gratings," in *Handbuch der Physik; Optische Instrumente*, S. Flügge (Ed.), Vol. 29, p. 503, by permission of Springer-Verlag.]

produces modulation of just that wavelength. Other wavelengths are either modulated less or not at all. The entire spectrum can be scanned by rotating the two gratings G_1 and G_2, and the frequencies and intensities are obtained directly by use of a demodulating amplifier similar to that employed in the Girard spectrometer. Luminosity gains up to 80 have been reported (Stroke, 1967).

The mock interferometer invented by Mertz (1965) is another form of multislit spectrometer (see also Ring and Selby, 1966; Selby, 1966; Stewart,

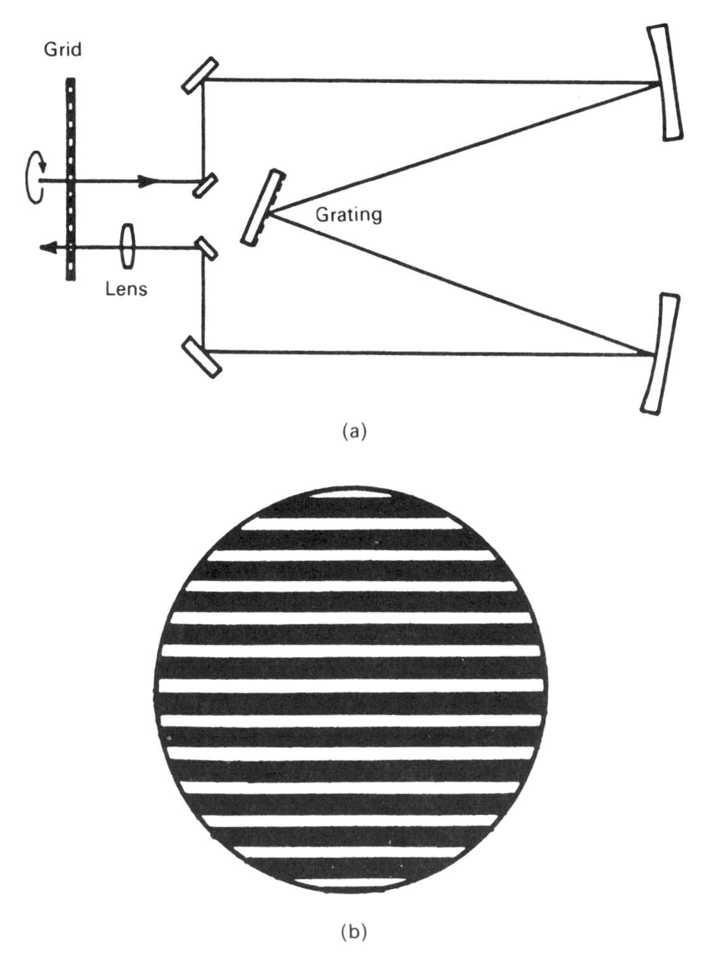

Figure 5.15. (*a*) Mertz's mock interferometer. (Reprinted from Stewart, 1970, p. 275, by courtesy of Marcel Dekker, Inc.) (*b*) Ronchi grid. (Reprinted from Stewart, 1970, p. 276, by courtesy of Marcel Dekker, Inc.)

1970). As shown in Figure 5.15*a*, the plane of the entrance and exit ports of the spectrometer contain a large Ronchi grid, a circular disk consisting of alternating open and closed slits (Figure 5.15*b*). Images of the entrance port appear at the exit port with each wavelength displaced by some distance determined by the dispersion of the grating. If the grid is rotated, each wavelength is chopped at a different frequency, producing a signal similar to that produced by a Michelson interferometer (Section 7.1.2). Different wavelengths can be discriminated using a tuned amplifier or by Fourier analysis.

Other methods that have been used to increase the luminosity of dispersive spectrometers have been reviewed in literature dealing with the design principles of spectrometers and spectrographs (Stroke and Stroke, 1963). Some of these techniques have also been used to enhance resolving powers attainable with gratings of given dimensions.

5.3. MULTIPLEXING

While all of the spectrometers described in Section 5.2 achieve a Jacquinot's advantage, except for the mock interferometer (Mertz, 1965) they still act as monochromators and monitor one wavelength channel at a time. Thus, only the question of resolution versus throughput has been treated, while the problem of SNR versus scanning time may still require the spectroscopist to select a set of compromise conditions that are not ideal for a particular experiment.

Under favorable circumstances, the need for compromise between scanning time and SNR can be eliminated by monitoring more than one wavelength channel at the same time (simultaneous multichannel detection or multiplexing). To understand why this occurs, we consider several ways of determining the weights of a collection of unknown objects (Yates, 1935). In the optical analogy, the weight of the object corresponds to the intensity associated with a particular wavelength channel, and the balance corresponds to the detector.

5.3.1. Conventional, One-at-a-Time Weighing

Following the development of Harwit (1978) and of Marshall and Comisarow (1978), we consider the problem of determining i unknown weights $x_1, x_2, \cdots, x_i$ using several weighing designs. The conventional procedure consists of weighing each of the unknowns, one at a time, on an analytical balance. Each of the resulting measurements yields an estimate $y_1, y_2, \cdots, y_i$ of the weights of the unknown objects directly, so no data reduction is necessary. Each object is weighed only once, and there will be a weighing error $e_1, e_2, \cdots, e_i$ associated with each determination. These errors are associated with the difference between the estimate y_i and the true weight x_i:

$$y_i = x_i + e_i \tag{5.18}$$

We assume that these errors are entirely random and associated primarily with the balance, that is, they are detector limited. In this case, the magnitude of e_i should not depend on the weight of the object, and the relative error r_{ei} associated with each measurement

$$r_{ei} = e_i / x_i \tag{5.19}$$

will tend, on the average, to be larger for objects that weigh less and smaller for objects that weigh more. If the relative error becomes too large, that is, the weight becomes comparable to the weighing error, then it will be impossible to weigh the object with any degree of precision.

The analogy between one-at-a-time weighing and conventional spectroscopy is straightforward. In conventional spectroscopy, each wavelength channel (object) is scanned (weighed) once, and the observed intensity in the spectrum is an estimate of the true intensity, which is proportional to the concentration of the chemical element in the sample. If the noise (error associated with each intensity measurement) is primarily due to the detector (balance), then the average noise level is relatively constant throughout the spectrum and does not depend on the intensity of the signal. Consequently, the SNR for each signal (inverse of the relative error) is less for more intense signals and greater for weaker signals. If the SNR becomes too small, that is, the intensity of the signal becomes comparable to the noise level, then this wavelength channel is not suitable for the analysis.

5.3.2. Multiple Weighings Using the Same Balance

Time averaging, or slow scanning, can generally be used to improve the SNR of a resolution element in conventional spectroscopy. In the weighing analogy, multiple determinations of individual weights are made using the same balance. Suppose object i is weighed n times. If the weighing error is entirely random, then over many determinations, the average of the error, $\langle e_i \rangle$, will approach zero, and the average of the estimates, $\langle y_i \rangle$, will approach the true value x_i.

The square of the error, e_i^2, is always nonzero, and σ, the root-mean-square error, or standard deviation, is often used to express the uncertainty associated with multiple determinations of the same quantity. For n separate weighings of object i,

$$\sigma_i = \langle e_i^2 \rangle^{1/2} = \left[\frac{1}{n}(e_{i1}^2 + e_{i2}^2 + \cdots + e_{in}^2) \right]^{1/2} \tag{5.20}$$

where the SNR of the measurement is

$$\mathrm{SNR} = 1/r_{ei} = x_i/\sigma_i \tag{5.21}$$

The SNR can be improved only by making more measurements and increasing the time required to complete the determination. If each of the i objects must be weighed n times on the same balance, then it requires approximately n times longer for all i weights to be determined. The same analogy holds for spectral measurements. The SNR for the ith signal is given by the true intensity of the signal, I_i, divided by the root-mean-square average of the error (noise) associated with the measurement:

$$\mathrm{SNR} = I_i/\sigma_i \tag{5.22}$$

Since an improvement in noise or error is proportional to the square root of n (Eqs. 5.20 and 5.22), continuing to increase the observation time yields proportionally less advantage in noise reduction, and the trade-off in scanning time eventually becomes experimentally unacceptable.

5.3.3. Multiple Balances

Another way to make multiple weighings of each object is to use more than one balance at a time. Suppose three objects are weighed on three separate balances. If all three balances are operated simultaneously, then a single determination of all three weights requires only one-third the time needed for conventional, one-at-a-time weighing using a single balance. Alternatively, all three objects could be weighed three times each on three separate balances in the same time period required for three single determinations using only one balance. Thus, multiple balances make it possible to either decrease measurement time while retaining the same SNR as in conventional weighing or improve SNR by keeping the same total observation time as required in conventional weighing.

A spectrometer (Figure 5.16) employing M detectors may be used to detect M separate wavelength channels simultaneously. Theoretically, all M channels could be monitored in approximately $1/M$th the time required by a conventional scanning spectrometer. A multichannel detector could, theoretically, also yield the desired M measurements with an improved SNR

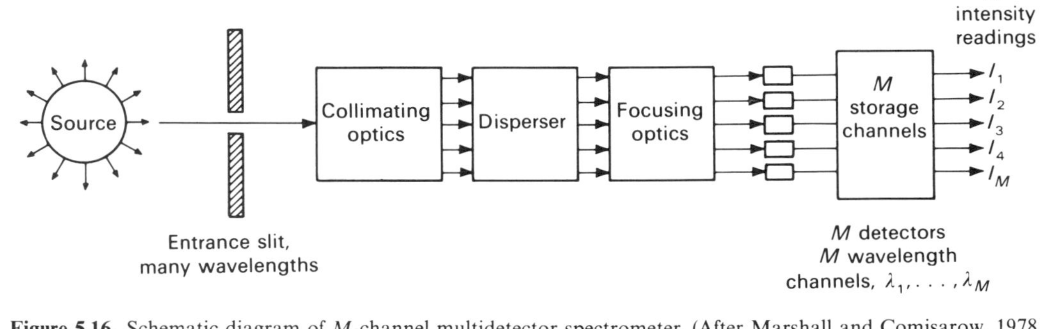

Figure 5.16. Schematic diagram of M-channel multidetector spectrometer. (After Marshall and Comisarow, 1978.)

proportional to $M^{1/2}$ in the same observation time required to scan M channels on a conventional spectrometer.

It should be noted that the advantages actually obtained from a multi-detector spectrometer depend on a number of factors, including the choice of detector. Conventional, UV–visible spectrometers generally use a photo-multiplier, which is a sensitive detector with rapid response time. Direct readers (Chapter 15) employing a bank of photomultipliers make it possible to monitor many energy channels simultaneously. The disadvantages of this arrangement include the large size and increased expense of the instrument. Alternatively, the bank of photomultipliers can be replaced by other types of multichannel detectors, such as the photographic plate or vidicon tube. Each type of detector has its own unique advantages and disadvantages. While the photographic plate (Chapter 8) is sensitive and has excellent resolution over a wide spectral range, it also requires considerable time and effort to develop, calibrate, and read. Vidicons (Chapter 15) respond more quickly but are useful only for applications requiring lower resolution and/or a limited spectral range.

5.3.4. Using the Weights in Combination: The Single-Balance, Single-Pan Design

With a little thought, it is possible to devise a method of weighing our unknown objects on a single balance so as to maintain the same observation time yet still perform multiple determinations of each object. Consider the specific case of three unknowns, x_1, x_2, and x_3. Suppose two of these objects are placed together on one pan of the balance and weighed simultaneously. Since the weights are being determined in combination, some data reduction will be required at the end of the analysis to transform the combinations, z_1, z_2, and z_3, into an estimate of the individual weights. With three unknowns, the following three linearly independent combinations of weights must be determined in order to estimate each weight individually,

$$\begin{aligned} z_1 &= x_1 + x_2 \phantom{{}+x_3} + e_1 \\ z_2 &= x_1 + \phantom{x_2+{}} x_3 + e_2 \\ z_3 &= \phantom{x_1+{}} x_2 + x_3 + e_3 \end{aligned} \tag{5.23}$$

where e_1, e_2, and e_3 are the errors associated with the three measurements.

Examination of the weighing design indicates that each object has been measured twice in the same time period as required for single determinations of each object using conventional, one-at-a-time weighing. For object 1, an estimate y_1 of its weight x_1 is given by

$$y_1 = \tfrac{1}{2}(z_1 + z_2 - z_3) = x_1 + \tfrac{1}{2}(e_1 + e_2 - e_3) \tag{5.24}$$

If error is detector limited (does not depend on the intensity of the signal), then the root-mean-square error for object 1 in the determination using this single-pan weighing design will be

$$\begin{aligned} \sigma_1 &= \{[\tfrac{1}{2}(e_1 + e_2 - e_3)]^2\}^{1/2} \\ &= [\tfrac{1}{4}(e_1^2 + e_2^2 + e_3^2)]^{1/2} \\ &\approx [\tfrac{1}{4}(3e_1^2)]^{1/2} \\ &= (3^{1/2}/2)e_1 \end{aligned} \tag{5.25}$$

Error cross products (e_1e_2, e_2e_3, etc.) disappear in step 2 of Eqs. 5.25 because each error (e_1, e_2, e_3) has an equal probability of being positive or negative, but the squares of any individual error are always positive. Step 3 is justified by assuming that error does not depend on signal intensity and so $e_1 = e_2 = e_3$ on the average.

Analysis of the weighing errors associated with objects 2 and 3 yields similar results. Therefore, by weighing the objects together using this experimental design, we have improved the SNR of each weight determination by a factor of $2/3^{1/2}$ over that obtained for any single determination (Eq. 5.18) without sacrificing anything in total observation time.

5.3.5. Using the Weights in Combination: The Single-Balance, Double-Pan Design

One additional weighing design will be considered. Suppose four unknowns must be determined, but this time the objects are distributed over both pans of the balance during each weighing. Again, combinations of weights will be measured, and four linearly independent equations are required to obtain an estimate of each weight separately:

$$\begin{aligned} z_1 &= x_1 + x_2 + x_3 + x_4 + e_1 \\ z_2 &= x_1 - x_2 - x_3 + x_4 + e_2 \\ z_3 &= x_1 - x_2 + x_3 - x_4 + e_3 \\ z_4 &= x_1 + x_2 - x_3 - x_4 + e_4 \end{aligned} \tag{5.26}$$

A negative sign in Eqs. 5.26 indicates that the object is placed on the right-hand pan of the balance, while a positive sign indicates that the object is placed on the left-hand pan of the balance.

Examination of this weighing design indicates that each object has been measured four times in the same time period required for single determinations of each object using conventional, one-at-a-time weighing. For object 1, an estimate y_1 of its weight x_1 is given by

$$\begin{aligned} y_1 &= \tfrac{1}{4}(z_1 + z_2 + z_3 + z_4) \\ &= x_1 + \tfrac{1}{4}(e_1 + e_2 + e_3 + e_4) \end{aligned} \tag{5.27}$$

If error is detector limited (does not depend on the intensity of the signal), then the root-mean-square of the error for object 1 in the determination using a double-pan weighing design will be

$$\begin{aligned} \sigma_1 &= \{[\tfrac{1}{4}(e_1 + e_2 + e_3 + e_4)]^2\}^{1/2} \approx [\tfrac{1}{16}(4e_1^2)]^{1/2} \\ &= \tfrac{1}{2}e_1 \end{aligned} \tag{5.28}$$

Analysis of the weighing errors associated with objects 2, 3, and 4 yields similar results. Therefore, by weighing the objects together in this fashion, we have improved the SNR of the weight determination by a factor of 2 over that obtained for any single weighing (Eq. 5.18) without sacrificing anything in total observation time.

5.4. SELECTING THE BEST MULTIPLEXING DESIGN

The best multiplexing design includes each individual object a maximum number of times in the least number of total observations. Generally, the total number of observations is never less than the number of objects being measured. The specific way of combining the objects so as to produce the optimum arrangement depends on how many objects are to be determined and whether the single- or double-pan design can be used. The following procedure for choosing a multiplexing design will be offered without proof. The reader may consult Harwit and Sloane (1979) for verification.

5.4.1. The *W*-Matrix

Any set of simultaneous, linear equations such as Eqs. 5.23 or 5.26 can be represented in matrix format:

$$\mathbf{z} = \mathbf{W}\mathbf{x} + \mathbf{e} \tag{5.29}$$

Table 5.4. *W*-Matrices for Three Weighing Designs

Combined Weights	Objects[a] (*a*)			Objects[b] (*b*)			Objects[c] (*c*)			
	x_1	x_2	x_3	x_1	x_2	x_3	x_1	x_2	x_3	x_4
z_1	1	0	0	1	1	0	1	1	1	1
z_2	0	1	0	1	0	1	1	−1	−1	1
z_3	0	0	1	0	1	1	1	−1	1	−1
z_4	—	—	—	—	—	—	1	1	−1	−1

[a] One-at-a-time weighing (Section 5.3.1).
[b] Single-pan, *S*-matrix design (Section 5.3.4).
[c] Double-pan, Hadamard matrix design (Section 5.3.5).

or

$$\begin{bmatrix} z_1 \\ z_2 \\ \vdots \\ z_i \end{bmatrix} = \begin{bmatrix} w_{11} & w_{12} & \cdots & w_{1i} \\ w_{21} & w_{22} & \cdots & w_{2i} \\ \vdots & & & \vdots \\ w_{i1} & w_{i2} & \cdots & w_{ii} \end{bmatrix} \begin{bmatrix} x_1 \\ x_2 \\ \vdots \\ x_i \end{bmatrix} + \begin{bmatrix} e_1 \\ e_2 \\ \vdots \\ e_i \end{bmatrix} \tag{5.30}$$

where **z**, **x**, and **e** are column matrices that contain the combined weight measurements, the true values for each individual weight, and the error associated with each measurement, respectively.

The $n \times n$ $(n = i)$ *W*-matrix is a representation of the weighing design in which each row corresponds to an experimental measurement and each column corresponds to a particular object, or unknown, whose weight is to be determined. The numbers in the matrix will be 1 if the object is placed on the left-hand pan of the balance, −1 if the object is placed on the right-hand pan of the balance, and zero if the object is not placed on either pan during that particular observation. Using this notation, we can represent our three weighing designs, (a) conventional one-at-a-time weighing, (b) multiple determinations of three objects using a single pan (Eqs. 5.23), and (c) multiple determinations of four objects using both pans (Eqs. 5.26), by the *W*-matrices shown in Table 5.4.

5.4.2. The Hadamard Matrix

Hadamard matrices make the best multiplexing designs when entries +1 and −1 can be used in a double-pan weighing design (Nelson and Fredman, 1971). A Hadamard matrix (Hadamard, 1893) is an $n \times n$ matrix of +1's and −1's with the property that the scalar product of any two rows is always zero, for example, $(w_{11})(w_{21})+(w_{12})(w_{22})+\cdots+(w_{1i})(w_{2i})=0$. (Such matrices are said to be orthogonal.) One simple way of constructing a Hadamard matrix is to begin by making all of the elements of the first row and first column of the matrix equal to $+1$. (Such Hadamard matrices are said to be normalized.) Since every remaining row has a zero scalar product with the first row, each row apart from the first must contain $+1$ in $\frac{1}{2}n$ places and -1 in $\frac{1}{2}n$ places. In addition, every remaining row also has a zero scalar product with every other remaining row. Therefore, for $n>2$, if i and j are any two distinct rows other than the first row, then when comparing row i to row j, there must always be exactly

$\frac{1}{4}n$ columns where row i is $+1$ and row j is $+1$,

$\frac{1}{4}n$ columns where row i is $+1$ and row j is -1,

$\frac{1}{4}n$ columns where row i is -1 and row j is $+1$, and

$\frac{1}{4}n$ columns where row i is -1 and row j is -1.

This means that Hadamard matrices exist only for orders (n values) of 1, 2, 4, and whole-number multiples of 4. The matrix labeled c in Table 5.4 is a 4×4 normalized Hadamard matrix. Other Hadamard matrices may be obtained from the normalized matrix by multiplying any of the rows or columns by -1.

5.4.3. The *S*-Matrix

S-Matrices make the best multiplexing designs when entries 0 and $+1$ are used in a single-pan weighing design (Hotelling, 1944; Sloane and Harwit, 1976). An *S*-matrix (Hadamard, 1893) is an $n\times n$ matrix obtained by taking a normalized Hadamard matrix, omitting the first row and column, and then changing all of the remaining $+1$'s to 0's and all of the remaining -1's to $+1$'s. This construction produces a matrix in which each row and each column contains $\frac{1}{2}(n-1)$ places occupied by 0's and $\frac{1}{2}(n+1)$ places occupied by $+1$'s. The name is derived from the fact that the patterns of 0's and 1's in the rows form a code, which in communication theory is known as a *simplex code*.

The matrix labeled b in Table 5.4 is a 3×3 *S*-matrix that has been derived from the Hadamard matrix in this same table using the procedure described in the preceding. Since Hadamard matrices only exist for orders 1, 2, 4, and whole-number multiples of 4, *S*-matrices only exist for orders 3, 7, 11, 15, $\cdots$, that is, for values of $4r-1$, where r is any positive integer.

5.4.4. Fellgett's Advantage

The benefits of using the optical analogs of these weighing designs is sometimes called *Fellgett's advantage* (Fellgett, 1951, 1958, 1971) or *the multiplex advantage* and may be stated as follows:

For optical measurements taken at equal resolution and in an equal measurement time with the same detector and on an instrument with the same optical throughput and efficiency, the SNR of a spectrum will be improved by a factor proportional to $n^{1/2}$, where n is the number of resolution elements or wavelength channels observed simultaneously.

The theoretical improvement in SNR depends on the weighing design and may be determined according to the following guidelines. It should be stressed that these statements apply *only* in the case of detector-limited noise.

1. If there are i unknowns and a Hadamard matrix of order $n = i$ can be constructed and used in a double-pan weighing design, then each object will be weighed n times, and the SNR will be improved by a factor of $n/n^{1/2}$ or $n^{1/2}$.
2. If there are i unknowns and an S-matrix of order $n = i$ can be constructed and used in a single-pan weighing design, then each object will be weighed $(n+1)/2$ times, and the SNR will be improved by a factor of $(n+1)/(2n^{1/2})$ or approximately $\frac{1}{2}(n^{1/2})$.

From the above, it can be seen that for comparable n values, Hadamard matrices give approximately twice the improvement in SNR compared to S-matrices. This is because each object is determined almost twice as often using a double-pan design as compared to a single-pan design.

Weighing designs other than those based on Hadamard or S-matrices can also be employed to obtain a multiplex advantage. However, in such cases all objects will not be weighed the same number of times, and the improvement in each SNR must be determined according to the specific design. For example, the following 4×4 W-matrix,

$$\mathbf{W} = \begin{bmatrix} 0 & 1 & 1 & 1 \\ 1 & 1 & 0 & 0 \\ 1 & 0 & 1 & 0 \\ 1 & 0 & 0 & 1 \end{bmatrix} \qquad (5.31)$$

represents a single-pan weighing design that does not employ an S-matrix. For object 1,

$$\begin{aligned}\sigma_1 &= \{[\tfrac{1}{3}(-e_1+e_2+e_3+e_4)]^2\}^{1/2} \\ &\approx [\tfrac{1}{9}(4e_1^2)]^{1/2} = \tfrac{2}{3}e_1 \end{aligned} \tag{5.32}$$

while for object 2,

$$\begin{aligned}\sigma_2 &= \{[\tfrac{1}{3}(e_1+2e_2-e_3-e_4)]^2\}^{1/2} \\ &= [\tfrac{1}{9}(e_1^2+4e_2^2+e_3^2+e_4^2)]^{1/2} \\ &\approx [\tfrac{1}{9}(7e_1^2)]^{1/2} = (7^{1/2}/3)e_1 \end{aligned} \tag{5.33}$$

Thus, the SNR for object 1 is improved by a factor of $\frac{3}{2}$, while the SNR for object 2 is improved by a factor of $3/7^{1/2}$. The reader may wish to verify that the SNRs for objects 2 and 3 are also improved by factors of $3/7^{1/2}$.

Fellgett's advantage can be particularly important for remote sensing (as in astronomical spectroscopy) and for other types of spectroscopic measurements that involve a weak source that can be monitored over a fairly lengthy time period. While measurements made in the same time yield an advantage in SNR, measurements at the same SNR can also be made n times faster using multiplexing methods. Therefore, if there is a limited time in which a large amount of data must be gathered, multiplexing may again offer some very substantial advantages over conventional spectroscopy.

5.5. HADAMARD TRANSFORM SPECTROMETERS

A singly encoded Hadamard transform (HT) spectrometer (Harwit, 1978; Harwit and Sloane, 1979) illustrates the spectroscopic use of multiplexing methods. The basic HT instrument, shown schematically in Figure 5.17, contains three essential components in addition to those of the conventional spectrometer (cf. Figure 5.1): (1) a mask located ahead of the detector, (2) postmask optics, and (3) a processor located after the detector. The purpose of the mask is not only to pass more than one resolution element at a time to the detector, but also to label (encode) each component in the spectrum. The postmask optics dedisperse the radiation transmitted by the mask (recombines the radiation into pseudowhite light) and focus the recombined intensities onto the detector. Since the detector always sees a combination of intensities from more than one resolution element, a processor is needed to analyze the combined radiation and reconstruct the intensity of each individual resolution element. Because there is only one entrance slit, the instrument

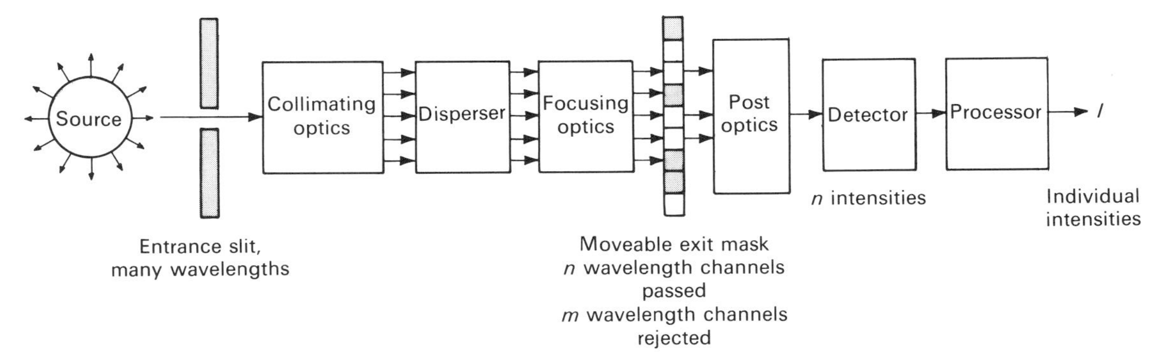

Figure 5.17. Schematic diagram of singly encoded Hadamard transform spectometer. (After Harwit and Decker, 1974.)

shown in Figure 5.17 does not have Jacquinot's advantage. However, since the intensities are measured in combination, there can be a multiplex advantage.

5.5.1. The Mask

An HT mask consists of a large encoding mask (Figure 5.18*a*) that contains a label for all resolution elements passed through the spectrometer and a smaller frame or blocking mask (Figure 5.18*b*) that permits radiation to reach only a certain portion of the encoding mask at any given time (Figures 5.18*c*, *d*). In practice, the encoding mask is simply an array of slits that are either open or shut. When shut, the light is blocked. When open, the slits transmit light to a single detector, which measures the sum of the intensities of the transmitted radiation in a (+1, 0) type of single-pan weighing design.

The combined intensities will yield linear measurements only if the detector employed in the spectrometer is a linear device, that is, double intensity produces double the signal. Fortunately, many detectors do behave linearly (Chapters 9–11) at low-intensity levels, conditions where multiplexing methods are most important.

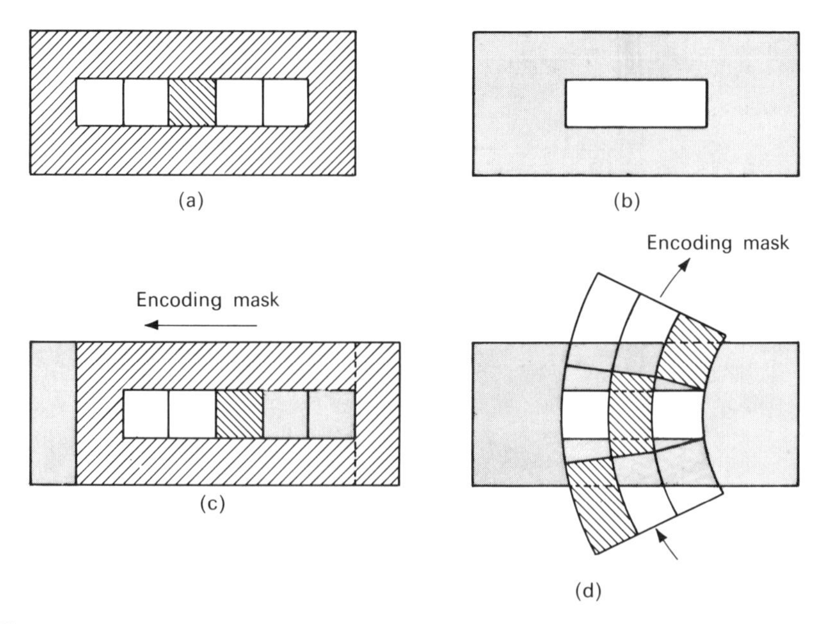

Figure 5.18. (*a*) Five-element encoding mask. (*b*) Framing mask. (*c*) Horizontally translating encoding mask. (*d*) Rotating encoding mask. (After Harwit and Decker, 1974.)

If the intensities of n different resolution elements are to be determined, then a minimum of n linearly independent measurements of the spectrum must be made. While it is possible to use a separate encoding mask for each measurement, it is much more efficient if all of the mask elements are placed on a single horizontal strip (Swift et al., 1972) or disk (Decker and Harwit, 1968; Hansen and Strong, 1972). During operation, the strip or disk can be moved in either a stepwise or continuous fashion past the frame, with each new position of the strip or disk corresponding to a new label or weighing design for the spectrum.

The best encoding design for the open or shut type of mask is the S-matrix. If this S-matrix is also cyclic, construction of the encoding mask is considerably facilitated. A left-circulant matrix has the property that each row is obtained by shifting the previous row one place to the left with any overflow from the left coming in on the right. (A right-circulant matrix shifts to the right). The matrix labeled b in Table 5.4 (Eq. 5.34) is a left-circulant S-matrix. Methods for constructing cyclic S-matrices are described in Harwit and Sloane (1979) and Sloane (1982). Cyclic S-matrices may be written for $n = 3, 7, 15, \cdots, 2r - 1$, where r is any positive integer:

$$\mathbf{S} = \begin{bmatrix} 1 & 1 & 0 \\ 1 & 0 & 1 \\ 0 & 1 & 1 \end{bmatrix} \tag{5.34}$$

The advantage of using a cyclic matrix is that instead of many separate masks (i.e., 110, 101, and 011 from Eq. 5.34), one single mask (11011) may be constructed. By placing the first three mask elements (110) over the frame, the first row of the S-matrix in Eq. 5.34 is obtained. By moving the mask one element to the left, the second row of the S-matrix (101) in Eq. 5.34 is obtained, and so forth. Thus, changing from one mask to another simply consists of moving a $2n - 1$ linear encoding mask across a frame of length n by one position per change. (The reader may wish to verify that the sequence 0101110010111 represents a 13-element mask. derived from a 7×7 cyclic S-matrix.)

Figure 5.19 shows how the encoding mask 11011 can be used in a stepwise fashion to obtain the three measurements required in the single-pan weighing design given by Eqs. 5.23. At T_1, the first three elements in the encoding mask are positioned over the frame opening, giving the pattern 110. At T_2, the encoding mask is moved up across the frame opening by one element, giving the pattern 101. At T_3, the exposed pattern is 011.

The size of the mask determines how many wavelength channels can be monitored simultaneously. If the mask could be made large enough, all

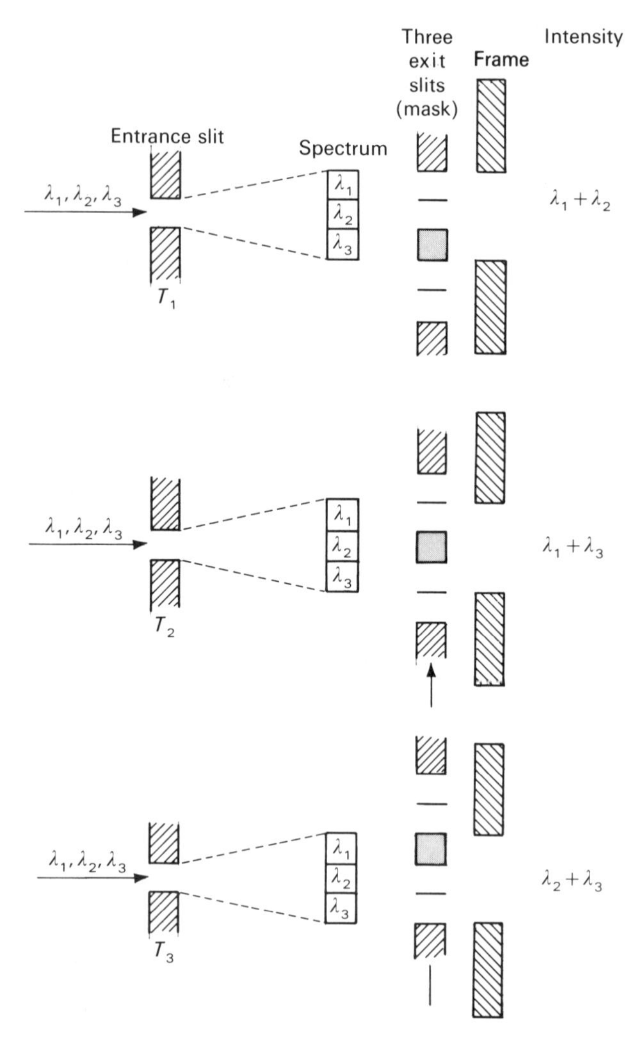

Figure 5.19. Channel intensities for singly encoded Hadamard transform spectrometer employing three entrance and three exit slits in *S*-matrix multiplexing design. Arrows show direction of mask motion.

desired wavelength channels could be monitored simultaneously without rotating the disperser. In theory, n resolution elements will require a mask containing a minimum of $2n - 1$ encoding elements. Placing these encoding elements on a rotating disk (Figure 5.18*d*) facilitates multiple determinations of each mask arrangement. This can be used to further enhance the SNR of the wavelength channels but only at the expense of additional measurement time.

5.5.2. The Processor

The function of the processor is to recover the original spectrum from the n combined intensity measurements made by the detector. This typically involves solving the n equations in n unknowns obtained from the weighing design (e.g., Eqs. 5.23), where the detector supplies the values for $z_1, z_2, \cdots, z_i$ and the values for the estimates $y_1, y_2, \cdots, y_i$ (the true intensity x_i plus the error e_i in measurement) must be calculated. Since

$$y_i = x_i + e_i \tag{5.18}$$

we may express Eqs. 5.23 in matrix form as

$$\begin{bmatrix} z_1 \\ z_2 \\ \vdots \\ z_i \end{bmatrix} = \begin{bmatrix} w_{11} & w_{12} & \cdots & w_{1i} \\ w_{21} & w_{22} & \cdots & w_{2i} \\ \vdots & & & \vdots \\ w_{i1} & w_{i2} & \cdots & w_{ii} \end{bmatrix} \begin{bmatrix} y_1 \\ y_2 \\ \vdots \\ y_i \end{bmatrix} \tag{5.35}$$

A matrix format makes the solutions for y_i particularly easy since

$$\mathbf{W}^{-1}\mathbf{W}\mathbf{y} = \mathbf{y} = \mathbf{W}^{-1}\mathbf{z} \tag{5.36}$$

where $\mathbf{W}^{-1}$ is the inverse of the W-matrix, which contains the weighing design. In other words, multiplying the **z** column matrix, which contains the combined intensity measurements made by the detector, by the inverse of the W-matrix yields the **y** column matrix, which contains all of the individual intensity estimates.

Obtaining the inverse of a Hadamard matrix or an S-matrix is particularly easy because all of the elements in the matrix are $+1$, 0, or -1. The inverse of an S-matrix of order n is readily generated by replacing each zero by -1 in the original W-matrix and then multiplying the result by $2/(n+1)$. To illustrate, the inverse of the S-matrix given by Eq. 5.34 is

$$\mathbf{S}^{-1} = \frac{2}{(n+1)} \begin{bmatrix} 1 & 1 & -1 \\ 1 & -1 & 1 \\ -1 & 1 & 1 \end{bmatrix} \tag{5.37}$$

where $n = 3$.

The inverse of a Hadamard matrix of order n may be found by taking the transpose of the original matrix and multiplying the result by $1/n$. If **H** is a

Hadamard matrix with elements a_{ij}, where i represents the row number and j represents the column number, then the transpose of **H** is a matrix whose (i, j)th entry is a_{ji}. In other words, to obtain the transpose of matrix **H**, all of the elements of the matrix are rotated 180° around the main diagonal, which runs from the top left corner to the bottom right corner of the matrix. (The main diagonal lies along the dotted line in Eq. 5.38.) To illustrate, if the Hadamard matrix is given by

$$\mathbf{H} = \begin{bmatrix} 1 & 1 & 1 & 1 \\ 1 & -1 & -1 & 1 \\ 1 & 1 & -1 & -1 \\ 1 & -1 & 1 & -1 \end{bmatrix} \tag{5.38}$$

then the inverse will be

$$\mathbf{H}^{-1} = \frac{1}{n} \begin{bmatrix} 1 & 1 & 1 & 1 \\ 1 & -1 & 1 & -1 \\ 1 & -1 & -1 & 1 \\ 1 & 1 & -1 & -1 \end{bmatrix} \tag{5.39}$$

where $n = 4$.

The use of matrices to obtain individual intensity estimates y_i from the combined intensity measurements z_i will be illustrated using a specific example. Suppose three combined intensity measurements with magnitudes of $z_1 = 3$, $z_2 = 5$, and $z_3 = 4$ units are measured by the detector using the weighing design given by the S-matrix in Eq. 5.34. Then

$$\begin{bmatrix} z_1 \\ z_2 \\ z_3 \end{bmatrix} = \begin{bmatrix} 3 \\ 5 \\ 4 \end{bmatrix} = \begin{bmatrix} 1 & 1 & 0 \\ 1 & 0 & 1 \\ 0 & 1 & 1 \end{bmatrix} \begin{bmatrix} y_1 \\ y_2 \\ y_3 \end{bmatrix} \tag{5.40}$$

From Eqs. 5.36 and 5.37,

$$\begin{bmatrix} y_1 \\ y_2 \\ y_3 \end{bmatrix} = \frac{2}{3+1} \begin{bmatrix} 1 & 1 & -1 \\ 1 & -1 & 1 \\ -1 & 1 & 1 \end{bmatrix} \begin{bmatrix} 3 \\ 5 \\ 4 \end{bmatrix} \tag{5.41}$$

Matrix multiplication is performed by multiplying rows into columns, that is,

$y_i = 2/(3+1)(w_{i1}z_1 + w_{i2}z_2 + \cdots + w_{ii}z_i)$, so

$$\begin{aligned} y_1 &= \tfrac{1}{2}(+3+5-4) = 2 \text{ units} \\ y_2 &= \tfrac{1}{2}(+3-5+4) = 1 \text{ unit} \\ y_3 &= \tfrac{1}{2}(-3+5+4) = 3 \text{ units} \end{aligned} \tag{5.42}$$

The usual method for multiplying a column vector by an $n \times n$ matrix requires about $n^2 - n$ additions and subtractions. This can be readily carried out by hand for small matrices or by a small microcomputer for larger matrices. Very large matrix multiplications can be carried out quite rapidly using a fast HT method (Nelson and Fredman, 1970; Harwit and Sloane, 1979), which reduces the number of steps to about $n \log_2 n$. (For $n = 8$, this amounts to 24 additions and subtractions, as opposed to 56 for direct evaluation.)

The possibility of realizing a multiplex advantage for grating spectrometers by optical encoding schemes based on Hadamard matrices was first pointed

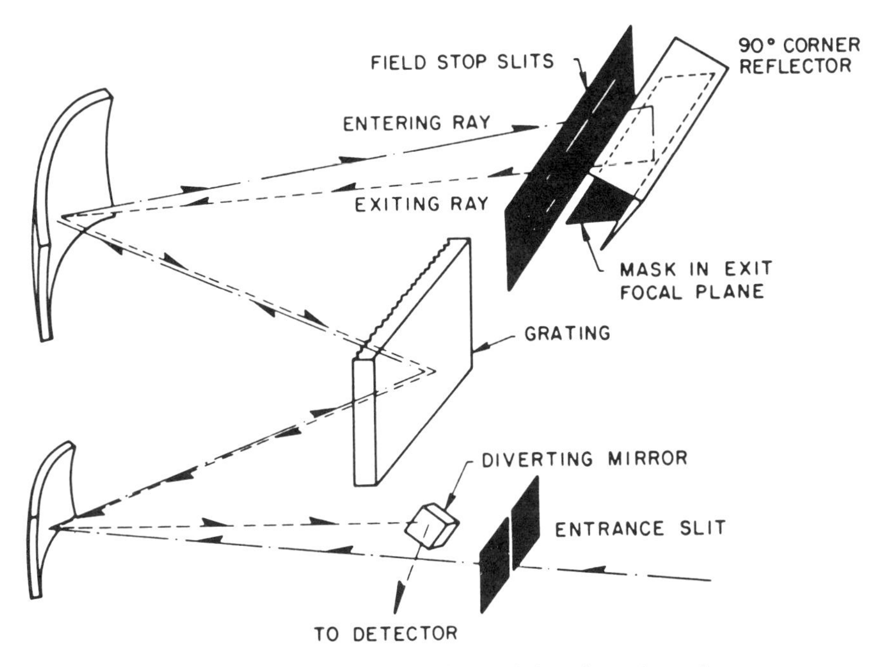

Figure 5.20. Experimental arrangement of singly encoded Hadamard transform spectrometer using Czerny–Turner arrangement of curved mirrors. [Reprinted with permission from M. Harwit and N. J. A. Sloane (1979), *Hadamard Transform Optics*, Academic, New York.]

out by Ibbett et al. (1968), and the related advantages of cyclic codes were discussed shortly afterward by Decker and Harwit (1968) with applications to multiplex spectrometry by Sloane et al. (1969). Construction and operation of a prototype HT multiplex grating spectrometer employing optical encoding at the exit focal plane was reported in 1969 (Decker and Harwit), and the first experimental verification of the multiplex advantage for singly encoded HT spectrometers was obtained by Decker (1971) using a commercially available 0.5-m Czerny–Turner spectrograph with a resolving power of about 1100. The encoding mask used a 255 cyclic S- matrix code with 509 total mask slots. The optical arrangement of this instrument is shown in Figure 5.20.

As shown in Figure 5.20, the instrument has one permanently open entrance slit and a wide exit aperture containing the mask. After passage through the mask, the direction of the rays is reversed by a corner reflector so that the light retraces its path back through the instrument. This postmask optical arrangement exactly recombines the multiplexed intensities into the image of the entrance slit. Due to the corner reflector, this image arrives slightly above or below the entrance slit, where it can be detected directly (Decker, 1971) or diverted to a detector (Figure 5.20) using a mirror.

Comparison spectra of the 1.5–1.7-μm emission spectrum of a mercury lamp were taken (Decker, 1971), operating the instrument in both the HT and scanning monochromator modes. Typical results are shown in Figure 5.21. The actual signal-to-noise gains agreed with the theoretical estimate to well within the standard deviation of the measurements.

5.5.3. Two-Dimensional Encoding Masks

In the usual HT arrangement, only one detector is employed, and the linear array of open and closed slits used as the mask is placed in the exit focal plane of the spectrometer. Instead of a straight line, the encoding slits can also be arranged in a rectangular array and used to produce an image of a two-dimensional object.

Figure 5.22 shows a schematic diagram of such an instrument (Gottlieb, 1968). In this representation, the detector responds to the total intensity of all radiation emitted by the object, and information concerning the intensities of individual wavelength channels is sacrificed to obtain spatial information. For purposes of illustration, the object shown in Figure 5.22 is a two-dimensional spectrum such as might be obtained from an Echelle spectrometer (Figure 4.20).

Suppose we divide the object into a 3×5 array of 15 separate sectors. Each of the 15 sectors has a radiation intensity associated with it, and if we determine these 15 intensities, we can form an image of the object. (While 15 is

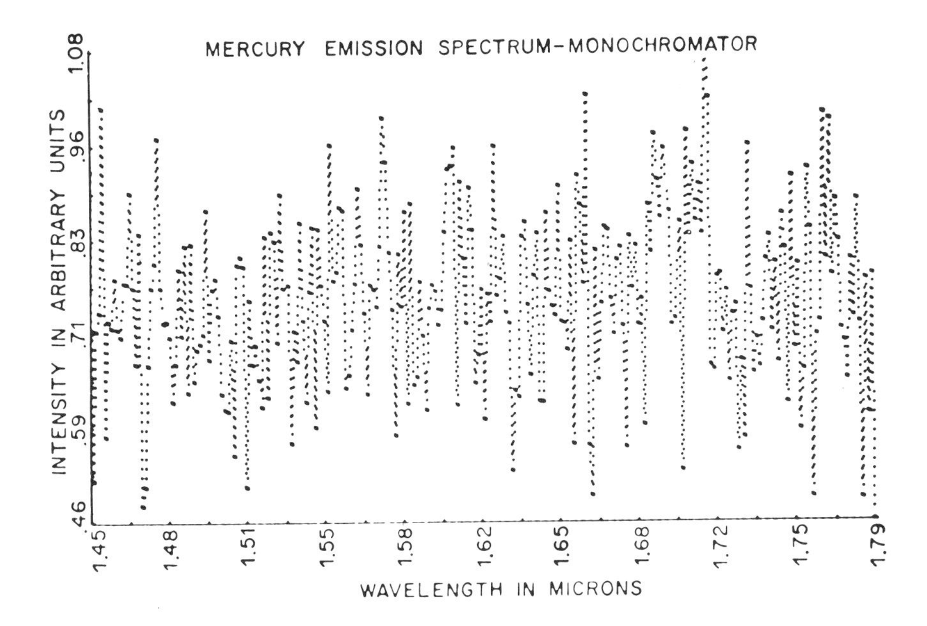

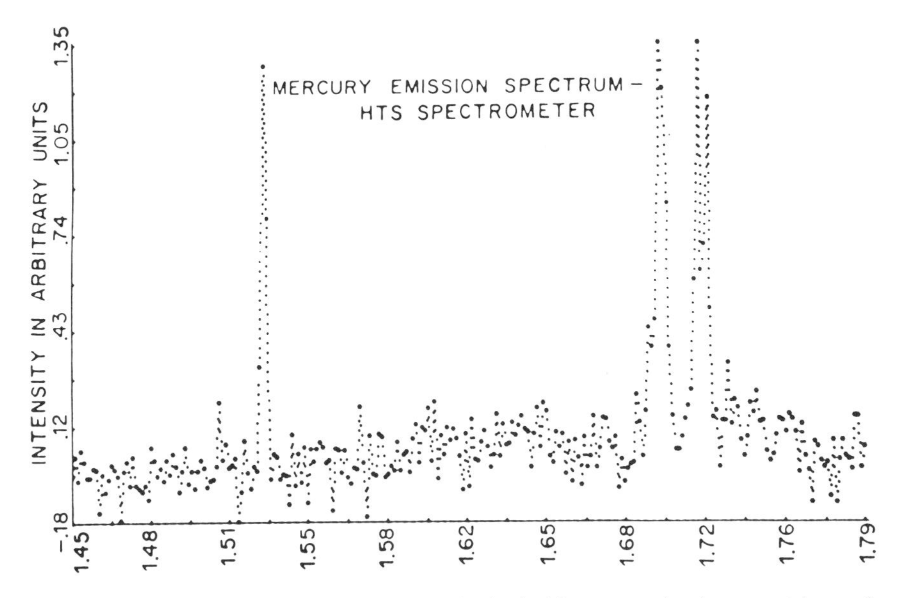

Figure 5.21. Mercury vapor emission spectra obtained with same grating instrument in equal time intervals: (*a*) using one exit slit; (*b*) using a 255-slit encoded mask. [Reprinted with permission from J. A. Decker, Jr., *Appl. Opt.*, *10*, 510. Copyright (1971) Optical Society of America.]

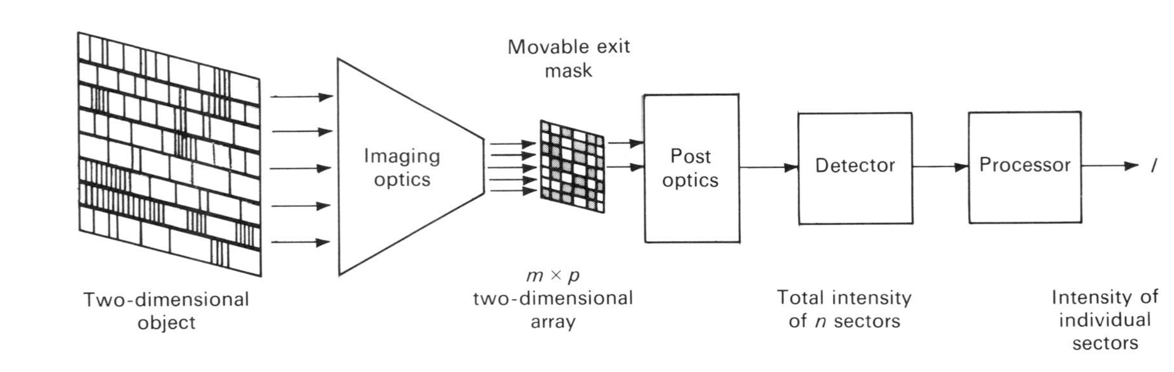

Figure 5.22. Schematic diagram of singly encoded Hadamard transform imaging spectrometer.

far too few to provide adequate detail about such a complex object as shown in Figure 5.22, 15 sectors will serve to illustrate the principle.)

If we have 15 unknown intensities, then we need 15 linearly independent mask arrangements, each with a 3×5 array of 15 slits that are either open or closed. An open or closed configuration is a $(+1, 0)$ type of weighing design, so the best mask arrangements will be derived from a 15×15 S-matrix such as

$$\mathbf{S} = \begin{bmatrix} 1&1&1&0&1&0&1&1&0&0&1&0&0&0&1 \\ 1&1&0&1&0&1&1&0&0&1&0&0&0&1&1 \\ 1&0&1&0&1&1&0&0&1&0&0&0&1&1&1 \\ 0&1&0&1&1&0&0&1&0&0&0&1&1&1&1 \\ 1&0&1&1&0&0&1&0&0&0&1&1&1&1&0 \\ 0&1&1&0&0&1&0&0&0&1&1&1&1&0&1 \\ 1&1&0&0&1&0&0&0&1&1&1&1&0&1&0 \\ 1&0&0&1&0&0&0&1&1&1&1&0&1&0&1 \\ 0&0&1&0&0&0&1&1&1&1&0&1&0&1&1 \\ 0&1&0&0&0&1&1&1&1&0&1&0&1&1&0 \\ 1&0&0&0&1&1&1&1&0&1&0&1&1&0&0 \\ 0&0&0&1&1&1&1&0&1&0&1&1&0&0&1 \\ 0&0&1&1&1&1&0&1&0&1&1&0&0&1&0 \\ 0&1&1&1&1&0&1&0&1&1&0&0&1&0&0 \\ 1&1&1&1&0&1&0&1&1&0&0&1&0&0&0 \end{bmatrix} \tag{5.43}$$

For a linear mask, each row of the matrix would be used to construct a 1×15 linear arrangement of open or closed slits. To form two-dimensional masks, the 15 elements in each row are folded, like words on a page, to produce 3×5 arrays. For example, the linear mask derived from the first row of our 15×15 matrix would be

$$s_{1,15} = \boxed{1\quad 1\quad 1\quad 0\quad 1\quad 0\quad 1\quad 1\quad 0\quad 0\quad 1\quad 0\quad 0\quad 0\quad 1} \tag{5.44}$$

while the corresponding 3×5 rectangular mask would be

$$s_{3,5} = \boxed{\begin{matrix} 1&1&1&0&1 \\ 0&1&1&0&0 \\ 1&0&0&0&1 \end{matrix}} \tag{5.45}$$

The subscripts in Eqs. 5.44 and 5.45 indicate the number of rows and columns in the mask, respectively. Each one of the 15 two-dimensional masks is obtained using this procedure. (Note that folding could also be used to produce any other $m \times p$ rectangular array as long as $m \times p = 15$.)

Instead of using 15 different masks, one larger encoding mask can be constructed by arranging these 3×5 row arrays side by side. If the original 15×15 S-matrix is cyclic (as in Eq. 5.43), all 15 folded rows from the S-matrix can be placed so that as few as 45 slits are required in the larger encoding mask. [In general, for an $m \times p$ picture array, the fewest number of slits required in the encoding mask will be $(2m-1)(2p-1)$.] Figure 5.23*a* shows how these folded rows can be arranged, where the letters, $a, b, \ldots, o$, correspond to the 1's and 0's in the first row of any 15×15 cyclic matrix. Figure 5.23*b* illustrates the specific case discussed here (Eq. 5.45).

During operation, a 3×5 framing mask (Figure 5.23*c*) is used to expose only one folded row at a time. For example, placing the frame over the slits in the upper left corner of Figure 5.23*b* yields the combined intensity readings given by the first row of the S-matrix (Eq. 5.43). The radiation passing through the mask is focused onto a single detector, and the combined intensities of all sectors passed by the mask is recorded. The encoding mask is then moved to a

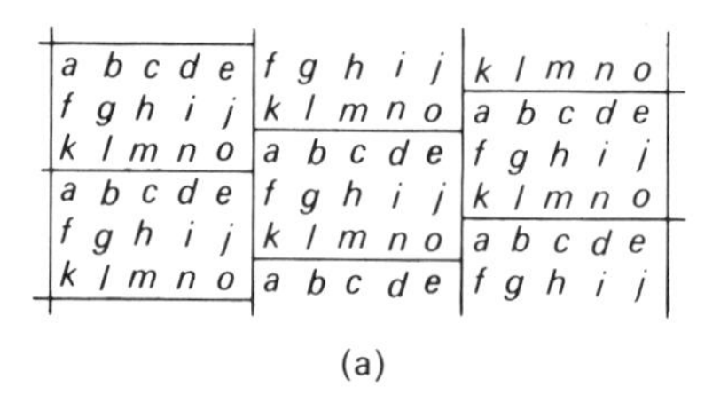

(a)

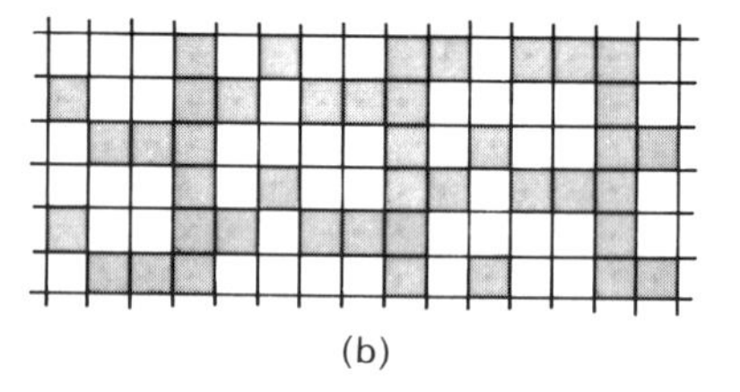

(b)

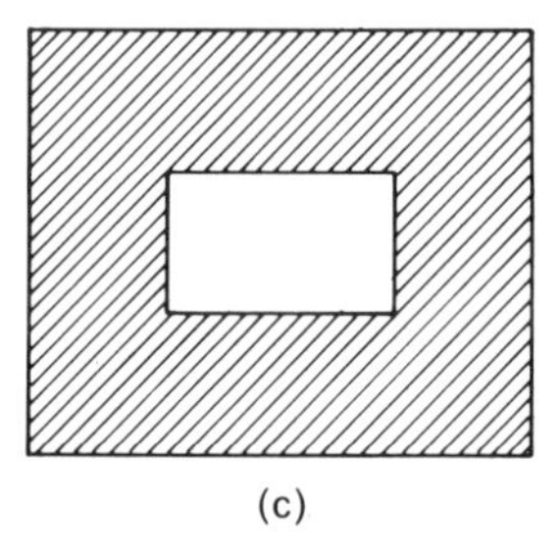

(c)

Figure 5.23. (*a*) Large array formed by repeating Eq. 5.45 in all directions. [Reprinted with permission from M. Harwit and N. J. A. Sloane (1979), *Hadamard Transform Optics*, Academic, New York.] (*b*) The 1's and 0's substituted for letters. (*c*) The 3×5 framing mask to be used with (*b*).

new position relative to the frame, exposing a different, linearly independent subarray of slits. For example, moving the encoding mask one sector to the left exposes the second row of the S-matrix in Eq. 5.43. The process is repeated until all 15 linearly independent row arrays have been examined. The processor then converts the combined sector intensities to 15 individual sector intensities, and the data is presented as a two-dimensional image. The theoretical improvement in SNR is $1/2\,(n^{1/2})$, where $n=15$. (Further information concerning two-dimensional mask designs may be found in Harwit and Sloane, 1979.)

5.5.4. Two-Detector Designs

Rather than using slits that are either open or shut, Harwit and Sloane (1979) have suggested that the mask might be constructed to contain elements that would either transmit light or reflect the light toward a second, reference detector (Figure 5.24). If the difference between the readings of the two detectors is recorded, then the wavelength channels will be measured in a $(+1, -1)$ type of pseudo-double-pan weighing design. The best encoding design for these masks is based on a Hadamard matrix.

Suppose the Hadamard matrix is given by

$$\mathbf{H}=\begin{bmatrix} 1 & 1 & 1 & 1 \\ 1 & -1 & -1 & 1 \\ 1 & 1 & -1 & -1 \\ 1 & -1 & 1 & -1 \end{bmatrix} \tag{5.38}$$

Then the four readings made by detector B are

$$\begin{aligned} z_1 &= y_1 + y_2 + y_3 + y_4 \\ z_2 &= y_1 \qquad\qquad\quad + y_4 \\ z_3 &= y_1 + y_2 \\ z_4 &= y_1 \qquad + y_3 \end{aligned} \tag{5.46}$$

where

$$\mathbf{W}_{\mathrm{B}}=\begin{bmatrix} 1 & 1 & 1 & 1 \\ 1 & 0 & 0 & 1 \\ 1 & 1 & 0 & 0 \\ 1 & 0 & 1 & 0 \end{bmatrix} \tag{5.47}$$

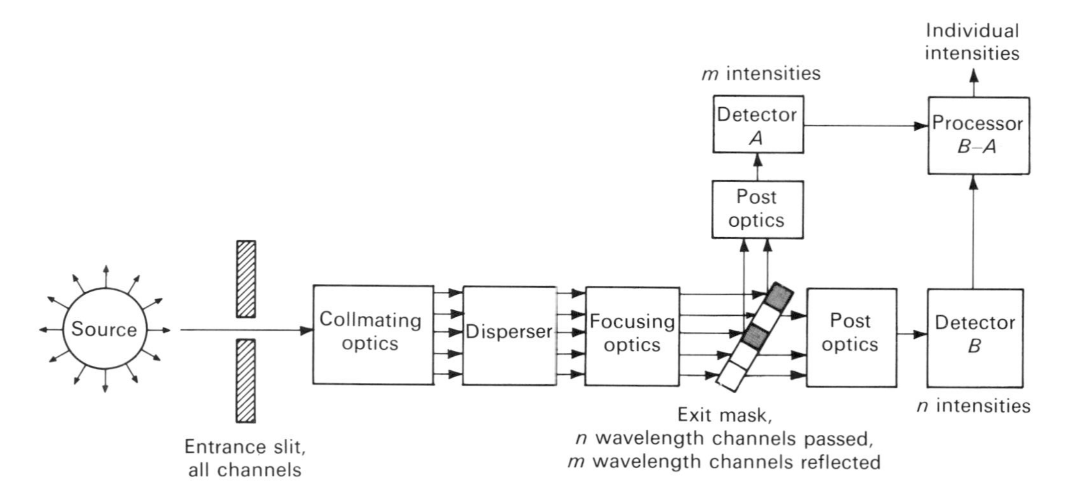

Figure 5.24. Schematic diagram of two-detector Hadamard transform spectrometer. (After Sloane and Harwit, 1976.)

and the four readings made by detector A are

$$\begin{aligned} z'_1 &= 0 \\ z'_2 &= \quad + y_2 + y_3 \\ z'_3 &= \qquad\quad + y_3 + y_4 \\ z'_4 &= \quad + y_2 \qquad + y_4 \end{aligned} \tag{5.48}$$

where

$$\mathbf{W}_A = \begin{bmatrix} 0 & 0 & 0 & 0 \\ 0 & 1 & 1 & 0 \\ 0 & 0 & 1 & 1 \\ 0 & 1 & 0 & 1 \end{bmatrix} \tag{5.49}$$

(The reader may wish to show that if the difference intensities obtained from the four mask arrangements given by Eq. 5.38 are $z''_1 = 11$ units, $z''_2 = -5$ units, $z''_3 = -3$ units, and $z''_4 = 1$ unit, where $z''_i = z_i - z'_i$, then use of Eqs. 5.36 and 5.39 will yield individual intensity measurements of $y_1 = 1$ unit, $y_2 = 3$ units, $y_3 = 5$ units, and $y_4 = 2$ units.)

An HT spectrometer based on this type of pseudo-double-pan arrangement is theoretically capable of achieving an improvement in SNR ratio equal to $(n/2)^{1/2}$ (Harwit and Sloane, 1979). This is about halfway between the ideal one-detector realization of the Hadamard matrix and the one-detector realization of the S-matrix.

5.5.5. Doubly Encoded HT Spectrometers

Hadamard transform spectrometers of this type encode radiation twice by using masks at both the entrance and exit apertures (Phillips and Harwit, 1971). These masks may be either one or two dimensional. The typical arrangement of components is shown schematically in Figure ·5.25.

Since multiple slits are employed at the entrance as well as the exit, doubly encoded HT instruments have the potential to achieve a Jacquinot's advantage in addition to a multiplex advantage (Harwit et al., 1970). However, depending on the mode of operation, both types of advantages may not always be achieved. In the example to be discussed, only a multiplex advantage is obtained because of the method used to process the combined intensity measurements. The instrument performs like a monochromator with a single entrance and exit slit, somewhat resembling Golay's static multislit spectrometer (Section 5.2.5). However, while Golay's instrument achieves a gain in luminosity due to Jacquinot's advantage, this doubly encoded HT

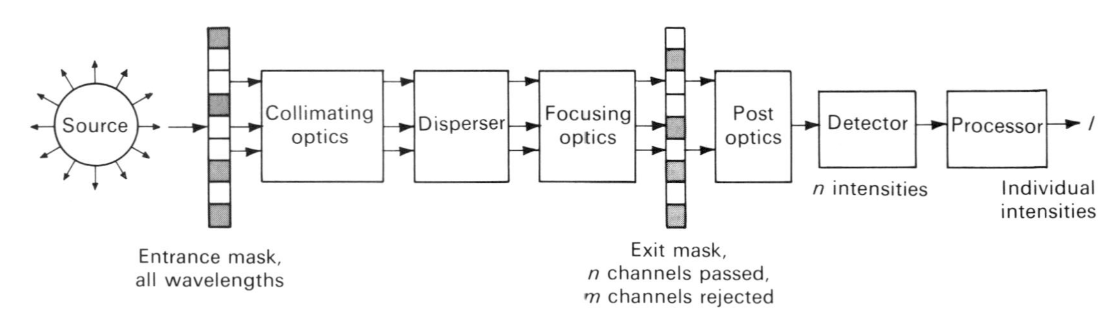

Figure 5.25. Schematic diagram of doubly encoded Hadamard transform spectrometer. (After Harwit, et al., 1970.)

spectrometer achieves a considerably increased SNR due to a multiplex advantage.

In this doubly encoded HT spectrometer, input radiation is first filtered to permit only a preselected number of wavelength channels to enter the instrument. The number of entrance slits n is made equal to the number of

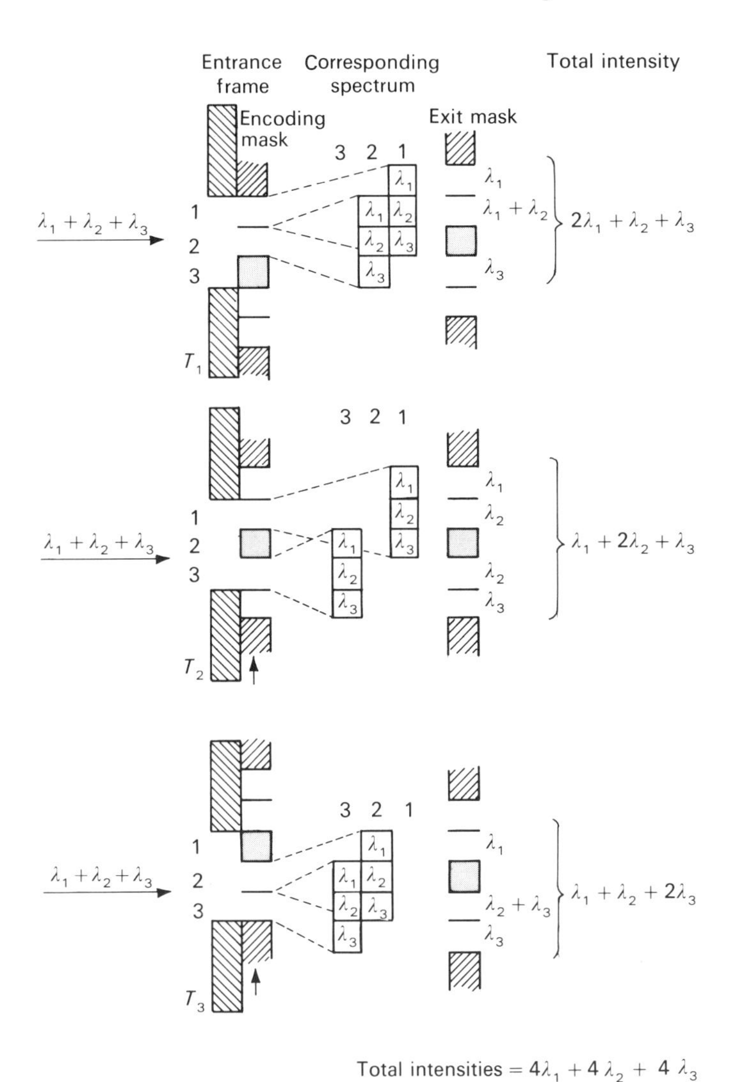

Figure 5.26. Channel intensities for mock monochromator derived from doubly encoded Hadamard transform spectrometer; T_1, T_2, and T_3 represent three different measurement time periods.

wavelength channels entering the instrument. The arrangement of exit slits must be such that all of the entering light would also pass through the exit framing mask if no exit encoding mask were present. Figure 5.26 illustrates this arrangement for $n=3$. There are three entrance slits to match the three wavelength channels λ_1, λ_2, and λ_3 entering the spectrometer. A total of $2n-1$, or 5, slits are placed in the exit focal plane to permit all of the overlapping images of λ_1, λ_2, and λ_3 to exit freely in the absence of the exit encoding mask.

Since the spectrometer measures radiation in a (0, 1) type of weighing design, the patterns for both masks are based on S-matrices. In this example (Figure 5.26), the entrance masks at T_1, T_2, and T_3 match the three rows of the left-circulant matrix

$$\mathbf{S}=\begin{bmatrix} 1 & 1 & 0 \\ 1 & 0 & 1 \\ 0 & 1 & 1 \end{bmatrix} \tag{5.34}$$

The entire entrance encoding mask contains $2n-1$ slits with the pattern 11011, but only three slits are exposed at any one time. To generate each row in Eq. 5.34, the entrance encoding mask is moved over one position (up in Figure 5.26) relative to the entrance frame. The exit mask is constructed from the same S-matrix, but the entire pattern 11011 is exposed during every measurement. Thus, in this instrument, only the entrance mask moves, while the exit mask is fixed.

All wavelength channels are measured during each of the three time periods T_1, T_2, and T_3, but as shown in Figure 5.26, a different channel receives preferential treatment during each period. Calculation of the final spectrum is particularly simple since no matrix algebra is necessary. If I_t equals the sum of all three intensity measurements,

$$I_t=4\lambda_1+4\lambda_2+4\lambda_3 \tag{5.50}$$

then subtracting $\frac{1}{4}I_t$ from each of the combined intensity measurements at T_1, T_2, and T_3 gives the intensity of each wavelength channel individually. In the general case, I_t is divided by $n+1$, this value is subtracted from the combined intensity measurements obtained during each of the n time periods, and multiplication by $4/(n+1)$ gives the spectrum.

Although subtraction is a particularly easy method for data processing, it has the disadvantage of eliminating any gain in luminosity resulting from the multiple entrance and exit slits. However, the multiplex advantage is still present and is theoretically equal to $\frac{1}{4}n$ (Harwit and Sloane, 1979), where n is the number of entrance slits utilized by the spectrometer (also the number of

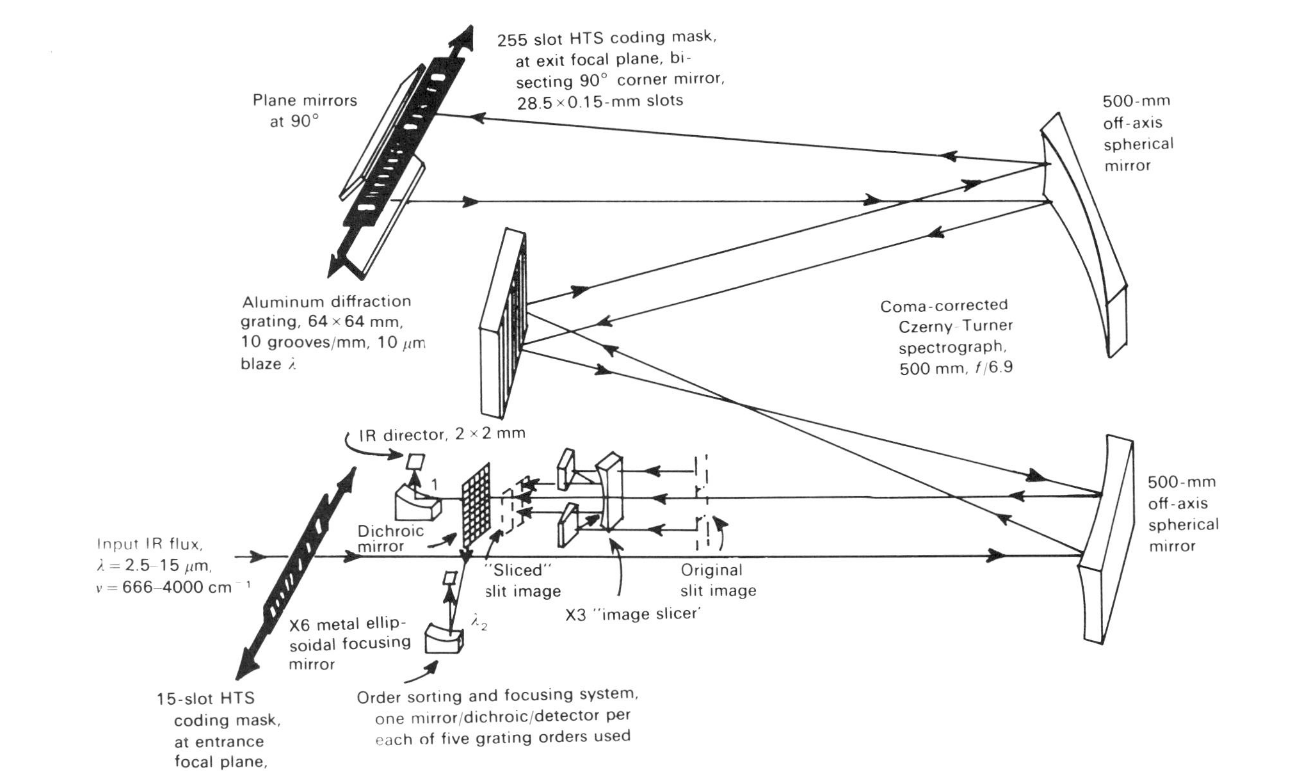

Figure 5.27. Optical diagram of spectral imaging Hadamard transform spectrometer. [Reprinted with permission from D. J. Lovell (1977), "Performance and Characteristics of Commercially Available Instruments," in *Spectrometric Techniques*, G. A. Vanasse (Ed.), Vol. 1, Academic, New York, pp. 331–348.]

combined intensity measurements made during one complete cycle). The reader may wish to verify that for the seven-membered S-matrix represented by the exit mask 1110100111010, a luminosity gain of 2 will be achieved for each wavelength channel provided multiplication by $4/(n+1)$ in the last step of the data processing is *not* carried out.

Imaging spectrometers have also been designed using the doubly encoded HT approach (Figure 5.27). In the singly encoded imaging spectrometer (Section 5.5.3), spectral information was sacrificed to obtain spatial information. In doubly encoded imagers, each additional mask dimension allows one additional parameter describing the radiation to be obtained. For example, a two-dimentional entrance mask combined with a one-dimensional exit mask makes it possible to obtain spectra for each image sector (as in a color photograph) or to obtain a series of two-dimensional images where each image is constructed using a different wavelength channel (Harwit, 1971, 1973).

5.6. NOISE CONSIDERATIONS

This chapter has approached multichannel advantages from a purely theoretical standpoint. Unless the spectroscopist understands the design and function of all of the components in a spectrometer, he or she may never be able to realize these advantages in practice. For example, a poor optical design will reduce throughput and offset the gain in luminosity obtained from multiple entrance and exit slits. An incorrect choice of detectors may not only reduce the improvement in SNR obtained from multiplexing but in some cases even produce a multiplex disadvantage. Optical design has been considered in Chapters 2–4. A brief introduction to noise and its role in the choice of detectors will be given at this point. A more detailed discussion of noise will be deferred until Chapters 9 and 13.

5.6.1. The Origins of Noise

The noise associated with a spectroscopic measurement may occur for a number of reasons, such as the presence of background radiation, variable transmission of radiation between the source and detector, detector characteristics, source characteristics, and so on. If this noise is systematic, it can usually be minimized by careful experimental design. However, random noise is inherent to the measurement and cannot be treated, except possibly through statistical procedures.

By determining how noise is related to the strength of the signal, it is possible to classify noise into three basic categories: photon shot-limited noise

N_p that is proportional to the square root of signal intensity, detector-limited noise N_d that is independent of signal intensity, and fluctuation noise N_f that varies directly with signal intensity. If each of these types of noise is statistically independent, the total experimental noise N_t will be related to the three independent types of noise according to the equation

$$N_t = (N_p^2 + N_d^2 + N_f^2)^{1/2} \tag{5.51}$$

Generally, one particular type of noise will be dominant, and the following discussion is restricted to such cases.

5.6.2. Photon Shot-Limited Noise

Consider the source radiation as a stream of photons arriving at the detector in a random fashion with time. If the energy of these photons is greater than kT, where T is the absolute temperature of the detector, then each individual photon will be counted with approximately equal sensitivity, and the error in the signal will be determined by the randomness of the photon stream (hence the name photon shot noise). If P photons interact with the detector in a given period, then the random signal fluctuations are on the order of $P^{1/2}$, and the noise N_p will be proportional to the square root of the signal,

$$N_p \propto P^{1/2} \propto s_m^{1/2} \tag{5.52}$$

where s_m represents the intensity of the mth wavelength channel. Photon shot-limited noise is typically encountered in UV-visible spectroscopy and in charged particle (photoelectron, ESCA, electron impact, etc.) spectroscopy. In this latter case, N_p is determined by the randomness of the charged particle stream.

5.6.3. Detector-Limited Noise

Throughout previous sections of this chapter, we have treated noise as "detector limited" by assuming that it is unrelated to the strength of the source, that is,

$$N_d = C \tag{5.53}$$

where C is a constant that may vary with wavelength. Detectors with high noise levels are commonly encountered in types of spectroscopy using radio-frequency, microwave, and infrared radiation because the photon energy associated with the radiation is less than kT, where T is the absolute temperature of the detector employed to make the measurement.

5.6.4. Fluctuation (Flicker, 1/*f*, Proportional, or Scintillation) Noise

Fluctuation noise is primarily encountered when signals are very intense or in remote sensing when low-level signals undergo unintended modulation on their way to the detector. Examples of situations producing fluctuation noise in remote sensing include the atmospheric variations in refractive index that make stars appear to flicker or wave patterns that make overhead light intensity seem to fluctuate when viewed from beneath the surface of a body of water. In analytical spectroscopy, fluctuation noise is often encountered when an intense signal is produced by a flickering flame. The magnitude of this type of noise depends directly on the strength of the source:

$$N_{\mathrm{f}} \propto s_m \tag{5.54}$$

5.6.5. Multichannel Advantages

Table 5.5 classifies the multichannel advantages theoretically obtainable for several different experimental arrangements according to the dominant type of noise present in the measurement. Values in columns 1–3 in Table 5.5 refer to the SNR expected for a multidetector spectrometer such as a direct reader (1) and a multiplexing spectrometer (2 and 3) divided by the SNR obtained *after the same total observation time* for a conventional scanning instrument having only one entrance and exit slit. (It should be noted that not all authors agree on the values in columns 2 and 3. See Section 14.1.1.) Values in column 4 refer to the SNR expected for a *single* measurement made on a multislit instrument with a luminosity gain of M divided by the SNR obtained for a conventional instrument with only one entrance and exit slit.

For an M-channel multidetector system (Section 5.3.3), each wavelength channel in the spectrum is monitored by its own detector. After time T, the time normally needed to scan the entire spectrum using a conventional spectrometer, each wavelength channel will have been observed M times. If photon noise dominates, noise accumulates as the square root of the number of measurements (Eq. 5.20), and the gain in SNR for repeated measurements will be $M^{1/2}$. If detector noise dominates, noise accumulates as the square root of the total observation time $T^{1/2}$. Consequently, the gain in SNR for repeated measurements will also be $M^{1/2}$. However, for fluctuation noise, noise accumulates in direct proportion to the signal, and signal averaging offers no advantage.

Table 5.5. Theoretically Expected Signal-to-Noise Enhancement Classified According to Dominant Type of Spectral Noise

Dominant Noise Type	M-Channel Multidetector Spectrometer[a, b] (1)	M-Channel Multiplexing Spectrometer[a, c] H-Matrix[f] (2)	S-Matrix[g] (3)	M-Channel Multislit Spectrometer[d, e] (4)
Detector-limited	$M^{1/2}$	$M^{1/2}$	$(1/2)M^{1/2}$	M
Photon-limited	$M^{1/2}$	1	$1/2^{1/2}$	$M^{1/2}$
Fluctuation	1	$1/M^{1/2}$	$1/M^{1/2}$	Variable

[a] SNR for multichannel spectrometer divided by SNR obtained by single-channel, scanning spectrometer after same total observation time.
[b] Signal-averaging effect (Section 5.3.3).
[c] Multiplex or Fellgett's advantage (Section 5.4.4); SNR affected by altering noise level.
[d] The SNR for single measurement obtained by multislit spectrometer with luminosity gain of M divided by SNR obtained by instrument with one entrance and exit slit.
[e] Jacquinot's advantage (Section 5.2.2); SNR affected by altering signal level; luminosity gain M is equivalent to number of corresponding pairs of entrance and exit slits open during measurement.
[f] Each channel measured M times.
[g] Each channel measured approximately $\frac{1}{2}M$ times. Also applies to Fourier transform–IR because half the intensity is lost.

For an M-channel multiplexing spectrometer using a Hadamard matrix design (Section 5.3.5), each channel will also be measured M times in the period normally required to scan the entire spectrum using a conventional spectrometer. However, in this case the signal S seen by the detector is the sum of M intensities, not individual intensities as in the multidetector case. Consequently, after time T,

$$S = s_1 + s_2 + \cdots + s_m \propto M \tag{5.55}$$

where $s_1, s_2, \cdots$ are the intensities of the individual wavelength channels being multiplexed. For the detector-limited case, noise is unaffected by the intensity of the signal, so N_d accumulates as the square root of the observation time, where time is proportional to M. Thus,

$$N_d \propto M^{1/2} \tag{5.56}$$

From Eqs. 5.55 and 5.56, $S/N_d \propto M/M^{1/2} = M^{1/2}$, yielding a multiplex advantage.

For photon shot-limited noise, N_p accumulates as the square root of M and is also increased by a factor proportional to the square root of the total signal intensity S (Eq. 5.52):

$$N_p \propto M^{1/2} S^{1/2} \propto M \quad (5.57)$$

From Eqs. 5.55 and 5.57, $S/N_p \propto M/M = 1$, showing that there is no advantage to be gained from multiplexing using the Hadamard design. In the case of fluctuation noise, the situation is even worse because

$$N_f \propto M^{1/2} S \propto M^{3/2} \quad (5.58)$$

From Eqs. 5.55 and 5.58, $S/N_f \propto M/M^{3/2} = M^{-1/2}$, which represents a multiplex disadvantage.

The results obtained from multiplexing with the S-matrix design (Section 5.3.4) can be determined using analogous reasoning. With the S-matrix design, each channel is measured for only about one-half of time T, so

$$S = \tfrac{1}{2}(s_1 + s_2 + \cdots + s_m) \propto \tfrac{1}{2} M \quad (5.59)$$

Detector-limited noise is unaffected by the strength of the signal, so

$$N_d \propto M^{1/2}, \quad (5.56)$$

while for photon shot-limited noise and fluctuation noise,

$$N_p \propto M^{1/2} S^{1/2} \propto (1/2)^{1/2} M \quad (5.60)$$

$$N_f \propto M^{1/2} S \propto (1/2) M^{3/2} \quad (5.61)$$

Use of Eq. 5.59 with Eqs. 5.56, 5.60, and 5.61 yields the entries in the S-matrix column (Table 5.5). There will be a multiplex advantage only in the case of detector-limited noise, while photon shot noise and fluctuation noise both yield multiplex disadvantages.

The last column in Table 5.5 shows the effect of Jacquinot's advantage on SNR. In the case of Jacquinot's advantage, SNR is affected by increasing the signal intensity, in contrast to multiplexing methods, which affect the SNR by decreasing the noise.

For multislit spectrometers, the gain in luminosity is governed by M, the number of pairs of corresponding entrance and exit slits (Section 5.2.2). Hence,

$$S \propto M \quad (5.62)$$

Detector-limited noise is unaffected by the strength of the signal, so

$$N_d = C \tag{5.53}$$

while for photon shot-limited noise and fluctuation noise,

$$N_p \propto S^{1/2} \propto M^{1/2} \tag{5.63}$$

$$N_f \propto S \propto M \tag{5.64}$$

Thus S/N_d will be improved by a factor of M and S/N_p will be improved by a factor of $M^{1/2}$ regardless of whether the signal is measured for a long or short period of time. For the case of fluctuation noise, multiple slits offer no advantage and may even be disadvantageous if noise grows faster than the signal, that is, the proportionality constant in Eq. 5.54 is greater than 1. (Such conditions are sometimes encountered when fluctuation noise results from atmospheric emission. See Decker, 1977.) Generally, the spectroscopist attempts to avoid circumstances where fluctuation noise dominates. If N_f varies in a predictable way, it may be possible to use a reference detector to subtract out the background fluctuations and reduce noise to a detector-limited or photon shot-limited noise case.

Columns 1–3 in Table 5.5 give the multichannel advantages in terms of SNR enhancement for constant total observation time compared to a single-channel, scanning spectrometer. It is also possible to state the multichannel advantage using a time savings argument by squaring the inverse of the SNR enhancement in each column. This gives the fraction of time needed to obtain the entire M-channel spectrum *with the same SNR* as compared to the conventional scanning instrument. For example, if photon shot-limited noise dominates, an M-channel multidetector can determine the spectrum in $1/M$th the time required by the single-channel detector, but it takes two times longer to determine the spectrum with the same SNR using an S-matrix multiplexing design.

REFERENCES

Connes, P. (1958), *J. Phys. Radium*, *19*, 262.

Connes, P. (1959), *Rev. Opt.*, *38*, 157, 416.

Connes, P. (1960), *Rev. Opt.*, *39*, 402.

Decker, Jr., J. A. (1971), *Appl. Opt.*, *10*, 24, 510.

Decker, Jr., J. A. (1977), "Hadamard-Transform Spectroscopy", in *Spectrometric Techniques*, Vol. 1, Vanassee, G. A. (Ed.), Academic, New York, Chapter 5, pp. 189–227.

Decker, Jr., J. A. and Harwit, M. (1968), *Appl. Opt.*, *7*, 2205.

Decker, Jr., J. A. and Harwit, M. (1969), *Appl. Opt.*, *8*, 2552.

Fellgett, P. B. (1951), Ph.D. Thesis, University of Cambridge.

Fellgett, P. B. (1958), *J. Phys. Radium*, *19*, 187, 237.

Fellgett, P. B. (1971), *Aspen International Conference on Fourier Spectroscopy*, Vanasse, G. A., Stair, A. T., and Baker, D. J. (Eds.), AFCRL-71-0019 Special Report No. 114, Aspen, Colorado, 1970, 139.

Girard, A. (1960), *Opt. Acta*, *7*, 81.

Girard, A. (1963a), *Appl. Opt.*, *2*, 79.

Girard, A. (1963b), *J. Phys. Radium*, *24*, 139.

Golay, M. J. E. (1949), *J. Opt. Soc. Am.*, *39*, 437.

Golay, M. J. E. (1951), *J. Opt. Soc. Am.*, *41*, 468.

Gottlieb, P. (1968), *IEEE Trans. Inform. Theor.*, *IT-14*, 428.

Hadamard, J. (1893), *Bull. Sci. Math.*, *17*, 240.

Hansen, P. and Strong, J. (1972), *Appl. Opt.*, *11*, 502.

Harwit, M. (1971), *Appl. Opt.*, *10*, 1415.

Harwit, M. (1973), *Appl. Opt.*, *12*, 285.

Harwit, M. (1978), "Hadamard Transform Analytical Systems," in *Transform Techniques in Chemistry*, Griffiths, P. R. (Ed.), Plenum, New York, Chapter 7, pp. 173–197.

Harwit, M. and Decker, J. A., Jr. (1974), "Modulation Techniques in Spectrometry," in *Progress in Optics*, Vol. 12, Wolf, E. (Ed.), North-Holland, Amsterdam, pp. 101–162.

Harwit, M., Phillips, P. G., Fine, T., and Sloane, N. J. A. (1970), *Appl. Opt.*, *9*, 1149.

Harwit, M. and Sloane, N. J. A. (1979), *Hadamard Transform Optics*, Academic, New York.

Hotelling, H. (1944), *Ann. Math. Stat.*, *15*, 297.

Ibbett, R. N., Aspinall, D., and Grainger, J. F. (1968), *Appl. Opt.*, *7*, 1089.

Jacquinot, P. (1959), *J. Opt. Soc. Am.*, *44*, 761.

Jacquinot, P. (1960), *Rep. Prog. Phys.*, *23*, 267.

Lovell, D. J. (1977), "Performance and Characteristics of Commercially Available Instruments," in *Spectrometric Techniques*, Vol. 1, Vanasse, G. A. (Ed.), Academic, New York, pp. 331–348.

Marshall, A. G. and Comisarow, M. B. (1978), "Multichannel Methods in Spectroscopy," in *Transform Techniques in Chemistry*, Griffiths, P. R. (Ed.), Plenum, New York, Chapter 3, pp. 39–68.

Mertz, L. (1965), *Transformations in Optics*, Wiley, New York.

Nelson, E. D. and Fredman, M. L. (1970), *J. Opt. Soc. Am.*, *60*, 1664.

Phillips, P. G. and Harwit, M. (1971), *Appl. Opt.*, *10*, 2780.

Ring, J. and Selby, M. J. (1966), *Infrared Phys.*, *6*, 33.

Selby, M. J. (1966), *Infrared Phys.*, *6*, 21.

Sloane, N. J. A. (1982), "Hadamard and Other Discrete Transforms," in *Fourier, Hadamard, and Hilbert Transformations in Chemistry*, Marshall, A. G. (Ed.), Plenum, New York, Chapter 2, pp. 45–67.

Sloane, N. J. A., Fine, T., Phillips, P. G., and Harwit, M. (1969), *Appl. Opt.*, *8*, 2103.

Sloane, N. J. A. and Harwit, M. (1976), *Appl. Opt.*, *15*, 107.

Stewart, J. E. (1970), *Infrared Spectroscopy: Experimental Methods and Techniques*, Marcel Dekker, New York.

Stroke, G. W. (1967), "Diffraction Gratings," in *Handbuck der Physik: Optische Instrumente*, Vol. 29, Flügge, S. (Ed.), Springer-Verlag, Berlin, pp. 426–754.

Stroke, G. W. and Stroke, H. H. (1963), *J. Opt. Soc. Am.*, *53*, 333.

Swift, R. D., Wattson, R. B., Decker, Jr., J. A., Paganetti, R., and Harwit, M. (1972), *Appl. Opt.*, *15*, 1596.

Yates, F. (1935), *J. Roy. Stat. Soc. Suppl.*, *2*, 181.

CHAPTER

6

PRINCIPLES OF INTERFERENCE

Chapter 5 has shown how multiple slits can be used to improve the optical throughput of a spectrometer (Jacquinot's advantage or luminosity gain) and how the signal-to-noise ratio or error involved in a spectroscopic measurement can be improved by measuring more than one resolution element at a time (Fellgett's or the multiplex advantage). While spectrometers that use masks make these advantages possible, such instruments are also subject to a number of problems that tend to limit their effectiveness. Many of these problems result from difficulties associated with fabrication of the mask and with mechanical operation of the mask during a series of spectroscopic measurements (Harwit and Sloane, 1979).

6.1. LIMITATIONS OF SPECTROMETERS THAT USE MASKS

The magnitude of Fellgett's advantage that is theoretically possible for any kind of a spectrometer is determined by the number of resolution elements that can be measured simultaneously. For a singly encoded, mask-type instrument, Fellgett's advantage will be directly proportional to the number of slits present in the exit mask pattern. To achieve Jacquinot's advantage, the spectrometer must be doubly encoded. Thus, even more slits are required because the instrument must have both an entrance and an exit mask.

The slits that make up the mask pattern are generally fabricated either by the deposition of metal according to a predetermined pattern or by the selective removal of metal from a thin film. In either case, some portions of the final pattern will generally be a bit larger or narrower than desired. Because the resolution elements are measured in combination, the error resulting from just one faulty slit will cause systematic errors in intensities throughout the entire spectrum. Thus, the larger the Jacquinot or Fellgett advantage that is possible for the instrument, the greater the chance for errors resulting from faulty mask construction.

During operation of the spectrometer, the encoding mask is moved, either stepwise or continuously, past the framing mask. Unless the encoding mask is exactly aligned with the frame, some of the slits will appear too wide or too narrow, and the effect on the spectrum will be similar to faulty mask construction as described previously. The gap between the encoding and framing mask is also a critical factor. If too narrow, the mask parts will catch. If too large, some of the radiation striking the blocking mask may pass into an encoding slit that was meant to be obscured and cause systematic errors throughout the entire spectrum.

Finally, each type of mask motion has its own associated problems. If continuous motion is employed, the velocity of the mask must remain strictly constant throughout the entire series of measurements. If a rotating disk is used, any slight radial eccentricity will also cause a radial motion of the mask, producing sinusoidal changes in intensity throughout the spectrum. In the case of stepwise motion, each step size must be exactly the same and must exactly match the encoding slit width. Even small differences in step size and slit width will accumulate over a large number of steps, producing substantial errors toward the end of a series of measurements. Again, the number of measurements and hence the size of the accumulated error will be directly proportional to the size of the Jacquinot or Fellgett advantage achieved. (See Section 14.3.2 for a discussion of a new type of encoding mask that uses an array of stationary liquid crystal electro-optic switches.)

6.2. SPECTROMETERS THAT USE INTERFERENCE

There is another, more commonly used spectroscopic technique that also produces large advantages of the Jacquinot and Fellgett type but does not require masks. Simply put, all of the emitted radiation is measured simultaneously as an interference pattern, and this interference pattern, or interferogram, which constitutes the spectral output, is analyzed for its component wavelengths and their associated intensities.

The key to the analysis of the interferogram lies in the independent behavior of each wavelength, that is, each wavelength produces a unique interference pattern independent of all other wavelengths. Consequently, the interferogram may be regarded simply as a sum of all of the interference patterns produced independently by each spectral component.

Even using this additivity principle, reconstruction of the original spectrum from the interferogram is a complex mathematical problem, generally requiring computer treatment of the data (see Foskett, 1978). Justification for employing such an indirect, mathematically complex technique lies in the many advantages of the interference spectrometer, or interferometer, as compared to the conventional grating spectrometer. These advantages include large gains in luminosity, greatly improved resolution, high wavelength accuracy, greatly reduced stray-light problems, very rapid scanning times, and a large wavelength range per scan. An interferometer also produces extremely reproducible spectra, making substantial enhancement of the signal-to-noise ratio possible through computer accumulation of multiple scans.

Finally, since the interferometer also detects all frequencies simultaneously, a large multiplex advantage is possible for the case of detector-limited noise. The magnitude of this multiplex advantage can be the same as that theoretically achievable using a mask-type spectrometer that employs a Hadamard matrix weighing design (Chapter 5), that is, $M^{1/2}$ (Table 5.5, column 2), where M is the total number of resolution elements that is monitored during formation of the interferogram. However, if some of the source radiation is not utilized by the interferometer, then the multiplex advantage will be reduced accordingly. For example, the Michelson interferometer (Section 7.1) has a multiplex advantage of only $\frac{1}{2}M^{1/2}$ because half of the radiation intensity entering the interferometer is lost at the half-silvered mirrors (Marshall and Comisarow, 1978).

Before discussing modern interferometry, Chapter 6 will consider some general principles governing the interference of light. Sections 6.3–6.5 deal with the factors that determine the appearance of the interference pattern. The remaining sections discuss how such patterns can be produced experimentally. Additional information on interference can be found in many texts dealing with physical optics and interferometry (Born and Wolf, 1986; Hariharan, 1985; Steel, 1983; Jenkins and White, 1976; Cook, 1971; Levi, 1968; Tolansky, 1955; Chandler, 1951). Many older volumes also contain much interesting material dealing with the history and experimental foundations of the wave theory of light (Wood, 1934; Michelson, 1903, 1927; Mach, 1921; Preston, 1912).

6.3. THE PRINCIPLE OF SUPERPOSITION

When two light waves coincide, or are superposed, the resultant displacement at any given point along the direction of wave propagation will be the sum of the displacements due to each wave separately. Consider the case of two light waves $y_1(t)$ and $y_2(t)$, each consisting of the same, single frequency f_1 (Figure 6.1). The variation of these signals with time t may be represented by cosine functions:

$$y_1(t) = a_1 \cos(2\pi f_1 t + \phi_1) \tag{6.1}$$

$$y_2(t) = a_2 \cos(2\pi f_1 t + \phi_2) \tag{6.2}$$

where a_1 and a_2 are constants that describe the amplitudes of the time-varying

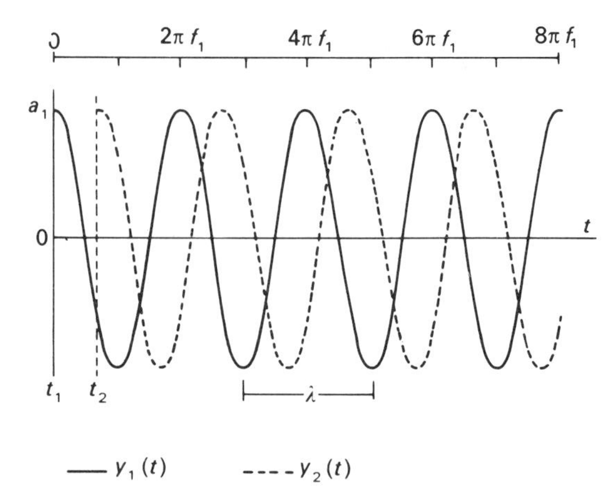

Figure 6.1. Two waves of same frequency and amplitude but with phase difference $\delta = 2\pi f_1(t_2 - t_1)$.

waves and ϕ_1 and ϕ_2 are the phase angles of the waves. Phase angle refers to the number of degrees or radians the cosine function has progressed from the origin. Thus, at time t_1 (Figure 6.1), the wave $y_1(t)$ is said to have zero phase.

The two light waves have a phase difference δ,

$$\delta = (\phi_2 - \phi_1) = 2\pi f_1(t_2 - t_1) = (2\pi/\lambda)(x_2 - x_1) \tag{6.3}$$

where $t_2 - t_1$ represents the time lag of one wave relative to the other. The wave $y_1(t)$ in Figure 6.1 is said to lead $y_2(t)$ by $t_2 - t_1$, or, conversely, $y_2(t)$ is said to lag $y_1(t)$ by $t_2 - t_1$. As indicated by Eq. 6.3, this phase difference is also given by 2π times the number of wavelengths in the path difference, $x_2 - x_1$, between $y_1(t)$ and $y_2(t)$ at any particular instant in time.

It should be noted that sine functions are equally valid representations of these light waves, that is,

$$y_1(t) = a_1 \sin(2\pi f_1 t + \phi_1') \tag{6.4}$$

Since sine functions lag cosine functions by 90°, the phase angle ϕ_1 differs from ϕ_1' by a factor of $\frac{1}{2}\pi$. Instead of using the cosine or sine functions separately, a third, mathematically equivalent representation is obtained through the use of exponential notation (Whittaker and Watson, 1935),

$$y_1(t) = a_1 \exp[i(2\pi f_1 t + \delta)] = a_1 \exp(i2\pi f_1 t)\exp(i\delta) \tag{6.5}$$

where δ represents a phase difference relative to some standard phase. According to Euler's relationship, $\exp(ix) = \cos(x) + i\sin(x)$. Hence, Eq. 6.5 is actually a way of expressing the sum of a cosine and a sine function.

In discussing interference patterns in Chapter 6, we generally choose to represent light waves as cosine functions (Eq. 6.1). However, the exponential form (Eq. 6.5) offers a number of mathematical advantages when dealing with spectroscopic applications. Consequently, the exponential form will be largely employed in Chapter 7.

According to the cosine representation, the resultant light wave $Y(t)$ produced by the superposition of $y_1(t)$ and $y_2(t)$ is found by summing Eqs. 6.1 and 6.2 and using the trigonometric identity $\cos(X+Y) = \cos X \cos Y - \sin X \sin Y$:

$$\begin{aligned} Y(t) &= a_1 \cos(2\pi f_1 t + \phi_1) + a_2 \cos(2\pi f_1 t + \phi_2) \\ &= a_1 [\cos(2\pi f_1 t)\cos\phi_1 - \sin(2\pi f_1 t)\sin\phi_1] \\ &\quad + a_2 [\cos(2\pi f_1 t)\cos\phi_2 - \sin(2\pi f_1 t)\sin\phi_2] \\ &= (a_1 \cos\phi_1 + a_2 \cos\phi_2)\cos(2\pi f_1 t) \\ &\quad - (a_1 \sin\phi_1 + a_2 \sin\phi_2)\sin(2\pi f_1 t) \end{aligned}$$

Since a_1, a_2, ϕ_1, and ϕ_2 are constants, we may write

$$a_1 \cos\phi_1 + a_2 \cos\phi_2 = A\cos\phi_3 \tag{6.6}$$

$$a_1 \sin\phi_1 + a_2 \sin\phi_2 = A\sin\phi_3 \tag{6.7}$$

Therefore,

$$\begin{aligned} Y(t) &= A\cos\phi_3 \cos(2\pi f_1 t) - A\sin\phi_3 \sin(2\pi f_1 t) \\ &= A\cos(2\pi f_1 t + \phi_3) \end{aligned} \tag{6.8}$$

The terms A and ϕ_3 in Eq. 6.8 can be evaluated by referring to Figure 6.2, where $y_1(t)$ and $y_2(t)$ are represented by two vectors of length a_1 and a_2, respectively (Goldman, 1948). These vectors rotate about point 0 with a constant angular velocity $2\pi f_1 t$ always maintaining the same angle $\delta = (\phi_2 - \phi_1)$ between them.

The amplitude A and phase ϕ_3 of the resultant vector are obtained by completing the vector parallelogram of a_1 and a_2. Thus, A is given by the sum of the projections of a_1 and a_2 on the x and y axes,

$$A = [(y_{a1} + y_{a2})^2 + (x_{a1} + x_{a2})^2]^{1/2} \tag{3.40}$$

$$= [(a_1 \sin\phi_1 + a_2 \sin\phi_2)^2 + (a_1 \cos\phi_1 + a_2 \cos\phi_2)^2]^{1/2} \tag{6.9}$$

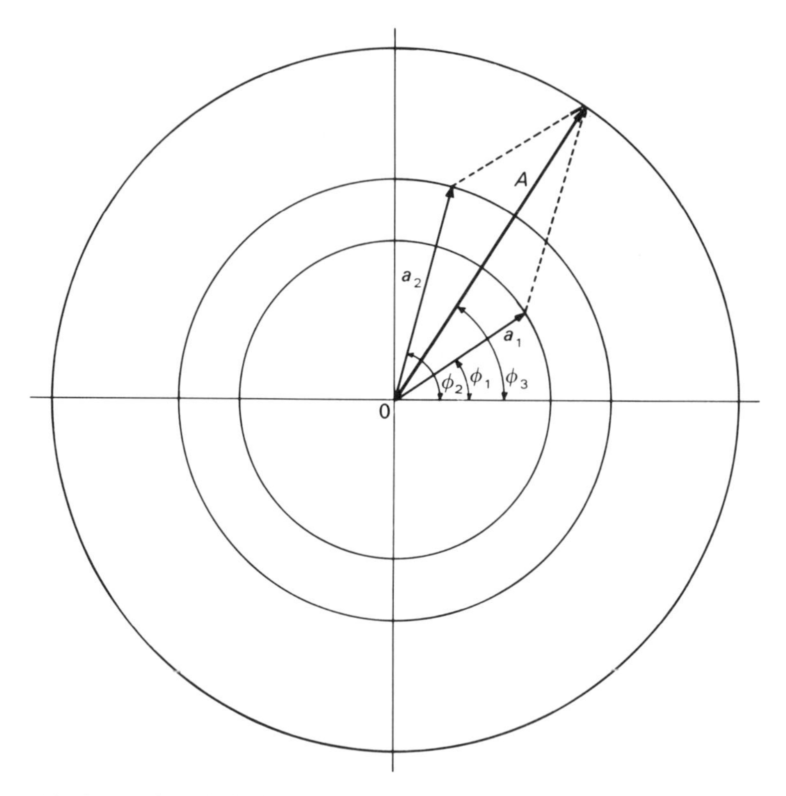

Figure 6.2. Geometric analysis of superposition of two simple harmonic vibrations having same frequency.

and the angle ϕ_3 is found from the trigonometric identity $\tan\theta = (\sin\theta/\cos\theta)$:

$$\tan\phi_3 = (a_1 \sin\phi_1 + a_2 \sin\phi_2)/(a_1 \cos\phi_1 + a_2 \cos\phi_2) \tag{6.10}$$

Equation 6.9 may be simplified by specifying that $\phi_1 = 0$. In this case, $\phi_2 = \delta$, the phase difference between the two superposed waves, and therefore,

$$A = (a_1^2 + a_2^2 + 2a_1 a_2 \cos\delta)^{1/2} \tag{6.11}$$

Equation 6.8 demonstrates that superposition of two light waves (cosine functions) of the same frequency produces another light wave (cosine function) of the same frequency but with a different amplitude A and phase angle ϕ_3. The addition of any number of cosine functions of the same frequency can be treated in the same manner since the displacements can be added successively (Figure 6.2), each addition producing an equation similar to 6.8.

Modifications of wave intensity obtained by superposition are referred to as interference effects. Since the intensity of a simple harmonic vibration is proportional to the square of the amplitude, from Eq. 6.11, the intensity (irradiance) I of the resultant wave will be

$$I \propto A^2 = a_1^2 + a_2^2 + 2a_1 a_2 \cos\delta \tag{6.12}$$

If the amplitudes of the original waves are the same, $a_1 = a_2$, and

$$A^2 = 2a_1^2(1+\cos\delta) = 4a_1^2\cos^2(\tfrac{1}{2}\delta) \tag{6.13}$$

Since $\cos\delta$ may vary from $+1$ to -1, A^2 may vary from $4a_1^2$ to zero. Thus, the intensity of $Y(t)$ will be a maximum whenever $\delta = \phi_2 - \phi_1 = 0, 2\pi, 4\pi, 6\pi, \cdots$, that is, when there is a whole-number difference in wavelengths between the two superposed waves. Similarly, the intensity will be a minimum whenever $\delta = \phi_2 - \phi_1 = \pi, 3\pi, 5\pi, \cdots$, that is, when there is a half-integer wavelength difference between the two superposed waves. These two extreme cases are illustrated in Figure 6.3 where $y_1(t)$ and $y_2(t)$ are represented by the light curves, and the resultant wave $Y(t)$ is represented by the heavy curve.

In Figure 6.3*a*, the phase difference between $y_1(t)$ and $y_2(t)$ is nearly zero, and the amplitude of the resultant wave is almost $2a_1$. In Figure 6.3*b*, the phase difference between the two superposed waves is close to $\frac{1}{2}\lambda$, and the

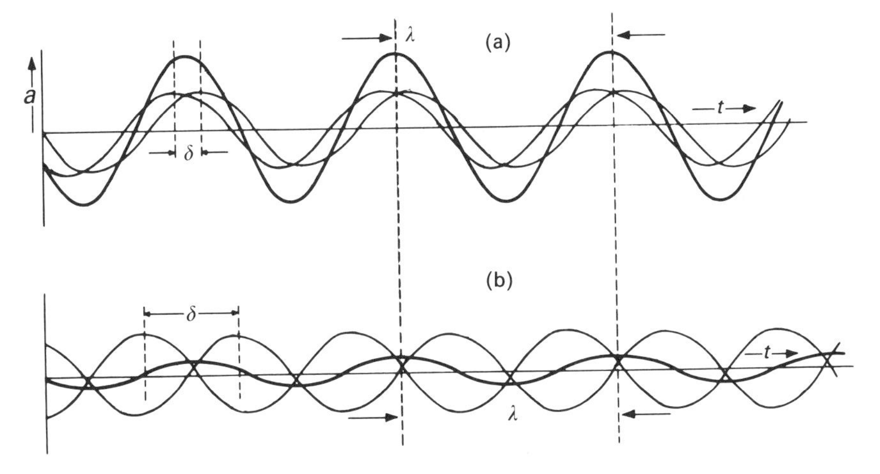

Figure 6.3. Superposition of two waves that are (*a*) almost in phase and (*b*) almost 180° out of phase. Light curves represent combining waves and heavy lines represent resultant waves. Both resultant and combining waves have same wavelength. (After Jenkins and White, 1976.)

amplitude of the resultant wave is practically zero. It should be noted that the ordinates of the heavy curves in Figure 6.3 can be obtained graphically by taking the algebraic sum of the ordinates of the light curves at any particular value of t.

Equation 6.8 has shown that superposition of simple cosine functions of the same frequency will give rise to another simple cosine function of the same frequency but with a different amplitude and phase angle. However, if waves of different frequencies are combined, the resulting wave is complex, that is, it is not a simple harmonic motion. Figure 6.4 illustrates several examples of this

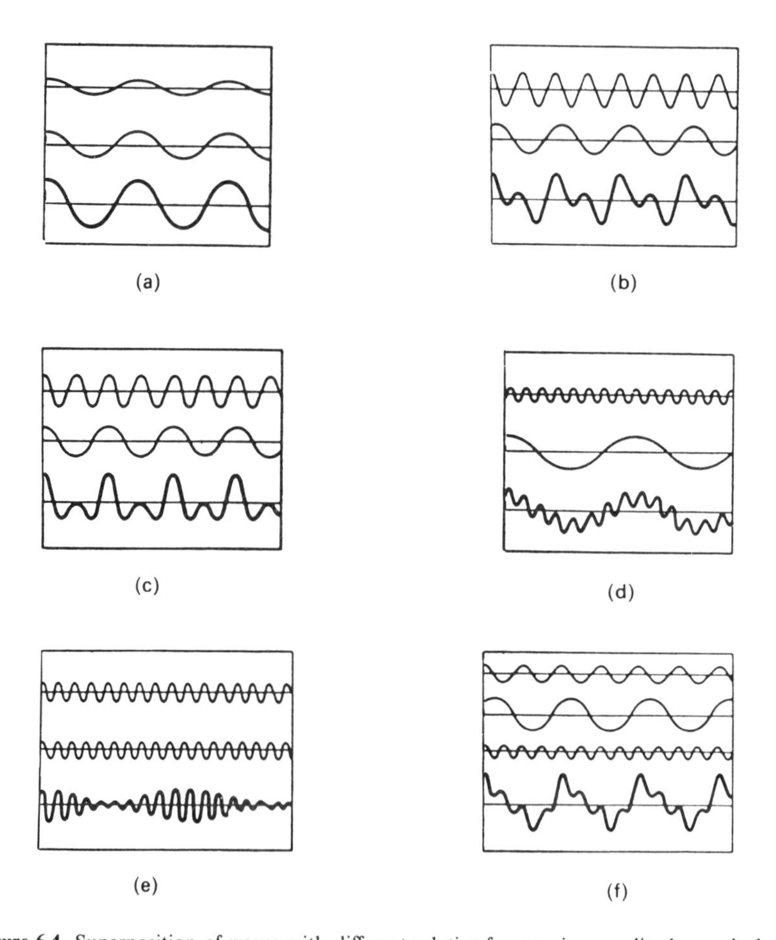

Figure 6.4. Superposition of waves with different relative frequencies, amplitudes, and phases. Light curves represent combining waves and heavy curves represent resultant waves. [Reprinted with permission from F. A. Jenkins and H. E. White (1976), *Fundamentals of Optics*, 4th ed., McGraw-Hill, New York.]

effect, the resultant curves (heavy lines) in each case being obtained by algebraic addition of the displacements due to the individual waves (light lines) at every point in time t.

For comparison, Figure 6.4*a* shows the addition of two light waves of the same frequency and phase but different amplitudes a_1 and a_2. The resultant is simple harmonic motion of the same frequency with amplitude $a_1 + a_2$. If n waves are superposed, each of the same frequency, amplitude and phase, the resultant amplitude will be simply the sum of the individual amplitudes, that is, $A = na$. Hence, the intensity at any given point along the direction of wave propagation will be n^2 times the intensity of any one wave at that point, that is, $A^2 = n^2 a^2$.

In Figure 6.4*b*, two waves of the same amplitude with frequencies in the ratio 2:1 are added. In this case, the resultant wave does not produce simple harmonic motion. Moreover, altering only the phase difference between the combining waves (Figure 6.4*c*) produces a resultant wave of very different form.

Figure 6.4*d* shows the effect of adding a wave of very high frequency to one of very low frequency. For the case of two waves of very similar frequencies (Figure 6.4*e*), the resultant motion has a wavelength that is an average of the wavelengths for the two superposed waves, but the amplitude is modulated to form groups. This behavior is best described using the trigonometric identity

$$\cos(2\pi f_1 t) + \cos(2\pi f_2 t) = 2\cos 2\pi t\,[\tfrac{1}{2}(f_1 + f_2)]\cos 2\pi t\,[\tfrac{1}{2}(f_1 - f_2)] \quad (6.14)$$

where $\frac{1}{2}(f_1 + f_2)$ is called the sum frequency and $\frac{1}{2}(f_1 - f_2)$ is called the difference or beat frequency. As f_1 and f_2 become more alike, the sum frequency approaches the individual frequencies while the beat frequency

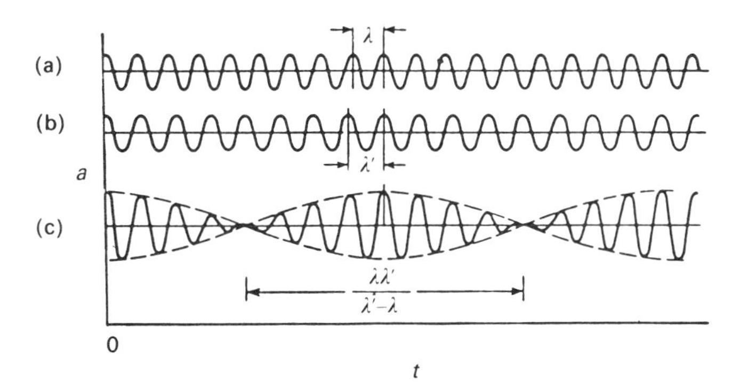

Figure 6.5. Superposition of two waves (*a*) and (*b*) of same amplitude but slightly different frequencies illustrating formation of groups. Beat frequency is shown in (*c*) by dashed lines while sum frequency is shown in (*c*) by solid line. (After Jenkins and White, 1976.)

becomes lower, eventually disappearing when $f_1 = f_2$. The combination of two waves of slightly different wavelengths is shown in more detail in Figure 6.5, where the envelope of modulations indicated by the broken curves (Figure 6.5*c*) represents the beat frequency and the solid line represents the sum frequency.

Figure 6.4*f* shows the extension of wave addition to three components having different frequencies, amplitudes, and phases. The resultant wave is now quite complex and very different from a simple cosine function. In the limit, the superposition of a very large number of waves having frequencies differing by only infinitesimal amounts produces white light, that is, a resultant wave with amplitude but no discernible modulation.

6.4. THE PRINCIPLE OF COHERENCE

Observation of the interference pattern resulting from the superposition of two or more waves requires that the phase angle ϕ_3 in Eq. 6.8 remain fixed for a discernible period of time. Light sources that emit waves with a constant, point-by-point phase relationship are said to be coherent, and the superposition of such waves will produce an interference pattern. However, when light beams from two independent (incoherent) sources combine, there is no fixed relationship between the phases of the beams, and stationary interference patterns cannot be formed.

The difficulty encountered when two independent sources are involved may be understood by considering the case of two separate lamps set side by side. The waves emitted from each lamp will undergo many phase changes over very short intervals of time, and these changes will occur independently of the behavior of the other lamp. While an interference pattern may exist during any particular instant in time, ϕ_3 will fluctuate so rapidly that the patterns will be impossible to detect.

The superposition of many waves of the same frequency but with random phases can be analyzed in a more rigorous fashion using vector addition of amplitudes, as developed in Section 6.3. The case of n waves of equal amplitude and frequency but with random phases is shown in Figure 6.6, where the phase ϕ for each wave may have any value between zero and 2π. The intensity of the resultant wave is found by squaring the sum of the projections of all vectors on the x axis, that is,

$$A_x^2 = [a(\cos\phi_1 + \cos\phi_2 + \cdots + \cos\phi_n)]^2 \tag{6.15}$$

and adding the results to the square of the corresponding sum for the y axis,

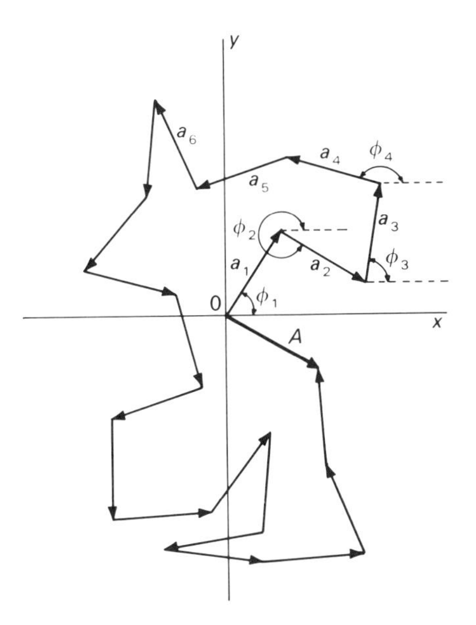

Figure 6.6. Resultant of 20 amplitude vectors with phases selected at random. All individual amplitudes, $a_1, a_2, \cdots, a_{20}$ are equal.

that is,

$$A_y^2 = [a(\sin\phi_1 + \sin\phi_2 + \cdots + \sin\phi_n)]^2 \tag{6.16}$$

When the quantities in the parentheses in Eqs. 6.15 and 6.16 are squared, terms of the form $\cos^2\phi_n$, $\sin^2\phi_n$, $\cos\phi_m\cos\phi_n$, and $\sin\phi_m\sin\phi_n$ are obtained. Since the phases are random, the cross-product terms will average to zero for large n, leaving only the squared terms. Consequently,

$$\begin{aligned} I \propto A^2 &= a^2(\cos^2\phi_1 + \cos^2\phi_2 + \cdots + \cos^2\phi_n) \\ &\quad + a^2(\sin^2\phi_1 + \sin^2\phi_2 + \cdots + \sin^2\phi_n) \\ &= a^2(\cos^2\phi_1 + \sin^2\phi_1) + a^2(\cos^2\phi_2 + \sin^2\phi_2) + \cdots \\ &\quad + a^2(\cos^2\phi_n + \sin^2\phi_n) \end{aligned} \tag{6.17}$$

Since $\cos^2 X + \sin^2 X = 1$, Eq. 6.17 reduces to

$$A^2 = a^2 n \tag{6.18}$$

Equation 6.18 shows that interference effects need not be considered when computing the average intensity resulting from the superposition of n waves with random phase. The resultant amplitude in Figure 6.6 will increase in proportion to $n^{1/2}$, and the average intensity is simply n times that due to any single wave.

To summarize, if light beams from incoherent sources arrive at point S, their intensities combine directly, and no interference effects will be observed. However, if light beams from coherent sources arrive at point S, resultant intensities must be computed according to the stationary interference pattern produced from the addition of wave amplitudes. These stationary wave patterns persist for discernible periods of time and appear as alternating bright and dark regions of light intensity called interference fringes.

6.5. THE INTERFERENCE PATTERN

The appearance of the interference pattern produced by superposition of two waves derived from coherent sources can be determined graphically using

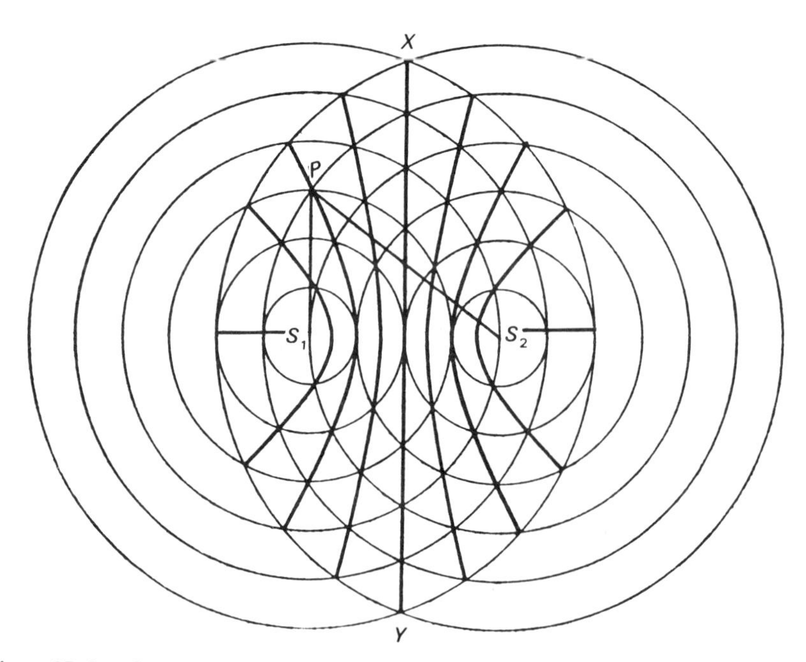

Figure 6.7. Interference produced by two-dimensional crossing of wavefronts from coherent sources S_1 and S_2 (light circles). Resultant light rays are indicated by heavy lines. Both sources emit monochromatic radiation of same frequency. (After Tolansky, 1955.)

Figure 6.7. For this discussion, we assume that sources S_1 and S_2 each emit radiation over an angle of 360° and that this radiation consists of the same single wavelength.

Let the circles in Figure 6.7 represent maxima that occur in the waves in the plane of the page. Where these circles intersect, as at point P, the condition $S_2P - S_1P = m\lambda$ ($m=0, 1, 2, 3, \cdots$) exists, producing a maximum in the interference pattern. The symbol m is called the order of interference, and $S_2P - S_1P$ is the difference in optical path from S_2 and S_1 for the two light beams.

Points for a specific value of m are interconnected with a heavy black line and represent the equation for a hyperbola, of which S_1 and S_2 are the foci. These thick curves show the location of light rays produced by the interference phenomenon, where each resultant light ray is characterized by a particular value of m. Along the line XY, the value of m is zero, indicating that the path difference for the interfering waves is zero. This is called the zero order. Successive orders (first, second, third, etc.) occur along the heavy lines where $m = 1, 2, 3, \cdots$. (See Figure 6.8)

To convert to the three-dimensional case, the plane waves of Figure 6.7 are transformed into spherical waves by rotating the pattern about the line S_1S_2. The heavy lines are now hyperboloids of revolution. If a screen is placed at position AB (Figure 6.8), that is, parallel to the line joining S_1 and S_2 and

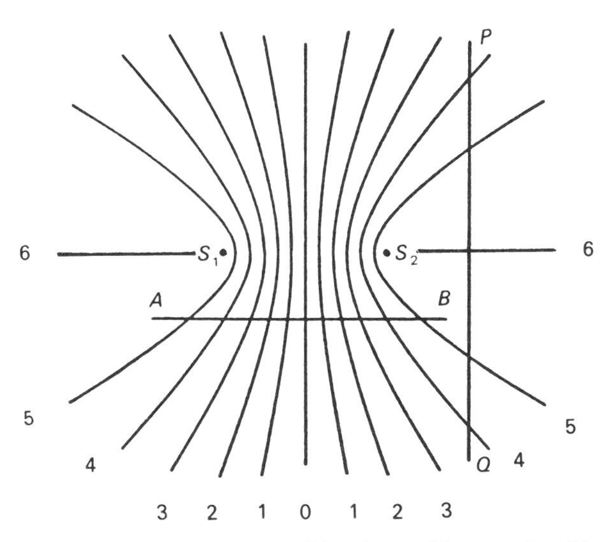

Figure 6.8. Screen placement for observation of interference fringes produced by two coherent sources S_1 and S_2 shown in Figure 6.7. Numbers correspond to order of interference m.

equidistant from both sources, the intersection of the hyperboloids with AB describes the interference pattern. If AB is placed at a distance that is comparable to the separation of S_1 from S_2, the successive patterns of light and dark regions (called interference fringes) will become more widely spaced toward each end of the screen. If the screen–source distance is much greater than the distance S_1S_2, the light and dark regions will appear as parallel, equidistant lines. These lines become more widely separated as the screen is moved back from the two sources.

If the screen in Figure 6.8 is moved to position PQ, perpendicular to the line S_1S_2, the interference fringes will appear as concentric circles with a bright spot in the center. If the screen AB is moved around to position PQ and then up to the top of the diagram where it is once again parallel to the line S_1S_2, the fringes on the screen (Figure 6.9) will pass successively through (a) straight lines, (b) arcs, (c) circles, (d) arcs, and finally back to (e) straight lines.

Aside from the general shape of the interference fringes, there is another fundamental difference between the pattern observed on the screen AB and that observed on PQ. At AB, the center of the interference pattern corresponds to the order $m = 0$, with successive lines occurring for values of m increasing by

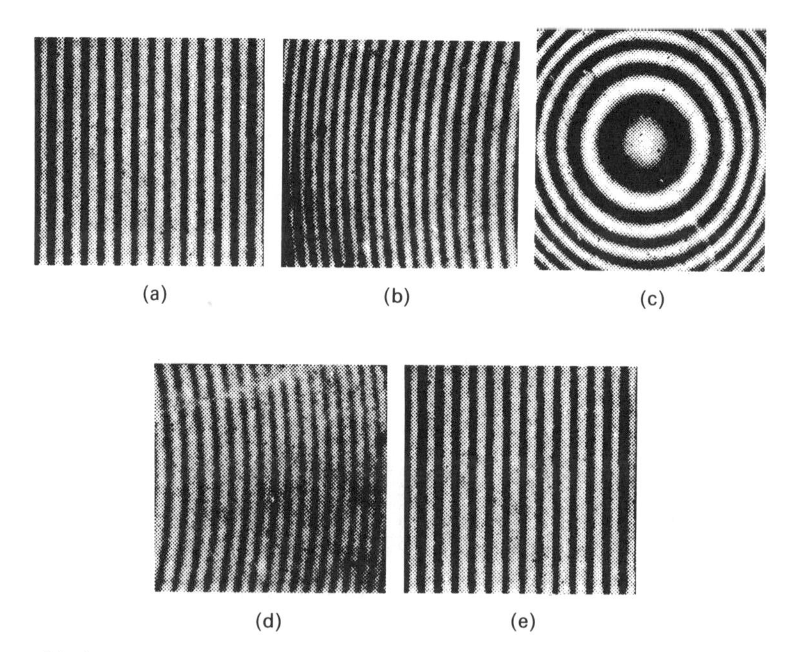

Figure 6.9. Appearance of interference fringes for various positions of screen (Figure 6.8). (After Jenkins and White, 1976.)

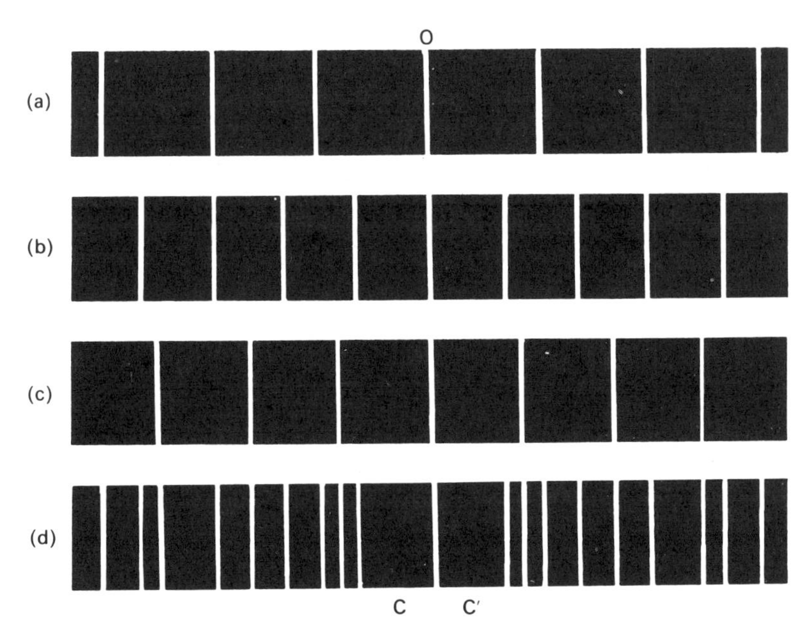

Figure 6.10. Interference fringes observed on screen AB (Figure 6.8) for two coherent sources. (*a*), (*b*), and (*c*) represent three different monochromatic cases. (*d*) S_1 and S_2 emit three wavelengths in (*a*), (*b*), and (*c*) simultaneously. The symbol O marks position of zero order, and C and C' mark positions of first dark fringes immediately to left and right of O.

1. At PQ, the center of the pattern (the "bulls-eye") corresponds to a maximum value of m, with successive rings occurring for values of m diminishing by 1.

If S_1 and S_2 represent polychromatic sources, then each wavelength will produce its own interference pattern. As a first approximation, we can simply say that each wavelength behaves independently of all others, and the final interference pattern will be the sum of the patterns for all wavelength components of the emitted spectrum. This feature is illustrated in Figure 6.10, where 6.10*a*–*c* represent the fringes obtained at screen position AB for three different monochromatic sources. Figure 6.10*d* shows that when S_1 and S_2 emit all three wavelengths simultaneously, the resulting interference fringes can be obtained by superposition of the three monochromatic patterns.

6.6. INTERFERENCE BY DIVISION OF WAVEFRONT: THE DOUBLE-SLIT EXPERIMENT

So far, nothing has been said about how two sources can be made coherent. The simplest way, and in some cases the only way, is to require that S_1 and S_2

both originate from a common source. In this manner, any phase fluctuations in S_1 will be exactly duplicated in S_2. Since both S_1 and S_2 retain a consistent phase relationship relative to each other, waves emitted by these sources can produce standing interference patterns.

There are basically two ways to produce two coherent sources from a common source. The first method, called division of wavefront, uses multiple slits, lenses, prisms, or mirrors to divide a single wavefront laterally to form two smaller segments that can interfere with each other. The second method, called division of amplitude, uses partial reflection at a beamsplitter to produce two wavefronts that maintain the same width but are of reduced amplitude. After following different optical paths, the two waves of reduced amplitude are recombined to produce an interference pattern.

Another way of distinguishing between the two methods of producing interference is to recognize that in division of wavefront, the interfering beams of radiation have left the source in different directions and some optical means is used to bring the beams back together. In division of amplitude, the interfering beams consist of radiation that has left the source in the same direction. This radiation is divided after leaving the source and later recombined to produce an interference pattern. While both methods of producing interference can be applied to spectroscopic measurements, division of wavefront is mainly useful with small sources (ideally originating from points or infinitely narrow slits). By contrast, division of amplitude can be used with extended sources, so the interference effects may be of greater intensity.

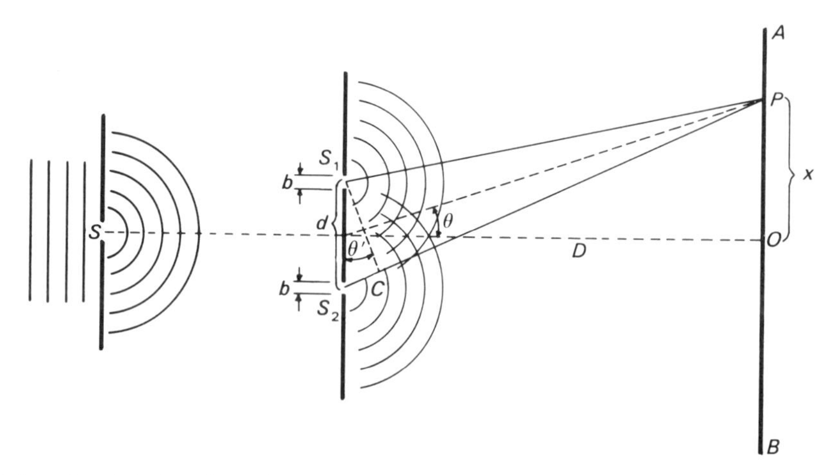

Figure 6.11. Division of wavefront using double-slit arrangement. Symbols S_1, S_2, A, B, and O correspond to those in Figures 6.7, 6.8, and 6.10.

Historically, division of wavefront was the earliest procedure used to demonstrate the interference of light. The experiment was first performed by Thomas Young in 1803 (Shamos, 1959; Young, 1803, 1804) using pinholes or very narrow slits arranged according to Figure 6.11. Diffraction of the parallel wavefront at the primary source, slit S, produces a single cylindrical wavefront, which can be divided into two cylindrical wavefronts by diffraction at slits S_1 and S_2. These two new wavefronts will be in phase provided that S_1 and S_2 are equidistant from S. If S_1 and S_2 are of equal width b and the separation d between them is small, the amplitudes of the two wavefronts will also be equal.

6.6.1. Position of the Interference Fringes

The double-slit experiment has been previously treated in Section 3.4 but will be reviewed here because it is an excellent example of interference by division of wavefront. At any point P on the screen, two waves arrive, one following the path S_1P and the second following the path S_2P. At point P, the phase difference in the two waves will be $\delta = (2\pi/\lambda)(S_2P - S_1P) = (2\pi/\lambda)(S_2C)$, where S_2C is the path difference between the two beams. The length S_2C is given by $d \sin \theta'$. If, as in the case of Young's experiment, the source–screen distance D is much greater than the slit separation d, then S_1CS_2 is approximately a right triangle, and $d \sin \theta' \approx d \sin \theta \approx d \tan \theta = d(x/D)$ The intensity of the interfering waves will be a maximum whenever $\delta = 0, 2\pi, 4\pi, 6\pi, \cdots$ and a minimum whenever $\delta = \pi, 3\pi, 5\pi, \cdots$. Hence the location of the interference fringes will be determined by Eqs. 6.19 and 6.20:

BRIGHT FRINGES

$$x = \pm m\lambda(D/d) \tag{6.19}$$

DARK FRINGES

$$x = \pm(m + \tfrac{1}{2})\lambda(D/d) \tag{6.20}$$

where λ is the wavelength of the interfering radiation, $m = 0, 1, 2, 3, \cdots$ is the order of interference, and x refers to the distance on the screen relative to point O, which marks the position of the bright fringe of zero order. (See Figures 6.9 and 6.10.)

6.6.2. Intensity Variations in the Interference Fringes

The pattern produced by the interference of the two wavefronts emerging from S_1 and S_2 (Figure 6.11) exists throughout the space behind the two slits, but the fringes are so close together that unless the source–screen distance D is rather large, a lens is generally required to observe them. Close examination of the bright fringes shows that they are not equal in intensity but display greatest brightness in the center, falling off in intensity at both sides. As discussed in Chapter 3, this variation in intensity occurs because the overall pattern is produced by a combination of two effects: diffraction and interference:

$$I \propto 4A_o^2(\sin^2 \beta/\beta^2)\cos^2(\tfrac{1}{2}\delta) \tag{3.38}$$

The term $\sin^2\beta/\beta^2$, where $\beta = (\pi b/\lambda)\sin\theta$, arises from diffraction by a single slit of width b, where θ is the angle of diffraction (Figure 6.11*b*). (These derivations are given in Chapter 3.) The term $\cos^2(\frac{1}{2}\delta)$, arises from the interference of two light beams of equal intensity having a phase difference of δ (Eq. 6.13). In the double-slit experiment, the phase difference δ is determined by the

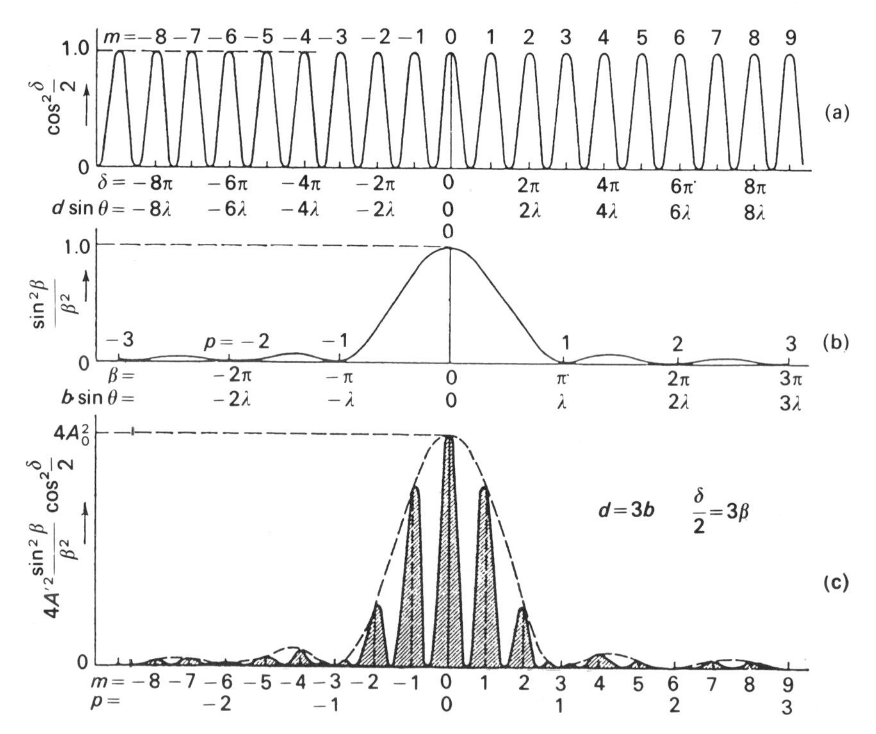

Figure 6.12. (*a*) Intensity variation for simple interference. (*b*) Intensity variation for simple diffraction. (*c*) Intensity variation obtained from double-slit experiment in Figure 6.11 for case $d = 3b$. [Reprinted with permission from F. A. Jenkins and H. E. White (1976), *Fundamentals of Optics*, 4th ed., McGraw-Hill, New York.]

path difference of the interfering beams leaving the double slit, that is, $\delta = (2\pi/\lambda)d \sin\theta$, as previously discussed.

Figure 6.12 shows the intensity (irradiance) variation expected from (a) simple interference from two beams of equal intensity and (b) simple diffraction for the case $d = 3b$. The intensity of the interference fringes in the double-slit experiment (c) is simply a superposition of the two intensity distributions due to both effects. For small θ, the amplitude in Figure 6.12*c* is now determined, not only by the amplitudes of the interfering waves, but also by b, the slit width, and D, the distance between the slits and the screen, that is, $A_o = A(b/D)$. (See Section 3.4.)

This discussion has made the following distinction between interference and diffraction. Interference is taken to mean the case in which the amplitude of a wave is modified by the superposition of a finite number of beams. Diffraction, on the other hand, is taken to mean the case in which the amplitude is determined by integration over all the infinitesimal elements of a wavefront (Figure 3.2). As Jenkins and White (1976) have pointed out, however, it is just as correct to consider diffraction as a more complicated case of interference, in which case the whole variation of intensity patterns shown in Figure 16.12*c* may be considered as interference effects.

6.6.3. Effect of Slit Width

The relationship between the two parameters b and d has a considerable effect on the appearance of the overall intensity pattern. As shown in Figure 6.13, the greater the ratio d/b, the greater the number of interference fringes contained within the central diffraction maximum. This effect can be understood from an examination of Eq. 3.38. If b is large compared to d, the $\sin^2\beta/\beta^2$ term dominates (Figure 6.13*a*), while if b is small compared to d, the $\cos^2(\frac{1}{2}\delta)$ term dominates (Figure 6.13*c*). If, as in the case of Young's experiment, the double slits are made very narrow and the source–screen distance is very large, the central maximum in the diffraction pattern will fill most of the viewing screen, and the intensities of the interference fringes within this central maximum will fall off very gradually as the order m increases.

As diffraction effects begin to dominate, some of the orders due to interference effects may appear to be missing. The phenomenon of missing orders occurs whenever the conditions for a maximum in the interference pattern ($d \sin\theta = 0, \lambda, 2\lambda, 3\lambda, \cdots = m\lambda$) and a minimum in the diffraction pattern ($b \sin\theta = \lambda, 2\lambda, 3\lambda, \cdots = p\lambda$; Eq. 3.28) occur for the same value of θ, that is, $d/b = m/p$. Since m and p are both integers, if d/b is in the ratio of two integers, then an order will be missing. Thus, if $d/b = 2$, as in Figure 16.13*a*, orders 2, 4, 6, $\cdots$ are not observed.

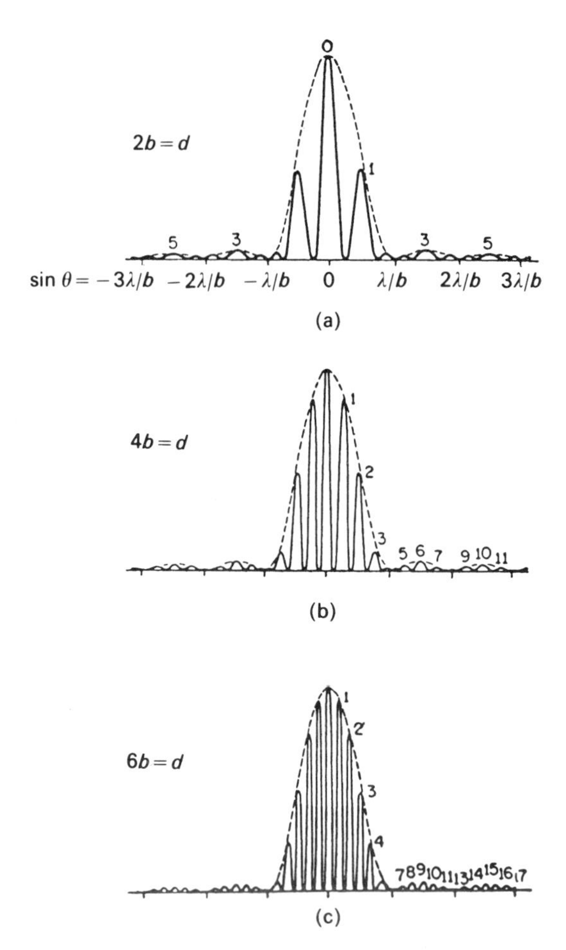

Figure 6.13. Intensity variations obtained for several double-slit experiments showing effect of changes in d/b: (a) $d = 2b$ with orders $m = 2, 4, 6, \cdots$ missing in interference pattern; (*b*) $d = 4b$ with orders $m = 4, 8, 12, \cdots$ missing in interference pattern; (*c*) $d = 6b$ with orders $m = 6, 12, 18, \cdots$ missing in interference pattern. [Reprinted with permission from F. A. Jenkins and H. E. White (1976), *Fundamentals of Optics*, 4th ed., McGraw-Hill, New York.]

The preceding discussion has assumed that the source S (Figure 6.11) is so narrow that only a single train of waves reaches the double slit. In actual practice, this condition would require that S be made infinitely thin. If S has finite width, a set of waves approaching S_1 and S_2 at slightly different angles will produce sets of interference patterns that are slightly shifted with respect to each other. In the simple case of two wave trains approaching the double slit from separate sources S' and S'' (Figure 6.14*a*), the positions of the central maxima of the interference patterns will be O' and O''. If the wavelength of the

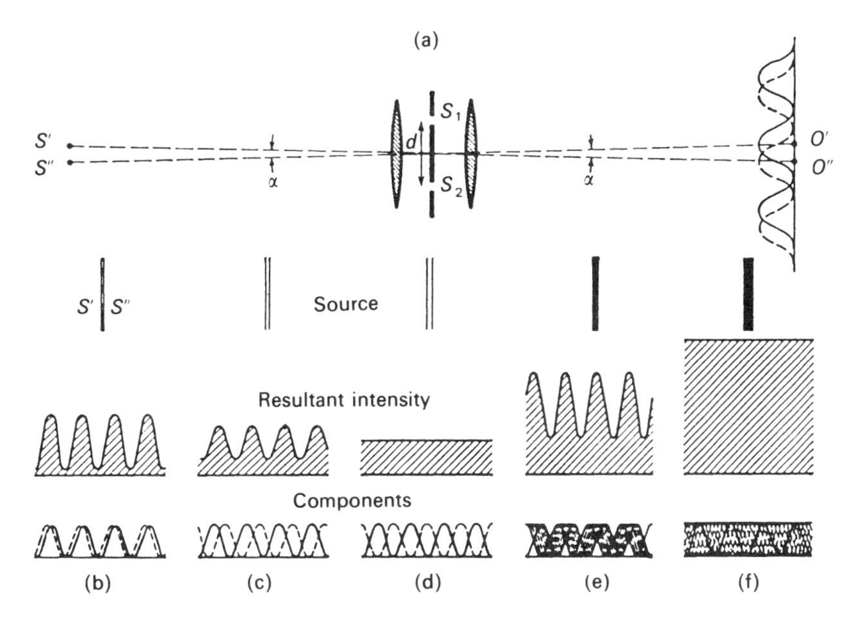

Figure 6.14. The effect of double source and wide source on double-slit experiment. [Reprinted with permission from F. A. Jenkins and H. E. White (1976), *Fundamentals of Optics*, 4th ed., McGraw-Hill, New York.]

interfering light, λ, is large with respect to the fringe displacement $O'O''$, the resultant intensity pattern will still resemble a $\cos^2(\frac{1}{2}\delta)$ curve but will not fall to zero at the minimum. Moreover, the positions of the maxima will be determined from the sum of the two individual interference patterns (Figure 6.14*b*).

As S' and S'' are moved farther apart, the distance $O'O''$ increases (Figure 6.14*c*). Eventually, the separate interference patterns become completely out of step, and the resultant intensity shows no fluctuation whatever (Figure 6.14*d*). This condition will occur whenever the distance $(S'S_1O' - S''S_1O') = (m + \frac{1}{2})\lambda$, that is, the source S'' is just a half integral wavelength farther from S_1 than the source S'. In general, the condition for disappearance is $\alpha = \lambda/2d$, $3\lambda/2d$, $5\lambda/2d, \cdots$, where α is the angle subtended by the two sources S' and S'' at the double slit.

If the source S is actually one wide slit (rather than two narrow slits as in Figure 6.14*a*), the resultant interference pattern will be the sum of a large number of interference patterns, each produced by a line element in the light strip of width $S'S''$ (Figure 6.14*e*). Intensity fluctuations in the interference pattern will be visible, except for the case in which the range covered by the component fringes extends over a whole fringe width (Figure 6.14*f*). In general, the condition for disappearance is $\alpha = \lambda/d$, $2\lambda/d$, $3\lambda/d, \cdots$, where α is the angle subtended by the width of the slit $S'S''$.

6.6.4. Polychromatic Sources

The previous discussion has assumed that source S (Figure 6.11) is monochromatic. If S is polychromatic, each wavelength will produce its own set of interference fringes according to Equations 6.19 and 6.20, and the superposition of these patterns will yield the overall effect. At zero order (O in Figures 6.10 and 6.11) all fringes overlap, producing a very bright central image flanked by two dark fringes C and C'. The central image at zero order will be the most intense image on the screen because it consists of contributions from every wavelength radiated by the source. Beyond the two dark fringes, bright fringes characteristic of each individual wavelength component begin to appear but eventually tend to run into each other and merge to form regions characteristic of more than one wavelength. In general, the greater the number of wavelengths present in the emitted radiation, the fewer the number of fringes characteristic of each individual wavelength that can be observed.

It should be noted that special lenses are required when dealing with polychromatic sources because the refractive index of all transparent media varies with wavelength. Therefore, a single lens will form a series of images, one for each wavelength of light present in the source. As shown in Figure 6.15*a*, a single lens acts like a prism, causing shorter wavelengths (in this case λ_1) to be brought to a focus nearer to the lens. The effect is called chromatic aberration.

One method of correcting for chromatic abberation is to employ a compound lens in which both a concave and a convex lens are cemented together. If both lenses have the same refractive index but the convex lens has a larger power, the combined focusing power will cause all wavelengths to be brought to approximately the same focus (Figure 6.15*b*).

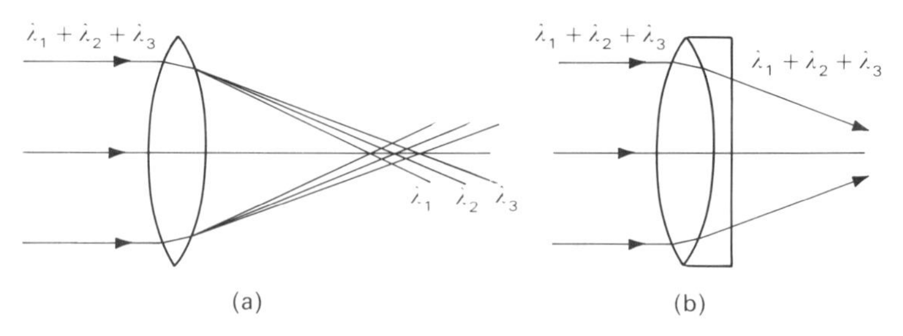

Figure 6.15. (*a*) Chromatic aberration of single lens. (*b*) Lens corrected for chromatic aberration. Relative wavelengths are $\lambda_1 < \lambda_2 < \lambda_3$.

6.6.5. The Rayleigh Interferometer

The "effective" optical path of a light wave is given by the product nl, where l is the distance traveled and n is the refractive index of the medium. When there are several segments, $l_1, l_2, l_3, \cdots$, of the light path in substances of different refractive indices $n_1, n_2, n_3, \cdots$, the optical path is the sum of the nl products (Eq. 3.6). Up to this point we have assumed that the interfering rays always traveled through air ($n = 1$) or else traveled comparable distances through media (e.g., lens) having the same index of refraction. However, if the interfering rays travel through media of different refractive indices before being combined, the difference $n_1 - n_2$ must be taken into account when determining the interference pattern.

We conclude Section 6.6 by illustrating how interference by division of wavefront using a double-slit arrangement can be used in a simple device to measure slight differences in the refractive indices of gases, liquids, and solutions with a high degree of precision. A schematic of this instrument, first developed by Lord Rayleigh (1896), is shown in Figure 6.16.

Monochromatic light from source S is collimated by lens L_1, then split into two beams using fairly wide slits S_1 and S_2. After first passing through tubes T_1 and T_2, each of the same length l, and then through the compensating plates G_1 and G_2, the beams are brought to focus at F by lens L_2.

With both tubes evacuated, the interferometer is adjusted to place the zero-order fringe in view. If two substances with refractive indices n_1 and n_2 are then introduced into T_1 and T_2, respectively, a path difference of $(n_1 - n_2)l$ will be created between the interfering rays, causing the fringes in the interference pattern to shift a number of orders such that $\Delta m\, \lambda = (n_1 - n_2)l$. The value of Δm can be determined by adjusting the compensator plates, and knowing the length of the two tubes allows the difference in the refractive indices of the two substances to be calculated.

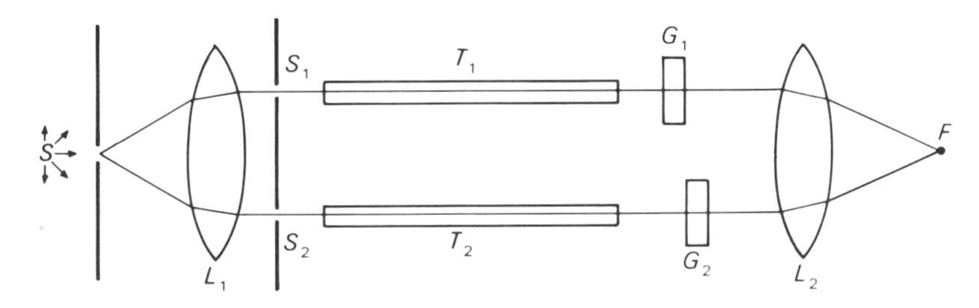

Figure 6.16. Rayleigh interferometer based on division of wavefront using double slits.

6.7. OTHER METHODS OF WAVEFRONT DIVISION

There are other ways of creating interference by division of wavefront, the earliest method being the double-slit experiment of Young. This section will examine some additional examples of wavefront division in order to demonstrate how interference can also be achieved using prisms, lenses, and mirrors. Some of these methods have found important practical applications. Lloyd's mirror (Figure 6.19*b*), for example, formed the basis of the first interferometer used in radio astronomy, the mirror there being the surface of the sea. Additional information on interference by division of wavefront can be found in many of the texts cited in the references, including Born and Wolf (1986), Jenkins and White (1976), and Tolansky (1955).

6.7.1. Dual Prisms

Fresnel's biprism (Figure 6.17) is formed from two equal prisms of small refracting angle placed together base to base. Using this device, radiation from source S can be divided by refraction into two beams of coherent radiation a_1b_1 and a_2b_2, which appear to originate from the virtual images S_1 and S_2, respectively. Since S_1 and S_2 are equidistant from the observation

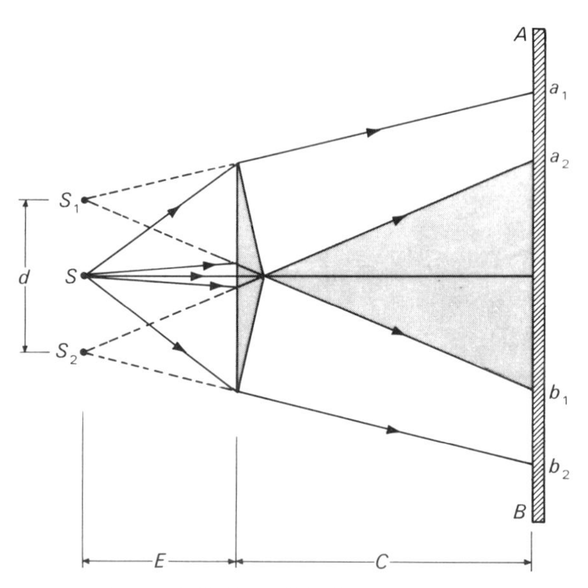

Figure 6.17. Fresnel biprism experiment.

screen AB, interference fringes will be observed as straight equidistant lines in the region a_2b_1.

The distance Δx between successive fringes on screen AB is given by

$$\Delta x = \lambda(E+C)/d \tag{6.21}$$

where E is the source–prism distance, C is the screen–prism distance, d is the distance between the virtual images S_1 and S_2, and λ is the wavelength of light emitted by the source S. To make the image separation d small, S must be placed close to the prism.

If the biprism is made from two identical pieces of glass, the intensity of the two virtual sources will be equal. However, if the biprism is made from two pieces of glass that have a different absorption for light, the two virtual sources S_1 and S_2 will have different intensities. For the case of unequal intensities,

$$I \propto A^2 = a_1^2 + a_2^2 + 2a_1a_2 \cos\delta \tag{6.12}$$

must be used without the simplifying assumption $a_1 = a_2$. Maxima and minima will still occur as $\cos\delta$ varies from $+1$ to -1, but the intensity will vary between a high of $(a_1+a_2)^2$ and a low of $(a_1-a_2)^2$. Thus, the cosine curve will not drop to zero at the minima, nor will it rise as high at the maxima.

6.7.2. Split Lenses

While Fresnel's biprism produces interference by creation of virtual images, the Billet split-lens device L shown in Figure 6.18*a* can be used to produce two real images S_1 and S_2 from a single source S. Again, unless both halves of the lens are made from the same glass, the intensities of the two sources will not be equal. The separation between S_1 and S_2 will depend on the separation of the two halves of the lens. The interference pattern can be observed as straight, equidistant lines at screen AB.

A variation of the Billet split lens (Meslin's experiment) is shown in Figure 6.18*b* (Meslin, 1893). When the two halves of the lens are displaced along the axis as indicated, the two sources S_1 and S_2 will appear behind each other and should produce concentric ring fringes when observed on screen PQ. However, because the region common to both beams is limited by a plane passing through the optical axis, the two beams only overlap in the region S_1S_2C, and the actual appearance is semicircular. The center of the pattern is an intensity minimum (rather than the maximum expected from Figure 6.9*c*) because the waves that appear to originate from the virtual source S_1 undergo a phase change of 180° on passage through the focus of the upper lens.

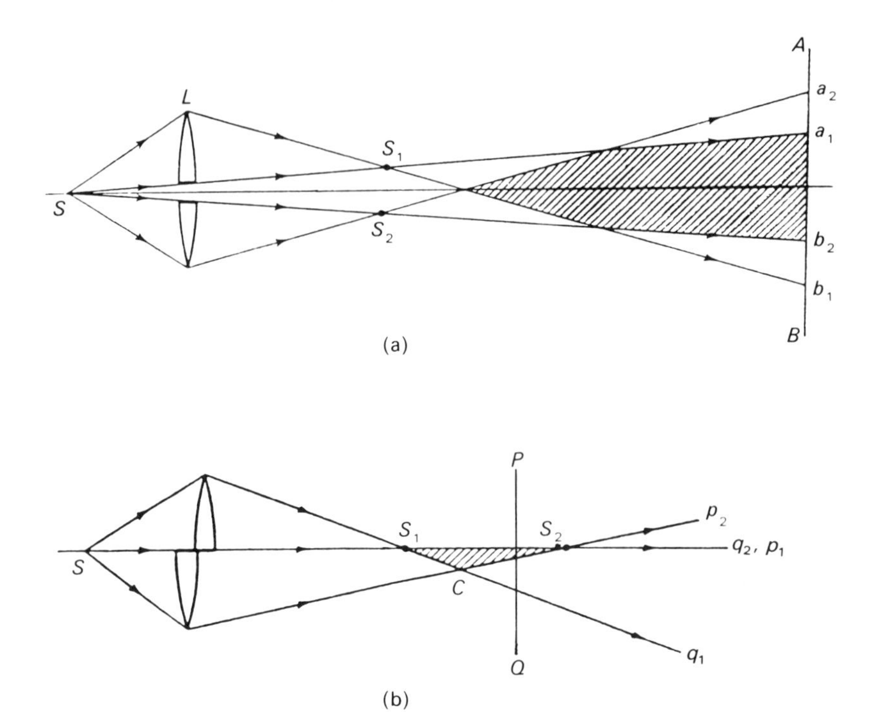

Figure 6.18. (*a*) Billet's split lens. (*b*) Meslin's experiment, variation of split lens. Relationship of screens AB and PQ to sources S_1 and S_2 shown in Figure 6.8. (After Born and Wolf, 1986.)

6.7.3. Mirrors

Figure 6.19*a* shows a double-mirror arrangement (Fresnel's mirrors) in which two light beams can be made to interfere. Light from slit S is reflected in two plane mirrors that are slightly inclined to each other at angle θ so as to produce two virtual images S_1 and S_2. Straight, equidistant interference fringes will be observed on screen AB in the region a_2b_2 where the reflected beams overlap.

An even simpler arrangement (Lloyd's mirror, Figure 6.19*b*) uses only one mirror to produce the same effect (Lloyd, 1837). The source S_1 is placed close to the plane containing the front-surface mirror. Light reflected at the mirror forms the virtual image S_2, which is close to and coherent with S_1. Interference is observed in the overlap region a_2b_2.

Since the distance S_2O and S_1O are equal in Figure 6.19*b*, the beams from the two different sources should be in phase at point O. However, if the observation screen is placed in contact with the end of the mirror, position $A'B'$, the edge O of the screen will contact a dark fringe instead of the bright

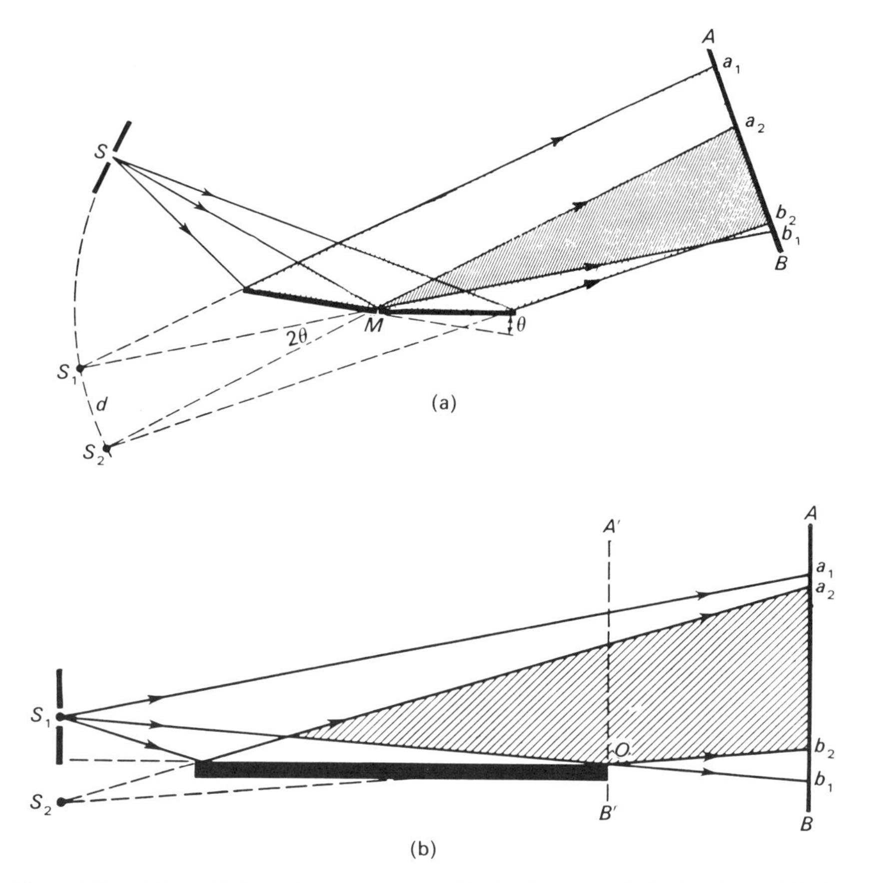

Figure 6.19. (*a*) Fresnel's two-mirror experiment. (*b*) Lloyd's mirror. (After Jenkins and White, 1976.)

one as expected. The observation of a dark fringe shows that there has been a phase change of π in one of the beams. Since the direct beam a_1b_1 cannot have undergone any alteration in phase, the reflected beam a_2b_2 must have undergone a phase shift of 180°. Such phase changes will be explored more fully in Section 6.8.1.

6.8. INTERFERENCE BY DIVISION OF AMPLITUDE

All the optical arrangements discussed in Sections 6.6 and 6.7 produced interference patterns by lateral division of a wavefront into segments generally having equal or nearly equal amplitudes. It is also possible to divide a wave by partial reflection at the interface between two media having different indices of refraction. The resulting waves are coherent, since they originate from the same source, but will be of reduced amplitudes. If these two waves follow different optical paths before being recombined, interference effects can be observed.

6.8.1. The Principle of Reversibility

Interference by division of amplitude is complicated by phase changes that sometimes occur when a wave encounters the interface between two media having different indices of refraction. For example, the dark fringe of zero order observed in the two-mirror experiment (Figure 6.19*b*) can be explained only if reflection at the mirror's surface produces a phase shift of 180°. To determine the generality of this phenomenon, we examine reflection using Stokes's principle of reversibility. According to this principle, in any ray-tracing procedure the reversal of all the rays must produce the original ray provided there is no absorption.

Figure 6.20*a* shows light ray A traveling through air, striking the surface of a glass plate at point P, where it is partially reflected. The reflected ray B has amplitude ar, while the transmitted ray has amplitude at. If a is the amplitude of the original ray A, then r and t are the fractions of the original ray that are reflected and transmitted at point P, respectively.

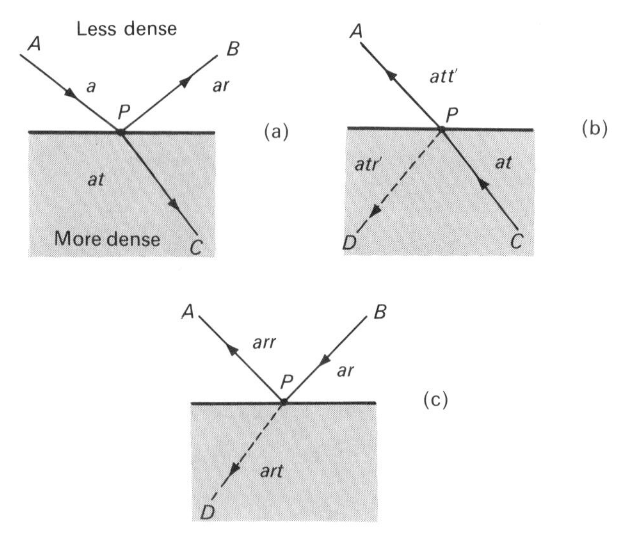

Figure 6.20. Reversibility principle. Wave AP (*a*) is reconstructed by reversing refracted wave PC (*b*) and reflected wave PB (*c*).

According to the principle of reversibility, A can be reconstructed by reversing rays B and C and adding the results. In Figure 6.20*b*, ray C strikes point P with amplitude at, where it is partially reflected. The reflected ray D will be of amplitude atr', while the transmitted ray A will be of amplitude att'. (The terms r' and t', the fractions of ray C reflected and transmitted at point P, are primed since it cannot be assumed that the reflectivity from glass is the same as that from air.) Figure 6.20*c* shows the results of reversing ray B. Ray B strikes point P with amplitude ar, where it produces reflected ray A of amplitude arr and transmitted ray D of amplitude art.

According to the principle of reversibility,

$$a = att' + atr' + arr + art$$

However, since ray D did not exist in the actual experiment,

$$atr' + art = 0$$

Therefore,

$$a = att' + arr$$

From these equations, the following relationships are obtained:

$$r' = -r \tag{6.22}$$

$$tt' = 1 - r^2 \tag{6.23}$$

The physical interpretation of Eq. 6.22 is that the fraction of intensity reflected at P is the same on both sides of the boundary, but rays B and D have a phase difference of 180°. Equation 6.22 does not indicate on which side of the boundary this phase shift occurs. However, the experiment with Lloyd's mirror (Section 6.7.3) has demonstrated that such a shift occurs when light waves approach a boundary from the side of higher velocity (i.e., from the side having the lower index of refraction). Therefore, ray B (Figure 6.20*a*), but not ray D (Figure 6.20*b*), undergoes a phase shift of 180°.

6.8.2. Division of Amplitude at Parallel Surfaces

To understand how interference can be produced by division of amplitude and to show how division of amplitude and division of wavefront can be distinguished from one another, we consider the two rays A_1 and A_2 in Figure 6.21. Both A_1 and A_2 are monochromatic and consist of the same wavelength

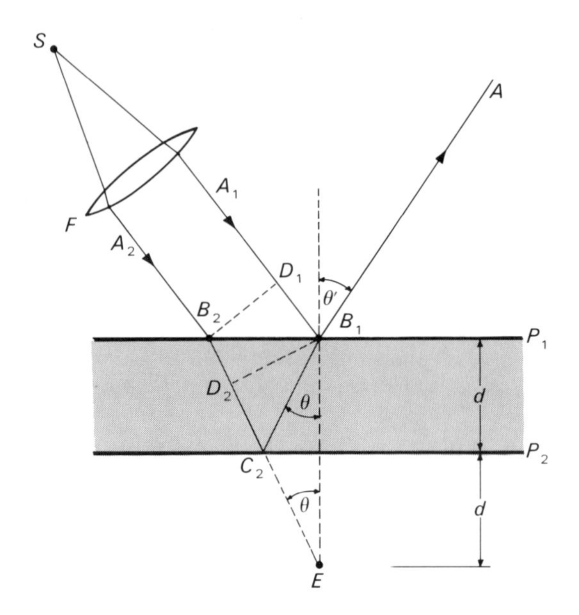

Figure 6.21. Optical path difference for two rays A_1 and A_2. Here, A_2 is refracted at surface P_1 and reflected at surface P_2, and A_1 is reflected at surface P_1.

λ. Both rays leave source S in different directions, are collimated by lens F, and strike a glass plate of thickness d in which the two surfaces P_1 and P_2 are strictly parallel. Ray A_1 strikes surface P_1 at point B_1 (θ' is the angle of incidence) and is totally reflected. Ray A_2 strikes surface P_1 at point B_2 and is completely transmitted through the air–glass interface (θ is the angle of refraction). After total reflection at point C_2, ray A_2 is superposed on ray A_1 at point B_1. Since both rays initially left the source in different directions, we would say that interference has occurred by division of wavefront.

The phase difference between A_1 and A_2 at point B_1 will depend on the optical path difference L of the two rays. If lines B_2D_1 and B_1D_2 represent perpendiculars from B_2 and D_2, respectively, then L will be given by $n(B_2C_2 + C_2B_1) - D_1B_1$, where n is the index of refraction in the glass. (The index of refraction of air is assumed to be 1.) Since $n > 1$, wave A_2 travels the shorter distance B_2D_2 in the same time period as wave A_1 travels the longer distance D_1B_1. Thus, $n(B_2D_2) = D_1B_1$, and $L = n(D_2C_2 + C_2B_1)$. If the isosceles triangle C_2EB_1 is constructed, $D_2C_2 + C_2B_1 = D_2E$, and

$$L = n2d \cos\theta \tag{6.24}$$

If L is a whole number of wavelengths, we would normally expect A_1 and A_2 to interfere constructively. However, A_1 has approached P_1 through air and undergoes a phase change of π on reflection at B_1. Ray A_2, on the other hand, has approached P_2 through glass and does not undergo a phase change at C_2. Thus, $\theta = 0, 2\pi, 4\pi, \cdots$, defines conditions producing destructive interference, while $\theta = \pi, 3\pi, 5\pi, \cdots$, defines conditions producing constructive interference.

To produce interference by division of amplitude, we simply reverse all of the rays in Figure 6.21. The single ray A (amplitude a) is partially reflected at point B_1 to form ray A_1 (amplitude ar), while the remainder of A is transmitted to surface P_2 where it undergoes total reflection to form ray A_2 (amplitude at). (The symbols t and r have the same meanings as in Figure 6.20*a*.) Since A_1 and A_2 originate from the same source, they are coherent and can be recombined using lens F to produce a stable interference pattern at point S. The conditions for destructive interference are given by Eq. 6.25, while those for constructive interference are given by Eq. 6.26:

MINIMA

$$n2d \cos \theta = m\lambda \tag{6.25}$$

MAXIMA

$$n2d \cos \theta = (m + \tfrac{1}{2})\lambda \tag{6.26}$$

where m is the order of the interference pattern ($m = 0, 1, 2, \cdots$). Since the two recombined rays have left the same source in the same direction and are of reduced amplitude, interference is said to occur by division of amplitude.

6.8.3. Interference Involving Multiple Reflections

Figure 6.22 shows a more complicated situation in which a single monochromatic ray A undergoes a series of multiple reflections and refractions within a glass plate where surfaces P_1 and P_2 are parallel. If the path difference $L = n2d \cos 2\theta = m\lambda$ (Eq. 6.25), A_1 will be out of phase with A_2. Since the geometry is the same, the path difference between A_2 and A_3 will also be given by $L = n2d \cos \theta$. However, neither A_2 nor A_3 have undergone a phase change upon reflection inside the glass, so A_2 and A_3 will interfere constructively. The same conditions hold for all succeeding pairs; that is, every pair except A_1A_2 is in phase. If all the rays A_1 through A_n are combined, the resultant amplitude will be

$$\begin{aligned} A &= ar - (atrt' + atr^3t' + atr^5t' + atr^7t' + \cdots) \\ &= ar - atrt'(1 + r^2 + r^4 + r^6 + \cdots) \end{aligned}$$

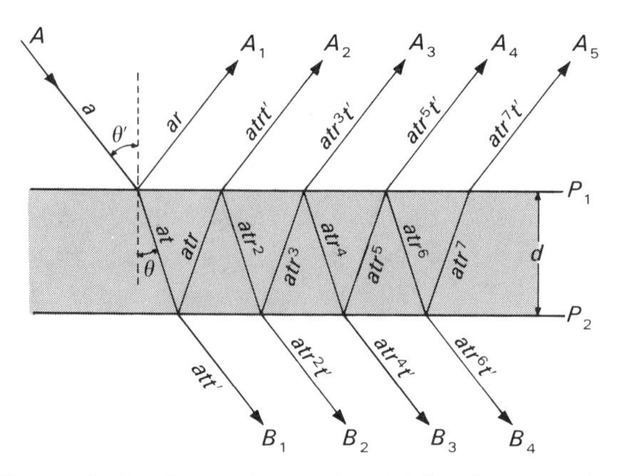

Figure 6.22. Amplitudes of successive rays in multiple reflections and refractions.

Since $r < 1$, the infinite geometric sum in parentheses has a finite value and can also be written as $(1 - r^2)^{-1}$, which, according to Eq. 6.23, is equivalent to $(tt')^{-1}$. Therefore,

$$A - ar - \frac{atrt'}{tt'} - ar - ar - 0$$

Therefore, provided there is no absorption, when $L = m\lambda$, the combination of all rays on the upper side of the glass surface P_1 results in complete destructive interference.

If the path difference $L = (m + \frac{1}{2})\lambda$ (Eq. 6.26), A_1 will in phase with A_2. However, A_2 will be out of phase with A_3, A_3 will be out of phase with A_4, and so forth. Thus, if all of the rays A_1 through A_n are combined, the resultant amplitude will be

$$A = ar + atrt' - atr^3t' + atr^5t' - atr^7t' + \cdots$$
$$= ar + atrt'(1 - r^2 + r^4 - r^6 + \cdots)$$

Since $r < 1$, the out-of-phase pairs do not completely cancel each other, so when $L = (m + \frac{1}{2})\lambda$, the combination of all of the rays on the upper surface of the glass produces a maximum of intensity.

The rays emerging from the lower side of the glass in Figure 6.22 can also be brought to interference using a lens. In this case, however, there are no phase changes at reflection for any of the rays. Maxima are produced when rays B_1, B_2, B_3, B_4, $\cdots$ are all in phase ($n2d \cos\theta = m\lambda$), while minima are produced when rays B_1, B_3, B_5, $\cdots$ are out of phase with B_2, B_4, B_6, $\cdots$ [$n2d \cos\theta$

$=(m+\frac{1}{2})\lambda]$. In the case of minima, B_1 is more intense than B_2, B_3 is more intense than B_4, and so on. Therefore, destructive interference does not produce complete cancellation, and the minima are not totally black.

Figure 6.23*a* compares the intensity contours for reflected and transmitted fringes from a glass plate having a reflectance r of 4%. The abscissa, δ, represents the phase difference between successive rays in the transmitted set or between all but the first pair in the reflected set, which from Eqs 6.3 and 6.24

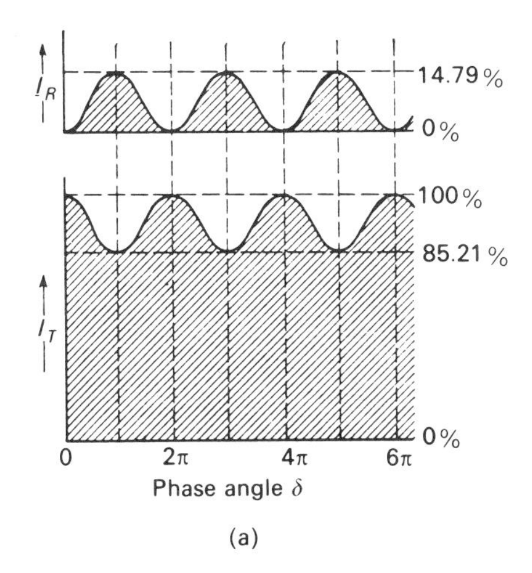

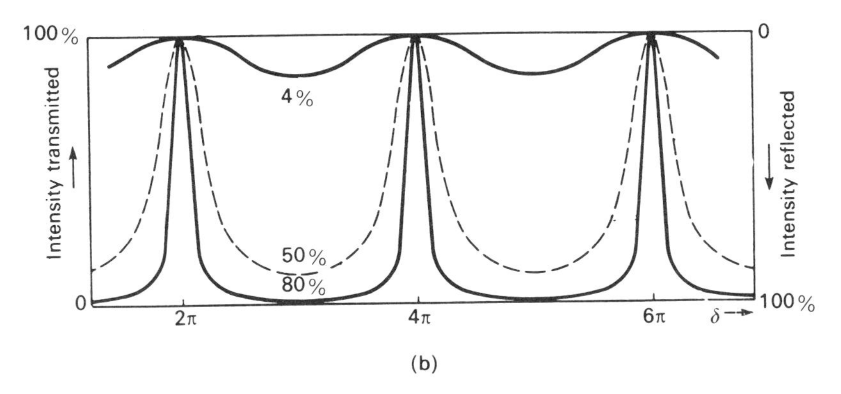

Figure 6.23. (*a*) Intensities of reflected (I_R) and transmitted (I_T) fringes from surface with reflectance of 4%. (*b*) Intensity of fringes due to multiple reflections showing variation of I_R and I_T for surfaces of different reflectance. Abscissas in both plots represent phase difference between successive rays in transmitted set or between all but first pair in reflected set. [Reprinted with permission from F. A. Jenkins and H. E. White (1976), *Fundamentals of Optics*, 4th ed., McGraw-Hill, New York.]

is given by

$$\delta = (2\pi/\lambda)L = (4\pi/\lambda)nd \cos\theta$$

The curves in Figure 6.23*a* resemble the $\cos^2$ contours obtained from the interference of two beams, except that the curve for transmitted light, I_T, does not fall to zero at the minima. Note that the two curves are out of phase with each other, that is, conditions that produce a minimum in the reflected light, I_R, produces a maximum in the transmitted light. This is to be expected since, for no absorption, the sum of the intensity of the transmitted light and the intensity of the reflected light must equal 100%. Thus, the two curves in Figure 6.23*a* are complementary, and one may be obtained from the other by turning the figure upside down.

Figure 6.23*b* shows the effect of increasing the reflectivity of the two surfaces P_1 and P_2 in Figure 6.22. (This can be accomplished experimentally by coating the surface with a reflecting film.) An increase in r causes the bright fringes formed by the transmitted light and, conversely, the dark fringes formed by the reflected light to become much narrower. The reason for this behavior can be understood by examining the equation describing the intensity of the combined rays $B_1, \cdots, B_n$ (Figure 6.22) as a function of r and δ. (The intensity of the combined rays $A_1, \cdots, A_n$ will be the complement of this equation.)

The derivation of the intensity equation describing the transmitted radiation will be considerably facilitated if the equations describing the rays (waves) are expressed in terms of complex functions rather than as cosines or sines. (See Section 3.4.) Using complex notation, at any instant in time a light wave may be represented as a vector $\mathbf{a}_1$ such that

$$\mathbf{a}_1 = a_1(\cos\phi_1 + i\sin\phi_1) = a_1 \exp(i\phi_1) \qquad (3.42)$$

where a_1 is the amplitude and ϕ_1 is the phase angle. For the combination of two waves, the resultant vector $\mathbf{A}$ will be

$$\mathbf{A} = A\exp(i\phi) = a_1 \exp(i\phi_1) + a_2 \exp(i\phi_2)$$

Therefore, the superposition of all transmitted waves in Figure 6.22 yields vector $\mathbf{A}_T$, where

$$\mathbf{A}_T = A\exp(i\phi) = att'\exp(i\phi_1) + atr^2t'\exp(i\phi_2)$$
$$+ atr^4t'\exp(i\phi_3) + atr^6t'\exp(i\phi_4) + \cdots$$

If the phase angles are expressed as phase differences relative to ϕ_1, then $\phi_2 - \phi_1 = \delta$, $\phi_3 - \phi_1 = 2\delta$, $\phi_4 - \phi_1 = 3\delta$, and so on,

$$\begin{aligned} \mathbf{A}_{\mathrm{T}} &= att' + atr^2t' \exp(i\delta) + atr^4t' \exp(i2\delta) + atr^6t' \exp(i3\delta) + \cdots \\ &= att'[1 + r^2 \exp(i\delta) + r^4 \exp(i2\delta) + r^6 \exp(i3\delta) + \cdots] \end{aligned} \quad (6.27)$$

Since $r^2 < 1$, the infinite geometric sum in brackets (Eq. 6.27) has a finite sum and can also be written as $[1 - r^2 \exp(i\delta)]^{-1}$. Therefore, Eq. 6.27 reduces to

$$\mathbf{A}_{\mathrm{T}} = A \exp(i\phi) = \frac{att'}{1 - r^2 \exp(i\delta)} \quad (6.28)$$

The intensity of the transmitted rays $B_1, \cdots, B_n$ is given by the square of the amplitude. For complex numbers, squaring is accomplished by multiplying the number by its complex conjugate, that is, $\mathbf{A}_{\mathrm{T}}^2 = \mathbf{A}_{\mathrm{T}} \mathbf{A}_{\mathrm{T}}^* = A \exp(i\phi) A \exp(-i\phi)$. Thus,

$$I_{\mathrm{T}} = \mathbf{A}_{\mathrm{T}}^2 = \frac{a^2(tt')^2}{1 - r^2[\exp(i\delta) + \exp(-i\delta)] + r^4}$$

From Euler's relationship, $[\exp(i\delta) + \exp(-i\delta)] = 2 \cos\delta$. Finally, if there is no absorption, we can set $a^2 = I_0$, where I_0 is the intensity of the incident beam A (Figure 6.22), and express tt' and r^2 in terms of intensity transmittance T and reflectance R, respectively. Therefore,

$$I_{\mathrm{T}} = \frac{I_0 T^2}{1 - 2R \cos\delta + R^2}$$

Using the trigonometric relationship $2 \sin^2(\delta/2) = (1 - \cos\delta)$,

$$I_{\mathrm{T}} = \frac{I_0 T^2}{(1 - R)^2 + 4R \sin^2(\delta/2)}$$

or alternatively,

$$I_{\mathrm{T}} = \frac{I_0 T^2}{(1 - R)^2} \left[1 + \frac{4R}{(1 - R)^2} \sin^2 \frac{\delta}{2} \right]^{-1} \quad (6.29)$$

The form of the intensity expression given in Eq. 6.29 is known as Airy's equation. From this, it is easy to obtain the reflected intensity I_{R} since $I_{\mathrm{R}} = 1 - I_{\mathrm{T}}$. The complete derivation is given in Tolansky (1955).

For transmitted light (Eq. 6.29), when $\delta = 0, 2\pi, 4\pi. \cdots$, $\sin^2(\delta/2) = 0$, and the intensity at the maximum will be

$$I_{max} = \frac{I_0 T^2}{(1-R)^2} = I_0 \tag{6.30}$$

Hence, provided there is no absorption, the transmitted fringe maxima have intensity equal to that of the incident light no matter what the reflecting coefficient may be. (See Figure 6.23b.) When $\delta = \pi, 3\pi, 5\pi, \cdots$, $\sin^2(\delta/2) = 1$, and the intensity at the minimum will be

$$I_{min} = \frac{I_0 T^2}{(1+R)^2} \tag{6.31}$$

This term never goes completely to zero so the minima will never be absolutely dark.

The width of the peaks in Figure 6.23*b* is determined by F, called the "coefficient of finesse," which appears in the denominator of Eq. 6.29:

$$F = \frac{4R}{(1-R)^2} \tag{6.32}$$

As r approaches 100%, F becomes increasingly important, and a small deviation of δ from $2\pi m$ ($m = 0, 1, 2, \cdots$) will result in a rapid drop in intensity, giving the narrow bright fringes shown in Figure 6.23*b*. For a large reflectance R, the fringe half-width, defined as the full width at half maximum (FWHM), can be shown (Born and Wolf, 1986; Stewart, 1970) to equal

$$\text{FWHM} \approx 2\frac{1-R}{R^{1/2}} = \frac{4}{F^{1/2}} \tag{6.33}$$

Thus, the sharpness of the fringes is greatly improved by increasing the reflectance of the two surfaces P_1 and P_2.

6.8.4. Fringes of Equal Thickness

The three variables that determine the order of interference m are given by Eqs. 6.25 and 6.26. These variables are n, the index of refraction d, the thickness of the plate, and θ, the angle of refraction. The index of refraction is generally constant for a given experiment, so variation in either d or θ provides a method for classifying the interference fringes produced by division of amplitude.

Figure 6.24 shows how the thickness of the plate, d, can be varied by using a wedge arrangement instead of two parallel plates. If the wedge angle is small, the basic equation derived for a parallel-sided plate will still apply. If the light striking the surface P_1 consists of parallel rays originating from a point source and the incidence is normal ($\cos\theta = 1$), dark fringes will be observed above P_1 at $nd = m\lambda/2$, where $m = 0, 1, 2, 3, \cdots$. Lines connecting all points on P_1 where m is a particular constant (AA', BB', CC', etc.) define locations where the wedge thickness is a constant. The resulting patterns are called fringes of equal thickness.

Fringes of equal thickness have long been used to test the surface quality of flat plates and lenses because the interference pattern produces a contour map of the wedge thickness. If either surface P_1 or P_2 is irregular, the fringe pattern will also be irregular, locating surface areas that require additional polishing to achieve the desired shape (Figure 6.25).

Figure 6.26*a* shows the experimental arrangement used to observe the circular fringes of equal thickness (called Newton's rings) produced in the air film formed between the convex surface of a lens and an optically flat glass

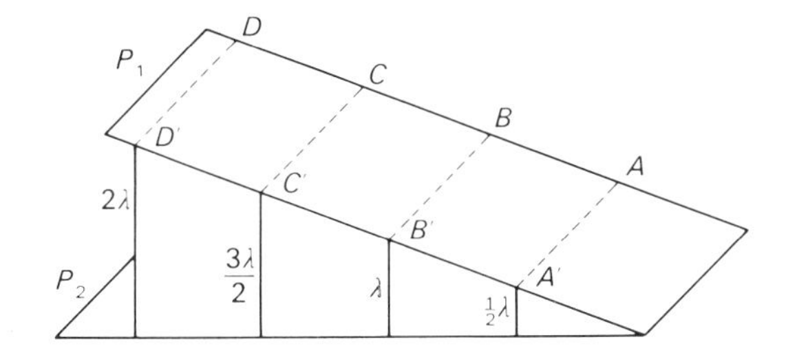

Figure 6.24. Fringes of equal thickness for $n = 1$.

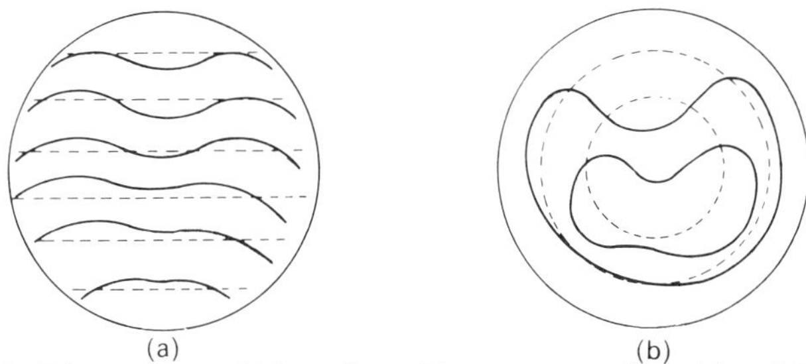

Figure 6.25. Fringes of equal thickness formed between two surfaces: (*a*) straight fringes formed for imperfect plane wedge; (*b*) circular fringes formed for imperfect lens. Dashed lines show ideal pattern.

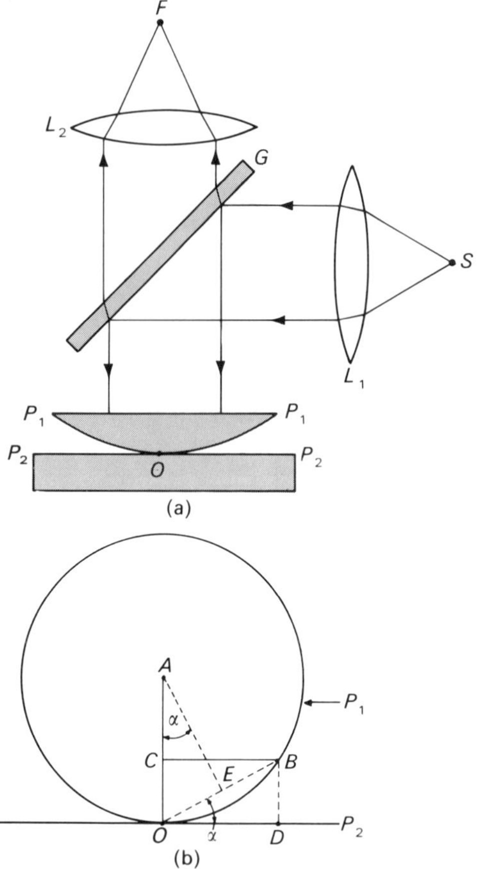

Figure 6.26. (*a*) Experimental arrangement for viewing Newton's rings. (*b*) Geometric relationship between radius of curvature $R = AO$ and radius of the interference fringes $r = OD$.

surface. In this arrangement, light from a monochromatic source S is collimated by lens L_1 and reflected by the glass plate G onto the back surface of the lens. The fringes are best observed if the angle of incidence is normal so that $\cos\theta = 1$. If the light is not incident normally, the rings become ellipses. After reflection within the air–glass wedge defined by P_1 and P_2, the light rays are transmitted through G and brought to interference at F by lens L_2.

The radius of curvature R of the lens is given by the line segment AO in Figure 6.26b, where O is the point at which the lens and plate just touch. If the radius of a ring in the interference pattern r is given by the line segment OD, then $\sin\alpha = DB/OB = OE/AO$. Since DB is the thickness of the air film where a fringe is produced, AO equals the radius of curvature of the lens, and $OB \approx OD$

$=r$ for small values of α, then the value of d for any ring of radius r will be given by

$$d = r^2/2R \tag{6.34}$$

The waves reflected from the lower surface P_2 undergo a phase change of 180° relative to those reflected at the upper surface P_1, so when the circular fringes are observed at F, the central spot corresponding to $m = 0$ is black. Since $2d = m\lambda$ for dark fringes and $2d = (m + \frac{1}{2})\lambda$ for bright fringes, it follows from Eq. 6.34 that the radii of the rings are proportional to the square root of the whole-number integers m, and the rings become more closely spaced toward the edges of the interference pattern as a result. No phase change occurs for the transmitted light, so if Newton's rings are observed beneath P_2, $2d = m\lambda$ defines conditions for constructive interference, and the central spot will be bright.

6.8.5. Fringes of Equal Inclination

While fringes of equal thickness are generally produced by parallel light beams originating from a point source, fringes of equal inclination require an extended source so that the angle of refraction θ can be varied. Figure 6.27 shows one way of using an extended source to produce an interference pattern.

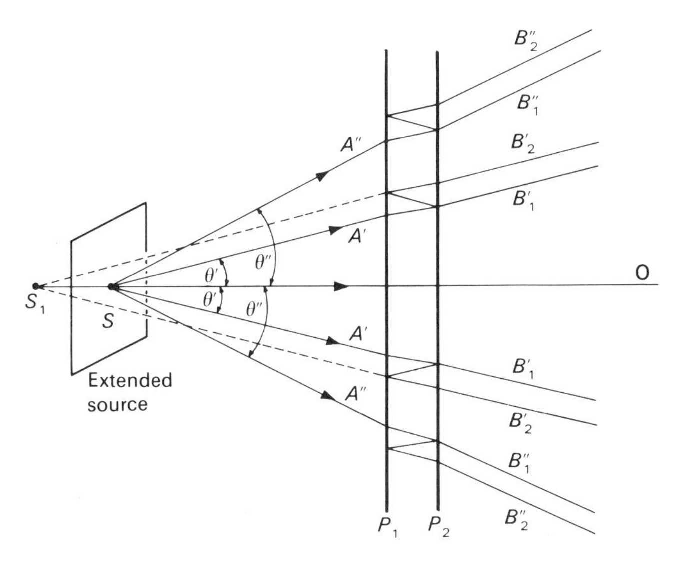

Figure 6.27. Fringes of equal inclination (Haidinger's rings).

If an extended source is aligned normal to a plate having parallel sides P_1 and P_2, light rays emitted from different parts of the source will strike P_1 with different angles of incidence and therefore have different angles of refraction. For example, the two rays A' originating from point S on the source strike surface P_1 at angle θ', while the two rays A'' from the same point strike P_1 at angle θ''. Since the angles of incidence (and therefore the angles of refraction) are different, the pairs of parallel beams B'_1 B'_2 and B''_1 B''_2 produced by division of amplitude will be out of phase by a different amount. When these parallel beams are recombined, each resultant wave will have a different intensity, and an interference pattern will be produced. Since there is no reversal of phase on reflection, maxima occur when the condition $m\lambda = 2nd \cos \theta$ is satisfied, while minima occur when the condition $(m + \frac{1}{2})\lambda = 2nd \cos \theta$ is satisfied. Since each maximum or minimum occurs for a specific value of refraction, the resulting interference pattern is called fringes of equal inclination.

The shape of the fringes produced by the arrangement depicted in Figure 6.27 will be circular. The presence of circular patterns, rather than straight lines or arcs, can be demonstrated by rotating Figure 6.27 about the line SO by $360°$ or by considering the position of the observer and referring to Figure 6.8. To the observer, it would appear that point S and its image S_1 (produced by reflection) are aligned normal to the line of sight. Therefore, the interference pattern will consist of concentric circles, with the maximum order m appearing at the center of the pattern. Consideration of the geometry involved in Figure 6.27 shows that, like Newton's rings, the radii of the fringes vary as the square root of the positive integers (Born and Wolf, 1986), causing the rings to become more closely spaced towards the edges of the interference pattern.

6.8.6. The Fabry–Perot Interferometer

Interferometers based on division of amplitude can be classified as two beam or multiple beam (Tolansky, 1948) depending on the number of beams that interfere. The most important two-beam device is the interferometer originally invented by Michelson (1891), some modification of which is utilized by almost all modern Fourier transform spectrometers. The Michelson interferometer produces circular fringes of equal inclination and has the potential for a large-throughput advantage because an extended source can be tolerated. Further details regarding the design and operation of the Michelson interferometer are reserved for Chapter 7, which deals with Fourier transform spectroscopy.

The Fabry-Perot interferometer (Fabry and Perot, 1901) also produces circular fringes of equal inclination (Figure 6.27) but falls into the multiple-beam category. The device consists of two plane–parallel plates (Figure 6.28) separated by a spacer of width d. Spacers typically range in value from about

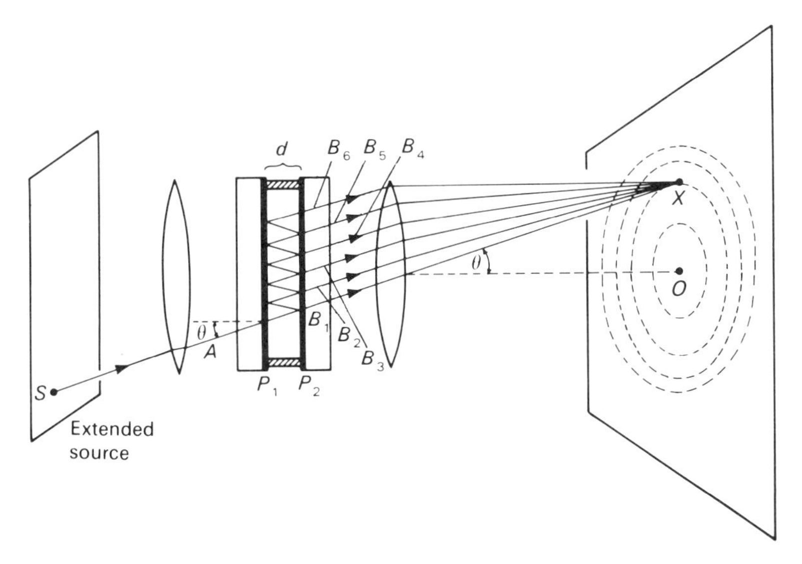

Figure 6.28. Fabry–Perot interferometer. Rays, A, B_1, B_2, $\cdots$ correspond to those shown in Figure 6.22.

0.1 to 20 cm. The surfaces P_1 and P_2 are covered with suitable reflecting film, and the space between them is generally filled with air. One pair of plates with a fixed spacer is sometimes called an *étalon*.

As shown in Figure 6.28, radiation from an extended source is partially reflected and partially transmitted by the two inner surfaces, such that each ray A from the source is decomposed into components $B_1, B_2, \cdots, B_n$ that have traversed the space between P_1 and P_2 a different number of times. The components are superposed by a lens, creating an interference pattern that appears as concentric rings. Again, the radii vary as the square root of the positive integers, and the maximum order appears at the center of the pattern.

Since all rays $B_2, \cdots, B_n$ go through an even number of reflections, when $m\lambda = n2d\cos\theta$, they will be in phase with each other and also with ray B_1. Therefore, maxima will be produced. When $(m + \frac{1}{2})\lambda = n2d\cos\theta$, rays B_1, B_3, $B_5, \cdots$ will be in phase with each other but exactly out of phase with rays B_2, B_4, $B_6, \cdots$. (See Section 6.8.3 for a more detailed discussion.) The Airy equation for transmitted intensity will apply;

$$I_{\mathrm{T}} = \frac{I_0 T^2}{(1-R)^2}\left[1 + \frac{4R}{(1-R)^2}\sin^2\frac{\delta}{2}\right]^{-1} \tag{6.29}$$

and when the reflectance R is large, very sharp fringes will be formed having a

FWHM given by

$$\mathrm{FWHM} \approx 2\frac{1-R}{R^{1/2}} = \frac{4}{F^{1/2}} \tag{6.33}$$

Another parameter frequently used to characterize the quality of the fringes obtained by a Fabry–Perot interferometer is finesse $\mathscr{F}$, the ratio of the separation of adjacent fringes and the FWHM. Since the separation of adjacent fringes corresponds to a change of 2π (Figure 6.23), the finesse is given by

$$\mathscr{F} = \frac{\pi F^{1/2}}{2} = \frac{\pi R^{1/2}}{1-R} \tag{6.35}$$

The formation of very sharp fringes permits the Fabry–Perot interferometer to achieve a very high resolving power, and prior to the development of the modern computer, this instrument was considered much superior to the Michelson interferometer for the study of hyperfine structure and linewidths in atomic spectra. If the spectrum was very complex, some auxiliary sorting of the wavelengths could be carried out using a low-power prism spectrograph, the method being illustrated in Figure 6.29. Light from source S_1 was collimated by lens L_1 before entering the Fabry–Perot

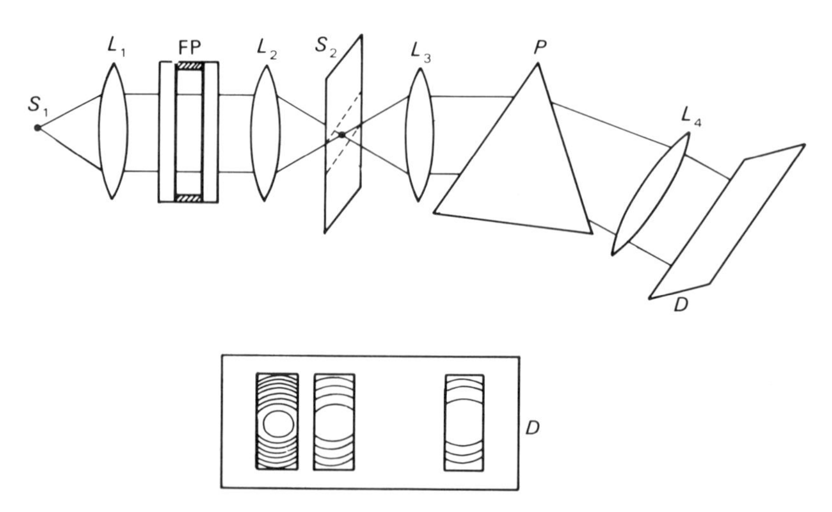

Figure 6.29. Fabry–Perot etalon–prism arrangement for separating ring systems produced by complex spectra. Symbols: S_1, source; S_2, slit; L, lenses; P, prism; D, photographic plate; FP, Fabry–Perot interferometer.

interferometer FP. The image of the fringes produced by the interferometer was then focused onto the slit S_2 of the spectrograph. The various wavelengths were separated by the prism and appeared on detector D (a photographic plate) as vertical sections of the full fringe system, the width of the section being determined by the width of slit S_2. If the spectrum had a very large number of closely spaced lines, the slit was made very narrow. Details of the procedure for converting such measurements into wavelengths and linewidths can be found in Tolansky (1955).

There are several potential problems associated with using the Fabry–Perot interferometer. First, unlike two-beam systems, the minima in multiple-beam interferometers never go completely to zero. Reduced fringe contrast can present problems when very weak spectral lines must be observed.

A second potential problem is limited spectral range between orders. Decreasing the plate separation in the *étalon* increases the free spectral range between successive orders but also adversely affects spectral resolution. To help overcome this problem, two or more *étalons* can be used in series, the plate separations being simple integral multiples of each other.

Figure 6.30*c* shows the fringe pattern passed by the combination of a thick *étalon* and thin *étalon* having exactly one third the plate separation of the first. The fringes passed by the thick *étalon* alone (*a*) are a third of the width of those

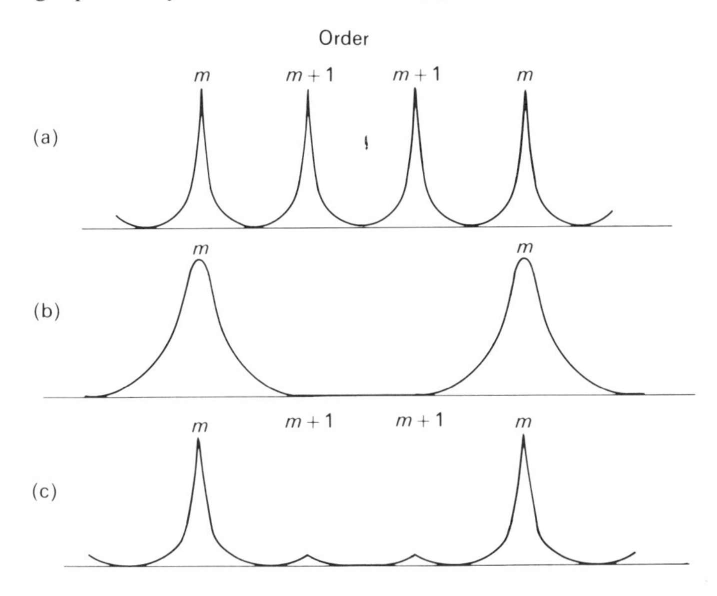

Figure 6.30. Interference fringes produced by single etalon (*a*) and second etalon (*b*) having one-third thickness of first. (*c*) Fringes formed as a result of compounding two etalons (*a*) and (*b*). [Reprinted with permission from S. Tolansky (1955), *An Introduction to Interferometry*, Longmans, England.]

passed by the thinner *étalon* (*b*), but they are also separated by only a third of the distance. The combination of the two *étalons* produces (*c*), a set of fringes having the same resolution as (*a*) but the free spectral range of (*b*). The intervening order in (*a*) is not completely eliminated because the minima do not go completely to zero. More information dealing with "compound Fabry–Perot interferometers" may be found in Born and Wolf (1986).

It is possible to make a scanning Fabry–Perot interferometer by varying d, that is, moving one of the reflecting plates relative to the other. However, in a variable-gap system it is very difficult to maintain the strictly parallel alignment of P_1 and P_2 required for good resolution. (Deviations of $\frac{1}{40}\lambda$ accumulate to half a wavelength after only 10 reflections! See Bradley, 1962.) Another scanning procedure involves changing the pressure inside the *étalon* (Jacquinot and Dufour, 1948), thereby altering n, the index of refraction of the medium between the two reflecting surfaces. Pressure scanning allows a wavelength interval of up to approximately 30 Å to be monitored over periods as short as 0.3 μsec (Greig and Cooper, 1968).

6.8.7. Interference Filters

If a Fabry–Perot *étalon* is placed in a parallel beam of white light, interference still occurs, but the fringes formed by the transmitted light will run together and cannot be observed unless dispersed by a prism or grating. The parallel, equally spaced pattern that results is sometimes called channeled spectra or Edsen–Butler fringes.

The maxima in these fringes are still determined by the equation $m\lambda = 2nd\cos\theta$, where m is any whole number and $\cos\theta = 1$. If d is large, there will

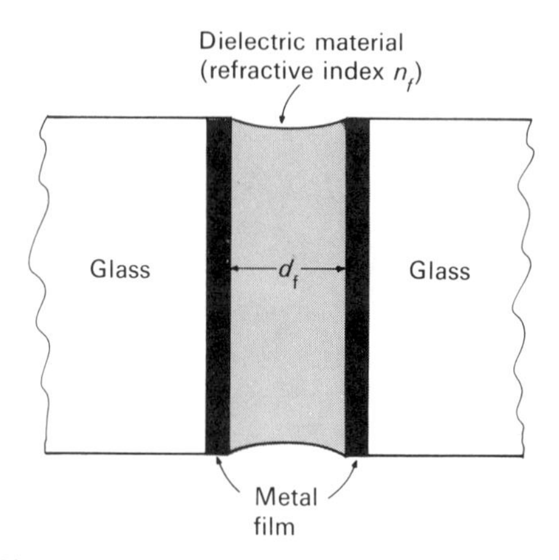

Figure 6.31. Schematic diagram of interference filter.

be many wavelengths represented in the fringe pattern, and high dispersion is needed to resolve them. However, if d is made very small, only a few wavelengths will be represented, and the *étalon* acts like a filter, passing nearly monochromatic light.

The intensity of the transmitted radiation is determined by Airy's equation (Eq. 6.29), and the ratio between the maximum and minimum values of the transmittance, called the filter rejection ratio $\mathscr{R}$, is determined from Eq. 6.30 and 6.31

$$\mathscr{R} = \frac{I_{\max}}{I_{\min}} = \frac{(1+R)^2}{(1-R)^2} \tag{6.36}$$

The value of the peak transmittance is called the optical efficiency of the filter.

To prepare an interference filter, shown schematically in Figure 6.31, a thin layer of some dielectric material is deposited by evaporation between two layers of partially reflecting metallic film supported on glass plates. The path difference in the filter is now determined by the thickness of the dielectric, d_f, and its refractive index n_f, with the wavelengths of maximum transmission for normal incidence given by the equation $\lambda = 2n_f d_f/m$. Filters with higher efficiency, coupled with higher rejection ratios, can be obtained by substituting multiple dielectric layers for the metallic reflection layers. Further specifics can be found in Levi (1980).

REFERENCES

Born, M. and Wolf, E. (1986), *Principles of Optics*, 6th (corrected) ed., Pergamon, New York.

Bradley, D. J. (1962), *J. Sci. Instrum.*, *39*, 41.

Chandler, C. (1951), *Modern Interferometers*, Hilger and Watts, Glasgow.

Cook, A. H. (1971), *Interference of Electromagnetic Waves*, Clarendon, Oxford.

Fabry, C. and Perot, A. (1901), *Ann. Chim. Phys.*, *22*, 564.

Foskett, C. T. (1978), "The Fourier Transform and Related Concepts: A First Look," in *Transform Techniques in Chemistry*, Griffiths, P. R. (Ed.), Plenum, New York, Chapter 2, pp. 11–37.

Goldman, S. (1948), *Frequency Analysis, Modulation, and Noise*, McGraw-Hill, New York.

Greig, J. R. and Cooper, J. (1968), *Appl. Opt.*, *7*, 2166.

Hariharan, P. (1985), *Optical Interferometry*, Academic, Sydney.

Harwit, M. and Sloane, N. J. A. (1979), *Hadamard Transform Optics*, Academic, New York.

Jacquinot, P. and Dufour, D. (1948), *J. Rech. Centre Nat. Rech. Sci. Lab. Bellevue (Paris)*, *6*, 1.

Jenkins, F. A. and White, H. E. (1976), *Fundamentals of Optics*, 4th ed., McGraw-Hill, New York.

Levi, L. (1968), *Applied Optics: A Guide to Optical System Design*, Vol. 1, Wiley, New York.

Levi, L. (1980), *Applied Optics: A Guide to Optical System Design*, Vol. 2, Wiley, New York.

Lloyd, H. (1837), *Trans. Roy. Irish Acad. 17*, 171. (Read, January 27, 1834.)

Mach, E. (1921), *Die Prinzipien Physikalischen Optik*, Barth, Leipzig.

Marshall, A. G. and Comisarow, M. B. (1978), "Multichannel Methods in Spectroscopy", in *Transform Techniques in Chemistry*, Griffiths, P. R. (Ed.), Plenum, New York, Chapter 3, pp. 39–68.

Meslin, G. (1893), *J. Phys.*, *2*, 205.

Michelson, A. A. (1891), *Phil. Mag.*, *31* (Series 5), 256.

Michelson, A. A. (1903), *Light Waves and Their Uses*, University of Chicago Press, Chicago.

Michelson, A. A. (1927), *Studies in Optics*, University of Chicago Press, Chicago.

Preston, T. (1912), *The Theory of Light*, 4th ed., Macmillan, London.

Rayleigh, J. W. S. (1896), *Proc. Roy. Soc.*, *59*, 201.

Shamos, M. H. (1959), *Great Experiments in Physics: Firsthand Accounts from Galileo to Einstein*, Dover, New York.

Steel, W. H. (1983), *Interferometry*, 2nd ed., Cambridge University Press, Cambridge.

Stewart, J. E. (1970), *Infrared Spectroscopy: Experimental Methods and Techniques*, Marcel Dekker, New York.

Tolansky, S. (1948), *Multiple-Beam Interferometry of Surfaces and Films*, Clarendon, Oxford.

Tolansky, S. (1955), *An Introduction to Interferometry*, Longmans Green, New York.

Whittaker, E. T. and Watson, G. N. (1935), *Modern Analysis*, Cambridge University Press, Cambridge, Chapter 1.

Wood, R. W. (1934), *Physical Optics*, Macmillan, New York; 1967, Dover, New York.

Young, T. (1803), "Experiments and Calculations Relative to Physical Optics," Bakerian Lecture, Royal Society of London, November 24.

Young, T. (1804), *Phil. Trans.*, 1–16.

CHAPTER

7

INTRODUCTION TO FOURIER TRANSFORM SPECTROSCOPY

Interference effects have been utilized in virtually every branch of science and have played a key role in many experiments of great fundamental significance. In addition to confirming the wave nature of light (Young, 1803, 1804; Shamos, 1959), experiments involving interference have resulted in the establishment of length standards (Michelson and Benoit, 1895) and the determination of the velocity of light with a high degree of accuracy (Michelson, 1927; Swenson, 1987). The demonstration by Michelson and Morley (1887) that the motion of the earth is not measurable (Swenson, 1972) marked the beginning of the theory of relativity, and confirmation of the wave nature of small particles (Davisson, 1937; Thomson, 1938) and determination of the fine structure in atomic spectra (Michelson, 1927) were of considerable importance in the development of modern quantum mechanics.

For the spectroscopist, interference effects provide a powerful method for the simultaneous measurement of spectral distributions with large resolving power, high wavenumber accuracy, large spectral range, fast scanning time, and much enhanced sensitivity over the more traditional techniques. Although a number of interferometers have been developed (Born and Wolf, 1986), the most successful to date for spectral analysis employ the basic Michelson design to produce the interference pattern and depend on Fourier transform (FT) methods to recover the original spectrum.

The purpose of this chapter is to provide a general introduction to FT spectroscopy that will enable the reader to appreciate both the instrumental design and mathematics associated with the technique. Applications of FT spectroscopy to spectrochemical analysis in the UV–visible will be discussed in Chapter 14.

7.1. THE MICHELSON INTERFEROMETER

The majority of interferometers in use today are based on a device originally designed by Michelson (1891a). Many features have been added to this basic design over the years to improve instrumental performance (Baker, 1977;

Griffiths, 1978; Sakai, 1977; Breckinridge and Schindler, 1981; Griffiths and deHaseth, 1986), but the fundamental means by which interference is produced has remained essentially unchanged.

7.1.1. Interferometer Design

The Michelson interferometer produces interference by division of amplitude (Section 6.8) and is shown schematically in Figure 7.1. Light beams from an extended source S are made parallel by lens D and strike mirror B (the beamsplitter) at 45°. Part of the light (ray A_1, ideally 50% of the incident radiation) is reflected at the half-silvered, back surface of B and, as shown by the dark line, directed toward mirror M_1. The other part of the light (ray A_2, shown by the light line) passes through B and is directed toward mirror M_2, which has been carefully adjusted to be perpendicular to M_1. Rays A_1 and A_2 are totally reflected from their respective mirrors (reflected rays shown slightly displaced for viewing ease) and return to B, where 50% of A_2 is reflected and 50% of A_1 is transmitted. Both rays are recombined at the focal point F of lens E, where they produce an interference pattern. (Note that half of the intensity originally emitted by S is directed out of the interferometer toward D and lost.)

The mirror M_1 is movable and can be translated back and forth along a path parallel to that of the approaching ray A_1. The presence of plate C (the

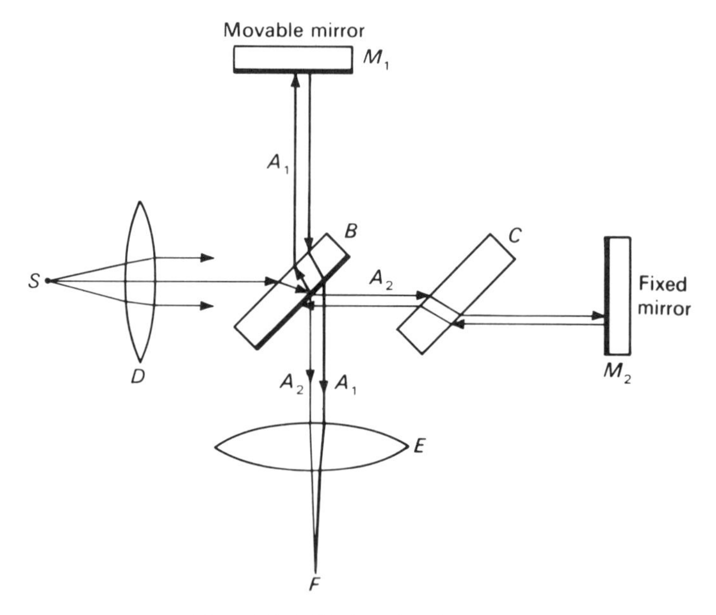

Figure 7.1. Michelson interferometer.

compensating plate) makes the path in glass of the two rays equal. (A compensator plate is not strictly necessary if S is monochromatic but becomes indispensable if S is polychromatic.) In this design, both A_1 and A_2 undergo two 180° phase changes due to reflection at air–silver and/or glass–silver interfaces, so when the optical paths of the two rays are equal, constructive interference occurs.

7.1.2. The Interferogram

The appearance of the interference fringes may be understood by examining Figure 7.2. When the mirrors M_1 and M_2 are perfectly adjusted, the real mirror M_2 is replaced by its virtual image M'_2, formed by reflection in the beamsplitter B. Mirror M'_2 is now parallel to M_1. The source S appears positioned at S', behind the observer, forming two virtual images S_1 and S_2 in M_1 and M'_2, respectively. If the mirror separation is d, the separation of the two virtual images will be $2d$. This is the optical path difference of the two rays A_1 and A_2.

When $2d$ equals an integral number of whole wavelengths, all of the light rays reflected normal to the mirrors will be in phase. Rays reflected at some angle θ, however, will have an extra path difference equal to $2d \cos\theta$ between points P' and P''. For those angles θ satisfying the relationship $2d\cos\theta = m\lambda$, maxima will be produced, while for conditions where $2d\cos\theta = (m + \frac{1}{2})\lambda$, minima will occur. These are fringes of equal inclination (Section 6.8.5) and will appear as concentric circles, with the radii of the rings decreasing toward

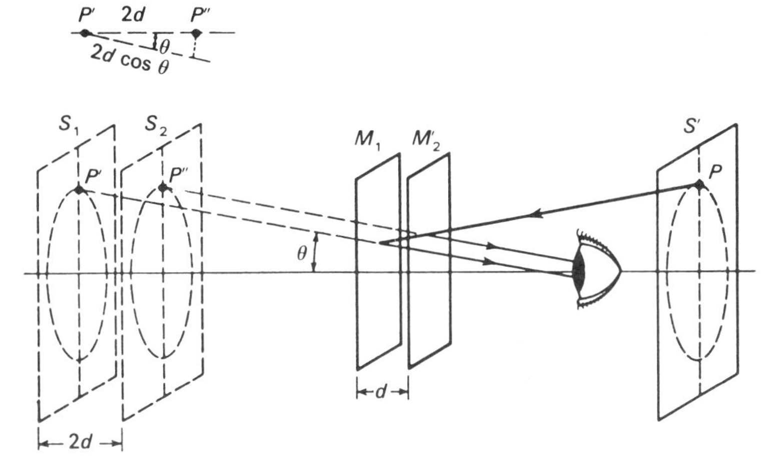

Figure 7.2. Formation of circular fringes of equal inclination in Michelson interferometer. [Reprinted with permission from F. A. Jenkins and H. E. White (1976), "Fundamentals of Optics," 4th ed., McGraw-Hill, New York.]

the edges of the pattern. For monochromatic light, the intensity distribution in the rings is

$$I = 4a_1^2 \cos^2(\tfrac{1}{2}\delta) \qquad (6.13)$$

and the phase difference δ is given by $(2\pi/\lambda)2d\cos\theta$.

Figure 7.3 shows how the circular fringes look under different conditions. If M_1 and M_2' are rather far apart, the pattern will consist of many rings that are very closely spaced. Since the center spot corresponds to the fringe of highest order, as M_1 moves toward M_2', the rings shrink and vanish into the center. At the center of the pattern, where $\cos\theta = 1$, the equation for constructive interference reduces to $2d = m\lambda$, indicating that a ring disappears each time $2d$ decreases by one wavelength. As the distance $2d$ continues to decrease, the pattern consists of fewer and fewer rings. When the two mirrors are finally in coincidence, $2d = 0$, and the central fringe spreads out to cover the entire field of view.

If S is strictly monochromatic and rays A_1 and A_2 are of equal amplitude, the intensities of the fringes will vary outward from the center of the interference pattern, as shown in Figure 7.4*a*. The ring of maximum order always appears in the center, and rings of successively lower orders appear closer together, their radii varying as the square roots of the positive integers. Because only one wavelength is present, the intensities at the maxima and minima do not change across the entire interference pattern.

In the very early days of interferometry, it was customary to view the entire set of interference fringes, the human eye often serving as the detector. With modern instruments, it is more typical to observe the intensity variation in one particular region of the interference pattern, generally at the center where

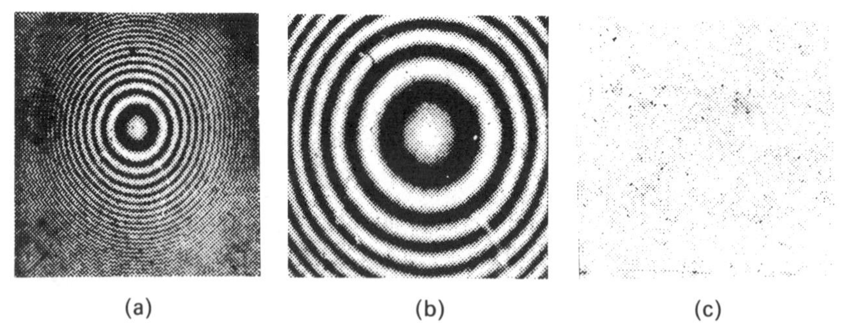

Figure 7.3. Appearance of circular fringes observed in Michelson interferometer as function of mirror separation d (Figure 7.2): (*a*) large d; (*b*) M_1 and M_2' nearly coincident; (*c*) M_1 and M_2' exactly coincident.

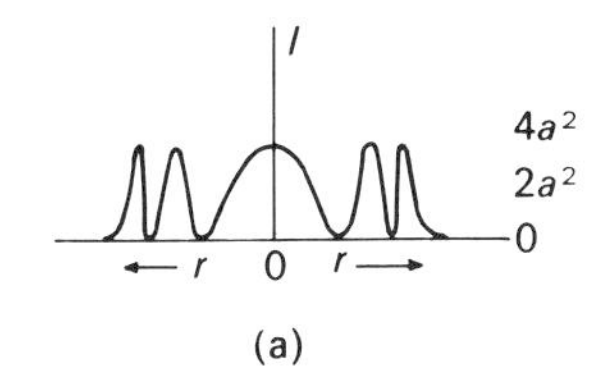

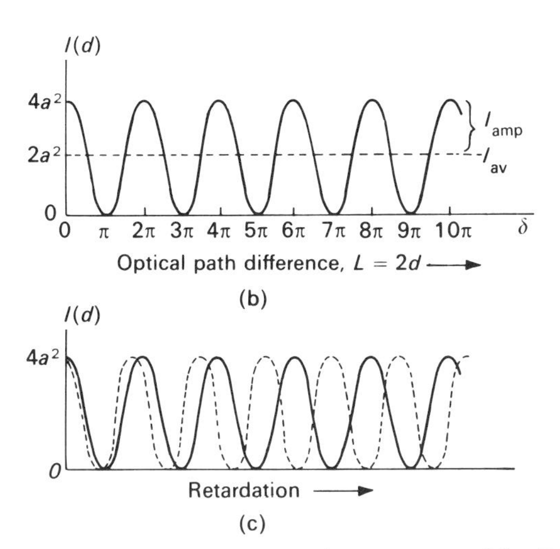

Figure 7.4. Variation of fringe intensities across interference pattern. (After Baker, 1977.) (*b*) Variation of intensity at center of interference pattern as function of retardation for monochromatic light. (*c*) Variation of intensity at center of interference pattern as function of retardation for two different wavelengths.

fringes of higher order first appear and where the ring has maximum thickness (Figure 7.4*a*). Figure 7.4*b* shows a plot of the intensity at the center of the pattern ($r = 0$) as a function of the optical path difference $L = 2d$ for the two rays A_1 and A_2. Since L determines the phase difference δ between rays A_1 and A_2, this figure is also a plot of intensity versus δ for the recombined rays.

The x axis in Figure 7.4*b* is also called retardation, and the entire plot is referred to as an interferogram. For a strictly monochromatic source of frequency f, the interferogram is a pure cosine wave,

$$I(L) = I_{\text{av}} + I_{\text{amp}} \cos(2\pi f L) \tag{7.1}$$

with an average intensity

$$I_{\text{av}} = \tfrac{1}{2}(I_{\max} + I_{\min}) \tag{7.2}$$

and an amplitude

$$I_{amp} = \tfrac{1}{2}(I_{max} - I_{min}) \tag{7.3}$$

where I_{max} and I_{min} are the maximum and minimum intensities observed during movement of the mirror. Since the amplitude of the intensity variation I_{amp} is given by $2a^2$, where a is the amplitude of the original light wave emitted by S, and the phase difference δ is determined by $(2\pi/\lambda)2d$, where λ is the wavelength of the light emitted by S, this interferogram contains all of the information needed to characterize the source radiation. (Characterization consists of determining both the frequency and intensity of the radiation.)

Polychromatic sources yield more complex interferograms. If S also emits radiation λ', a wavelength of very similar frequency and identical amplitude to λ, the motion of the mirror will generate a cosine function for each wavelength. The two cosine functions will have maxima and minima at different retardation values (Figure 7.4*c*), and the interferogram will be the sum of both functions. The result of adding two cosine waves of the same amplitude and similar frequencies has already been shown in Figure 6.5. In order to reconstruct the spectrum emitted by the source, this resultant interferogram could be analyzed mathematically according to Eq. 6.14. (Optical path difference L and time t are interrelated through mirror velocity.)

In the general sense, where S radiates many frequencies with a wide variety of amplitudes, the interferogram becomes so complex that computer analysis is required to characterize the spectral distribution. The sum of a large number of cosine waves may be expressed as a Fourier series, and recovery of the spectral distribution (wavelengths with their associated intensities) is accomplished by a mathematical procedure known as Fourier transformation.

7.1.3. Resolution

Before beginning a discussion of Fourier transformation, we consider some of the ways in which spectrometers based on the Michelson interferometer differ from dispersive spectrometers. In dispersive spectrometers, resolution depends on the slit width and the ruling of the grating. The Michelson interferometer has no slits or gratings, and to a first approximation resolution is determined by the maximum optical path difference achieved by the interferometer during a scan. This relationship is easy to understand on a purely intuitive basis.

Suppose a spectrum consists of a doublet at λ and λ', each component of this doublet having equal intensity. The motion of the mirror will produce two cosine waves, one for each component of the doublet, and the interferogram will be the superposition of these two cosine waves. (Consult Figure 6.5.) Our criterion for resolution is that we must see at least one full beat frequency in order to completely resolve the doublet. Therefore, the optical path difference L must be at least

$$L = \frac{\lambda\lambda'}{\lambda - \lambda'} = (\Delta\bar{\nu})^{-1}$$

for λ and λ' to be resolved. Here, $\Delta\bar{\nu} = \bar{\nu} - \bar{\nu}'$ is the difference in wavenumbers between the two spectral lines. The narrower the separation of the doublet, the greater the distance that must be traveled by the mirror in order to go through one complete period of beat frequency. Thus, intuitively, if the maximum optical path difference in the interferometer is L_{max}, the best resolution we could obtain using this interferometer is

$$\Delta\bar{\nu} = (L_{\mathrm{max}})^{-1} \tag{7.4}$$

The relationship given in Eq. 7.4 places a very fundamental limit on the resolution that can be obtained from the Michelson interferometer. In practice, resolution is also determined by the mathematical function used to truncate the interferogram at the end of the scan (the apodization function), the sampling interval used to digitize the interferogram, and the diameter of the aperture in the interferometer. Instability in the mirror drive system as well as a number of other instrument-specific factors will also play a role. Many excellent references may be consulted for additional information on this topic (Hirschfeld, 1979; Griffiths and deHaseth, 1986). The effect of apodization, sampling, and finite aperture will be discussed briefly in Section 7.6.

7.1.4. Throughput

The maximum throughput Θ_{I} of the interferometer is given by the product of the maximum solid angle of the beam of radiation passing through the interferometer (Eq. 7.56) and the area A_{M} of the mirrors being illuminated,

$$\begin{aligned} \Theta_{\mathrm{I}} &= 2\pi A_{\mathrm{M}}(\Delta\bar{\nu}/\bar{\nu}_{\mathrm{max}}) \qquad \mathrm{cm}^2\ \mathrm{sr} \\ &= 2\pi A_{\mathrm{M}}/R_{\mathrm{I}} \end{aligned} \tag{7.5}$$

where $\Delta\bar{\nu}$ is the resolution, $\bar{\nu}_{\mathrm{max}}$ is the maximum wavenumber in the spectrum, and R_{I} is the resolving power of the interferometer. For a grating instrument, throughput is given by

$$\Theta_G = Wl/F^2 \approx (l/F)(A_G/R_G) \tag{7.6}$$

where W is the slit width, l is the slit length, F is the focal length of the collimator, A_G is the slit area, and R_G is the resolving power. Assuming the same areas, the same focal lengths for the collimators, and the same resolving powers for the two instruments, the ratio of the interferometer and grating throughputs will be

$$\frac{\Theta_I}{\Theta_G} = 2\pi(F/l) \tag{7.7}$$

Since the ratio (F/l) is generally much greater than 1 (typically on the order of 30 for an infrared instrument), considerably more power can be put through the interferometer than through the grating spectrometer. (See Bell, 1972, for a more detailed discussion.)

7.1.5. Sensitivity

The theoretical increase in sensitivity for the interferometer is found by multiplying the values of the throughput advantage (Eq. 7.7) and the multiplex advantage. The magnitude of the multiplex advantage depends on the total number of frequencies being monitored (M in Table 5.5) and the type of noise dominating the measurement. (Derivation of the entries in Table 5.5 has already been discussed.) The entries under "Multiplexing, S-Matrix" apply to both the Hadamard spectrometer and Fourier transform spectrometers that employ the Michelson interferometer. The factor of $\frac{1}{2}$ in Hadamard transform instruments arises because each resolution element is detected for only half of the total time needed to complete the measurement. Using the Michelson interferometer, all of the resolution elements are monitored all of the time, but half of the spectral intensity is lost at the half-silvered mirror (Marshall and Comisarow, 1978).

In practice, the actual sensitivity of the FT spectrometer may be limited by many factors including the dynamic range of the detector and the efficiency of the beamsplitter. More details may be found in the references. (See, e.g., Griffiths and deHaseth, 1986.)

7.2. FOURIER TRANSFORM SPECTROSCOPY: A HISTORICAL PERSPECTIVE

Fourier transform spectroscopy has progressed gradually over a history of a hundred or more years as advances in both science and technology have made the technique, once experimentally tedious and theoretically difficult to interpret, into something so useful that FT spectrometers are now found in virtually every modern laboratory. The following account of the development of FT methods in spectral analysis is far from complete but does serve to show how the ultimate success of any new spectroscopic method can generally be attributed to a combination of three essential factors: (1) a need by the spectroscopist to make measurements that are outside the capabilities of available instrumentation, (2) the ability of the new method to make such measurements possible, and (3) the existence of the technology required to make the new method experimentally feasible. Many excellent histories of the development of FT spectroscopy exist elsewhere (Loewenstein, 1966; Mertz, 1971; Connes, 1969; Bell, 1972; Chamberlain, 1979) and may be consulted for additional information.

Although interference effects had been known since Young's experiment in 1803 (Shamos, 1959), the first application to spectral analysis did not occur until 1862 (Fizeau). Using an experimental arrangement similar to that shown in Figure 6.26*a*, Fizeau studied Newton's rings produced by the yellow light of a sodium flame and observed that the intensity or distinctness of the interference fringes was not uniform. When the lens and plate were in contact, the interference fringes became almost invisible near the position of the 490th ring, but regained their original distinctness near the 980th ring. By moving the lens and plate apart, Fizeau was able to follow these periodic variations of distinctness through 52 cycles, each cycle containing approximately 980 rings! From these observations, Fizeau correctly concluded that the yellow sodium light has two wavelength components that are different by a factor of about 1 part in 980. This conclusion was later confirmed by direct observation using a prism spectroscope.

Beginning in 1881, Michelson made more elaborate attempts at spectral analysis (1881, 1891a, b, 1892) using the interferometer pictured in Figure 7.5 (Baly, 1927; Hardy and Perrin, 1932) and described in Section 7.1. Recognizing that the intensity distribution of the circular interference fringes produced by his interferometer depended on the spectral distribution of the radiation transmitted by the instrument, Michelson attempted to analyze the visible spectra emitted by excited atoms, a subject of considerable interest during the last decade of the nineteenth century.

In general, if the entire emission spectrum was allowed to pass through the interferometer, fringes could not be observed, except at extremely small mirror separations (parameter d in Figures 7.2 and 7.4). This occurred because the spectrum usually contained a large number of wavelengths, each of which produced a unique set of interference fringes that coincided only at zero order ($d = 0$). As d increased and higher orders became visible, fringes from different wavelengths began to separate out, blurring the overall pattern and producing an apparently uniform field of illumination. In order to observe fringes

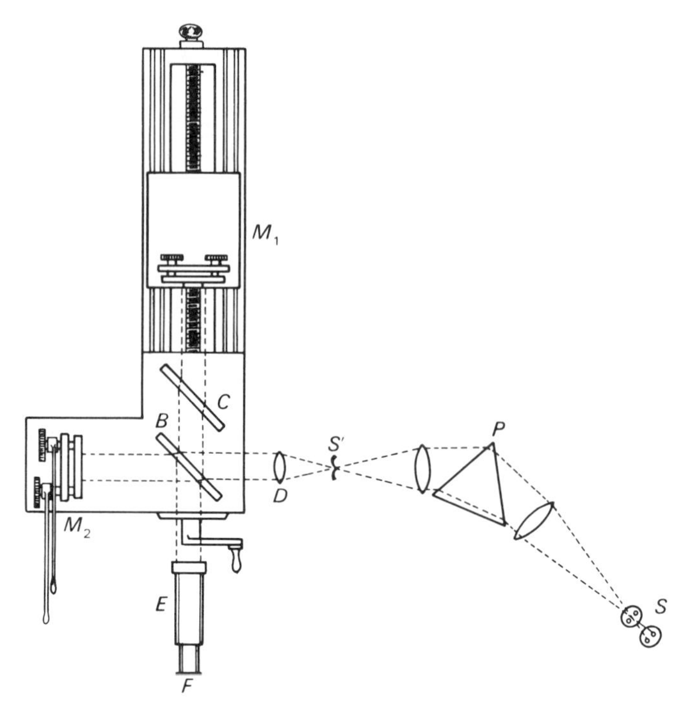

Figure 7.5. Early Michelson interferometer. (After Baly, 1927.)

over mirror separations of several centimeters, Michelson simplified his interference patterns by placing a prism predisperser in front of the interferometer and selecting out quasi-monochromatic portions of atomic spectra for study.

Since no detector was available, other than the human eye, Michelson was unable to determine the entire interference function produced by his instrument. Rather than counting fringes as Fizeau had done, Michelson compared the intensities of adjacent bright and dark fringes in the pattern, $I_{\max}$ and $I_{\min}$, respectively, and reported the "fringe visibility",

$$V(d) = \frac{I_{\max}(d) - I_{\min}(d)}{I_{\max}(d) + I_{\min}(d)} \tag{7.8}$$

a quantity that represents the distinctness with which the fringes appear to the eye. This variation of fringe visibility with mirror separation is presented in the form of a visibility plot (Figure 7.6), with the zero of position corresponding to the spot at which the two mirrors coincide.

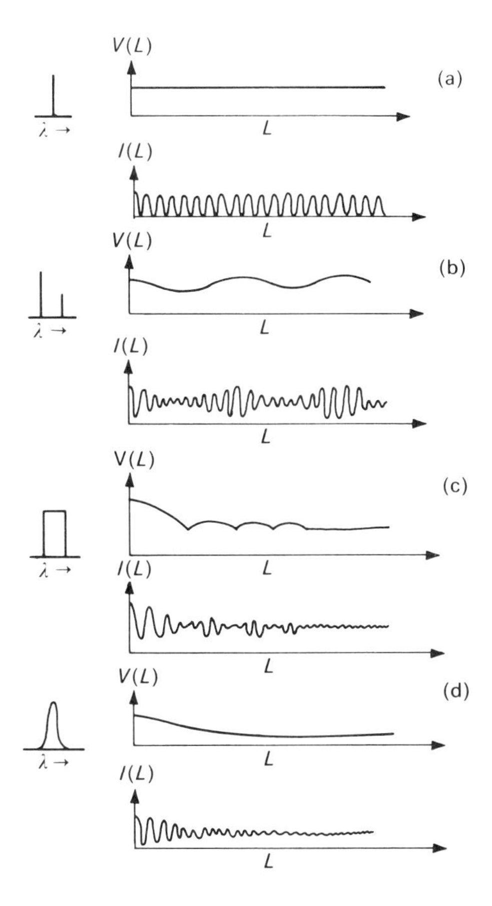

Figure 7.6. Visibility plots $V(L)$ and corresponding interference functions $I(L)$ for four simple spectral distributions. (After James and Sternberg, 1969.)

Determination of the visibility plot was a tedious, time-consuming process, requiring visual estimation by the experimenter using a telescope. (See Baly, 1927, for details.) During an actual experiment, the fringe visibilities would be determined at the center of the interference pattern for every 1-mm change in path length of the interferometer. The intensity data was then corrected for personal error using correction factors determined by visual inspection of a set of fringes having known visibility.

Without computer facilities, interpretation of the visibility plots presented yet another problem. Michelson began by calculating the visibility curves that would result from various types of single, double, and multiple lines. Figure 7.6 shows examples of such plots obtained for four simple examples: a perfectly monochromatic spectrum (Figure 7.6*a*), a spectrum consisting of two infinitely narrow lines of unequal intensity (Figure 7.6*b*), a single line spectrum consisting of a narrow bandwidth of frequencies all of the same amplitude [also called a boxcar function (Figure 7.6*c*)], and a single line spectrum consisting of a narrow bandwidth of frequencies whose amplitudes follow a Gaussian distribution (Figure 7.6*d*). Comparison with the full interference functions shows that the visibility curves, as determined by Michelson, represent the upper envelopes of the interferogram. Thus, Michelson's visibility procedure gave only beat or difference frequencies, while the interferogram gives the entire variation or sum frequency for the combining waves.

To assist in analyzing his results, Michelson also invented a mechanical harmonic synthesizer in which up to 80 harmonic motions could be superimposed and used to drive a pen that plotted a visibility profile (Michelson, 1903). In this manner, the spectral distribution of light transmitted by the interferometer could be deduced by trial and error rather than by direct calculation using Fourier transformation.

From his visibility plots, Michelson was able to determine the spectral profiles of a number of well-known lines in atomic spectra that were not resolvable by existing grating or prism instruments. True singlets showed only a gradual decrease in visibility with increasing mirror separation, the falloff in visibility occurring more rapidly the broader the line. (Compare Figures 7.6*a* and 7.6*d*.) Multiplets gave modulated visibility curves (Figure 7.6*b*), the decrease and increase in visibility occurring less rapidly the more closely spaced the lines. From such studies, Michelson concluded that the red line of cadmium (644 nm) was a true singlet (Figure 7.7*a*) but the red Balmer line of hydrogen (656 nm, transition from principal quantum number $n = 3$ to $n = 2$) was actually a doublet (Figure 7.7*b*). For such observations, as well as for other applications involving his interferometer, Michelson was awarded the Nobel Prize in physics in 1907.

Visibility curves provide only partial characterization of the spectrum, and although the spectral distribution in Figure 7.7*b* shows the weaker line of the hydrogen doublet at longer wavelength, beat frequencies alone are insufficient to determine on which side of a principal line the weaker component lies (Lord Rayleigh, 1892). (Beat frequencies give only the wavelength *difference* between the two lines.) Moreover, for spectra containing a large number of lines, there will be many different intensity distributions, all of which will give the same visibility plot. Thus, except for a few simple cases, Michelson was not able to recover a unique spectral distribution.

Because of the experimental difficulties associated with Michelson's early attempts at spectral analysis, the visibility method was little used outside of his own laboratory. Moreover, with the invention of the Fabry–Perot

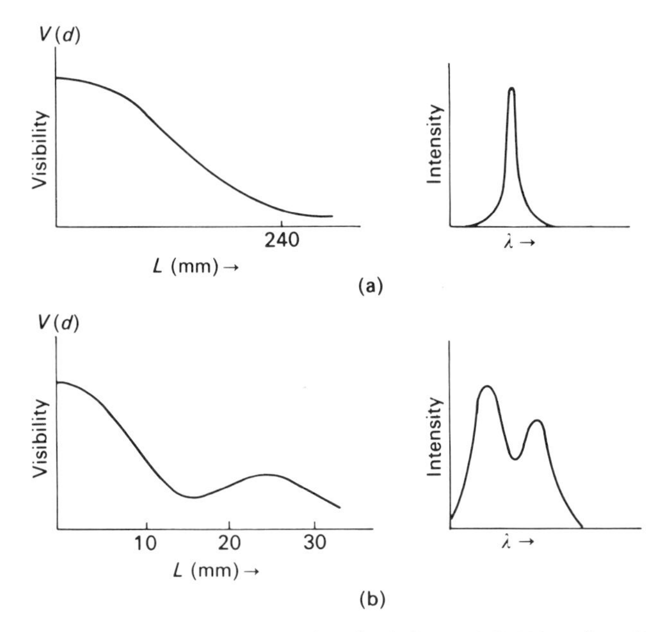

Figure 7.7. Visibility plots for (*a*) 644-nm line of cadmium and (*b*) 656-nm line of hydrogen. (After Michelson, 1927.)

interferometer in 1901, the Michelson dual-beam interferometer was superseded by improved multiple-beam methods that produced much sharper fringes (Section 6.8.6) and were capable of resolving closely spaced multiplets directly, without the use of visibility plots.

The difference in appearance of the fringes resulting from the two interferometers is illustrated in Figure 7.8, where the circular interference patterns produced by a single spectral line are compared. If a second line was present, it would only reduce the visibility in Figure 7.8*a* but show as a separate set of rings in Figure 7.8*b*. Taking advantage of the greater resolving power of their instrument, Fabry and Perot studied the fine structure of a number of spectral lines, obtaining results the Michelson instrument was incapable of yielding.

The first extension of interference techniques to the infrared occurred during the early part of the twentieth century (Rubens and Wood, 1911; Wood, 1934). These experiments are particularly significant because this region of the electromagnetic spectrum required a detector other than the human eye. Use of a thermal sensing device, a detector that measured total incident flux, allowed the entire interferogram to be measured for the first time.

Rubens and Wood did not use a Michelson interferometer but employed two thin, crystalline quartz plates that were mounted parallel to each other

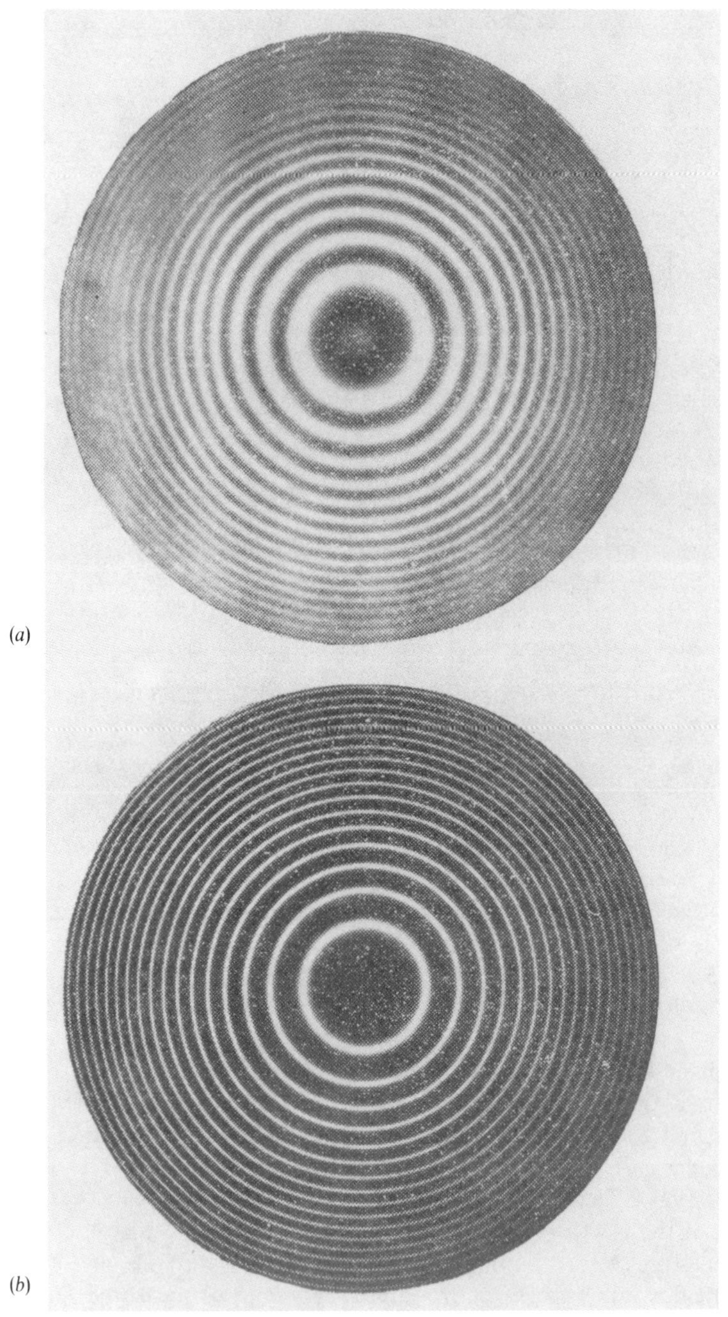

Figure 7.8. Comparison of fringes obtained using (*a*) Michelson interferometer and (*b*) Fabry–Perot interferometer with surfaces having reflectance of 0.80. [Reprinted with permission from F. A. Jenkins and H. E. White (1976), "Fundamentals of Optics," 4th ed., McGraw-Hill, New York.]

and could be moved apart to create variable path differences between the interfering beams. The interferogram was obtained by selecting a path difference, focusing the center portion of the interference pattern onto the detector, taking a reading, moving to a new path difference, and repeating the process.

The curve obtained by the Rubens and Wood procedure is shown in Figure 7.9 and represents the first true interferogram (as opposed to visibility plots obtained by Michelson). Although the maxima and minima are quite pronounced, the interferogram is rapidly damped out due to the large number of wavelengths transmitted by the instrument. Again, the spectral distribution was not determined by direct Fourier transformation, but from the distance between adjacent maxima the average wavelength of the radiation was estimated to be about 107 μm.

Although Michelson and Rubens had shown that the interferometer could be used for the analysis of spectral distributions, little additional work was carried out in this area for almost 40 years. Fellgett (1967) noted that much of the reason for this lack of interest could be attributed to a prevailing misconception of the time:

> It was regarded as self-evident that, although the method might work for discrete emission lines, all the information would be so jumbled up and smeared out that it would be irretrievably lost in the interferogram of a continuous source! (p. 167)

This prejudice was so strong that Fellgett (1967) also stated: "My practical

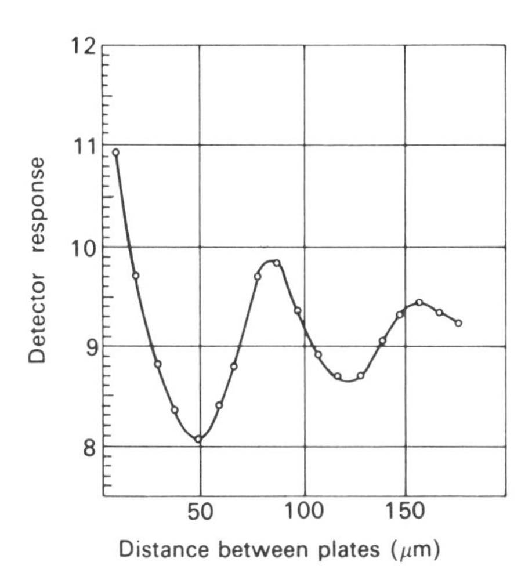

Figure 7.9. The first interferogram. [Reprinted from H. Rubens and R. W. Wood (1911), *Phil. Mag.*, *21*, 249. By permission of Taylor and Francis, London.]

demonstration [that such a source could be studied by interference methods (1951)] caused practically no modification of this opinion" (p. 167).

During the first half of the 1950s, interest in interference spectroscopy was revived by the recognition that the method offered two major theoretical advantages over dispersive methods of spectral analysis. The first advantage, pointed out by Fellgett (1951), arises as a direct consequence of the way in which the spectral information is acquired. Monitoring the sinusoidal fluctuations of intensity at the center of the interference pattern allows the *entire* spectral distribution to be measured throughout the duration of the experiment. Dispersive spectrometers, on the other hand, can only observe each spectral component sequentially over a short period, the sum of these periods giving the total time of the experiment. Depending on the predominant type of noise present in the experiment (Table 5.5), interference methods could experience a gain in signal-to-noise ratio up to as much as $M^{1/2}$, where M is the number of spectral lines being measured. This is the multiplex or Fellgett advantage discussed in Chapter 5.

Using a laboratory interferometer employing an air wedge formed by two sheets of glass (Figure 7.10), Fellgett (1951, 1958a) demonstrated that the spectrum of a continuous source could be obtained by interference methods. The emission spectrum from a mercury lamp L was reflected onto the wedge W by mirror M_1. The wedge was fixed, and the path difference was introduced by rotating the turntable T that carried the two mirrors M_1 and M_2. For the first time, the spectrum (Figure 7.11*b*) was obtained by direct Fourier transformation of the interferogram (Figure 7.11*a*) using numerical methods. The results were in good agreement with the spectrum obtained using a grating instrument (Figure 7.11*c*).

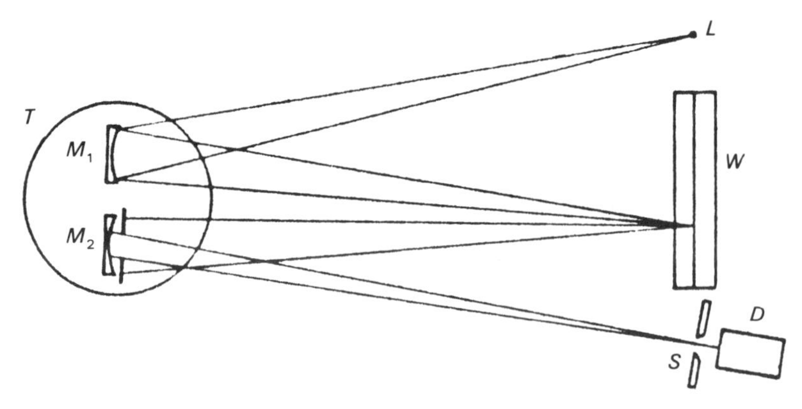

Figure 7.10. Interferometer used by Fellgett to determine spectral distribution of continuous source. [Reprinted from P. Fellgett (1958), *J. Phys. Radium*, *19*, 237, with permission from the Société Française de Physique, Paris, France.]

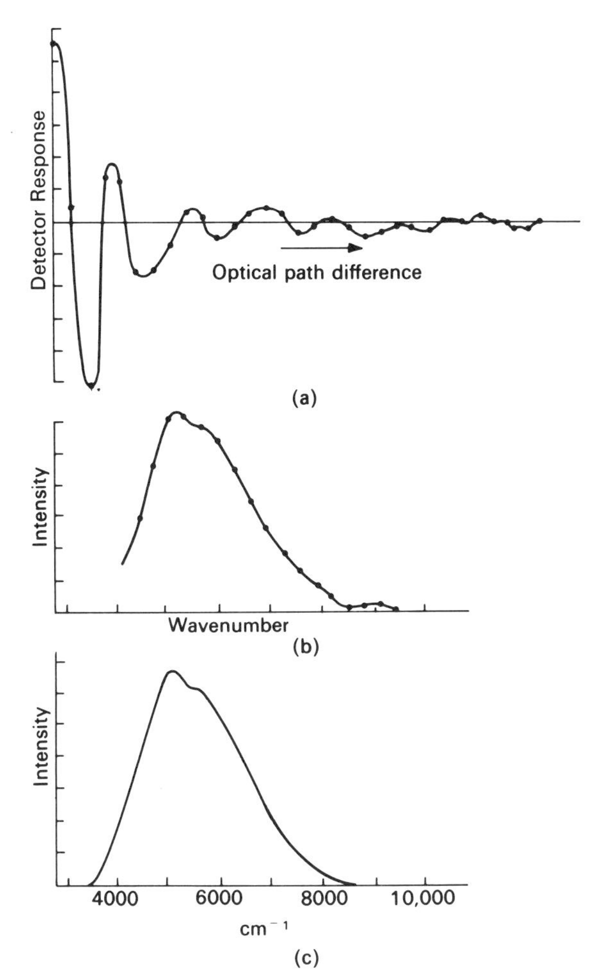

Figure 7.11. (*a*) Interferogram and (*b*) Fourier transformation of radiation emitted by mercury lamp. (*c*) Spectrum of same source obtained by conventional grating instrument. [Reprinted from P. Fellgett (1958), *J. Phys. Radium 19*, 237, with permission from the Société Française de Physique, Paris, France.]

The second advantage offered by interferometric methods was pointed out by Jacquinot (1954). Unlike the dispersing instrument, the interferometer depends on differences in path length, not on slit width, for its resolving power. Consequently, for the same resolving power, the radiant throughput of the interference spectrometer should be much higher than that of either a grating or a prism instrument (Figure 1.10). The Michelson interferometer produced much broader fringes and much larger path length differences than the Fabry–Perot device. As a result, interest in the Michelson interferometer was revived because as long as spectral information could be determined in the form of an interferogram, the Michelson design made it possible to study a much broader spectral range with greater throughput and resolution than the Fabry–Perot device.

Fellgett recognized that the superiority of interference methods was especially significant for astronomy, where measurements are severely limited by low light levels. He subsequently constructed a Michelson instrument of improved design (Fellgett, 1958a) and used it to study stellar emissions in the near-infrared region. Use of a lead sulfide detector gave a signal whose noise was detector limited, making it possible to achieve the full multiplex as well as throughput advantages. A resolving power (R) of about 60 in the region 4000–8000 cm^{-1} was obtained (Figure 7.12).

The second half of the 1950s saw the development of a variety of new interference spectrometers as well as many improvements in the basic Michelson design. The first conference dealing with progress in interference spectroscopy was held in France (the 1957 Bellevue Colloquium, published in

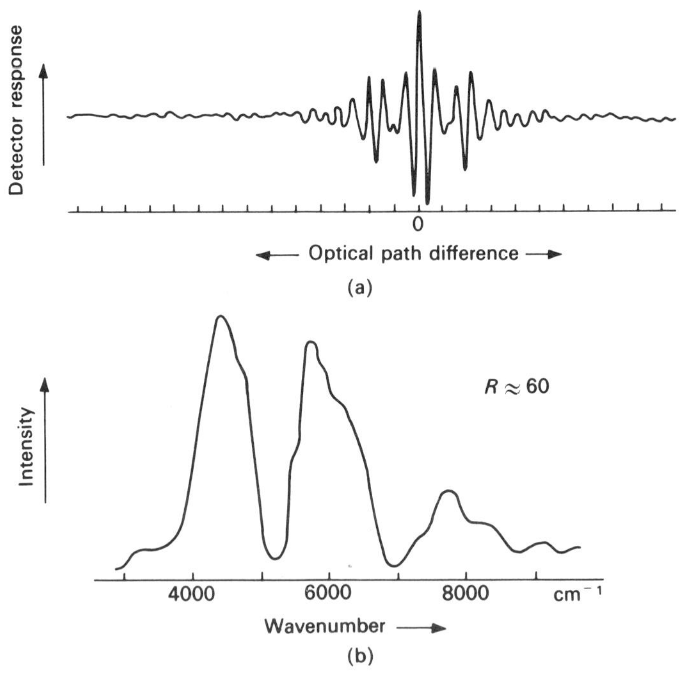

Figure 7.12. (*a*) Interferogram and (*b*) corresponding spectrum emitted by α-Herculis (magnitude 3.5). [Reprinted from P. Fellgett (1958), *J. Phys. Radium 19*, 237, with permission from the Société Française de Physique, Paris, France.]

1958 in volume 19 of the *Journal de Physique et le Radium*) and contributed much to bringing the various ideas together. The success of this colloquium prompted a second one in Orsay, France, whose proceedings were published in 1967 as a supplement to the *Journal de Physique*. Papers presented at a third international conference, held in Aspen, Colorado, in 1970 (The Aspen International Conference on Fourier Spectroscopy) were published in 1971 (AFCRL Special Publication; Vanasse, Stair, and Baker, 1971).

Following Fellgett's lead, many of the studies conducted during the 1950s and 1960s involved astronomical applications in the infrared (1–4 μm PbS region). This portion of the spectrum was rich in scientific potential because it included the absorption bands of many molecules, it contained many relatively bright astronomical sources, and it did not have severe telluric absorptions for ground-base work. Many astronomical spectra never before observed in this region were obtained as a result of utilizing both the multiplex and throughput advantages gained from interference methods. (See Fink and Larson, 1979, for a review.)

Beginning in the middle of the 1950s, both French (Terrien and Hamon, 1954; Terrien and Masui, 1956; Terrien, 1958) and American (Harrison and Stroke, 1955; Mertz, 1958, 1973; Stroke, 1958) investigators extended interference methods to the UV–visible region by employing a photomultiplier as a detector. Since the photomultiplier gave a signal that was photon shot noise limited, only a throughput advantage was still possible. Following a number of improvements in instrumentation, including the discovery that mylar could be used as a beamsplitter, spectra in a wavelength region from about 5.5 to 50 μm could also be satisfactorily determined using interference methods (Richards, 1964).

Today, when nearly every laboratory has access to a high-speed computer, it is easy to forget that Fourier transforms of even a few thousand points presented major difficulties for early workers in this field. In 1951, Fellgett was forced to transform his interference patterns numerically using Beevers–Lipson strips (Beevers and Lipson, 1936). The technique was originally developed as an aid to Fourier calculations in X-ray crystallography but could also be applied in the more general case to the analysis of any set of periodic functions (Ross, 1943; Fellgett, 1958b). Strips of paper giving the amplitudes of each sinusoid in the interferogram were used to facilitate calculation of resultant amplitudes, with the whole process of adding and checking carried through for 29 harmonics in about 2 hr (Ross, 1943)! The analysis of a crystal structure consisting of 80 F values summed over 1800 points (a total of 124,000 terms) required about 8 hr to complete (Beevers and Lipson, 1936).

In 1956, Gebbie and Vanasse published the first digitally computed far-infrared spectrum (Figure 7.13), showing water vapor absoption lines that

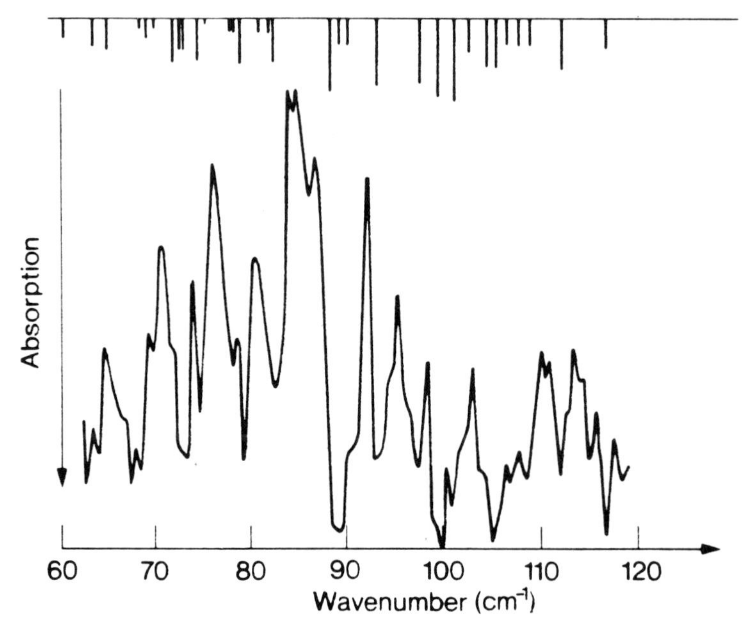

Figure 7.13. First spectrum obtained by digital transformation (Gebbie and Vanasse, 1956) of atmospheric interferogram. Theoretical (line) spectrum of water vapor is shown at top for comparison. [Reprinted from J. Chamberlain (1979), *The Principles of Interferometric Spectroscopy*. By permission of John Wiley and Sons, New York.]

were in satisfactory agreement with theory. Improvement in resolution, however, required that the number of input observations be increased in direct proportion to the observed frequency. For example, to obtain a resolution of 0.1 cm^{-1} in the 50-cm^{-1} region (200 μm, the far infrared), only 500 input observations were required. To achieve the same resolution at 2000 cm^{-1} (5 μm), a total of 20,000 input points was needed. If there were N data points and it was necessary to determine the amplitude of N separate sinusoids, then computation time was proportional to N^2, the number of multiplications performed by the computer. According to Connes (1971), the longest transform carried out with this computational technique was 12,000 points and took 12 hr to complete.

Because of apparent computational limitations as well as many unsolved optical, mechanical, and electronic problems associated with achieving large path differences, early interferometers were restricted to low resolution, which in many cases was inferior or at most equal to spectra obtained by conventional grating spectrometers. As Connes (1971) put it, "Just as the gain of Fourier spectroscopy becomes high, the difficulties become very large" (p. 83). Doubts about the potential of the method were voiced, and it seemed that perhaps the technique might remain the province of a few investigators who, like the astronomers, could not obtain their information in any other manner.

The demonstration that Fourier spectroscopy not only works but gives vastly superior results over conventional spectroscopy can be largely attributed to the efforts of the Conneses. By 1959, Connes and Gush had obtained a resolving power of 1000 in a single observation (Figure 7.14) using a two-beam Michelson interferometer. For comparison, using a grating instrument, it had been necessary to take the average of 10 spectra in order to obtain an effective resolving power of 150. This improvement enabled rotational fine structure within the vibrational bands to be clearly resolved for the first time.

Following this demonstration, Connes (1961) carefully analyzed the problems associated with obtaining high-quality spectra, and during a visit to the Jet Propulsion Laboratory, P. Connes built an interferometer that incorporated most of these requirements. The results obtained with this instrument (Connes and Connes, 1966) changed the cautious attitude toward Fourier spectroscopy and demonstrated that the method could be widely useful.

Figure 7.15 puts these results into proper perspective. Figure 7.15*a* shows a portion of a Venus spectrum considered to be the limit of the art with a grating spectrometer and best available detector (Kuiper, 1962). These measurements, taken with the McDonald 82-in. telescope, have a resolution of 8 cm^{-1} and show the CO_2 bands on Venus quite clearly. Figure 7.15*b* shows the first astronomical results obtained with the Connes interferometer at the University of Arizona Steward Observatory 36-in. telescope (Connes and Connes, 1966). The increase in resolving power is so great that only one of the CO_2 bands can be covered, and the rotational structure is now fully resolved. Using

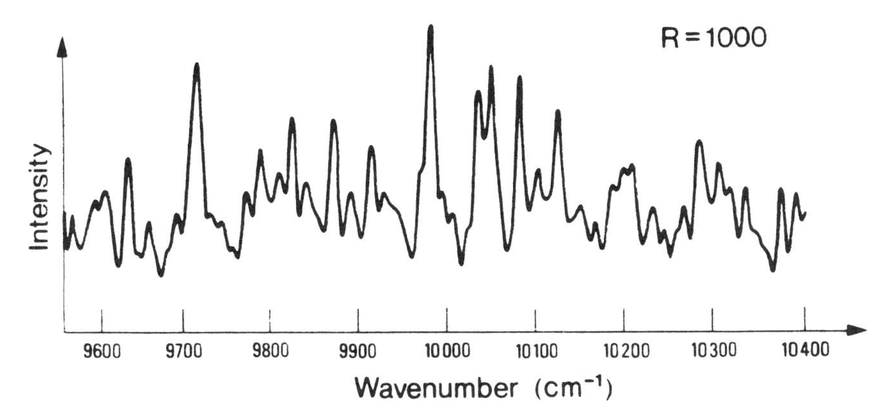

Figure 7.14. Part of night-sky spectrum observed by Connes and Gush (1959, 1960). [Reprinted from J. Chamberlain (1979), *The Principles of Interferometric Spectroscopy*. By permission of John Wiley and Sons, New York.]

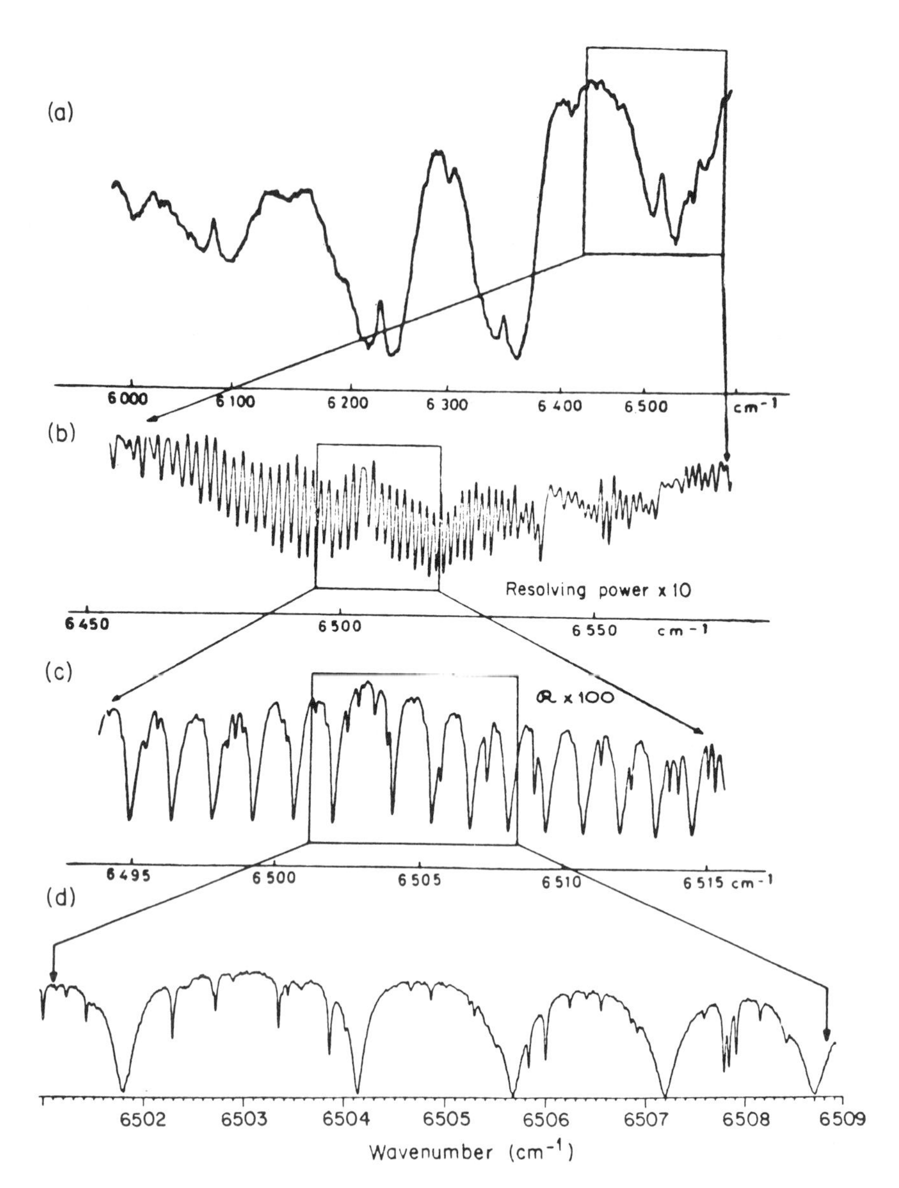

Figure 7.15. Sections of Venus spectrum centered on 30°1–00°0 (30°1 I) CO_2 band showing progress of FT–IR spectroscopy. (*a*) Grating spectrum obtained before advent of FT–IR spectroscopy (Kuiper, 1962). (*b*) First good FT–IR spectrum (Connes and Connes, 1966). (*c*) Excerpt of Connes et al. (1969) atlas. (*d*) Spectrum taken with third-generation interferometer at 500-cm Mt. Palomar telescope showing resolving power of about 500,000 (Connes and Michel, 1975). [Reprinted with permission from U. Fink and H. P. Larson (1979), "Astronomy: Planetary Atmospheres", in *Fourier Transform Infrared Spectroscopy: Applications to Chemical Systems*, J. R. Ferraro and L. J. Basile (Eds.), Vol. 2, Academic, New York, Chapter 7, pp. 243–314.]

a larger telescope (Saint-Michel, 193 in.) and longer observation time, the same interferometer was able to produce the spectrum given in Figure 7.15*c*, showing a further gain in quality and resolving power (Connes et al., 1969).

About the same time that the Conneses were demonstrating the practicality and effectiveness of Fourier spectroscopy, the problem of computing large transforms was solved with the publication of the fast Fourier transform (FFT) algorithm (Cooley and Tukey, 1965). The story behind the development of this algorithm is given by Brigham (1974, p. 8):

> During a meeting of the President's Scientific Advisory Committee, Richard L. Garwin noted that John W. Tukey was writing Fourier transforms (Cooley, Garwin, Rader, Bogert, and Stockham, 1969). Garwin, who in his own research was in desperate need of a fast means to compute the Fourier transform, questioned Tukey as to his knowledge of techniques to compute the Fourier transform. Tukey outlined to Garwin essentially what has led to the famous Cooley–Tukey algorithm.
>
> Garwin went to the computing center at IBM Research in Yorktown Heights to have the technique programmed. James W. Cooley was a relatively new member of the staff at IBM Research and by his own admission was given the problem to work on because he was the only one with nothing important to do (Cooley, Garwin, Rader, Bogert, and Stockham, 1969). At Garwin's insistence, Cooley quickly worked out a computer program and returned to his own work with the expectation that this project was over and could be forgotten. However, requests for copies of the program and a writeup began accumulating, and Cooley was asked to write a paper on the algorithm. In 1965, Cooley and Tukey published the now famous "An Algorithm for the Machine Calculation of Complex Fourier Series" in the *Mathematics of Computation*. [Reprinted with permission from E. O. Brigham (1974), *The Fast Fourier Transform*, Prentice-Hall, Englewood Cliffs, NJ.]

Fast Fourier transform reduced the computer time needed to obtain a spectrum from a time proportional to N^2 to a time proportional to $N \log_2 N$. The calculation that once took 12 hr on an IBM 7040 (Connes, 1971) could be done in about 20 min using the new algorithm and is further shortened to about 5 sec using the more powerful computers available today. Increase in computing speed due to the development of this algorithm completely revolutionized FT spectroscopy, making possible the transformation of interferograms containing more than a million data points. Within three years of the publication of FFT, commercial, stand-alone FT–IR (infrared) and FT–NMR (nuclear magnetic resonance) spectrometers were being produced and marketed for general laboratory use.

Realizing that resolving power was now limited only by instrumental design, the Conneses constructed a second and third generation of spectrometers that produced spectra with path differences of up to 2 m and containing over a million sample points. A section of the Venus spectrum (Figure 7.15*d*) obtained at the 200-in. Mt. Palomar telescope shows a resolving power of over 500,000 (Connes and Michel, 1975). The line shape of the rotational structure of the CO_2 bands is now apparent, and many additional weak lines lying between the major CO_2 lines have become visible. It should be noted that both spectra 7.15*a* and 7.15*d* were obtained using a PbS detector with approximately the same detector sensitivity. Spectrum 7.15*d* was obtained using a larger telescope; however, most of the difference in quality is due to Fourier spectroscopy and the multiplex advantage. These data are perhaps the clearest demonstration of the great power of the technique.

Fourier transform spectroscopy has now become routine for most applications involving IR (Griffiths and deHaseth, 1986; Theophanides, 1984; Ferraro and Basile, 1978, 1979, 1982; Vanasse, 1977, 1981) and NMR spectroscopy (Farrar, 1978; Cooper, 1978). Applications of transform methods to other types of experiments [i.e., nuclear quadrupole resonance (Klainer, Hirschfeld, and Marino, 1982), microwave spectroscopy (Flygare, 1982), two-dimensional NMR (Morris, 1982), ion cyclotron resonance (Comisarow, 1982), and Raman spectroscopy (Chase, 1987)] have required additional theoretical and/or technical developments and thus occurred later.

Because the general utility of the FT method for spectrochemical analysis in the UV–visible region has yet to be determined, rapid commercialization of FT instrumentation for UV–visible spectroscopy has not yet occurred. Ultraviolet–visible FT spectroscopy is still being done in only a few laboratories with specialized, one-of-a-kind instrumentation (Nordstrom, 1982; Horlick, Hall, and Yuen, 1982; Griffiths and deHaseth, 1986). The development of FT spectroscopy in the UV–visible spectral region and its applications to spectrochemical analysis will be discussed in Chapter 14.

7.3 FOURIER TRANSFORMATION: A MATHEMATICAL INTRODUCTION

The Fourier transform is an important analytical tool in many fields familiar to the chemist, such as optics, probability theory, and quantum mechanics. In optics, the Fourier transform not only provides a method for treating one-dimensional spectra but also forms the basis for the analysis of two- and three-dimensional image formation and processing with lens systems. Synthetic chemists encounter three-dimensional Fourier transformation routinely in the interpretation of X-ray crystallographic data. Less familiar applications are found in the communication and information science of electrical engineering, including signal analysis that is not optical in origin (Bracewell, 1989).

In order to facilitate the discussion of the experimental aspects of FT spectroscopy, it will be helpful to have a general understanding of some of the mathematical concepts involved in the procedure. The development of these concepts will be partly rigorous and partly intuitive in nature, with the use of pictorial methods to supplement the equations whenever possible. More formal treatments of the Fourier transform and its applications, especially in the field of optics, may be found in Bracewell (1965), Bell (1972), Brigham (1974), Duffieux (1983), and Steward (1987).

7.3.1. Transform Analysis

Fourier transformation is but one example of transform analysis, a procedure that involves two steps. In the first step, information is converted to a new form; in the second step, a related operation recovers the information in its original form. The use of logarithms provides a familiar example. Data in the form of real numbers can be transformed into logarithms (step 1) and reconverted back into real numbers by the use of antilogarithms (step 2). Transform analysis is not generally used unless there is some advantage to be gained from the procedure. For example, in the absence of a calculator, it may be much easier to evaluate x^y by taking the antilogarithm of $(y \log x)$ than trying to determine x^y directly.

Hadamard transform (HT) spectroscopy (Chapter 5) involves *spatial transformation.* Light, consisting of M pieces of amplitude–frequency information, is dispersed spatially by the grating in a spectrometer. Rather than measuring each of the M components individually, a mask is used to transform the information into sums of the M components. These sums are transformed back into pieces of individual amplitude–frequency information by solving a set of linearly independent equations. In this example, the mask performs the first step in the transform analysis, while step 2 is performed mathematically as part of the data analysis. Justification for using such an indirect method is the multiplex advantage that results from measuring the frequencies in combination.

Fourier transform spectroscopy involves *frequency transformation.* In step 1, the interferometer transforms the M pieces of amplitude–frequency information into an amplitude–time form. This change occurs because the motion of the mirror converts each frequency component into a cosine function whose rate of oscillation depends on two factors: (1) the frequency of the incoming radiation and (2) the velocity with which the mirror is moved. Thus, every optical frequency in the input signal is uniquely represented in the form of a time-varying wave.

In step 2 of the FT analysis, the summation of all cosine oscillations resulting from the input radiation is converted back into the amplitude–frequency form by mathematical manipulation. As in HT spectroscopy, FT spectroscopy uses an instrument to perform the first step in the transformation, while the second step is performed during final data analysis. A substantial increase in resolving power, wavenumber accuracy, scanning time, and sometimes signal-to-noise ratio provides justification for the procedure.

7.3.2. The Fourier Transform

The mathematical equations that accomplish amplitude–time and amplitude–frequency transformations are

$$I(t) = \int_{-\infty}^{\infty} A(f)\exp(i2\pi ft)df \quad \text{(frequency to time)} \tag{7.9}$$

$$A(f) = \int_{-\infty}^{\infty} I(t)\exp(-i2\pi ft)dt \quad \text{(time to frequency)} \tag{7.10}$$

where the variables in the two integrals have reciprocal dimensions: seconds for time t and reciprocal seconds for frequency f. The range of each of these variables is said to form a domain. In Eq. 7.9, integration is carried out over the frequency domain (all frequencies monitored during the experiment), while in Eq. 7.10, integration is carried out over the time domain (the duration of the experiment or the scan time).

The preexponential functions in the integrals, $A(f)$ and $I(t)$, are called "kernels" and may be regarded as the Fourier transforms of each other, or as FT pairs. In FT spectroscopy, FT pairs represent the experimental information: $I(t)$ is the interferogram and $A(f)$ is the spectrum.

The form of Eq. 7.9 can be deduced somewhat intuitively from the following considerations. Suppose, in the simplest possible case, the source spectrum (input signal) is composed of just one frequency, f, with its associated amplitude a. The output signal from the interferometer will be a pure cosine function (Figure 7.4*b*) whose form is given by

$$I(L) = I_{av} + I_{amp}\cos(2\pi fL) \tag{7.1}$$

where I_{av} (Eq. 7.2) $= I_{amp}$ (Eq. 7.3) $= 2a^2$ and L is the optical path difference for the two interfering rays. Since the variable L depends on the velocity with which the mirror is moved, Eq. 7.1 is also a function of the time during which the experiment is conducted, and we may equally well write

$$I(t) = 2a^2 + 2a^2\cos(2\pi ft) \tag{7.11}$$

Inspection of Eq. 7.11 shows that it consists of two components, a constant, or dc signal, and an oscillation, or ac signal. The dc portion of the signal contains only amplitude information, which is also present in the ac signal. Consequently, we can exclude the constant term from further consideration. The ac signal, which contains all of the amplitude–frequency information, is the experimentally determined interferogram. (Some authors call the total record of intensity vs. time the interference function and reserve the term *interferogram* for the ac component of Eq. 7.11.)

If the input signal is polychromatic, the interferogram will represent the summation of all of the cosine oscillations produced from each of the M input frequencies:

$$I(t) = \sum^{M} a_M(f) \cos(2\pi f_M t)$$

where $a_M(f) = 2a_M^2$, or in the more general case,

$$I(t) = \int_0^\infty a(f) \cos(2\pi ft) df \tag{7.12}$$

Equation 7.12 is simply the integral of the ac portion of Eq. 7.11 (the monochromatic case), so $a(f)$ now represents the intensity of the source as a function of frequency, that is, the original spectrum.

If the interferogram is perfectly symmetrical about $t=0$ (defined as $L=0$ when the two mirrors are coincident), it is said to be an even function. Even functions have the property that reflection through the y axis does not change either the value or the sign of the function (Figure 7.16*a*) For the interferogram, this property of evenness may be expressed mathematically as

$$I_e(t) = I_e(-t) \tag{7.13}$$

that is, the value of the even function I_e at $+t$ is the same as the value of the function at $-t$.

Any perfectly even harmonic function may be expressed as the sum of cosine functions, which are themselves perfectly even harmonic functions. The generation of an even square wave from the summation of an infinite number of cosine functions, $s_e(t) = \cos(2\pi ft) - \frac{1}{3}\cos(6\pi ft) + \frac{1}{5}\cos(10\pi ft) - \frac{1}{7}\cos(14\pi ft) + \cdots$ provides an interesting example of this fact (Figure 7.17). If the interferogram is perfectly symmetrical about $t=0$, all of its components will consist of cosines.

An odd function has the property that reflection through the y axis changes the sign but not the magnitude of the function (Figure 7.16*b*):

$$I_o(t) = -I_o(-t) \tag{7.14}$$

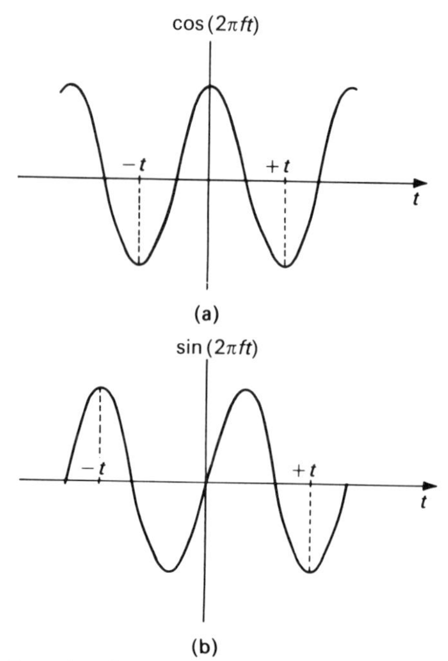

Figure 7.16. Examples of (*a*) purely even and (*b*) purely odd functions.

Any purely odd harmonic function may be expressed as the sum of sine functions, which are themselves purely odd harmonic functions. Shifting the even square wave in Figure 7.17 by 90° produces an odd square-wave function that can be generated by the infinite sum, $s_o(t) = \sin(2\pi ft) + \frac{1}{3}\sin(6\pi ft) + \frac{1}{5}\sin(10\pi ft) + \frac{1}{7}\sin(14\pi ft) + \cdots$. If the interferogram is odd, all of its components will consist of sines.

Real interferograms are generally unsymmetric and cannot be represented by either even or odd functions alone. Unsymmetrical functions can, however, be broken down into sums of components that are themselves either even or odd,

$$I_u(t) = \sum I_e(t) + \sum I_o(t) \tag{7.15}$$

A simple example of the resolution of an unsymmetrical interferogram into even (cosine) and odd (sine) components is shown in Figure 7.18.

Another way of rationalizing the even and odd properties of unsymmetrical interferograms is to remember that $I(t)$ is a summation of sines and cosines, and this combination may be represented as $I(t) = \Sigma a' \cos(2\pi ft + \phi)$, where ϕ is the phase angle. When $\phi = 0, 2\pi, 4\pi, \cdots$, $I(t)$ consists only of pure cosines and so is completely even. When $\phi = \pi, 3\pi, 5\pi, \cdots$, $I(t)$ consists only of pure

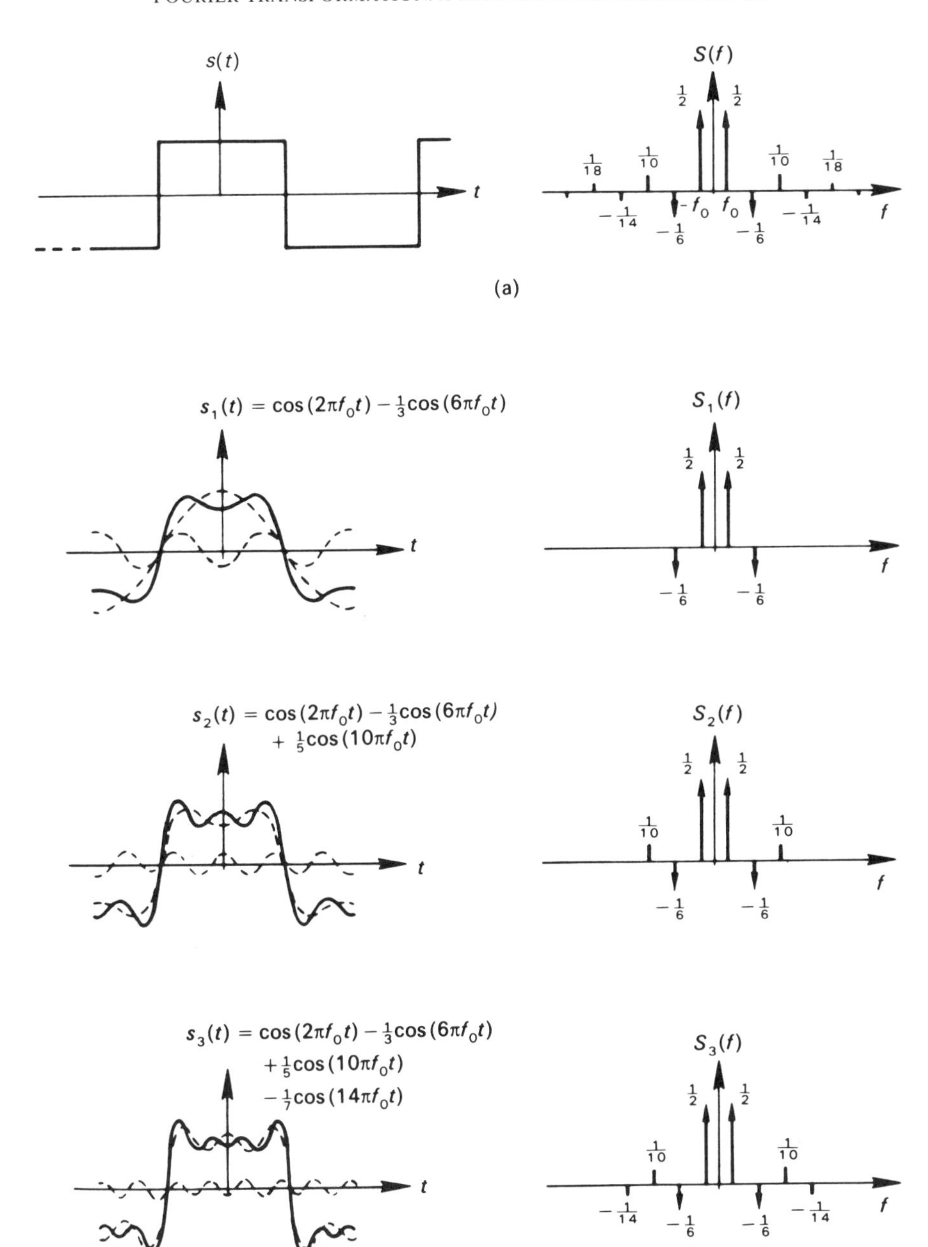

Figure 7.17. Construction of purely even square wave (*a*) from infinite series of cosine waves. Fourier transforms are shown on right. [Reprinted with permission from E. O. Brigham (1974), *The Fast Fourier Transform*, Prentice-Hall, Englewood Cliffs, NJ.]

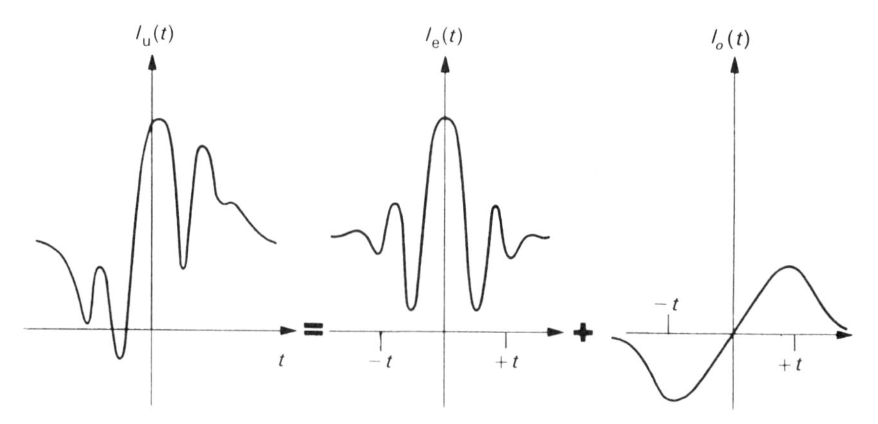

Figure 7.18. Unsymmetrical interferogram resolved into even and odd components. [Reprinted from J. Chamberlain (1979), *The Principles of Interferometric Spectroscopy*. By permission of John Wiley and Sons, New York.]

sines and so is completely odd. Therefore, the presence of nonzero phase angles in some of the components of the interferogram will generally require the inclusion of both even (cosine) and odd (sine) terms in the summation representing $I(t)$:

$$I(t)=\sum^{M} a_M(f)\cos(2\pi f_M t)+\sum^{N} b_N(f)\sin(2\pi f_N t) \tag{7.16}$$

where $a_M(f)$ represents the intensities of the frequencies present in the ideal spectrum and $b_N(f)$ represents the intensities of the frequencies introduced as a result of the presence of nonzero phase angles.

Sums of cosines and sines can be conveniently represented as complex functions using Euler's relationship, that is, $\exp(i2\pi ft)=\cos(2\pi ft)+i\sin(2\pi ft)$ and $\exp(-i2\pi ft)=\cos(2\pi ft)-i\sin(2\pi ft)$. Hence, modification of Eq. 7.12 to include both even and odd terms can also be accomplished in the following manner:

$$I(t)=\int_{-\infty}^{\infty} A(f)\exp(i2\pi ft)df \tag{7.9}$$

where $A(f)$ represents the intensities of all of the frequencies present in the spectrum. This now includes frequencies from the original spectrum as well as frequencies introduced deliberately or inadvertently by nonidealities present in the experiment.

The two sums in Eq. 7.16 are taken over all positive frequency values. The use of complex notation, however, requires that integration in Eq. 7.9 be taken over the limits from minus infinity to plus infinity. Physically, of course, frequency is confined to positive values, but inclusion of negative frequencies allows us to distinguish between cosine and sine functions when complex notation is employed. Another way of looking at the problem is that we need two pieces of information to describe each element of the interferogram: the numerical value of f and its parity (evenness or oddness). Parity is determined by the symmetry (evenness or oddness) of the function with respect to the ordinate, and unless this is explicitly indicated in the equation, we can determine this only by looking at the behavior of the function at both negative and positive values of f.

Equation 7.10 (transformation from amplitude–time to amplitude–frequency) is found by writing the two successive transformations as a repeated integral:

$$A(f)=\int_{-\infty}^{\infty}\left(\int_{-\infty}^{\infty} A(f)\exp(i2\pi ft)df\right)\exp(-i2\pi ft)dt$$

$$=\int_{-\infty}^{\infty} I(t)\exp(-i2\pi ft)dt \qquad (7.10)$$

Since sine and cosine functions are orthogonal, Eq. 7.10 is equivalent to writing

$$A(f)=\int_{-\infty}^{\infty} I_e(t)\cos(2\pi ft)dt - i\int_{-\infty}^{\infty} I_o(t)\sin(2\pi ft)dt \qquad (7.17)$$

Therefore, the Fourier transform is a complex quantity,

$$A(f)=A_{re}(f)+A_{im}(f)=A_e(f)-iA_o(f) \qquad (7.18)$$

consisting of a real part $A_{re}(f)$ and an imaginary part $A_{im}(f)$. (See Figure 7.19.)

Examination of Eq. 7.17 reveals an additional advantage of using complex notation. All of the even components of the interferogram, $I_e(t)$, are preserved in the real portion of the transform, while all of the odd components of the interferogram, $I_o(t)$, are segregated in the imaginary portion (Figure 7.19). This provides a convenient way of sorting out the two types of frequency contributions: those originating in the actual spectrum and those resulting from the conditions of the experiment.

7.3.3. Terms in the Fourier Transform

For those who are uncomfortable with complex functions, we attempt to illustrate the various terms contained within the Fourier transform using graphical methods (Brigham, 1974). Suppose that the interferogram is rep-

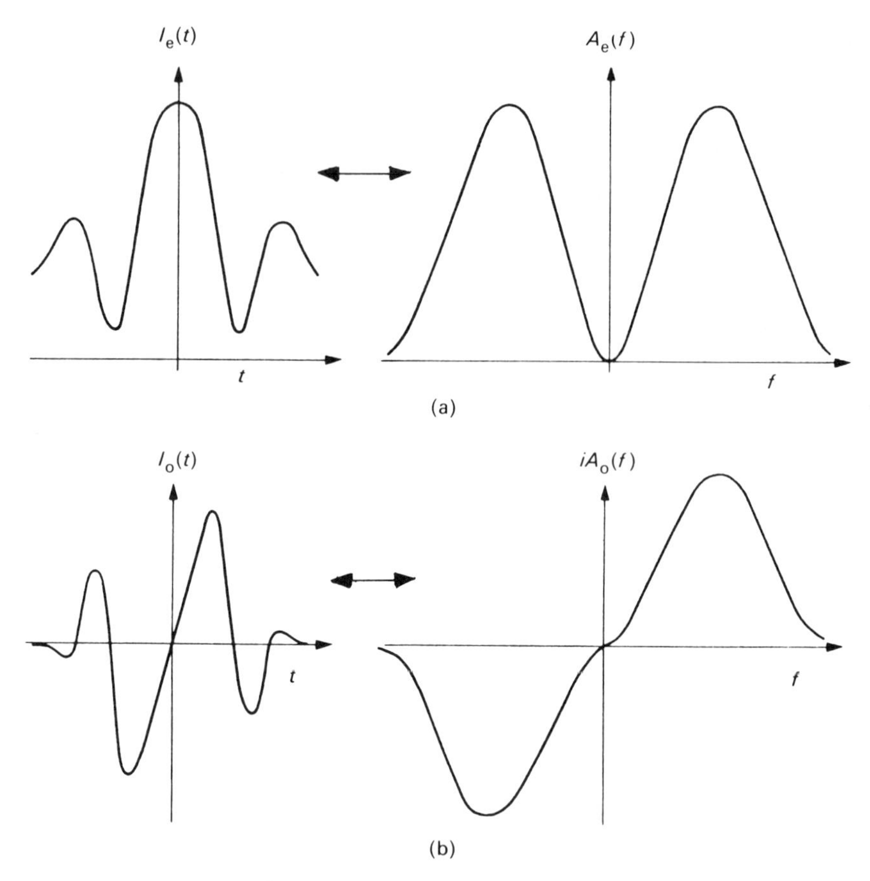

Figure 7.19. (*a*) Fourier transformation of real, even function yields real, even function. (*b*) Fourier transformation of real, odd function yields imaginary, odd function. Fourier transformation of real, unsymmetrical function yields complex function (Figure 7.20). [Reprinted from J. Chamberlain (1979). *The Principles of Interferometric Spectroscopy*. By permission of John Wiley and Sons, New York.]

resented by the equation

$$I(t) = \begin{cases} \beta \exp(-\alpha t), & t > 0 \\ 0, & t < 0 \end{cases} \tag{7.19}$$

as shown in Figure 7.20*a*. This function has been chosen because it is unsymmetrical and has both even and odd components. Therefore, the Fourier transform will be complex.

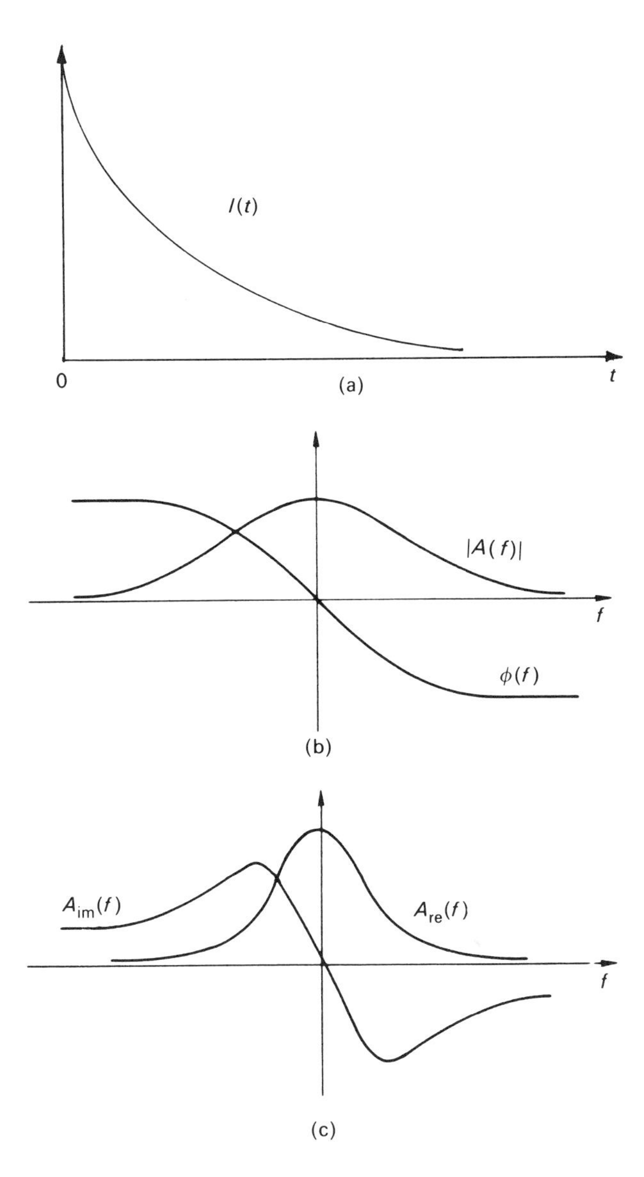

Figure 7.20. Real, $A_e(f)$, imaginary, $iA_o(f)$, magnitude, $|A(f)|$, and phase, $\phi(f)$, portions of Fourier transform of unsymmetrical interferogram, $I(t)$. [Reprinted with permission from E. O. Brigham (1974), *The Fast Fourier Transform*, Prentice-Hall, Englewood Cliffs, NJ.]

The spectrum $A(f)$ is recovered from the interferogram $I(t)$ using Eq. 7.10:

$$A(f) = \int_0^\infty \beta \exp(-\alpha t) \exp(-i2\pi f t)\, dt = \beta \int_0^\infty \exp[-(\alpha + i2\pi f)t]\, dt$$

$$= -\frac{\beta}{\alpha + i2\pi f} \exp[-(\alpha + i2\pi f)t] \Big|_0^\infty = \frac{\beta}{\alpha + i2\pi f}$$

After multiplying and dividing by $\alpha - i2\pi f$, we obtain

$$A(f) = \frac{\beta\alpha}{\alpha^2 + (2\pi f)^2} - i\frac{2\pi f \beta}{\alpha^2 + (2\pi f)^2} \tag{7.20}$$

Hence, the real and imaginary components of $A(f)$ are

$$A_{re}(f) = \beta\alpha / [\alpha^2 + (2\pi f)^2] \tag{7.21}$$

$$A_{im}(f) = -i2\pi f \beta / [\alpha^2 + (2\pi f)^2] \tag{7.22}$$

These two portions of $A(f)$ are plotted individually in Figure 7.20*c*.

In order to plot the entire function $A(f)$, we represent each component of the complex function as a vector in a rectangular coordinate system with orthogonal axes *a* and *b*, where *b* is the imaginary axis and *a* is the real axis. (A vector treatment of complex numbers has also been used in Chapter 3.) Therefore, the real and imaginary components of $A(f)$ may be represented in three-dimensional space by plotting $A_{re}(f)$ in the *af* plane and $A_{im}(f)$ in the *bf* plane (Figure 7.21*a*). This treatment has the advantage of graphically demonstrating the orthogonality of the even (cosine) and odd (sine) portions of $A(f)$ since multiplication by *i* causes the imaginary portion of the function to be rotated 90° out of the *af* plane.

The magnitude (modulus) of the vector representing a complex number must be real and is given by

$$|A(f)| = [A_{re}^2(f) + A_{im}^2(f)]^{1/2} \tag{7.23}$$

where $A_{im}^2(f) = A_{im}(f) A_{im}(f)^*$ and $A_{im}(f)$ and $A_{im}(f)^*$ are complex conjugates of each other. The phase angle $\phi(f)$ describes the twist or rotation of the vector out of the real plane:

$$\phi(f) = \tan^{-1}[A_o(f)/A_e(f)] \tag{7.24}$$

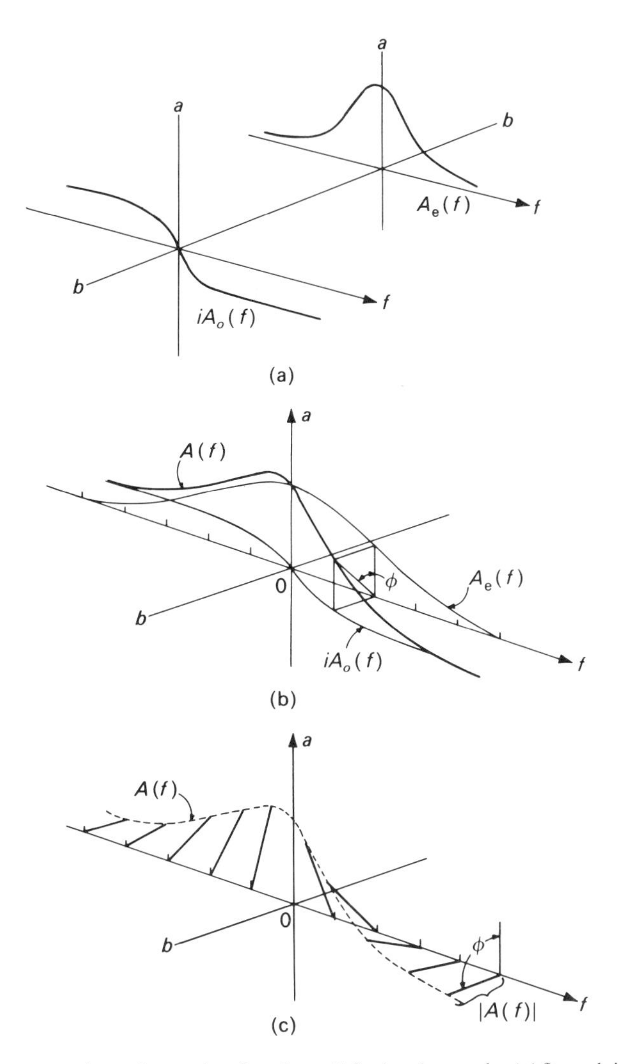

Figure 7.21. Representation of complex function $A(f)$ showing real, $A_e(f)$, and imaginary, $iA_o(f)$, components graphed on orthogonal planes. [Reprinted from P. M. Duffieux (1983). "The Fourier Transform and Its Application to Optics," 2nd ed. By permission of John Wiley and Sons, New York.]

(Compare Eqs. 7.23 and 7.24 with Eqs. 6.9 and 6.10, which represent the two-dimensional case.)

For the interferogram described by Eq. 7.19,

$$|A(f)| = \beta/[\alpha^2 + (2\pi f)^2]^{1/2} \tag{7.25}$$

and

$$\phi(f) = \tan^{-1}[(-2\pi f)/\alpha] \tag{7.26}$$

The magnitude and phase angle of $A(f)$ are plotted in Figure 7.20*c*.

Since length and direction are sufficient to describe a vector, the magnitude and phase angle can also be used to define the complex function

$$A(f) = |A(f)|\exp[i\phi(f)] \tag{7.27}$$

(See Eq. 3.42 for the two-dimensional case.) Therefore, in addition to Eq. 7.20, we have the alternative representation

$$A(f) = \{\beta/[\alpha^2 + (2\pi f)^2]^{1/2}\}\exp[i\tan^{-1}(-2\pi f/\alpha)] \tag{7.28}$$

Figure 7.21 shows the relationship between the real, imaginary, magnitude, and phase portions of the FT using a three-dimensional graphical representation. Note that the even portion of the transform appears in the real (af) plane, while the odd portion of the transform appears in the imaginary (bf) plane (Figure 7.21*a*). Vector addition of the real and imaginary components yields the transform $A(f)$ shown by the dark line in Figure 7.21*b*.

Some of the individual vectors defining the resultant function are shown in Figure 7.21*c*. At any given value of f, the magnitude of the transform is given by the modulus (magnitude) of the vector at that frequency, while the phase angle of the transform is given by the angle of rotation of that vector out of the real plane.

7.4. SPECTRAL LINE SHAPES AND THEIR INTERFEROGRAMS: SOME USEFUL FOURIER TRANSFORM PAIRS

Section 7.3 has described the various components present in a Fourier transform and shown how these components can be represented graphically. This section will deal with the Fourier transformation of several simple functions that are used by spectroscopists to describe spectral profiles.

7.4.1. The Delta Function

The delta function $\delta(x)$ has the property that it is zero everywhere except at $x = 0$, and at that point it is infinite:

$$\delta(x) = 0, \qquad x \neq 0 \tag{7.29}$$

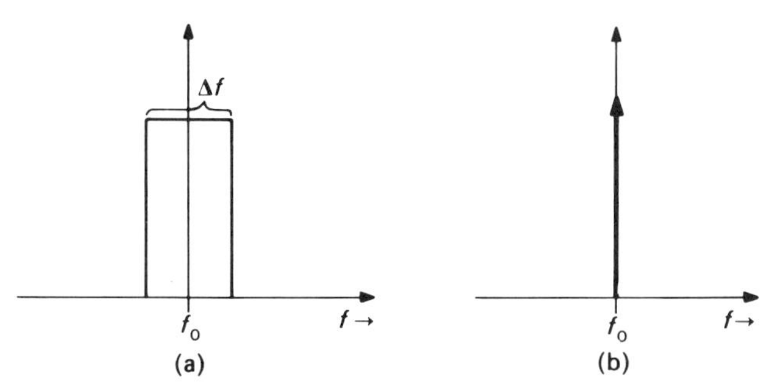

Figure 7.22. Delta function. Narrow pulse of unit area (*a*) becomes delta function of infinitesimal width and infinite height (*b*) as bandwidth Δf goes to zero.

However, since its width at x is infinitely narrow, the area under the function is unity:

$$\int_{-\infty}^{\infty} \delta(x)\, dx = 1 \tag{7.30}$$

In spectroscopy, the delta function represents ideal monochromatic radiation, a spectral profile having infinitely narrow bandwidth and infinitely large amplitude (Figure 7.22).

Delta functions at x_0, not at the origin, are designated as $\delta(x-x_0)$ for locations on the $+x$ axis and as $\delta(x+x_0)$ for locations on the $-x$ axis. When $x=|x_0|$, the condition of Eq. 7.29 is satisfied, allowing the delta function to have a nonzero value at x_0.

Delta functions have the following further interesting property:

$$\int_{-\infty}^{\infty} F(x)\delta(x-x_0)\, dx = F(x_0) \tag{7.31}$$

This is called the 'sifting" property of the function since operation of $\delta(x-x_0)$ on $F(x)$ sifts out the value of $F(x)$ at the delta function location. This sifting property is depicted graphically in Figure 7.23 for a narrow impulse signal centered at f_0. As the bandwidth of the signal goes to zero, the area under the curve approaches the value of $F(f_0)$.

One important example of a FT pair is found by taking the Fourier transform of a delta function located at the origin of the *frequency* domain. As a consequence of the sifting property of the delta function,

$$I(t) = \int_{-\infty}^{\infty} \delta(f)\exp(i2\pi ft)\, df = \exp(0) = 1 \tag{7.32}$$

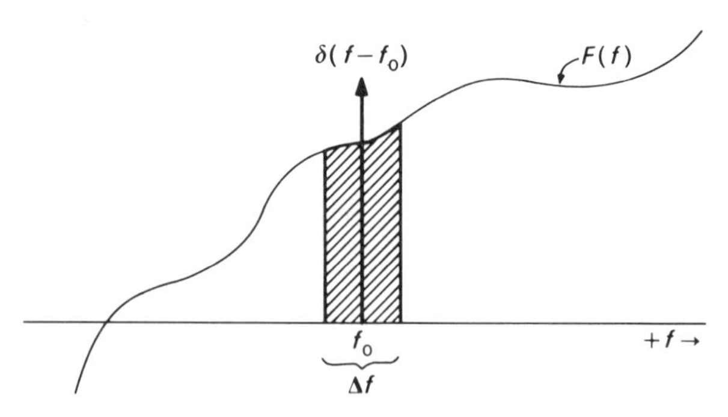

Figure 7.23. Shifting property of delta function. Shaded area approaches $F(f_0)$ as Δf goes to zero. (After Bracewell, 1986.)

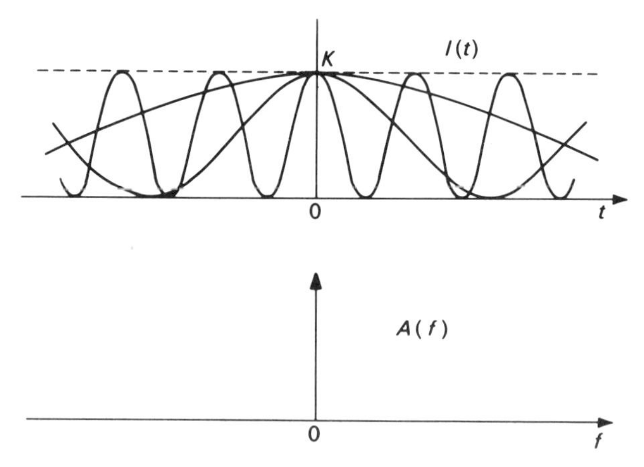

Figure 7.24. Fourier transform of $\delta(f)$. As f_0 decreases, modulation occurs more slowly. At $f_0=0$, interferogram assumes constant value for all mirror positions.

Figure 7.24 illustrates the physical interpretation of this result. A single frequency in the spectrum is modulated by the interferometer to produce a time-varying wave. As the value of the spectral frequency decreases, an increasingly long scan time is required to resolve this modulation. At $f_0=0$, scan time becomes infinitely long, and the interferogram assumes a constant value for all mirror positions. This result has considerable significance because it bears directly on the question of spectral resolution (section 7.5.3).

Fourier transform pairs are complementary; either one may be said to be the Fourier transform of the other. This is true regardless of which way we choose to perform the transformation or which domain we choose to examine.

Thus, if the delta function is located at the origin of the *time* domain, Fourier transformation produces a constant in the frequency domain:

$$A(f)=\int_{-\infty}^{\infty}\delta(t)\exp(-i2\pi ft)\,dt=\exp(0)=1 \tag{7.33}$$

The physical significance of Equation 7.33 is also important. A constant value for $A(f)$ represents white light, a spectrum consisting of *all* frequencies. The only order common to the interference patterns from an infinite number of frequencies is zero. Zero order will be detected when the two mirrors in the interferometer are coincident, that is, when $t=0$. Therefore, the interferogram of a true continuum is a single sharp spike located at the origin. Figure 7.25 depicts the formation of this delta function in the interferogram graphically.

For a delta function situated at f_0, a position other than the origin,

$$I(t)=\int_{-\infty}^{\infty}\delta(f-f_0)\exp(i2\pi ft)\,df=\exp(i2\pi f_0t)$$

$$=\cos(2\pi f_0t)+i\sin(2\pi f_0t) \tag{7.34}$$

$$=\int_{-\infty}^{\infty}\delta(f+f_0)\exp(i2\pi ft)\,df=\exp(-i2\pi f_0t)$$

$$=\cos(2\pi f_0t)-i\sin(2\pi f_0t) \tag{7.35}$$

By adding Eqs. 7.34 and 7.35, we obtain

$$2\cos(2\pi f_0t)=\exp(i2\pi f_0t)+\exp(-i2\pi f_0t)$$

and therefore,

$$\cos(2\pi f_0t)=1/2\,[\mathscr{F}\,\delta(f-f_0)+\mathscr{F}\,\delta(f+f_0)] \tag{7.36}$$

where the symbol $\mathscr{F}$ is used to mean "Fourier transform of." Since Fourier transformation is a linear operation (Section 7.5.2), the Fourier transform of a sum is equal to the sum of the Fourier transforms. Therefore,

$$\cos(2\pi f_0t)=1/2\mathscr{F}\,[\delta(f-f_0)+\delta(f+f_0)] \tag{7.37}$$

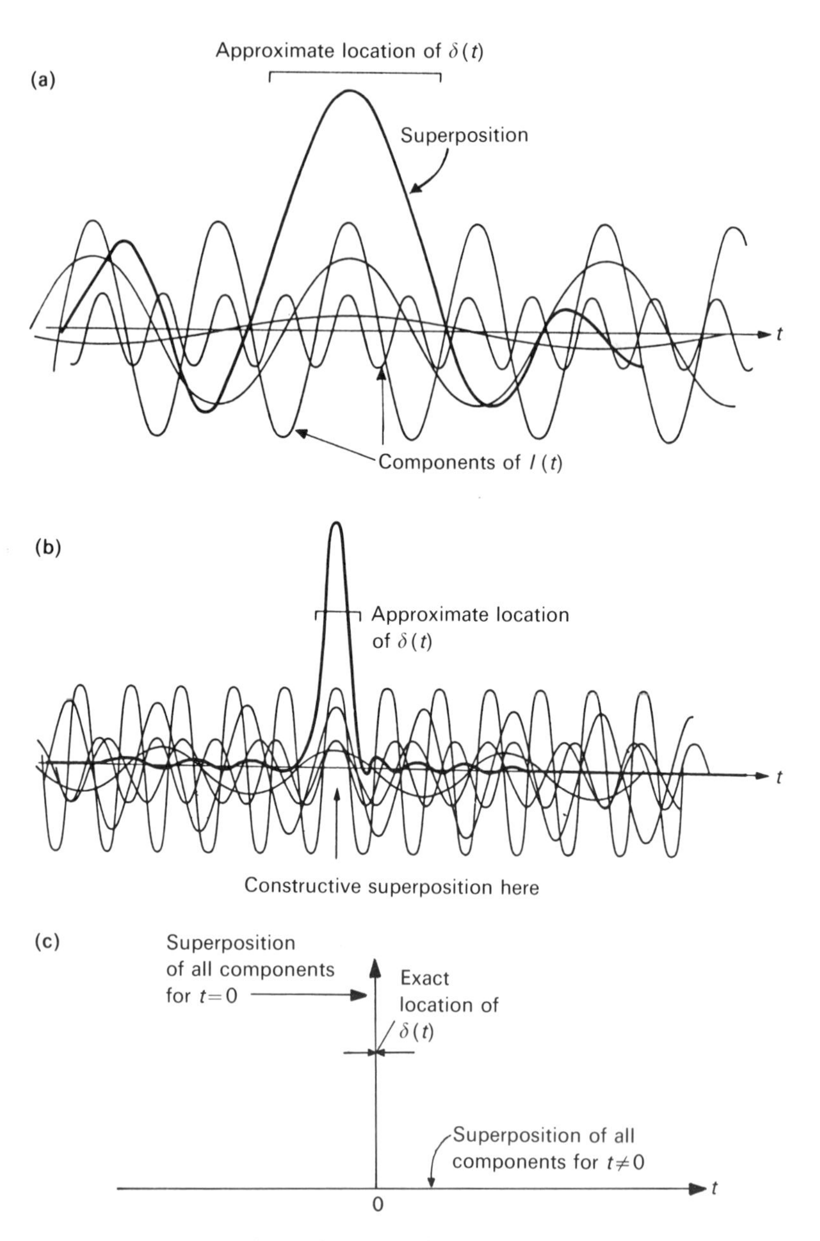

Figure 7.25. Construction of delta function in interferogram of spectrum consisting of large number of frequencies. As number of frequencies increases, number of cosine waves in interferogram increases. When number of cosine waves becomes infinite, ac signal is localized at $t = 0$. Superposition of cosine components of $I(t)$ shown by dark lines. [Adapted from Figure 13.14, Atkins, P. W. (1986), *Physical Chemistry*, W. H. Freeman, New York.]

Subtracting Eqs. 7.34 and 7.35 yields

$$2i\sin(2\pi f_0 t)=\exp(i2\pi f_0 t)-\exp(-i2\pi f_0 t)$$

$$\sin(2\pi f_0 t)=\frac{1}{i2}[\mathscr{F}\,\delta(f-f_0)-\mathscr{F}\,\delta(f+f_0)]$$

Since $i^{-1}=-i$,

$$\sin(2\pi f_0 t)=\frac{i}{2}\mathscr{F}\,[\delta(f+f_0)-\delta(f-f_0)] \tag{7.38}$$

Equations 7.37 and 7.38 define two more sets of Fourier transform pairs (Figure 7.26) and confirm what we have already deduced intuitively. A simple sinusoidal modulation of intensity in the interferogram indicates the presence of a monochromatic spectrum ($f_0 \neq 0$). The single frequency in this spectrum

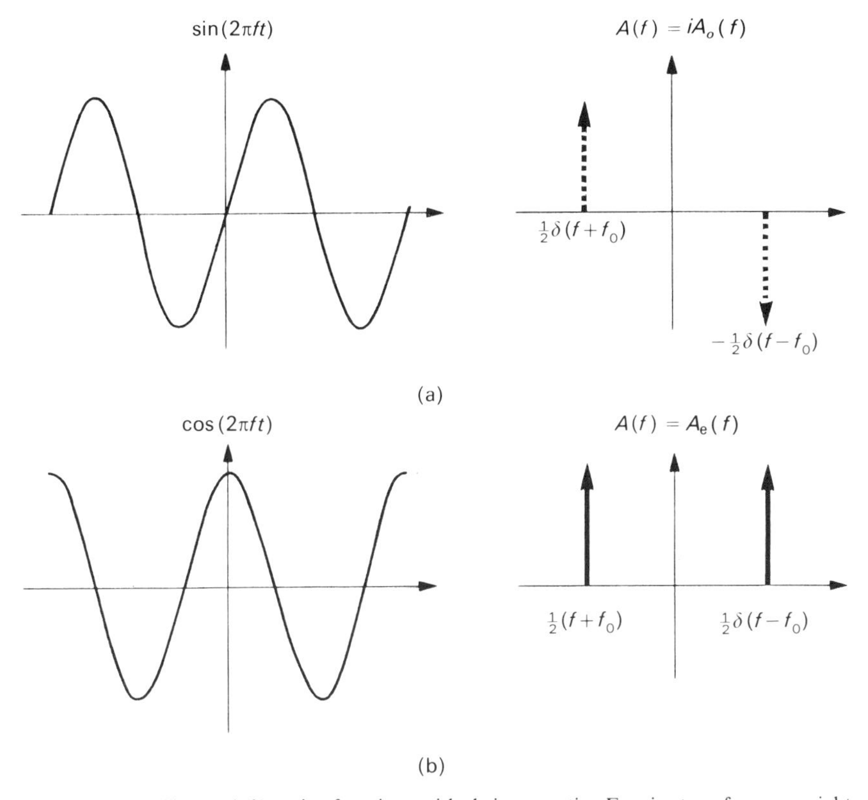

Figure 7.26. (*a*) Sine and (*b*) cosine functions with their respective Fourier transforms on right. Broken arrows indicate that plot lies in imaginary plane.

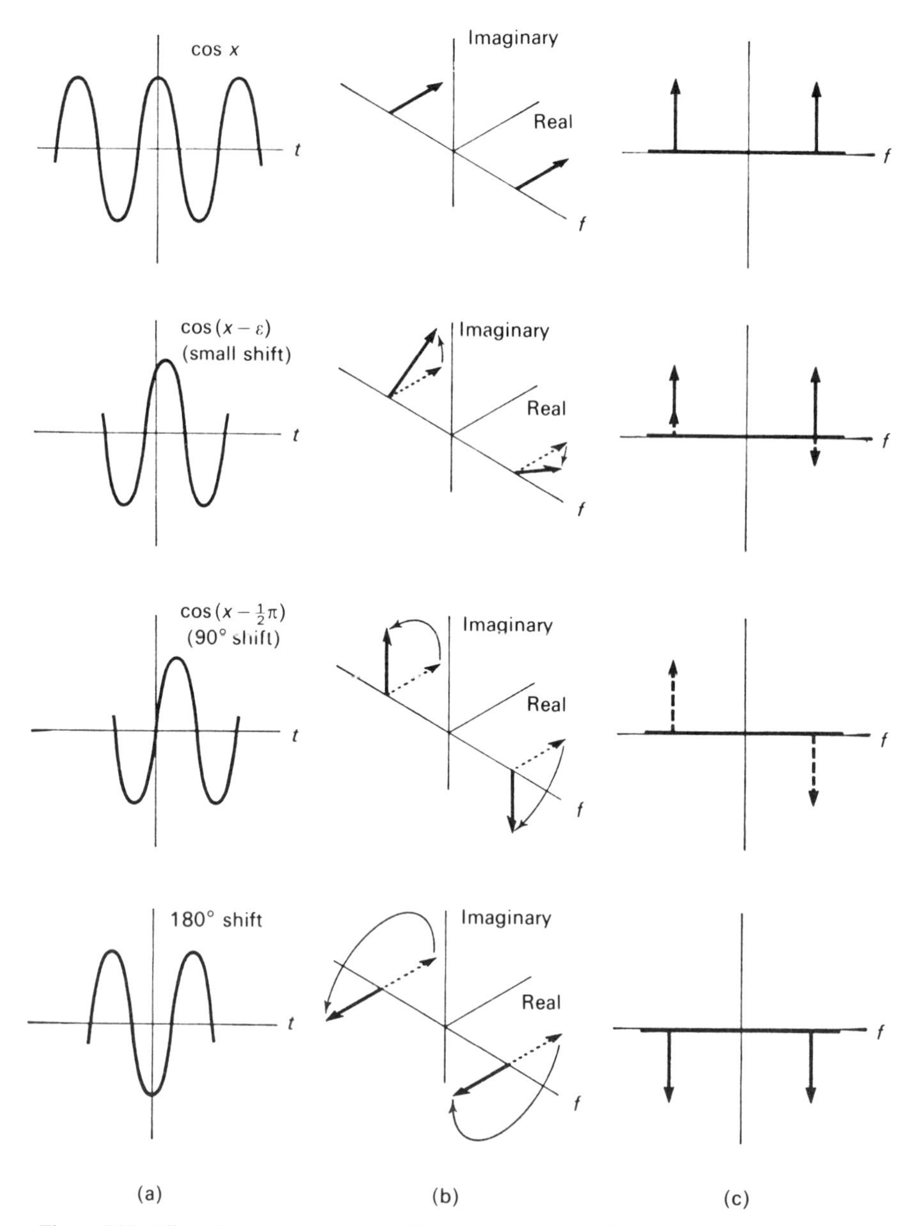

Figure 7.27. Effect of introducing phase shift. Fourier transform of (*a*) shown in (*b*) and (*c*). In (*b*), real and imaginary components are graphed on orthogonal planes. In (*c*), real components are shown by solid lines, imaginary components by broken lines. [Reprinted with permission from R. N. Bracewell (1986), *The Fourier Transform and Its Applications*, 2nd (revised) ed., McGraw-Hill, New York.]

can be determined from the location of the delta function, and the phase is determined by observing the behavior of the delta function on both sides of the ordinate.

If the interferogram is neither even (cosine) nor odd (sine), the transform of $I(t)$ will require both even and odd components. Figure 7.27 shows the effect of shifting a cosine function. A small shift requires the introduction of a small imaginary part in the Fourier transform. At $\frac{1}{2}\pi$ there is no real component left, and at π the real components have undergone a full reversal of phase.

7.4.2. The Boxcar Function

The boxcar function (Figure 7.28) represents an idealized, narrow "white" band of radiation and in the frequency domain is defined as

$$A(f) = \begin{cases} a, & |f| < f_0 \\ \frac{1}{2}a, & f = \pm f_0 \\ 0, & |f| > f_0 \end{cases} \tag{7.39}$$

Since $A(f)$ is real and even, the interferogram of this spectrum must also be

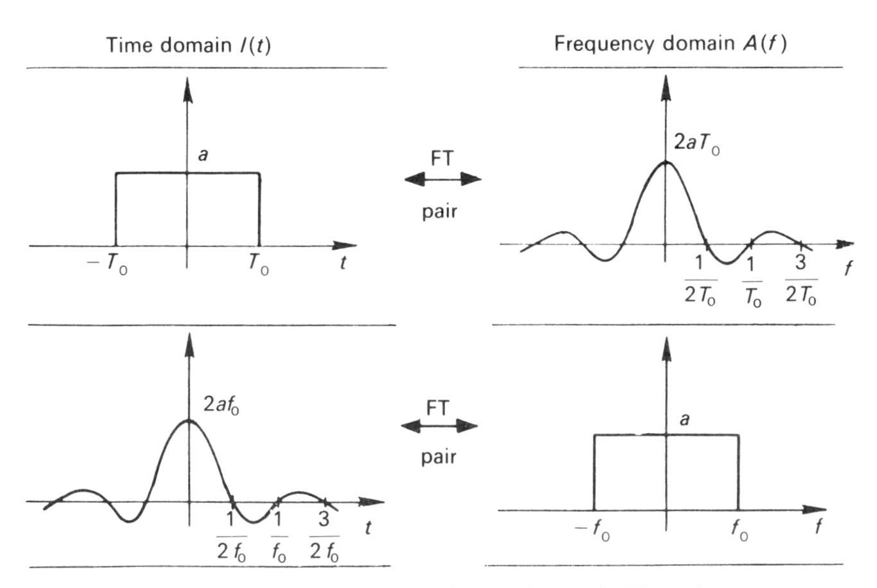

Figure 7.28. Box car and sinc functions—Fourier transform pair. Figure demonstrates symmetric nature of two functions and shows reciprocal relationship between frequency and time domains.

real and even:

$$I(t)=\int_{-\infty}^{\infty} a \exp(i2\pi ft)\, df$$

$$=a\int_{-\infty}^{\infty} \cos(2\pi ft)\, df + ia\int_{-\infty}^{\infty} \sin(2\pi ft)\, df$$

The second integral is zero since the integrand is odd. Therefore,

$$I(t)=[a/(2\pi t)]\sin(2\pi ft)\Big|_{-f_0}^{f_0}$$

$$=\frac{a}{2\pi t}[\sin(2\pi f_0 t)-\sin(-2\pi f_0 t)]$$

$$=\frac{a}{2\pi t}[2\sin(2\pi f_0 t)]$$

which, after rearrangement, yields

$$I(t)=2af_0[\sin(2\pi f_0 t)]/(2\pi f_0 t) \tag{7.40}$$

Functions of the form $(\sin x)/x$ are called sinc functions, so Eq. 7.40 can also be written as

$$I(t)=2af_0 \operatorname{sinc}(2\pi f_0 t) \tag{7.41}$$

A plot of the sinc function is shown in Figure 7.28.

Sinc functions have been encountered previously in the discussion of Fraunhofer diffraction through a single slit (Chapter 3). A single slit may be described by a boxcar function, and its diffraction pattern obeys a sinc function. (Amplitude is represented by the sinc function, while intensity obeys a sinc^2 function.) The slit width has units of length, while the scale of the diffraction pattern has the dimensions of reciprocal length. We are therefore led to the concept of describing a Fraunhofer diffraction pattern as the Fourier transform of the slit function. Diffraction patterns produced by multiple apertures and gratings may be described as sampled sinc functions and are also amenable to treatment by Fourier methods. Fourier transformation has many more applications than simple recovery of a spectrum from an interferogram: It pervades the entire subject of optics and spectroscopy!

7.4.3. Gaussian and Lorentzian Profiles

We now turn to more realistic examples of spectral profiles. There are two main classes: the Gaussian (Doppler) shape in which the width is due to

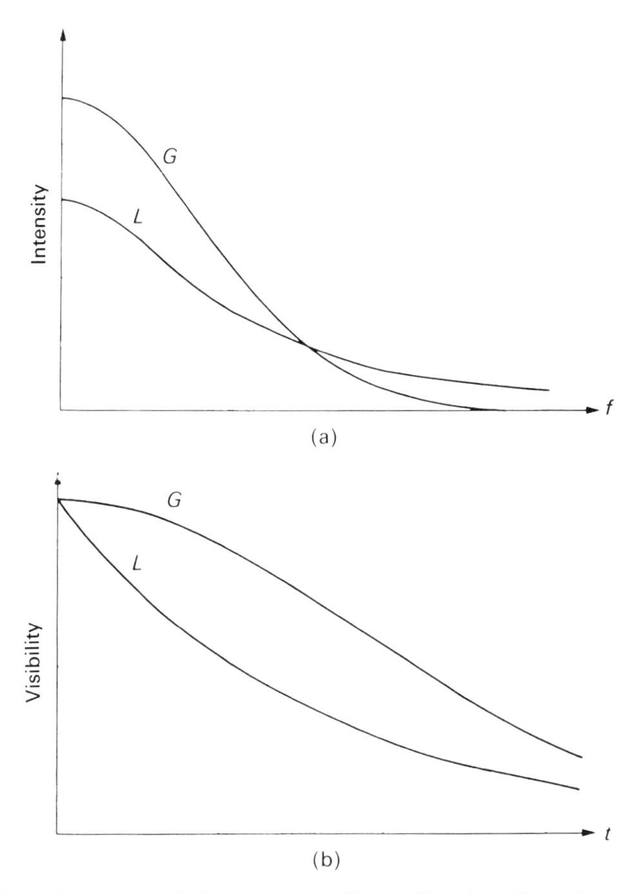

Figure 7.29. (*a*) Spectral and (*b*) interference profiles for Gaussian (G) and Lorentzian (L) line shapes. Only one-half of the spectral plot is shown. Both $I(t)$ and $A(f)$ are symmetric with respect to ordinate. (After Chamberlain, 1979.)

thermal agitation and the Lorentzian (resonance) shape in which broadening is due to interatomic collisions. Figure 7.29*a* compares these profiles under conditions where both curves have the same area. The line shapes are very similar near the peak, the major differences occurring in the wings where the signal is relatively small.

The Lorentzian line shape obeys the general equation

$$A(f)=\alpha^2/[\alpha^2+(2\pi f)^2] \tag{7.42}$$

where α represents the product of a number of constants. This function is even, and therefore its transform will be real and even. Comparison of Eqs. 7.21 and

7.42 shows that 7.42 is, in fact, the real portion of the Fourier transform of the function

$$I(t)=\begin{cases}\beta\exp(-\alpha t), & t>0\\ 0, & t<0\end{cases} \tag{7.19}$$

for the special case in which $\beta=\alpha$. Therefore, the Fourier transform of Eq. 7.42 must have a form similar to Eq. 7.19 but be even (symmetric with respect to the ordinate):

$$I(t)=\tfrac{1}{2}[\alpha\exp(-\alpha|t|)] \tag{7.43}$$

(The factor of $\frac{1}{2}$ compensates for the increase in area resulting from consideration of both positive and negative values of t.) A more rigorous treatment of this problem may be found in Brigham (1974).

A Gaussian line shape obeys the general equation

$$A(f)=\exp[(-\pi^2 f^2)/\alpha] \tag{7.44}$$

where α again represents the product of a number of constants. Since $A(f)$ is even, $I(t)$ will be real and even, so it is only necessary to consider the cosine portion of the transform,

$$I(t)=\int_{-\infty}^{\infty}\exp[(-\pi^2 f^2)/\alpha]\cos(2\pi ft)\,df \tag{7.45}$$

Since α is a positive number, Eq. 7.45 is of a form to be readily evaluated from a table of integrals,

$$2\int_{0}^{\infty}\exp(-a^2x^2)\cos(bx)\,dx=2(\pi^{1/2}/2a)\exp[-b^2/(4a^2)] \tag{7.46}$$

Substituting $x=f$, $a^2=\pi^2/\alpha$, and $b=2\pi t$ into Eq. 7.46 yields

$$I(t)=(\alpha/\pi)^{1/2}\exp(-\alpha t^2) \tag{7.47}$$

Thus, the Fourier transform of a Gaussian is another Gaussian!

Figure 7.29*b* also shows the profiles of $I(t)$ (the visibility curves) resulting from the Lorentzian and Gaussian line shapes. Since the Lorentzian curve is the broader of the two in the wings, it gives rise to an interferogram that falls off more rapidly than in the Gaussian case. This rapid falloff is readily detected close to $t=0$ and enables the distinction between the spectral profiles to be perceived more easily in the interferogram than in the actual spectrum.

Analysis of spectral line shapes is often simplified when the visibility curve is used instead of the actual spectrum. This is the case even when the spectral profile is a mixture of the Lorentzian and Gaussian forms (Voigt profiles) or shows self-reversal. (Self-reversal occurs when cold gas absorbs the radiation emitted from a hotter region of the same gas and is often a problem in spectroscopic sources.) Interested readers are referred to Chamberlain (1979) for further details regarding this subject.

7.5. PROPERTIES OF THE FOURIER TRANSFORM

We conclude our introduction to the mathematics of Fourier transformation by discussing some important properties of the transform. Graphical examples will be used to illustrate relevant applications for spectroscopy. Mathematical proofs of these properties will not be given but can be found in Brigham (1974), Bracewell (1986), as well as many other texts on the subject.

7.5.1. Symmetry

Interchanging the parameters t and f in a Fourier transform pair has no effect other than interchanging the domains of the two functions. This property is illustrated in Figure 7.28. The Fourier transform of a boxcar function in the frequency domain yields a sinc function in the time domain. If we interchange f and t, we still produce the same Fourier pair, that is, the Fourier transform of a sinc function in the frequency domain yields a boxcar function in the time domain. The transformation works either way. If $I(t)$ and $A(f)$ are a Fourier pair, then $I(t)$ yields $A(f)$, and $A(f)$ yields $I(t)$ under Fourier transformation.

7.5.2. Linearity

The Fourier transform of a sum of functions is the sum of the Fourier transforms of each individual function in the sum. This property has already been used in the derivation of Eqs. 7.37 and 7.38 and is further illustrated in Figure 7.17. The Fourier transform of a sum of cosine functions is the sum of the Fourier transforms of all the individual cosine functions. Therefore, the Fourier transform of an infinite number of cosine functions is an infinite number of delta functions.

7.5.3. Scaling

Time Scaling: If $I(t)$ has the Fourier transform $A(f)$, then $I(kt)$, k a real constant greater than zero, has the Fourier transform $|k|^{-1}A(f/k)$. The time-scaling property states that time-scale expansion corresponds to frequency-scale compression. This is illustrated in Figure 7.30 using the boxcar function. Note that as the time scale expands, the frequency scale not only contracts but also increases in amplitude so as to keep the area under the curve constant. In the limit, when the boxcar function becomes infinitely long, $A(f)$ becomes a delta function (Figure 7.24). The spectroscopic implications of this property are quite profound; resolution in the frequency domain is improved by

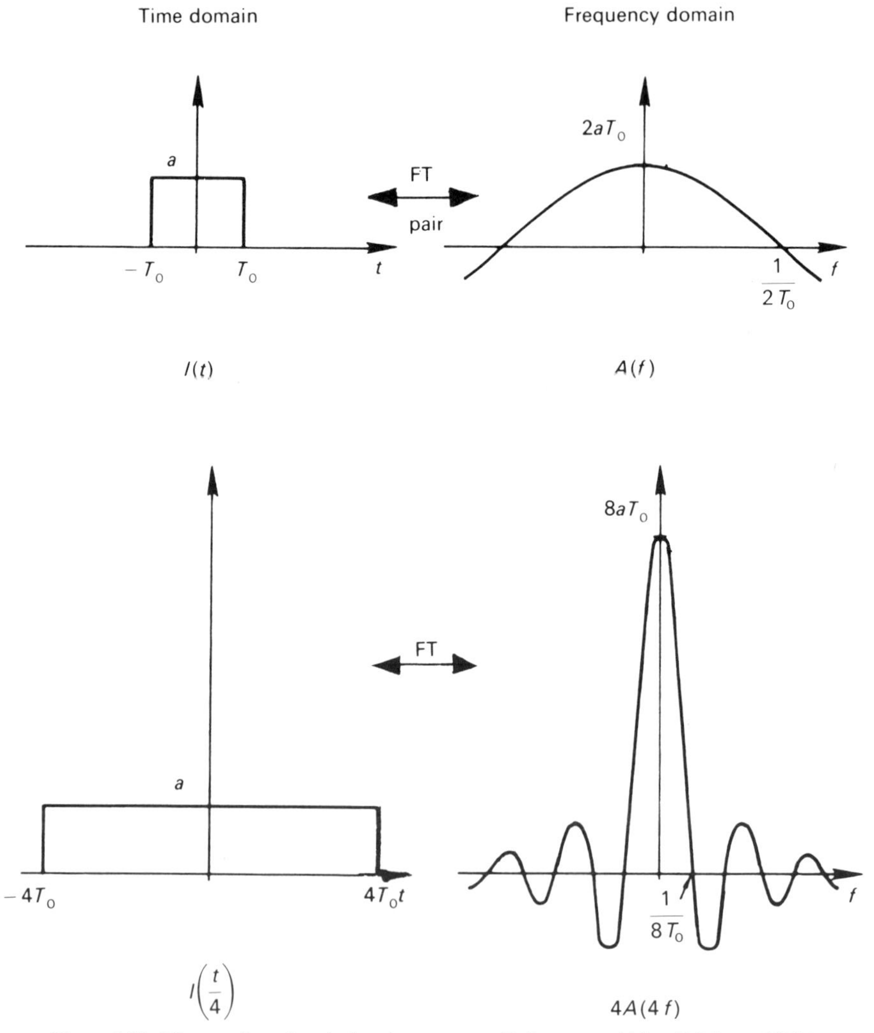

Figure 7.30. Time scaling. Resolution improves as T_0 increases. (After Brigham, 1974.)

expanding the time scale (maximum optical path difference in the interferometer).

Frequency Scaling: If $A(f)$ has the Fourier transform $I(t)$, then $A(kf)$, k a real constant greater than zero, has the Fourier transform $|k|^{-1}I(t/k)$. This statement reflects the symmetry property of Fourier transform pairs. Analogous to time scaling, frequency scale expansion results in a contraction of the time scale (Figure 7.31). An increase in $A(f)$ means that more frequencies are

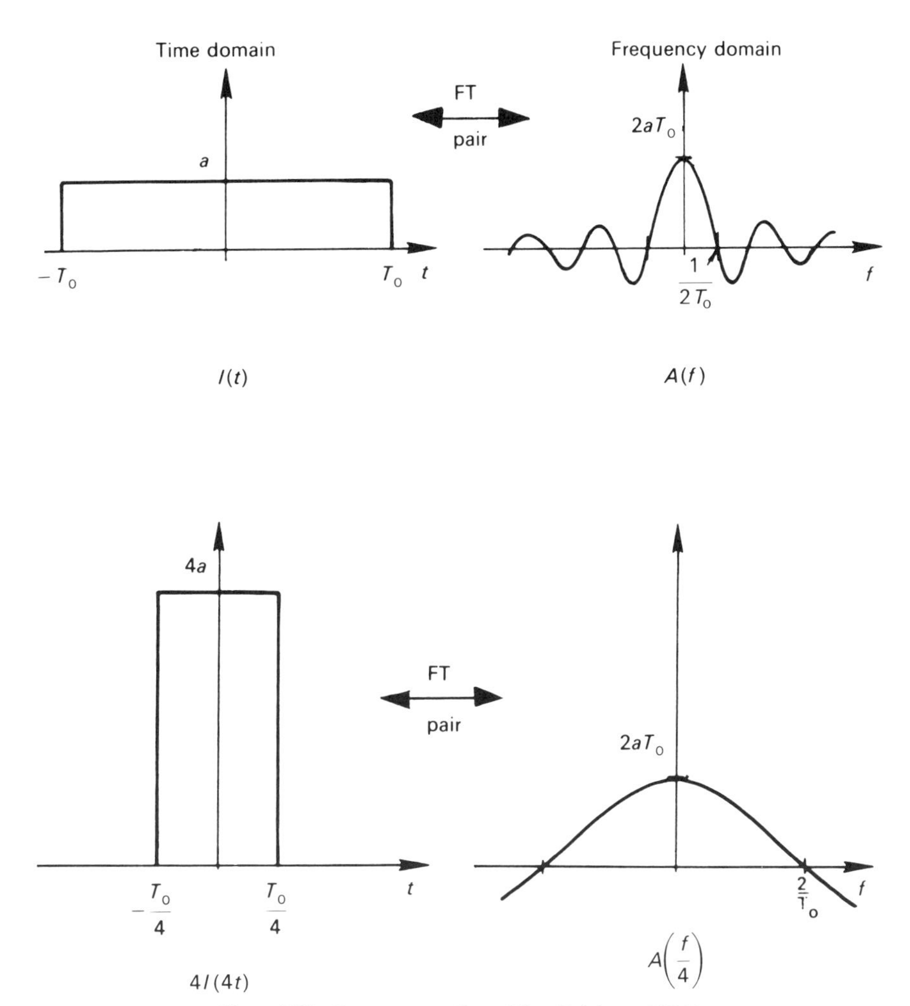

Figure 7.31. Frequency scaling. (After Brigham, 1974.)

present in the spectrum. In the limit, when $A(f)$ becomes infinite, the spectrum is a continuum, and $I(t)$ becomes a delta function. (See Figure 7.25.)

Taken together, the two scaling properties represent a form of the Heisenberg uncertainty principle,

$$\Delta t\ \Delta f \geq h$$

which states that the uncertainty in frequency (Δf) and the uncertainty in time (Δt) are inversely related. Thus, it takes an infinitely long time to precisely locate one frequency, and it requires an infinite number of frequencies to precisely locate one instant in time. (Particle momentum and position exhibit a similar relationship and can also be related through the Fourier transform.)

7.5.4. Shifting

Time Shifting: If $I(t)$ is shifted by a constant t_0, the only effect on $A(f)$ will be multiplication by $\exp(-i2\pi f t_0)$.

Since we can express $A(f)$ by the equation

$$A(f) = |A(f)| \exp[i\phi(f)] \tag{7.27}$$

where $|A(f)|$ is the magnitude of the frequency vector and ϕ is its phase, the time-shifting theorem is equivalent to saying that a shift in the time domain causes no Fourier component changes in magnitude but results in a phase shift of $\exp(-i2\pi f t_0)$. Since f is the frequency of each spectral component, the higher the frequency, the greater the change in phase angle. Figure 7.27 illustrates the effect of introducing a phase shift in $I(t)$. Note that only the direction, but not the magnitude, of the *resultant* vectors are affected.

Frequency Shifting: If $A(f)$ is shifted by a constant f_0, the only effect on $I(t)$ will be multiplication by $\exp(i2\pi f_0 t)$. Again, this statement reflects the symmetry property of FT pairs. If $A(f)$ is real, frequency shifting amounts to multiplication of $I(t)$ by $\cos(2\pi f_0 t)$. Given $I(t)$, we may picture $I(t)\cos(2\pi f_0 t)$ as an oscillation lying within the envelope $I(t)$ and $-I(t)$. The result, called modulation, is shown in Figure 7.32.

7.5.5. Convolution

To find the Fourier transform of a product of two functions, convolve their individual Fourier transforms. The convolution theorem describes what happens under Fourier transformation to the product of two functions,

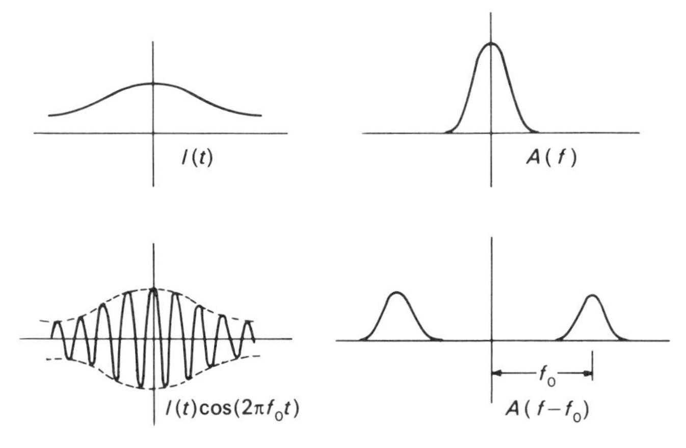

Figure 7.32. Frequency shifting or modulation. (After Bracewell, 1986.)

that is,

$$A(f)=\int_{-\infty}^{\infty} f(t)g(t)\exp(-i2\pi ft)\,dt \tag{7.48}$$

The theorem can be used to simplify the derivation of many FT pairs. It is also important to the spectroscopist as it can be used to analyze the effect of instrumental factors (apertures, gratings, lenses, etc.) on the input signal in order to determine the shape of the output, that is, the "spectral response" of the instrument. The convolution of two functions may be thought of as the net effect of distributing one function in accordance with a law specified by the other function. In a sense, then, the response of the system is due to one of the functions smeared out as a result of the effects of the second function.

The concept of convolution has already been introduced in Section 4.1.3, where the effect of the exit slit on the image of the entrance slit was determined. Both the entrance slit image and the exit slit were represented as boxcar functions, and the spectral response was given by a triangular apparatus function (Figure 4.6).

Convolution is easiest to visualize using a graphical approach. For any two functions $f(x)$ and $g(x)$, the procedure involves the following sequence of steps:

1. *Folding*: Select one of the two functions, say, $g(x)$, and reflect it through the ordinate. [Selecting $f(x)$ for reflection does not alter the final result.]
2. *Displacement*: Shift the reflected image $g_r(x)$ by an amount x' along the abscissa.

3. *Multiplication*: Multiply the shifted function $g_r(x-x')$ by the unshifted function $f(x)$.
4. *Integration*: Determine the area under the product curve by integration. This area is the value of the convolution of the two functions at position x'.

To determine the apparatus function, we add one more step:

5. *Graphing*: Repeat steps 3 and 4 for other values of x'. Present all the convolution values in the form of a plot.

This sequence of steps is demonstrated in Figure 4.6. First the image of the entrance slit (1), which overlaps the exit slit, is reflected through the ordinate (taken as the left edge of the exit slit). Next (1–9), the reflected image is moved over the exit slit. The integrated intensity of the radiation passing through the exit slit at a given image position represents the convolution of the two boxcar functions. A plot of this intensity (convolution) versus position of the entrance slit describes the spectral response to scanning the image over the exit slit. This response is the triangular apparatus function shown in Figure 4.6.

The power of the convolution theorem is revealed when we try to determine the Fourier transform of the triangular apparatus function (Figure 7.33). Since a sinc function is the Fourier transform of a boxcar function, then the Fourier transform of the triangular function (convolution of two boxcars) must be the product of two sinc functions, that is, a sinc squared function.

It is worth reiterating the relationship between Figure 7.33 and some basic optical principles discussed in Chapters 3 and 4. In the top half of Figure 7.33, we see that the effect of the exit slit on the image of the entrance slit (convolution of two boxcar functions) produces a triangular slit profile. The right- and left-hand portions of Figure 7.33 show that Fraunhofer diffraction (Fourier transformation) of a monochromatic beam of light through a single slit modulates the amplitude of the radiation according to a sinc function. The bottom half of Figure 7.33 shows that the intensity of this radiation, found by squaring the amplitude, obeys a sinc squared function. Therefore, by the convolution theorem, the Fourier transform of a sinc squared function will be the triangular slit function shown in the middle of Figure 7.33.

7.6. SOME FUNDAMENTAL PHYSICAL LIMITATIONS

Up to this point we have discussed Fourier transformation from a purely theoretical standpoint. In practice, many factors limit the ability of the technique to produce an accurate representation of the true spectrum. Some factors are instrument specific; others are common to all FT spectrometers.

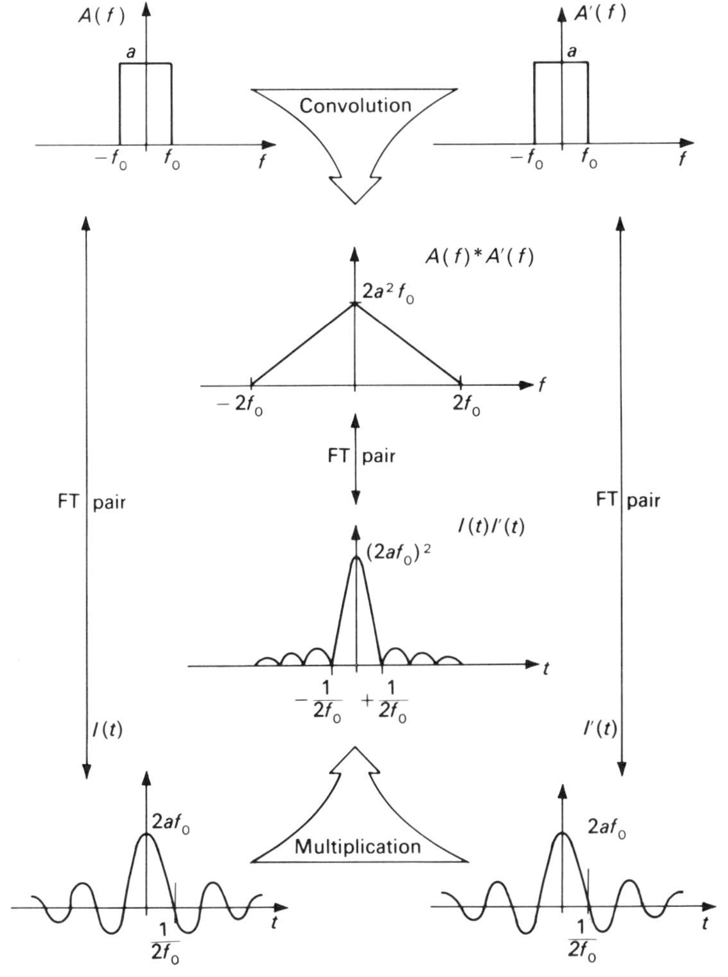

Figure 7.33. Graphical example of convolution theorem. Refer to Figure 4.6 for determination of triangular function $A(f)*A'(f)$. [Reprinted with permission from E. O. Brigham (1974), *The Fast Fourier Transform*, Prentice-Hall, Englewood Cliffs, NJ.]

We conclude this introduction by discussing the effect of three fundamental physical limitations on Fourier transformation and show how these limitations alter the instrument profile.

7.6.1. Finite Optical Path Difference

The interferogram of a single monochromatic beam of radiation with frequency f can be represented by a simple cosine function. If the distance traveled by the mirror could be made infinitely long, the cosine function would also be infinitely long, and Fourier transformation would yield two delta functions symmetrically placed in the real plane (Figure 7.26*b*). These delta functions would define the frequency f exactly, and spectral resolving power would be infinite.

In practice, the cosine function must be truncated at the end of each scan. The effect of truncation can be described mathematically as the product of an

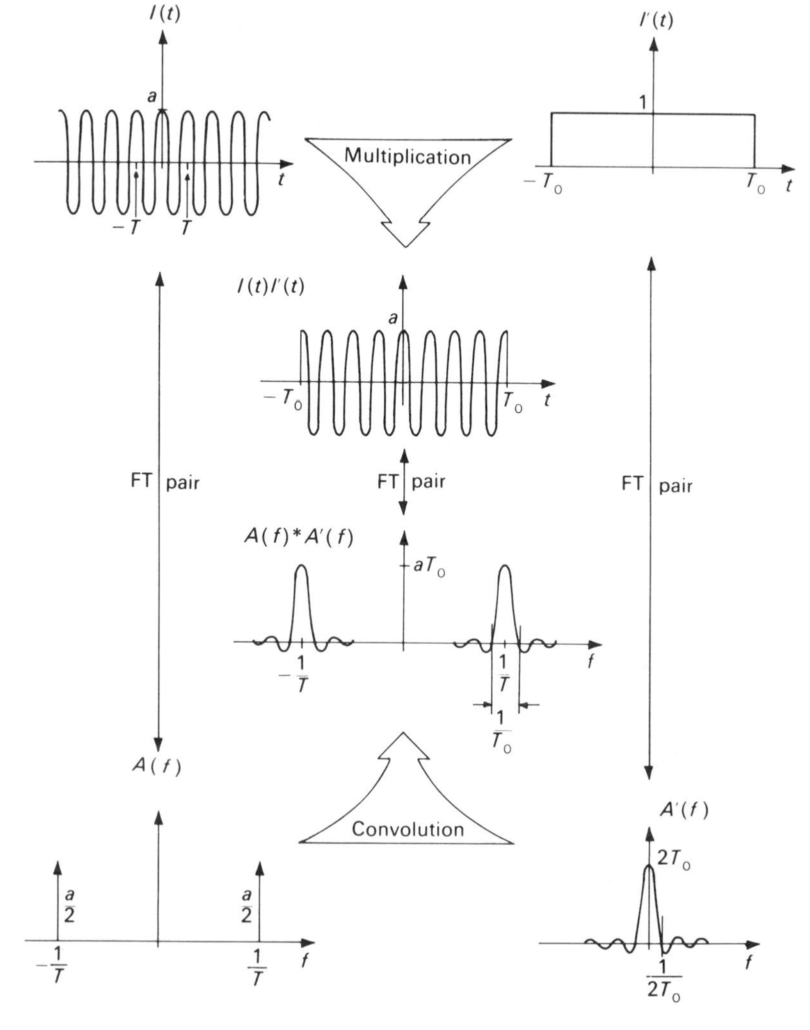

Figure 7.34. Effect of finite retardation on instrumental line shape. [Reprinted with permission from E. O. Brigham (1974), *The Fast Fourier Transform*, Prentice-Hall, Englewood Cliffs, NJ.]

infinitely long cosine wave and a boxcar function (top of Figure 7.34). The height of this box car is unity, and the length is determined by the maximum retardation or optical path difference, L_{max}, produced by the interferometer.

Using the convolution theorem discussed in Section 7.5.5, we can find the Fourier transform of a truncated cosine function by convolving the Fourier transform of an infinitely long cosine function with the Fourier transform of a boxcar function. Figure 7.34 shows that the instrument profile will obey a sinc function. The sinc function still defines frequency f, but its width is now finite. Moreover, much of the amplitude is spread out into side lobes. These side lobes, also called ringing, produce false sources of energy at nearby frequencies in the transformed spectrum and have two important consequences. First, side lobes may be mistaken for real lines, especially if the spectrum is a bit noisy. Second, for complex spectra, side lobes are blended into nearby peaks, affecting both the strengths and positions of neighboring lines.

In principle, the most effective way of dealing with both the resolution and side-lobe problem is to increase the maximum optical path difference in the interferometer. As L_{max} increases, resolution improves (Figure 7.30), and eventually the side lobes become more and more localized around the central peak. However, depending on the spectrum, the required path length may be outside the capabilities of the instrument or too costly in terms of scan time. Another common solution is to apodize the interferogram using a different weighting function.

In signal-processing nomenclature, the boxcar function is called a "weighting" function. Other weighting functions besides the boxcar can be used, the usual procedure being to multiply the interferogram by the chosen weighting function just prior to final Fourier transformation. Modification of the interferogram in this fashion causes the delta function (representing the true spectrum) to become convolved with the Fourier transform of the new weighting function, which therefore determines the profile present in the final spectrum. If side lobes are suppressed, the process is called *apodization*, after a Greek word meaning "without feet." Functions that weight the interferogram for this purpose are called apodization functions.

Any even function that has a value of unity at zero order and decreases with increasing path difference can be used as an apodization function. The triangular function

$$A(t) = \begin{cases} 1 - \dfrac{L}{L_{max}}, & L \leqslant L_{max} \\ 0, & L > L_{max} \end{cases} \tag{7.49}$$

provides a simple example. We already know from Figure 7.33 that the Fourier transform of a triangular function is a sinc squared function.

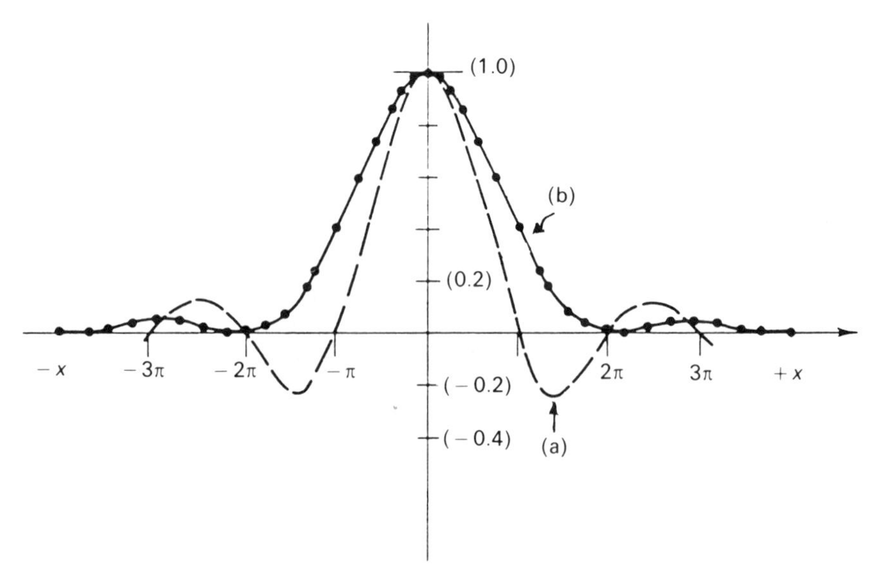

Figure 7.35. Plots of (*a*) sinc x and (*b*) sinc^2 x functions. [Reprinted with permission from R. J. Bell (1972), *Introductory Fourier Transform Spectroscopy*, Academic, New York.]

Therefore, the instrument profile resulting from a triangular apodization function will also obey a sinc squared function.

Figure 7.35 compares the instrument profile obtained for a monochromatic source modified by the same L_{max} using the boxcar and triangular weighting functions. In the sinc function, the central peak is reasonably narrow (FWHM equal to $0.604/L_{max}$), but the first side lobe drops about 21% below zero. In the sinc squared function, negative intensities are removed, and the side-lobe excursions are reduced by a factor of 4. The FWHM is $0.886/L_{max}$, so resolution is degraded somewhat.

Ideally, we would like to have an apodization function that not only removes side lobes but also improves resolution. Several studies on apodization functions (Filler, 1964; Norton and Beer, 1976, 1977; and Kauppinen et al., 1981) have been carried out. As a general rule, the narrower the central line, the greater the relative magnitude of the strongest side lobes. Several examples of apodization functions and the resulting instrument profiles are given in Figure 7.36. In each case, the maximum optical path difference in the interferometer is identical. The resultant FWHM and side-lobe excursion are given in Table 7.1. Of these functions, the Gaussian is not far from optimum and is often used.

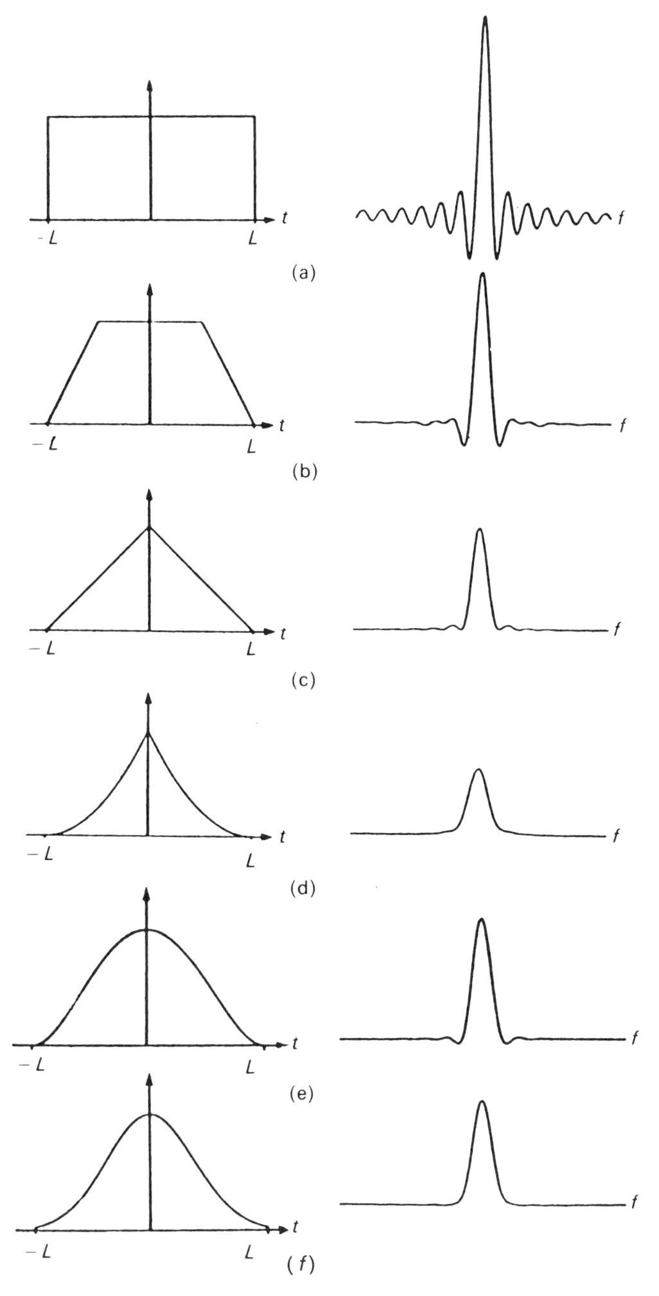

Figure 7.36. Series of weighting functions and their corresponding instrumental profile: (*a*) boxcar; (*b*) trapezoidal; (*c*) triangular; (*d*) triangular squared; (*e*) Bessel; (*f*) Gaussian. (After Kauppinen et al., 1981.)

Table 7.1. Apodization Functions and Resultant Line Shape Parameters

Apodization Function	FWHM,[a] L_{max}	Side-Lobe Amplitude[b] (%)
Boxcar	0.604	−21
Trapezoidal	0.773	−15
Triangular	0.886	+4.5
Triangular squared	1.180	+0.7
Bessel	0.952	−4.1
Gaussian	1.015	−0.45

[a] Full width at half maximum divided by maximum optical path difference achieved by interferometer.
[b] Amplitude of largest side lobe expressed as percentage of amplitude of central peak.

7.6.2. Sampling

The modern digital computer and such algorithms as the FFT are absolutely essential to obtain high-resolution spectra from Fourier transformation. But *digital* implies discrete, so an essential part of the method for carrying out the numerical Fourier transformation of interferograms consists of sampling the continuous interferogram at equal intervals of path difference.

Sampling at equal intervals can be described mathematically by the Shah function $\mathrm{III}(t)$, an infinite sequence of unit impulses spaced at regular intervals T along the time axis:

$$\mathrm{III}(t) = \sum_{n=-\infty}^{n=\infty} \delta(t - nT) \tag{7.50}$$

A Shah function is shown at the top right-hand corner of Figure 7.37. As can be seen from this figure, a Shah function (also called a Dirac comb) resembles an array of Dirac delta functions except that the impulse height is unity rather than infinity. Multiplication of the function $I(t)$, representing the continuous interferogram, by $\mathrm{III}(t)$ causes the interferogram to be sampled at equal intervals.

For simplicity, consider that we have a monochromatic source, and the interferogram $I(t)$ is a simple cosine function. The product $I(t)\,\mathrm{III}(t)$ is the sampled cosine function shown in the middle of Figure 7.37. To determine the effect of sampling on the instrument profile, we once more make use of the convolution theorem. The Fourier transform of the product $I(t)\,\mathrm{III}(t)$ is the convolution of the Fourier transform of the individual functions $I(t)$ and $\mathrm{III}(t)$.

The Fourier transform of the cosine function is given by two symmetrically placed delta functions in the real plane (left side, Figure 7.37). (We are ignoring

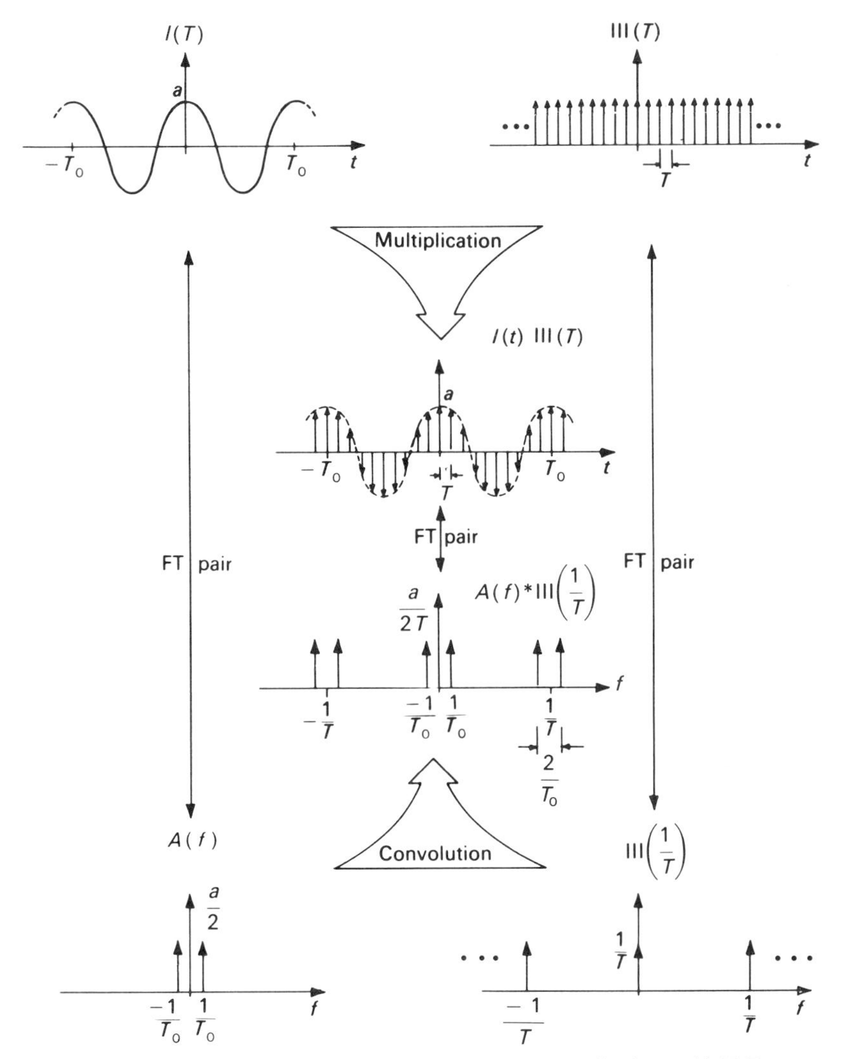

Figure 7.37. Effect of sampling on transformed interferogram. Sampling interval is T. Frequency of radiation is $1/T_0$. (After Brigham, 1974.)

the effect of finite path length, which for boxcar truncation would cause the line to appear as sinc functions.) The Fourier transform of a Shah function is another Shah function (right side, Figure 7.37). A graphical proof of this relationship is shown in Figure 7.38. As indicated in this figure, a Shah function can be written as the sum of an infinite number of cosine functions, each cosine transforming into a pair of symmetrically placed delta functions in the real plane. As the number of cosines in the sum approaches infinity, the

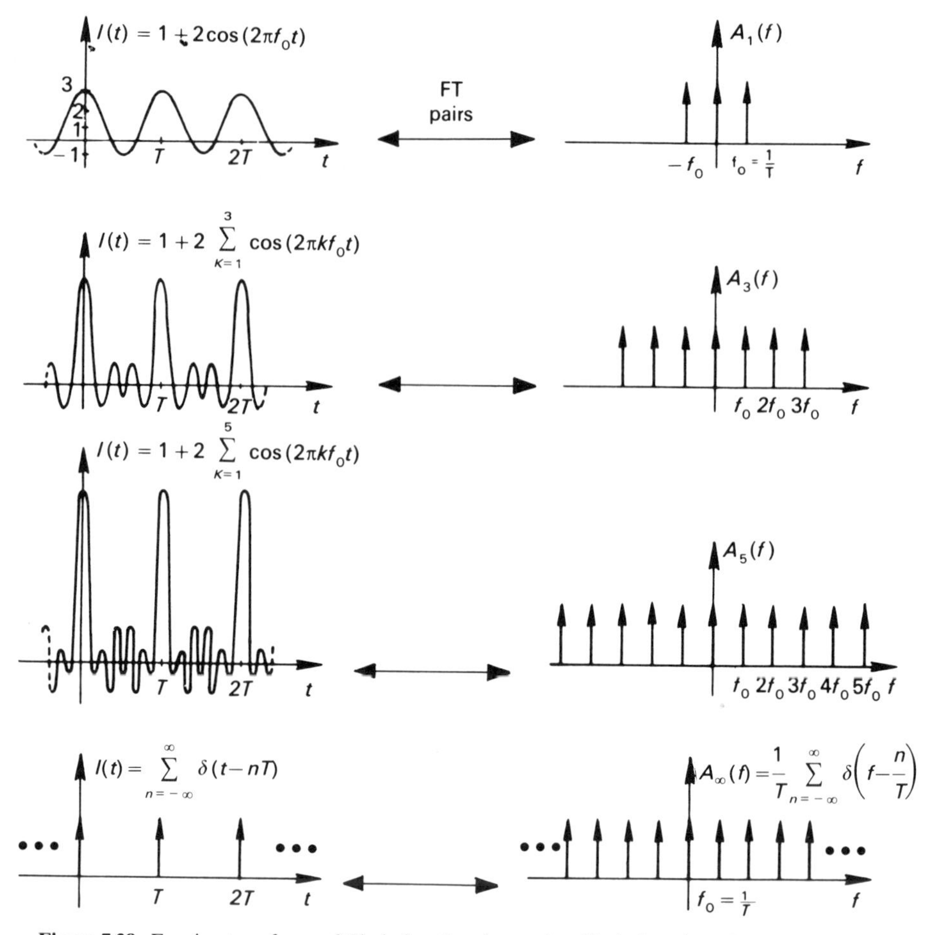

Figure 7.38. Fourier transform of Shah function is another Shah function. Graphical development. [Reprinted with permission from E. O. Brigham (1974), *The Fast Fourier Transform*, Prentice-Hall, Englewood Cliffs, NJ.]

superposition of these cosines and their transforms approach an infinite sequence of impulses spaced at regular intervals along the time and frequency axes. (See Papoulis, 1962, for a more formal derivation.)

The Fourier transform of the digitized interferogram, $I(t)\,\text{III}(t)$, is shown in the center of Figure 7.37. As can be seen from the figure, the discrete Fourier transform contains the same information as the continuous Fourier transform, but sampling has caused the information to be replicated. As long as these replicas can be distinguished from one another, the Fourier transform of the digitized interferogram will provide a wavelength-accurate representation of the spectrum, $A(f)$.

The sampling interval T is highly critical for proper reconstruction of $A(f)$. Figure 7.39 shows the effect of oversampling (c, d) and undersampling (g, h) the interferogram of a spectrum in which the maximum frequency is f. In each case, the sampled interferogram produces a version of the true spectrum (b), but only when sampling occurs at a "critical" value (e, f) is the transform fully specified without gaps or overlap. The critical sampling rate $T_c = (2f)^{-1}$ is known as the Nyquist frequency (Lathi, 1968).

Restated, the Nyquist sampling theorem says that the highest frequency present in the spectrum must be sampled at a rate of at least two points per cycle to be represented accurately. If less than two points per cycle are used,

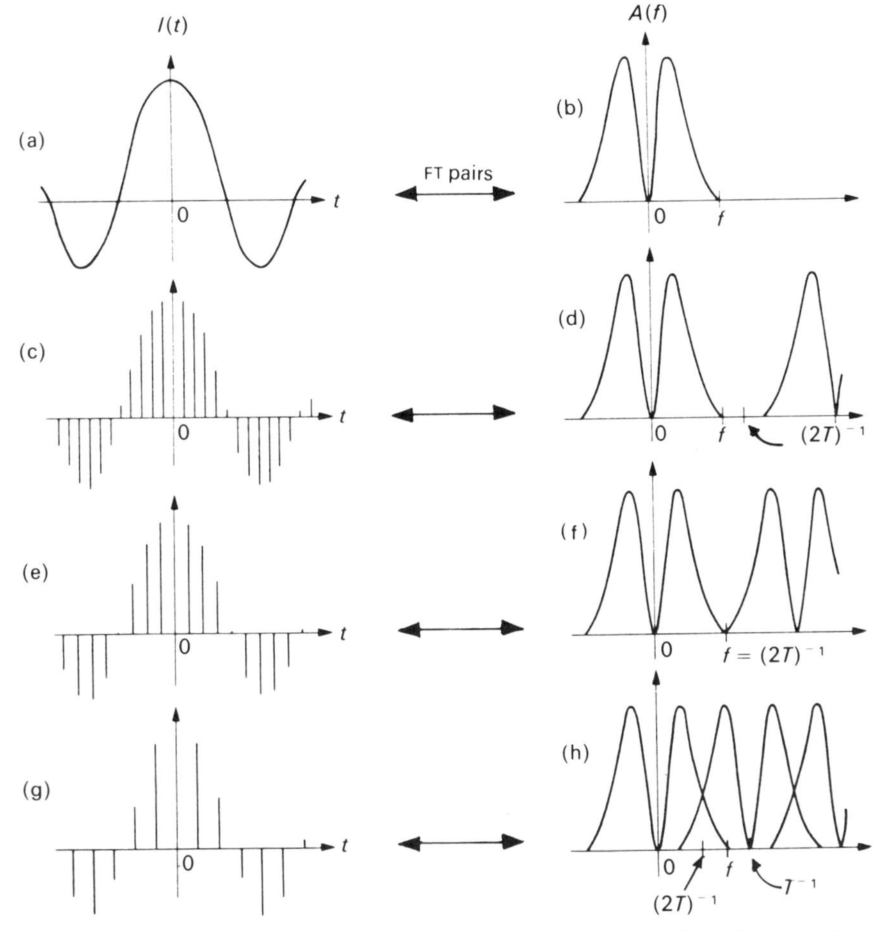

Figure 7.39. Effect of (c) subcritical, (e) critical, and (g) supercritical values of sampling interval on transformed interferogram. (a) Analog version of interferogram with (b) its true transform. [Reprinted from J. Chamberlain (1979), *The Principles of Interferometric Spectroscopy*. By permission of John Wiley and Sons, New York.]

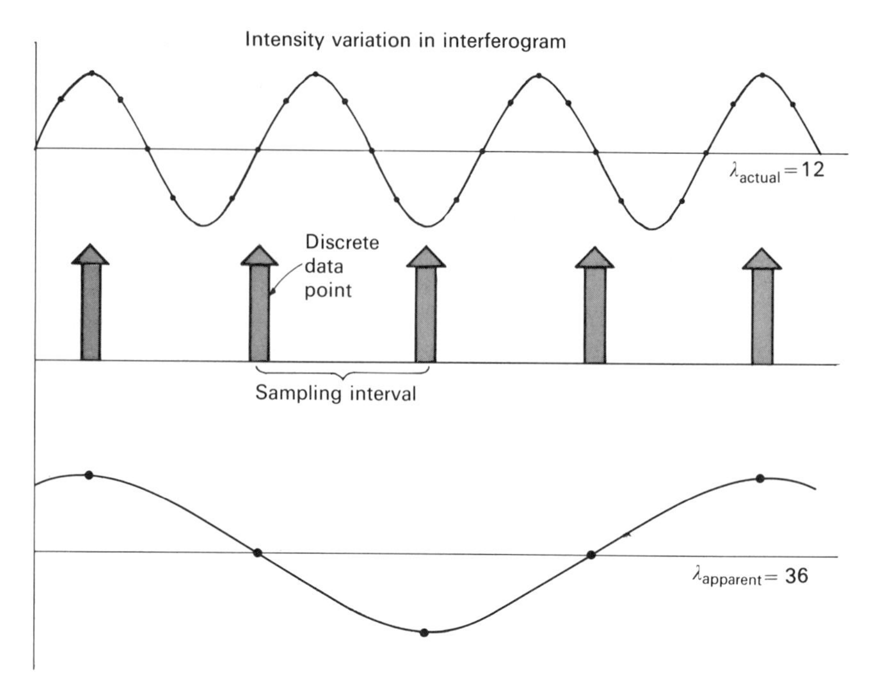

Figure 7.40. Intensity variation in sampled interferogram showing effect of undersampling.

the computer will see this frequency as being less than the actual frequency (Figure 7.40). Spectral components of higher frequency than the Nyquist frequency are "folded back" into the spectrum at a frequency exactly as much lower than the Nyquist frequency as the original frequency was higher than the Nyquist frequency. This phenomenon is also known as aliasing (Horlick and Malmstadt, 1970; Malmstadt et al., 1974).

For broadband continuous spectra, it is essential to avoid aliasing because of the overlapping spectral replicas that result. However, for relatively dilute line spectra, aliasing may sometimes be used to interleave these replicas without any actual overlap occurring (Horlick and Yuen, 1975; Yuen and Horlick, 1977). The advantage is a reduction in the amount of computer memory required to represent the interferogram and to make the transform. This is particularly important for work in the UV–visible region because short-wavelength spectra require a very high sampling rate. A major disadvantage of aliasing is the increase in spectral noise, the signal-to-noise ratio of the spectrum being degraded by a factor of $2^{1/2}$ each time the spectrum is folded.

To illustrate the phenomenon of aliasing, we consider the consequence of using a He–Ne laser referencing system for our interferometer. The basic

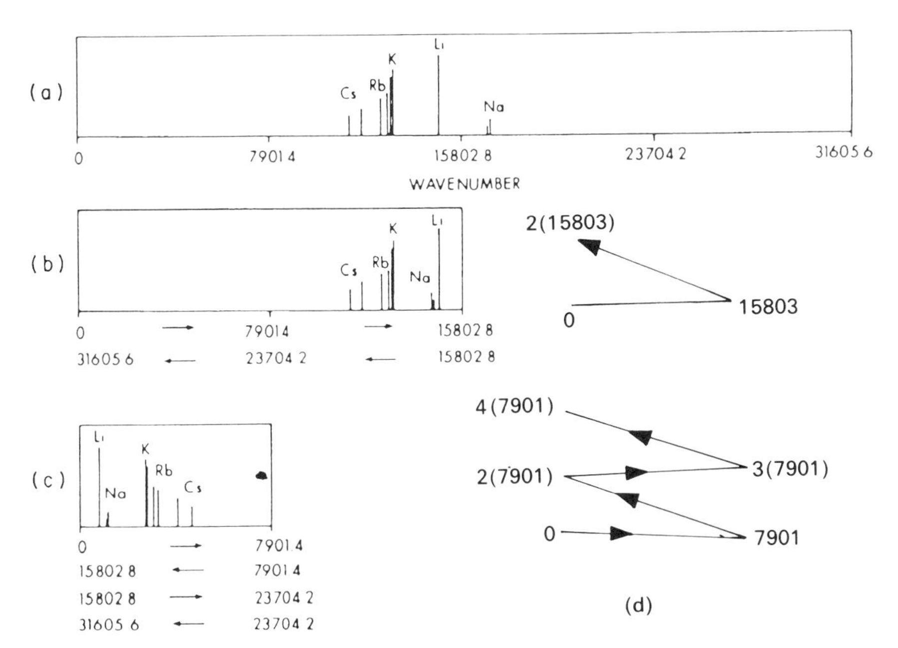

Figure 7.41. (*a*) Schematic spectrum of Li, Na, K, Rb, and Cs. (*b*) One-level aliasing. (*c*) Two-level aliasing. (*d*) Schematic folding of actual spectrum in one- and two-level aliasing. [Reprinted with permission from G. Horlick and W. K. Yuen (1982), "Atomic Emission Spectrochemical Measurements with a Fourier Transform Spectrometer," in *Fourier Transform Infrared Spectroscopy: Applications to Chemical Systems*, J. R. Ferraro and L. J. Basile (Eds.), Vol. 2, Academic, New York, Chapter 6, pp. 193–242.]

sampling interval for this laser is 0.6328 μm, so 1.266 μm (7901.4 cm^{-1}) is the shortest wavelength that can be sampled without foldover. Figure 7.41*a* shows a schematic flame emission spectrum of the alkali metals. All major lines fall in the near-infrared and visible spectral regions, well below 1.266 μm. In order to measure this spectrum without aliasing, it would be necessary to sample at about twice the normal rate of the He–Ne reference clock, that is, at 0.3164-μm intervals. This would achieve an unaliased spectrum down to 0.6238 μm (15802.8 cm^{-1}), sufficient for all lines except those of sodium.

Figure 7.41*b* shows the spectrum that would result from sampling at 0.3164-μm intervals. Foldover occurs at 15,803 cm^{-1}, the 15,803–31,606-cm^{-1} region being aliased into the 0–15,803-cm^{-1} region. This positions the sodium doublet between the potassium and lithium lines, so no spectral interference has been created.

In Figure 7.41*c*, the sampling interval has been reduced to 0.6328 μm, the basic He–Ne sampling rate. Now sodium is aliased twice and the remaining lines only once. The observed spectrum is the sum of all these replicas over the 0–7901-cm^{-1} interval. One simple way to think of this sum is to imagine that the original spectrum (7.41*a*) has been plotted on Z-fold paper with the folds at n(7901) cm^{-1} ($n=1, 2, 3, 4$). The paper is folded up so that when n is even, the spectrum is viewed normally, but for odd values of n the spectrum is reversed. Even with three layers of aliasing present, the lines do not overlap exactly, so interpretation of the final spectrum is still possible.

7.6.3. Beam Divergence

In Section 7.1, all radiation entering the Michelson interferometer was assumed to be perfectly collimated so that intersection of the beam with the two mirrors always occurred at right angles. In actual practice, the source is extended (astronomical sources being a possible exception), so radiation passing through the interferometer will diverge somewhat. Beam divergence limits the resolution of the interferometer and also shifts the wavelength of a computed spectral line from its true value.

To show the effect of beam divergence on resolution, we begin by considering that the source radiation is monochromatic. Figure 7.42 shows that at zero retardation both the on-axis ray (7.42*a*) and the off-axis rays (7.42*b*) experience constructive interference. However, when the movable mirror M_1 is displaced a distance d, the on-axis ray (7.42*c*) experiences an optical path difference $L=2d$, while the off-axis rays (7.42*d*) experience an optical path difference $L=2d/\cos\alpha$, where α represents the half angle of divergence. Therefore, a path difference Δx is created between the rays, where

$$\Delta x=\frac{2d}{\cos\alpha}-2d=2d\left(\frac{1-\cos\alpha}{\cos\alpha}\right) \tag{7.51}$$

Since α is small, we can truncate the series expansion representing the cosine function after two terms, that is,

$$\cos\alpha\approx 1-\tfrac{1}{2}\alpha^2$$

so,

$$\Delta x=2d\left(\frac{\alpha^2}{2\cos\alpha}\right)=d\alpha^2 \tag{7.52}$$

In modern interferometers, only the intensity variation at the center of the fringe pattern is monitored by the detector. During a typical scan, the movable mirror is driven from zero to some maximum path difference,

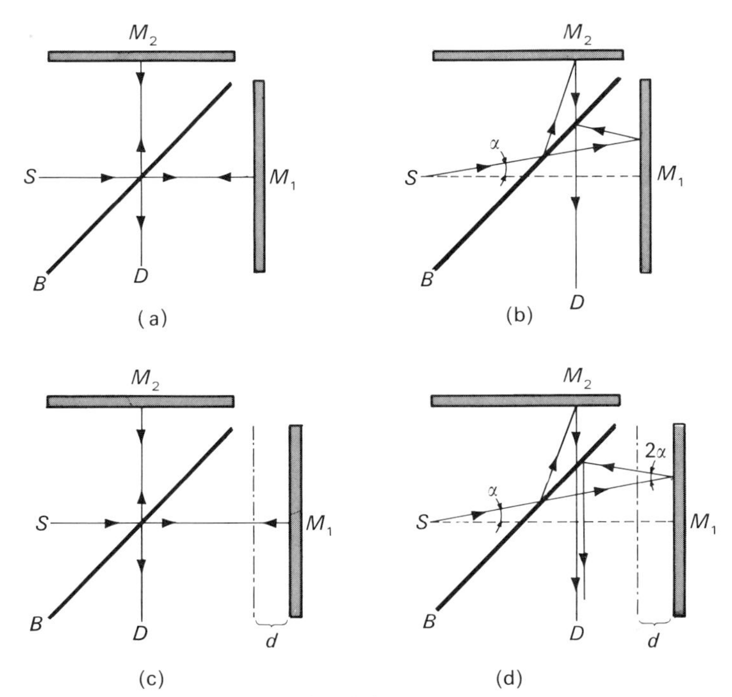

Figure 7.42. Schematic representation of radiation passing through Michelson interferometer. On-axis ray at (*a*) zero retardation and (*c*) mirror displacement *d*. Off-axis ray with divergence half angle α at (*b*) zero retardation and (*d*) mirror displacement *d*. Symbols: M_2, fixed mirror; M_1, movable mirror; *B*, beamsplitter; *S*, source; *D*, focusing lens and detector.

causing the center ring to decrease from one of infinite size (Figure 7.3*c*) to one of smaller (Figure 7.3*b*) and smaller (Figure 7.3*a*) diameter. If the detector sees too many of the rings in this pattern, the integrated intensity of the bright and dark fringes will average to a constant value, causing the cosine function to damp out (Figure 7.43). Therefore, it is important to match the size of the detector to the size of the center ring.

As the scan progresses from zero retardation, the extreme off-axis ray and the on-axis ray will first come into destructive interference when

$$\Delta x = \tfrac{1}{2}\lambda = (2\bar{\nu})^{-1} = d\alpha_{\max}^2 \tag{7.53}$$

Thus, beam divergence limits the size of the center ring, and for constant $\alpha_{\max}$, the extent of this limitation is directly proportional to wavelength.

For the case of polychromatic radiation, the detector must be matched to the radius of the center ring obtained at the maximum optical path difference,

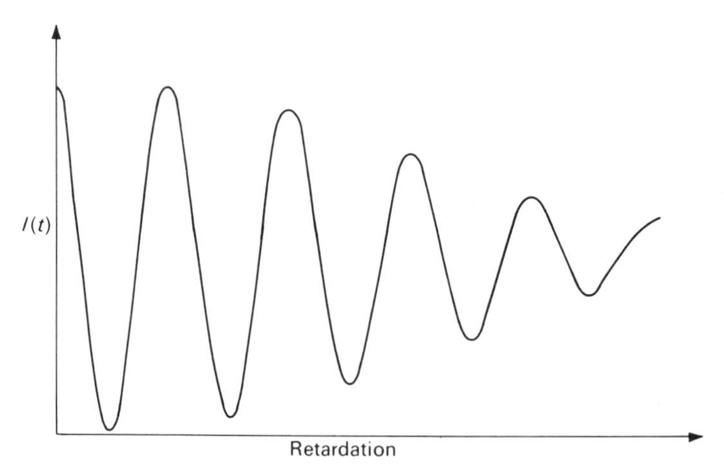

Figure 7.43. Monochromatic interferogram showing effect of damping.

$L_{max} = 2d_{max}$, required to resolve the shortest wavelength $(\bar{\nu}_{max}^{-1})$ present in the spectrum. Therefore,

$$\Delta x = (2\bar{\nu}_{max})^{-1} = d_{max}\alpha_{max}^2 \tag{7.54}$$

gives the conditions that effectively limit the size of the center ring, and

$$2d_{max} = (\Delta\bar{\nu})^{-1} \tag{7.4}$$

defines the optical path difference required to achieve the desired resolution $\Delta\bar{\nu}$.

Combining Eqs. 7.54 and 7.4 gives the greatest half angle divergence, α_{max}, that can be tolerated in the interferometer:

$$\alpha_{max} = (2d_{max}\bar{\nu}_{max})^{-1/2} = (R_I)^{-1/2} \tag{7.55}$$

where R_I is the resolving power $(R_I = \bar{\nu}_{max}/\Delta\bar{\nu})$. Therefore, the maximum solid angle Ω_{max} that can be accepted by the interferometer is

$$\Omega_{max} = 2\pi\alpha_{max}^2 = 2\pi/R_I \tag{7.56}$$

and the maximum throughput of the interferometer, Θ_I, cannot exceed

$$\Theta = 2\pi A_M/R_I \tag{7.5}$$

where A_M is the area of the mirrors being illuminated.

In Section 7.6.1 it appeared that the only limitation imposed on resolution was due to the finite length of the interferogram. In actual practice, an infinitely long path difference would never be practical because the interferogram is self-truncating. Whatever detector size is chosen, a critical value of d_{max} will finally be reached, too many fringes will be observed, and the interferogram will damp out. Damping out can be delayed with a corresponding increase in spectral resolution by decreasing the entrance aperture of the interferometer. But a decrease in aperture decreases the throughput. While interferometers are capable of achieving a throughput that is at least an order of magnitude greater than that of a dispersive instrument, in the end, the fundamental limitations of the two types of instruments are quite similar.

The effect of beam divergence on wavelength can be treated intuitively by recognizing that the path difference Δx between the on-axis and off-axis rays causes the off-axis rays to experience an increased retardation of $L+\Delta x$. Therefore, the true wavelength λ will be changed to an apparent wavelength λ'. The greater α_{max}, the greater the spread in values obtained for λ'. For monochromatic radiation, the effect makes a line source appear like a boxcar function, the width of the boxcar determined by α_{max}.

To a first approximation, the wavenumber of the line computed from the interferogram, $\bar{\nu}_{av}$, will be the mean of the wavenumbers represented by the on-axis, $\bar{\nu}$, and extreme off-axis rays. To obtain the exact solution,

$$\bar{\nu}_{av} = \bar{\nu}\left(\frac{1-\Omega_{max}}{4\pi}\right) \tag{7.57}$$

these wavenumbers must also be weighted according to their relative contributions to the total signal. Mathematically, this amounts to convolving each frequency in the transformed spectrum with a boxcar function that is exactly one resolution wide. The complete derivation may be found in Brault (1985) and Chamberlain (1979).

As can be seen from Eq. 7.57, $\bar{\nu}_{av}$ is wavelength dependent and is also determined by the field of view of the interferometer. It is possible to compensate for wavelength shifts by using a scaling factor during computation of the final spectrum. A variety of optical methods for field compensation can also be employed (Baker, 1977).

REFERENCES

Baker, D. (1977), "Field-Widened Interferometers for Fourier Spectrometers," in *Spectrometric Techniques*, Vol. 1, Vanasse, G. A. (Ed.), Academic, New York, Chapter 2, pp. 71–106.

Baly, E. C. C. (1927). *Spectroscopy*, Vol. 2, 3rd ed., Longmans Green, New York, Chapter 1.

Beevers, C. A. and Lipson, H. (1936), *Nature, 137*, 825.

Bell, R. J. (1972), *Introductory Fourier Transform Spectroscopy*, Academic, New York.

Born, M. and Wolf, E. (1986), *Principles of Optics*, 6th (corrected) ed., Pergamon, New York.

Bracewell, R. (1986), *The Fourier Transform and Its Applications*, 2nd (revised) ed., McGraw-Hill, New York.

Bracewell, R. N. (1989), *Scientific American, 260*, p. 86.

Brault, J. W. (1985), "Fourier Transform Spectrometry." in *High Resolution in Astronomy*, Benz, A., Huber, M., and Mayor, M. (Eds.), 15th Advanced Course, Swiss Society of Astronomy and Astrophysics, March 25–30, 1985 (Valais, Switzerland), Geneva Observatory, Chemin des Maillettes 51, CH-1290, Sauverny, Switzerland.

Breckinridge, J. B. and Schindler, R. A. (1981), "First Order Optical Design for Fourier Spectrometers," in *Spectrometric Techniques*, Vol. 2, Vanasse, G. A. (Ed.), Academic, New York, Chapter 2, pp. 63–160.

Brigham, E. O. (1974), *The Fast Fourier Transform*, Prentice-Hall, Englewood Cliffs, NJ.

Chamberlain, J. (1979), *The Principles of Interferometric Spectroscopy* (completed, collated, and edited by Chantry, G. W. and Stone, N. W. B.), Wiley, New York.

Chase, B. (1987), *Anal. Chem., 59*, 881A.

Comisarow, M. B. (1982), "Fourier Transform Ion Cyclotron Resonance Spectroscopy," in *Fourier, Hadamard, and Hilbert Transformations in Chemistry*, Marshall, A. G. (Ed.), Plenum, New York, Chapter 5, pp. 125–146.

Connes, J. (1961), Thesis, University of Paris, October 7, 1960, subsequently published, *Rev. Opt. Theor. Instrum., 40*, 45, 116, 171, 231.

Connes, J. (1971), *Aspen International Conference on Fourier Spectroscopy*, Vanasse, G. A., Stair, A. T., and Baker, D. J. (Eds.), AFCRL-71-0019 Special Report No. 114, Aspen, CO, 1970, p. 83, National Technical Information Service, Springfield, MA.

Connes, J. and Connes, P. (1966), *J. Opt. Soc. Am., 56*, 896.

Connes, J., Connes, P., and Maillard, J. P. (1969), *Atlas of Near Infrared Spectra of Venus, Mars, Jupiter and Saturn*, CNRS, Paris.

Connes, J. and Gush, H. P. (1959), *J. Phys. Radium, 20*, 915.

Connes, J. and Gush, H. P. (1960), *J. Phys. Radium, 21*, 645.

Connes, P. (1969), *Lasers and Light*, Freeman, San Francisco.

Connes, P. and Michel, G. (1975), *Appl. Opt., 14*, 2067.

Cooley, J. W. and Tukey, J. W. (1965), *Math. Comput., 19*, 297.

Cooley, J. W., Garwin, R. L., Rader, C. M., Bogert, B. P., and Stockham, T. C. (1969), in "The 1968 Arden House Workshop on Fast Fourier Transform Processing," *IEEE Trans. Audio Electroacoustics, AU-17* (2), pp. 66–76.

Cooper, J. W. (1978), "Advanced Techniques in Fourier Transform NMR," in

Transform Techniques in Chemistry, Griffiths, P. R. (Ed.), Plenum, New York, Chapter 9, pp. 227–255.

Davisson, C. J. (1937), "The Discovery of Electron Waves," in *Nobel Lectures in Physics, 1922–1941*, Elsevier, Amsterdam, pp. 381–394. (Read, December 13, 1937.)

Duffieux, P. M. (1983), *The Fourier Transform and Its Application in Optics*, 2nd ed., Wiley-Interscience, New York.

Fabry, C. and Perot, A. (1901), *Ann. Chim. Phys., 22*, 564.

Farrar, T. C. (1978), "Pulsed and Fourier Transform NMR Spectroscopy," in *Transform Techniques in Chemistry*, Griffiths, P. R. (Ed.), Plenum, New York, Chapter 8, pp. 199–226.

Fellgett, P. B. (1951), Ph.D. Thesis, University of Cambridge.

Fellgett, P. B. (1958a), *J. Phys. Radium, 19*, 237.

Fellgett, P. B. (1958b), *J. Scient. Instrum.*, *35*, 257.

Fellgett, P. B. (1967), *J. Phys.*, *28*, 165.

Ferraro, J. R. and Basile, L. J. (Eds.) (1978, 1979, 1982), *Fourier Transform Infrared Spectroscopy: Applications to Chemical Systems*, Vols. I, II, and III, Academic, New York.

Filler, A. H. (1964), *J. Opt. Soc. Am.*, *54*, 762.

Fink, U. and Larson, H. P. (1979), "Astronomy: Planetary Atmospheres," in *Fourier Transform Infrared Spectroscopy: Applications to Chemical Systems*, Vol. 2, Ferraro, J. R. and Basile, L. J. (Eds.), Academic, New York, Chapter 7, pp. 243–314.

Fizeau, H. (1862), *Comp. Rend.*, *54*, 1237.

Flygare, W. H. (1982), "Pulsed Fourier Transform Microwave Spectroscopy," in *Fourier, Hadamard, and Hilbert Transformations in Chemistry*, Marshall, A. G. (Ed.), Plenum, New York, Chapter 8, pp. 207–270.

Gebbie, H. A. and Vanasse, G. A. (1956), *Nature*, *178*, 432.

Griffiths, P. R. (1978), "Fourier Transform Infrared Spectrometry: Theory and Instrumentation," in *Transform Techniques in Chemistry*, Griffiths, P. R. (Ed.), Plenum, New York, Chapter 5, pp. 109–139.

Griffiths, P. R. and deHaseth, J. A. (1986), *Fourier Transform Infrared Spectrometry*, Wiley, New York.

Hardy, A. C. and Perrin, F. H. (1932), *The Principles of Optics*, 1st ed., McGraw-Hill, New York.

Harrison, G. R. and Stroke, G. W. (1955), *J. Opt. Soc. Am.*, *45*, 112.

Hirschfeld, T. (1979), "Quantitative FT–IR: A Detailed Look at the Problems Involved," in *Fourier Transform Infrared Spectroscopy: Applications to Chemical Systems*, Vol. 2, Ferraro, J. R. and Basile, L. J. (Eds.), Academic, New York, Chapter 6, pp. 193–242.

Horlick, G., Hall, R. H., and Yuen, W. K. (1982), "Atomic Emission Spectrochemical Measurements with a Fourier Transform Spectrometer," in *Fourier Transform Infrared Spectroscopy: Applications to Chemical Systems*, Vol. 3, Ferraro, J. R. and Basile, L. J. (Eds.), Academic, New York, Chapter 2, pp. 37–81.

Horlick, G. and Malmstadt, H. V. (1970), *Anal. Chem.*, *42*, 1361.

Horlick, G. and Yuen, W. K. (1975), *Anal. Chem.*, *47*, 775A.

Jacquinot, P. (1954), *J. Opt. Soc. Am.*, *44*, 761.

James, J. F. and Sternberg, R. S. (1969), *The Design of Optical Spectrometers*, Chapman & Hall, London.

Jenkins, F. A. and White, H. E. (1976), *Fundamentals of Optics*, 4th ed., McGraw-Hill, New York.

Kauppinen, J. K., Moffatt, D. J., Cameron, D. G., and Mantsch, H. H. (1981), *Appl. Opt.*, *20*, 1866.

Klainer, S. M., Hirschfeld, T. B., and Marino, R. A. (1982), "Fourier Transform Nuclear Quadrupole Resonance Spectroscopy," in *Fourier, Hadamard, and Hilbert Transformations in Chemistry*, Marshall, A. G. (Ed.), Plenum, New York, Chapter 6, pp. 147–182.

Kuiper, G. P. (1962), *Commun. Lunar Planet. Lab.*, *1*, 83.

Lathi, B. P. (1968), *Communications Systems*, Wiley, New York, p. 89.

Loewenstein, E. V. (1966), *Appl. Opt.*, *5*, 845.

Malmstadt, H. V., Enke, C. G., Crouch, S. R., and Horlick, G. (1974), *Optimization of Electronic Measurements*, Benjamin, New York.

Marshall, A. G. and Comisarow, M. B. (1978), "Multichannel Methods in Spectroscopy," in *Transform Techniques in Chemistry*, Griffiths, P. R. (Ed.), Plenum, New York, Chapter 3, pp. 39–68.

Mertz, L. (1958), *J. Phys. Radium*, *19*, 233.

Mertz, L. (1971), "Fourier Spectroscopy: Past, Present, and Future," in *Aspen International Conference on Fourier Spectroscopy*, Vanasse, G. A., Stair, A. T., and Baker, D. J. (Eds.), AFCRL-71-0019 Special Report No. 114, Aspen, CO, 1970, National Technical Information Service, Springfield, MA.

Mertz, L. (1973), *Opt. Commun.*, *6*, 354.

Michelson, A. A. (1881), *Am. J. Sci.*, *22*, 1920.

Michelson, A. A. (1891a), *Phil. Mag.*, *31*, 256.

Michelson, A. A. (1891b), *Phil. Mag.*, *31*, 338.

Michelson, A. A. (1892), *Phil. Mag.*, *34*, 280.

Michelson, A. A. (1903), *Light Waves and Their Uses*, University of Chicago Press, Chicago.

Michelson, A. A. (1927), *Studies in Optics*, University of Chicago Press, Chicago.

Michelson, A. A. and Benoit, R. (1895), *Trav. Mem. Bur. Poids Mes.*, *XI*.

Michelson, A. A. and Morley, E. W. (1887), *Phil. Mag.*, *24*, 449.

Morris, G. A. (1982), "Two-Dimensional Fourier Transform NMR Spectroscopy," in *Fourier, Hadamard, and Hilbert Transformations in Chemistry*, Marshall, A. G. (Ed.), Plenum, New York, Chapter 9, pp. 271–306.

Nordstrom, R. J. (1982), "Aspects of Fourier Transform Visible/UV Spectroscopy," in *Fourier, Hadamard, and Hilbert Transformations In Chemistry*, Marshall, A. G. (Ed.), Plenum, New York, Chapter 14, pp. 421–452.

Norton, R. H. and Beer, R. (1976), *J. Opt. Soc. Am.*, *66*, 259.

Norton, R. H. and Beer, R. (1977), *J. Opt. Soc. Am.*, *67*, 419.

Papoulis, A. (1962), *The Fourier Integral and Its Applications*, McGraw-Hill, New York, p. 44.

Rayleigh, J. W. S. (1892), *Phil. Mag.*, *34*, 407.

Richards, P. L. (1964), *J. Opt. Soc. Am.*, *54*, 1474.

Ross, M. A. S. (1943), *Nature*, *152*, 302.

Rubens, H. and Wood, R. W. (1911), *Phil. Mag.*, *21*, 249.

Sakai, H. (1977), "High Resolving Power Fourier Spectroscopy," in *Spectrometric Techniques*, Vol. 1, Vanasse, G. A. (Ed.), Academic, New York, Chapter 1, pp. 1–70.

Shamos, M. H. (1959), *Great Experiments in Physics: Firsthand Accounts from Galileo to Einstein*, Dover, New York.

Steward, E. G. (1987), *Fourier Optics: An Introduction*, 2nd ed., Halsted, New York.

Stroke, G. W. (1958), *J. Phys. Radium*, *19*, 415.

Swenson, Jr., L. S. (1972), *The Ethereal Aether: A History of the Michelson–Morley–Miller Aether-Drift Experiments, 1880–1930*, University of Texas Press, Austin.

Swenson, Jr., L. S. (1987), *Invent. Tech.*, *3*, 43.

Terrien, J. (1958), *J. Phys. Radium*, *19*, 390.

Terrien, J. and Hamon, J. (1954), *C. R. Acad. Sci.*, *239*, 586.

Terrien, J. and Masui, T. (1956), *C. R. Acad. Sci.*, *243*, 776.

Theophanides, T., Ed. (1984), *Fourier Transform Infrared Spectroscopy*, Reidel, Dordrecht.

Thomson, G. P. (1938), "Electron Waves," in *Nobel Lectures in Physics, 1922–1941*, Elsevier, Amsterdam, pp. 397–403. (Read, June 7, 1938.)

Vanasse, G. A., Ed. (1977, 1981), *Spectrometric Techniques*, Vols. I and II, Academic, New York.

Vanasse, G. A., Stair, A. T., and Baker, D. J., Eds. (1971), *Proc. Aspen Int. Conf. Fourier Spectrosc.*, AFCRL 71-0019 Special Report No. 114, Aspen, CO, 1970, National Technical Information Service, Springfield, MA.

Wood, R. W. (1934), *Physical Optics*, 3rd (revised) ed., MacMillan, New York.

Young, T. (1803), "Experiments and Calculations Relative to Physical Optics," Bakerian Lecture, Royal Society of London, November 24.

Young, T. (1804), *Phil. Trans.*, pp. 1–16.

Yuen, W. K. and Horlick, G. (1977), *Anal. Chem.*, *49*, 1446.

CHAPTER

8

PRINCIPLES OF PHOTOGRAPHIC DETECTION

So far, we have discussed the various ways in which polychromatic light can be separated into its component wavelengths but have not discussed the detection process whereby this radiation is converted to some readily observable and hopefully quantifiable parameter. A wide variety of parameters have been employed as measures of the amount of radiation incident on a detector surface. In the electrical data domain, transducers that convert radiant energy into an electrical quantity such as conductance, voltage, current, capacitance, and resistance are all well known (Pinson, 1985) and will be discussed in later chapters. However, no book on detection systems for multielement analysis would be complete without some discussion of photographic detection.

A photographic plate used in spectrochemical work typically consists of a glass plate upon which has been deposited a photosensitive layer or *emulsion*. The emulsion typically consists of a suspension of finely divided silver bromide crystals (~ 1–$5\ \mu$m in diameter) in a gelatin matrix. Gelatin is a nearly ideal matrix to hold the silver bromide crystals in position because it is transparent to a wide range of wavelengths and its swelling–permeability characteristics facilitate solution chemical processing of the exposed photographic plate (Sawyer, 1963). In addition, when properly dried, the gelatin layer returns to its initial form and position after processing.

Before discussing the photographic process, it is worthwhile to summarize some of the advantages of photographic detection in multielement analysis. The three most important advantages of photographic detection for spectrochemical analysis are (Harrison, Lord, and Loofbourow, 1948) (1) all wavelengths shorter than about 1.3 μm are accessible; (2) spectral features covering broad regions of the spectrum can be monitored simultaneously; and (3) even relatively weak spectral sources can be employed if a sufficiently long exposure time is possible because the photographic emulsion is an integrating detector responding to the total energy incident upon it rather than the radiant power. In addition, the photographic plate represents a permanent record of the spectrum, which can be stored for future reference. In contrast to the various image detectors that will be discussed later in this book, there is no difficulty in fabricating photographic plates that can cover wide wavelength

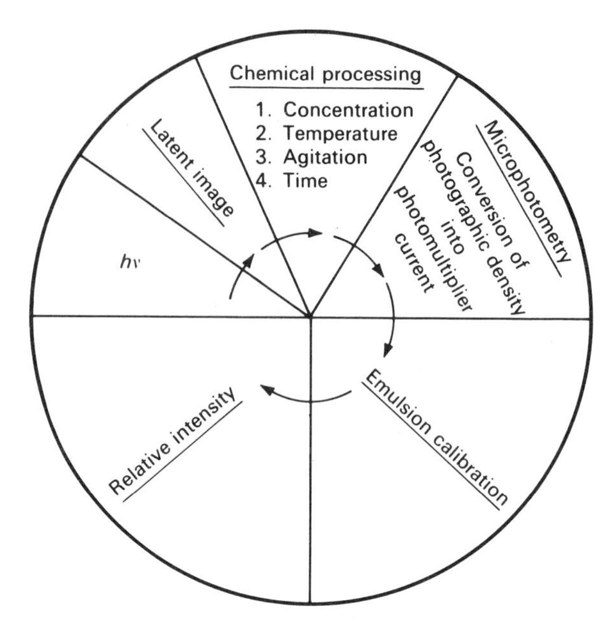

Figure 8.1. Data domain conversions involved in photographic photometry.

regions of the spectrum when used in a typical spectrograph (i.e., plates 4 × 10 in. are readily available). Finally, photographic emulsions are capable of high resolution under proper conditions.

In spite of the aforementioned advantages, photographic detection is not without disadvantages as a means of radiation detection. Chief among these is the need for chemical processing prior to readout of the stored information. This processing not only requires additional facilities (i.e., a darkroom and related equipment) but introduces a host of chemically related variables that must be controlled during the development process. As a result of the time required for chemical processing, photographic detection is not a real-time means of detection. Even after the latent image present on the exposed plate has been transformed into a visible form by developing, the information stored on the plate in terms of varying degrees of blackening must be retrieved by means of microphotometry before the intensity information can be obtained in numerical form. Figure 8.1 shows the various data domain conversions involved in photographic photometry.

8.1. THE PHOTOGRAPHIC PROCESS

When a photographic plate is exposed to light, a *latent image* that is not directly observable is formed in the emulsion. This latent image is due to small changes in the crystal lattice of the silver halide. According to the Gurney–Mott mechanism (Levi, 1980), the absorption of a photon results in the promotion of an electron into the conduction band of the crystal. Defects within the crystal structure then act as traps for the mobile electrons produced by photon absorption. Traps containing these electrons become negatively charged, thereby attracting mobile interstitial silver ions. These silver ions attracted to the trap are reduced to metallic silver. This process can then repeat itself until a small grain of metallic silver is formed at the trap. This grain then becomes what is known as a *development center*. It is the formation of the tiny metallic silver speck in the latent image that acts as a catalyst during the development process, causing the entire silver bromide grain to be reduced to metallic silver.

Development of the latent image formed by exposure of the plate to light is a three-step chemical process. In the first step, the plate is immersed in a bath containing a reducing agent known as the *developer*. When conducted under proper conditions of concentration, temperature, time, and agitation, this treatment will reduce those silver halide grains that were sensitized by exposure to light but will not affect similar grains not so exposed. The chemical reduction process results in the production of metallic silver particles that are stable to further exposure to light. If development is continued too long, even grains that have not been exposed to light begin to react, producing what is known as *fog*.

The second step in the process is to immerse the plate briefly in a dilute solution of acetic acid known as the *stop bath*. As the name implies, this bath prevents further action of the developer.

The final treatment is required to remove the remaining unexposed grains of silver bromide from the gelatin matrix. This is done with a chemical complexing agent known as the *fixer*. The reagent commonly used for this purpose is a solution of sodium thiosulfate, which is a strong complexing agent for silver ion. Solutions of sodium thiosulfate are sometimes referred to in the photographic literature as *hypo*.

After the three chemical treatments have been conducted for the proper periods of time, the plate is washed for a period of about 30 min in running water to remove the remaining chemicals. The final step in the process is to dry the plate.

The overall high sensitivity of the photographic process is the result of a chemical amplification process quite analogous to the concept of electrical amplification in that it is characterized by a *gain*. This gain results from the fact that the absorption of a mere 20 photons can sensitize the entire crystal by the formation of a development center so that 10^9 silver ions are converted to metallic silver during development (Levi, 1980).

8.2. RESPONSE OF THE EMULSION TO LIGHT

The photographic response of a plate toward exposure to light is, after development, observable as a blackening of the emulsion due to the tiny grains of metallic silver that have been formed. For quantitative purposes the degree of blackening must be measured by determining the amount of light transmitted through the photographic image. The degree of blackening is expressed in this process in terms of a quantity known as *density*, defined as

$$D=\log(I_0/I)=\log O=\log(1/T) \tag{8.1}$$

where I_0 is the amount of light transmitted through an unexposed portion of the plate and I is the amount of light transmitted through some exposed portion of the plate. The ratio I_0/I is known as the opacity O, while the inverse ratio is known as the transmission T.

If the density, as defined in the preceding discussion, is measured as a function of the exposure, an *emulsion characteristic curve* can be prepared. The exposure H is defined as

$$H=\int_0^t E(t)\,dt \tag{8.2}$$

where $E(t)$ is the variation of irradiance with time. For a constant irradiance, $H=E^*t$. If the density D is plotted versus log H, a characteristic curve known as a *Hurter and Driffield* (*HD*) *plot* is obtained.

The characteristic curve obtained with an HD plot is sigmoidal and can be divided into several regions, as shown in Figure 8.2. In this figure, the irradiance is colloquially referred to as the intensity. The initial region for low exposure is known as the *toe* (Slavin, 1971) and corresponds to underexposure of the plate. The region of correct exposure is characterized by the straight-line portion of the characteristic curve. The *shoulder* region where the response begins to fall off corresponds to overexposure of the plate. The region where the response begins to decrease with increasing exposure corresponds to exposures that are so intense that factors that destroy the development centers begin to dominate. This phenomenon is known as *solarization.*

The linear region of the curve is referred to as the *latitude* and is characterized by a slope referred to as *gamma.*

$$\gamma=\Delta D/\Delta \log H=\tan^{-1}\theta \tag{8.3}$$

In photographic terminology, gamma represents the *contrast* of the photographic material. If the linear portion of the curve is extrapolated back to the

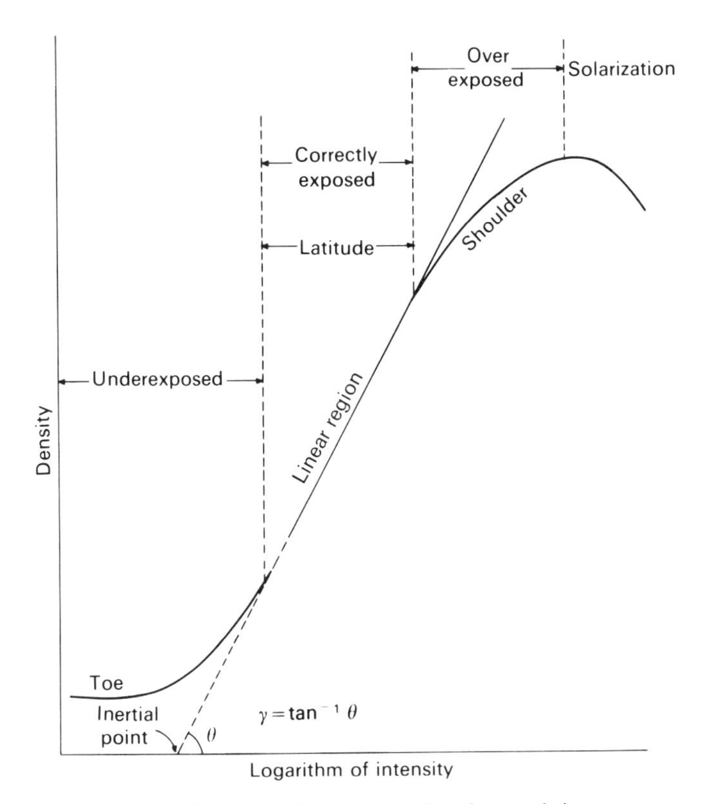

Figure 8.2. Hurter and Driffield emulsion characteristic curve.

abscissa, the point where it intersects the abscissa is known variously as the emulsion *inertia* or the *inertial point*. While most authors in this area refer to this point as inertia, Slavin (1971) refers to it as the inertial point.

From an analytical standpoint, the HD plot is significant because it represents the calibration curve for the emulsion. Since this calibration curve is a function of wavelength, individual calibrations must be prepared every 25 Å along the plate in careful spectrochemical work. Since it is convenient for purposes of linear regression to have linear calibration curves, various ways of linearizing the HD plot have been proposed. Probably the most familiar is the so-called *Seidel transformation* in which Δ is plotted versus $\log H$, where Δ is defined by

$$\Delta = \log[(1/T) - 1] \tag{8.4}$$

Figure 8.3 shows a plot of Seidel density versus $\log H$.

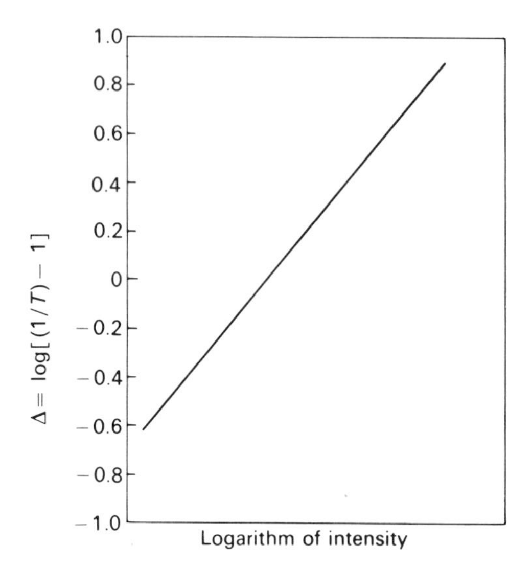

Figure 8.3. Seidel emulsion characteristic curve.

8.3. MISCELLANEOUS EMULSION PROPERTIES

Several additional emulsion properties deserve mention. The first of these is the *Bunsen–Roscoe reciprocity law* (Levi, 1980), which states that the photographic response depends only on the total energy absorbed and is independent of the time interval of the exposure. According to the reciprocity relationship,

$$H = E_1 t_1 = E_2 t_2 = \text{const.} \tag{8.5}$$

and the same exposure should result for different irradiances (E) and times as long as their product is constant. Although this law is reasonably reliable for intermediate exposure times, deviations are observed when very short and very long exposure times are involved. Figure 8.4 is a typical reciprocity curve, which is a plot of log H versus log E. This effect is not a function of wavelength.

A second, related property is the so-called *intermittency effect*. If a photographic plate is exposed to a periodically interrupted exposure, it is often observed that for certain chopping frequencies the interrupted exposure does not produce the same photographic response as an equivalent continuous exposure.

Both of the aforementioned effects can be explained in terms of the Gurney–Mott mechanism. In this mechanism, photoelectrons promoted to the

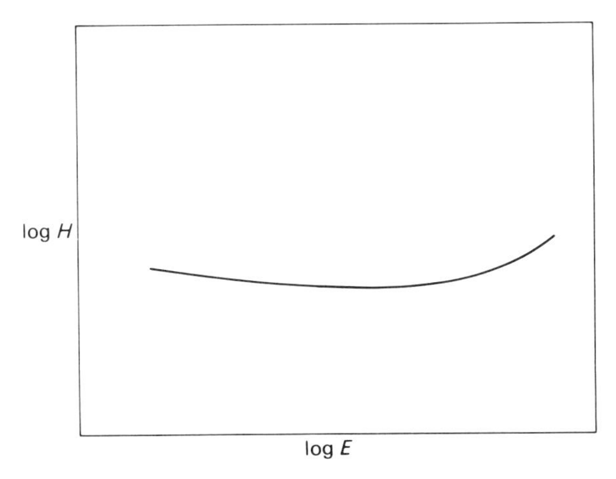

Figure 8.4. Typical photographic reciprocity behavior.

conduction band become localized in traps where they capture silver ions, reducing them to metallic silver. According to the mechanism, the first silver ion trapped is not tightly bound to the site and may escape unless another silver ion is trapped within a short time period. As a result, when the irradiance on the plate is low, escape of the initial silver ion from the sensitivity site diminishes the photographic effect. Under conditions of high irradiance, a second photoelectron may arrive at the trap before the first has had time to capture a silver ion. In this case, the second photoelectron is repelled from the trap and lost, producing a diminution of the photographic response. In the case of intermittent exposure, if each silver halide crystal receives an average of one photon per modulation cycle, the time duration of the off period has no effect on the photographic response. The photographic response produced is therefore equal to that obtainable with a continuous exposure at the mean irradiance. Thus at intermediate irradiance levels, the exposure is found to be independent of chopping frequency (Levi, 1980).

Two other effects often observed in spectrochemical work come under the category of *adjacency effects*. An adjacency effect is said to occur when one point on the image affects the processing of a neighboring point. Adjacency effects arise during the developing process as a result of diffusional limitations by processing chemicals within the gelatin.

Figure 8.5 shows two effects known respectively as the *Eberhardt effect* and the *Kotinsky effect*. Both of these effects arise at the border between a region of high exposure and one of low exposure. At the border between the two regions diffusion of fresh and spent developer takes place laterally within the emulsion. Thus, fresh developer is diffusing laterally from the low-exposure

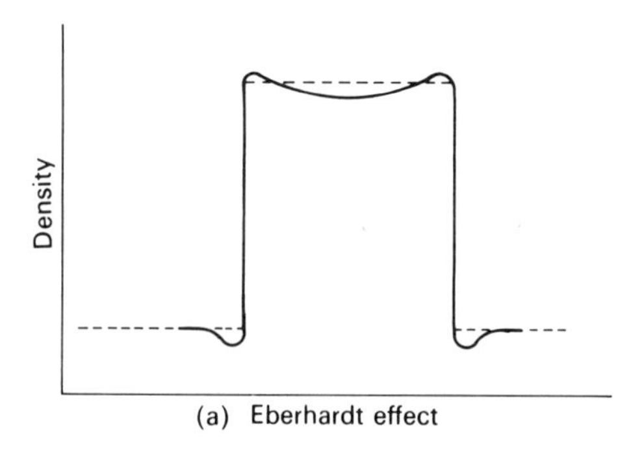

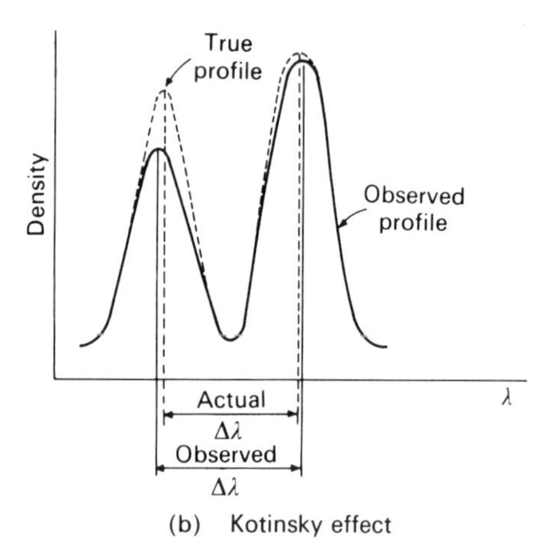

Figure 8.5. Photographic adjacency effects: (*a*) Eberhardt effect; (*b*) Kotinsky effect.

region to the high-exposure region, producing a slight overdevelopment of the image at the edge. At the same time, spent developer is diffusing from the high-exposure region to the low-exposure region, retarding the development process in the low-exposure region slightly. These processes result in a distortion of the developed image compared with the true image.

In the Eberhardt effect, the results of diffusional processes are most noticeable for a wide spectral line. In this case, diffusion of fresh developer at the edge of the line results in increased density, while the center of the line experiences decreased density from being in contact with partially exhausted developer. In the Kotinsky effect, the line profiles of two closely spaced spectral lines are distorted during development so that the distance between the centers of maximum density is altered (Mika and Török, 1974). This effect can produce errors in the determination of wavelength.

8.4. PHOTOGRAPHIC PHOTOMETRY IN SPECTROCHEMICAL ANALYSIS

At this point, it is worthwhile to describe the procedure used in applying photographic photometry to spectrochemical analysis. To illustrate the procedure, we describe the steps involved in a typical spectrochemical analysis. Data for this illustration have been taken from two sources (Ahrens and Taylor, 1961; Feldman, 1953) and combined. This example will illustrate the determination of potassium in a geological sample using a dc arc as the excitation source. In the determination, rubidium will be used as an internal standard. The analytical line that has been selected for potassium is 404.7 nm, while the internal standard line employed for rubidium is 420.2 nm. Since the purpose of this discussion is to illustrate the photometric principles involved in a determination, other analytical considerations such as line selection, selection of an internal standard, sample preparation and excitation, as well as matrix effects will not be discussed. For a complete discussion of the principles of spectrochemical analysis, the reader is referred to the standard references on this subject (Ahrens and Taylor, 1961; Feldman, 1953; Harvey, 1950; Mika and Török, 1974; Nachtrieb, 1950; Slavin, 1971).

The first step in the photometric procedure involves the preparation of the emulsion calibration curve in the form of either the HD plot or its Seidel transformation. In either case, this procedure involves plotting some function of the plate density versus the logarithm of the relative intensity (i.e., irradiance actually).

Plate density measurements are made by determining the transmission properties of the plate using a device known as a *microphotometer*, commonly referred to in this country as a *densitometer*. For a complete discussion of the various types of densitometers available and their use, the reader is referred to the standard spectroscopy references (Harrison et al., 1948; Sawyer, 1963). For our present purposes we assume that a functioning densitometer is available for making transmission measurements on spectrographic plates.

8.4.1. Methods for Producing a Graded Series of Exposures

The establishment of a scale of relative intensity is somewhat less straightforward. A variety of methods for obtaining a series of accurately known relative intensities have been proposed, none of which is entirely satisfactory. For a discussion of the pros and cons of the various techniques, the reader is referred to Slavin (1971), Ahrens and Taylor (1961), and Harrison et al. (1948).

The ideal situation would be to vary the plate exposure by varying the irradiance of the light source for a series of exposures of fixed duration. This procedure most nearly approximates what is done in a quantitative analysis where a series of different standards is photographed with a fixed exposure time. Emulsion calibrations prepared by varying the intensity of the light source for a fixed exposure period are referred to as *intensity-scale characteristics*. An alternative procedure, albeit somewhat less satisfactory, is to vary the plate exposure by keeping the source intensity constant and varying the exposure time. This procedure results in what is commonly referred to as a *time-scale characteristic*. This method of plate calibration will be satisfactory as long as the reciprocity law holds and intermittency effects are avoided.

Intensity-scale measurements are commonly made using a graded neutral density filter known as a *step filter*. Figure 8.6 shows a typical two-step filter for spectrochemical work. The filter consists of a quartz plate, part of which has been coated with a thin film of metal that has been deposited by sputtering or evaporation. Typical metallic films are prepared from platinum, palladium, and rhodium (Mika and Török, 1974). Since perfectly neutral filters (i.e., filters whose transmission does not vary with wavelength) cannot be prepared, the neutrality of any filter used should be verified spectrophotometrically over the wavelength interval of interest before use. In a two-step filter, the clear quartz portion would have a relative transmission of 100 units while the coated portion might have a relative transmission of 60 or 80 units.

In use, the filter is mounted immediately in front of the spectrograph slit. If the filter is mounted as shown in Figure 8.6 so that the steps are perpendicular to the slit, then the slit image will be correspondingly attenuated as long as the spectrograph is stigmatic (see Chapter 4). Naturally, in order for the intensity attenuation along the spectral line to be related to the transmission properties of the filter, it is necessary that the entire slit be uniformly illuminated. This may not be easy to achieve in practice if a tall slit is employed (see Chapter 4).

Another way to produce exposure variations along a spectral line is by using a so-called *rotating step sector*. A step sector consists of a disk with steps

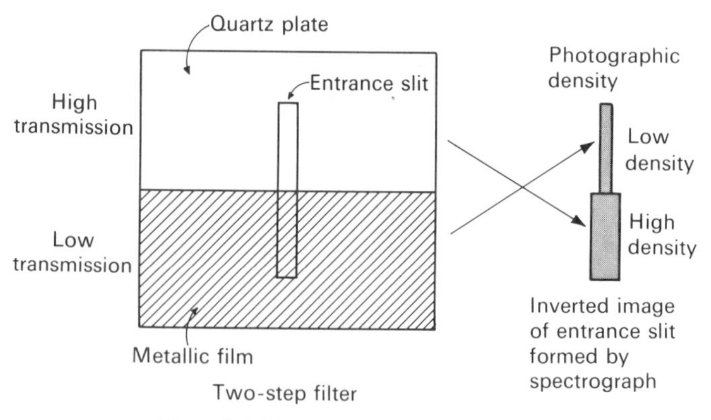

Figure 8.6. Two-step neutral density filter.

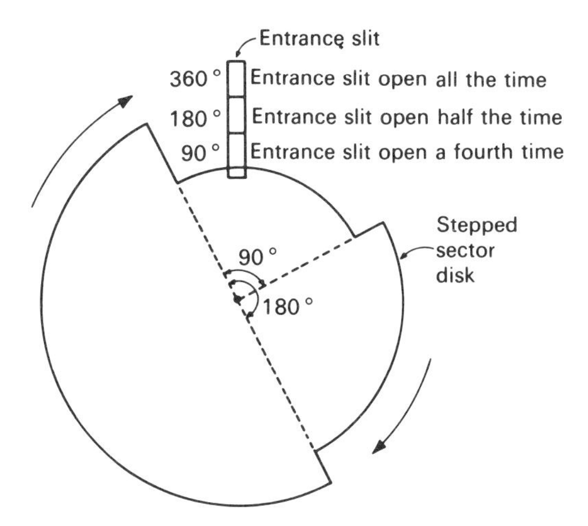

Figure 8.7. Rotating step sector.

cut along the edge as shown in Figure 8.7. When the disk is positioned at the stigmatic point immediately in front of the spectrograph slit and rotated at some constant speed, the exposure along the spectral line is varied by varying the exposure time in a known fashion. This form of calibration results in a time-scale characteristic.

As with the step filter, use of the rotating step sector requires that the slit be uniformly illuminated. To avoid errors due to intermittency effects, the disk must be rotated at a minimum of several hundred revolutions per minute (rpm). Webb (1933) recommends a speed of 1800 or more rpm for accurate results. It should also be noted that a time-scale characteristic curve may have a somewhat different slope than an intensity-scale characteristic due to deviations from true reciprocity (Ahrens and Taylor, 1961).

8.4.2. Photometric Procedure

With the preliminary details behind us, we are now in a position to discuss the photometric steps involved in the spectrographic determination of potassium in a geological sample. The procedure will be illustrated using the so-called *two-step method* (Feldman, 1953). To perform the determination, a series of exposures will be made on a single photographic plate. The plate will be displaced vertically between exposures to produce an array of spectra arranged one above the other on the plate.

For emulsion calibration purposes, the spectrum of an iron arc is photographed using either a two-step filter or a rotating step sector that has been stopped down to two steps by adjusting the slit height. An iron arc is usually selected because it is rich in spectral lines and therefore permits plate calibration over a wide wavelength range at closely spaced intervals. For this purpose, Slavin (1971) recommends an iron-bead arc. This arc is formed when a small amount of high-purity iron (200–400 mg) is melted by a low-current arc in the cavity of a graphite-cup electrode. A graphite rod is used as the cathode.

After the calibration spectrum has been photographed, a series of analytical standards is prepared containing different amounts of the element to be determined. Each powered sample is loaded into the cavity of a graphite-cup electrode. Each electrode is then burned in the arc for a fixed exposure time of, say, 30 sec, and the photographic plate is displaced vertically between exposures. Finally, a sample of unknown material is prepared in an identical fashion to that used for the standards, and the sample is burned under identical conditions for the same exposure period used with the standards. The plate is then removed from the spectrograph and developed.

When the plate has dried, a group of iron lines whose wavelengths are close to that of the analytical line and the internal standard line is selected. This group of lines should cover an adequate range of densities. If the analytical line and the internal standard line are widely separated in terms of wavelength, it may be necessary to prepare two emulsion calibration curves in order to take into account the variation of photographic response with wavelength.

With the plate properly positioned in the densitometer, the densities of the selected group of iron lines are read. Since a two-step procedure has been used in this determination, each spectral line will have two densities, which are labeled weak and strong for convenience. Table 8.1 shows the results obtained for a group of four lines. If this information is plotted on an HD plot, a series of overlapping line segments is produced, as shown in Figure 8.8. Although a smooth HD plot could conceivably be prepared by the horizontal displace-

Table 8.1. Photographic Densities Obtained with Two-Step Method for Four Lines

Line	Weak Density	Strong Density
1	0.602	1.20
2	0.299	0.780
3	0.097	0.354
4	0.029	0.130

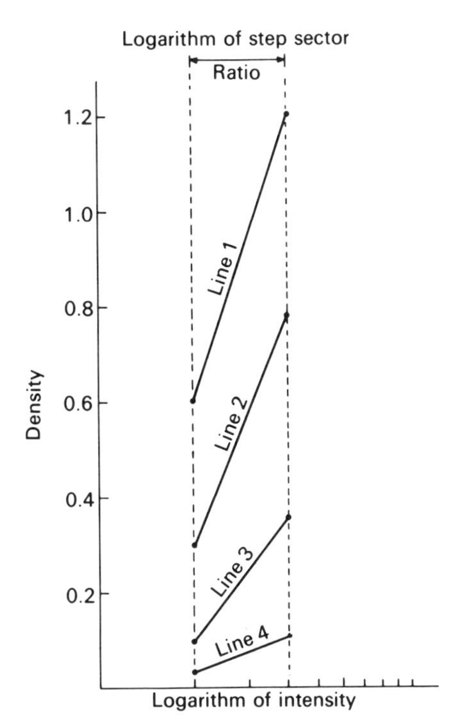

Figure 8.8 Optical density versus logarithm of intensity for four iron lines.

ment of the four line segments shown in Figure 8.8, a more elegant procedure is to use the so-called *preliminary curve method.*

With the preliminary curve method, the data from Table 8.1 are plotted in terms of density on a coordinate system where the abscissa represents the strong density for a given line and the ordinate is the weak density for the same line. Figure 8.9 shows the preliminary curve obtained from the data in Table 8.1. Since four lines have been selected, there are four points on the curve. The curious bow-shaped curvature shown in Figure 8.9 is typical. For work requiring the highest accuracy, a larger number of lines should be employed.

Having prepared the preliminary curve, we are now in a position to prepare the HD curve from it. Starting with the highest point on the preliminary curve (point A in Figure 8.9), we obtain a weak density D_w of 0.60 for this line after passing through the attenuating portion of the step modulator. If we had been fortuitous enough to have picked a line in the calibrating group with a strong density D_s of 0.60, it would have had a corresponding weak density of 0.20, as shown by point C in Figure 8.9. In fact, there is no actual line measurement at

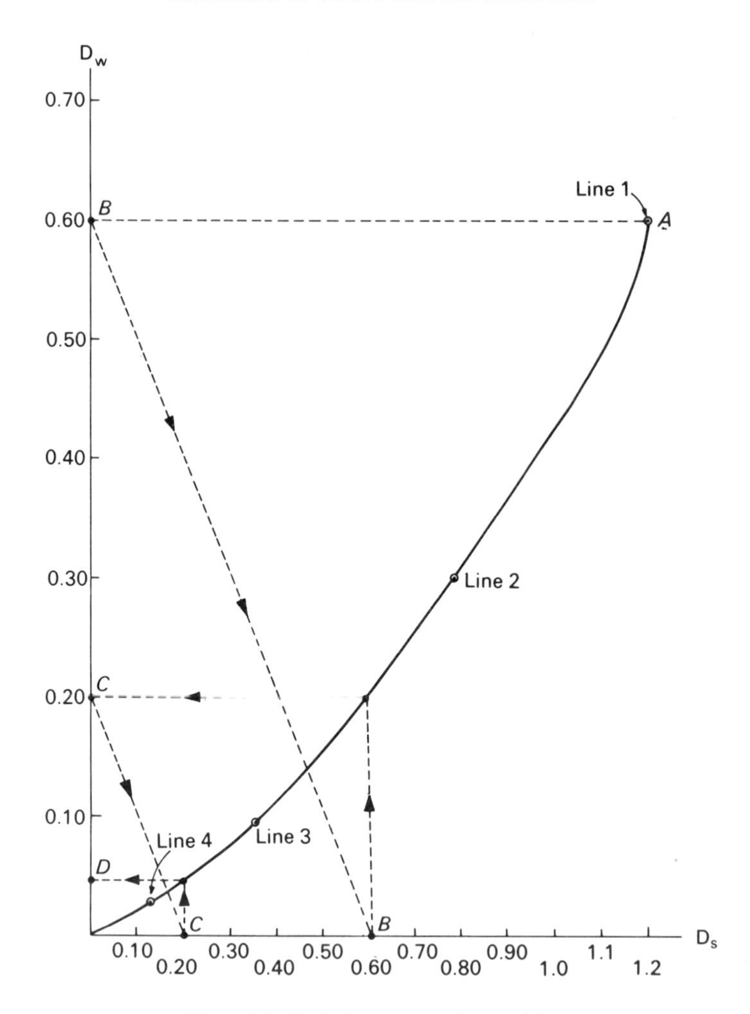

Figure 8.9. Preliminary curve for emulsion.

this point, but because we have the preliminary curve, we can read off what the weak density would have been for such a line anyway. By using this procedure, we can construct a continuous HD curve from successive points (*A*, *B*, *C*, and *D*) shown in Figure 8.9.

Figure 8.10 shows the HD curve prepared from points *A*, *B*, *C*, and *D* of Figure 8.9. Starting with point *A*, each point is shifted to the left by the logarithm of the step sector (or filter) ratio. In this example, the step sector ratio was 2.

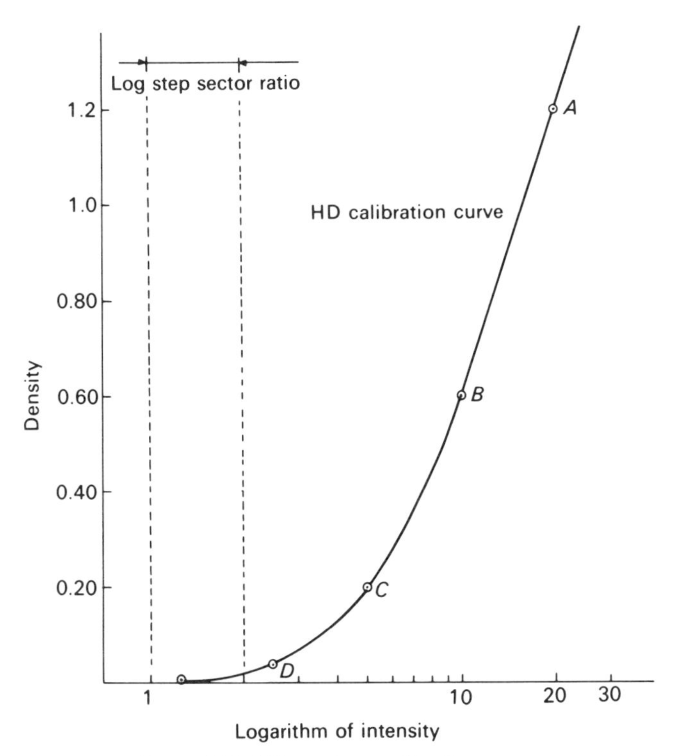

Figure 8.10. HD emulsion calibration curve.

Table 8.2 shows the data obtained for a series of six analytical standards containing various percentages of K_2O. The second column gives the opacity data obtained from the plate for potassium and rubidium. The third column gives the corresponding density data for each standard. The fourth column

Table 8.2. Analytical Data

Percentage of K_2O	Opacity		Density		Relative Intensity		I_K/I_{Rb}
	K	Rb	K	Rb	K	Rb	
5.0	25.6	15.6	1.41	1.19	25	20	1.25
2.5	10.4	9.13	1.02	0.96	16	15	1.07
1.5	5.1	7.7	0.71	0.89	11.3	14	0.81
1.0	3.0	7.1	0.48	0.85	8.5	13.3	0.64
0.4	2.1	14.0	0.32	1.15	6.5	18.5	0.35
0.1	1.2	9.8	0.08	0.99	3.2	15.9	0.20

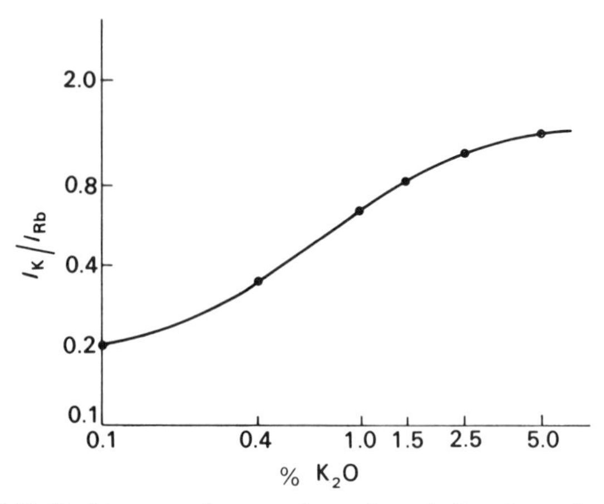

Figure 8.11 Working curve for potassium using rubidium internal standard.

gives the relative intensities obtained from the emulsion calibration curve (Figure 8.10). The fifth column gives the ratio of the intensities of the analytical line (K 404.7 nm) to the internal standard line (Rb 420.2 nm).

Figure 8.11 is a plot of the actual analytical curve or *working curve* for this determination on a log–log format. From a plot of this type, the percentage of K_2O in an unknown can readily be determined once the intensity ratio for the lines has been determined.

From the preceding discussion, it can be seen that spectroscopic analysis using a photographic plate as a means of detection represents a significant amount of effort. For this reason, much research in spectroscopy has been devoted to the development of electronic substitutes for the photographic plate. A description of possible substitutes will be the subject of the next few chapters.

REFERENCES

Ahrens, L. H. and Taylor, S. R. (1961), *Spectrochemical Analysis*, Addison-Wesley, Reading, MA.

Feldman, C. (1953), *Methods for Emission Spectrochemical Analysis*, American Society for Testing Materials, Philadelphia, PA.

Harrison, G. R., Lord, R. C., and Loofbourow, J. R. (1948), *Practical Spectroscopy*, Prentice-Hall, Englewood Cliffs, NJ.

Harvey, C. E. (1950), *Spectrochemical Procedures*, Applied Research Laboratories, Glendale, CA.

Levi, L. (1980), *Applied Optics*, Vol. 2, Wiley, New York.

Mika, J. and Török, T. (1974), *Analytical Emission Spectroscopy*, Crane, Russak & Company, New York.

Nachtrieb, N. H. (1950), *Principles and Practice of Spectrochemical Analysis*, McGraw-Hill, New York.

Pinson, L. J. (1985), *Electro-Optics*, Wiley, New York.

Sawyer, R. A. (1963), *Experimental Spectroscopy*, Dover, New York.

Slavin, M. (1971), *Emission Spectrochemical Analysis*, Wiley, New York.

Webb, J. H. (1933), *J. Opt. Soc. Am.*, *23*, 157, 316.

CHAPTER

9

INTRODUCTION TO PHOTOELECTRONIC DETECTION

Research in quantitative analytical spectroscopy has tended to proceed along two general lines. One avenue of investigation has been the development of new excitation sources such as the induction-coupled plasma (Fassell and Kniseley, 1974a,b) capable of detecting trace and ultratrace levels of analyte in samples. Coupled with the development of new spectral sources has been the development of a quantitative theory of spectrochemical excitation (Boumans, 1966). The other major avenue of investigation has centered around the development of instrumentation for optical radiation detection.

For a long time in the development of quantitative spectroscopy, photographic detection with its many variables and tedious calibration procedures was the only means for determining relative intensities. With the advent of photomultiplier tubes capable of sensing very low light levels, spectroscopists (Margoshes, 1970a–c) began to wonder whether it would be possible to develop alternative detection systems based on electronic principles that could replace the photographic plate. In fact, the development of a completely satisfactory electronic analog to the photographic plate suitable for spectroscopic work has not yet been accomplished. Nevertheless, developments in the detection of optical and infrared radiation have led to a wide variety of detection systems that have potential analytical application.

The renaissance in optics and electro-optic detection has been fueled by concomitant developments in semiconductors, integrated circuits, computers, and lasers. Most of the work on new radiation sensors has been for purposes other than spectroscopic applications such as television broadcasting and military applications, and for this reason many of the detection systems that have been produced are not ideally suited for spectroscopic work.

In general, research in detector development and fabrication involves high technology and falls outside the scope of the average research spectroscopist. As a result, the role of the research spectroscopist interested in advances in instrumentation for multiwavelength detection of optical radiation has been primarily of an evaluative nature. Recently, commercial spectrometers employing photodiode array detectors have become available, and this has prompted the American Society for Testing and Materials (ASTM) to form a committee whose on-going goal will be to formulate performance standards and criteria for these instruments as they are developed. One potential benefit of these discussions could be the interaction and possible collaboration of electro-optic engineers with research spectroscopists and analytical chemists leading to the eventual development of detection systems specifically designed for spectroscopic work.

For the moment, however, analytical spectroscopists must be content with evaluating rather than fabricating new sensors. This, in itself, is no small task. Numerous electro-optic detection systems have been described in the literature with some having more analytical potential than others. To select sensors with good analytical potential for a particular application, the research spectroscopist needs to be familiar with the performance specifications for the new detectors. Unfortunately, performance specifications for these new detectors are all specified in the language of the electro-optic engineer. Since much of the terminology used in electro-optics is unfamiliar to the average analytical spectroscopist, technology transfer in this area has been slow.

The purpose of this chapter will be to introduce the terminology of electro-optics at a level intelligible to research spectroscopists with no prior experience in the area. References to the original literature will be kept to a minimum at this point since the major goal will be one of discussing terminology and concepts. This course of action is also prudent in terms of the long-term value of a monograph of this type. Electro-optics is a rapidly developing area, and original literature citations that seem significant today may quickly go out of date. In contrast, concepts generally change rather slowly, if at all. It is hoped that this chapter will provide sufficient depth and background to enable the reader to understand and appreciate the orginal literature. The reader who wishes a more extensive introduction to electro-optics should consult Keyes (1980), Boyd (1983), Kingston (1978), Budde (1983), Pinson (1985), and Levi (1968, 1980).

9.1. CLASSIFICATION OF DETECTOR TYPES

The process of optical detection refers to the transformation of the energy present in a radiation field into some physical (or chemical as discussed in Chapter 8) effect. In this chapter, we are concerned with detectors that transform the energy into some electrical form such as current, voltage, conductivity, capacitance, or resistance. In terms of a classification system, there are two major categories of radiation detectors. One type is known as *thermal detectors*. With thermal detectors, the energy of the radiation field is absorbed and converted into heat. The heat generated within the device may alter a junction potential, as in the *Seebeck effect*, which occurs in thermocouples, or it may cause a change in the resistance of a material, as in the thermistor bolometer. In general, since the incident photons do not directly produce charge carriers as a result of absorption (rather the electrical effects are secondary effects due to changes in temperature), thermal detectors are nonselective with respect to wavelength and respond over a wide wavelength range. In spite of this wide wavelength range, thermal detectors are used primarily in the infrared region of the spectrum where photon energies are too small for the second major category of detectors, namely, *photon* or *quantum detectors*.

Photons absorbed by photon detectors interact directly with the atoms and molecules of the detector without prior thermalizing. Compared with thermal detectors, quantum detectors have a higher sensitivity and a faster response time and are therefore used preferentially in the UV–visible region of the spectrum. Since quantum detectors involve a direct rather than indirect interaction with the atoms or molecules composing the detector material, quantum detectors show a more pronounced variation of response with different wavelengths. Since the wavelengths of interest in multielement analysis fall in the UV–visible region of the spectrum, our discussion of electro-optic detectors will be limited to quantum detectors.

9.2. INTRODUCTION TO SOLID-STATE CHEMISTRY

To understand the principles involved in photon detection, one must first understand something of the phenomena involved. These phenomena are, in turn, described in the language of solid-state chemistry.

9.2.1. Band Theory of Solids

Elementary chemistry tells us that the electronic energy levels of free atoms are arranged in the form of discrete energies according to the laws of quantum mechanics. When a large number of atoms combine together, however, to form a crystal lattice, the discrete electronic energy levels associated with the free atoms tend to rearrange themselves into a series of very closely spaced levels known as *bands*. This rearrangement process leads to the formation of two bands separated by a forbidden gap devoid of levels known as the *band gap*. Figure 9.1 shows a schematic diagram of the band structure for a perfect crystalline solid. The band with the lower energy is known as the *valence band* and contains electrons that are localized about individual nuclei and are therefore unable to participate in conduction. The band with the higher energy is known as the *conduction band*. Electrons in the upper band are free to move about in response to applied fields.

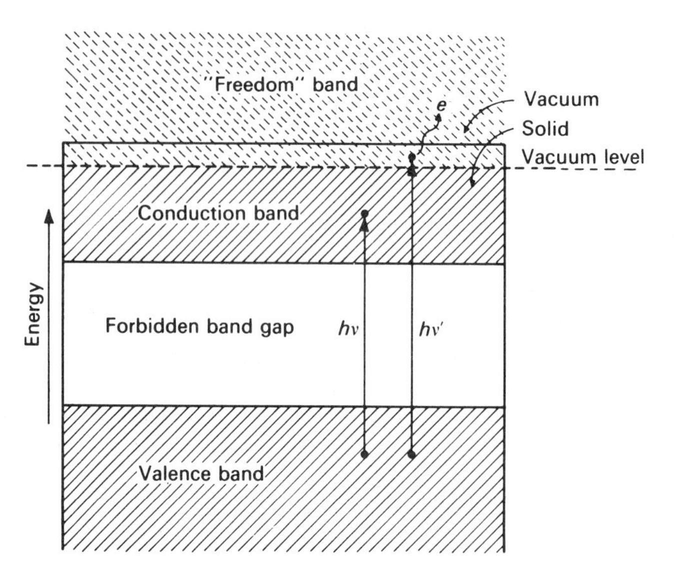

Figure 9.1. Band structure for perfect crystalline solid.

Beyond the conduction band (i.e., at still higher energies) lies a continuum of levels where electrons are no longer bound to the solid. This so-called *freedom band* (Levi, 1980) partially overlaps the conduction band. The lower level of the freedom band is known as the *vacuum level*.

In general, materials may be classified into three major categories on the basis of their conductivities. Materials with extremely small or no conductivity are referred to as *insulators*. In terms of the *band theory* of solids described in the preceding discussion, insulators are crystals where the band gap is relatively large and the valence band is not completely filled with electrons. This means that it takes a relatively large amount of energy to promote an electron from the valence band into the conduction band (i.e., everything becomes a conductor if enough energy is available). In addition, any electrons introduced into the insulating material can occupy unfilled levels in the valence band, thereby becoming lost to conduction. Substances that have electrons in the conduction band that are free to move about in response to applied fields are known as *conductors*. Between these two extremes is an important class of materials known as *semiconductors*. Semiconductors are materials where the valence band is just filled with electrons and the band gap is not too large.

9.2.2. Semiconductors

Semiconductor materials can be classified into two basic types. *Intrinsic semiconductors* are pure materials with band gaps that can be crossed by the thermal promotion of an electron from the valence band to the conduction band. The process of promoting an electron from the valence band to the conduction band gives rise to a positive center in the valence band. This positive center is known as a *hole* and is capable of moving through the valence band. In contrast to metallic conduction where current is carried entirely by electrons, current flow in semiconductors is carried by both electron and hole flow. For this reason, electrons and holes are referred to as *carriers. Extrinsic semiconductors* have crystal defects deliberately introduced into the crystal lattice. When crystal defects due to chemical impurities or crystal irregularities are present in a crystal, they generally lead to the formation of allowed energy levels within the forbidden band gap. The presence of these allowed levels within the band gap facilitates conduction as described in what follows.

In fabricating extrinsic semiconductors, two types of chemical impurities are often introduced into a pure intrinsic semiconductor material in a controlled manner by the process of *doping*. If the chemical impurity has more valence electrons than the intrinsic semiconductor material it replaces, it has more valence electrons for bonding than it actually needs to fit into the intrinsic semiconductor lattice. The extra valence electron (or electrons), which does not participate in bonding, occupies a level (or levels) in the upper portion of the band gap known as a *donor level*. Extrinsic semiconductors fabricated in this manner are known as *n-type semiconductors* because the majority carriers in this type of material are electrons that have been promoted to the conduction band from the nearby donor levels. It should not be inferred from the foregoing discussion that the *n*-type material is negatively charged in some manner, as this is not the case.

If the chemical impurity has less valence electrons than the intrinsic semiconductor material it replaces, it has less valence electrons for bonding than it needs to fit into the crystal lattice. This results in the formation of a *vacant* energy level known as an *acceptor level*. The acceptor levels lie within the band gap just above the valence band. These acceptor levels readily accept electrons from the valence band, producing mobile holes. Since the majority charge carriers in this type of material are holes, this type of extrinsic semiconductor is referred to as a *p-type semiconductor*.

9.2.3. The *pn* Junction

Figure 9.2 shows the energy level structure for *p*- and *n*-type semiconductors. The dotted line labeled E_F shows the location of the so-called *Fermi level*. The location of the Fermi level on the diagram provides an indication of the electron occupancy with respect to energy. Levels with energies lower than the Fermi level tend to be completely filled with electrons, while those with energies higher than the Fermi level tend to be empty. The Fermi level itself

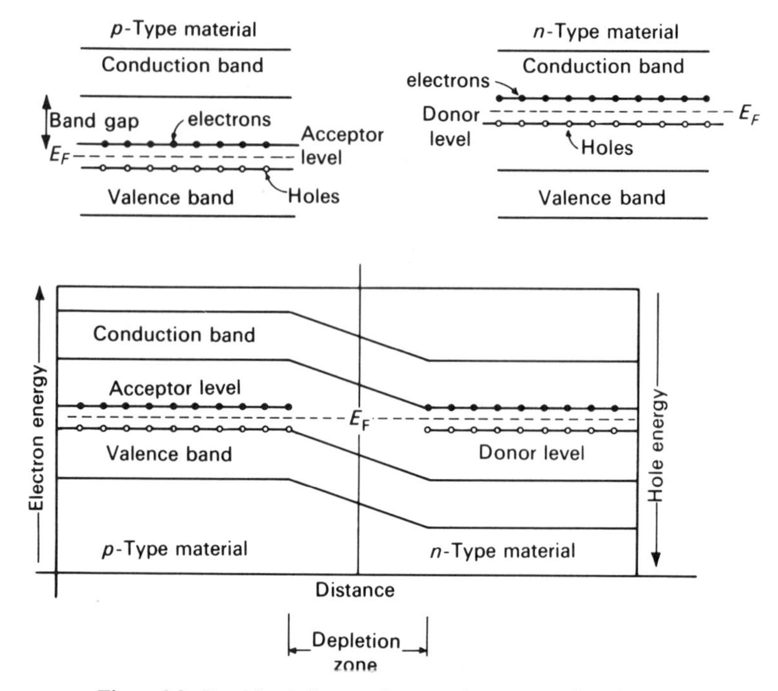

Figure 9.2. Band level diagram for *p*- and *n*-type semiconductors.

represents the energy where, based on probability considerations and the Pauli exclusion principle, a level exactly half filled with electrons would be located. From the figure, it can be seen that in each case the Fermi level is located near its respective impurity level. When materials with different Fermi levels are brought in contact, electrons will tend to flow from the material with the higher Fermi level into the material with the lower Fermi level. This transfer raises the lower Fermi level and lowers the higher one. At equilibrium, the Fermi levels will be equal.

With this in mind, let us consider what happens when a sample of *p*-type material is brought in contact with some *n*-type material. When the materials are brought into electrical contact, charge carriers from both materials immediately begin to diffuse across the junction in response to the concentration gradient. Since the holes are the majority carriers in the *p*-type material, their concentration is much higher in the *p*-type material than in the *n*-type material (where because of their low concentration they are known as *minority carriers*), and they therefore diffuse from the *p*-type material into the *n*-type material. Similarly, electrons are the minority carriers in the *p*-type material, and consequently they diffuse from the *n*-type material (where they are

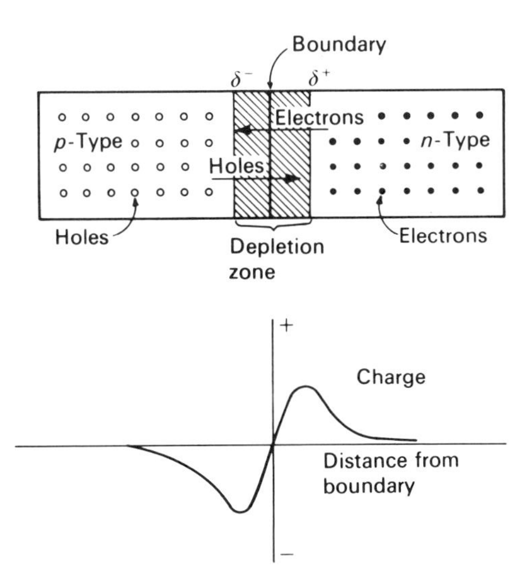

Figure 9.3. Carrier diffusion and charge in vicinity of *pn* junction.

majority carriers) into the *p*-type material. This diffusion process results in the formation of a junction potential whose electric field opposes the further diffusion of additional majority carriers across the junction. The polarity of the junction potential is shown in Figure 9.3.

Since electrons have diffused into the *p*-type material and holes have diffused into the *n*-type material, the *p*-type material adjacent to the junction becomes negatively charged with respect to the *n*-type material. Although the formation of the junction potential tends to retard the futher diffusion of majority carriers across the junction, it tends to facilitate the diffusion of minority carriers. Thus, minority carriers are encouraged to diffuse across the junction regardless of their energy, whereas only majority carriers with enough energy can cross the barrier imposed by the junction potential. At equilibrium, the flow of minority carriers across the junction is just balanced by the flow of majority carriers in the opposite direction. As a result of these processes, current flow across the *pn* junction depends significantly on the height of the energy barrier established at the *pn* junction.

The device created by joining the *p*- and *n*-type materials is known as a semiconductor, or solid-state, *diode*. Another phenomenon also occurs at the *pn* junction that deserves mention, namely, the formation of the *depletion layer* or *depletion zone*. Figure 9.3 shows the location of the depletion zone at the *pn* junction. It should be remembered that a hole is merely a vacancy or unfilled level in the valence band. As a result, whenever a hole and an electron meet, the electron can fill the vacancy in the valence band, thereby annihilating both carriers. This process is known as *recombination.* At the junction, because of the diffusion of carriers across the boundary, the probability of recombination increases. Since recombination reduces the number of charge carriers, the region where it occurs is depleted in charge carriers and is therefore known as the depletion zone. Since the depletion zone contains relatively few extrinsic charge carriers, its conductivity is considerably less than the surrounding material, approaching the conductivity of the undoped intrinsic semiconductor material. As a result, the resistance of the depletion zone is large.

Both the size of the depletion zone and the current through the device vary markedly with the application of an external bias potential, as shown in Figure 9.4. As shown in the figure, the size of the depletion zone increases when a reverse-bias potential is applied to the diode. Application of a reverse-bias potential also increases the size of the charge barrier carriers must cross.

Figure 9.5 shows the effect of an external bias voltage on the band structure of the semiconductor. Current through the diode is the sum of four components, two *diffusion currents* and two *generation currents.* The diffusion currents are due to the flow of majority carriers across the junction. Since

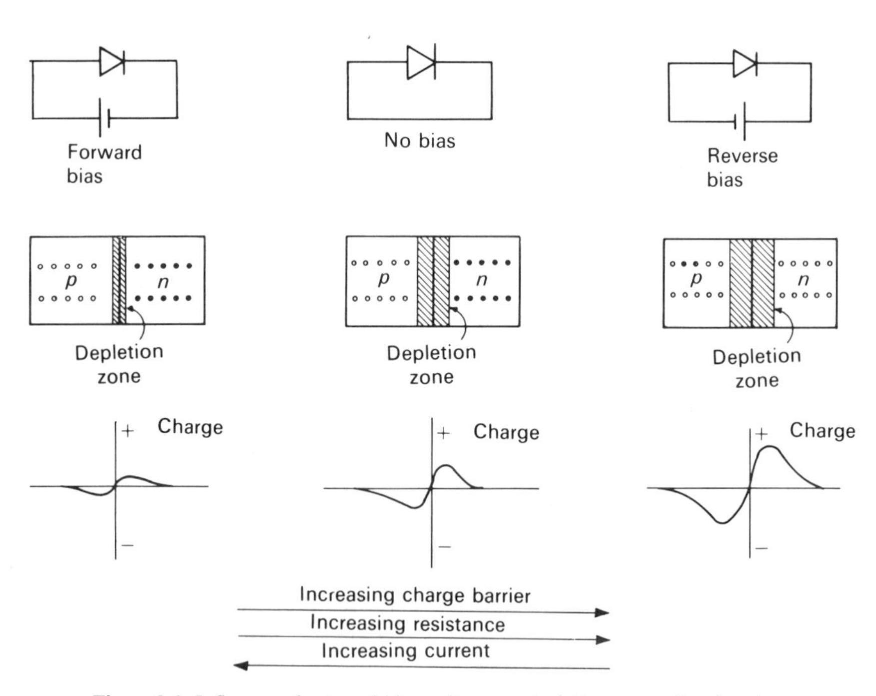

Figure 9.4. Influence of external bias voltage on depletion zone of *pn* junction.

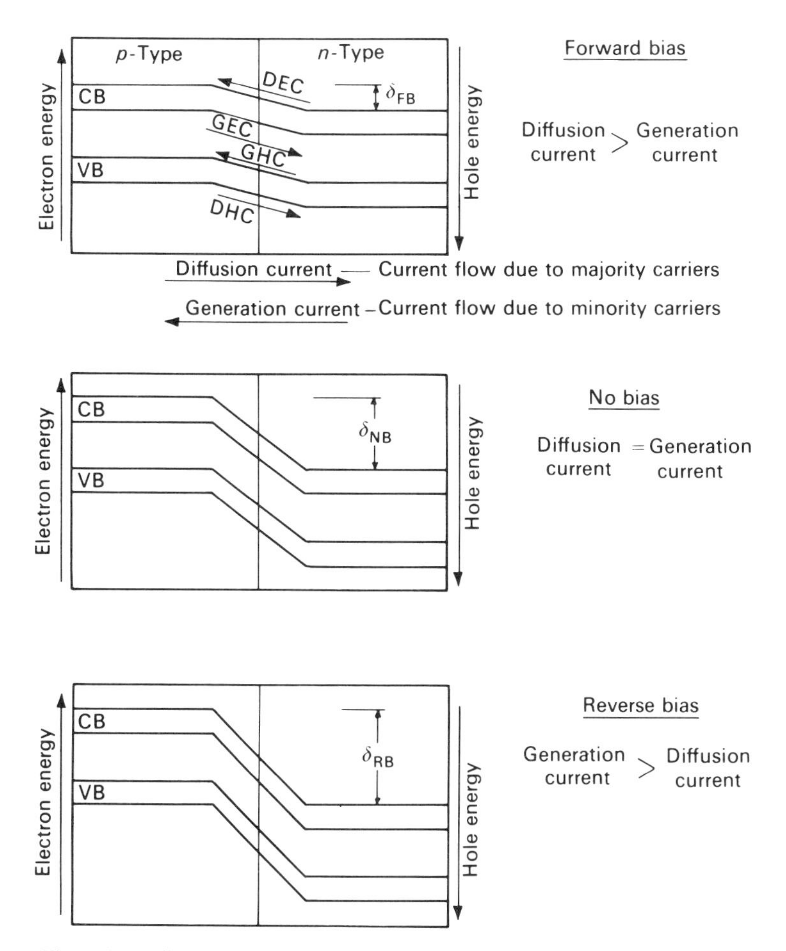

Figure 9.5. Effect of external bias voltage on band structure of semiconductor.

current in a semiconductor is carried by electrons and holes, there are two diffusion currents. The diffusion electron current (DEC) is due to electrons flowing from the *n*-type material into the *p*-type material. The diffusion hole current (DHC) is due to holes flowing from the *p*-type material into the *n*-type material. Although these two diffusion currents flow in opposite directions, they result in a net current from left to right that is the sum of both components since the direction of the current is taken as the direction of positive charge flow. Both of these diffusional processes involve a flow against an energy gradient, as indicated in the figure. It should be noted that the energy gradient for holes is opposite to that for electrons (i.e., hole energies increase in the downward direction, whereas electron energies increase in the upward direction). As a result of the energy gradient, the DEC and DHC are quite sensitive to the magnitude of the energy barrier δ. Only that fraction of carriers with energies greater than δ can cross the junction. Since the reverse-biased δ (δ_{RB}) is greater than the forward-biased δ (δ_{FB}), the diffusion current is greatly reduced under reverse-bias conditions. From the figure it can be seen that $\delta_{RB} > \delta_{NB} > \delta_{FB}$.

The other component of the current, namely, the generation current, flows in the opposite direction to the diffusion current and consists of two carrier flows in opposite directions. The generation current refers to the current flow due to minority carriers that have been thermally generated. Thus, one portion of the generation current consists of thermally generated electrons formed in the *p*-type material and that flow into the *n*-type material. The other portion of the generation current, flowing in the opposite direction, is due to thermally generated holes formed in the *n*-type material. By analogy with the diffusion current, both components of the generation current combine together to produce a net generation current flowing from right to left, as shown in the figure. Both of these components to the generation current involve a flow with the energy gradient as indicated in the figure. As a result, all of the thermally generated carriers can cross the junction regardless of the magnitude of δ. For this reason, the generation current is not affected by the bias voltage but remains constant for a given temperature. Since the generation current is due to thermal excitation, its magnitude is much smaller than the diffusion current that can flow under forward-bias conditions (see Figure 9.10).

9.3. PRINCIPLES OF PHOTON DETECTION

The absorption of light by an appropriate photosensitive material may, under the right conditions, result in the promotion of an electron from the valence band into the conduction band. For this to occur, the energy of the photon must be at least as great as the band gap present in the material. At this point, two effects can be distinguished based on the energy of the photon. If the energy of the photon is very large, the electron may gain sufficient energy to be promoted from its location in the valence band to a level in the conduction band above the vacuum level. Electrons with these energies have a high probability of escaping the solid, whereupon they can be collected by suitably designed electrodes, producing a photocurrent. This process is an example of the *external photoelectric effect* or the *photoemissive effect* (Levi, 1980) and is the principle upon which photomultiplier tubes are based. If the energy absorbed is such that the electron is merely promoted from the valence band into the conduction band without escaping from the material, the process alters the conductivity of the material. As a result, this effect is known as the *photoconductive effect* (Levi, 1980) or the *internal photoelectric effect*.

9.3.1. The External Photoelectric Effect

The minimum energy needed to eject an electron from a material is known as the *work function* ϕ. If we take the Fermi level as an indication of the highest filled level in the material (see Figure 9.1) and the vacuum level as the minimum energy required to eject an electron from the surface, then the work function is simply the energy difference between the vacuum and Fermi levels. Since the Fermi level often occurs in the band gap, the work function under these conditions becomes equal to the difference between the vaccum level and the top of the valence band. Since the energy of a photon is given by the Planck relationship,

$$E = h\nu = hc/\lambda \tag{9.1}$$

where E is the energy of the photon, h is Planck's constant, and c is the speed of light, the threshold wavelength may be determined by substituting the work function for E,

$$\lambda_{\text{th}} = hc/\phi \tag{9.2}$$

The threshold wavelength is the longest wavelength that can be detected with the photoemissive detector.

To extend the long-wavelength response of photoemissive detectors, it is important to seek ways to reduce the work function. To understand how this might be accomplished, it is necessary to look more closely at the work function. The work function can be considered to be composed of the sum of two contributions (Levi, 1980; Boyd, 1983), the band gap energy previously described and the electron affinity. The electron affinity is the energy difference between the bottom of the conduction band and the vacuum level. Thus, one way to improve the long-wavelength response of photoemissive detectors is to find ways to reduce the electron affinity of the material. This has recently become possible (Boyd, 1983) through a process known as *band bending*, involving surface treatment and doping. Figure 9.6 shows the effect of band bending on the location of the vacuum level near the surface. By proper treatment, photoemissive materials with negative electron affinities (E_a) have been prepared. Materials with these characteristics suggest that even electrons promoted to the bottom of the conduction band should, in principle, have enough energy to escape the solid. Using this procedure, threshold wave-

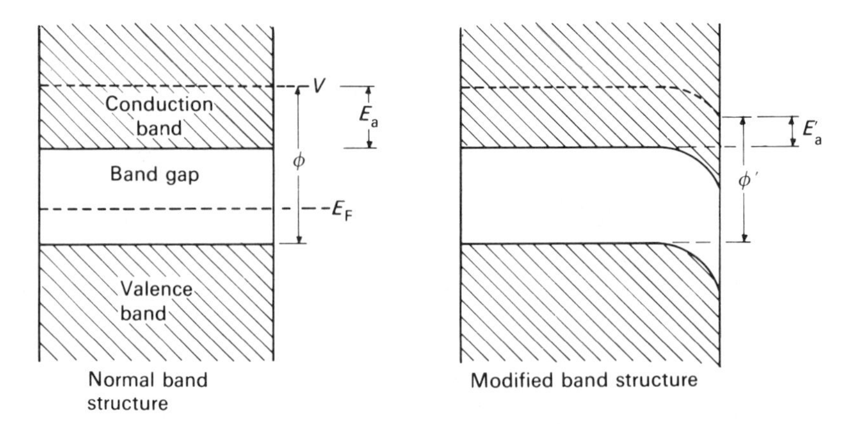

Figure 9.6. Effect of band bending on location of vacuum level near surface.

lengths for photoemissive detectors have been extended out to 1.5 μm (Boyd, 1983).

The short-wavelength response of photoemissive detectors is generally limited by a combination of factors, the most important of which is the transmission properties of the detector envelope itself. However, even in the absence of optical losses in window materials, short-wavelength response is observed to decrease because the absorption of photons in the detector material becomes less likely as the difference between photon energy and work function increases (Levi, 1980). As the probability of absorption decreases, photons penetrate more deeply into the material before being absorbed. Photoelectrons produced deep within the material have further to travel to reach the surface and are less likely to be able to overcome the electron affinity upon reaching it. Both of these factors, decreased absorptivity and photoelectron losses, intrinsically limit the short-wavelength response of photoemissive detectors.

It should be noted that not all incident photons that strike the detector are effective in releasing photoelectrons. The ratio of the number of photons that strike the surface to the number that produce the desired effect (in this case, photoemission of electrons) is known as the *quantum efficiency* η.

9.3.2. The Internal Photoelectric Effect

A variety of solid-state semiconductor detectors make use of the internal photoelectric effect as a means of radiation detection. These detectors can be classified in two ways. One classification scheme employs the terms *photoconductive devices* and *photovoltaic devices*. One disadvantage with this method of classification is that photovoltaic devices can be employed in a photoconductive mode. This leads to a certain degree of confusion as pointed out by Budde (1983).

A somewhat less ambiguous method of classification (Budde, 1983) is based on the presence or absence of a junction in the semiconductor material. According to this scheme, *bulk detectors* are homogeneous detectors composed of a uniform semiconductor material. In contrast, *junction semiconductor detectors* are not uniform but possess a *pn* junction or some modification of it.

Bulk detectors may be either intrinsic or extrinsic semiconductors and are based on the observation that the electrical conductivity of many semiconductor materials increases when the material is exposed to light. Since photoconductivity is the only mode of operation for bulk detectors, all bulk detectors fall in the category of photoconductive detectors.

The explanation for the increase in conductivity upon exposure to light (i.e., photoconductivity) may be understood by reference to Figure 9.1. If a photon of energy $h\nu$ is absorbed by the semiconductor, an electron can be promoted from the valence band into the conduction band. Since this process produces a hole in the valence band and an electron in the conduction band, the conductivity of the material increases. In contrast to the external photoelectric effect, the electron is not promoted to the vacuum level, where it would be lost from the solid, but instead remains in the conduction band. As shown in the figure, a photon must have an energy equal to or greater than the band gap to be absorbed. This sets an upper limit on the long-wavelength response of the detector given by an equation analogous to Eq. 9.2.

$$\lambda_{\text{th}} = hc/E_{\text{bg}} \tag{9.3}$$

where E_{bg} is the band gap energy.

A variety of approaches have been employed to extend the long-wavelength cutoff of photoconductive detectors. One procedure that has been used with intrinsic semiconductors has been to use solid solutions produced by mixtures of two semiconductors (Boyd, 1983). Such mixtures are observed to have band gap energies that depend on the mole fraction of the components.

Another approach for extending long-wavelength response is to use extrinsic semiconductor materials instead of intrinsic ones. Figure 9.2 shows the location of acceptor levels in the *p*-type material and donor levels in the *n*-type material. A homogeneous portion of either material may be used as a photoconductive detector. In either case, the material must be cooled so that the impurities are not ionized by thermal effects (i.e., in the absence of light for an ideal material there are no electrons in the acceptor level if *p*-type material is used or holes in the donor level if *n*-type material is considered). Photons with energies greater than the ionization energy of the impurity will, upon being absorbed, produce charge carriers within the material, thereby altering its conductivity. Since the ionization energy of the impurity is less than the band gap energy of the corresponding intrinsic semiconductor, the long-wavelength response is extended compared with undoped semiconductor material. Figure 9.7 presents a comparison of the use of intrinsic and extrinsic semiconductors.

Figure 9.8 shows a schematic diagram of a typical experimental arrangement using a photoconductive detector. The most common arrangement is to orient the detector so that the incident radiation is perpendicular to the direction of current flow (Keyes, 1980). This arrangement is referred to as *transverse geometry*. Several configurations are possible for signal detection.

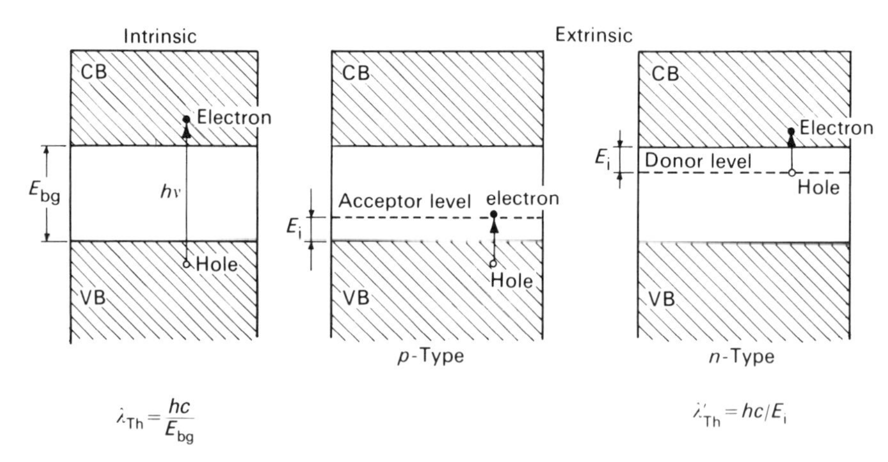

Figure 9.7. Long-wavelength response of intrinsic and extrinsic semiconductor material showing $\lambda'_{Th} > \lambda_{Th}$.

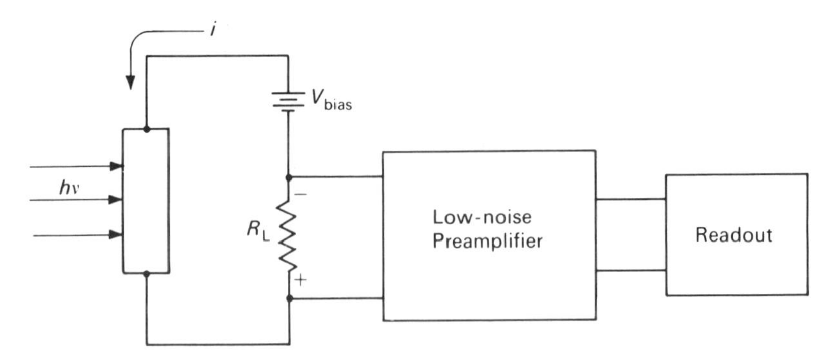

Figure 9.8. Signal detection with intrinsic photoconductor.

In the simplest, the change in current through the circuit is detected as a change in voltage across a load resistor in series with a bias voltage source and the detector, as shown in the figure. Alternatively, the change in voltage drop across the detector itself can be used. In all cases, a bias voltage is required to induce the flow of charge carriers in the detector. In contrast to junction semiconductor detectors, the polarity of the applied bias voltage has no effect on the detector circuit other than to change the direction of current flow.

Junction photodetectors in the form of silicon photodiodes are becoming widely used as detectors in the UV–visible region of the spectrum because they have a wavelength range of 200–1100 nm (Budde, 1983). In operation, light is made to strike the depletion zone of a *pn* junction, as shown in Figure

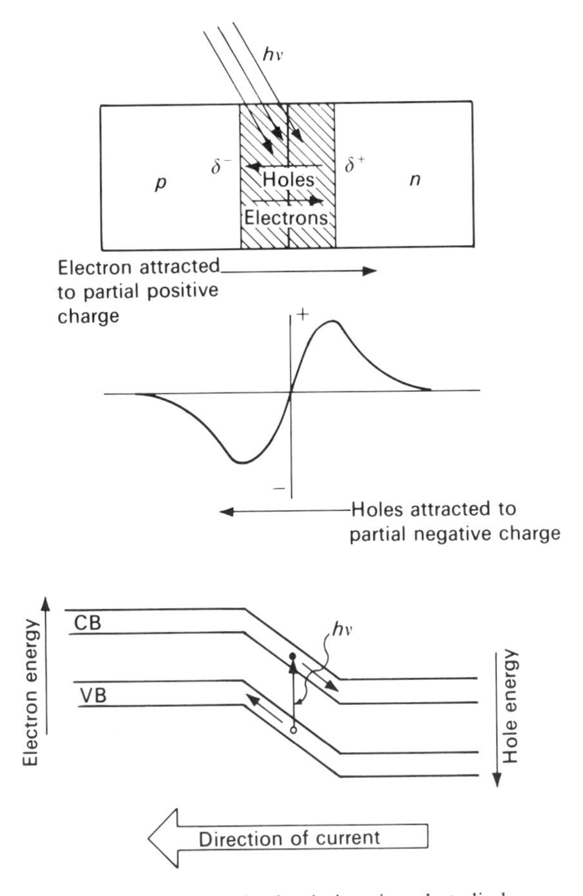

Figure 9.9. Photoionization in junction photodiodes.

9.9. If the incident photons have energies greater than the band gap energy, they can be absorbed, producing holes and electrons.

The presence of these photoinduced carriers may be monitored in a number of ways depending on whether a bias voltage is employed. The various modes of operation can be illustrated by reference to Figure 9.10, which shows a typical diode current–voltage characteristic curve. In the absence of an external bias, the photoinduced charge carriers are accelerated across the junction by the built-in electric field present at the junction as a result of majority carrier diffusion (see Figure 9.3). Thus, holes are accelerated toward the *p*-type material and electrons toward the *n*-type material. This flow of charge carriers can be detected as a current in the *short-circuit mode* of operation. Alternatively, in the *open-circuit mode* of operation, current from

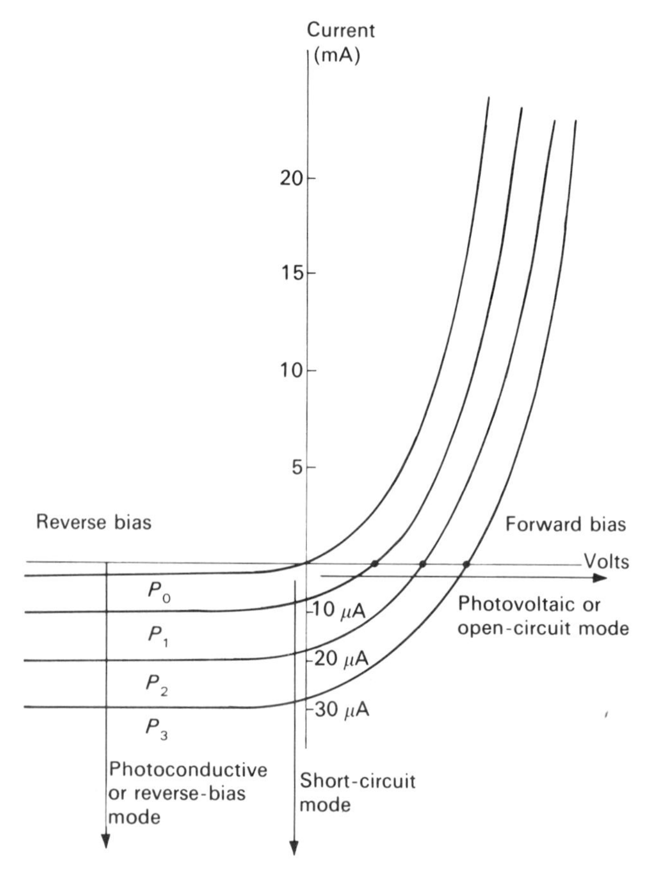

Figure 9.10. Typical photodiode current–voltage characteristic curve. P_0–P_3 represent increasing incident radiant power striking the detector.

photoinduced charge carriers is allowed to flow until the resulting voltage produced prevents any further flow of current. The zero-current voltage that results is an example of the *photovoltaic effect* from which these detectors derive their name.

Since the photovoltaic effect is nonlinear with respect to incident light intensity, junction photodetectors are often used in a *photoconduction mode* where an external reverse-bias voltage is applied to the diode. As shown in Figure 9.10, the current under reverse-bias conditions is a linear function of

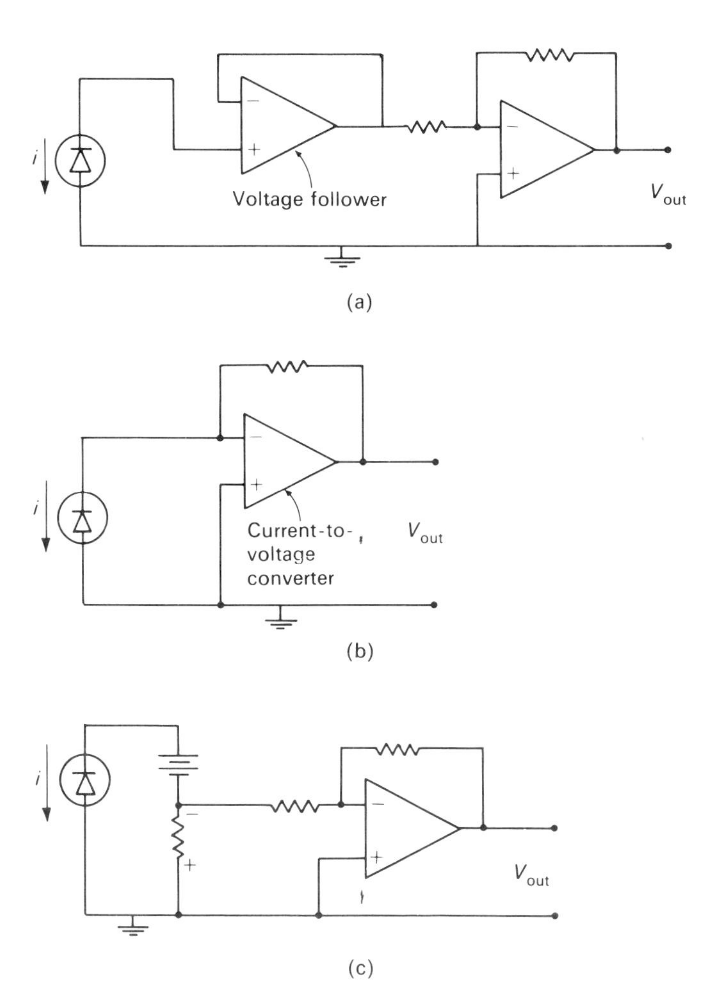

Figure 9.11. Three possible modes of using junction photodiodes: (*a*) photovoltaic mode (open-circuit voltage), no bias; (*b*) photocurrent mode (short-circuit current), no bias; (c) photoconductive mode, reverse bias.

incident power striking the detector. Figure 9.11 shows three possible ways in which junction photodiodes may be employed.

Reverse-biased photodiodes may also be used in a *storage* or *light-integrating* mode. The storage mode of operation is based on the capacitance produced at the *pn* junction. Such capacitance is inherent in any *pn* junction since the diode consists of two conducting materials separated by a non-conducting depletion zone. When the diode is reverse biased, the depletion zone becomes larger and the effective capacitance of the diode increases. Figure 9.12 shows the equivalent circuit for a reverse-biased photodiode. When the switch is closed, a charging current will flow through R_L until the diode capacitance is charged to the bias potential. If the switch is now opened, the diode capacitance will remain charged for a period of time determined by the magnitude of the discharge mechanism.

Three discharge mechanisms exist as shown in the figure. One discharge mechanism is due to *leakage current* given by

$$i_L = V_b/R_i \tag{9.4}$$

where i_L is the leakage current, V_b is the applied reverse-bias voltage, and R_i is the equivalent internal resistance of the reverse-biased diode. Another mechanism is due to the *dark current*. Since the dark current is due primarily to thermally generated charge carriers, the magnitude of the dark current in

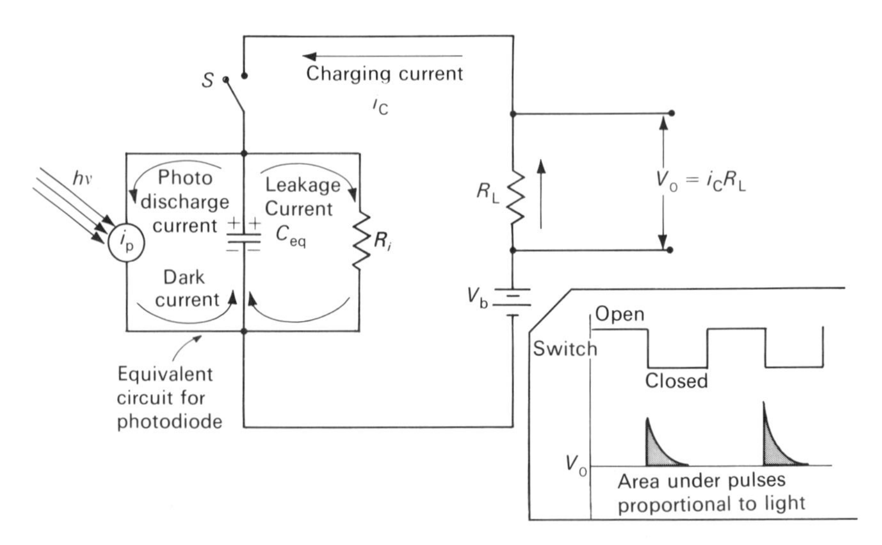

Figure 9.12. Equivalent circuit for reverse-biased junction photodiode.

photodiodes can be substantially reduced by cooling. The final discharge mechanism is the photocurrent produced by photon-generated carriers. If the parasitic effects due to leakage and dark current can be made reasonably small, the extent of discharge over a given time period can be related to the integrated intensity striking the detector, according to the following equation:

$$i_p(t) = k\,I(t) = dq/dt \tag{9.5}$$

$$q = \int_0^t i_p(t)\,dt = k\int_0^t I(t)\,dt \tag{9.6}$$

In these equations, $i_p(t)$ is the photocurrent as a function of time produced by the time-varying intensity $I(t)$. Equation 9.6 shows that the extent of discharge represented by q is proportional to the integral of the intensity over the time period t. When the switch is closed after the integration period, a charging current will flow through R_L until the missing charge on the capacitor is replaced.

The integrating behavior of photodiodes is an important characteristic that is used in scanning detectors consisting of arrays of photoelements (Levi, 1980). With these systems, each photoelement is charged on a periodic basis and the integrated light intensity since the last recharge is monitored.

Before leaving the topic of junction photodetectors, two additional types of junction detectors deserve mention. These are the *avalanche photodiode* and the *pin photodiode*. As mentioned in the preceding discussion, all photodiode junctions exhibit a certain amount of capacitance that ultimately limits the speed with which these detectors can respond. To decrease the response time, *pin* diodes employ a lightly doped intrinsic region between the *p* and *n* materials. This intrinsic layer between the two semiconductor materials reduces the capacitance, thereby decreasing the response time.

Avalanche photodiodes are used solely in the reverse-bias mode. If the reverse-bias voltage is large enough, photogenerated charge carriers originating in the depletion zone will be accelerated by the bias voltage to energies where they produce additional carriers by impact ionization (Budde, 1983). The production of secondary charge carriers as a result of impact ionization results in a photodiode with gain (Levi, 1980). Application of too high a bias voltage results in a condition known as *breakdown*, resulting in uncontrolled current. For this reason, selection of the appropriate bias voltage necessary to achieve high gain while avoiding breakdown deserves careful consideration, especially since the breakdown voltage is a function of temperature. Because of their speed and gains of several hundred (Budde, 1983), avalanche photodiodes are most commonly employed where weak, fast pulses are to be detected.

9.4. NOISE IN THE DETECTION PROCESS

The smallest light level a given detector can sense is ultimately limited by small, random fluctuations known as *noise*. These small, random fluctuations obey the mathematical laws of chance and may arise from fluctuations inherent in the light source itself, fluctuations produced in the detector, or fluctuations arising from any associated electronics such as power supplies and amplifiers. In principle, it is possible through careful design to minimize the noise arising from the electronics (Motchenbacher and Fitchen, 1973). By contrast, noise arising from the quantum nature of matter and light represents a fundamental limit that cannot be avoided or circumvented by clever design. If noise arising from within the detector can be made negligible, the only remaining source of noise is the light source itself, and this situation is described as *photon noise limited*. An *ideal photon detector* is one where no output is produced in the absence of incident radiant power and no noise source exists other than the random generation of charge carriers due to photon absorption. At the other extreme, it is frequently observed with certain classes of detectors (i.e., thermal detectors) that noise arising from within the detector itself is much larger than the photon noise. In this case, the photon noise limit is not reached, and the system is described as *detector limited.*

Even under photon-noise-limited conditions (also known as *source limited*), two situations are possible. The ultimate performance possible is attained when an ideal photon detector is employed in such a way that no extraneous background strikes the detector. This situation is described as being *signal fluctuation limited.* At the other extreme, if the predominant source of noise is photon noise originating from the background, the system is said to be *background noise limited.*

9.4.1. Types of Noise

In 1918, Schottky studied the fluctuation in the current of a vacuum tube due to the combined effect of large numbers of independently emitted electrons. Since this noise reminded him of the sound produced by a hail of shot striking a target, he termed this form of noise *shot noise* (Goldman, 1948). Since the statistical fluctuations in the emission of photoelectrons from the cathode of a photoemissive detector is similar to that observed from the hot cathode of a vacuum tube, the *shot effect* is a good description of the fluctuations present in photoemissive detectors such as photomultipliers. In fact, since the shot effect is a good description of any process resulting from the combined effect of large numbers of charge carriers crossing an energy barrier, the concept can be extended to solid-state detectors. Thus the noise associated with the photoinduced current flow in a reverse-biased solid-state photodiode is classified as shot noise.

Schottky showed that the root-mean-square (rms) value of the fluctuating current produced by the combined effect of the emitted electrons was given by

$$\langle i_N \rangle = [2e\langle i \rangle \Delta f]^{1/2} \tag{9.7}$$

where $\langle i_N \rangle$ is the rms noise current, e is the charge on the electron (1.6×10^{-19} C), $\langle i \rangle$ is the average current, and Δf is the electrical bandwidth of the system. If the electrical system has a flat response described by a Bode diagram, the frequency response bandwidth will correspond to the 3-dB point on the diagram (see appendix at the end of this chapter for a discussion of Bode diagrams). Although the rms shot noise current is a function of the frequency response bandwidth Δf, it is not a function of the frequency itself. As a result, if shot noise power is plotted versus frequency, a constant power is observed regardless of the frequency considered. Such a *noise power spectrum* (i.e., noise power vs. frequency) is described as *white* by analogy with optical radiation, where "white" light is composed of all colors (i.e., frequencies).

Another form of noise described in 1928 by Johnson (Goldman, 1948) is known as *thermal noise*. Johnson showed that this noise was the result of tiny fluctuations caused by the thermal motion of electrons in resistive components. On the basis of statistical thermodynamics, in 1928 Nyquist showed (Goldman, 1948) that noise generated in a resistive component by thermal effects was given by

$$\langle e_{rms} \rangle = [4RkT\Delta f]^{1/2} \tag{9.8}$$

where $\langle e_{rms} \rangle$ is the rms thermal noise voltage, R is the resistance in ohms, k is Boltzmann's constant (1.37×10^{-23} w-sec/degree), T is the absolute temperature, and Δf is the frequency response bandwidth already discussed. Like shot noise, thermal noise power is independent of frequency and is therefore a form of white noise. Thermal noise is also referred to as *Johnson noise* or *Nyquist noise*. The thermal noise may be expressed as a current by dividing Eq. 9.8 by R,

$$\langle i_{rms} \rangle = [(4kT\Delta f/R]^{1/2} \tag{9.9}$$

where $\langle i_{rms} \rangle$ is the rms value of the current in amperes. On the basis of Eqs. 9.8 and 9.9, it is interesting to note that the thermal noise power ($P = ie$) is independent of resistance.

In contrast to Johnson noise and shot noise, *flicker noise power* is a function of frequency. Since flicker noise has a power spectrum that can be approximately characterized as being inversely proportional to frequency, flicker noise is commonly referred to as *1/f noise*. In spectroscopy there are two sources of flicker or *fluctuation noise*. In the electrical domain, $1/f$ noise is often associated with potential barriers at electrical contacts. In the optical domain, flicker noise is often referred to as *source flicker noise*, *modulation noise*, or *scintillation noise*. These radiation noise components arise, in part, because of convection currents and refractive index changes that cause the intensity of the radiation transmitted from the source to fluctuate at low frequency. Since the exact origin of flicker noise is not known, the usual expression is an empirical one of the form

$$\langle i_{\mathrm{F}} \rangle = [K i^{\alpha} \Delta f / f^{\beta}]^{1/2} \tag{9.10}$$

where $\langle i_{\mathrm{F}} \rangle$ is the rms flicker current, K is a constant, i is the current, Δf is the frequency response bandwidth, f is the frequency, and α and β are approximately 2 and 1, respectively.

A form of shot noise peculiar to solid-state semiconductor detectors is known as *generation–recombination noise*, or *gr noise*. Generation–recombination noise is due to the statistical fluctuation in the number of charge carriers in the semiconductor material. In this respect, it arises from considerations not unlike those described for a photoemissive device. In contrast to the photoemissive situation, where the current fluctuation is due to the

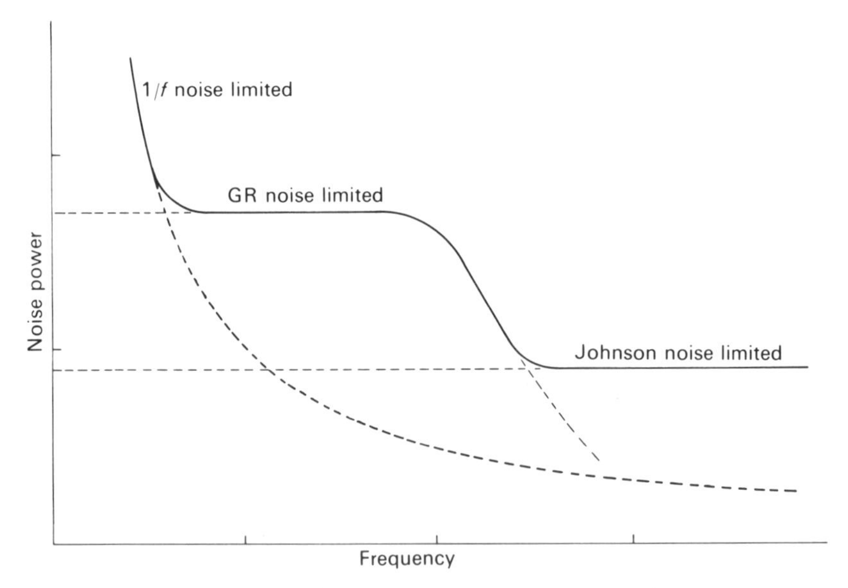

Figure 9.13. Idealized noise power spectrum for photoconductive detector.

random generation of photoelectrons, in semiconductors there are two sources of charge carrier fluctuation as a result of *both* generation *and* recombination. For a simple cooled extrinsic semiconductor with a gain of 1, the *gr* noise is given by an equation of the form (Keyes, 1980)

$$\langle i_{gr} \rangle = \{4\langle i \rangle e\, \Delta f/[1+(2\pi f \tau)^2]\}^{1/2} \tag{9.11}$$

where $\langle i_{gr} \rangle$ is the rms generation–recombination noise, $\langle i \rangle$ is the mean current through the photoconductor, e is the charge on the electron, f is the frequency, Δf is the frequency response bandwidth, and τ is the lifetime of free carriers. For low frequencies where $f \ll 1/2\pi\tau$, Eq. 9.11 reduces to

$$\langle i_{gr} \rangle = [4\langle i \rangle e\, \Delta f]^{1/2} \tag{9.12}$$

Comparison of Eq. 9.7 with Eq. 9.12 reveals that a *gr*-noise-limited detector is expected to be $2^{1/2}$ times noisier than a shot-noise-limited detector. Figure 9.13 shows an idealized noise power spectrum for a photoconductive detector.

9.4.2. Characteristics of Noise

At this point, it is worthwhile to mention some of the properties of random noise. If two or more independent noise sources are present in a system, the total noise power is the sum of the individual noise powers. Since the noise power is $\langle i \rangle^2 R$, the total noise from two sources will be

$$\langle i_T \rangle^2 R = \langle i_1 \rangle^2 R + \langle i_2 \rangle^2 R \tag{9.13}$$

$$\langle i_T \rangle^2 = \langle i_1 \rangle^2 + \langle i_2 \rangle^2 \tag{9.14}$$

$$\langle i_T \rangle = [\langle i_1 \rangle^2 + \langle i_2 \rangle^2]^{1/2} \tag{9.15}$$

If the noise is considered from a statistical standpoint, the rms noise corresponds to the standard deviation of the signal, while the square of the rms noise represents the variance of the quantity. Statistical theory reveals that the total variance arising from independent sources is simply the sum of the individual variances. Thus the law of noise power additivity is simply a consequence of the statistical nature of random noise.

Another consequence of the statistical nature of noise is the possibility of noise reduction by averaging. According to statistical theory, the magnitude of the random error associated with a given measurement is inversely proportional to the square root of the number of independent measurements. Since the random error is considered to be equally likely to be positive or negative, the fluctuation of an average value is less than the fluctuation observed for individual measurements because in the case of the average value the random errors tend to average to zero.

The manner in which a given signal is averaged can take on two different forms. If the signal is averaged over a period of time, the process is referred to as *integration*. If the waveform is periodic, the average of a large number of identical signals can be obtained. This process is known as *signal averaging*. Integration is an example of *time averaging*, while signal averaging is an example of *ensemble averaging*. If the system behaves *ergodically*, ensemble averaging will give the same result as time averaging.

Noise reduction by integration can be considered as a form of bandwidth reduction. It can be shown (Boyd, 1983) that for systems that integrate the signal over a sampling period T, $\Delta f = 1/2T$. Thus an increase in T is equivalent to a reduction in electrical bandwidth, Δf. Although bandwidth reduction (i.e., integration) is an effective means of noise reduction for white-noise sources such as shot noise and Johnson noise (see Eqs. 9.7 and 9.9), it is not effective as a means of noise reduction for systems that are flicker noise limited.

The distinction between flicker noise and the other two can be seen by comparing Eqs. 9.7, 9.9, and 9.10. At first glance, it might appear that all three noise sources could be reduced by bandwidth reduction since all three noise powers are directly proportional to Δf. In the case of flicker noise, however, the noise power is proportional to $\Delta f/f$. It should be realized that Δf and f are not truly independent quantities because f must fall within the electrical bandwidth Δf. Thus as the sampling period is increased, f is reduced along with Δf.

The effect of bandwidth reduction is illustrated schematically in Figure 9.14 for a $1/f$ noise power spectrum. The total noise power transmitted by the system is equal to the integrated area under the curve corresponding to the given frequency response bandwidth. As shown in the figure, as the electrical bandwidth is reduced, f is also reduced, and the area under the curve remains constant. For this reason, bandwidth reduction has no effect on reducing flicker noise. By the ergodic principle, the same result holds for signal averaging as well.

Finally, it is worthwhile to compare thermal noise, shot noise, and flicker noise in terms of their dependence on signal. From Eqs. 9.7, 9.9, and 9.10, it can be seen that (1) thermal noise power is independent of signal current; (2) shot noise power is proportional to signal current; and (3) flicker noise power is proportional to the square of the signal current.

In terms of spectroscopy, most systems employing quantum or photon detectors are shot noise limited in the middle of their range. At the upper extreme (i.e., high intensities) where the signal current becomes large, the system may become flicker noise limited. As the intensity is decreased, the system eventually becomes *dark-current noise limited*.

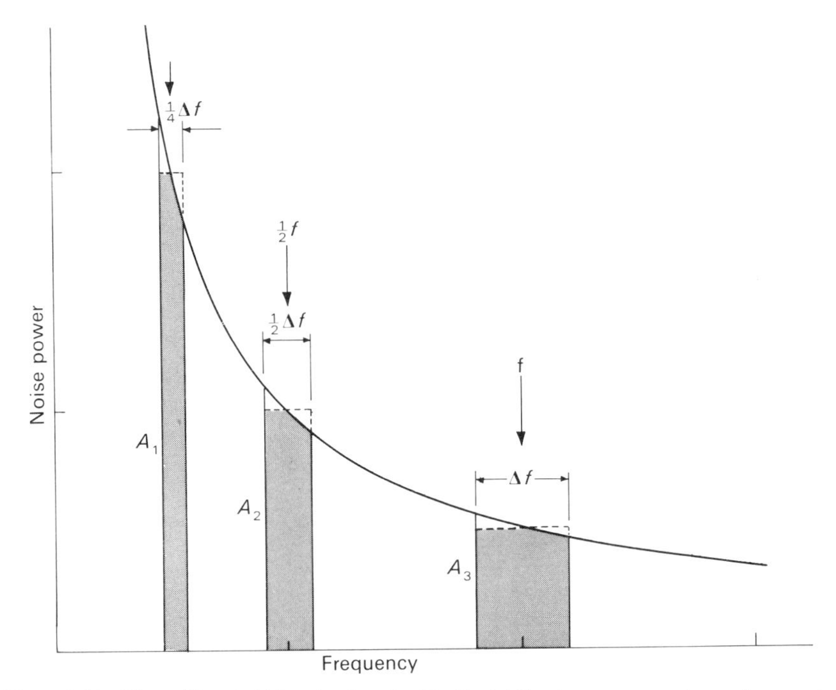

Figure 9.14. Effect of bandwidth reduction for idealized $1/f$ noise power spectrum. Area under curves gives noise power transmitted by system ($A_1 = A_2 = A_3$).

The dark current in a detector is the small residual signal observed from the detector in the absence of light. In photoemissive detectors, dark current is primarily due to the thermal emission or *thermionic emission* of electrons from the cathode and dynodes. Other sources of dark current in photoemissive detectors such as photomultipliers include field effects, residual gas ionization within the tube, ohmic leakage, and fluorescence (Levi, 1980). In solid-state detectors, dark current is due to the thermal generation of charge carriers. In either case (photoemissive or solid state), dark current may be reduced by cooling the detector. The rms dark-current noise will be due either to shot noise (Eq. 9.7) in the case of the photoemissive detector or *gr* noise (Eq. 9.11) for a semiconductor.

9.5. DETECTOR PERFORMANCE PARAMETERS

The overall picture of detector performance can be obtained from measurements of signal output versus optical input. A plot of signal output versus

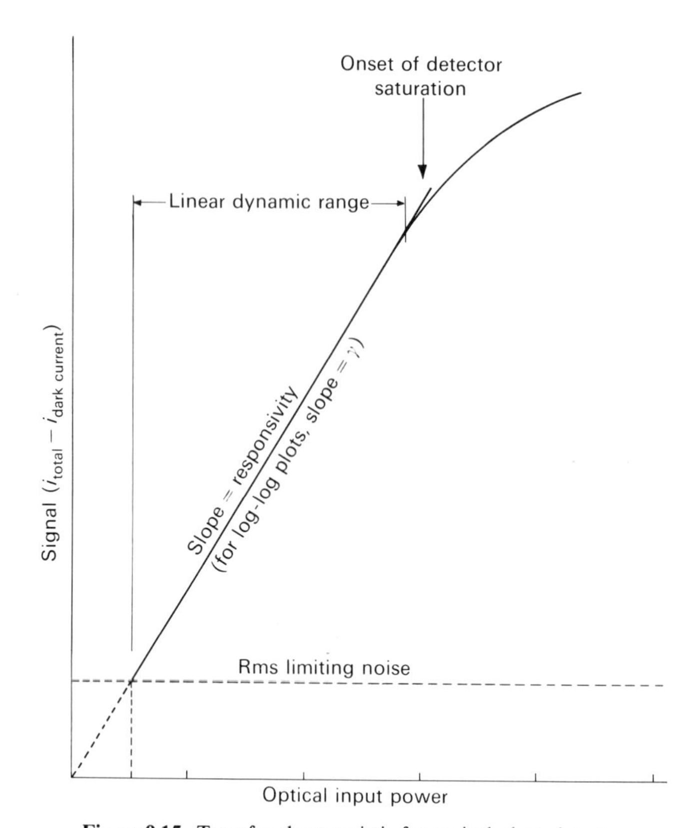

Figure 9.15. Transfer characteristic for typical photodetector.

optical input is known as a *transfer characteristic*. Figure 9.15 shows the transfer characteristic for a typical detector. This curve has many similarities with the emulsion characteristic curve (see Figure 8.2) discussed previously. It should be noted in Figure 9.15 that the signal that is plotted has been corrected for dark current. Thus in the following discussion, the signal will be taken as $i_s = i_{total} - i_{dark}$ current.

As in the case of the emulsion characteristic curve, the characteristic curve for a photoelectronic detector is divided into a number of regions. At the upper extreme (i.e., high values of optical input), the detector saturates and the response falls off with increasing optical input. At the other extreme (i.e., very low levels of optical input), the detector response is ultimately limited by noise considerations. In between the lower noise limit and the saturation level is a region where the signal output varies linearly with optical input.

9.5.1. Responsivity

The slope of the transfer characteristic is a measure of the ability of the detector to sense small changes in optical input and is known as the *responsivity* R. Thus, the responsivity is given by

$$R = \lim [\Delta(\text{signal output})/\Delta(\text{optical input})] \tag{9.16}$$

If the transfer characteristic is plotted on log–log coordinates, the slope of the curve is referred to as gamma (γ) by analogy with the case of the photographic emulsion (see Section 8.2):

$$\gamma = \lim [\Delta(\log \text{signal output})/\Delta(\log \text{optical input})] \tag{9.17}$$

In comparing responsivities for a group of detectors, care must be taken to be sure that the units of responsivity are the same for all members of the group. Particular attention must be paid to the form of the optical input since responsivity can be reported for monochromatic as well as polychromatic radiation. If, for example, the output is a current, the responsivity may be reported in units of amperes per watt if the total incident flux is measured. If the optical input is an irradiance, the units of responsivity will be ampere meters squared per watt since irradiance has units of watts per meters squared.

If the output is the photocurrent produced by incident monochromatic radiation of a given wavelength, the responsivity will be a function of wavelength and will be referred to as the *spectral responsivity*. If polychromatic radiation is employed, the *total responsivity* is the ratio of the total output current for all wavelengths divided by the total optical flux from all wavelengths. If the spectral responsivity is defined as

$$R(\lambda) = i(\lambda)/F(\lambda) \tag{9.18}$$

where $i(\lambda)$ is the photocurrent produced by an incident monochromatic optical flux $F(\lambda)$, the total responsivity will be given by (Budde, 1983),

$$R = \frac{\int_0^\infty [F(\lambda) R(\lambda) d\lambda]}{\int_0^\infty [F(\lambda) d\lambda]} \tag{9.19}$$

where the numerator represents the sum of all the current from the different wavelengths and the denominator is the total optical flux. For infrared detectors, the total responsivity is frequently reported for a blackbody source with a temperature of 500 K (Keyes, 1980). For visible detectors, a blackbody temperature of 2870 K is often used as a reference.

Since all photoelectronic detection systems are characterized by a response time, the responsivity will be a function of the temporal characteristics of the light input,

$$R(\lambda, f) = i(\lambda, f)/F(\lambda, f) \tag{9.20}$$

where $R(\lambda, f)$ is the responsivity as a function of both wavelength and chopping frequency, i is the output current at the chopping frequency f, and F is the input flux also chopped at f. In general, the temporal responsivity decreases as the chopping frequency increases.

From a spectroscopic point of view, the spectral responsivity and the temporal responsivity are important parameters. The former is important because the spectroscopist is more interested in response at specific wavelengths than broadband performance. The latter is important because optical sources are frequently chopped (i.e., *modulated*) to reduce the effect of electronic drift and background radiation. If the background radiation is not chopped while the source radiation is, the signal of interest will be an ac signal, while the background radiation will result in a dc signal. The signal of interest can be recovered from the total signal by use of an ac amplifier.

The spectral responsivity can be related to the quantum efficiency by performing a simple dimensional analysis. From the definition of spectral responsivity (Eq. 9.18), it can be seen that two terms are involved, $i(\lambda)$ and $F(\lambda)$. The photocurrent $i(\lambda)$ can be analyzed dimensionally into the product of two components,

$$i(\lambda) = \begin{cases} \text{(photoelectrons/sec) (charge/photoelectron)} & (9.21) \\ \text{(photoelectrons/sec)}q & (9.22) \end{cases}$$

Similarly, the optical flux can be resolved into the product of two components,

$$F(\lambda) = \begin{cases} \text{(photons/sec) (energy/photon)} & (9.23) \\ \text{(photons/sec)}h\nu & (9.24) \end{cases}$$

Since the quantum efficiency is the number of photoevents per photon,

$$\eta(\lambda) = \frac{i(\lambda)/q}{F(\lambda)/h\nu} = \frac{R(\lambda)h\nu}{q} = \frac{R(\lambda)hc}{\lambda q} \tag{9.25}$$

Equation 9.25 permits spectral responsivity data to be transformed into quantum efficiency as a function of wavelength.

For an in-depth discussion of responsivity and how it is measured, the reader is referred to Budde (1983).

9.5.2. Noise Equivalent Power

Although the responsivity allows us to predict the change in signal output from the detector in response to small variations in the optical input, it does not allow us to determine the lowest light level to which the detector can respond. One way to specify the lowest light level to which the detector can respond is in terms of the *noise equivalent power* (*NEP*). The NEP refers to the incident radiant power that will produce a signal just equal to the rms noise. In other words, the NEP is the radiant power that will produce a signal-to-noise ratio of 1.

To get some idea of the factors that affect NEP, let us consider an ideal photon detector that is shot noise limited. If the detector is irradiated by an incident monochromatic flux whose power is P_s watts, the resulting photocurrent will be

$$i_s(\lambda) = R(\lambda)P_s(\lambda) = \eta(\lambda)qP_s(\lambda)/hv \tag{9.26}$$

according to Eqs. 9.18 and 9.25. If the detector is shot noise limited,

$$\langle i_s(\lambda)\rangle = \begin{cases} [2qi_s(\lambda)\Delta f]^{1/2} & (9.27) \\ [2q^2\eta(\lambda)P_s(\lambda)\Delta f/hv]^{1/2} & (9.28) \end{cases}$$

Dividing Eq. 9.26 by Eq. 9.28 gives the signal-to-noise ratio (S/N)

$$S/N = [\eta(\lambda)P_s(\lambda)/2hv\Delta f]^{1/2} \tag{9.29}$$

When the signal-to-noise ratio is unity,

$$P_s(\lambda) = 2hv\,\Delta f/\eta(\lambda) = 2hc\,\Delta f/\lambda\eta(\lambda) = \text{NEP} \tag{9.30}$$

Equation 9.30 is actually an oversimplification of the situation under shot-noise-limited conditions. Keyes (1980) has shown that a more realistic equation can be obtained on the basis of Poisson statistics assuming that at least one photon has a high probability of arriving during each observation period. On the basis of these assumptions, the minimum power required to produce a 99% probability of photon detection is

$$P_{\min} = 9.22hc\,\Delta f/\lambda\eta(\lambda) \tag{9.31}$$

Regardless of the derivation used, NEP is inversely proportional to both wavelength and quantum efficiency and directly proportional to frequency response bandwidth. Thus to improve the performance of the detector, one must seek ways to minimize NEP. For a given detector and wavelength of interest, bandwidth reduction is the only practical means of reducing NEP. Since bandwidth is inversely proportional to observation time ($\Delta f = 1/2T$), this is equivalent to increasing the observation period.

9.5.3. Detectivity

Another way in which the lower performance limit of a detector is commonly specified is in terms of its *detectivity D*. The detectivity is defined as the inverse of NEP. Since NEP is the radiant power required to give a signal just equal to the noise, it can be written in a generalized form as

$$i_s(\lambda) = R(\lambda)P(\lambda) = \langle i_s(\lambda) \rangle \tag{9.32}$$

$$P(\lambda) = \text{NEP} = \langle i_s(\lambda) \rangle / R(\lambda) \tag{9.33}$$

Equation 9.33 shows that NEP is inversely proportional to the responsivity $R(\lambda)$ and directly proportional to the limiting noise $\langle i_s(\lambda) \rangle$. From Eq. 9.33, it follows that the detectivity will be given by

$$D(\lambda) = R(\lambda) / \langle i_s(\lambda) \rangle \tag{9.34}$$

The advantage of the detectivity as a figure of merit over the NEP is that improved performance is associated with an increase in the former whereas it is related to a decrease in the latter. Although it is not specifically indicated in Eqs. 9.33 and 9.34, it should be noted that both NEP and detectivity are functions of the frequency response bandwidth as a result of their dependence on the rms limiting noise. For a white-noise-limited situation, this functional dependence will take the form $(\Delta f)^{1/2}$ as discussed earlier. In the absence of any information regarding the magnitude of Δf, it is assumed that Δf corresponds to 1 Hz.

One of the primary purposes of a figure of merit such as D is to permit intercomparison between different types of detectors. For some types of detectors, the noise is proportional to $(A\,\Delta f)^{1/2}$, where A is the area of the detector and Δf is the bandwidth. A good example of such a detector is the photomultiplier. In this case, the limiting noise will be due to the shot noise imposed on the dark current. Since the dark current is proportional to the amount of thermionic emission and the amount of thermionic emission is proportional to the area of the cathode, the shot noise, according to Eq. 9.7, will be proportional to $(A\,\Delta f)^{1/2}$.

Since the detector area and frequency response bandwidth are likely to depend on the experimental arrangement, a normalized figure of merit is often used known as the *specific detectivity*, D^*, where D^* is obtained by multiplying the detectivity by $(A\,\Delta f)^{1/2}$ to eliminate the area and bandwidth dependence associated with D. Thus, D^* is given by

$$D^*(\lambda) = D(\lambda)\,(A\,\Delta f)^{1/2} = R(\lambda)\,(A\,\Delta f)^{1/2}/\langle i_s(\lambda)\rangle \tag{9.35}$$

Finally, it should be noted that although D^* is a good figure of merit for comparing different types of detectors, it is the actual NEP that must be considered once a given type of detector has been selected.

APPENDIX 9A

A *Bode diagram* is a plot of the gain of a circuit in decibels versus the logarithm of the frequency. Plotting the frequency response of a circuit in this form is especially convenient because it results in a graph consisting of two straight-line segments that intersect at a point that can be easily calculated. To illustrate the basis for the Bode diagram, consider the low-pass filter shown in Figure 9A.1.

From elementary electronics, the gain of this circuit is given by

$$G = (e_{out}/e_{in}) = X_C/(R^2 + X_C^2)^{1/2} \tag{9A.1}$$

In this equation, R is the resistance in ohms and X_C is the capacitive reactance given by $X_C = 1/2\pi fC$, where f is the frequency and C is the capacitance. If f is in hertz and C is in farads, the capacitive reactance will be in ohms. To express

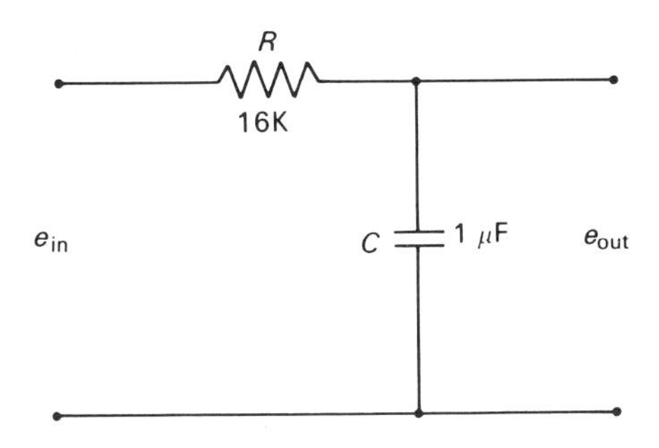

Figure 9A.1. Low-pass *RC* filter.

the gain of this circuit in decibels, we must take the logarithm of the gain and multiply by 20,

$$G(\mathrm{dB}) = 20 \log(e_{\mathrm{out}}/e_{\mathrm{in}}) \tag{9A.2}$$

Applying Eq. 9A.2 to Eq. 9A.1 gives the following result:

$$G(\mathrm{dB}) = 20 \log X_C - 10 \log(R^2 + X_C^2) \tag{9A.3}$$

Using Eq. 9A.3, three conditions can be identified depending on the frequency. From Eq. 9A.3, it can be seen that when $X_C \gg R$, the gain in decibels will be constant and equal to zero. Since this condition exists for low frequencies, the low-frequency portion of the Bode diagram is a horizontal line at 0 dB.

When $X_C = R$, Eq. 9A.3 reveals that the gain in decibels will be -3 dB. This condition will occur when $f = 1/2\pi RC$, where RC is the RC time constant of the filter (i.e., the product of R and C) in seconds. This point on a Bode diagram has a variety of names such as *3-dB point*, *half-power frequency*, *breakpoint frequency*, and *frequency response bandwidth*.

At still higher frequencies, $X_C \ll R$ and Eq. 9A.3 reduces to

$$G(\mathrm{dB}) = \begin{cases} 20 \log X_C - 20 \log R & (9A.4) \\ -20 \log(2\pi f C) - 20 \log R & (9A.5) \end{cases}$$

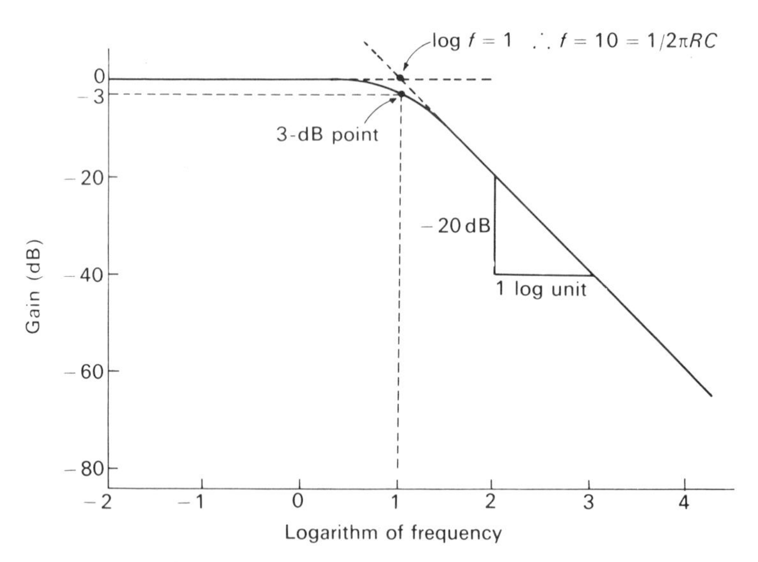

Figure 9A.2. Bode diagram for low-pass *RC* filter.

Equation 9A.5 reveals that for high frequencies the gain of the filter in decibels is a linear function of frequency with a slope of -20 dB per logarithmic unit of frequency.

Figure 9A.2 shows the complete Bode diagram for the low-pass filter. As shown in the figure, the diagram consists of two straight-line segments that intersect at the 3-dB point. The slope of the high-frequency portion of the graph is sometimes referred to as the *roll-off*.

REFERENCES

Boumans, P. W. J. M. (1966), *The Theory of Spectrochemical Excitation*, Plenum, New York.

Boyd, R. W. (1983), *Radiometry and the Detection of Optical Radiation*, Wiley, New York.

Budde, W. (1983), *Optical Radiation Measurements, Vol. 4, Physical Detectors of Optical Radiation*, Academic, New York.

Fassel, V. and Kniseley, R. N. (1974a), *Anal. Chem.*, *46*, 1110A.

Fassel, V. and Kniseley, R. N. (1974b), *Anal. Chem.*, *46*, 1155A.

Goldman, S. (1948), *Frequency Analysis, Modulation and Noise*, McGraw-Hill, New York.

Johnson, J. B. (1928), *Phys. Rev.*, *32*, 97.

Keyes, R. J., Ed. (1980), *Optical and Infrared Detectors*, Springer-Verlag, New York.

Kingston, R. H. (1978), *Detection of Optical and Infrared Radiation*, Springer-Verlag, New York.

Levi, L. (1968), *Applied Optics—A Guide to Optical System Design*, Vol. 1, Wiley, New York.

Levi, L. (1980), *Applied Optics—A Guide to Optical System Design*, Vol. 2, Wiley, New York.

Margoshes, M. (1970a), Pittsburgh Conference on Analytical Chemistry and Applied Spectroscopy, Cleveland, OH, March, Paper No. 99.

Margoshes, M. (1970b), *Opt. Spectra*, *4*, 26.

Margoshes, M. (1970c), *Spectrochim. Acta, Part B*, *25*, 113.

Motchenbacher, C. D. and Fitchen, F. C. (1973), *Low-noise Electronic Design*, Wiley, New York.

Nyquist, H. (1928), *Phys. Rev.*, *32*, 110.

Pinson, L. J. (1985), *Electro-Optics*, Wiley, New York.

Schottky, W. (1918), *Ann. Phys.*, *57*, 541.

CHAPTER

10

INTRODUCTION TO IMAGE DETECTORS

In the previous chapter, detectors that respond to the total optical flux incident on their light-sensitive surfaces were discussed. When used in conjunction with some form of optical or mechanical scanning, such detectors can provide information about the flux distribution in an image. By contrast, detectors that can respond inherently to the distribution of flux in an image without the need for additional optical or mechanical components are known as *image detectors*. Image detectors can be classified into two categories depending on the form of image transduction. If the output of the device is an optical image, the device is known as an *image intensifier* if it is used to increase the luminance of the image or an *image converter* if it is used to change the wavelength of the image (i.e., convert incident radiation from the infrared or ultraviolet region of the spectrum to visible radiation). On the other hand, if the output of the device is an electrical signal, the device is known as an *image pickup tube*. Since the latter category of image detector is more appropriate for spectroscopic applications, the remainder of this chapter will be devoted to optical–electrical image transducers.

10.1. THE DEVELOPMENT OF IMAGE DETECTORS

Before discussing the material needed to understand modern-day image detectors, it is worthwhile to review some of the more important milestones in the development of image detectors. Since the early devices use the same general principles as modern-day image detectors, an appreciation of the early work will be useful as an introduction to understanding current technology.

The first attempt at developing a rapid scanning detection system for the electrical transmission of an image was the so-called *Nipkow disk* invented by Nipkow in 1884. Light from the scene to be transmitted was focused on a rotating disk provided with a spiral of holes around the circumference, as shown in Figure 10.1. A photodetector placed behind the disk was used to monitor the light transmitted through a given hole. Only one hole transmitted light from the scene to the photodetector at any given time. As one hole rotated out of view of the detector, the next hole, which was displaced

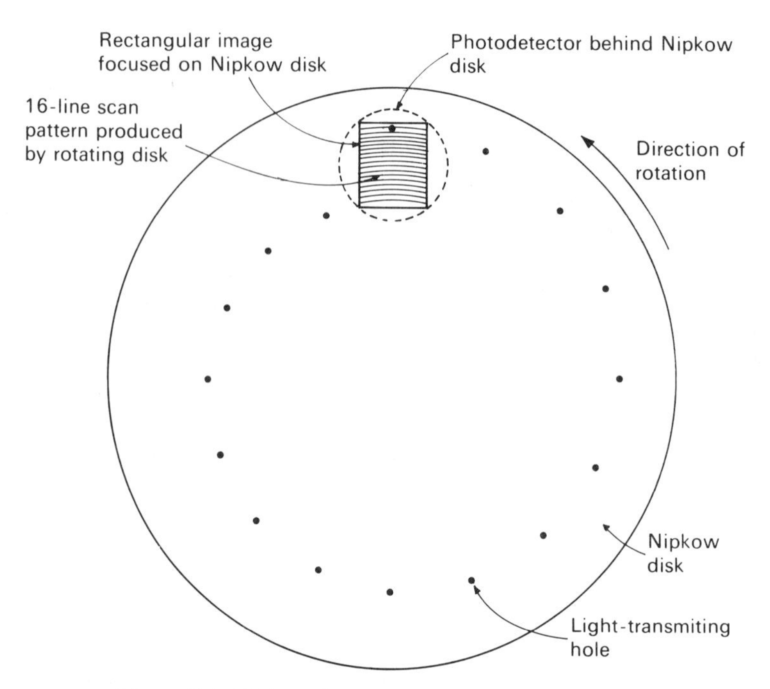

Figure 10.1. Nipkow disk for electrical image transmission.

vertically from the previous one, rotated into position. In this way, the scene was divided up into a series of horizontal lines that were then transmitted in serial fashion by the photodetector. The number of horizontal lines in the picture transmitted depended on the number of holes around the circumference of the disk. The major limitation of the Nipkow disk system, insofar as its application to conventional television transmission is concerned, was the small amount of light transmitted through the hole in the rotating disk and the small sampling period for each resolution element in the picture.

To overcome these difficulties, Ekström (1910) proposed a reverse optical system that used the Nipkow rotating disk. In Ekström's system, a bright arc was placed behind the Nipkow disk so that light from any given hole could be focused on the object as a small moving spot. As the disk rotated, the intense light spot would scan the object in a pattern of vertically displaced horizontal lines. By employing a photodetector housed in a large reflector, light reflected by the "flying spot" off of different objects in the scene was converted to an electrical signal.

These preliminary experiments made possible the first practical transmission of images by photoelectrical means. On June 14, 1923, Jenkins performed the first laboratory demonstration of "radio vision" in his Washington laboratory (Jenkins, 1925). On January 27, 1926, Baird gave a public demonstration of his "televisor" in London (Dinsdale, 1926). In April of the following year, Bell Telephone Laboratories unveiled their system of television and published an account of their experiments in an article in the *Bell System Technical Journal* (Vol. 6, October 1927).

With the development of the principles of electron optics in the 1920s, the possibility of developing detectors that employed electronic scanning was considered (Meyers, 1939). In 1931, the American inventor Philo T. Farnsworth described a detector he called a "dissector tube" in an article in *Television News*. Figure 10.2 shows a schematic of the system devised by Farnsworth. The tube consisted of a photocathode (C) parallel to and in close proximity to a screen mesh that served as the anode (A). At the opposite end of the tube, a target electrode (T) with a small active area was positioned to receive the electron flux from the photocathode. The operation of the tube was described by Farnsworth (1931, pp. 48) as follows:

> This tube, considered broadly, is a photoelectric cell wherein provision is made for forming an "electron image" of an optical image focused on its cathode surface. By "electron image" it is meant that, if a fluorescent screen were placed in the plane of the electron image, the original optical image would be reproduced. The condition necessary for the formation of this electron image is that all the electrons emitted from any single point on the cathode surface shall meet again in a corresponding point in the plane of the electron image.
>
> An image of the object to be transmitted is focused upon the cathode, and the photoelectrons emitted therefrom are accelerated by a potential of the order of 500 volts between the cathode and the anode screen. Most of them are projected into the region between the screen and the target and, by means of the focusing magnetic field combine to give an electron image in the plane of the target. This electron image, made up as it is of a prism of moving electrons, can be shifted by a magnetic field at right angles to the tube. By this means, the image is moved over the scanning aperture in the target shield.

The next major development in the evolution of television pickup tubes was the development of the charge storage tube by Zworykin in 1934. Use of a charge storage principle produced a gain in sensitivity of the order of 10^5 compared with real-time devices such as the Nipkow disk and the Farnsworth dissector. This gain in sensitivity was a result of the integrating nature of the charge storage tube.

To appreciate the origin of the sensitivity gain, we must look more closely at the readout process of the two types of tubes. With real-time detectors, only 10^{-7} sec is spent interrogating any given picture element when the tube is operated at conventional transmission frequencies, and the photocurrent

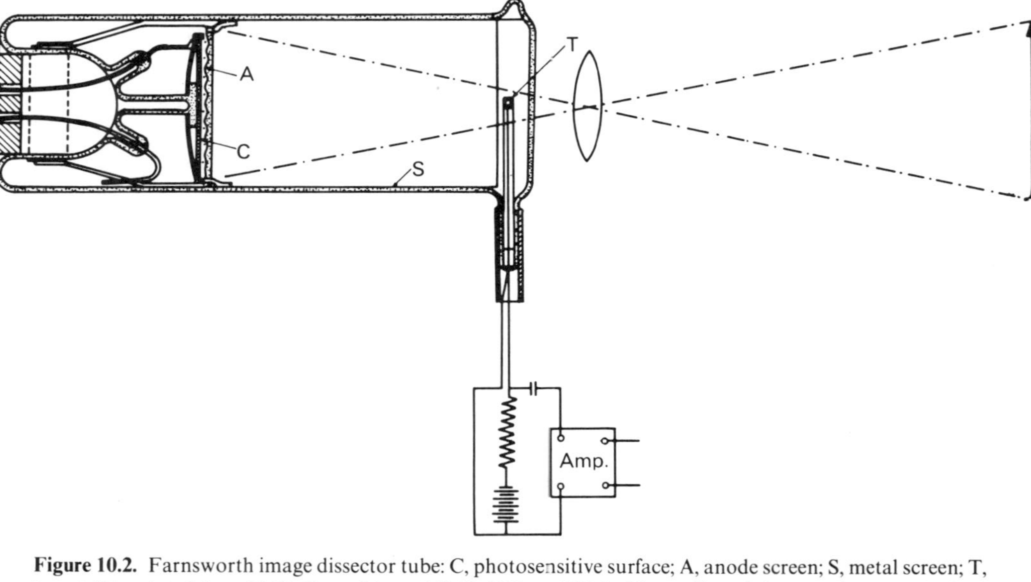

Figure 10.2. Farnsworth image dissector tube: C, photosensitive surface; A, anode screen; S, metal screen; T, target. [Reprinted from V. K. Zworykin and E. D. Wilson (1934), *Photocells and their Applications*, 2nd ed. By permission of John Wiley & Sons, New York.]

produced during this 10^{-7} sec period must serve as the signal. By comparison, if an integrating detector is used, the signal produced is not due to the instantaneous signal accumulated during the period the given resolution element is interrogated but, in contrast, represents the signal that has accumulated since the last time the particular resolution element was interrogated. At conventional transmission frequencies, the time period between interrogation of individual resolution elements is typically about $\frac{1}{30}$ sec. Thus, if a scene with a given brightness is monitored by both a dissector tube and a charge storage tube, a picture element in a charge storage tube will have a chance to respond to 10^5 more photons during readout than a picture element in a real-time detector.

Figure 10.3 shows a schematic diagram of the *iconoscope* (from the Greek word *eikon*, which means "image") developed by Zworykin (1934). The tube consisted of an evacuated glass bulb containing a photosensitive mosaic and an electron gun located opposite the mosaic and inclined 30° to the normal passing through the middle of the mosaic. The photosensitive mosaic consisted of a metal plate (known as the *signal plate*) covered with a large number of individual miniature photoelectric cells that were insulated from each other and from the plate, as shown in Figure 10.4.

The photosensitive mosaic was produced by starting with a thin sheet of the insulating natural material mica, having a uniform thickness and serving as the insulating support for the mosaic. The signal plate was then formed by

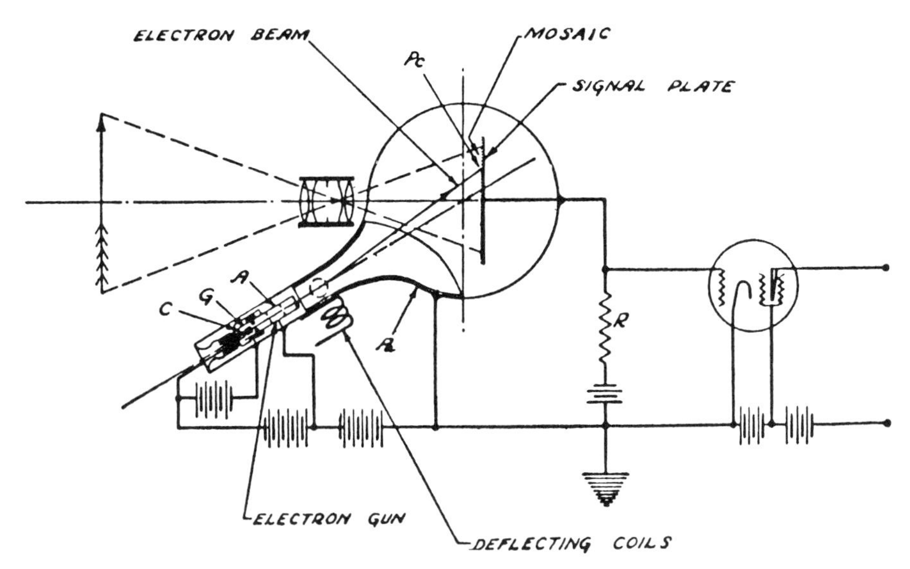

Figure 10.3. Schematic diagram of iconoscope. [Reprinted from V. K. Zworykin and E. D. Wilson (1934), *Photocells and their Applications*, 2nd ed. By permission of John Wiley & Sons, New York.]

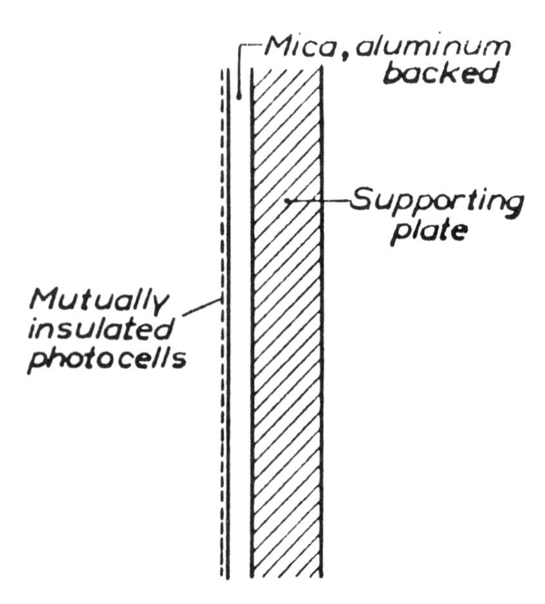

Figure 10.4. Photosensitive target or signal plate of iconoscope. [Reprinted from H. Bruining (1954), *Physics and Applications of Secondary Electron Emission*. By permission of McGraw-Hill Book Company, New York.]

coating one side of the mica sheet with a metallic layer. The mosaic was formed on the opposite side of the mica sheet by vacuum evaporation of a photoelectric metal such as an alkali metal onto the mica. Oddly enough, this procedure produced a mosaic of photosensitive elements rather than a continuous photosensitive element because of the thinness of the layer. When the evaporated film of photosensitive metal is extremely thin, it is not continuous but consists of a uniformly distributed conglomeration of minute spots or globules, each spot being insulated from its neighbors and the signal plate by the intervening mica.

The photosensitive target produced in this manner consisted of a mosaic of tiny capacitors where the capacitance of each individual element of the array was determined by the thickness and dielectric constant of the insulating mica layer between the individual elements and the common signal plate.

Figure 10.5 shows an equivalent circuit for an individual element in the photosensitive mosaic. Light striking the photosensitive side of the mica target would produce photoelectrons that could be drawn to an electrode known as the *collector* where they would be drained to ground. As a result of the release of photoelectrons, a positive residual charge would form and build up on each element of the array in proportion to the amount of light absorbed by the individual photoelements. This charge pattern that developed on the target could be measured and neutralized at the same time by scanning the

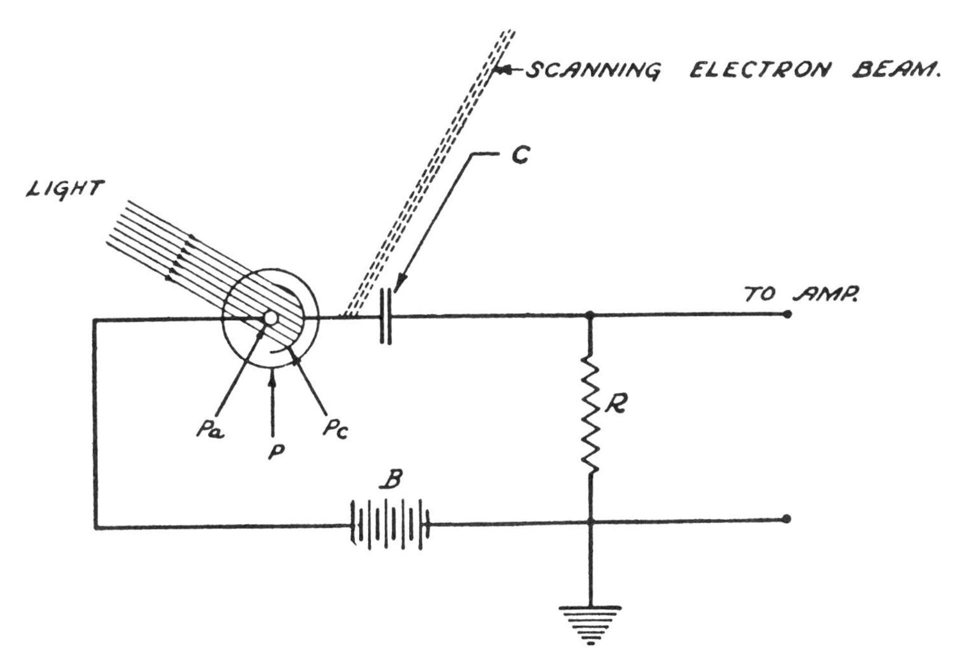

Figure 10.5. Schematic diagram of equivalent circuit of single element in mosaic of iconoscope. [Reprinted from V. K. Zworykin and E. D. Wilson (1934), *Photocells and their Applications*, 2nd ed. By permission of John Wiley & Sons, New York.]

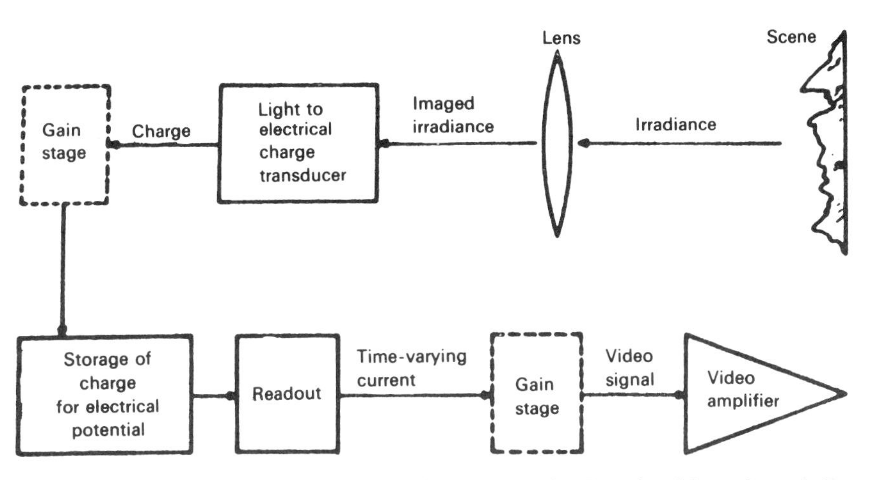

Figure 10.6. Block diagram of modern television camera tube. [Reprinted from Cope, A. D., Gray, S., and Hutter, E. C. (1974), "The Television Camera Tube as a System Component," in *Photoelectronic Imaging Devices*, Vol. 2, Biberman, L. M. and Nudelman, S. (Eds.), Plenum, New York.]

surface of the target line by line with an electron beam from the electron gun. The discharge current produced when the electron beam replaces the negative charge lost between two scans flows from the signal plate through the input resistance of a wide-band amplifier known as a *video amplifier*.

The basic principles behind the operation of modern-day image detectors have not changed since the development of the iconoscope in 1934. Figure 10.6 shows a block diagram of the functions of a modern television camera tube. As illustrated in the figure, the basic principle of operation is the transduction of a light image into a charge pattern that can be stored on a target for subsequent readout by a scanning electron beam. The dramatic improvements in the sensitivity and performance of modern image tubes has been the result of improvements in target fabrication and scanning technology.

The remainder of this chapter will be devoted to a discussion of the principles governing the operation of the various components that make up an image detector. The chapter will conclude with a discussion of four common image detectors.

10.2. ELECTRON MULTIPLIER

Before discussing examples of various image pickup tubes, it is appropriate to discuss some basic aspects of electron multiplication and electron optics since both processes are commonly used in these devices.

Of the various forms of amplification available, electron multiplication introduces the least amount of noise and for this reason is frequently employed in low-light-level detectors such as photomultipliers (i.e., more correctly known as *multiplier phototubes*). The process of electron multiplication is based on the observation that *secondary electrons* are released from a solid as a result of electron impact by energetic electrons (Bruining, 1954). The number of secondary electrons produced depends on the energy of the bombarding electrons (the primary electrons) and the material being bombarded. The efficiency of the process is described in terms of the *secondary emission ratio*, which is the ratio of the total number of electrons (both secondary and scattered primary) emitted from the surface to the number of incident electrons. Regardless of the energy of the primary electrons, the energy of the secondary electrons produced is about 2 eV (Levi, 1980).

If the secondary emission ratio is measured as a function of the energy of the primary electrons, it is found to be small for very low incident energies where electron absorption appears to be the predominant process. As the energy of the incident electrons increases, the secondary emission ratio increases to a maximum value whereupon it begins to decrease. This decrease in secondary electron production is believed to be due to the increased depth of penetration of the primary electrons into the material. When the primary electron is absorbed deep within the material, the secondary electrons must travel farther to reach the surface and frequently arrive with insufficient energy to escape. To minimize this parasitic effect, the angle of incidence of the primary electrons is frequently increased to minimize penetration depth of the primary electrons.

To make use of the secondary emission of electrons from solids, two arrangements are possible. Both rely on a potential difference to accelerate electrons to the required energy and both require an evacuated envelope. The first uses a series of discrete electrodes known as *dynodes* fabricated from a good photoelectric emitter material. Various dynode geometries are employed in fabricating a dynode chain (Figure 10.7). The particular type of dynode geometry to use depends on the given application (RCA, 1970). For high-speed operation, a focused dynode is the fastest because it reduces the spread in electron transit times. With proper care, transit times of 17 nsec are possible (Levi, 1980).

The second arrangement is known as a *continuous-channel electron multiplier* orginally described by Farnsworth (1930), or *Channeltron* (registered trademark of Galileo Electro-Optics Corp). The Channeltron consists of a narrow glass tube approximately 1 mm in diameter the inside of which has been coated with an appropriate photoemissive material. When a 2-kV potential is applied across the length of the tube, electrons traveling down the tube strike the wall along the way, producing secondary electrons in the process. These secondary electrons in turn may strike the wall, producing still more secondary electrons. Figure 10.8 shows a schematic diagram of a continuous-channel electron multiplier. To reduce ion feedback, most channeltrons are curved (Kurz, 1979). Either of the two approaches are capable of producing gains on the order of 10^6.

The overall gain obtained with a discrete dynode chain depends on the individual gains obtained at each stage. If an electron multiplier with 12 stages is used and each has an average gain (i.e., secondary emission ratio) of 4, the overall gain of the tube will be

$$G = \delta^n = 4^{12} = 16.7 \times 10^6 \tag{10.1}$$

Since the secondary emission ratio is a function of the kinetic energy of the primary electrons and the kinetic energy in turn is a function of the accelerating voltage between stages, precise control of the overall gain G requires careful control of the interdynode voltages. If the secondary emission ratio is related to the interdynode voltages by the equation (Budde, 1983)

$$\delta = kV^{0.7} \tag{10.2}$$

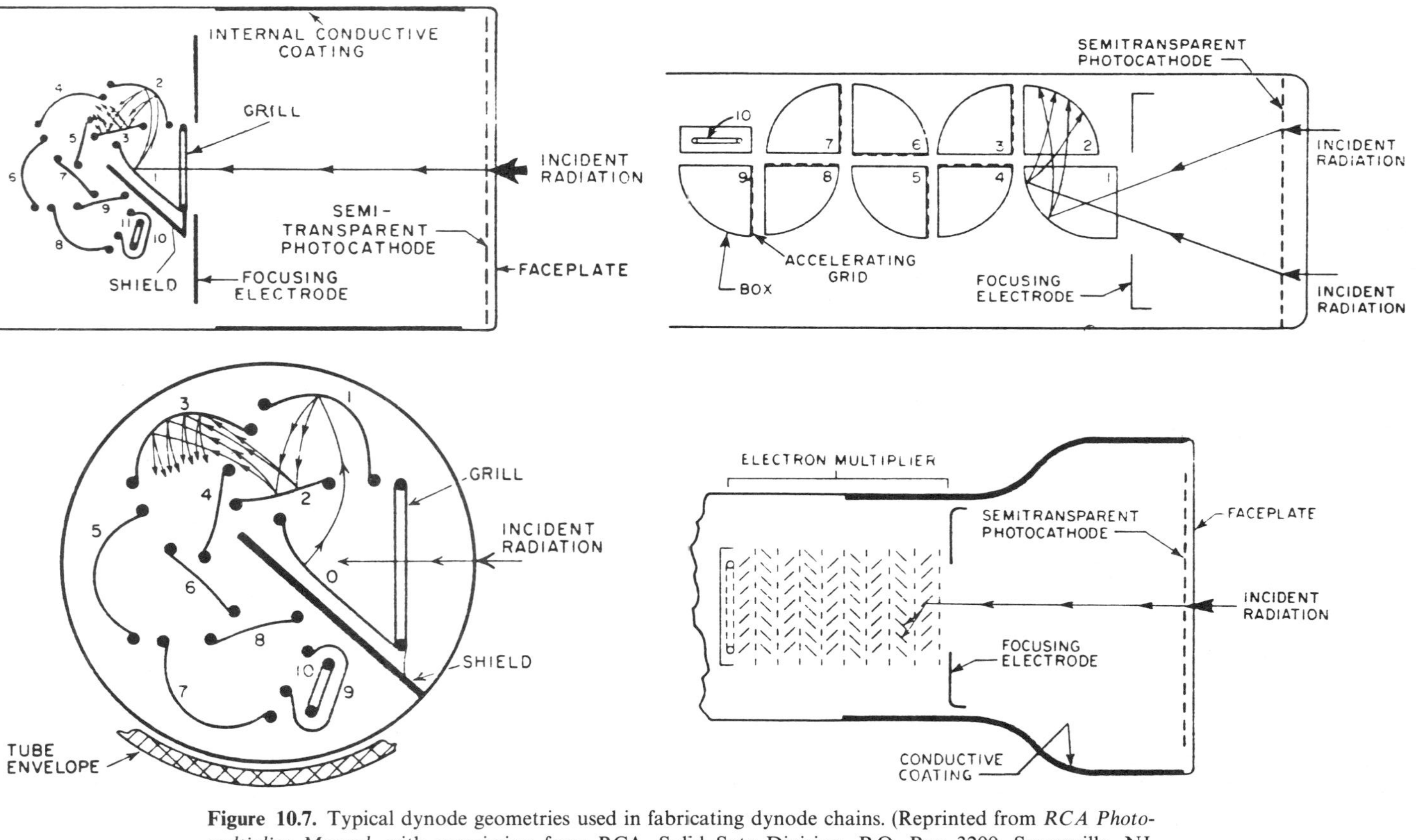

Figure 10.7. Typical dynode geometries used in fabricating dynode chains. (Reprinted from *RCA Photomultiplier Manual*, with permission from RCA, Solid Sate Division, P.O. Box 3200, Somerville, NJ. Copyright 1970 RCA Corporation.)

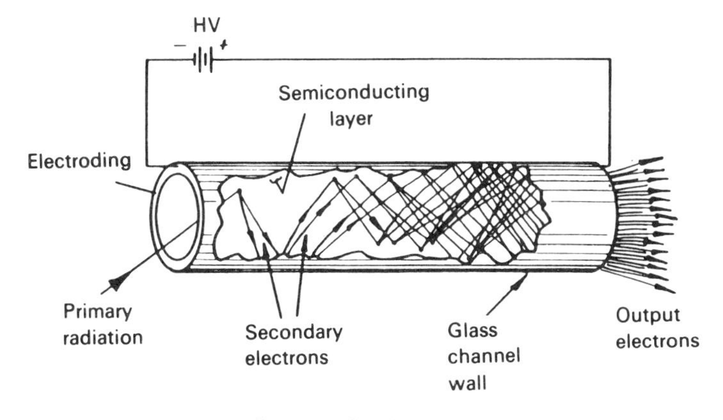

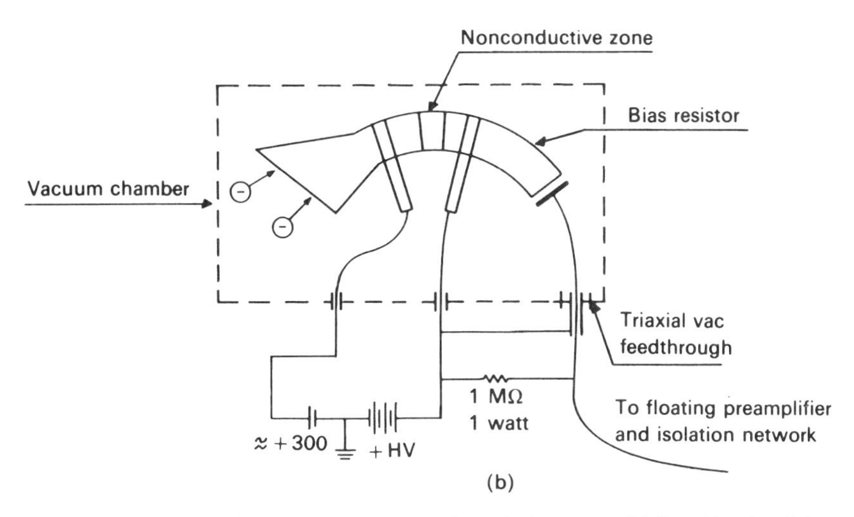

Figure 10.8. Schematic diagram of continuous-channel electron multiplier. (Reprinted from *American Laboratory*, Vol. 11, No. 3, pp. 67–82, 1979. Copyright 1979 by International Scientific Communications, Inc.)

the overall gain according to Eq. 10.1 will be given by

$$G = k'V^{0.7n} \tag{10.3}$$

The change in gain with interdynode voltage will therefore be

$$dG = 0.7\,nk'V^{0.7n-1}dV \tag{10.4}$$

The relative change in gain will be

$$dG/G = 0.7\, nk' V^{0.7n-1} dV/G = 0.7n(dV/V) \tag{10.5}$$

Since the total voltage across the entire chain is linearly related to the voltage between stages, Eq. 10.5 is true for the total voltage as well as the interdynode voltages. Equation 10.5 can be used to obtain some idea of the stability required of the power supply used to provide the interdynode voltages. If a 10-stage tube is used, the coefficient of the relative voltage change from Eq. 10.5 is 7, which gives rise to the rule of thumb, namely, the stability of the power supply must be better than 10 times the desired output stability. Thus, a desired precision of 0.1% in the output of the electron multiplier requires a 0.01% precision in the voltage power supply.

Figure 10.9 shows a typical voltage divider circuit for biasing the dynode chain. In a typical installation, the positive terminal of the power supply is grounded and the negative terminal is connected to the cathode. For proper stability, the current through the voltage divider should be about 10 times greater than the maximum photoanodic current. Since the maximum anodic current is typically about 1 mA when an electron multiplier is used in a

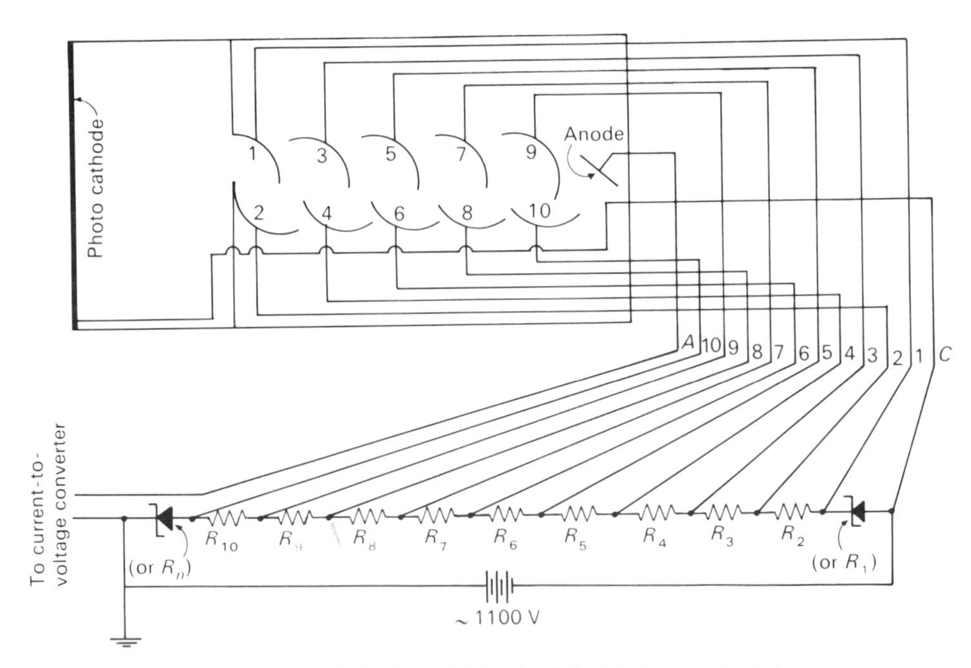

Figure 10.9. Typical voltage divider circuit for biasing dynode chain.

photoemissive detector, the current through the voltage divider should be at least 10 mA in this application.

Even with a well-regulated power supply nonlinearity can be introduced whenever bursts of electrons travel down the electron multiplier. Such bursts draw current from the dynode chain, thereby altering the interdynode voltages. The effect is particularly important for the first and last dynode, and for this reason R_1 and R_n are often replaced with zener diodes, as shown in Figure 10.9. For a comprehensive discussion of the factors involved in the design and fabrication of a dynode voltage divider for an electron multiplier, the reader should consult Budde (1983) and the *RCA Photomultiplier Handbook* (RCA, 1980).

Although the photoanodic current is often converted to a voltage output by means of a large load resistor, a more desirable current-to-voltage conversion can be obtained with an operational amplifier as shown in Figure 10.10. In the figure, the shunt capacitance is introduced as a by-pass for high-frequency noise components. If the output from the current-to-voltage converter is fed to a voltage-to-frequency converter followed by a counter, the circuit will function as an integrating digital voltmeter (Budde, 1983). Using this circuit, signal-to-noise ratios may be improved by increasing the observation time (i.e., bandwidth reduction).

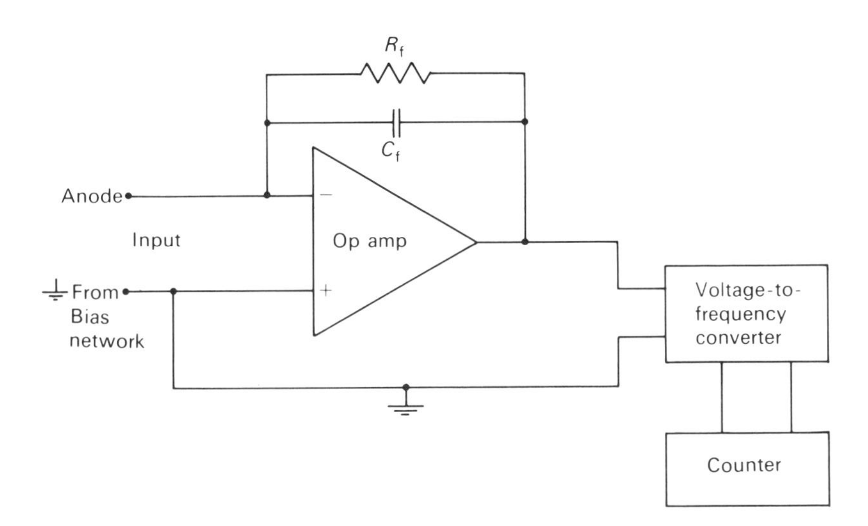

Figure 10.10. Transimpedance amplifier configuration for current-to-voltage conversion.

10.3. ELECTRON OPTICS

The ability to deflect and focus electron beams using electrical and magnetic fields is in many ways analogous to the manipulations that occur in ordinary light optics and for this reason is known as electron optics. The birth of electron optics is said to have occurred in 1926 when Busch (1926, 1927) showed that the effect of a short axially symmetric magnetic field was analogous to the action of a glass lens on light. In geometric optics with light rays, the rays are deflected by encountering changes in the refractive index, according to Snell's law (Eq. 2.9). In electron optics, it can be shown (Levi, 1980; Klemperer, 1953) that the square root of the potential of the applied electric field is analogous to the refractive index in light optics. Thus changes in the electric field can cause deflections in electron beams in the same way that changes in the refractive index can cause deflection in light rays. In contrast to ordinary optics, where deflections occur at abrupt discontinuities in the refractive index (i.e., at the interface between two materials), in electron optics it is possible to vary the extent of deflection continuously by gradually varying the electric field strength. This is analogous to producing a lens with a continuously varying refractive index.

10.3.1. The Electron Lens

The deflection and focusing of electron beams can be accomplished using either electric or magnetic fields or both. In fact, any system of superimposed electric and magnetic fields will act as an electron lens as long as it is axially symmetric. To illustrate this fact, consider an electrostatic potential in an axially symmetric field. Such a potential may be represented by a series of the following form:

$$V(r, z) = V(0, z) - (r^2/4)V''(z) + (r^4/64)V''''(z) - \cdots \tag{10.6}$$

where r is the radial distance from the axis, z is the distance along the axis, and $V''(z)$ and $V''''(z)$ are the second and fourth derivatives of the potential along the axis (Vine, 1974). The radial component of the electric field can be obtained from Eq. 10.6 by differentiating with respect to r,

$$-E(r, z) = -(\tfrac{1}{2}r)V''(z) + (\tfrac{1}{16}r^3)V''''(z) \tag{10.7}$$

If we consider only the first term in Eq. 10.7, we see that an electron traveling close to the axis will experience a radial force component that is proportional to its radial distance from the axis. Therefore, as long as the electron path makes only a small angle with the axis, the deflection due to the electric field will be proportional to its radial distance from the axis. This is the same criterion (i.e., deviation proportional to radial distance) that operates with paraxial light rays incident on a thin lens. Thus an axially symmetric electric field will behave as a thin lens as long as the electrons are close to the axis and the angle that the electron path makes with the axis is small so that the sine of the angle is equal to the angle itself. Thus for electrons that satisfy these conditions (i.e., *paraxial electrons*), only the first term in Eq. 10.7 is significant and the electron lens is said to follow *first-order theory*. As electrons deviate from the preceding conditions, the second term in Eq. 10.7 becomes important, giving rise to what are known as *third-order aberrations*.

Aberrations in electron optics are categorized in ways analogous to light optics. Aberrations due to deficiencies in the shape of the focusing field give rise to what are known as *geometric aberrations*. *Chromatic aberrations* arise as a result of velocity differences among electrons in the beam.

An *electrostatic electron lens* usually consists of a series of coaxial cylinders or apertures maintained at different potentials. A good example of this type of lens is the so-called *three-cylinder lens* shown in Figure 10.11. In this arrangement, the first and last cylinders are maintained at the same potential. Focusing is accomplished by varying the potential of the center cylinder. This is particularly convenient because adjusting the focus does not alter the potential in object and image space. Figure 10.12 shows the equipotential surfaces formed by a typical three-cylinder lens along with their effect on a paraxial bundle of electrons. Figure 10.13 shows the focusing properties of a three-cylinder lens. Image formation by this lens is analogous with the

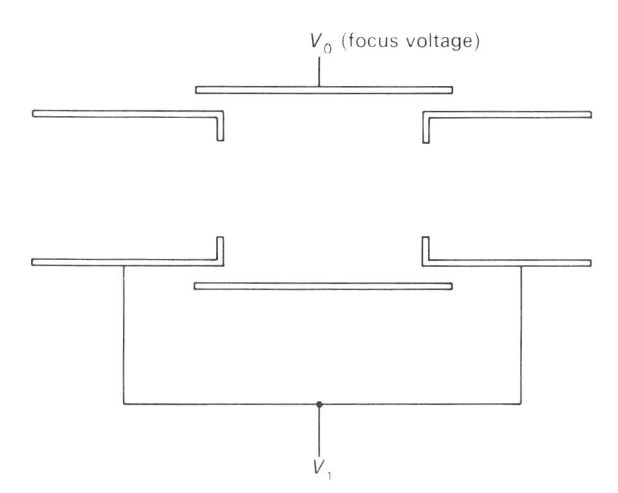

Figure 10.11. Three-cyclinder electrostatic electron lens.

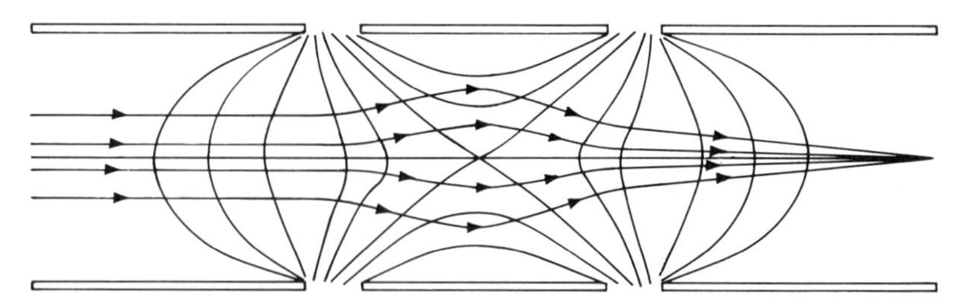

Figure 10.12. Equipotential surfaces formed by typical three-cylinder lens. Arrows show effect of potentials on paraxial bundle of electrons.

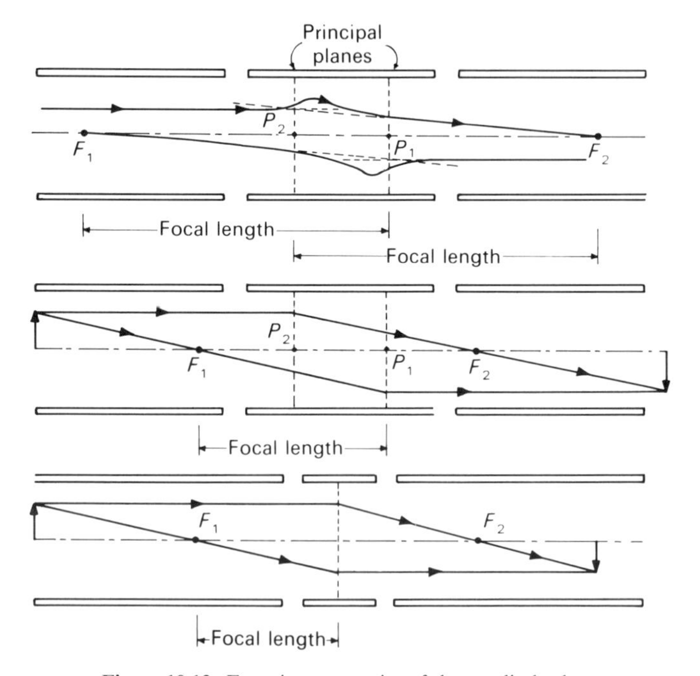

Figure 10.13. Focusing properties of three-cylinder lens.

situation described in Section 2.3.1 for a thick lens. Thus, the intersection of the projections of the incident and emergent rays occurs at the two principal planes, and the points where these planes cut the axis are the principal points. The distance between a given principal point and the corresponding focal point is the focal length. As the length of the central cylinder is reduced, the principal planes approach one another and eventually merge when the lens becomes a thin lens. As shown in the figure, the three-cylinder electrostatic lens forms an inverted image of the object and obeys the thin-lens formula (Eq. 2.29) when the central cylinder is short enough.

10.3.2. The Electrostatic Image Lens

In a typical image tube, the three-cylinder lens is commonly used to focus the electron beam emitted from the electron gun as a small spot on the target of the tube. Another operation that is commonly required in image tubes is the formation of an electron image from the photoelectrons emitted by a photocathode. This requires a somewhat different electrostatic lens. Figure 10.14 shows the situation that results when a spherical anode is concentrically located at the center of a larger spherical cathode. With this arrangement, it can be shown (Vine, 1974) that the image formed will be located on another concentric sphere whose radius is a function of the radii of the cathode and anode,

$$R_i = R_a R_c / (R_c - 2R_a) \tag{10.8}$$

where R_i is the radius of the image sphere, and R_a and R_c are the radii of the anode and cathode, respectively. One serious limitation with the arrangement shown in Figure 10.14 is that it requires an anode that is transparent to

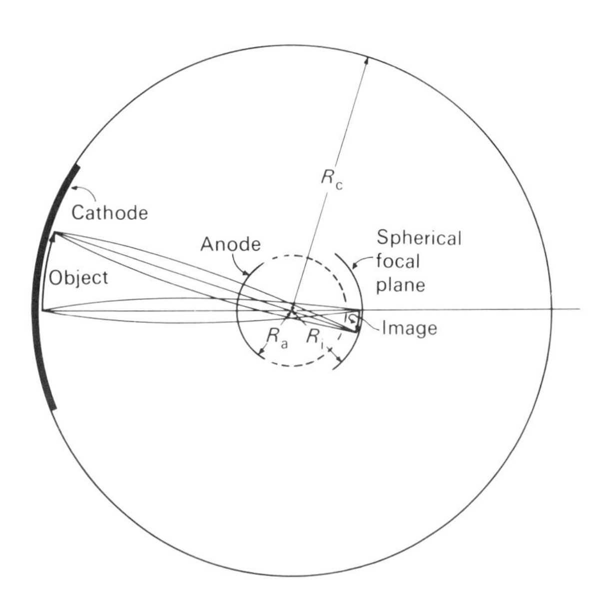

Figure 10.14. Image formation by electrostatic image lens.

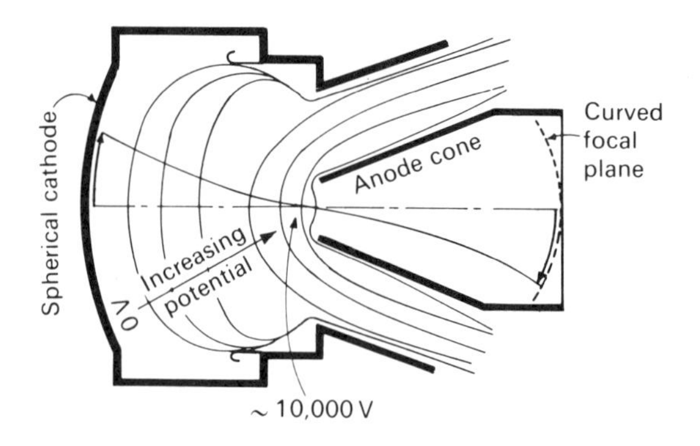

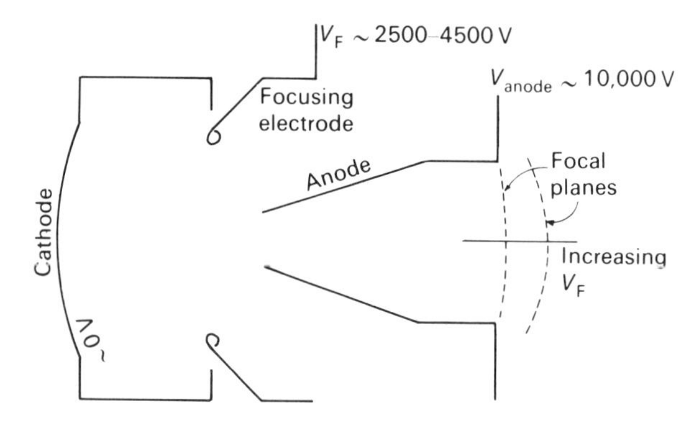

Figure 10.15. Schematic diagram of typical electrostatic image lens.

electrons, something that is not easy to achieve in practice. To overcome this limitation, an alternative arrangement is used that produces equipotential surfaces that are approximately spherical, as shown in Figure 10.15. In this arrangement, the cathode surface is still spherical, while the anode is cone shaped with an aperture to admit the electrons from the cathode.

By analogy with the three-cylinder lens, focusing of the electrostatic image lens can be achieved by introducing a focusing electrode between the cathode and the anode, the potential of which can be varied independently of the other two. Such a configuration is known as a triode arrangement. If the image formed by the lens is focused on a flat target, a certain amount of defocusing will occur at the edges of the image because the focal plane is spherical.

10.3.3. Magnetic Focusing

Focusing of electron beams can also be achieved magnetically as shown in Figure 10.16. This action is generally obtained by producing a uniform

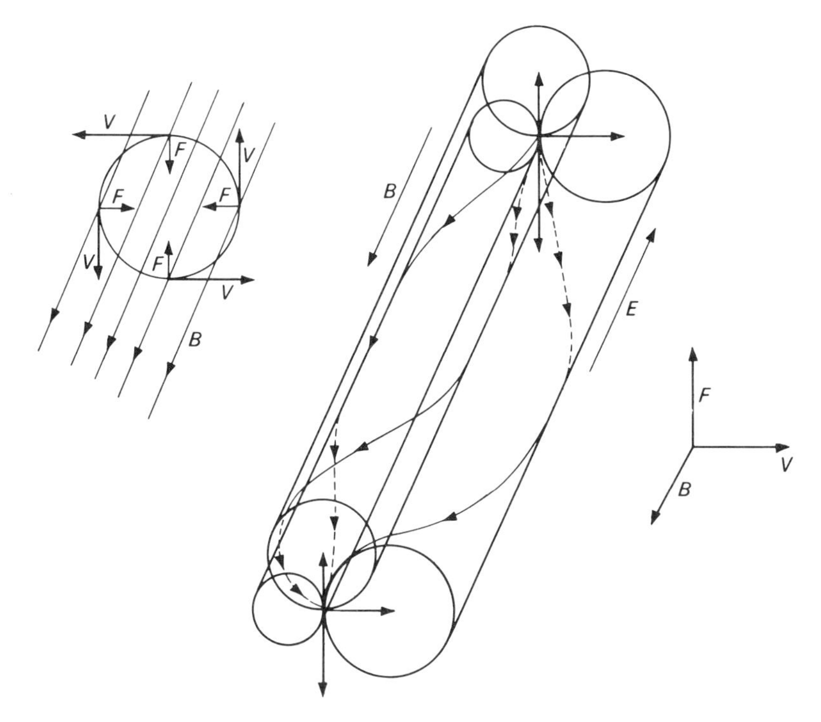

Figure 10.16. Electron focusing with uniform magnetic field.

magnetic field by means of a solenoid and for this reason is known as *solenoidal magnetic focusing* (Vine, 1974).

To understand the principle behind magnetic focusing, we must understand the influence of a magnetic field on a moving charge. Figure 10.16 shows the direction of the force exerted on a *negative* charge moving perpendicular to a uniform magnetic field. (*Note*: For a positive charge, the direction of the force is reversed.) From the figure, it can be seen that the directions of the force, velocity, and magnetic field are all mutually perpendicular. The relationship between the three quantities may be expressed mathematically by

$$\mathbf{F} = e\mathbf{V} \times \mathbf{B} \tag{10.9}$$

where the force $\mathbf{F}$ is the vector cross product between the velocity $\mathbf{V}$ and the magnetic induction $\mathbf{B}$ and e is the charge on the electron. In SI units, the force will be in newtons when the charge is in coulombs, the velocity in meters per second, and the magnetic induction in tesla, where $1\,T = 10{,}000\,\mathrm{G}$. Equation 10.9 indicates that only the velocity component orthogonal to the magnetic field can interact to produce a force. This means that if a uniform magnetic field is oriented along some axis, only the radial velocity (i.e., the velocity component orthogonal to the magnetic field) can interact with the field to produce a force. Thus, electrons moving parallel to the axis of the field are unaffected. Those moving at some angle to the axis, however, will experience a force proportional to the radial component of their velocity.

Since the force produced by the interaction of the velocity and the magnetic field is orthogonal to the velocity, it will not affect the magnitude of the velocity but merely alter its direction. Figure 10.16 illustrates this point for a magnetic field directed out of the plane of the paper. From the figure, it can be seen that regardless of the direction of the radial velocity component, the magnitude of the force remains unchanged and the direction of the force always points to the same spot. The orbit of the electron, therefore, is a circle as a result of the centripetal force directed toward the center. If the direction of the magnetic field is taken as coming out of the plane of the paper, as shown in Figure 10.16, the electrons will rotate about circular orbits in a clockwise direction.

Since the centripetal force is equal to the force produced by the magnetic field,

$$F = eVB = mV^2/r \tag{10.10}$$

The radius of the circular orbit in Eq. 10.10 is given by

$$r = mV/eB \tag{10.11}$$

where r is the radius in meters, m is the mass of the electron in kilograms, V is the radial velocity component in meters per second, e is the charge on the electron in coulombs, and B is the magnetic induction in tesla. Since the period for a complete revolution is simply the distance divided by the velocity, the period will be given by

$$T = 2\pi r/V = 2\pi m/eB \tag{10.12}$$

The important point to realize from Eq. 10.12 is that the period for a complete revolution is *independent of the radial velocity component of the electron.* Since the period of revolution is independent of the radial velocity component, it follows that all electrons emitted from a given point will return to a second point regardless of their initial radial velocity component as long as they all have the same axial velocity. A focus will be obtained, therefore, if the transit time from object to image is an integral multiple of the period of revolution. For a constant axial velocity, the distance from the object to any of the successive image points will be

$$x_n = nV_xT \tag{10.13}$$

where x_n is the axial distance from object to image to reach the nth image point, V_x is the axial velocity, T is the period for a complete revolution, and n is the number of loops completed in reaching the nth image point.

In magnetic focusing, the uniform axial electron velocity is usually achieved by means of an axial electric field. In the presence of the electric field, the electrons will undergo an axial translation in addition to the rotation caused by the magnetic field. The resultant motion from the combined forces will be that the electrons spiral down the axis in a helical path with a constant pitch. This results in a sequence of focal points along the axis separated by the distance traveled during successive loop times. Equating the kinetic energy of the electrons to the energy obtained from the accelerating voltage gives

$$V_x = (2eE/m)^{1/2} \tag{10.14}$$

where V_x is the axial velocity, e is the charge on the electron, m is the mass of the electron, and E is the accelerating voltage. Substituting V_x into Eq. 10.13 gives

$$x_1 = (2\pi/B)[2mE/e]^{1/2} \tag{10.15}$$

According to Eq. 10.15, the first focal point will be located 4.7 cm from the object assuming an accelerating voltage of 500 V and field strength of 100 G.

The focusing action of a magnetic field is illustrated in Figure 10.16 for the case of an object immersed in a uniform magnetic and electric field. The figure shows the paths followed by three electrons emitted from a common point on the object with different radial velocities as indicated by the vectors. For each electron, a circular orbit is drawn tangent to the vector in such a way that the electron will rotate in a clockwise direction. The radius of the circle is proportional to the radial velocity (i.e., the length of the vector) according to Eq. 10.11. An electron emitted from the object will follow a path along the surface of a cylinder whose radius is determined by the radial velocity of the electron. If the lengths of each cylinder are exactly equal to the axial distance traveled by the electron during one revolution (Eq. 10.15), all three electrons will arrive at the same image point at the same time in spite of the fact that they have each followed different paths. Since each has completed one revolution, the radial velocity vectors will be oriented in the same directions as drawn initially. If the fields are long enough, the process will repeat itself, producing a sequence of focal points as indicated by Eq. 10.13.

Although variations in radial velocity do not influence image formation as discussed previously, variations in axial velocity produce variations in transit times that lead to chromatic aberration. As a result, the energy spread of the emitted photoelectrons must be kept small with respect to the accelerating voltage to achieve good resolution.

By contrast with the electrostatic lens, the magnetic lens results in an upright image at unit magnification. With a uniform magnetic field and the proper accelerating voltage, distortion effects are very low and a good image is formed over the entire screen (RCA, 1974).

10.3.4. Proximity Focusing

A third form of focusing used in image detectors is known as *proximity focusing*. In this form of "focusing," the photocathode is placed parallel to the intended target plane so that the two are separated by only a very short distance. If a high electric field is established between the two plates, electrons accelerated from the photocathode to the target will not have much opportunity to deviate from trajectories parallel to the tube axis. Although this form of focusing is conceptually simpler than electrostatic or magnetic focusing, the image produced is not as good. In addition, the presence of the high electric field in close proximity to the photocathode tends to increase the dark current as a result of field effects (Levi, 1980).

10.3.5. Electron Deflection and Raster Scanning

In addition to focusing, electron beams may be deflected either magnetically or electrostatically. This is accomplished by orienting the electric or magnetic field so that it is transverse to the electron beam. Figure 10.17 shows the effects of an electric and magnetic field on the direction of an electron beam.

Electrostatic deflection is accomplished by pairs of plate-shaped electrodes. A pair of plates oriented vertically will deflect a beam of electrons in the horizontal plane and is therefore known as the *horizontal-deflection plates*. Conversely, a pair of deflection plates oriented in the horizontal plane will cause a beam of electrons to be deflected in the vertical plane and is therefore known as the *vertical-deflection plates*.

Magnetic deflection is usually accomplished by means of a coil known as the *magnetic deflection yoke*. As indicated in Section 10.3.3, electrons travel in circular orbits when in a magnetic field. From Figure 10.17, it can be seen that the extent of the deflection can be readily calculated knowing the velocity of the electron, the strength of the magnetic field, and the extent of the magnetic field (Levi, 1980). The sine of the angle of deflection is equal to L/r, where L is

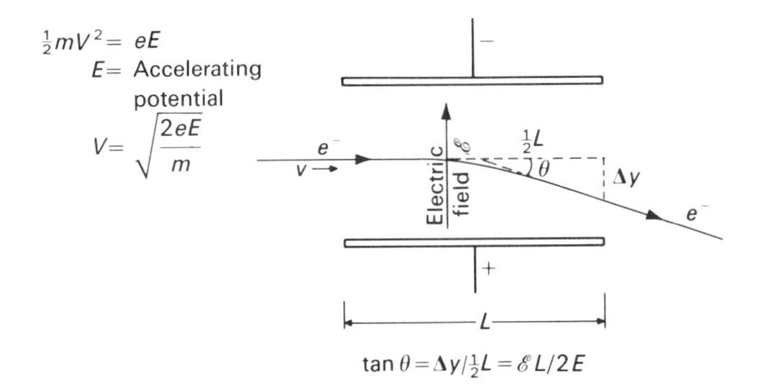

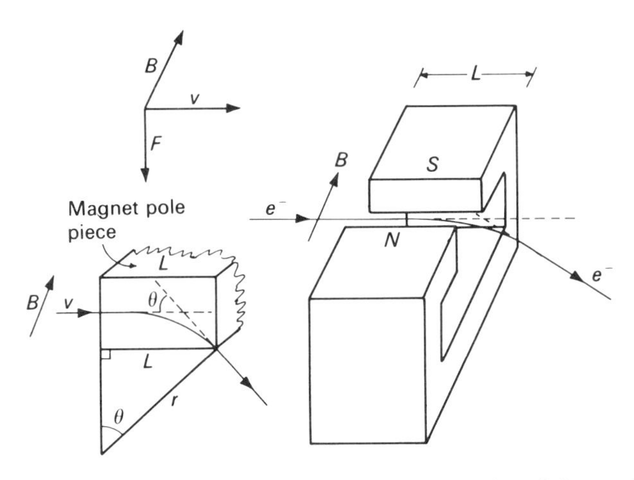

Figure 10.17. Effect of electric and magnetic fields on direction of electron beam.

the length of the magnetic field and r is given by Eq. 10.11,

$$\sin\theta = LBe/mV \tag{10.16}$$

Equation 10.16 shows that for small angles where $\sin\theta \sim \theta$, the extent of deflection will be directly proportional to the magnetic induction for a given electron velocity and magnet geometry. Thus the extent of deflection of an electron beam can be varied at will by varying the current in the deflection yoke.

In image detectors it is frequently necessary to interrogate a target surface with an electron beam that has been focused down to a tiny spot. As discussed previously, this can be accomplished either electrostatically or magnetically by varying either the potential applied to deflection plates or the current through a deflection yoke. If the target is interrogated in a nonsequential fashion, the mode of interrogation is known as *random-access interrogation.* Frequently the target is interrogated sequentially so that any given portion of the target is accessed with a fixed period. This form of interrogation is usually accomplished by what is called *raster scanning.* A completed raster is known as a *frame,* in analogy with cinematography.

Raster scanning is accomplished by applying a saw-tooth waveform with a given period from a *sweep generator* to one set of deflection plates (or coils). At the same time, a second sweep generator applies a much higher frequency waveform to the other set of deflection plates. The net result of both waveforms is shown in Figure 10.18. The period of the higher frequency waveform determines the *linescan period,* while the period of the lower frequency waveform determines the *framescan period.* In conventional television applications, a framescan is typically on the order of $\frac{1}{30}$ sec, or 33 msec. To scan 500 lines, for example, in this time period requires a linescan period of about

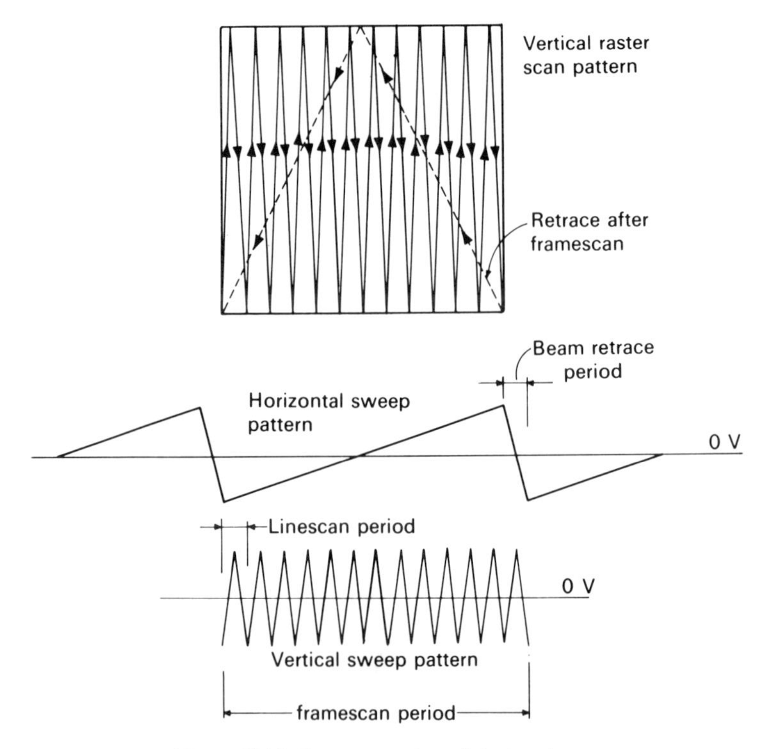

Figure 10.18. Raster scanning of electron beam.

65 μsec. Thus the period of a linescan is dependent on the framescan period, and the maximum length of the framescan period is set by the dark current of the detector.

10.4. MICROCHANNEL PLATES

Besides various electron optical components, another component frequently found in image detectors is the *microchannel plate*, or *mcp*. An mcp is an array of miniature channel electron multipliers oriented parallel to one another. The array, which can contain as many as 10^4–10^7 channel electron multipliers, is usually fabricated from a lead glass wafer containing channels that have been treated to optimize secondary electron emission. Typical channel diameters range from 10 to 100 μm. Metallic coating of the front and rear surfaces of the mcp provides electrical contact with each channel. By applying a potential across the plate so that the input side is connected to the negative terminal of the power supply and the output side is connected to the positive terminal, each channel will behave as a continuous dynode structure. Because the total resistance between the front and rear electrodes is on the order of $10^9\ \Omega$, the plate resistance acts as its own dynode resistor chain.

A key feature concerning the design of mcp's is the fact that the electrical performance is a function of the length-to-diameter ratio (α) rather than a function of either parameter individually. This makes drastic size reductions possible without altering the performance of the device. Figure 10.19 shows a schematic diagram of an mcp.

The overall gain of an mcp is given by

$$G=\left(\frac{AV}{2\alpha\sqrt{V_0}}\right)^{4V_0\alpha^2/V} \tag{10.17}$$

assuming that secondary emission occurs normal to the channel walls (Wiza, 1979). In Eq. 10.17, A is a proportionality constant, V is the total voltage across the channel, V_0 is the initial energy of an emitted secondary electron (~ 1 eV), and α is the length-to-diameter ratio.

With straight-channel configurations, maximum gains obtainable with mcp's are limited to 10^3–10^5 because of the onset of *ion feedback*. Ion feedback occurs as a result of electron collisions with residual gas molecules and desorbed gas molecules from the walls. Ions produced as a result of these electronic collisions are positive and are therefore accelerated by the electric field toward the channel input. This results in ion after-pulses and, if a photocathode is located in close proximity to the input side of the mcp, can result in regenerative feedback due to ion impact with the photocathode. Such

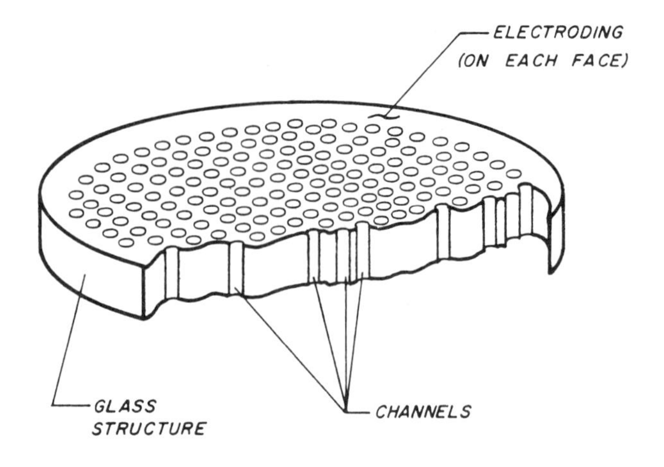

Figure 10.19. Schematic diagram of microchannel plate. [Reprinted with permission from J. L.Wiza (1979), *Nuclear Instruments and Methods 162*, 587. Copyright 1979 Elsevier Science Publishers.]

behavior results in performance instability and in extreme cases, where a photocathode is involved, can result in the destruction of the photocathode.

Suppression of ion feedback in mcp's is accomplished in an analogous manner to the way the problem was solved in electron channel multipliers, namely, by introducing some curvature to the channel (Kurz, 1979). To understand the principle of channel curvature as a means of ion feedback suppression, it is necessary to look at ion production and its effect more closely. Ion production is a result of electronic collisions with residual gas molecules and is most probable near the output end of the channel where the secondary electron density is the highest. As the ions produced by electron impact are accelerated toward the input, they may strike the walls of the channel, producing spurious secondary electrons that are amplified by the channel. Since the extent of the amplification these spurious secondary electrons receive increases for ion–wall impacts nearer the input end of the channel, the goal of ion feedback suppression is to prevent ion–wall collisions near the input end of the channel. By curving the channel, the distance toward the input that can be traversed by the ions before an ion–wall collision occurs is limited. As a result, the gain received by the spurious secondary electrons from ion–wall collisions is small compared with the overall gain of the entire channel.

Since mcp's are usually relatively thin (~ 0.5 mm), channel curvature is difficult to achieve. Instead of channel curvature, a chevron arrangement of two mcp's is commonly used to suppress ion feedback. This configuration is shown in Figure 10.20. The angle the channels make with the plate normal is known as the *channel bias angle*. Typical bias angle combinations for the mcp pair are $8°–8°$ or $0°–15°$. Either combination of angles provides a sufficient change in direction so as to prevent ions produced at the output from reaching the input. Gains up to 10^7 have been achieved by this approach with plate separations of 50–150 μm.

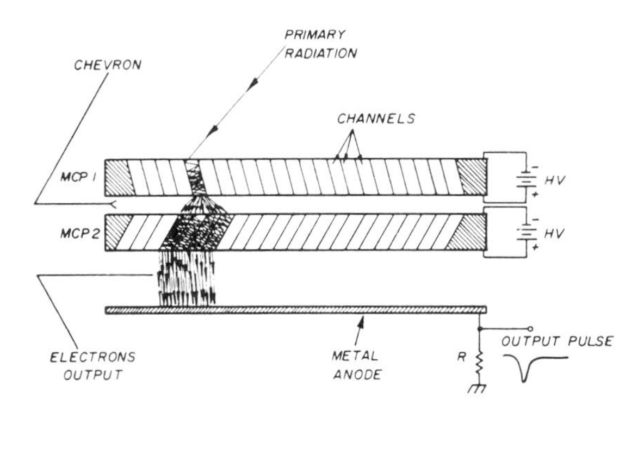

Figure 10.20. Chevron configuration of two microchannel plates for suppression of ion feedback. [Reprinted with permission from J. L. Wiza (1979), *Nuclear Instruments and Methods*, *162*, 587. Copyright 1979 Elsevier Science Publishers.]

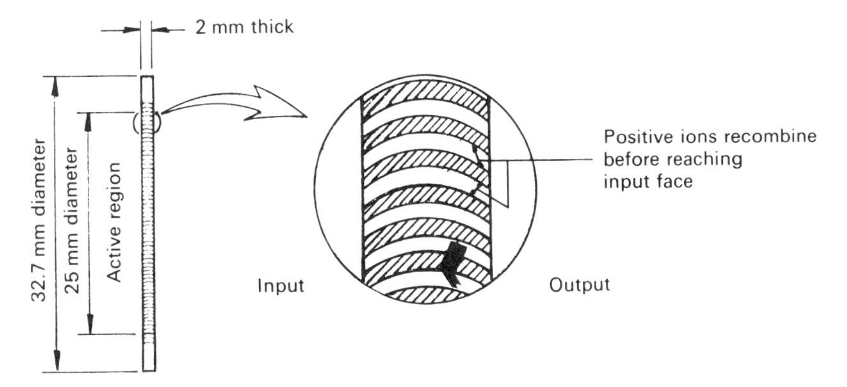

Figure 10.21. Curved-channel microchannel plate for suppression of ion feedback. [Courtesy of Galileo Electro-Optics Corporation.]

Recently, procedures for fabricating curved channel mcp's (Figure 10.21) have been developed at the Laboratoires d'Electronique et de Physique Appliquee in Limeil Brevannes, France (Boutot et al., 1974, 1976). The performance of curved channel mcp's has been studied by Timothy (1974; Timothy and Bybee, 1977) for mcp's with 25-mm plate diameters. Gains in the low 10^6 range were obtained with channel diameters of 12–40 μm and $\alpha=80$.

The high degree of dimensional uniformity obtained in fabricating mcp's is shown in Figure 10.22. This uniformity is obtained by a clever manufacturing technology based on highly reliable fiber-drawing techniques (Wiza, 1979). The process starts with a solid glass fiber drawn from a rod consisting of a soluble glass core surrounded by an insoluble lead glass cladding. Hexagonal arrays of these fibers are packed together and drawn to produce hexagonal multifibers. These hexagonal multifibers are in turn stacked together and fused to form rods that are sliced into thin wafers at an angle of 8°–15° from the normal to the channel axis. The wafers are then treated chemically to dissolve the soluble glass cores, thereby producing the channels. The final step in the manufacture involves a furnace treatment during which the lead oxide on the surface of the channel is converted to agglomerated semiconducting lead particles, forming a continuous-channel electron multiplier.

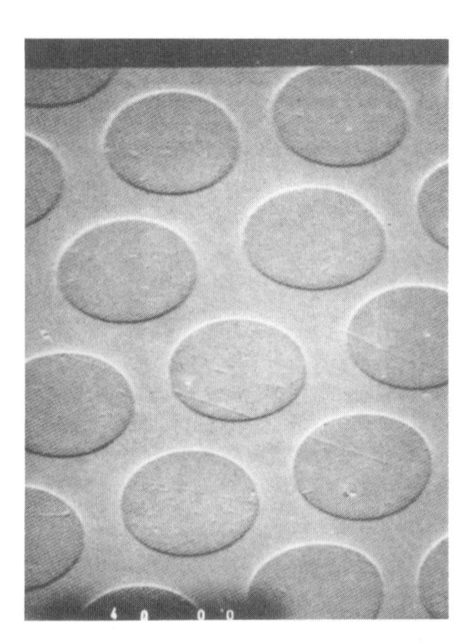

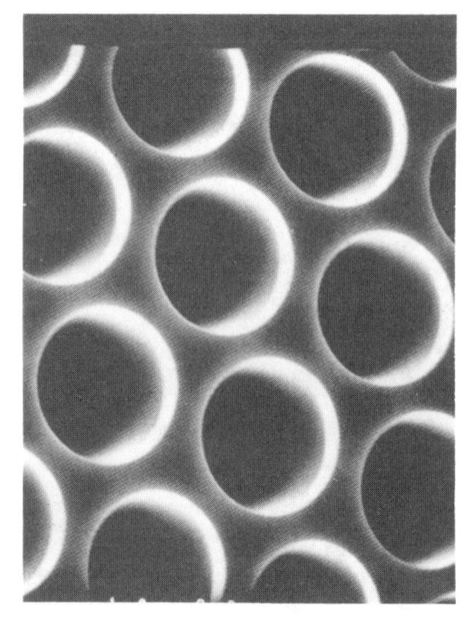

Figure 10.22. Photomicrograph of microchannel plate showing high degree of dimensional uniformity obtained in fabrication. [Reprinted with permission from J. L. Wiza (1979), *Nuclear Instruments and Methods*, *162*, 587. Copyright 1979 Elsevier Science Publishers.]

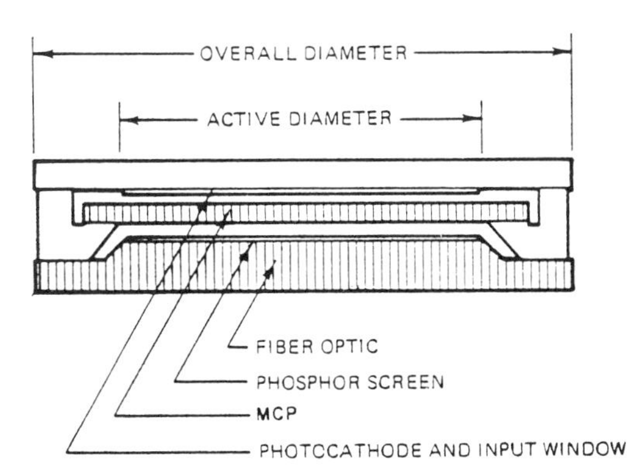

Figure 10.23. Proximity-focused image intensifier employing microchannel plate. [Reprinted with permission from J. L. Wiza (1979), *Nuclear Instruments and Methods*, *162*, 587. Copyright 1979 Elsevier Science Publishers.]

Although originally developed for use in image intensifiers, mcp's have been used to detect various charged particles as well as vacuum ultraviolet (VUV) radiation and hard and soft X-rays. Figure 10.23 shows a typical application of an mcp in a proximity-focused image intensifier. In this detector, photoelectrons emitted by the photocathode are proximity focused by a 300-V potential onto the input of the mcp. Output electrons from the mcp are subsequently proximity focused by a 6-kV accelerating potential onto a phosphor screen. Overall gain produced by this device is the product of the photocathode quantum efficiency, the gain of the mcp, and the optical gain of the phosphor due to energetic electron impact.

Such a direct-viewing device is capable of emitting more visible photons than the number of photons absorbed and for this reason behaves as a light amplifier. The amount of amplification obtained can be increased by cascading a series of image intensifier sections, although the increase in gain is obtained at the expense of a decrease in resolution. The use of mcp's with proximity focusing makes possible the fabrication of image intensifiers having considerably shortened dimensions.

10.5. SPECTROSCOPIC INTERLUDE

In the next several sections, we discuss a variety of image detectors. The majority of these detectors have been developed for purposes other than spectroscopy, such as television applications, where light levels are typically relatively high. To get some idea of the potential utility of an image detector for spectroscopic applications, we need an estimate of the radiance emitted from a typical spectral source. The radiance emitted from an excitation source in thermal equilibrium is given by (Mavrodineanu and Boiteux, 1965)

$$L = A_T h\nu D g_u N e^{(-h\nu/kT)}/4\pi B(T) \tag{10.18}$$

where L is the radiance in ergs/sec cm^{2}sr, A_T is the Einstein transition probability in reciprocal seconds, h is Planck's constant (6.6×10^{-27} erg sec), ν is the frequency of the emitted radiation in reciprocal seconds, D is the optical depth in centimeters of the excitation source along the optical axis, g_u is the statistical weight of the upper level, N is the number density of emitting atoms in cm^{-3}, k is Boltzmann's constant (1.38×10^{-16} ergs K^{-1}), T is the temperature in kelvin, and $B(T)$ is the electronic partition function given by

$$B(T) = g_0 + g_1 e^{(-E_1/kT)} + g_2 e^{(-E_2/kT)} + \cdots \tag{10.19}$$

In Eq. 10.19, g_0 is the statistical weight of the ground state and the E_i's are the energies of the various excited states. If the transition under consideration is a *resonance transition* (i.e., one involving the ground state) where the energy of the upper level is more than 1.5–2.0 eV, Eq. 10.19 can be taken as $B(T) = g_0$ since the other terms will be small. Substituting for $B(T)$ in Eq. 10.18 gives

$$L = A_T h\nu D g_u N e^{(-h\nu/kT)}/4\pi g_0 \tag{10.20}$$

Equation 10.20 is a function of the atomic concentration and the temperature of the excitation source. To obtain a meaningful estimate of the radiance expected from an excitation source for a given concentration of analyte in the sample, we need a relationship between analyte solution concentration and atomic number density in the source. Winefordner and Vickers (1964) have shown that the relationship between analyte concentration and atomic number density for a combustion flame is given by

$$N = 3 \times 10^{21} n_R \phi \varepsilon \beta C / n_T T Q \tag{10.21}$$

where n_R and n_T are the numbers of moles of reactants and products, respectively, ϕ is the rate of sample aspiration in milliliters per minute, ε is the aspiration efficiency of the nebulizer (typically about 0.1), β is the free atom fraction that accounts for incomplete dissociation and ionization, C is the concentration of analyte in moles per liter, T is the temperature of the flame in kelvin, and Q is the flow rate of unburnt flame gases in milliliters per second introduced at 1 atm pressure and room temperature into the flame.

For purposes of illustration, let us consider an excitation source with a temperature of 2500 K that is being fed with a solution containing 10 ppm

Na. To obtain an estimate of N, we need to make some assumptions. For purposes of this calculation, we assume that $\phi = 5$ mL/min, $\varepsilon = 0.10$, $\beta = 1$, $Q = 70$ mL/sec, and $n_R \sim n_T$. If the analyte concentration is 10 ppm, C will be 4.3×10^{-4} mol/L. Substituting these values into Eq. 10.21 gives $N \sim 4 \times 10^{12}$ cm^{-3}.

The radiance emitted from the source can now be estimated by substituting N into Eq. 10.20. For this purpose, we assume that the source has an optical depth of 1 cm, and the transition of interest is the 589-nm resonance line with a transition probability of 9×10^7 sec^{-1}, where $g_u/g_0 = 2$. Substituting these parameters into Eq. 10.20 gives a radiance of 1×10^4 ergs/sec cm^{2}sr, or 1×10^{-3} W/cm^2 sr.

If a 1:1 image of this source is focused on the entrance slit of a 0.5-m, $f/8.7$ spectrometer so that the entrance aperture is filled with light, the irradiance incident on a detector at the focal plane can be estimated by multiplying the radiance by the solid angle collected by the spectrometer. The solid angle collected by the instrument is given by

$$\Omega = A/F^2 \tag{10.22}$$

where A is the area of the collimating mirror and F is its focal length. From the f-number of the instrument, the diameter of the mirror is estimated to be 57 mm, corresponding to an equivalent area of 2592 mm^2. Substituting this result in Eq. 10.22 gives a solid angle of 0.01 sr. Using this solid angle, the irradiance incident at the focal plane will be on the order of 10^{-5} W/cm^2 assuming that all the light incident on the entrance slit is transmitted to the image (something that is not true for a grating spectrometer producing multiple orders).

It should be remembered that the value we have calculated for the irradiance at the focal plane produced by aspirating 10 ppm of sodium into a flame is only an estimate based on numerous assumptions. Still, it does provide an order of magnitude that may be useful in predicting the analytical potential of various image detectors.

10.6. SIGNAL-GENERATING IMAGE DEVICES

At this point, we turn our attention to signal-generating image devices or image pickup devices. These devices fall into two general categories. The first category includes devices that make use of scanning electron beams. Since these devices are a form of vacuum tube, they are referred to as signal-generating image tubes or *camera tubes* since they are generally used for television applications. The second category of signal-generating image device does not employ electron beams. This form of image device consists of a mosaic of solid-state sensors scanned by an arrangement of multiplex switches. Since these devices do not require a scanning electron beam, they are referred to as *self-scanned array detectors.* Self-scanned array detectors, as well as other solid-state image detectors, will be discussed in detail in Chapter 11.

Regardless of the category of image device considered, signal generation is accomplished by converting the light pattern incident on the device into a corresponding charge pattern that is detected electrically and used to generate an electrical signal. The electrical signal generated as a result of the charge pattern is then sent to a high-speed amplifier (i.e., wide bandwidth) known as a *video amplifier* for amplification prior to further signal processing.

10.6.1. The Image Dissector

Perhaps the oldest and simplest camera tube is the image dissector originally described by P. T. Farnsworth (1931, 1934). Figure 10.24 shows a schematic diagram of an image dissector, which, in principle, is simply a form of scanning photomultiplier tube. In the device, electrons emitted from a photocathode are accelerated and focused on an aperture plate that separates the photocathode from the electron multiplier. Only photoelectrons that can pass through the opening in the aperture plate can be amplified by the electron multiplier. By using magnetic deflection coils to provide orthogonal deflection, the photoelectron image on the aperture plate can be made to pass through the opening in a sequential raster manner, thereby dissecting the image into a sequence of photoanodic currents. The instantaneous photoanodic current observed at the output of the electron multiplier is proportional to the instantaneous irradiation on that portion of the photocathode that corresponds to the photoelectron image passed by the aperture.

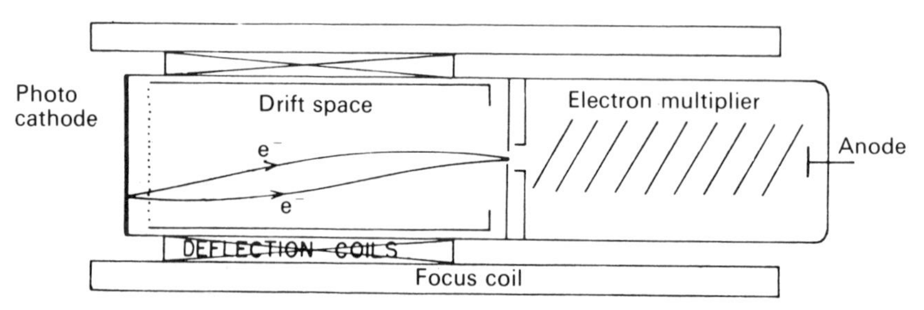

Figure 10.24. Schematic representation of image dissector tube. [Reprinted with permission from H. L. Felkel and H. L. Pardue (1979), "Simultaneous Multielement Determinations by Atomic Absorption and Atomic Emission with a Computerized Echelle Spectrometer/Imaging Detector System," Chapter 3 in *Multichannel Image Detectors*, Y. Talmi (Ed.), ACS Symposium Series 102, American Chemical Society, Washington, DC, pp. 59–96. Copyright 1979 American Chemical Society.]

To get some idea of the magnitude of the signal-to-noise ratio that might be expected for a spectroscopic application (such as discussed in Section 10.5) with an image dissector, let us consider a tube with an S-20 photocathode and a circular aperture with a diameter of 0.008 cm. From the *RCA Photomultiplier Handbook* (RCA, 1980), the responsivity of an S-20 photocathode at 589 nm will be about 43 mA/W. For purposes of this calculation, let us assume that an area of the photocathode is uniformly illuminated by an irradiance E. If the electron optics used in the tube provide unity magnification, the area of the aperture in the aperture plate will correspond to an equivalent area on the photocathode. The signal current transmitted to the electron multiplier will be given by

$$i_s = REa \tag{10.23}$$

where R is the responsivity of the photocathode in amperes per watt, E is the irradiance of the photocathode surface in watts per square centimeter, and a is the area of the aperture in square centimeters.

Under shot-noise-limited conditions, the noise associated with the signal current transmittted to the electron multiplier will be due to the random emission of photoelectrons from the photocathode and will be given by Eq. 9.7,

$$i_n = (2ei_s \Delta f)^{1/2} \tag{10.24}$$

The signal-to-noise ratio under these conditions will be given by

$$S/N = (REa/2e\,\Delta f)^{1/2} \tag{10.25}$$

If we assume that a wide entrance slit is used so that the slit image formed on the photocathode is wider than the aperture opening in the aperture plate of the image dissector, the signal will be limited by the area of the aperture (in this case, 5×10^{-5} cm^2). The only remaining parameter that needs to be specified in order to calculate the signal-to-noise ratio is the bandwidth of the video amplifier.

For television applications using raster scanning, a high-speed video amplifier is usually employed. Such an amplifier might have a bandwidth of 3.5 MHz. If this is the case, the estimated signal-to-noise ratio for the spectroscopic application discussed in Section 10.5 will be

$$S/N = \frac{[(4.3 \times 10^{-2}\ \text{A/W})(5 \times 10^{-5}\ \text{cm}^2)(10^{-5}\ \text{W/cm}^2)]^{1/2}}{[2(1.6 \times 10^{-19}\ \text{C})(3.5 \times 10^{6}\ \text{Hz})]^{1/2}} \tag{10.26}$$

$$= 19$$

Considering the relatively large analyte concentration (10 ppm), this signal-to-noise ratio estimate is relatively small, suggesting a relatively poor detection limit. The situation can be improved by two factors, as indicated by Eq. 10.25. Equation 10.25 shows that for a given irradiance the signal-to-noise ratio is a function of the square root of the aperture area and the electrical bandwidth. On this basis, it should be possible to improve the signal-to-noise performance by increasing the aperture area and decreasing the electrical bandwidth. These modifications are not without concomitant trade-offs, however. For example, it should be realized that the resolution of the detector is determined by the size of the opening in the aperture plate. Although it is apparent from Eq. 10.25 that the signal-to-noise ratio is directly proportional to the diameter of the aperture, increasing the size of the aperture will degrade the resolution obtainable with the detector. Thus, a compromise exists between signal-to-noise ratio and resolution with an image dissector.

A similar compromise exists between electrical bandwidth and the speed with which the image can be scanned. If the entire image is to be scanned in a raster pattern, a fast scan time will be required. This is the situation inherent in television applications. In spectroscopic work, however, it is often the case that large regions of the image do not contain information (i.e., spectral lines) of interest. If this is the case, random-access scanning (Section 10.3.5) may be used in place of raster scanning to interrogate only those locations on the image that contain intensity information of interest. Since the entire image is not being scanned, the scan rate can be reduced to any desired level with an image dissector. If a bandwidth of 10 kHz is selected, the observation period (Section 9.4.2) (known as the *dwell time* of the image dissector) will be $t_{\text{dwell}} = 1/2\,\Delta f = 50\ \mu\text{sec}$. This bandwidth will give a signal-to-noise ratio of 82, an approximately fourfold improvement.

One final feature of the image dissector deserves mention, namely, the real-time nature of the device. Since the image dissector is nothing more than a scanning photomultiplier, the same effect could be achieved with a mechanical system (i.e., a scanning monochromator) and an ordinary photomultiplier. The only real difference between the two systems is the increased convenience, speed, and reliability of electron scanning compared with mechanical scanning. Thus, the image dissector is essentially a single-channel scanning device that only monitors one resolution element at a time. For this reason, it is not possible to achieve any form of multichannel advantage with this detector. On the positive side, since the image dissector is not a storage device, it does not suffer from lag or blooming (see Chapter 11).

10.6.2. Charge Storage Tubes

We now turn our attention to those image tubes that offer the possibility of a multichannel advantage. Image tubes that continuously monitor all the resolution elements regardless of the location of the scanning electron beam are known as *charge storage tubes.* Charge storage tubes may be classified into two basic types. In the first type, a single element functions as both primary radiation sensor and charge storage element. In the second type, the functions of radiation sensing and charge storage are accomplished by different components within the tube.

Figure 10.25 shows a schematic diagram of a typical storage tube belonging to the second category. With these devices, signal integration (i.e., storage) is accomplished by means of an integrating *target* that divides the tube into two sections known as the *write section* and the *read section.*

The write section typically consists of a photocathode and an electron imaging stage whose function is to focus an electron image on the leading surface of the target so that the electron image produced corresponds to the optical image focused on the photocathode. Photoelectrons that strike the leading surface of the target produce an alteration in the potential of the other side of the target by some means. The accelerating potential associated with the electron imaging stage increases the energy of the emitted photoelectrons, frequently providing a certain amount of gain when they strike the target. This increase is known as *prescanning intensification* or *target gain.*

The read section of the tube consists of an electron gun with an associated electron lens to focus the electron beam emitted from the gun as a tiny spot ($\sim 20\ \mu m$) on the target. Deflection plates or coils are used to scan the electron beam across the target in the desired pattern. When the target is scanned by the electron beam, the charge pattern that has developed on the back side of the target is monitored. The means by which the charge pattern is developed on the back side of the target and subsequently sensed by the electron beam

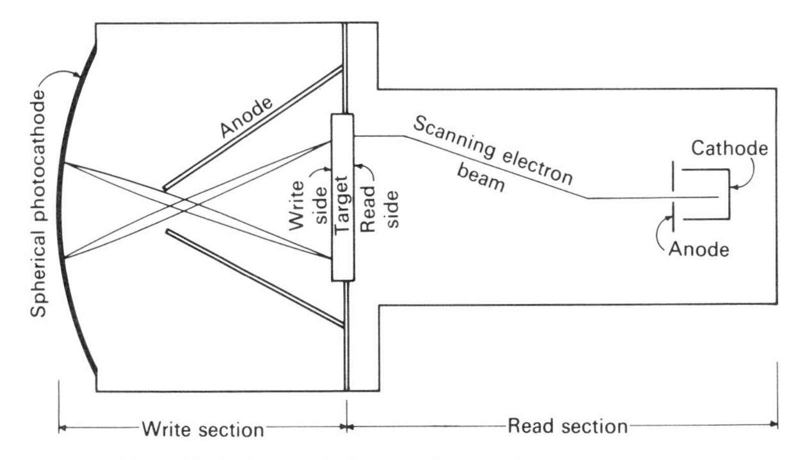

Figure 10.25. Schematic diagram of typical charge storage tube.

varies, and this has given rise to a variety of storage tubes. The following discussion will present a brief survey of three common storage tubes.

10.6.3. Image Orthicon

We begin our discussion with the *image orthicon* shown in Figure 10.26 and discussed in detail by Redington (1974). The storage target, which divides the tube into the image section and read section, is a thin layer of material composed of either soda lime glass (as in conventional tubes) or MgO (as in newer tubes). During operation, the scanning electron beam charges the read side of the target down to the potential of the cathode of the electron gun (typically ground potential). At the same time, the image side of the target is charged to zero volts due to current leakage through the target.

In tubes with glass targets, current is carried by mobile sodium ions. In tubes with MgO targets, current is carried by electrons. The MgO targets are generally superior because they have a high secondary emission ratio (~ 15) and are not subject to *burn-in* from continued use. Burn-in is a form of permanent image retention that occurs in glass targets because the ion drift is always in the same direction, permanently altering the resistivity of the glass after several hundred hours of operation.

Since the photocathode is maintained at -500 V, photoelectrons emitted by the photocathode are accelerated toward the target and simultaneously focused by a uniform axial magnetic field. Upon striking the target, the primary photoelectrons produce secondary electrons that are collected by a mesh electrode known as the *target mesh electrode*. The target mesh electrode is maintained at $+2$ V and is located in close proximity to the image side of the target.

Since secondary electrons are being removed from the image side of the target, the target material must have sufficient conductivity to permit charge from the read side to have enough time to flow through the target and make up the loss. To accomplish this within a typical frame time requires a resistivity of 10^{11} Ω-cm for materials with dielectric constants from 1–10 (Redington, 1974).

When the electron beam returns to any given portion of the target after a framescan period, it replaces the charge that was lost during the frame period, thereby charging the target at that point back down to the potential of the electron gun. The charging process continues as long as the electron beam is energetically capable of depositing electrons on the surface. The amount of charge deposited on the target at a particular location is determined indirectly by measuring the change in the total reflected return beam current. The advantage of the *return beam mode* of operation results from the ability to use an electron multiplier to amplify the return beam current. Since a typical tube

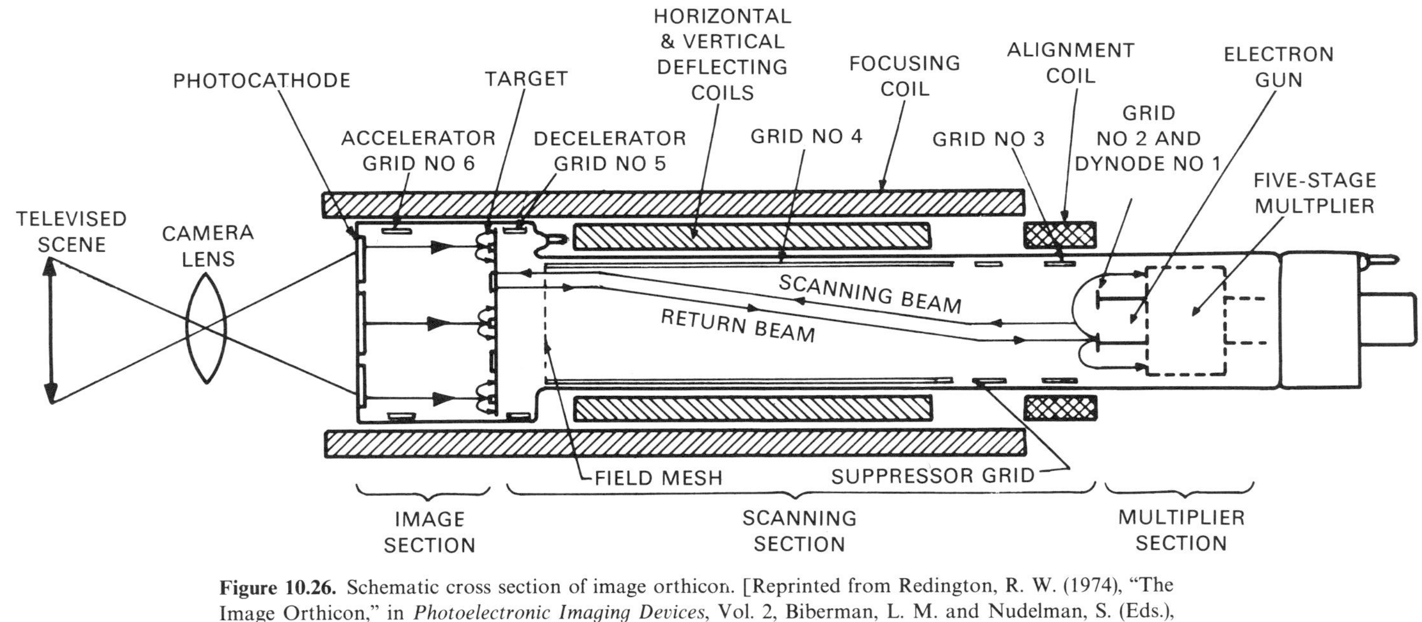

Figure 10.26. Schematic cross section of image orthicon. [Reprinted from Redington, R. W. (1974), "The Image Orthicon," in *Photoelectronic Imaging Devices*, Vol. 2, Biberman, L. M. and Nudelman, S. (Eds.), Plenum, New York.]

uses a five-stage electron multiplier with an average secondary emission ratio of 4, gains of about 1000 are possible.

The presence of the target mesh stabilizes the detector by limiting the amount of charge that can be removed from the target (Bruining, 1954). If the target mesh is maintained at +2 V, secondary electrons emitted from the target will be collected by the target mesh electrode until the potential of the target reaches +2 V. When a point on the target reaches the target mesh potential, secondary electrons emitted from that point will be reflected back to the target rather than being collected. This prevents the target from becoming positively charged to the point where the reading electron beam begins to produce secondary electrons.

Because of the presence of the target mesh electrode, the video signal is contained in a potential range from zero to +2 V. In order to read this signal with the scanning electron beam, extreme uniformity in beam-landing characteristics is required if the target is to be reproducibly recharged to the potential of the electron gun. This requirement places restrictions on the manner in which the target is scanned. For example, if the path of the scanning electron beam between the electron gun and the target were a straight one, the landing potential of the electron beam would be different near the edge of the target from that at the center by about 1% (Redington, 1974). Since the accelerating potential of the drift space is typically 200–300 V, this would result in a 2–3-V variation in the potential across the target. Such variations are known as *beam-landing errors*. Clearly this much variation is unacceptable since the expected video signal variation is only 2 V. To avoid this problem, the deflection of the electron beam is arranged so that the beam always arrives perpendicular to the target, as shown in Figure 10.26. To minimize the deflection angles required, a relatively long electron gun is employed (~20 cm), which limits the minimum size of the tube.

One of the problems with the image orthicon is the inverse relationship that exists between target illumination and noise. This arises because the tube monitors the total reflected beam current, which is simply the difference between the incident electron beam current and the current required to recharge the target. Since the shot noise on the total reflected electron beam is proportional to the square root of the beam current (Eq. 9.7) and since the total reflected beam is greatest under conditions of minimum illumination, it follows that the maximum noise will occur in the dark portions of the image. This limitation is eliminated in the *image isocon* discussed in detail by Musselman (1974).

10.6.4. Image Isocon

The write section of the image isocon from the photocathode to the storage target is identical to that in the image orthicon. The major difference between the image orthicon and the image isocon is the manner in which the reflected electron beam is sampled. To understand the distinction between the two tubes, the various interactions between the electron beam and the target must be analyzed. There are three ways in which the incident electron beam can interact with the target. One interaction is the deposition of electrons on the target, resulting in a charging current. Along with the deposition of electrons on the target, a certain portion of the electrons that strike the target are scattered. The final component of the electron beam does not have sufficient energy to strike the target and is simply reflected. Equating these three components with the incident electron beam gives

$$i_{\text{inc}} = i_c + i_s + i_r \tag{10.27}$$

where i_{inc} is the incident beam current, i_c is the target charging current, i_s is the current due to scattering electrons from the target and is proportional to i_c ($i_s = ai_c$, where a is the *scatter gain* of the target), and i_r is the current due to specularly reflected electrons.

Fom Eq. 10.27, it can be seen that the total return beam current is composed of two components, the scattered electron beam and the reflected electron beam. With the image orthicon, the total return beam current is monitored, whereas with the image isocon only the scattered electron beam is used to produce the signal. This difference results in superior dark-noise performance with the image isocon. Since the scattered electron beam current is directly proportional to the charging current, the signal current decreases with decreasing illumination levels. By contrast with the image orthicon, the shot noise on the signal current decreases with decreasing illumination in the case of the image isocon.

To isolate the scattered electron beam current from the total return beam, some form of beam separation is required. This is accomplished by means of a small aperture in the first dynode of the electron multiplier, as shown in Figure 10.27. If the incident electron beam strikes the target orthogonally, the reflected beam will return over the same general path. Since the energy spread in the scattered electron beam will always be larger than that in either the incident beam or the reflected beam, the scattered electron beam will have a larger circular cross section than the other two beams. Thus, by careful design, it is possible to locate an aperture in the first dynode that will pass both the incident and reflected beams while intercepting the majority of the scattered beam. For efficient separation, the aperture must be large enough to pass both the incident and reflected beams while being small enough to collect most of the scattered beam. Since the signal produced by the image isocon is given by

$$i_{\text{isocon}} = fGai_c \tag{10.28}$$

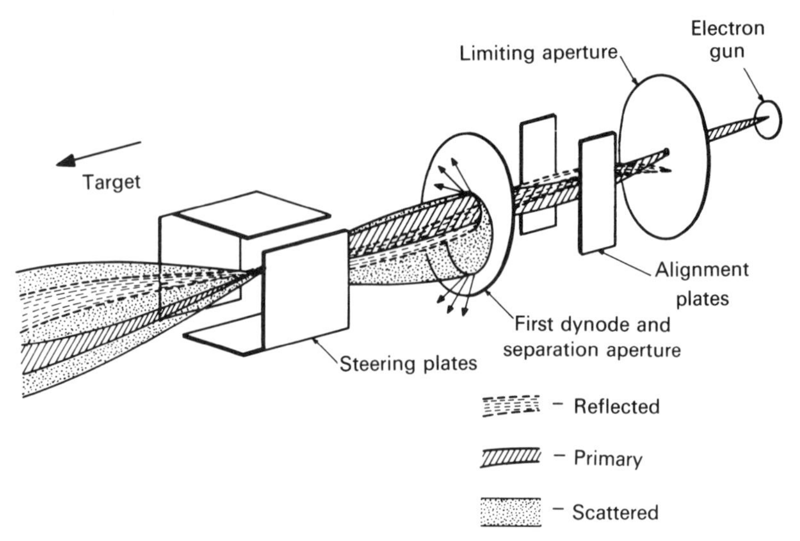

Figure 10.27. Image isocon beam separation system. [Reprinted from Musselman, E. M. (1974), "The New Image Isocon-Its Performance Compared to the Image Orthicon" in *Photoelectronic Imaging Devices*, Vol. 2, Biberman, L. M. and Nudelman, S. (Eds.), Plenum, New York.]

where f is the fraction of the scattered beam collected by the first dynode and G is the gain of the electron multiplier, it is desirable to keep the size of the separation aperture as small as possible.

10.6.5. Secondary-Electron Conduction Camera Tubes

The final image tube we discuss in this chapter is the secondary-electron conduction tube discussed in detail by Goetze and Laponsky (1974). The development of the secondary-electron conduction (SEC) camera tube was an outgrowth of work originally aimed at producing transmission secondary-emission dynodes for use in photoelectronic imaging devices. Although this research was ultimately successful in that it resulted in the development of high-gain transmission dynodes (Sternglass, 1955), their use was limited by the charging that took place at the exit surface. Although the surface charging effect observed with the low-density layers that were developed limited their use as transmission secondary-emission dynodes, it did provide an opportunity for the development of a charge storage target. For a complete discussion of the SEC process and its applications, the reader should consult *Advances in Electronics and Electron Physics*, Vols. 22 and 28, published by Academic Press.

Figure 10.28 shows a schematic diagram of an SEC camera tube. During operation, the photocathode is maintained 8000–10,000 V negative with respect to ground. Primary photoelectrons emitted from the photocathode

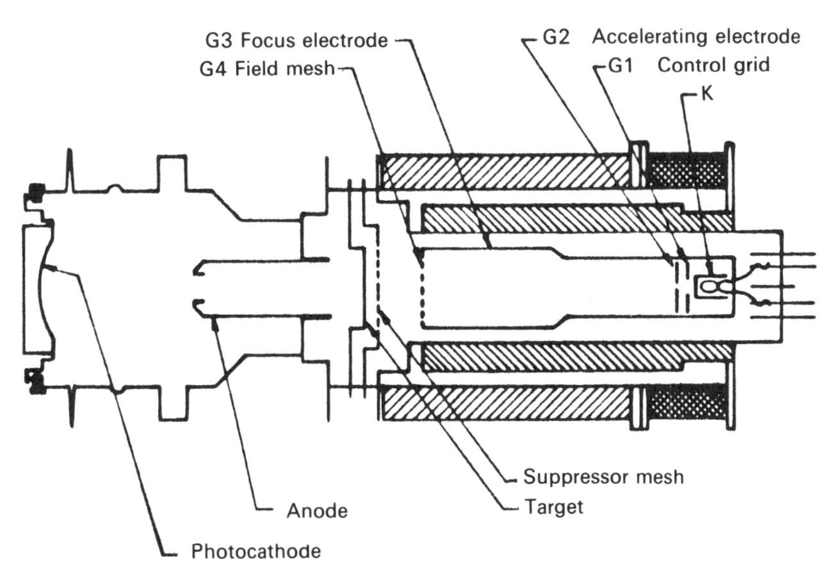

Figure 10.28. Cross section of SEC camera tube. [Reprinted from Hall, J. A. (1974), "Evaluation of Signal-Generating Image Tubes" in *Photoelectronic Imaging Devices*, Vol. 2, Biberman, L. M. and Nudelman, S. (Eds.), Plenum, New York.]

are focused electrostatically so as to form an electron image on the SEC target. These primary photoelectrons strike the target with energies of 8–10 keV, releasing secondary electrons that are conducted to ground through the signal plate. The positive charge pattern left on the target as a result of this process is neutralized by periodically scanning the target with a low-energy electron beam. To understand the principle of operation of the SEC tube in more detail, it is necessary to look more closely at the structure of the target.

Figure 10.29 shows a cross section of a typical SEC target. As shown in the figure, the target is composed of three layers. The *supporting layer* is normally an aluminum oxide disk 700 Å thick. The *conducting layer*, which forms the signal plate for the tube, is generally a 700-Å-thick layer of aluminum that is deposited on the supporting layer. The final layer is an *insulating layer* approximately 20 μm thick that is formed on the aluminum conducting layer by vacuum evaporation of the material in an inert-gas atmosphere at a pressure of about 2 Torr. Although other materials have been employed, the most common material used to produce the insulating layer is potassium chloride. Deposited in the manner described previously, the insulating layer has a density of only 1–2% of the normal density and consists of a fibrous network composed of 100-Å-thick fibers.

As a result of the fibrous nature of the low-density insulating layer, approximately 99% of the layer volume is void volume or vacuum. When struck by a 10-keV primary photoelectron, approximately 2keV energy is lost

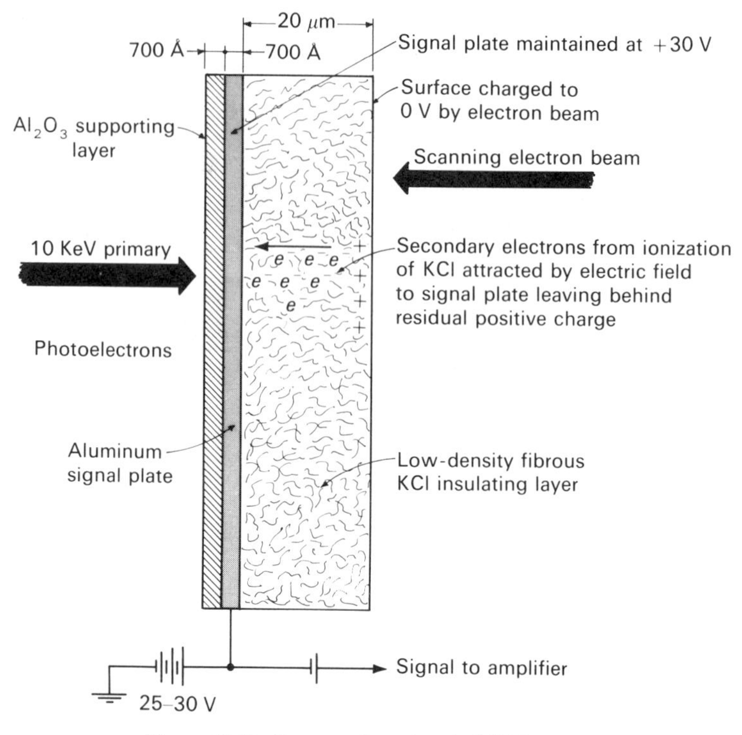

Figure 10.29. Cross section of typical SEC target.

in penetrating the support layer and the conducting layer, and the remaining 8 keV is dissipated in ionizing the KCl layer. Because of the fibrous nature of the target, secondary electrons produced by incident primary photoelectrons have a high probability of escaping the thin KCl fibers into the vacuum interstices of the layer. Because the signal plate is normally maintained at a small positive potential and the insulating layer is maintained at ground potential by the scanning electron beam, an electric field is established that causes the secondary electrons to be conducted through the target voids to the signal plate. Since each secondary electron produced requires an energy of about 30 eV, approximately 250 secondary electrons will be released for every primary photoelectron that strikes the supporting layer with an energy of 10 keV.

The local discharge of the target caused by the loss of the secondary electrons can be replaced by periodic scanning of the target with the low-energy electron beam. The recharging of the layer by the electron beam creates a current in the signal plate that is used as the video signal. Because many secondary electrons are produced for every primary photoelectron, the SEC tube has a *target gain* of about 250.

Aside from the target gain, one of the most attractive features of the SEC tube in terms of spectroscopic applications is its ability to integrate signal for long periods of time. Because the target has a resistivity in excess of 10^{17} Ω-cm (Goetze and Laponsky, 1974), the amount of leakage across the target is extremely small, and a reciprocity similar to the Bunsen–Roscoe photographic reciprocity law (see Chapter 8) exists. This means that very low level static light sources can be monitored for several hours with the electron beam turned off to allow the charge pattern to develop on the target to a satisfactory extent before readout. In this way, the SEC tube resembles a photographic plate, and for this reason it has been employed extensively in astronomical observations where long exposures are often needed. Even greater sensitivities can be obtained with an intensified SEC tube. This tube has an image intensifier stage coupled to the SEC tube so that there are a total of two image sections prior to the target.

REFERENCES

Boutot, J. P., Eschard, G., Polaert, R., and Duchenois, V. (1974), presented at *6th Symp. Photoelectr. Im. Dev.*, London.

Boutot, J. P., Eschard, G., Polaert, R., and Duchenois, V. (1976), *Adv. Electron. Electronic Phys.*, *40A*, 103.

Bruining, H. (1954), *Physics and Applications of Secondary Electron Emission*, McGraw-Hill, New York.

Budde, W. (1983), *Physical Detectors of Optical Radiation*, Academic, New York.

Busch, H. (1926), *Ann. Phys., Lpz.*, *81*, 974.

Busch, H. (1927), *Arch. Elektrotech*, *18*, 583.

Dinsdale, A. (1926), *Television*, Pitman, London.

Ekström, A. (1910), Swedish Patent No. 32,200, January 24.

Farnsworth, P. T. (1930), U.S. Patent No. 1,969,399.

Farnsworth, P. T. (1931), *Television News*, March–April, p. 48.

Farnsworth, P. T. (1934), *J. Franklin Inst.*, *218*, 411.

Goetze, G. W. and Laponsky, A. B. (1974), "Camera Tubes Employing High-Gain Electron-Imaging Charge-Storage Targets," in *Photoelectronic Imaging Devices*, Vol. 2, Biberman, L. M. and Nudelman, S. (Eds.), Plenum, New York.

Jenkins, C. F. (1925), *Vision by Radio*, Jenkins Laboratories, Washington, DC.

Klemperer, O. (1953), *Electron Optics*, Cambridge University Press, Cambridge,

Kurz, E. A. (1979), *American Laboratory*, March, p. 67.

Levi, L. (1980), *Applied Optics*, Vol. 2, Wiley, New York.

Mavrodineanu, R. and Boiteux, H. (1965), *Flame Spectroscopy*, Wiley, New York.

Meyers, L.M. (1939), *Electron Optics*, Chapman, London.

Musselman, E. M. (1974), "The New Image Isocon—Its Performance Compared to the Image Orthicon," in *Photoelectronic Imaging Devices*, Vol. 2, Biberman, L. M. and Nudelman, S. (Eds.), Plenum, New York.

Nipkow, P. (1884), German Patent No. 30,105, January 6.

RCA (1970), *Photomultiplier Manual*, RCA Technical Series PT-61, RCA Solid State Division, Lancaster, PA.

RCA (1974), *Electro-Optics Handbook*, RCA Technical Series OEH-11, RCA Solid State Division, Lancaster, PA.

RCA (1980), *RCA Photomultiplier Handbook*, RCA Technical Series PMT-62, RCA Solid State Division, Lancaster, PA.

Redington, R. W. (1974), "The Image Orthicon," in *Photoelectronic Imaging Devices*, Vol. 2, Biberman, L. M. and Nudelman, S. (Eds.), Plenum, New York.

Sternglass, E. J. (1955), *Rev. Sci. Instr.*, *26*, 1202.

Timothy, J. G. (1974), *Rev. Sci. Instr.*, *45*, 834.

Timothy, J. G. and Bybee, R. L. (1977), *Proc. SPIE Int. Tech. Symp.*, San Diego, CA, August.

Vine, J. (1974), "Electron Optics," in *Photoelectronic Imaging Devices*, Vol. 1, Biberman, L. M. and Nudelman, S. (Eds.), Plenum, New York.

Winefordner, J. D. and Vickers, T. J. (1964), *Anal. Chem.*, *36*, 1939.

Wiza, J. L. (1979), *Nucl. Instr. Meth. 162*, 587.

Zworykin, V. K. (1934), *J. Franklin Inst.*, *217*, 1.

CHAPTER

11

SOLID-STATE IMAGE DETECTORS

One of the primary goals of spectroscopic research is the development of ways to increase the rate of information transmission in spectroscopic systems. It was not long after the general adoption of photomultipliers as spectroscopic detectors that it was realized that the convenience and sensitivity of these detectors was offset, at least in part, by the trade-off between scan speed and signal-to-noise ratio. The attainment of a satisfactory signal-to-noise ratio often dictated an intolerably slow scan rate. Use of a slow scan rate is particularly unsatisfactory with transient sources or if the spectral source under observation is subject to slow drift. The presence of source drift during a scan precludes the accurate measurement of relative intensities of widely separated spectral features because the measured intensities are a function of both wavelength and time (where the time dependence is random). In addition, a conventional scanning system, because it measures only one resolution element at a time, spends only a very small fraction of the total scan time observing any one resolution element. Thus, if the spectral source is subject also to higher frequency random variations, the particular instant the given resolution element is sampled will affect the outcome of the measurement.

These drawbacks to the conventional scanning monochromator led spectroscopists to wonder whether such limitations associated with electronic detection could be overcome. Two basic approaches have been proposed to ease the dilemma caused by the scan speed/signal-to-noise trade-off. In the infrared region, where systems are frequently detector noise limited, it was realized that the simultaneous monitoring of several resolution elements could improve the signal-to-noise ratio. This approach led to the development of the transform techniques discussed in Chapters 5, 7, and 14. Even with transform techniques, however, source variations such as low-frequency drift and photon shot noise exert a deleterious influence and can easily nullify the sought-for multiplex advantage.

The other avenue of investigation into ways of increasing the rate of information transmission involves the use of multichannel image detectors (Kazan and Knoll, 1968). The ultimate goal of this research is the development of an electronic analog to the photographic plate. The ideal image detector for spectroscopic application should combine the multichannel advantage of a parallel detector with the convenience of electronic readout. It should be realized at the outset, however, that the majority of image detectors commercially available today has been developed for applications other than spectroscopic detection. As a result, the application of image detectors in spectroscopy is not without trade-offs and compromises.

The purpose of this chapter is to provide an introduction to the important performance characteristics of the more common solid-state image detectors available today. The discussion will focus on the general principles of operation rather than a comparison of actual specifications. Image detector development is an active area of research, and specific performance figures quickly go out of date as improvements are made. General principles, on the other hand, tend to change only slightly.

11.1. PERFORMANCE CHARACTERISTICS OF IMAGE DETECTORS

Before discussing the various solid-state image detectors available today, it is essential to understand the terminology used to describe the performance of these devices. The following terms will be defined briefly in what follows, and the more important ones will be discussed in more detail later.

11.1.1. Pixel

A pixel is a term that refers to a picture element. The whole image is made up of a very large number of pixels of different intensity.

11.1.2. Fixed-Pattern Noise

Fixed-pattern noise is a term used to describe a characteristic dark-response variation that occurs from one pixel to another producing the same variation in response (or pattern) across the target from scan to scan under similar operating conditions. One common source of fixed-pattern noise is the variation of diode leakage current from one diode to another across the target. Although it appears random like a true noise, the variation in response from one pixel to another is actually nonrandom in nature since it is due to variation in the dark current across the target. These dark-current variations (or variations in diode leakage current) are a characteristic of the individual diodes making up the array and produce a reproducible effect under similar conditions. As a result, this response variation from what should be nominally identical sensors produces a fixed pattern that, because it is actually an offset signal, can be removed by subtraction of the dark current from the signal on a diode-by-diode basis.

11.1.3. Transfer Characteristic

Image detectors typically produce a signal related to some power of the detector irradiance (i.e., the detector is often nonlinear), as given by

$$S = kI^{\gamma} \tag{11.1}$$

where S is the signal, k is a proportionality constant, I is the irradiance on the detector, and γ is a parameter characteristic of the detector. The power law relationship with a particular γ generally holds over several orders of magnitude of irradiance so that a plot of the logarithm of the signal versus the

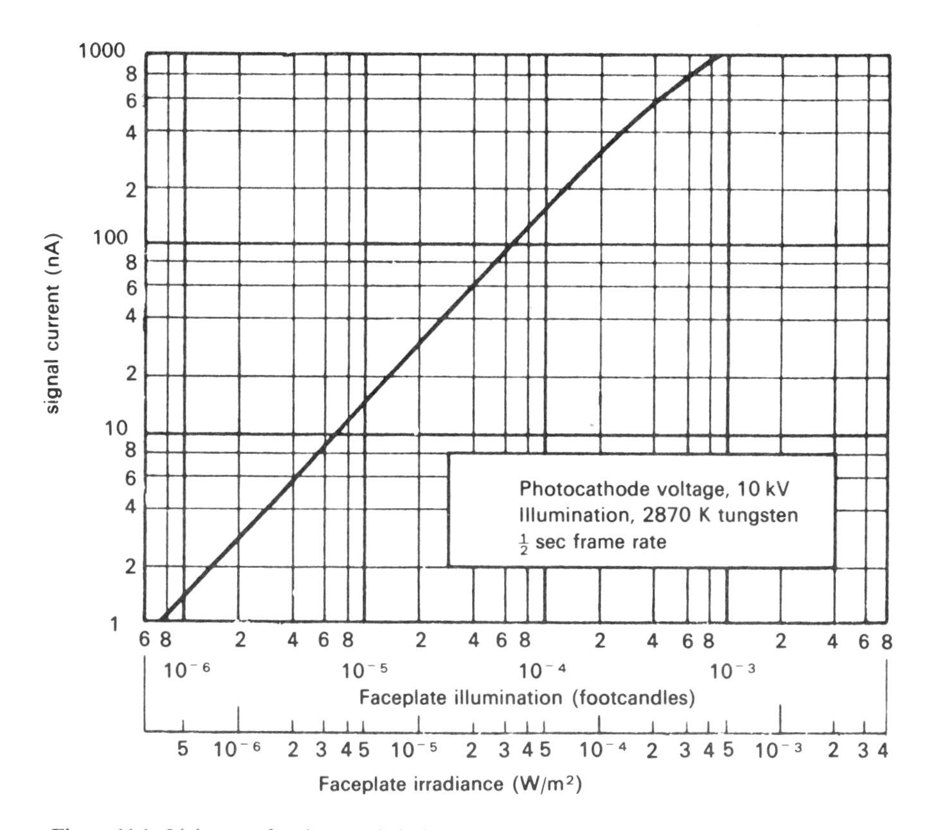

Figure 11.1. Light transfer characteristic for silicon diode array target camera tube (based on WL-30691). [Reprinted from Goetze, G. W. and Laponsky, A. B. (1974), "Early Stages in the Development of Camera Tubes Employing the Silicon-Diode Array as an Electron-Imaging Charge-Storage Tube" in *Photoelectronic Imaging Devices*, Vol. 2, Biberman, L. M. and Nudelman, S. (Eds.), Plenum, New York.]

logarithm of irradiance (known as the transfer characteristic) has a straight-line portion, as shown in Figure 11.1. The slope of the transfer characteristic is the change in signal for a small change in light level and is a measure of the ability of the detector to respond to small variations in light intensity. By analogy with the photographic emulsion characteristic curve (Chapter 8), the slope of the transfer characteristic is called the *gamma* of the tube and represents the contrast of the detector.

11.1.4. Dynamic Range

The dynamic range is the ratio of the maximum irradiance level that can be stored by the target to the minimum irradiance that can be detected over the noise. The maximum irradiance that can be stored depends on the target charge storage capability CV, where C is the target capacitance and V is the maximum allowable voltage excursion of the target. By contrast, the minimum signal that can be detected over the noise is a function of the readout process. For a return-beam tube like the image orthicon, the shot noise in the reading beam will determine the minimum detectable light level. For a direct readout tube like the silicon vidicon (discussed later in this chapter), preamplifier noise will determine the minimum detectable signal.

11.1.5. Lag

The term *lag* refers to the persistence of the image produced by a camera tube on the charge storage target. The terminology has been borrowed from phosphor research where the decay of phosphorescence is characterized by a *time lag*.

In conventional television transmission, each pixel is read out every $\frac{1}{30}$ sec. Ideally, the image stored during this 33-msec framescan period should be completely erased during target readout. This is not the case in actual practice, however, and a certain portion of the image from one frame often appears in a subsequent frame. The residual image due to the incomplete erasure of information between subsequent framescans is a result of the lag of the charge storage portion of the detector. The magnitude of the lag depends on the rate at which charge can be replaced on the target by the electron beam after the target has been exposed to light. As a result, it depends on the RC time constant of the electron beam–target combination, where R is the effective resistance of the electron beam and C is the target capacitance.

The magnitude of the lag is commonly expressed as a percentage of the original signal level on the target. *Third-field lag* is the amount of the original image remaining on the target three field periods after the optical image was removed.

To understand this definition of lag, the term *field* must be defined. In Chapter 10, a simple form of raster scanning was described where each line was scanned in sequence (i.e., 1, 2, 3, 4, · · ·). In most television applications, a more complicated form of scanning, known as *interlaced scanning* (Zworykin and Morton, 1940), is employed to reduce the amount of flicker perceived by the viewer in the picture tube phosphor. Although a variety of interlacing schemes is possible, the most common is an even–odd arrangement. Under this form of interlaced scanning, the odd-numbered lines of the raster, which compose the *odd field*, are scanned first (i.e., 1, 3, 5, 7, · · ·). When all of the odd-numbered lines of the raster have been scanned, the even-numbered lines of the raster, which compose the *even field*, are scanned (i.e., 2, 4, 6, 8, · · ·). With this form of scanning, the *field frequency* is twice the framescan frequency. Thus the period of a field is one-half the framescan period. If the typical framescan period is taken as $\frac{1}{30}$ sec, then third-field lag refers to the percentage of the signal still remaining on the target $\frac{3}{60}$ sec after the image was removed.

11.1.6. Blooming

"Blooming" is a term used to describe the spillover of signal charge on the target into adjacent pixels as a result of sensor saturation. Blooming occurs in high-light-level regions of the scene because of the lateral diffusion of charge carriers on the target. Lateral diffusion occurs because of the mutual repulsion of like-charge carriers.

11.1.7. Integration Time

In conventional television, the typical framescan period is $\frac{1}{30}$ sec, and the signal for any pixel is the charge that has accumulated since the last time the picture element was interrogated. Used in this manner, solid-state image devices are integrating detectors like a photographic plate with an exposure of $\frac{1}{30}$ sec. In the real-time mode of operation, the detector transmits an updated framescan every $\frac{1}{30}$ sec. It should be mentioned, however, that while the integration period is the same for all photosites of a typical image detector, all photosites are not integrating incident radiation at precisely the same time (i.e., the reset interval is shifted in time for different portions of the image).

For spectroscopic work, the period of a framescan can be increased to allow a larger amount of charge to build up on the target prior to readout. The limitation to the length of signal integration on the target itself is determined by the magnitude of the dark current (i.e., diode leakage current) and the tendency for lateral charge spreading on the target with time. Since most image tubes that are available commercially today were designed for television application, they were not made with the goal of lengthy integration periods in mind, and consequently, large gains in sensitivity should not be expected from this mode of operation. The SEC tube is an exception to this rule, however, and integration periods of several hours have been employed successfully in astronomical studies (Eccles, Sim, and Tritton, 1983). With devices based on silicon solid-state technology, cooling can be used to reduce the dark current and increase the available integration time.

11.1.8. Resolution

One of the most important characteristics of an image tube is its ability to reproduce faithfully the light pattern incident on the detector. The ability of any optical component to reproduce faithfully an image of an object can be expressed in terms of its *modulation transfer function* (MTF). The modulation transfer function (sometimes referred to as the *sine wave response*) is simply the ratio of the amount of modulation present in an image formed by the optical system to the amount of modulation present in the object being observed. The amount or extent of modulation (referred to as the *depth of modulation* in electronics) is simply the ratio of the amplitude of the modulation to the average value of the signal. The depth of modulation (often referred to in optics as the *contrast*) is given by

$$M = \frac{(I_{\max} - I_{\min})/2}{(I_{\max} + I_{\min})/2} \tag{11.2}$$

$$= \frac{I_{\max} - I_{\min}}{I_{\max} + I_{\min}}$$

where the numerator of the first equation represents the amplitude of the modulation and the denominator is the average value. If the amplitude of the modulation is equal to the average value of the signal, the contrast or amount of modulation is 100%. Figure 11.2 shows how the depth of modulation is related to the minimum and maximum intensities $I_{\min}$ and $I_{\max}$.

If a suitable object with a pattern having 100% modulation is imaged by an optical system that produces an image having a modulation of 100%, the modulation transfer function (i.e., the ability of the optical system to transfer the modulation in the object to the image) will be 1. Naturally, the ability of the optical system to reproduce the modulation in the object will be a function of the spatial frequency in the test pattern under observation. As the spatial frequency of the pattern increases, the extent of modulation in the image will decrease, and the MTF will decrease toward zero.

Figure 11.3 shows a test pattern consisting of a series of light and dark bars whose intensity varies in a sinusoidal fashion and whose spatial frequency

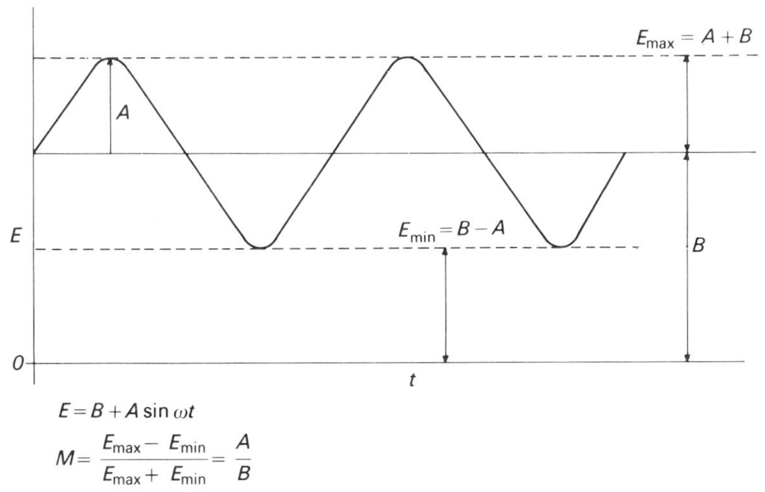

Figure 11.2. Graphical construction showing relationship between depth of modulation and minimum and maximum amplitudes of sine wave.

increases from left to right. The spatial frequency of the test pattern is expressed as the number of *line pairs* per millimeter where a line pair consists of one black bar and one white bar. Figure 11.3*b* shows the negative obtained when the test pattern is photographed. Figure 11.3*c* shows the densitometer tracing obtained when the negative is scanned. It is clear from Figure 11.3*c* that the depth of modulation in the image is a definite function of the spatial frequency of the test pattern. At low values of the spatial frequency, the optical system has no difficulty in reproducing the extent of modulation in the test pattern, and the modulation transfer function is close to 1. As the spatial frequency in the test pattern increases, however, diffraction and aberrations cause some of the light from the white portions of the target to spill over into the dark portions. As this occurs, the light portions become less bright and the dark portions become brighter until the contrast between the two becomes equal. At this point, all resolution is lost.

Figure 11.4 shows the MTF for a hypothetical photographic camera. From the figure, it can be seen that each component acts as a low-pass spatial frequency filter with a particular roll-off. The advantage of using the MTF as a means of expressing optical performance is due to the ability to cascade the modulation transfer functions of several components to obtain the overall modulation transfer function of the entire system.

Consider for a moment the case of a two-component optical system. Suppose that a test pattern with a given modulation represented as $M_{object1}$ is imaged by a lens with a modulation transfer function MTF_{lens1}. Suppose that

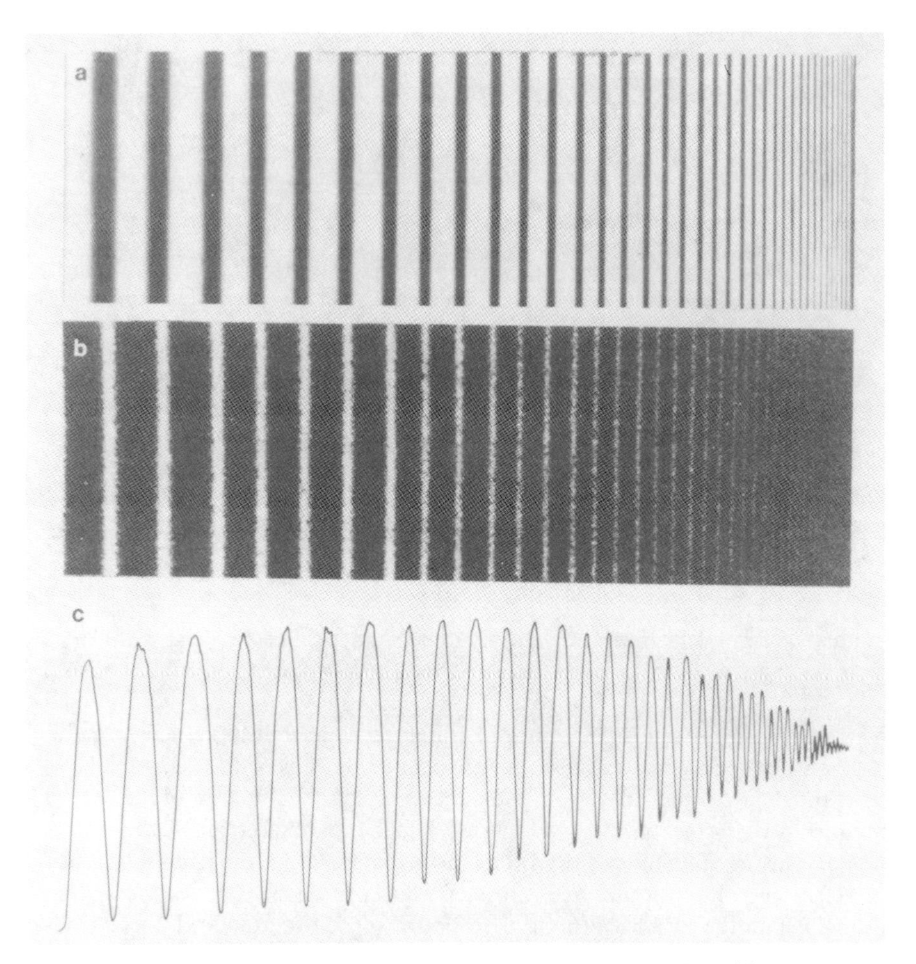

Figure 11.3. Experimental measurement of modulation transfer function with test pattern: (*a*) intensity variation in test pattern; (*b*) negative obtained when test pattern is photographed; (*c*) densitometer tracing obtained when negative is scanned. [Reprinted with permission from R. Kingslake (1983), *Optical System Design*, Academic, New York. Copyright 1983 Academic Press.]

the image formed by the first lens is focused on a screen by a second lens with a modulation transfer function $\mathrm{MTF}_{\mathrm{lens}2}$. Clearly, the object for the second lens is the image produced by the first lens,

$$M_{\mathrm{object}1} \times (M_{\mathrm{lens}1}/M_{\mathrm{obj}}) = M_{\mathrm{object}2} \tag{11.3}$$

where $M_{\mathrm{lens}1}/M_{\mathrm{obj}}$ is the MTF of the first lens, and $M_{\mathrm{object}2}$ is the depth of

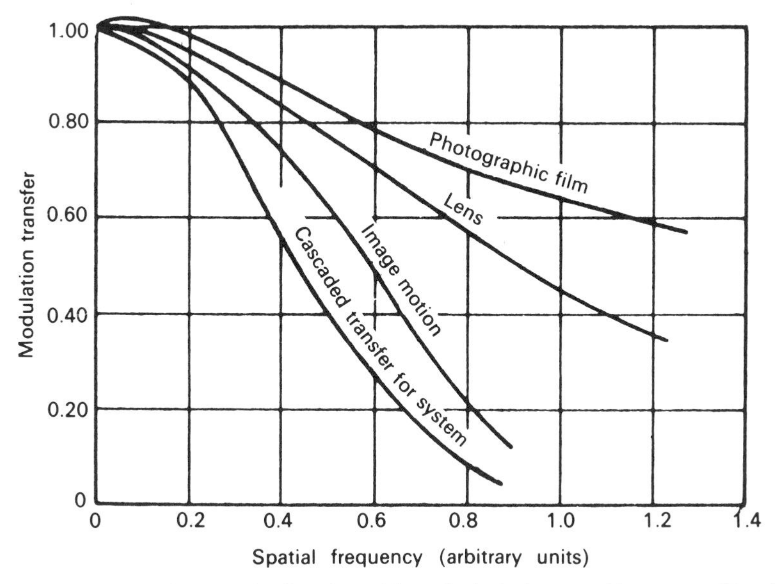

Figure 11.4. Modulation transfer function of hypothetical photographic camera. [Reprinted from Cope, A. D., Gray, S., and Hutter, E. C. (1974), "The Television Camera Tube as a System Component," in *Photoelectronic Imaging Devices*, Vol. 2, Biberman, L. M. and Nudelman, S. (Eds.), Plenum, New York.]

modulation in the image produced by the first lens. The extent of modulation in the final image produced on the screen by the second lens will be

$$M_{\text{final image}} = M_{\text{object2}} \times (M_{\text{lens2}}/M_{\text{obj}}) \tag{11.4}$$

where $M_{\text{lens2}}/M_{\text{obj}}$ is the MTF of lens 2. From Eqs. 11.3 and 11.4, it can be seen that the overall MTF will be the product of the individual MTFs:

$$\text{MTF}_{\text{overall}} = M_{\text{final image}}/M_{\text{object1}} = \text{MTF}_{\text{lens1}} \times \text{MTF}_{\text{lens2}} \tag{11.5}$$

In Figure 11.4, the overall MTF for the photographic camera is obtained from the product of the individual MTFs for a given spatial frequency. The limiting resolution the eye can detect with an optical system is generally taken as a MTF of 3–5%.

Because of the difficulty in producing test patterns with a sinusoidal intensity variation as discussed previously, test patterns with a square-wave spatial frequency variation are often used for convenience. Resolution measurements made with a square-wave test pattern result in a transfer function known as the *contrast transfer function.* Although the contrast transfer function may be more convenient to measure, it cannot be cascaded the way the modulation transfer function can to obtain the overall system performance. Fortunately, the contrast transfer function can be converted conveniently into the MTF by means of a simple calculation. The following equation gives the relationship between the contrast transfer function and the MTF:

$$\mathrm{MTF}(N) = \tfrac{1}{4}[\mathrm{C(N)} + \tfrac{1}{3}\mathrm{C(3N)} + \tfrac{1}{5}\mathrm{C(5N)} + \tfrac{1}{7}\mathrm{C(7N)} + \tfrac{1}{11}\mathrm{C(11N)} - \tfrac{1}{13}\mathrm{C(13N)} - \tfrac{1}{15}\mathrm{C(15N)} - \tfrac{1}{17}\mathrm{C(17N)} \cdots] \tag{11.6}$$

where C is the contrast transfer function and MTF(N) is the MTF for a given spatial frequency N. The ninth-order term is missing in Eq. 11.6, and the expression for the general term is somewhat complicated (Coltman, 1954).

11.2. SPECIFYING IMAGE DETECTOR RESOLUTION IN SPECTROSCOPY

The resolution obtained when an image device is used as a detector in spectroscopy can be specified in a number of different ways. Because of this, care must be used in comparing published values for resolution in image detector spectrometers. Most image detectors used in spectroscopy monitor the intensity of light in a series of slit-shaped resolution elements or channels. Thus, one way in which the resolution of the instrument has been specified in the literature is to report the channel width in wavelength units (i.e., the channel width times the reciprocal linear dispersion). This method of reporting resolution, however, does not convey much information. For example, it does not reveal how close two spectral lines can be before they cannot be resolved.

11.2.1. Rayleigh Criterion and the Modulation Transfer Function

To determine the resolving power of an image detector spectrometer, more information is needed about the intensity profile of the spectral line as a function of channel position. This is illustrated in Figure 11.5. If the line profile is such that 40% of the peak intensity is located in the two adjacent channels, then the half-intensity bandwidth (FWHM) is approximately two channels wide, and lines that are separated by this amount—two channels—will be just resolved with approximately a 20% valley between them. This corresponds roughly to the Rayleigh criterion for resolving power. In wavelength units, this amount of separation corresponds to 2 times the channel width times the reciprocal linear dispersion. To obtain baseline resolution for the two lines, a minimum of four-channel peak separation is

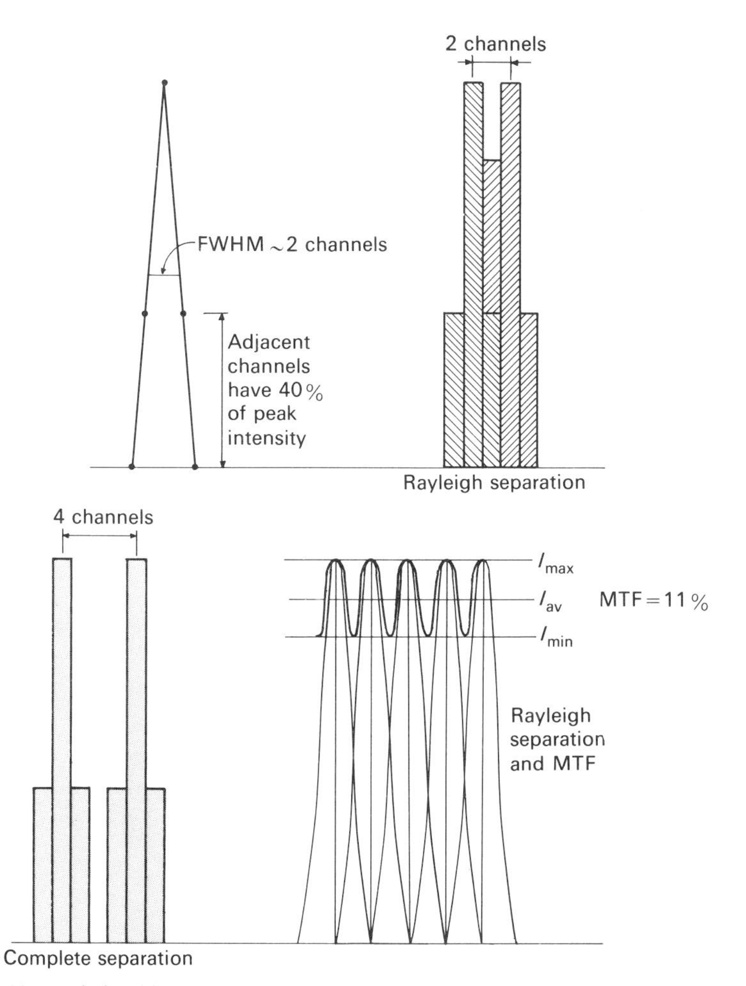

Figure 11.5. Relationship between Rayleigh criterion for resolving power and modulation transfer function.

required, as shown in Figure 11.5. Figure 11.5 also shows that this "Rayleigh" criterion corresponds to a modulation transfer function of 11%.

Assuming that the detector is the component that limits the resolution (i.e., the MTFs of all the other components of the spectrometer are close to 1.0), the spatial frequency corresponding to a detector MTF of 11% can be used to estimate the wavelength separation anticipated for the Rayleigh criterion of resolving power. This wavelength separation is given by

$$\Delta\lambda = (\text{line pair/mm})^{-1} \times R_D \tag{11.7}$$

where $\Delta\lambda$ is the wavelength separation, R_D is the reciprocal linear dispersion, and the spatial frequency used corresponds to a MTF of 11% for the detector. If the MTFs of the other components of the spectrometer are known, the overall MTF can be determined as described previously and used to give a more reliable estimate of system performance.

Finally, it should be noted that the MTF of an image detector is a function of detector irradiance, and a compromise exists between limiting resolution and light level. The decrease in modulation observed at low light levels is due to the corresponding decrease in signal-to-noise ratio as the irradiance decreases. For this reason, care should be taken when using published MTF data taken at high irradiance levels to predict the performance of an image detector at low light levels.

11.2.2. Aliasing

All image detectors having photosites spaced a finite distance apart sample the optical signal at a particular spatial frequency governed by the pixel spacing. Because the data collection process with this type of image detector involves periodic spatial sampling as opposed to continuous spatial sampling, the system must satisfy the *Nyquist sampling theorem* to avoid the loss of

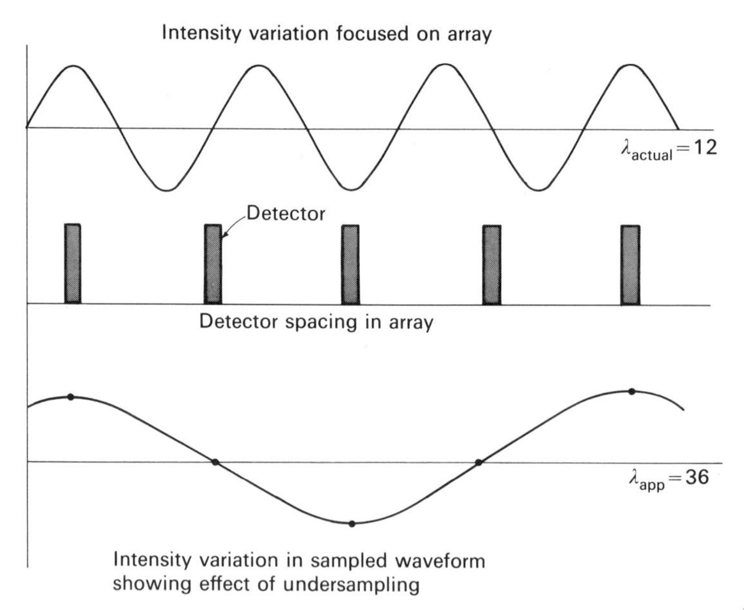

Figure 11.6. Effect of undersampling periodic spatial intensity variation with array detector with fixed pixel spacing. λ_{app}, apparent wavelength.

information. According to the Nyquist sampling theorem, a waveform composed of a mixture of frequencies must be sampled at a frequency at least twice the highest frequency present to avoid loss of information. Figure 11.6 shows the effect of undersampling on a periodic spatial intensity variation that has been focused on an array detector with a fixed pixel spacing in the array. Because the detector does not sample the incident waveform at least twice during a period of the incident sine wave, the Nyquist sampling theorem is not satisfied. As a result, a loss of information occurs, and the incident waveform cannot be unambiguously determined. It is apparent from Figure 11.6 that undersampling the incident waveform produces an apparent waveform with a lower frequency than the incident waveform. This production of a spurious lower frequency waveform as a result of undersampling is known as *aliasing*. To avoid aliasing with an array detector, the maximum spatial frequency component of an incident intensity pattern cannot exceed one-half the spatial frequency corresponding to the photosite spacing.

11.3. THE IMPORTANCE OF DYNAMIC RANGE IN SPECTROSCOPY

Aside from adequate resolution, the two next most important characteristics of image detector performance as far as spectroscopic application is concerned are sensitivity and dynamic range. The term sensitivity as used here refers to the irradiance level that will produce an adequate signal-to-noise level. As discussed in Chapter 10, the radiance produced by 10 ppm (not a particularly trace concentration by any means) of sodium introduced into a flame is $1 \times 10^{-3}\ \mathrm{W\,cm^{-2}\,sr^{-1}}$. If an image detector is used in conjunction with an $f/8.7$ 0.5-m polychromator, the solid angle collected by the spectrometer will be 0.01 sr, and the irradiance produced in the focal plane will be about $10^{-5}\ \mathrm{W\,cm^{-2}}$ assuming complete optical transfer within the spectrometer. Thus to be satisfactory for use in analytical atomic spectroscopy, an image detector must be able to produce a satisfactory signal-to-noise ratio at irradiance levels equal to $10^{-5}\ \mathrm{W\,cm^{-2}}$ and, hopefully, several orders of magnitude lower.

Aside from adequate sensitivity, an image detector suitable for use in analytical atomic spectroscopy must possess an adequate dynamic range. The need for a large dynamic range arises because of the wide range of intensities of spectral lines encountered in a typical emission spectrum. Figure 11.7 shows a typical multielement flame emission spectrum for a sample containing eight elements. The wide range of intensities observed is due to the different excitation energies of the various lines as well as the range of concentrations of the various elements present in the sample. To be completely satisfactory for use in analytical atomic spectroscopy, an image

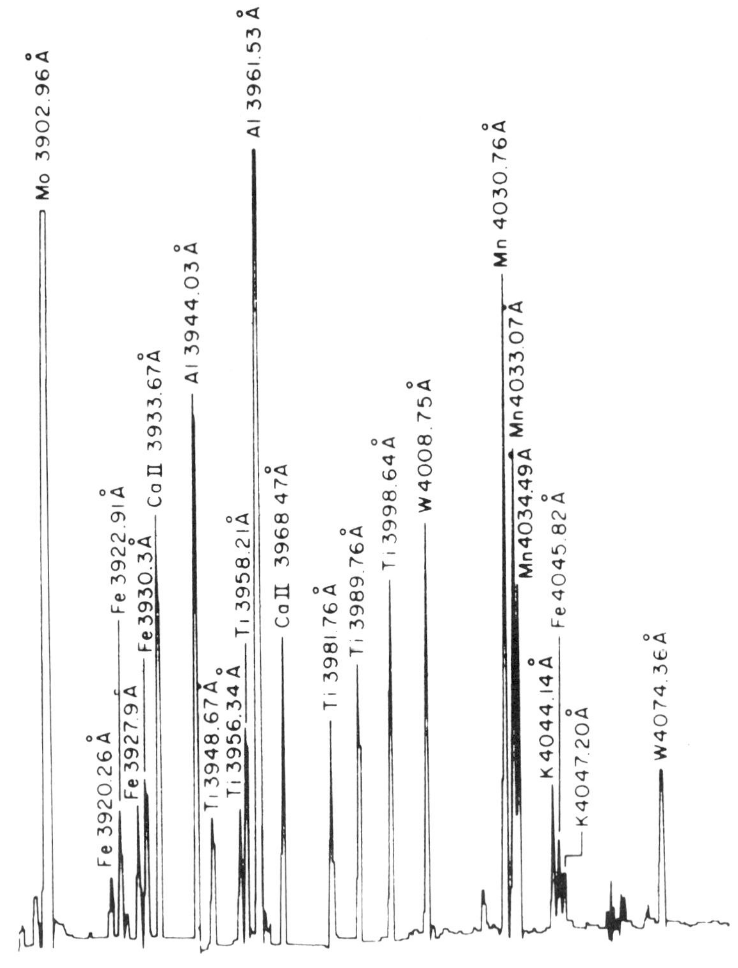

Figure 11.7. Multielement flame emission spectrum from 388.6 to 408.6 nm. [Reprinted with permission from K. W. Busch, N. G. Howell and G. H. Morrison (1974), *Anal. Chem.*, **46,** 575. Copyright 1974 American Chemical Society.]

detector needs to be able to monitor the intensity of a weak trace-element line that is in close proximity to a strong major-component line. For a typical image tube, the maximum integrated irradiance the target can store during a framescan period is limited by the charge that can be deposited by the electron gun over the area of a channel during the linescan period. For areas of the target exposed to relatively high levels of irradiance, target blooming characteristics become a significant parameter. If the blooming is particularly severe, photoinduced charge carriers will spill over into adjacent channels effectively wiping out information over large areas of the target.

For most television applications for which image detectors were designed, the dynamic range of the device is perfectly adequate to monitor the variations of intensity found in a typical scene (i.e., the *intrascenic dynamic range*). For spectroscopic work, however, the range of irradiance levels that need to be monitored simultaneously frequently exceeds the dynamic range available in the image detector. To be useful as detectors in analytical atomic spectroscopy, image detectors must be used in such a way that detector saturation and blooming are avoided for the intense major-component lines, while adequate sensitivity is maintained to detect the weak trace-element lines. Since the tube itself frequently does not have the necessary dynamic range, various optical and electronic target-scanning means often are employed to modify the range of intensities to match the dynamic range of the particular detector (see Chapter 15).

11.4. SILICON VIDICON DETECTORS

A wide variety of image detectors is available today, ranging from image tubes that employ electron beam readout to self-scanned solid-state array detectors. This survey will be limited to a discussion of the principles and general characteristics of the more common solid-state image detectors. The term *solid state* as used here refers to detectors based on silicon chip technology.

The term *vidicon* was used first (Weimer, Forgue, and Goodrich, 1950, 1951) to describe a television camera tube based on a photoconductive principle. The *silicon vidicon*, which employs a target consisting of an array of photodiodes, was discussed first in a series of papers beginning in 1967 (Buck et al., 1968); Crowell et al., 1967; Crowell and Labuda, 1969; Gordon, 1967; Gordon and Crowell, 1968; Wendland, 1967). This work culminated in the introduction of a diode array tube by a Bell Telephone Laboratories. The silicon vidicon tube was developed by Bell Laboratories as a simple, relatively rugged image detector that could be used in a "picture-phone" system. Since its introduction, a variety of silicon vidicon detectors has been developed and is commercially available.

11.4.1. Fabrication of the Silicon Vidicon

A key factor contributing to the success of the silicon vidicon is the ease with which the diode array can be fabricated using standard integrated circuit (IC) technology. In fact, the production of a diode array target is, in many ways, less difficult than an ordinary IC chip because only diodes are used and there are no internal connections. On the other hand, since the ultimate resolution of the tube depends on the diode density, high diode density is necessary, which, in turn, requires a lower leakage current than is commonly required in standard IC fabrication.

To prepare a target, a 0.010-in.-thick slice of silicon known as the *substrate* is oxidized, and windows are opened in the oxide layer on one side of the slice using standard photolithographic procedures. Then *pn* junctions are formed through the array of open windows by diffusion of boron. Typical center-to-center diode spacing is 20 μm with an oxide hole diameter of 8 μm. The opposite side of the slice is then treated to form an n^+ layer. An n^+ region is simply one that is more heavily doped than usual. The purpose of the n^+ layer is to establish an electric field at this face that is strong enough to separate the photon-induced holes from the electrons before they can recombine (Eccles et al., 1983). This is especially important at the surface where the majority of charge carrier production occurs because photons do not penetrate very far into the target. The final step is the addition of a thin film of material over the diode side of the substrate to prevent excessive charging of the SiO_2 insulating film. The prevention of excessive SiO_2 charging is necessary to avoid electron beam landing errors and can be accomplished in a number of ways, the most common being (1) the deposition of a thin film of highly resistive material known as a *resistive sea* or (2) the formation of isolated metallic pads over the

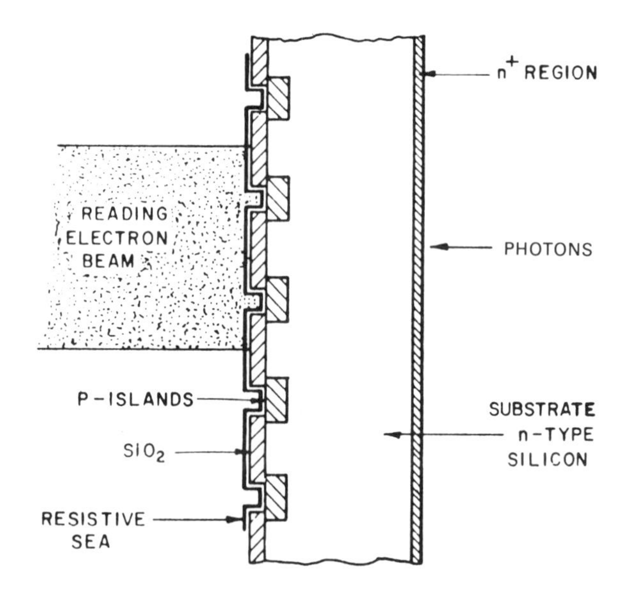

Figure 11.8. Schematic of silicon diode array target. [Reprinted from Goetze, G. W. and Laponsky, A. B. (1974), "Early Stages in the Development of Camera Tubes Employing the Silicon-Diode Array as an Electron-Imaging Charge-Storage Tube" in *Photoelectronic Imaging Devices*, Vol. 2, Biberman, L. M. and Nudelman, S. (Eds.), Plenum, New York.]

diodes. For a complete discussion of the modifications to the basic target structure that have been made to avoid SiO_2 charging and improve performance, the reader is referred to Crowell and Labuda (1974). Figure 11.8 shows a cross section of a typical diode array vidicon target.

The diode array produced in this manner is relatively stable chemically and can easily withstand the 400°C back-out given the tube during the final stages of manufacture to prolong its life. In addition, the silicon vidicon television tube can withstand brief exposures to intense light sources without suffering permanent damage.

11.4.2. Principle of Operation

Figure 11.9 shows a schematic diagram of a vidicon television pickup tube. During operation, the diode side of the wafer, which faces the electron gun, is scanned by the electron beam. To eliminate registration problems between the beam and the mosaic, the beam diameter is generally larger than the diode spacing in the array. As the beam scans the surface in a raster pattern, it deposits electrons until the resulting surface charge that builds up becomes strong enough to prevent any further deposition. This point is reached when the surface acquires a potential very close to the cathode potential of the electron gun (~ 0 V).

At the same time, the n-type silicon wafer (i.e., the substrate) is maintained at a positive potential of 10 V relative to the cathode potential (i.e., the *target potential*). The combination of the positive potential applied to the substrate and the charging effect of the scanning electron beam effectively reverse biases the diodes of the array by 10 V. As discussed in Chapter 9, a pn junction is reverse biased when the n-type material is made positive relative to the p-type

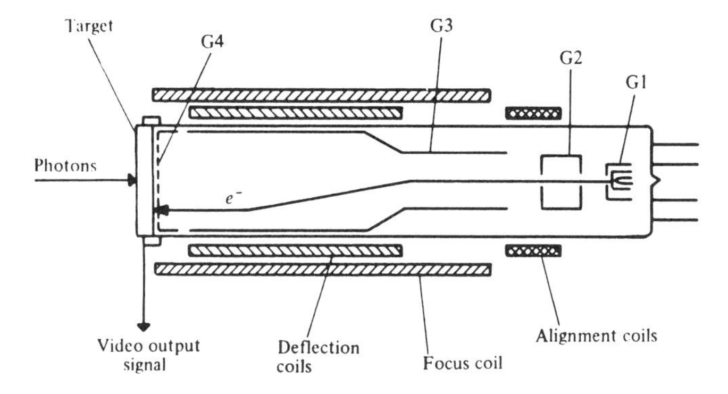

Figure 11.9. Schematic of vidicon television pickup tube. [Reprinted from Eccles, M. J., Sim, M. E., and Tritton, K. P. (1983), *Low-Light Level Detectors in Astronomy*, Cambridge University Press, Cambridge.]

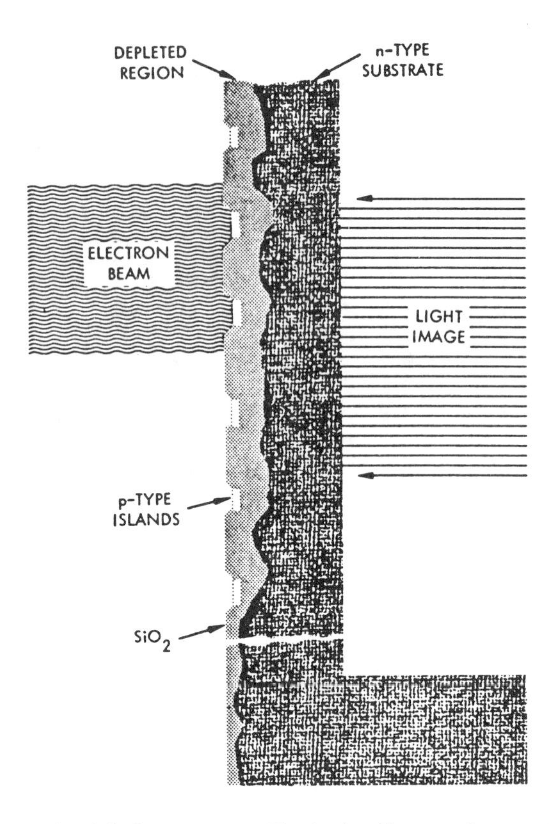

Figure 11.10. Schematic of diode array target. To obtain self-supporting structure, perimeter of wafer is left much thicker than substrate in area of diode array. [Reprinted from Crowell, M. H., and Labuda, E. F. (1974), "The Silicon-Diode-Array Camera Tube" in *Photoelectronic Imaging Devices*, Vol. 2, Biberman, L. M. and Nudelman, S. (Eds.), Plenum, New York.]

material. One of the primary effects of reverse biasing a *pn* junction is the formation of an enlarged depletion layer (i.e., a region having few charge carriers and therefore behaving like an insulator) in the vicinity of the junction, as shown in Figure 11.10. For a 10-Ω-cm substrate under a 10-V reverse-bias potential, the width of the depletion zone is typically 5 μm (Crowell and Labuda, 1974). Formation of the depletion zone (i.e., insulating layer) converts the diode array into a mosaic of equivalent tiny capacitors, each capable of storing the charge deposited by the electron beam.

The amount of charge that can be stored is determined by the target junction capacitance (typically 2000 pF cm^{-2}) and the maximum permissible voltage allowed to develop on the diode side of the array. If the voltage on the diode side of the array is allowed to become too large (i.e., positive), it can result in deflection of the low-energy scanning electron beam, producing beam-bending errors in the raster scanning process. On the other hand, the maximum junction capacitance that can be tolerated will depend on the ability of the electron beam to recharge each target element back down to the potential of the cathode of the electron gun during the interrogation time allotted to each picture element by the scanning circuitry. If the target capacitance is too great, the electron beam will not be able to deposit enough charge during the scanning time, and some of the image will not be "erased". As a result, some of the image from a previous frame will appear in a subsequent frame, resulting in lag.

If the resistivity of the reverse-biased array is high enough (i.e., the diode leakage current is very small), the diodes will remain fully reverse biased during the period of the framescan as long as they have not been exposed to light. If the *n*-type substrate is exposed to light as shown in Figure 11.10, electron–hole pairs will be produced in the *n*-type material for every photon absorbed. Since silicon has a rather high absorption coefficient for radiation between 200 and 1100 nm, most of the radiation will be unable to penetrate very far into the wafer and will be absorbed near the illuminated surface of the substrate. This will result in the production of an excess of minority carriers (i.e., holes) in the *n*-type material. These holes will migrate (under the combined influence of the electric field across the target and the concentration gradient established at the surface) toward the reverse-biased junctions on the opposite side of the wafer. If the lifetime of the holes is sufficiently long so that they do not suffer recombination during their journey, a large fraction of the holes produced by photon absorption will reach the *pn* junction by diffusion. Upon reaching the junction, the holes cross the depletion layer under the influence of the electric field, thereby reducing the negative charge on the *p*-type islands (i.e., making them more positive) in the vicinity where light was absorbed in the *n*-type wafer. As long as the array is illuminated and the diodes remain reverse biased, a junction current will continue to flow, discharging the tiny "capacitors" in those regions of the target exposed to light.

When the electron beam returns to a given pixel after a framescan period, it deposits electrons on the surface of those diodes that have lost charge and whose potential is now more positive than 0 V (i.e., the cathode potential). The charging current produced as the charge on each picture element is restored is the video signal.

11.4.3. Electron-Bombardment-Induced Response

Although similar in principle to the SEC tube, the ordinary silicon vidicon does not possess the target gain of 100 produced by the SEC target (see Chapter 10). As a result, the ordinary silicon vidicon is not especially suited for low-light-level applications such as encountered in emission spectroscopy. To improve the low-light-level response of the silicon diode array target, *electron-bombardment-induced response* (EBIR) is used in place of direct photon excitation. This improvement in low-light-level response was first described by Gordon and Crowell in 1968. Electron bombardment response is achieved by use of an electrostatically focused image intensifier stage (see following discussion) prior to the silicon target. Such intensified vidicons are known by a variety of names. In the United States, they are commonly referred to as SIT vidicons, a name adopted by RCA to describe their detector and which stands for silicon-intensified target. Although similar in principle to the SEC tube discussed in Chapter 10, the SIT vidicon is generally 15–25 times more sensitive than the SEC tube (Goetze and Laponsky, 1974).

Figure 11.11 shows a schematic of an SIT vidicon. Primary photoelectrons emitted by a photocathode are accelerated and focused electrostatically onto the reverse-biased silicon diode array target where they strike with energies of about 10 keV. Since approximately 3.4 eV are required to produce one electron–hole pair in the target, one 10-keV photoelectron is capable theoretically of producing about 3000 electron–hole pairs (Goetze and Laponsky, 1974). In actual practice, however, the gains realized are somewhat less than predicted theoretically due to hole recombination during the journey to the *p*-type islands. If the target is sufficiently thin, collection efficiencies of 70% are possible, resulting in target gains of about 2000.

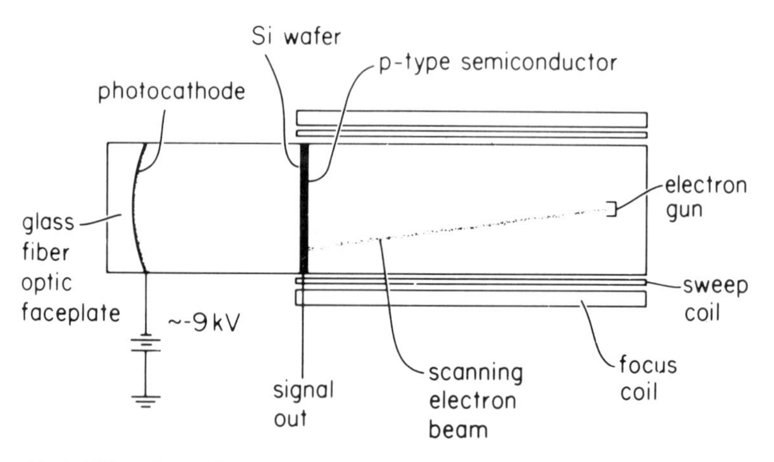

Figure 11.11. Silicon-intensified target vidicon detector. [Reprinted with permission from K. W. Busch, N. G. Howell, and G. H. Morrison (1974), *Anal. Chem.*, *46*, 1231. Copyright 1974 American Chemical Society.]

11.4.4. Development of Cascade Image Intersifiers

Before discussing the intensified SIT (ISIT) tube in the next section, it is worthwhile to discuss some of the developments that have led to the availability of modern cascaded modular image intensifiers such as used in the ISIT. The idea of light amplification by employing a series of image intensifiers was first suggested by von Ardenne (1936). Although serious interest in the idea of image intensification by cascading a series of image intensifier stages can be traced to the mid-1930s (Barthelemy and Leithine, 1936; Meyers, 1939; Morton, 1964), it was not until 1949 that a two-stage intensifier was built by RCA for the U.S. Navy to be used with a Schmidt telescope (Schnitzler and Morton, 1974). The lengthy delay between conception and actual realization of the first image intensifier was due to a combination of factors.

The basic principle behind image intensification is the use of a phosphor screen in conjunction with a photocathode to produce a two-dimensional electron current amplifier. Image intensification (i.e., amplification) was to be accomplished by accelerating the primary photoelectrons through a potential difference of several thousand volts so that, upon striking the phosphor layer, numerous electron–hole pairs would be produced that, upon recombination, would emit a great number of photons. When the idea was first tried in the 1930s, it was unsuccessful. Using the best available photocathode [which at the time was the Ag–O–Cs (S-1) photoemitter developed by Koller (1929)] and the best phosphor available, the amount of light generated at the output of the intensifier was found to be less than that incident on the photocathode! It was not until the development of the *cesium antimony* photocathode by Görlich (1936, 1938, 1941) as well as advances in phosphor technology that the first successful single-stage image intensifier was developed.

Although the goal of still higher gains obtained by cascading a series of image intensifier stages was now technically feasible, problems associated with coupling the stages together still needed to be solved. Early attempts at cascading a series of single-stage tubes were unsuccessful because of the lateral diffusion of light that occurred in passing from the phosphor of one tube through the intervening glass windows and onto the photocathode of the second tube. The first solution to this problem was the fabrication of multistage image intensifiers in a single tube envelope. These devices employed special dynode stages that consisted of a phosphor and a photoemitter deposited on opposite sides of a thin mica sheet. Although this configuration did reduce the amount of lateral light spreading, it was expensive to fabricate these tubes, and magnetic focusing was required because the dynodes were flat. Use of magnetic focusing added to the size and weight of the device because of the solenoid coil and associated constant-current circuitry.

An alternative technique for internal amplification within an image intensifier that was investigated was the use of transmission secondary-electron dynodes referred to in Chapter 10 (Section 10.6.5). The basic concept that was investigated was the feasibility of developing a magnetically focused image-forming photomultiplier by using a series of plane-parallel thin-film dynodes (Sternglass, 1955; Wachtel, Doughty, Goetze, et al., 1960; Wilcock, Emberson, and Weekly, 1960; Wachtel, Doughty, and Anderson, 1960). Like the target of the SEC tube, the thin-film dynodes consisted of an aluminum oxide supporting layer on which was deposited a thin layer of aluminum and a secondary emitter such as low-density KCl (see Figure 10.29). Secondary electrons emitted from the first dynode were focused by a magnetic field onto the second dynode, and so on until the desired degree of amplification was achieved. Assuming an electron gain of 7 per stage and a phosphor gain of 25, total gains on the order of 50,000 were possible with a four-stage tube (Hiltner, 1962). Figure 11.12 shows a schematic of a secondary-emission image intensifier. In spite of the potential high gains obtainable with secondary-emission image intensifiers, the employment of these dynodes was found to be limited because of the effects of positive charging at the exit surface where the secondary electrons were emitted (Goetze and Laponsky, 1974).

Further dramatic improvements in the development of true modular cascaded image intensifiers that could be fabricated from single-stage tubes took place during the 1950s with the development of more sensitive photoemitters such as the *tri-alkali antimony* photocathode described by Sommer (1955, 1968). The most significant factor in the development of modular cascaded image intensifiers, however, was the availability of fiber-optic plates developed in the 1950s (Potter and Hopkins, 1958). These fiber-optic plates could be shaped to any configuration and permitted the use of electrostatic focusing for the first time. Electrostatic focusing was possible because the inside surface, on which the photocathode or phosphor was deposited, could now be made with the proper curved shape to meet the requirements of electrostatic focusing (see Section 10.3.2). The outside surface, on the other hand, could be flat to allow the efficient optical coupling of several stages by placing the fiber-optic plates in contact, as shown in Figure 11.13.

Since the development of the SIT and ISIT would not have been possible without the availability of fiber-optic faceplates as optical couplers, it is worthwhile to say something about the origin and principle of operation of these image couplers. An individual optical fiber is a long thread of material (typically glass, although other materials have been used) with a circular cross section that transmits light along its length by repeated total internal reflection at the surface. To function as an effective optical waveguide, the core of the fiber must a have high transmission and a higher refractive index than the material that surrounds the fiber. Thus, according to Snell's law (Eq. 2.28),

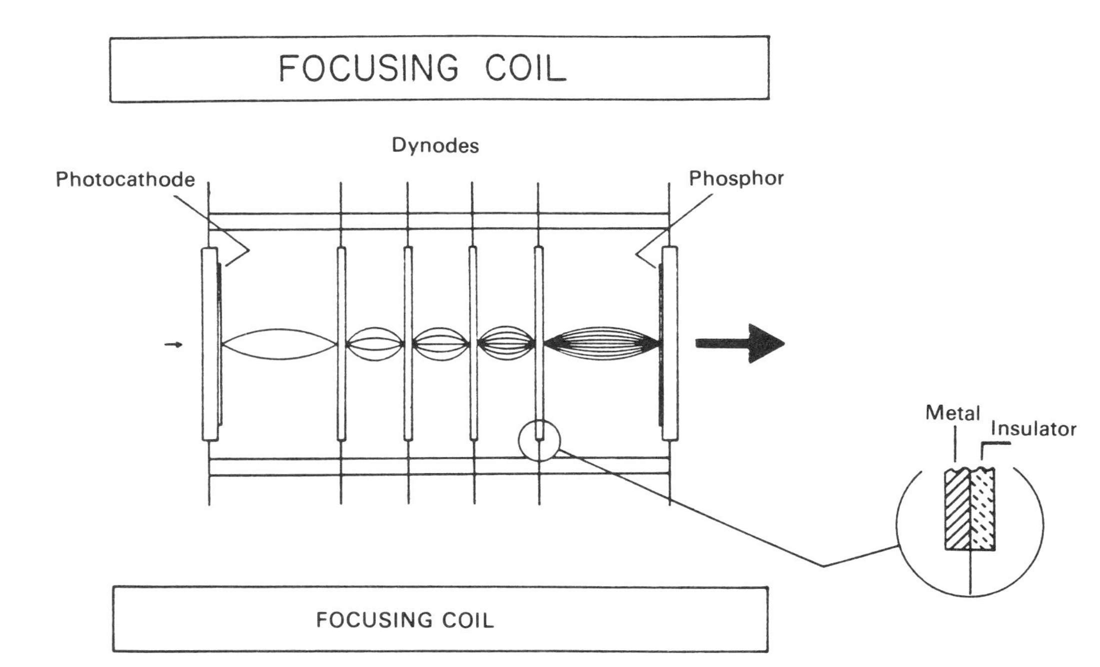

Figure 11.12. Schematic diagram of image converter with secondary-electron multiplication. [Reprinted from W. A. Hiltner (1962), "Image Converters for Astronomical Photography," in *Astronomical Techniques*, W. A. Hiltner (Ed.), University of Chicago Press, Chicago. Copyright 1962 University of Chicago Press.]

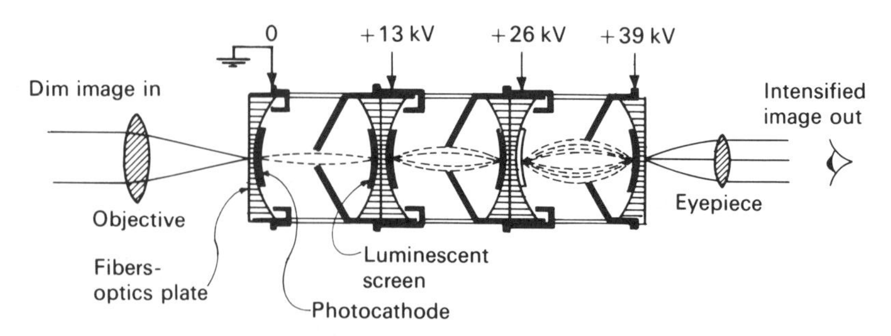

Figure 11.13. Schematic diagram of modular-type cascade image intensifier. [Reprinted from Morton, G. A., and Schnitzler, A. D. (1974), "Cascade Image Intensifiers" in *Photoelectronic Imaging Devices*, Vol. 2, Biberman, L. M. and Nudelman, S. (Eds.), Plenum, New York.]

the critical angle for total internal reflection will be given by

$$i_c = \sin^{-1}(n/n') \quad (11.8)$$

where i_c is the critical angle and n and n' are the refractive indices of the material surrounding the fiber (i.e., the *cladding*) and the fiber itself, respectively. Thus, as long as light within the fiber strikes the walls with an angle greater than i_c, it will be reflected back into the fiber without any loss.

To achieve the essential refractive index requirement for total internal reflection, a number of approaches have been taken (Levi, 1980). The simplest, most straightforward method of optical fiber manufacture is based on coating the fiber with a well-defined sheath of lower refractive index material. This results in a *stepped-index fiber.* A *graded-index fiber* may be prepared by diffusing suitable materials into the glass fiber as it is pulled at elevated temperatures so that a gradual reduction in refractive index occurs with distance from the fiber axis. Finally, *doubly clad fibers* are made by coating the glass fiber with a lower refractive index cladding that is then itself surrounded by another cladding having a refractive index intermediate between the glass core and the initial cladding layer.

To be useful for optical coupling with image detectors, individual optical fibers are combined to form *fiber bundles* that can be either *incoherent* or *coherent*. When the information transmitted does not depend on the relative positions of the various fibers comprising the bundle, the bundle is referred to as an incoherent bundle, or simply a *light guide*, and is not capable of transmitting spatial information. When spatial information is important, as in image detector applications, a coherent bundle must be employed. In the coherent bundle, the relative positions of the various fibers comprising the bundle are controlled. Short, rigid coherent bundles are referred to as *image conduits*. If the bundle length is short in comparison to its diameter, the image conduit is called a *fused-fiber plate.*

When used in image intensifiers, fused-fiber plates can still result in a certain amount of light spreading into adjacent fibers. This is due to the refractive index of the phosphor being much greater than the refractive indices of available optical glasses. As a result, a certain fraction of the light emitted by the phosphor strikes the fiber at an angle less than the critical angle and refracts out of the fiber into adjacent fibers. Use of doubly clad fibers coated with a second sheath of absorbing material reduces the amount of light spreading into adjacent fibers and improves the resolution of the image detector.

11.4.5. The Intensified SIT Tube

Still greater low-light performance can be obtained (albeit at the expense of decreased resolution) by adding an additional image intensifier stage to the SIT to produce an intensified SIT or ISIT. Figure 11.14 shows a schematic diagram of an ISIT. Because electrostatic focusing is employed to form the electron image on the subsequent stage, both the SIT and the ISIT employ curved (i.e., spherical) photocathodes to achieve the necessary geometry of an electrostatic lens. To achieve the required curvature of the photocathode while at the same time providing a flat faceplate on which to focus an optical image, a curved fiber-optic coupler between the faceplate and the photocathode emissive layer is employed. Since most fiber-optic components available today are made of glass, the short-wavelength spectral response of the SIT and ISIT is limited by the transmission properties of glass to wavelengths greater than about 350 nm. The short-wavelength response of the SIT and ISIT can be extended by spraying the faceplate of the tube with a suitable fluorescent material that can absorb short-wavelength radiation and fluoresce at a wavelength greater than 350 nm. The long-wavelength response, on the other hand, is limited by the long-wavelength cutoff of the photoemissive layer. Figure 11.15 shows the spectral response of the silicon vidicon and the SIT vidicon.

11.4.6. Performance Characteristics of the SIT and ISIT Detectors

For low-light-level applications such as encountered in emission spectroscopy, the SIT and ISIT tubes are clearly superior to the ordinary silicon vidicon pickup tube in terms of sensitivity. At very low light levels, where the photocurrents are at their lowest, tubes with the highest internal gain have the advantage because preamplifier noise becomes negligible. In fact, the gain obtained with the ISIT tube is so large that the tube performance is photoelectron noise limited (Rosell, 1974), and the overall gain of the device is more than sufficient to make preamplifier and other system noises negligible.

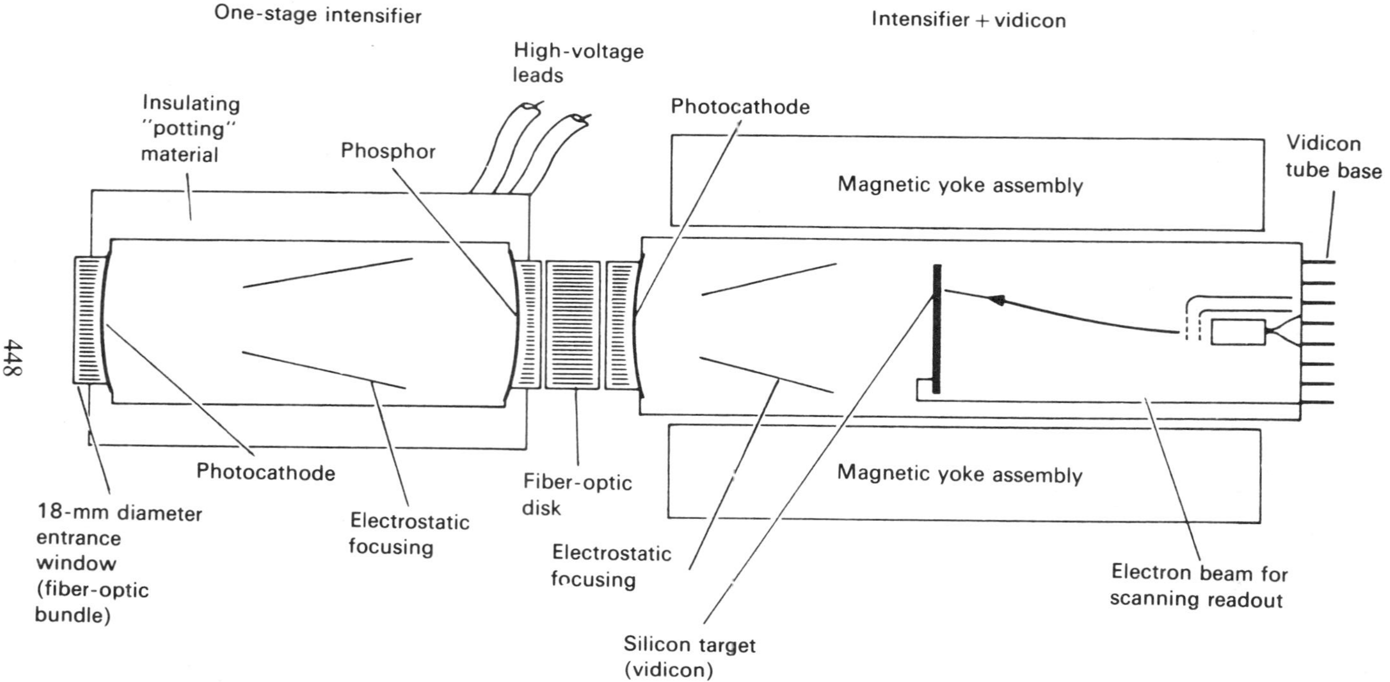

Figure 11.14. Intensified silicon-intensified target vidicon. [Reprinted from Horowitz, P., and Hill, W. (1980), *The Art of Electronics*, Cambridge University Press, Cambridge.]

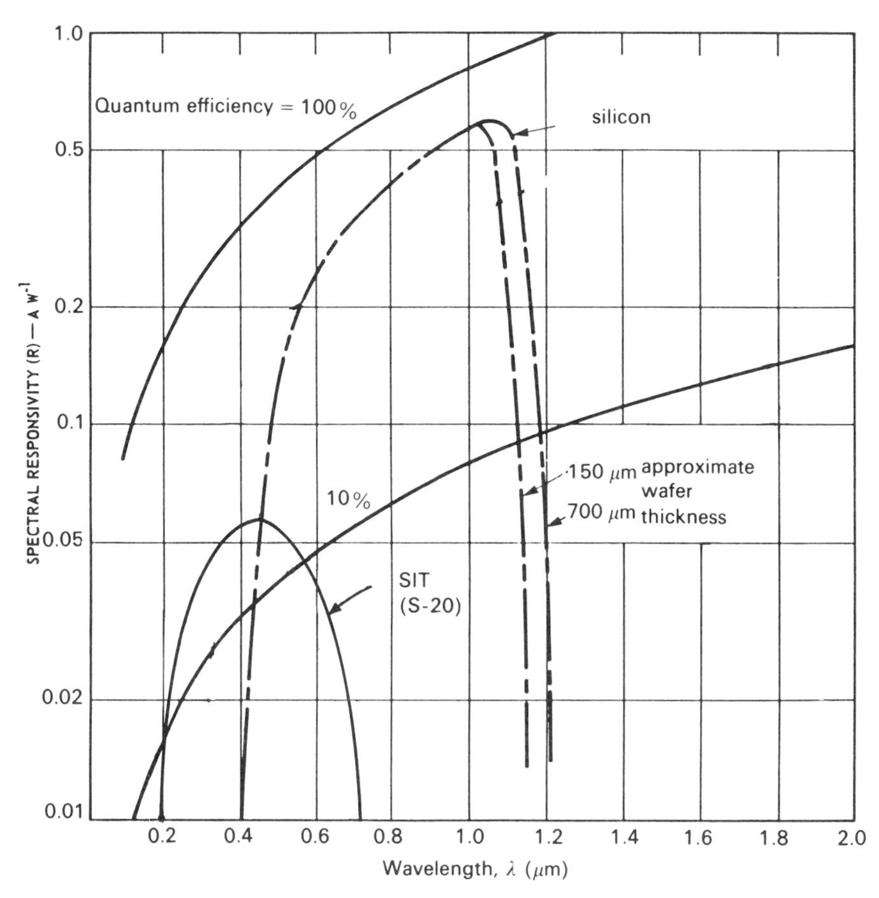

Figure 11.15. Spectral response of silicon vidicon and SIT vidicon. (Reprinted from *RCA Electro-Optics Handbook*, with permission from RCA, Solid State Division, P.O. Box 3200, Somerville, NJ. Copyright 1974 RCA Corporation.)

By contrast, the ordinary silicon vidicon pickup tube has no internal means of amplification, and as a result, preamplifier noise is the limiting noise. Figure 11.16 shows a plot of signal current versus photocathode irradiance in watts per square meter for tubes with various photocathode diameters. These irradiance levels can be converted into watts per square centimeter by multiplying by 10^{-4}. Thus, an irradiance of 10^{-6} W/m^2 is equivalent to 10^{-10} W/cm^2. Comparing the irradiance levels shown in Figure 11.16 with the 10^{-5}-W/cm^2 figure we have been using as a typical spectroscopic irradiance level, it is clear that both the SIT and ISIT tubes possess the requisite sensitivity for spectroscopic application.

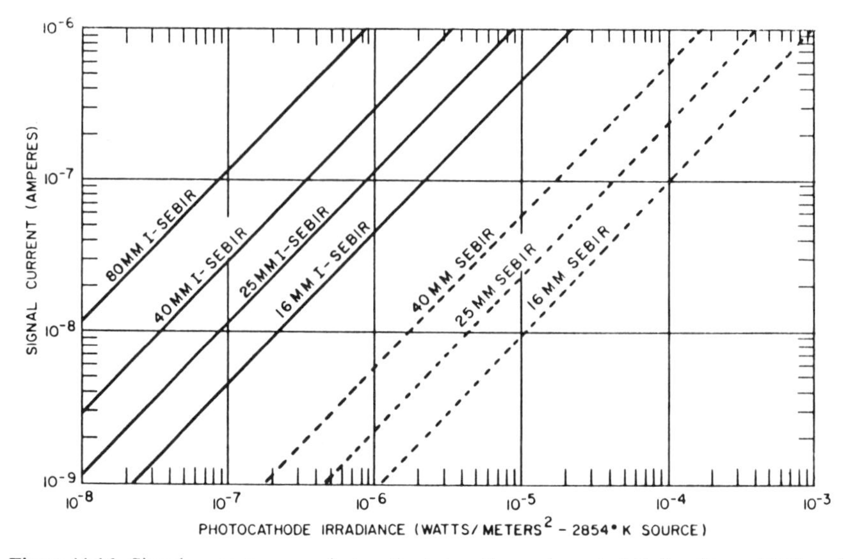

Figure 11.16. Signal current versus photocathode irradiance characteristic for silicon–EBIR and intensifier silicon–EBIR cameras for various input photocathode diameters. [Reprinted from Rosell, F. A. (1974), "Television Camera Tube Performance Data and Calculations" in *Photo-electronic Imaging Devices*, Vol. 2, Biberman, L. M. and Nudelman, S. (Eds.), Plenum, New York.]

Figure 11.17 shows the square-wave response (contrast transfer function) and sine wave response (modulation transfer function) as a function of the spatial frequency in line pairs per millimeter obtained with various intensified vidicons. From the figure, it can be seen, not unexpectedly, that the resolution of the ISIT tube is less than the resolution of the SIT tube because of the presence of the additional intensifier stage. A 25-mm diameter SIT tube has a limiting resolution (11% MTF) of approximately 16 line pairs per millimeter compared with about 13 line pairs per millimeter for the same diameter ISIT tube. Because of the use of electrostatic focusing, both tubes suffer from pincushion distortion (see Figure 2.28), and as a result, the resolution is degraded toward the edges of the target compared with the center.

In terms of performance, both tubes are relatively rugged but are affected by the presence of magnetic fields and vibrations. They can tolerate brief exposure to moderate light levels, but the photocathode must be protected against damage.

Because of the use of scanning electron beams for target readout, both tubes can employ a random-access mode of interrogation rather than conventional raster scanning. By computer control of the electron beam, any portion of the target can be erased while skipping other regions. This allows signal charge to build up for different periods of time on the target before being erased by the electron beam. Random-access scanning can be used in spectroscopy as a

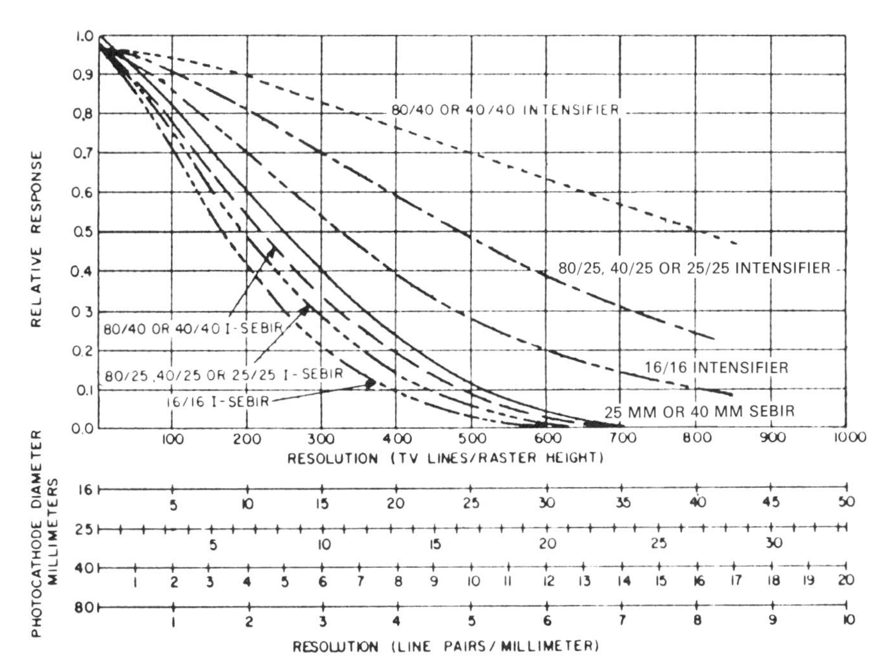

Figure 11.17. Uncompensated horizontal square or sine wave response for silicon–EBIR and intensifier silicon–EBIR cameras. (Intensifier responses are sine wave; I-SEBIR and SEBIR responses are square wave.) [Reprinted from Rosell, F. A. (1974), "Television Camera Tube Performance Data and Calculations" in *Photoelectronic Imaging Devices*, Vol. 2, Biberman, L. M. and Nudelman, S. (Eds.), Plenum, New York.]

means of extending the apparent dynamic range of the detector electronically. By using random-access scanning, very intense spectral lines can be erased more frequently than weak lines. This allows the signal charge associated with the weak lines to build up on the target while avoiding saturation and blooming, which would invariably occur if the intense lines were allowed to integrate for the same time period. In essence, random-access scanning is a means of controlling and varying the integration time for different regions of the target.

Another feature that is often useful for spectroscopic work is the ability to turn the detector on and off at high speed in synchronization with a pulsing light source, for example. The use of intermittent detection, known as *gating*, ensures that the detector is on only when light information of interest is present. By using gating, noise from other sources can often be avoided because the detection system is turned off except when information of significance is expected. Since the accelerating voltage used in the image section of the SIT tube and ISIT tube can be turned on and off rapidly, it can be used like the grid in a triode vacuum tube is used to control the plate current. By turning the 10-keV accelerating voltage off, photoelectrons from the photocathode are not accelerated toward the target, and most fail to reach it. Those that do reach the target do not have enough energy to create many electron–hole pairs. The effectiveness of gating is expressed by the *on–off gating ratio*, which is the ratio of the signal obtained when the tube is gated on to that obtained when the tube is gated off. Typical values for the gating ratio of the SIT tube are 10^4–10^6.

For a complete discussion of the performance characteristics of SIT and ISIT tubes, the reader should consult the chapter on this topic by Talmi and Busch (1983).

11.5. PHOTODIODE ARRAY DETECTORS

The advances in IC technology that occurred in the 1960s and were ultimately responsible for the development of the technology needed to produce the target of the silicon vidicon also offered the possibility of producing a *self-scanned mosaic sensor* that would not require an electron beam (Weimer et al., 1967). Elimination of the electron beam, although a formidable goal because of the large number of pixels making up a normal television scene (i.e., on the order of 250,000), would remove the need for a vacuum envelope as well as the heavy solenoid used for magnetic focusing and deflection. In addition, it would reduce the normally relatively high power consumption associated with producing the uniform axial magnetic field used for focusing in vacuum

Table 11.1. Comparison of the Vidicon and Solid-State Array Detector

Vidicon	Solid-State Array Detector
Fragile	Rugged
Medium power dissipation (1 W)	Low power dissipation (< 1 W)
High-voltage (about 500 V)	Low voltage (about 15 V)
Some image lag	No image lag
Low dynamic range (250:1)	High dynamic range (about 1000:1)
Some geometric distortion	No geometric distortion
Can be damaged by image burning	No damage by image burning
Good antiblooming characteristics	Can be made to have good antiblooming characteristics
Affected by magnetic fields	Not affected by magnetic fields
Continuous analog signal output	Sampled analog signal output

Source: Reprinted from Prettyjohns (1987).

image detectors (See Section 10.3.3). The potential advantages of a self-scanned solid-state array detector compared with a vidicon detector employing electron beam scanning would be reduced power consumption, compactness, reduced weight, and increased ruggedness (because no vacuum envelope was needed). Table 11.1 gives a comparison of a vidicon with a typical solid-state imaging array. In addition, use of digital scanning should provide a geometric accuracy of interrogation and flexibility of addressing not possible with electron beams (Weimer et al., 1974). With these goals in mind, solid-state engineers began the search for an array of photosensitive elements that could be addressed by on-chip scanning circuitry and connected to on-chip video-coupling circuits. By 1968, a variety of *monolithic* (i.e., on a single silicon chip) solid-state sensors had been developed, and an entire issue of the *IEEE Transactions on Electron Devices* (1968) was devoted to papers on self-scanned array detectors.

11.5.1. Morphology of the Photodiode Array

One of the results of this research was the development of the modern *photodiode array detector*, available today in standard IC packages (such as

Figure 11.18. Photograph of three modern photodiode array detectors. (Courtesy of Hammamatsu Corporation.)

the 22-lead or 28-lead, dual in-line pin, or DIP package), as shown in Figures 11.18 and 11.19, having either glass, quartz, or fiber-optic windows. In addition to the actual array of photodiodes (which may be obtained in a variety of formats including one- or two-dimensional layouts), the sensor chip also contains a digital shift register and an array of switching transistors (sometimes referred to as multiplex switches) so that there is one *pnp* switch for each photodiode, as shown in Figure 11.20.

The photodiode array is prepared by oxidizing the surface of the *n*-type substrate to produce a silicon oxide layer having a thickness of 0.4 μm. Bar-shaped windows are then opened in the oxide layer by standard photolithographic procedures, and the *pn* junctions are formed by the diffusion of an appropriate acceptor impurity (i.e., *p*-type impurity). Figure 11.21 shows a

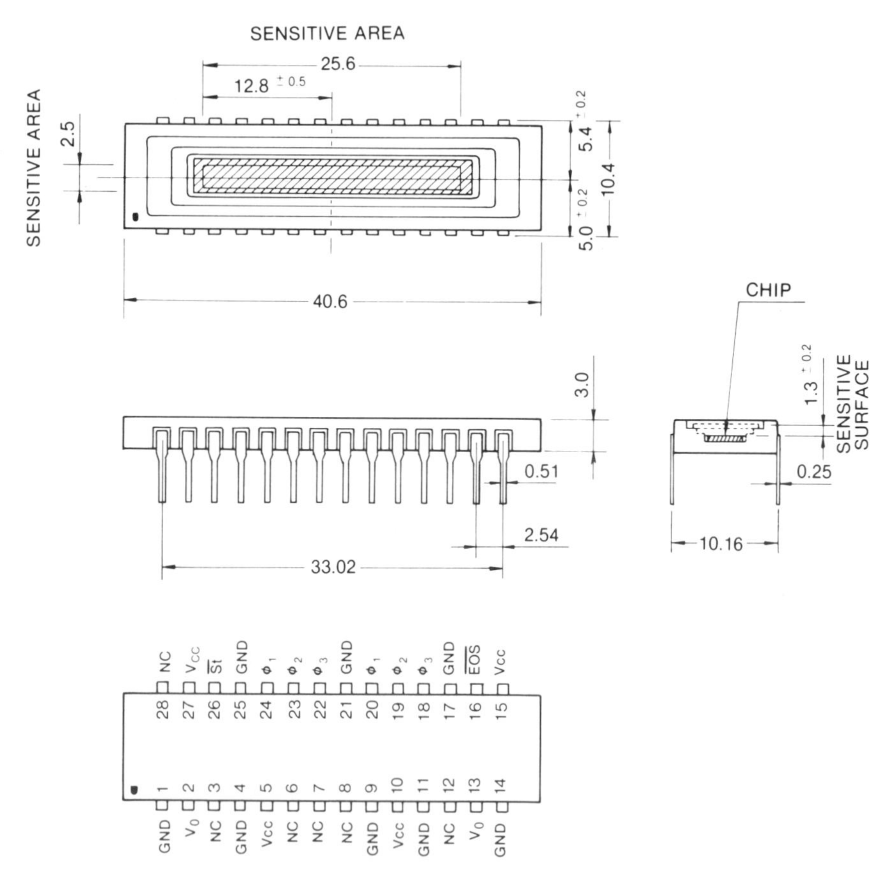

Figure 11.19. Schematic of photodiode array detector showing pin connections for dual-in-line pin package. (Courtesy of Hammamatsu Corporation.)

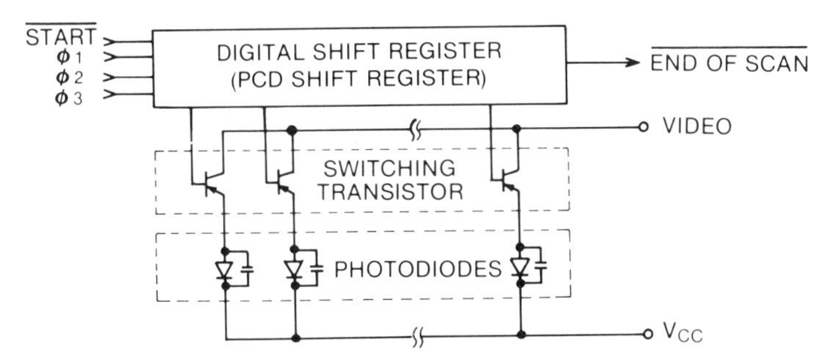

Figure 11.20. Schematic of photodiode array detector showing on-chip digital shift register and array of multiplex switching transistors. (Courtesy of Hammamatsu Corporation.)

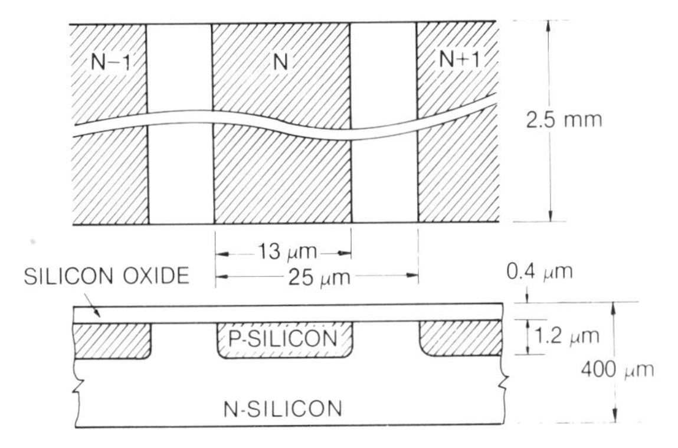

Figure 11.21. Schematic of silicon chip used in photodiode array detector showing sensor geometry. (Courtesy of Hammamatsu Corporation.)

cross section of a silicon chip in which the individual *p*-type islands have been formed as described previously. Figure 11.21 also shows the typical sensor geometry. The so-called *S-series* linear array, available from EG & G Reticon (see EG & G Reticon *Image Sensing Products*) and Hamamatsu in arrays with up to 1024 diodes, has been designed specifically for spectroscopic applications and, as a result, has a slit–like geometry with a diode height of 2.5 mm and a center-to-center diode spacing of 25 μm. Other arrays with up to 4096 diodes are available commercially today but with a reduced aspect ratio of 40:1.

11.5.2. Principle of Operation

In operation, each photodiode is reverse biased by V_{cc} (typically +5 V) as shown in Figure 11.20. This results in the formation of a depletion zone at the *pn* junction, causing each diode to behave as a tiny storage capacitor in the same manner as discussed in conjunction with the silicon vidicon. It should be noted that the "capacitors" shown in the equivalent circuit of Figure 11.20 represent the equivalent diode junction capacitance and are not actual additional components in the circuit. Photons absorbed by the *p*-type islands produce holes and electrons in the usual manner if the energy of the photon is greater than the bandgap. These charge carriers are attracted across the *pn* junction by the presence of the electric field across the diode so that the electrons are attracted to the positive terminal of the reverse-biased diode (i.e., the *n*-type substrate) and the holes are attracted to the negative terminal of the reverse-biased diode (i.e., the *p*-type islands). As a result of the diffusion of photon and thermally generated charge carriers, the reverse-bias potential across the diodes decreases with time. Photons can also be absorbed by the *n*-type substrate, producing charge carriers that also tend to discharge the tiny capacitors. Charge generated by photon absorption in the substrate between two *p*-type islands will be divided between the adjacent diodes to give a response function similar to that shown in Figure 11.22.

In the real-time mode of operation, a negative start-scan pulse is applied to the digital shift register on a periodic basis determined by the desired integration time, and this pulse is shifted to the successive flip-flops in the shift register after each new clock pulse. As this negative pulse shifts from one flip-flop to the next in the shift register, it is applied to the base of successive switching transistors, as shown in Figure 11.20. The application of a negative pulse to the base of the *pnp* transistor addressed by the digitial shift register forward biases the particular switching transistor, allowing a charging current to flow from V_{cc} until the photodiode is once again fully reverse biased. This charging current flows through the photodiode and switching transistor and appears on the video output line. In the Hamamatsu version of the linear diode array, a three-phase clock is used to drive the negative start-scan pulse through the digital shift register. The video signal from successive photodiodes is fed to an operational amplifier integrator with an analog reset, as

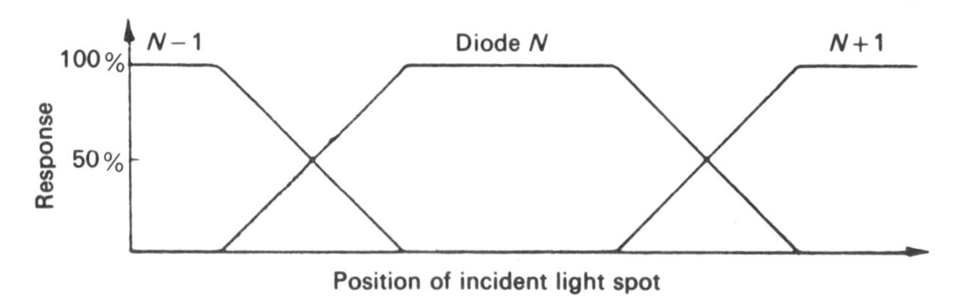

Figure 11.22. Response function between adjacent diodes. [Reprinted with permission of EG & G Reticon.]

shown in Figure 11.23. Figure 11.24 shows the timing diagram for the start-scan pulse, the three-phase clock, and the analog reset. Because of the presence of the integrator and the reset pulses at the end of each sampling period, the output waveform has a boxcar shape. When all of the diodes in the array have been read out serially, another start-scan pulse can be applied to the shift register to start the process over again. The signal from each diode is the charge integral from photon absorption and thermal generation since the last time the diode was interrogated by the shift register.

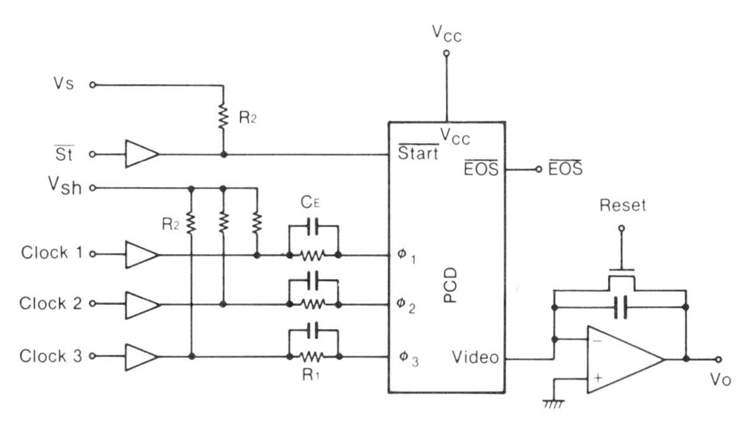

Figure 11.23. Clocking and signal processing used in Hammamatsu version of linear photodiode array. (Courtesy of Hammamatsu Corporation.)

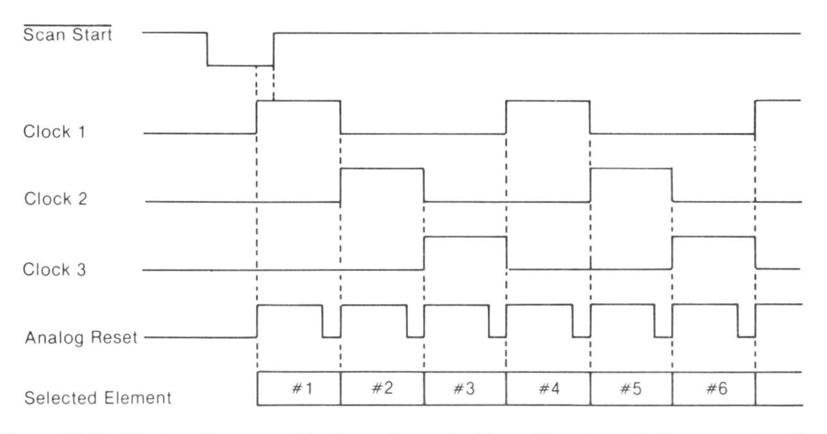

Figure 11.24. Timing diagram with three-phase clocking. (Courtesy of Hammamatsu Corporation.)

11.5.3. Performance Characteristics

One of the basic problems associated with self-scanned mosaic arrays is the appearance of high-frequency switching transients on the video line due to feed-through of clock pulses by the multiplex switches. Since these switching transients are coupled to the video line capacitively, a great deal of effort has been devoted to reducing the internal capacitance of the solid-state components associated with the shift register, the multiplex switches, and the video amplifier.

Photodiode arrays are also subject to a form of reset noise that arises as a result of having to charge the equivalent photodiode junction capacitance to a certain level through a switch. Whenever a capacitor is charged through a resistance, the voltage across the capacitor will be subject to the Johnson noise voltage fluctuations that appear across the resistor. These voltage fluctuations are given by Eq. 9.8:

$$\langle V_{\mathrm{n}} \rangle^2 = 4kTR\,\Delta f = \frac{1}{2\pi} 4kTR \int_0^{\infty} [1 + (\omega RC)^2]^{-1/2}\, d\omega \tag{11.9}$$

where the effective bandwidth Δf is given by

$$\Delta f = \frac{1}{2\pi} \int_0^{\infty} [1 + (\omega RC)^2]^{-1/2}\, d\omega = \frac{1}{2RC\pi} \tag{11.10}$$

where ω is the angular frequency, R is the resistance, C is the capacitance, and the other parameters in Eq. 11.9 have been defined previously. Substituting the effective bandwidth from Eq. 11.10 into Eq. 11.9 gives the magnitude of the noise voltage across the capacitor as

$$\langle V_{\mathrm{n}} \rangle^2 = 2kT/3C \tag{11.11}$$

which interestingly enough is independent of the magnitude of the resistance. If this noise is expressed in terms of the number of electrons rather than a voltage, the fluctuation in the number of electrons stored by the capacitor will be

$$\Delta n = C\langle V_{\mathrm{n}} \rangle / q = (1/q)(\tfrac{2}{3}kTC)^{1/2} \tag{11.12}$$

which equals

$$\Delta n = 328[C(\mathrm{pF})]^{1/2} \tag{11.13}$$

at room temperature. Because of the form of Eq. 11.12, this form of noise is known as *kTC noise.*

EG & G Reticon has reduced the effect of the switching transients as well as the effect of dark current by diode leakage by using two identical linear arrays of photodiodes. One array is exposed to light from the sensor window while the other, known as the *dummy array*, is maintained in the dark. The shift register is connected to the corresponding diode of each array by MOSFET multiplex switches. Thus, when the *n*th diode of the active array is interrogated by the shift register, the *n*th diode of the dummy array is also switched on at the same time. By reading the active video line and the dummy video line differentially, both the switching transients and the diode dark current are eliminated by subtraction.

Although digital signal subtraction can remove the fixed-pattern contribution of the dark current, it cannot remove the shot noise superimposed on the dark current. This can be reduced, however, by cooling the array to $-25°C$ or lower with a Peltier effect cooler or other cryogenic source. Figure 11.25 shows the effect of temperature on the dark current for a typical photodiode array. Since the shot noise associated with the dark current is proportional to

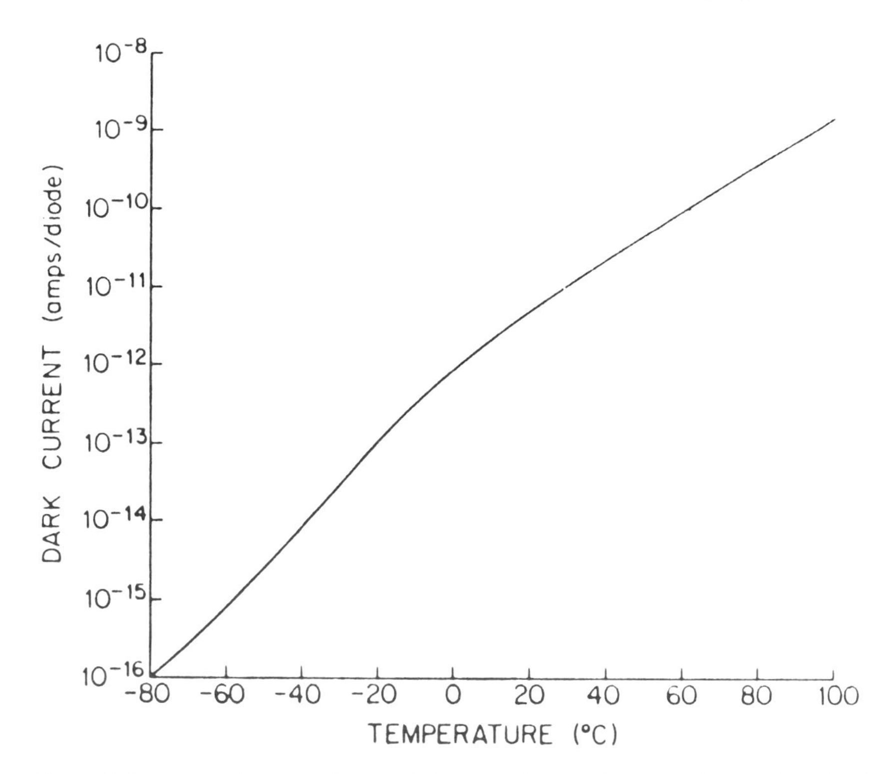

Figure 11.25. Temperature dependence of dark current for typical photodiode array. [Reprinted with permission of EG & G Reticon.]

the square root of the dark current (see Eq. 9.7), cooling is an effective means of reducing this source of noise. For a more comprehensive discussion of dark-current noise as well as other noise sources in photodiode arrays, the reader is referred to the papers by Simpson (1979) and Talmi and Simpson (1980).

Figure 11.26 shows the output charge produced as a function of exposure H (see Eq. 8.2) obtained by irradiating the sensor with 750 nm radiation. The exposure in nanojoules per square centimeter is obtained by multiplying the irradiance in microwatts per square centimeter by the integration time in milliseconds. An exposure of 25 nJ/cm^2 will be accumulated in 25 msec with an irradiance of 1×10^{-6} W/cm^2. As discussed earlier (see Section 10.5), the irradiance produced in the focal plane of a 0.5-m polychromator when 10 ppm sodium is introduced into a flame is on the order of 10^{-5} W/cm^2. This means that the photodiode array can detect the low light levels typically encountered in atomic spectroscopy; however, some form of signal enhancement may be necessary to detect signals from ultratrace samples.

On-chip integration can be used to improve the signal-to-noise ratio in a given application if the source can be observed for a sufficiently long time period. For very low light levels, where the system is likely to be readout noise limited, on-chip integration is preferable to signal averaging a number of framescans because, in the case of detector integration, the signal-to-noise

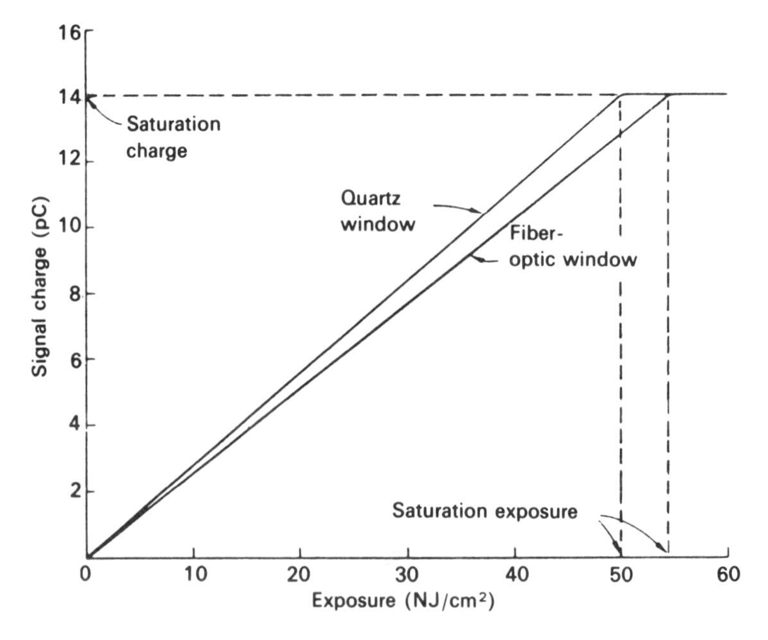

Figure 11.26. Signal charge versus exposure at 750 nm wavelength. [Reprinted with permission of EG & G Reticon.]

ratio is directly proportional to integration time, whereas with ensemble averaging the signal-to-noise ratio improves only as the square root of the observation time. In the case of ensemble averaging, the observation time is taken as the number of framescans accumulated in memory times the framescan period. At higher light levels, where the system is likely to be photon noise limited, both forms of signal accumulation give equivalent results. For a photodiode array, the cell-to-cell sampling rate is determined by the clock frequency (typically several megahertz), while the integration time is the interval between start pulses.

The duration of signal integration on the detector is limited by several factors including the maximum charge that can be stored on the diode junction capacitance (typically 14 pC, which corresponds to an exposure of 50 nJ/cm^2), the magnitude of the dark current, and the speed and size of the analog-to-digital (A/D) converter used to convert the analog video signal into digital form. Since the thermally generated dark charge competes with the photon-generated charge, the presence of excessive dark current can severely limit the duration of signal integration. In the absence of dark current, the maximum signal that can be accumulated on the sensor is frequently limited by the speed of the A/D converter, which must be able to transform the analog video signal into digital form in the time span allowed by the clock frequency. In the absence of speed limitations, the size (i.e., number of bits) of the A/D converter ultimately determines the largest number that can be produced. For a 12-bit A/D converter, the largest number that can be produced is 4095 $(10^{12} - 1)$.

One drawback in the performance of the linear photodiode array is the lack of true random-access capability. In comparison with electron beam detectors such as the SIT and ISIT, where the target can be interrogated in a *random-access* mode (i.e., nonsequential), the diodes in a linear photodiode array must be scanned sequentially because of the use of the digital shift register as the scanning circuitry. True random-access scanning is possible when selective portions of the target can be interrogated nonsequentially as desired without interrogating other portions of the target that are not of interest.

A form of selective scanning, known as *pseudo-random-access readout*, can be employed with the linear photodiode array by increasing the clock frequency to 2 MHz for those portions of the array that do not contain relevant information. When portions of the spectrum that are of interest are reached, the clock speed can be returned to the normal readout frequency (60 kHz).

11.5.4. Intensified Photodiode Arrays

As with conventional television pickup tubes, the low-light-level performance of photodiode arrays can be extended by use of various forms of image intensification. Thus, for true low-light-level applications, some form of image intensification is required to produce signal levels that exceed the readout and dark noise of the diode array. Any image intensifier, such as the three-stage cascaded unit shown in Figure 11.13 or the proximity-focused mcp image intensifier shown in Figure 10.23, is suitable for this purpose. The fiber-optic outputs of either of these intensifiers are easily coupled to any photodiode array that has a fiber-optic input window. Luminous gains from about 200 to 10^4 can be obtained by varying the high voltage applied to the image intensifier stage. Care must be taken in the design and use of the intensifier, however, to prevent the formation of a corona discharge within the device. This factor frequently limits the maximum high voltage that can be applied.

Intensified photodiode arrays are available commercially from a number of suppliers, including Tracor Northern (TN-1710-21 IDA), EG&G Reticon (RL-512SF), and Princeton Instruments (IRY-512). The Tracer Northern device uses a Reticon 512-element diode array at the output of an electrostatically focused image-inverting mcp intensifier with a maximum luminous gain of 35,000. The photocathode of the intensifier has an extended-red response (S-20ER) and is deposited on the output of a spherical fiber-optic faceplate similar to the one shown in Figure 11.11 for the SIT.

With proximity-focused mcp intensifiers, the resolution across the array is practically constant, producing a spectral line profile with a full-width at half maximum (FWHM) of three to four diodes. With electrostatically focused image intensifiers, however, the resolution is not uniform across the target and worsens toward the edges as a result of *pincushion distortion* (See Figure 2.28), characteristic of electrostatic image sections.

For a complete discussion and comparison of the performance characteristics of the SIT, the ISIT, the self-scanned photodiode array, and the intensified photodiode array, the interested reader should consult the chapter by Talmi and Busch (1983).

11.6. CHARGE TRANSFER DEVICES

At about the same time that self-scanned arrays using photodiodes were being developed, workers at Bell Laboratories were searching for an equivalent to the magnetic-bubble memory that could be fabricated with the already well-developed silicon technology. This work culminated in the invention of the *charge-coupled device* (CCD) in 1969 by Boyle and Smith (1970). In essence, a CCD is an array of closely spaced *metal–insulator–semiconductor* (*MIS*) diodes formed on a wafer of semiconductor material. Figure 11.27 shows a cross section of one MIS diode composing the array. If the insulating layer is an oxide, the diodes are known as *metal–oxide–semiconductor diodes*, or *MOS*

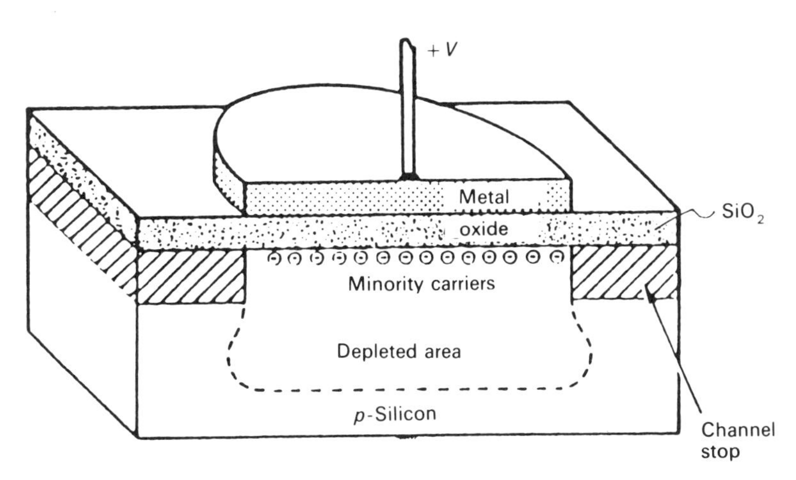

Figure 11.27. Cross section of one MIS diode in an array. [Reprinted from W. S. Chan (1988), *Infrared Detection, Focal Plane Technology and Systems*, Vol. 1, Electro-Optek Corporation, Torrance, CA. Copyright 1988 Electro-Optek Corporation.]

diodes, where the term *MOS* indicates the order in which the three materials making up the device are arranged with respect to one another. If a proper sequence of clock voltage pulses is applied to the array of gates formed by the metal contacts shown in Figure 11.27, the charge transfer device can move electrical charge across the array in a controlled manner. Soon after the conception of the CCD, it was realized that the principle of charge coupling could be applied in the development of imaging systems, and the first actual CCD produced was used as a simple line imager (Amelio et al., 1970).

Since that time, CCDs have had a major impact in the field of imaging because of their freedom from fixed-pattern noise (Barbe, 1980) and other advantages. Because of the numerous potential applications of CCD technology, interest in this area developed rapidly, and by 1976 most of the fundamental physics governing the behavior of CCD devices had been developed. In 1975, Sequin and Tompsett published a book entitled *Charge Transfer Devices*, which summarized most of the basic theory of CCD performance. Also in 1975, Barbe published a paper on the use of CCDs as imaging devices. In 1977, Barbe and Campana wrote a chapter entitled "Imaging Arrays Using the Charge-Coupled Concept," which appeared in the book *Advances in Image Pickup and Display*. In that same year, the IEEE press published a compilation of important papers on the charge-coupled concept (Melen and Buss, 1977). Since that time, numerous articles and monographs have appeared, including the works by Barbe (1980), Howes and Morgan (1979), Beyon and Lamb (1980), Hobson (1978), and Nicollian and Brews (1982).

11.6.1. Introduction to MOS Technology

A complete discussion of the physics of MOS diodes is beyond the scope of this book (see Sze, 1981, for more details). In essence, however, a CCD is simply an array of MOS diodes that function as an array of densely packed capacitors capable of storing charge and moving it in a controlled fashion across the array. Figure 11.27 shows a cross section of a so-called *n-channel device* revealing the sandwich structure consisting of a metal gate that has been deposited on a thin insulating layer of silicon dioxide (i.e. the oxide layer), which covers a single crystal of silicon substrate (i.e., the semiconductor). The silicon crystal making up the substrate has been lightly doped with acceptor impurities to form a *p*-type material. Since the metal gate electrode is insulated from the semiconductor by the presence of the oxide layer, the device behaves as a capacitor. A *p-channel device* is fabricated in the same fashion except that the semiconductor substrate is made from *n*-type material.

On a qualitative basis, if a negative potential is applied to the gate of an *n*-channel device, free holes (i.e., the majority carriers in *p*-type material) will be attracted to the silicon dioxide–semiconductor interface by the presence of the electric field. By the same token, if a positive potential is applied to the gate of an *n*-channel device, the holes will be repelled from the oxide–semiconductor interface, producing a depletion region in the vicinity of the gate. If the positive potential is increased further, a point is reached where the electric field becomes so strong that minority carriers (i.e., electrons in this case) are attracted to the oxide–semiconductor interface.

Figure 11.28 shows the energy band arrangement that exists at the surface of a *p*-type semiconductor when a positive voltage is applied to the metallic gate electrode. In the figure, E_V is the top of the valence band, E_C is the bottom of the conduction band, and E_F is the Fermi level (see Chapter 9). The intrinsic Fermi level E_i shows the location of the Fermi level in undoped (i.e., intrinsic) silicon. Because of the definition of the Fermi level (see Chapter 9), the intrinsic level is located $\frac{1}{2}E_g$ above the top of the valence band, where E_g is the bandgap. The potential ψ at any point in the diagram is measured with respect to the intrinsic Fermi level E_i. In the figure ψ_s is the surface potential and ψ_B is the potential difference between the Fermi level E_F and the intrinsic Fermi level E_i. The surface potential that exists at the oxide–semiconductor interface is a function of the applied gate voltage, the oxide layer thickness, and the doping concentration of the silicon substrate. For an ideal MOS diode, no current flows in the device when a potential is applied, so the Fermi level E_F remains constant in the semiconductor, as shown in Figure 11.28.

When a positive potential is applied to the gate of a *p*-type MOS diode, the bands in the semiconductor bend downward, and the surface potential is

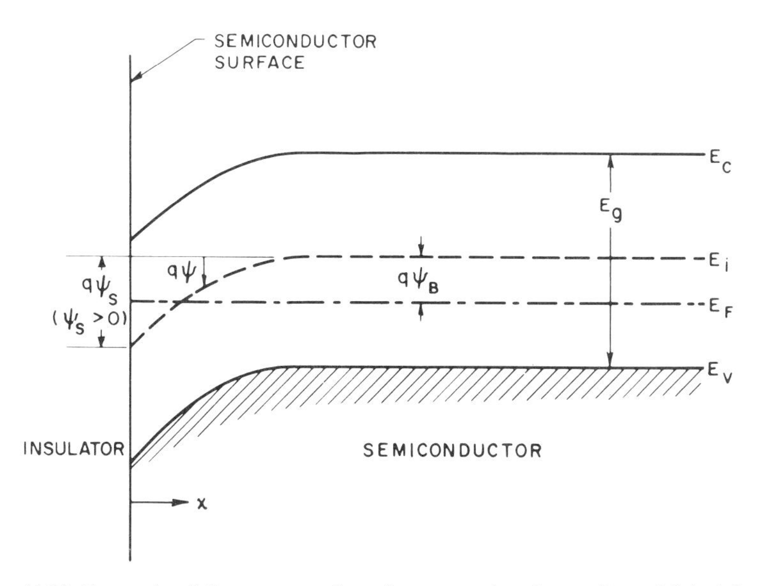

Figure 11.28. Energy band diagram at surface of *p*-type semiconductor. Potential ψ, defined as zero in bulk, is measured with respect to intrinsic Fermi level E_i. Surface potential ψ_s is positive as shown. (*a*) Accumulation occurs when $\psi_s < 0$. (*b*) Depletion occurs when $\psi_B > \psi_s > 0$. (*c*) Inversion occurs when $\psi_s > \psi_B$. [Reprinted from Sze, S. M. (1981), *Physics of Semiconductor Devices*, 2nd ed. By permission of John Wiley & Sons, New York.]

taken to be positive, as shown in Figure 11.28. Figure 11.29 shows the energy band diagrams for ideal *p*- and *n*-type MIS diodes for different applied voltages.

Since the surface electron concentration and surface hole concentration depend on the surface potential ψ_s that exists at the semiconductor–insulator interface, they are given by

$$n_s = n_{p0} \exp(\beta\psi_s) \qquad p_s = p_{p0} \exp(-\beta\psi_s) \tag{11.14}$$

where n_{p0} and p_{p0} are the equilibrium densities of electrons and holes, respectively, in the bulk semiconductor and $\beta = q/kT$.

At this point, five different conditions can be identified depending on the magnitude of ψ_s. For a *p*-type semiconductor MIS diode where a negative potential is applied to the gate, the bands bend upward at the semiconductor–insulator interface, and $\psi_s < 0$. Under these conditions, according to Eq. 11.14, the surface concentration of holes increases, and majority carriers (i.e., holes) accumulate at the surface under the gate. This state of the device is known as the *accumulation state*. When $\psi_s = 0$, no band bending occurs, and the device is said to be in the *flat-band* condition. If

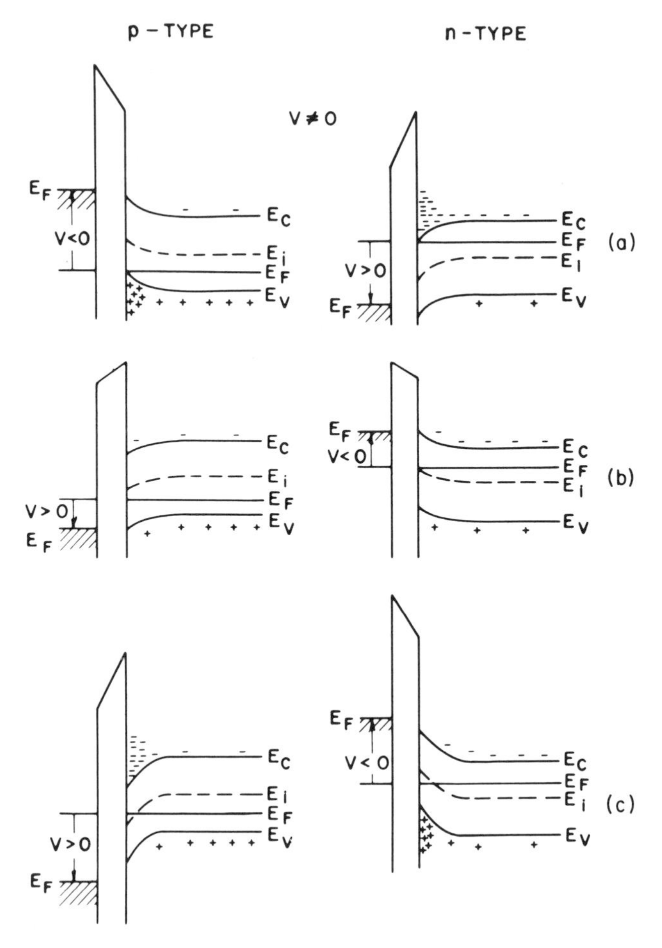

Figure 11.29. Energy band diagrams for ideal MIS diodes when $V \neq 0$ for (*a*) accumulation; (*b*) depletion; (*c*) inversion. [Reprinted from Sze, S. M. (1981), *Physics of Semiconductor Devices*, 2nd ed. By permission of John Wiley & Sons, New York.]

$\psi_B > \psi_s > 0$, the bands bend downward, and the surface concentration of majority carriers (i.e., holes) is reduced according to Eq. 11.14. This state is referred to as the *depletion state.* If $\psi_s = \psi_B$, the surface concentration of charge carriers (both electrons and holes) is equal to the intrinsic semiconductor concentration (see Chapter 9).

If a still larger positive voltage ($\psi_s > \psi_B$) is applied to the device, the bands continue to bend downward so that the intrinsic level at the surface, E_{is}, drops below the Fermi level. When the intrinsic level at the surface crosses over the Fermi level, the number of minority carriers (i.e., electrons in this case) at the surface becomes larger than the number of holes at the surface. Since the conditions at the surface have been inverted, this state is known as the *inversion state*. For an ideal MOS device, strong inversion begins when the surface potential is equal to approximately $2\psi_B$.

As long as a sufficiently positive potential is maintained on the gate, the device will remain in the inversion state, and electrons can be collected and stored at the oxide–semiconductor interface without danger of recombination with holes because the formation of the depletion zone in the p-type substrate has forced the holes away from the interface. The downward bending of the bands in the vicinity of the gate interface results in the formation of a *potential well* that can store charge. Electrons generated in the depletion zone, in an effort to attain the lowest energy, will collect in this well until it is full. As the electrons accumulate in the potential well, however, the bands tend to flatten out in the vicinity of the gate interface until the well is full and cannot accommodate any more electrons.

To avoid recombination of the minority carriers (i.e., the useful signal) stored in the potential wells, the depletion regions must be maintained so that majority carriers do not enter the potential well regions. The need to maintain the depletion region determines the amount of full-well charge that can be stored by the device. The full-well charge that can be stored by the device can be estimated from the condition that the depletion region be just on the point of disappearing. Under this condition (Hobson, 1987), the continuity of the electric displacement at the oxide–semiconductor interface requires that

$$-\rho_{\mathrm{nfull}} = -\varepsilon_{\mathrm{I}}\varepsilon_0 E_{\mathrm{Ifull}} + \rho_{\mathrm{I}} \tag{11.15}$$

where ρ_{nfull} is the maximum possible minority carrier charge per unit area, ρ_{I} is the fixed charge per unit area at the interface due to lattice imperfections, ε_{I} is the relative permittivity of the oxide insulation, ε_0 is the permittivity of free space, and E_{Ifull} is the electric field within the oxide insulation under full-well conditions.

The maximum possible minority carrier charge that can be stored under a given gate will be determined by the maximum bias voltage that can be applied to maintain the depletion zone. The maximum bias voltage that can be applied, however, is determined by the maximum electric field (see Eq. 11.15) the insulator can tolerate before dielectric breakdown occurs.

If a constant bias voltage is maintained on the gate electrode in the *absence of any stored charge*, the electric field will be redistributed so that the maximum field will occur at the semiconductor interface. To avoid avalanche breakdown in the semiconductor itself, doping levels within the semiconductor must be kept as low as possible. Even with this precaution, however, the lower breakdown voltage in the semiconductor means that the maximum electric field that can be tolerated across the insulator (E_I) in the absence of any stored charge is much less than can be tolerated under full-well conditions (Eq. 11.15). The large variation of E_I required from empty to full-well conditions can be satisfied only if the majority of the bias voltage exists across the semiconductor when the well is empty. To meet this requirement, the oxide insulator layer must be made as thin as possible without the onset of pinhole leakage. Oxide layer thicknesses of 0.1 μm are typical of current MOS technology.

A device that meets the maximum allowable doping level and minimum allowable oxide thickness will have a depletion layer width of 5.8 μm under a bias voltage of 17 V (Hobson, 1987). Such a device will be able to store a saturation charge of 3.8 pC (24×10^6 electrons) under an electrode area of 10×100 μm.

To illustrate the charge transfer process with CCDs, the potential wells under each gate are often represented as shown in Figure 11.30. With regard to the potential well concept, it should be stressed that the bottom of the well

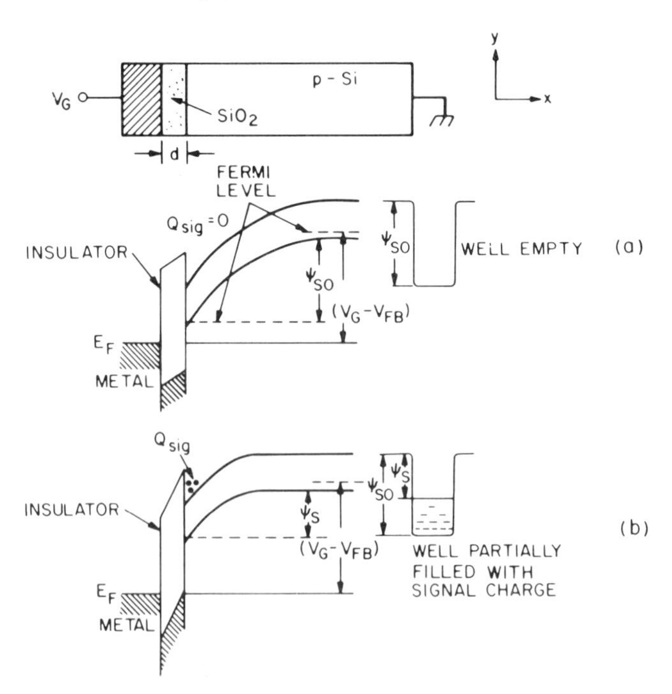

Figure 11.30. Energy band diagrams for surface channel MOS diode: V_G, gate voltage; V_{FB}, flat-band voltage. (*a*) Band bending at deep depletion and empty potential well. (*b*) Band bending at Si–SiO_2 interface and partially filled potential well representation. [Reprinted from Sze, S. M. (1981), *Physics of Semiconductor Devices*, 2nd ed. By permission of John Wiley & Sons, New York.]

is the region immediately adjacent to the interface, and the well fills from this point outward into the substrate. When potential wells are represented pictorially as in Figure 11.30, the well is drawn as a depression under the gate and is shown as filling from the bottom up in analogy with water filling a well. As long as the true charge storage mechanism is understood, there is no harm in the way potential wells are drawn, and this method of representing the electron storage mechanism of CCDs is particularly convenient in showing how charge transfer from one gate to another is accomplished.

The electrons that accumulate under the gates when the capacitors are maintained in strong inversion arise from thermally generated electron–hole pairs as well as electron–hole pairs created as a result of photon absorption in the depletion zone. In both cases, the electrons generated are attracted toward the potential well at the interface by the presence of the electric field while the holes are repelled and flow out into the substrate. In this way, an MOS device can be used to store a charge pattern of minority carriers produced by imaging a light pattern on the device.

11.6.2. Implementation of Charge Transfer

As indicated by the previous discussion, charge-transfer devices (CTDs) store charge at successive minima of potential energy, known as *potential wells*, which are created in a semiconductor material as a result of the presence of an electric field applied under an electrode. By appropriately manipulating the potential energy minima (i.e., the gate electrode voltages), isolated charge packets can be transferred from one adjacent potential minimum to another. The analog signal transmitted across the array is carried by *minority carriers* that are stored in potential minima. Currently, many schemes have been developed to implement charge transfer.

All forms of charge transfer within a CTD are based on two principles: (1) charge will collect in the deepest potential well and (2) potential barriers can be formed to prevent charge packets from mixing. In addition to charge packet isolation, *unidirectional* charge transfer is also necessary.

Perhaps the most common and simplest method of charge transfer involves a three-phase arrangement where every third electrode in the array is connected to a common potential source, as shown in Figure 11.31. With this arrangement, bias voltages ϕ_1, ϕ_2, and ϕ_3 can be applied by means of a three-phase clock so that they are staggered in time, as shown in Figure 11.31. Notice that for efficient unidirectional charge transfer, clock waveforms must overlap in time. This is accomplished by having the falling edges of the clock pulses decrease slowly to zero *after* the rising edge of the next phase has occurred. Slowly collapsing the previous potential well encourages charge to flow into the adjacent well. With three-phase clocking, at least one of the three

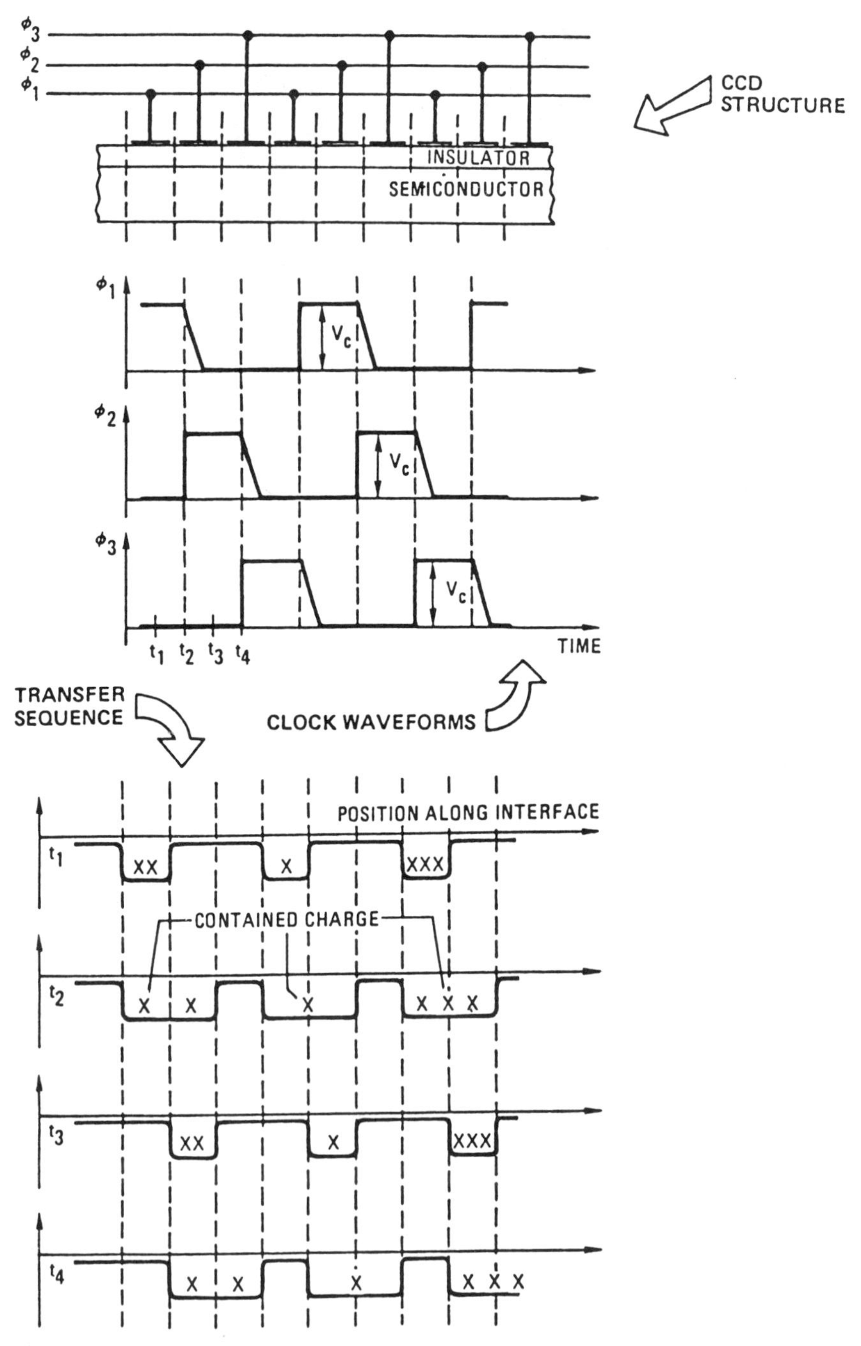

Figure 11.31. Three-phase charge transfer process in typical CCD. [Reprinted from W. S. Chan (1988), *Infrared Detection, Focal Plane Technology and Systems*, Vol.1, Electro-Optek Corporation, Torrance, CA. Copyright 1988 Electro-Optek Corporation.]

potential well regions must be collapsed at any time to provide a barrier to separate charge packets from adjacent detector elements.

For purposes of discussion, let us assume that the clock voltages are connected to an *n*-channel CCD. This means that whenever positive potentials are applied to the gate electrodes, minority carriers (i.e., electrons) will be stored in potential wells whose depth depends on the magnitude of the positive potential applied. As shown in Figure 11.31, at time t_1, ϕ_1 is applied to every third electrode while the other electrodes connected to ϕ_2 and ϕ_3 are held at zero bias. As a result, charge is stored under electrodes connected to ϕ_1. At time t_2, electrodes connected to ϕ_2 are biased positively, causing the charge that was formerly localized under the ϕ_1 electrodes to be distributed within the single potential well created under the ϕ_1 and ϕ_2 electrodes. Soon after biasing the ϕ_2 electrodes, the bias under the ϕ_1 electrodes is removed slowly so that the charge spills into the potential wells established under the ϕ_2 electrodes. At time t_4, a single potential well is created under the ϕ_2 and ϕ_3 electrodes. As the process repeats itself, charge packets are transferred sequentially to the right.

Various other potential well profiles can be employed that involve either different phase clocking schemes or different electrode arrangements. A four-phase clocking scheme has been developed for high-frequency applications because it is easier to produce clock waveforms with a 90° phase difference than to produce the 120° phase shift required with three-phase clocking.

Clearly, *charge transfer efficiency* is an important consideration in CCD design because any charge residual that is not transferred during a particular clock sequence ends up in adjacent charge packets. This form of distortion is called *charge smearing*. Since up to 3000 charge transfers will be required with three-phase clocking to clock an image from the distal side of a 1000-element array, the maximum allowable fraction of incomplete charge transfer must not exceed 10^{-3}–10^{-4}.

Maintaining a high charge transfer efficiency becomes more difficult as the clock frequency increases. In addition, the presence of surface states (see following discussion) can lead to trapping of signal charge. Finally, the presence of potential humps in the region between two electrodes can cause some charge loss during transfer.

11.6.3. Morphology of the Charge Transfer Device

A problem encountered with the MOS structure described in the preceding discussion (i.e., a so-called *surface channel CCD*) is the formation of *surface states*. Surface states arise as a result of the disruption of the periodic lattice structure at the SiO_2–Si interface. This lattice disruption leads to a high density of energy states near the interface that lie within the forbidden gap of the material. Since the surface states lie within the forbidden gap, they behave as traps for charge carriers. Since the traps are more numerous in the vicinity of the interface and since the potential wells fill from the interface outward into the substrate, their presence is most serious for low light levels. The presence of the traps reduces the efficiency of charge transfer because trapped charge is immobile and is effectively lost because it cannot be transferred.

Since the trapping effect of surface states is more pronounced for low light levels, its presence introduces a certain amount of nonlinearity into the light transfer characteristic of the device. This nonlinearity can be removed by pre–illuminating the device with a low-level light source prior to the exposure of interest. This procedure ensures that the surface state traps are filled prior to the exposure of interest and introduces an artificial bias known as *fat zero*.

To avoid the effect of surface traps, a modified MOS structure has been developed that is capable of storing charge away from the interface (Boyle and Smith, 1974; Walden et al., 1972). These *buried-channel* CCDs are made by adding a thin *n*-type layer under the SiO_2 layer by means of ion implantation, as shown in Figure 11.32 for the *n*-channel device. When a positive potential is applied to the *n*-type layer immediately beneath the oxide layer, a depletion zone is produced at the junction because the *pn* junction between the thin *n*-type layer and the *p*-type substrate is reverse biased. If the bias potential is made sufficiently positive, the depletion layer expands until the shallow *n*-type layer is completely depleted. When this occurs, the device is said to be in a *punch-through condition.* In this state, the potential underneath a given gate becomes independent of the positive bias potential and is determined solely by the applied gate voltage. Figure 11.33 shows the band structure of a buried-channel CCD in a punch-through state. As illustrated in the figure, the potential well occurs at the *pn* junction and, as a result, the channel is shifted away from the interface with the insulating oxide layer. As the amount of charge stored in the potential well increases, the carriers are distributed closer

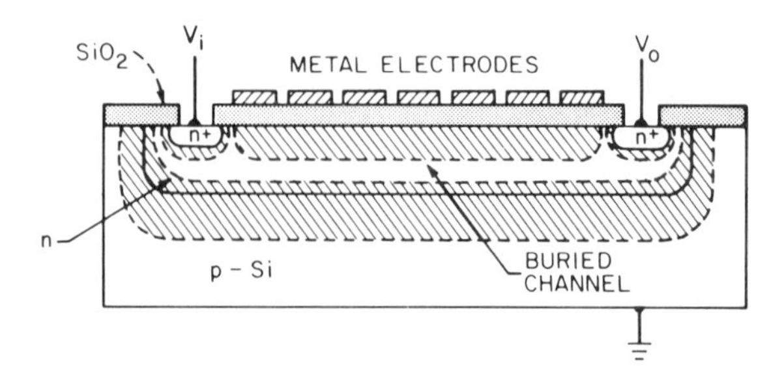

Figure 11.32. Cross-sectional view of BCCD. [Reprinted from Sze, S. M. (1981), *Physics of Semiconductor Devices*, 2nd ed. By permission of John Wiley & Sons, New York.]

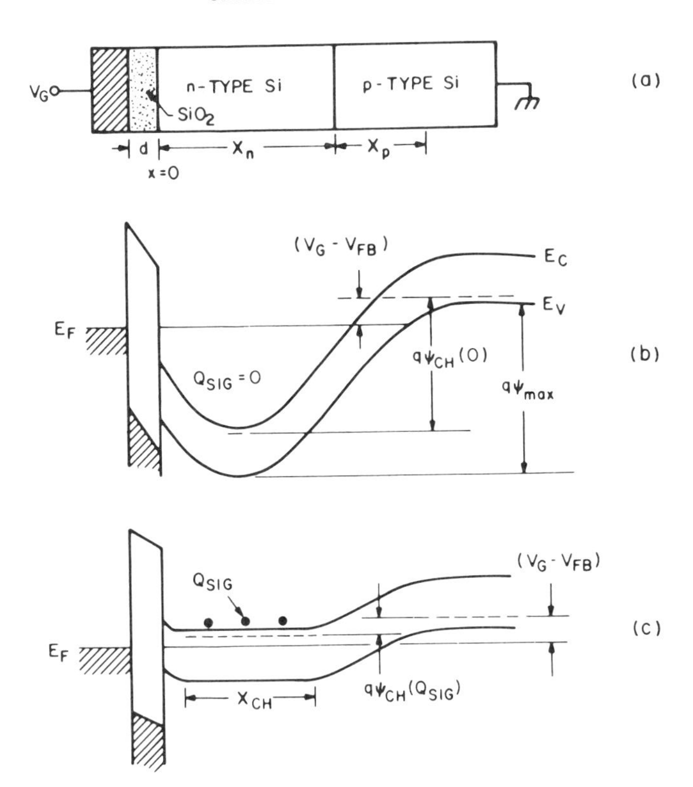

Figure 11.33. (*a*) A BCCD. (*b*) Energy band for empty well. (*c*) Energy band when signal packet is present. [Reprinted from Sze, S. M. (1981), *Physics of Semiconductor Devices*, 2nd ed. By permission of John Wiley & Sons, New York.]

to the interface and the potential barrier between the channel and the interface decreases.

Clocking requirements for the buried-channel device are similar to the surface channel configuration with the exception of the need for the positive offset bias voltage applied to the *n*-type layer. When a small number of electrons has been generated in the material (by thermal effects or photon absorption), they are stored at the potential well formed at the *pn* junction that is located away from the SiO_2–Si interface, as already described. These charge carriers can be transferred to the next electrode by external clocking, as previously discussed.

Compared with the *surface channel* CCD described, the buried-channel device offers two important advantages. The first advantage is a higher transfer efficiency as a result of the elimination of the influence of surface trapping. This increase in charge transfer efficiency in turn improves the overall linearity in light response. The second advantage obtained with the buried-channel configuration is an increase in operating frequency. The increase is possible because the orientation of the electric fields used for charge transfer is more effective and also because the bulk mobility of charge carriers is about twice the surface charge mobility.

Another advantage of the buried-channel structure, which should not be overlooked, is the dramatic improvement in the noise level. This reduction in noise is due to the elimination of *trapping noise* that results because of the interaction of charge carriers with the surface state traps. Since trapping and emission of charge carriers by the surface states are a random process, their presence results in a random fluctuation of charge carriers (i.e., noise). By avoiding the area where the traps are located, the noise performance is improved.

One disadvantage of the buried-channel configuration is its small maximum-charge-handling capacity. However, since the minimum-size charge packet that can be transferred is also reduced, a wide dynamic range is still possible if the device is connected to a high-performance, low-noise output circuit. Table 11.2 gives a comparison of the characteristics of buried-channel and surface channel CCDs.

Another problem encountered with the use of CCDs as imaging devices involves the effect the mosaic of clock electrodes on the surface of the device has on the spectral response of the detector. In order to penetrate into the substrate of the detector and be absorbed, incident photons must have sufficient energy to pass through the gate electrode and the underlying oxide layer. When metallic electrodes are employed with a front-illuminated CCD, the quantum efficiency is generally limited by multiple reflection losses arising from the metallic electrodes. For some wavelengths, these metallic electrodes are essentially opaque to the incident radiation, and as a result, photons can enter the device only in the transparent SiO_2 regions surrounding the electrodes. Thus, the use of metallic electrodes not only limits the quantum

Table 11.2. Comparison of Buried Channel CCD with Surface Channel CCD

Buried Channel CCD	Surface Channel CCD
Higher charge transfer efficiency	Higher signal-handling capability
Higher clock frequency operation	Easier fabrication
No "fat zero" required	Able to operate below 77 K
Lower intrinsic noise	Better linearity

Source: Reprinted from Prettyjohns (1987).

efficiency by reflection losses, but it also reduces the effective sensing area of the detector.

To counteract the disadvantages caused by the use of metallic electrodes, two approaches have been taken. One approach involves the use of gate electrodes made of polysilicon, a form of silicon having metallic properties. Since polysilicon gate electrodes have good transmission properties for optical wavelengths down to 450 nm, their use can improve the light collection efficiency of a CCD in the visible region of the spectrum.

To improve the blue response of CCDs even further, *backside-illuminated* CCDs have been developed. With this approach, the substrate side of the CCD is used as the photosensitive portion of the array. By this means, incident photons can be absorbed directly in the substrate without having to pass through the electrodes or the oxide layer. Since short-wavelength radiation cannot penetrate very deeply into the substrate before it is absorbed, the substrate must be thinned to ensure that the photogenerated charge can be collected in the depletion regions under the electrodes. For this reason, backside-illuminated CCDs are generally thinned to a substrate thickness of approximately 10 μm, which makes them somewhat more delicate than their front-illuminated counterparts. Compared with typical front-illuminated CCDs (with a quantum efficiency of about 0.2), back-illuminated devices have a quantum efficiency on the order of 0.8 (Howes and Morgan, 1979, p. 267).

11.6.4. Charge Readout

If a two-dimensional CCD array is used as an image detector, some provision must be made to avoid the inevitable image smearing that would occur during readout of the array as a result of the continuous generation of charge carriers by the incident radiation. One way to accomplish this is to use a shutter to block the incident radiation from striking the array during readout.

A continuous readout at 30 frames per second, however, can be achieved by employing various masking techniques. The two masking procedures that have been employed are the *frame transfer CCD* and the *interline transfer CCD*. Figure 11.34 shows a schematic diagram of a conventional frame transfer architecture used in normal television applications. With this arrangement, one-half of the CCD is designated as the *image area*, while the other half is covered with an opaque mask and is designated as the *storage area*. Adjacent to the storage area is a high-speed serial shift register. The storage area has separate clock connections from the imaging area so that the two halves may be clocked independently.

In operation, both clocks are stopped initially, and one phase of the gates in the image area is biased to form isolated potential wells under that series of gates. During this period, charge generated by photon absorption is inte-

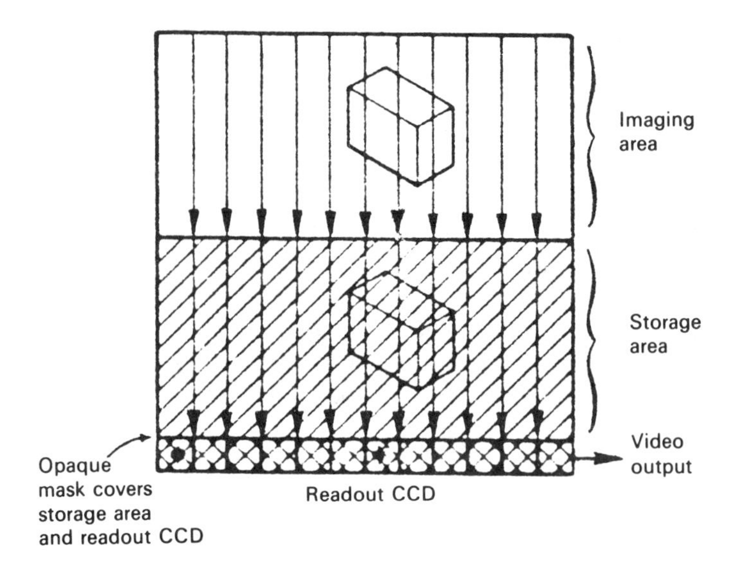

Figure 11.34. Frame transfer (FT) CCD architecture. [Reprinted with permission from K. N. Prettyjohns (1987), "Imaging Arrays, Solid State," in *Encyclopedia of Physical Science and Technology*, Vol. 6, Academic, New York, p. 570. Copyright 1987 Academic Press]

grated by the imaging section. At the end of the integration period, the clocks are activated, and the charge is transferred to the masked storage area. When this has been accomplished, the clocks are stopped and the image area gates are biased once again to produce charge-collecting potential wells. In the meantime, the storage area is clocked periodically to transfer successive horizontal lines of the image to the high-speed shift register. When the entire storage area has been read one line at a time by the high-speed shift register, the imaging area is ready to transfer the next frame of integrated photogenerated charge to the storage area. If the time required to transfer the charge from the image area to the storage area can be kept short by high-speed clocking of the CCD, the amount of charge smearing from continuous irradiation of the array can be kept small.

Figure 11.35 shows a schematic diagram of an interline transfer CCD. With this architecture, alternate columns of the array are covered with opaque masks so that the array is divided into imaging columns and readout columns. A horizontal shift register is connected to the readout columns at the bottom of the array, as shown in Figure 11.35.

At the end of an integration period, photogenerated charge that has collected in the individual photosites of the imaging columns is transferred simultaneously to the adjacent masked readout columns. Once the image

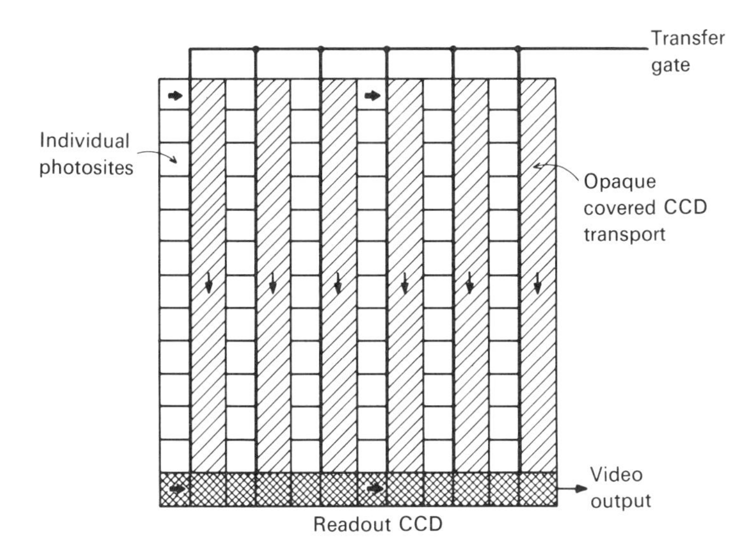

Figure 11.35. Interline transfer (ILT) CCD architecture. [Reprinted with permission from K. N. Prettyjohns (1987), "Imaging Arrays, Solid State," in *Encyclopedia of Physical Science and Technology*, Vol. 6, Academic, New York, p. 570. Copyright 1987 Academic Press.]

charge has been transferred to the respective vertical readout columns, the readout columns are all clocked simultaneously to transfer the first horizontal line of the image to the horizontal shift register that transmits the image line sequentially to the video output at high speed. When this has been accomplished, the next row of charge from the readout columns is clocked into the horizontal shift register. The process continues until the entire array has been read. When the entire array has been read, the photogenerated charge from the next frame is transferred from the imaging columns to the readout columns and the process begins again.

The primary advantage of the interline transfer architecture is the elimination of charge-smearing effects that result when light is incident on the array during readout. Since each imaging column is adjacent to a readout column, the entire image can be transferred to the opaque readout columns with one clock pulse. Once the image has been transferred to the readout columns, charge smearing is eliminated because all charge transfers involved in readout are conducted in the absence of incident light. The principal disadvantage to this architecture is that image sampling is reduced by the presence of the readout columns. The loss of image information due to the presence of the readout columns increases the likelihood of aliasing and reduces the ultimate image resolution that is available with the array.

At this point, it is worthwhile to discuss the actual readout process in more detail. The quantity of charge contained in a given charge packet is sensed by a reverse-biased output diode (sometimes referred to as a *floating diffusion*) that converts the output charge to an output voltage. The single output node (i.e., floating diffusion), which is located at the edge of a one- or two-dimensional array, is used to convert signal charge into voltage prior to detection by an on-chip MOS amplifier that senses the potential change. When a charge packet is transferred to the floating diffusion, it flows into the nodal capacitance produced by the reverse-biased *pn* junction of the output diode, thereby producing a change in voltage that is detected by the MOS amplifier. Between successive charge outputs, the capacitor voltage is reset to a defined voltage with a reset switch composed of a reset gate and reset drain, as shown in Figure 11.36.

Since the total capacitance of the output (the combined capacitance of the *pn* junction and the input gate of the MOS amplifier) is extremely low, readout noise in the form of *kTC* noise (see Eq. 11.12) is extremely low. It is the presence of this extra-low output capacitance that gives the CCD its ultra-low read noise and distinguishes it from other solid-state arrays in terms of performance.

In spite of the low readout noise of the CCD, the charge transfer fluctuations that result during charge transfer across the array are still much less than the *kTC* output fluctuations (high-frequency switching transients

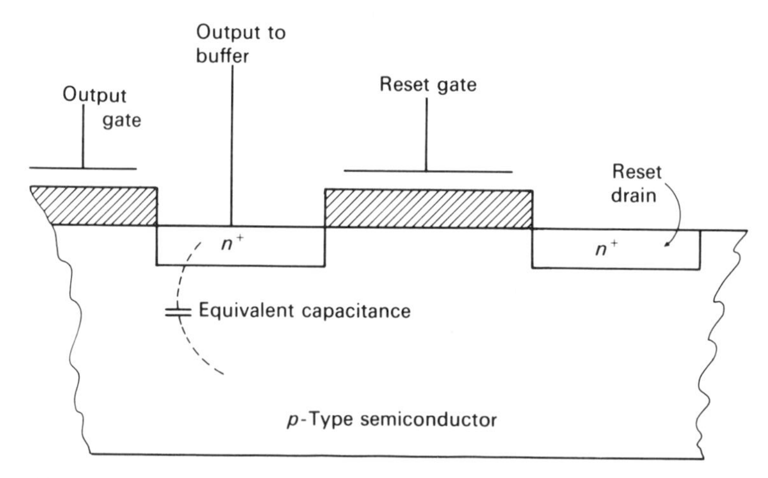

Figure 11.36. Schematic of CCD output showing floating diffusion and reset gate. [Reprinted with permission from G. S. Hobson (1987), "Charge-Transfer Devices," in *Encyclopedia of Physical Science and Technology*, Vol. 2, Academic, New York, p. 622. Copyright 1987 Academic Press.]

due to reset switching breakthrough) that result during readout. Actually, if the charge transfer process were 100% efficient, there would be no noise component associated with charge transfer because there would be nowhere to share the charge fluctuation. Since the only charge component that can fluctuate during the charge transfer process is smeared charge (which is typically only 10^{-4} of the signal charge), charge transfer noise is extremely small.

To take advantage of the extremely low charge transfer noise inherent in CCDs, an output technique with lower noise, known as *correlated double sampling*, has been devised. In essence, correlated double sampling is a technique that samples the output voltage before and after reception of the signal charge in the output diode and takes the difference between the voltage levels. If the reset noise voltage in the output circuit does not change significantly during the sampling process, the difference in voltages should eliminate the switching transients.

The actual process of correlated double sampling is shown schematically in Figure 11.37. The voltage follower amplifier shown in the figure measures the voltage across capacitor C_2, which changes in response to the opening and closing of switch S_2. Switch S_1 periodically resets the voltage at point B to ground at time periods $t_1, t_3, t_5, \cdots$ after the reset transient has occurred. Once the reset transient has passed, S_1 opens and S_2 closes to allow capacitor C_2 to charge to the output voltage *difference* that exists between times t_1 and t_2. The output of the voltage follower amplifier responds to changes in the voltage level across C_2, which changes only at $t_2, t_4, t_6, \cdots$ In essence, the CCD output waveform (V_A) is sampled at t_1/t_2, t_3/t_4, t_5/t_6, and so on, and the respective voltage differences that exist at these times are amplified by the system. The net effect of this procedure is to remove the reset pulses from the output waveform.

11.6.5. Charge Injection Devices

Another form of charge-transfer device pioneered by the General Electric Company is the *charge injection device* (CID). The charge injection device shown in Figure 11.38 consists of an array of individual MOS photosites formed on an n-doped layer that has been epitaxially grown on a p-doped silicon wafer. In contrast to the CCD, however, each photosite consists of *two* polysilicon gate electrodes, where one gate (i.e., the row electrode) is connected to a particular row address and the other (i.e., the column electrode) is connected to a particular column address. A thin silicon oxide (or silicon nitride) insulating layer separates the polysilicon gate electrodes from the n-doped silicon region. Since the depletion region is formed in the n-doped

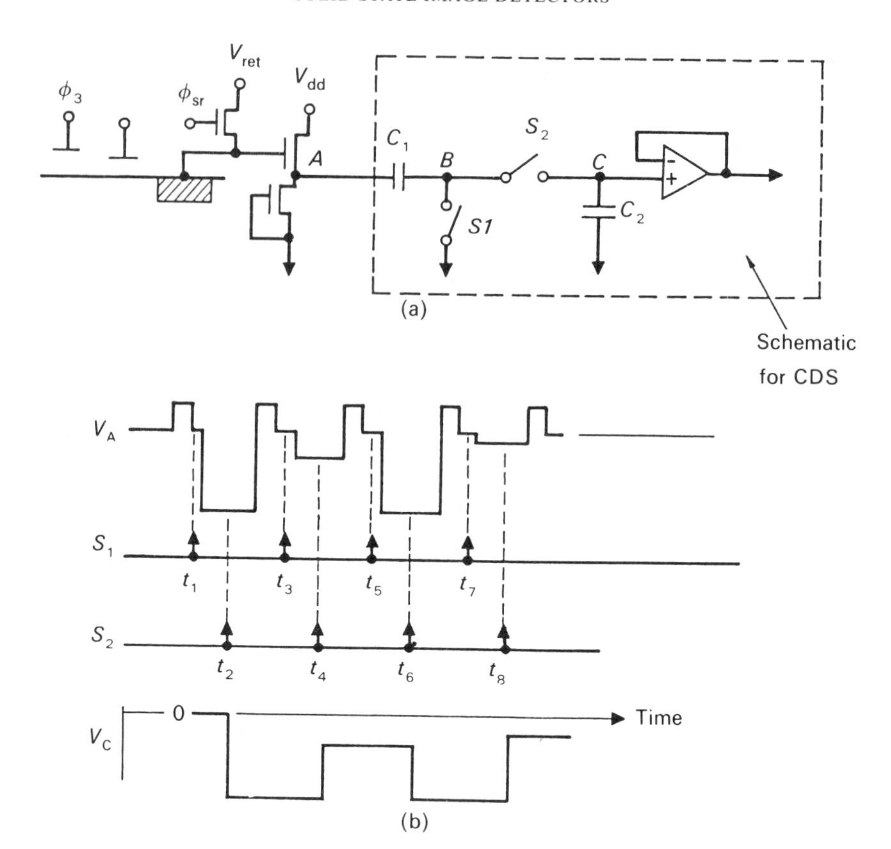

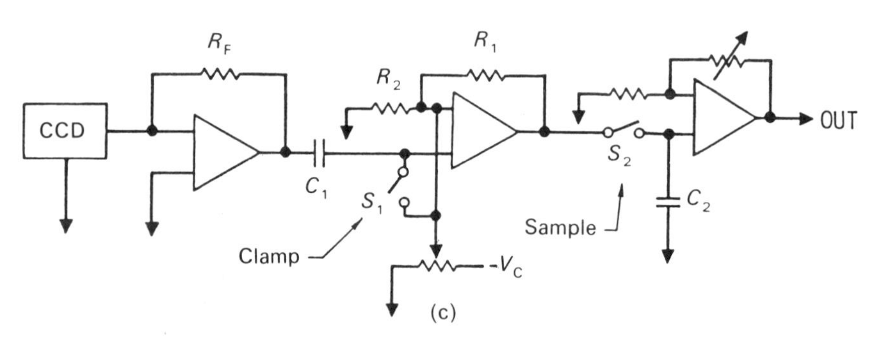

Figure 11.37. Correlated double sampling: (*a*) Schematic for correlated double sampler; (*b*) timing diagram for switches S_1 and S_2 in correlated double sampler; (*c*) typical circuit for implementing correlated double sampling. [Reprinted from W. S. Chan (1988), *Infrared Detection, Focal Plane Technology and Sytems*, Vol. 1., Electro-Optek Corporation, Torrance, CA. Copyright 1988 Electro-Optek Corporation.]

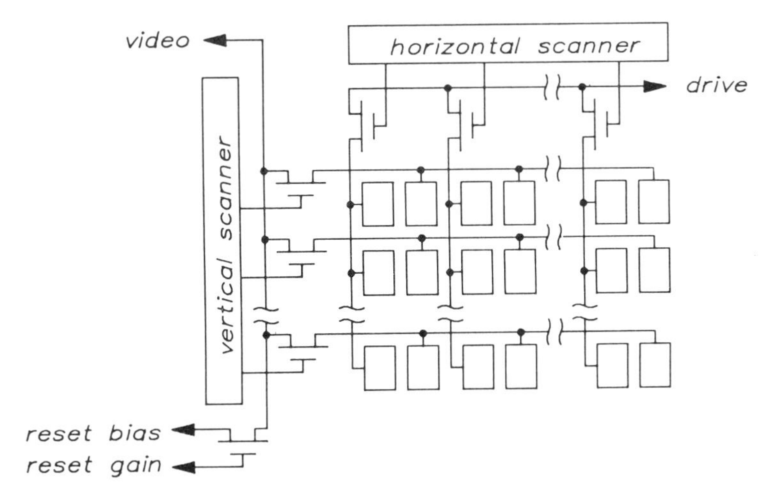

Figure 11.38. Schematic of typical two-dimensional array based on charge injection principle. [Reprinted from R. B. Bilhorn, J. V. Sweedler, P. M. Epperson, and M. B. Denton (1987), *Appl. Spectrosc.*, *41*, 1114, with permission from the Society for Applied Spectroscopy.]

semiconductor, photogenerated holes (i.e., the minority carriers) are stored in the potential wells created by biasing the gate electrodes negatively.

Column and row electrodes associated with each photosite allow access to any detector element within the array. Although a number of readout schemes have been devised for CIDs, all involve manipulation of the electrode potentials within a given photosite. The simplest form of CID readout is to address sequentially the various rows and columns and sequentially collapse the potential wells under each photosite by biasing simultaneously both the column and row electrodes for that particular site to zero. By collapsing the potential wells of both photosite electrodes simultaneously, the photogenerated charged that has been collected is *injected* into the substrate, and the resulting current produced can be measured by means of a load resistor connected to the substrate. This form of readout is called *destructive readout* because the charge is destroyed in the process of measurement.

Another form of readout available with CIDs is *nondestructive readout* (NDRO). Nondestructive readout can be accomplished by movement of charge from one electrode (the collection electrode) to another (the sensing electrode). The movement of charge within a given detector element has been called *intracell charge transfer* (Billhorn, Sweedler et al., 1987).

With intracell charge transfer (see Billhorn, Sweedler et al., 1987), photogenerated charge is collected under the collection electrode by biasing the collection electrode with a more negative potential than the sensing electrode, as shown in Figure 11.39. To read the charge collected after a given

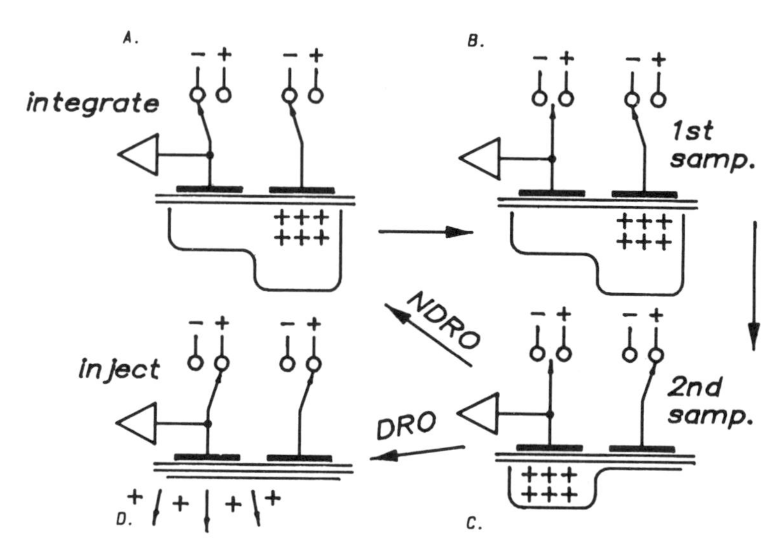

Figure 11.39. Readout scheme used to quantify amount of photogenerated charge in individual detector element of CID. (*a*) The CID is in charge integration mode. (*b*) First of two measurements of potential at charge amplifier is performed. (*c*) Second measurement of potential after charge has been moved to row electrode is performed. Readout sequence can be completed by shifting charge back to column electrode as shown in (*a*), by a nondestructive readout (NDRO), or by injecting charge into substrate as shown in (*d*), a destructive readout (DRO). [Reprinted from R. B. Bilhorn, J. V. Sweedler, P. M. Epperson, and M. B. Denton (1987), *Appl. Spectrosc.*, *41*, 1114, with permission from the Society for Applied Spectroscopy.]

integration period, the potential well under the collection electrode is collapsed by applying a positive voltage. This causes the charge that has been collected under the collection electrode to migrate to the potential well under the adjacent sensing electrode. The movement of charge from the collection electrode to the sensing electrode results in a potential change given by

$$dV = dQ/C \tag{11.16}$$

where dV is the potential change induced by the movement of charge from one electrode to the other, dQ is the amount of charge transferred, and C is the capacitance of the MOS capacitor. By measuring the potential prior to charge movement and after the charge has been transferred, a voltage difference propotional to the quantity of stored charge can be measured.

Intracell charge transfer is inherently nondestructive because the process is reversible (i.e., no net change was made in the quantity or location of the photogenerated charge). After the potential difference given by Eq. 11.16 has been measured by the output amplifier, it is a simple matter to reestablish the

Table 11.3. Comparisons of CCD and CID Imagers

Characteristic	Frame Transfer CCD	Interline Transfer CCD	CID
Versatility	Front- or black-surface illumination; can be used for frame-to-frame integration for low noise	Front-surface illumination only; separate sensors make video signal processing possible	Front-surface illumination only; more complicated injection sites for back-surface illumination; random access readout possible
Spectral sensitivity	Large fraction of theoretical silicon response for back-surface illumination; front-surface layers cause optical interference modifications to spectral response	Shielded front surface and complicated deposition patterns reduce sensitivity more than two-fold	Good front-surface collection efficiency in low-density arrays; large fraction of silicon response in back-surface illumination
Use of silicon chip area	50% available for imaging	50% available for imaging	90% available for imaging
Antiblooming control	Accumulation around cells or overflow drains give good blooming performance	Overflow drains can be placed between columns, giving good blooming performance	Accumulation around cells provides excellent antiblooming control
Low-light and low-noise performance	Cooled buried-channel devices and low-capacity collection nodes give good noise performance and good low-light performance	Similar to frame transfer CCD	Higher capacity charge collection nodes and reset noise give rise to problem that is offset by ability to integrate over many frame cycles with nondestructive readout
Resolution	Overlapping pixels on interlace	Good	Good
Infrared performance	Suitable for silicon-based charge transfer	Suitable for silicon-based charge transfer	Suitable for compound semiconductors owing to independence of charge transfer efficiency
Special problems	High-speed vertical transfer	Complex cell	Fixed pattern noise

Source: Reprinted from Hobson (1987).

Table 11.4. Formats of Representative Charge Transfer Devices

Manufacturer	Imager (Model No.)	Dimensions (Active Elements)	Detector Size (μm)	Overall Photoactive Area (mm)
		CIDs		
General Electric	CID11B	244 × 248	47 × 35	11.5 × 8.7
General Electric	CID17B	244 × 378	27.2 × 23.3	6.5 × 8.7
General Electric	CID35	512 × 512	14 × 14	7.2 × 7.2
General Electric	CID75	1 × 1	1000 × 1000	1 × 1
General Electric	CID62	512 × 32	29 × 127	14.8 × 4.1
		CCDs		
Texas Instruments	VP800R	800 × 800	15 × 15	12 × 12
Texas Instruments	VP1M	1024 × 1024	18 × 18	18.4 × 18.4
Texas Instruments	TC104	3456 × 1	10.3 × 10.3	35.6 × 0.01
RCA	SID501EX	512 × 320	30 × 30	15.4 × 9.6
RCA	SID504DD	256 × 403	19 × 16	4.8 × 6.5
Tektronix	TK512M	512 × 512	27 × 27	13.8 × 13.8
Tektronix	TK2048M	2048 × 2048	27 × 27	55.3 × 55.3
Thomson-CSF	TH7882CDA	384 × 576	23 × 23	8.8 × 13.2
Kodak	M1A	1320 × 1035	6.8 × 6.8	9.0 × 7.0

Source: Reprinted from Bilhorn, Sweedler et al. (1987).

Table 11.5. Electro-optical Characteristics of Selected CTDs

Device	Architecture	Dimensions (Elements)	Peak QE (%) Wavelength	Range over Which QE Exceeds 10% (nm)	Full Well (Carriers)	Noise (Carriers)	Notes
RCA[a] SID501EX	CCD, three-phase back-side	512 × 320	95% at 500 nm	200[b]–950	4×10^5	50	AR coating for 500 nm
Tektronix[c] TK512M-011	CCD, three-phase front-side	512 × 512	35% at 750 nm	450–950	9×10^5	6	—
Texas Instruments[d]	CCD, three-phase back side	800 × 800	90% at 550 nm	0.1[b]–1050	5×10^4	5	Optimized process, prototype AR coating for 550 nm
Texas Instruments[a] TC104-1	CCD, uniphase front side	3456 × 1	95% at 400 nm	200[b]–1000	3×10^5	80	No gates over photo-active region, AR coated for 400 nm
General Electric[e] CID17-B	CID	244 × 388	47% at 550 nm	210–850	6×10^5	60	
General Electric[f] CID75	CID	1 × 1	35% at 250 nm	200[b]–850	1×10^8	80	Prototype devices

[a] Measurements made by Denton and co-workers.
[b] Point at which QE measurements stopped. QE may exceed 10% beyond this point.
[c] From Epperson et al. (1987).
[d] Janesick, J., Campbell, D., Elliot, T., Daud, T., and Ottley, P. (1986), "Flash Technology for Imaging in the UV," in *UV Technology*, Huffman, R. (Ed.), *Proc. SPIE, 687*, p. 36, Bellingham, WA.
[e] Bilhorn, R. B., and Denton, M. B., unpublished results.
[f] From Sweedler et al. (1987).
Source: Reprinted from Bilhorn, Sweedler et al. (1987).

potential well under the collection electrode and cause the charge to migrate back to its initial position. Since the device can be restored to its initial condition, computer summation of a number of such nondestructive readouts can be averaged to reduce read noise. If the read noise is white, it will be reduced by a factor proportional to the square root of the number of NDROs accumulated in the average. Sims and Denton (1987a) have shown that read noise can be reduced by over a factor of 10 by signal averaging multiple NDROs. When sufficient NDROs have been obtained, the charge can be removed by collapsing the potential wells under both sets of electrodes simultaneously, as shown in Figure 11.39, thereby injecting the charge into the substrate.

11.6.6. Performance Characteristics of Charge Transfer Devices

The majority of recent work on the spectroscopic application of charge transfer devices has been done by Denton and co-workers at the University of Arizona (Bilhorn, Sweedler et al., 1987; Bilhorn, Epperson et al., 1987; Sweedler et al., 1988; Epperson et al., 1988; Aikens, Epperson, and Denton, 1984; Sims and Denton, 1987a, b; Epperson et al., 1987; Sweedler et al., 1987). Of all the solid-state image detectors available today, charge transfer devices are the most useful for scientific applications (see Janesick et al., 1984). As a class, they offer high sensitivity, broad wavelength response, and minimal dark current. Table 11.3 summarizes some of the characteristics of CCDs and CIDs.

Charge injection devices as a class are very resistant to migration of excess charge to adjacent detector elements (i.e., blooming) and, because of their architecture, can be accessed randomly. The possibility of nondestructive readout can be used to advantage to reduce readout noise. Tables 11.4 and 11.5 give the formats and electro-optic characteristics of representative CIDs.

Charge-coupled device technology has reached a point where very high levels of performance are available. For example, backside-illuminated CCDs with special antireflection coatings (thin flash-grown oxide layers) result in CCDs with quantum efficiencies of nearly 100% for wavelengths out to 800 nm (Bilhorn, Sweedler et al., 1987). Three- and four-phase clocking schemes are available that give charge transfer efficiencies of 0.999995. On-chip amplifiers are available that produce read noise of under 5 electrons. Fixed potential barriers or *channel stops* have been incorporated into CCD arrays to prevent charge migration. Finally, the dark count rate with cooled CCDs is in the range of < 0.001 to 0.03 electrons per detector element per second, so extremely long integration times are possible.

REFERENCES

Aikens, R. S., Epperson, P. M., and Denton, M. B. (1984), "Techniques for Operating Charge-Coupled Devices (CCD's) in Very High Speed Framing Mode," in *State-of-the-Art Imaging Arrays and Their Applications*, Vol. 501, Society of Photo-Optical Instrumentation Engineers, Bellingham, WA, pp. 49–54.

Amelio, G. F., Tompsett, M. F., and Smith, G. E. (1970), *Bell Syst. Tech. J., 49*, 593.

von Ardenne, M. (1936), *Electr. Nachr. Tech., 13*, 230.

Barbe, D. F. (1975), *Proc. IEEE, 63*, 38.

Barbe, D. F., Ed. (1980), *Charge-Coupled Devices*, Springer-Verlag, Berlin.

Barbe, D. F. and Campana, S. B. (1977), "Imaging Arrays Using the Charge-Coupled Concept," in *Advances in Image Pickup and Display*, Vol. 3, Kazan, B. (Ed.), Academic, New York.

Barthelemy, R. and Leithine (1936), French Patent No. 802244, August 31.

Beynon, J. D. E. and Lamb, D. R. (1980), *Charge-Coupled Devices and Their Applications*, McGraw-Hill, New York.

Bilhorn, R. B., Epperson, P. M., Sweedler, J. V., and Denton, M. B. (1987), *Appl. Spectrosc., 41*, 1125.

Bilhorn, R. B., Sweedler, J. V., Epperson, P. M., and Denton, M. B. (1987), *Appl. Spectrosc., 41*, 1114.

Boyle, W. S. and Smith, G. E. (1970), *Bell Sys. Tech. J., Briefs, 49* (4), 587.

Boyle, W. S. and Smith, G. E. (1974), U.S. Patent 3,792,322.

Buck, T. M., Casey, Jr., H. C., Dalton, J. V., and Yamin, M. (1968), *Bell Sys. Tech. J., 47* (9), 1827.

Coltman, J. W. (1954), *J. Opt. Soc. Am., 44*, 468.

Crowell, M. H., Buck, T. M., Labuda, E. F., Dalton, J. V., and Walsh, E. J. (1967), *Bell Sys. Tech. J., 46*(2), 491.

Crowell, M. H. and Labuda, E. F. (1969), *Bell Sys. Tech. J., 48*(5) 1481.

Crowell, M. H. and Labuda, E. F. (1974), "The Silicon-Diode-Array Camera Tube," in *Photoelectronic Imaging Devices*, Vol. 2, Biberman, L. M. and Nudelman, S. (Eds.), Plenum, New York.

Eccles, M. J., Sim, M. E., and Tritton, K. P. (1983), *Low-Light Level Detectors in Astronomy*, Cambridge University Press, Cambridge.

Epperson, P. M., Sweedler, J. V., Bilhorn, R. B., Sims, G. R., and Denton, M. B. (1988), *Anal. Chem., 60*, 327A.

Epperson, P. M., Sweedler, J. V., Denton, M. B., Sims, G. R., McCurnin, T. W., and Aikens, R. S. (1987), *Opt. Eng., 26*, 715.

Goetze, G. W. and Laponsky, A. B. (1974), "Early Stages in the Development of Camera Tubes Employing the Silicon-Diode Array as an Electron-Imaging Charge-Storage Target," in *Photoelectronic Imaging Devices*, Vol. 2, Biberman, L. M. and Nudelman, S. (Eds.), Plenum, New York.

Gordon, E. J. (1967), *Bell Lab. Rec., 45*(6) 174.

Gordon, E. J. and Crowell, M. H. (1968), *Bell Sys. Tech. J.*, *47*(9), 1855.

Görlich, P. (1936), *Z. Physik*, *101*, 335.

Görlich, P. (1938), *Z. Physik*, *106*, 374.

Görlich, P. (1941), *J. Opt. Soc. Am.*, *31*, 504.

Hiltner, W. A. (1962), "Image Converters for Astronomical Photography," in *Astronomical Techniques*, Hiltner, W. A. (Ed.), University of Chicago Press, Chicago.

Hobson, G. S. (1978), *Charge-Transfer Devices*, Arnold, London.

Hobson, G. S. (1987), "Charge-Transfer Devices," in *Encyclopedia of Physical Science and Technology*, Vol. 2, Academic, New York, p. 622.

Howes, M. J. and Morgan, D. V., Eds. (1979), *Charge-Coupled Devices and Systems*, Wiley, New York.

Janesick, J. R., Elliot, T., Collins, S., Marsh, H., Blouke, M. and Freeman, J. (1984). "The Future Scientific CCD," in *State-of-the-Art Imaging Arrays and Their Applications*, Prettyjohns, K. (Ed.), *Proc. SPIE*, *501*, p. 2, Bellingham, WA.

Kazan, B. and Knoll, M. (1968), *Electronic Image Storage*, Academic, New York.

Koller, L. R. (1929), *J. Opt. Soc. Am.*, *19*, 135.

Levi, L. (1980), *Applied Optics*, Vol. 2, Wiley, New York.

Melen, R. and Buss, D. (1977), *Charge-Coupled Devices: Technology and Applications*, IEEE, New York.

Meyers, L. M. (1939), *Electron Optics*, Chapman, London.

Morton, G. A. (1964), *Appl. Opt.*, *3*, 651.

Nicollian, E. H. and Brews, J. R. (1982), *MOS Physics and Technology*, Wiley, New York.

Potter, R. J. and Hopkins, R. E. (1958), presented at *Image Intensifier Symposium*, Ft. Belvoir, VA (Oct. 6–7).

Prettyjohns, K. N. (1987), "Imaging Arrays, Solid State," in *Encyclopedia of Physical Science and Technology*, Vol. 6, Academic, New York, p. 570.

Rosell, F. A. (1974), "Television Camera Tube Performance Data and Calculations," in *Photoelectronic Imaging Devices*, Vol. 2, Biberman, L. M. and Nudelman, S. (Eds.), Plenum, New York.

Schnitzler, A. D. and Morton, G. A. (1974), "Cascade Image Intensifiers," in *Photoelectronic Imaging Devices*, Vol. 2, Biberman, L. M. and Nudelman, S. (Eds.), Plenum, New York.

Sequin, C. H. and Tompsett, M. F. (1975), *Charge Transfer Devices*, Academic, New York.

Simpson, R. W. (1979), *Rev. Sci. Instrum.*, *50*, 730.

Sims, G. R. and Denton, M. B. (1987a), *Opt. Eng.*, *26*, 999.

Sims, G. R. and Denton, M. B. (1987b), *Opt. Eng.*, *26*, 1008.

Sommer, A. H. (1955), *Rev. Sci. Instrum.*, *26*, 725.

Sommer, A. H. (1968), *Photoemissive Materials—Preparation, Properties, and Uses*, Wiley, New York.

Sternglass, E. J. (1955), *Rev. Sci. Instrum.*, *26*, 1202.

Sweedler, J. V., Bilhorn, R. B., Epperson, P. M., Sims, G. R., and Denton, M. B. (1988), *Anal. Chem.*, *60*, 282A.

Sweedler, J. V., Denton, M. B., Sims, G. R., and Aikens, R. S. (1987), *Opt. Eng.*, *26*, 1020.

Sze, S. M. (1981), *Physics of Semiconductor Devices*, Wiley, New York.

Talmi, Y. and Busch, K. W. (1983), "Guidelines for the Selection of Four Optoelectronic Image Detectors for Low-Light Level Applications," in *Multichannel Image Detectors*, Vol. 2, Talmi, Y. (Ed.), American Chemical Society, Washington, DC.

Talmi, Y. and Simpson, R. W. (1980), *Appl. Opt.*, *19*, 1401.

Wachtel, M. M., Doughty, D. D., and Anderson, A. E. (1960), in *Photo-electronic Image Devices*, McGee, J. D. and Wilcock, W. L. (Eds.), Academic, New York, p. 59.

Wachtel, M. M., Doughty, D. D., Goetze, G., Anderson, A. E., and Sternglass, E. J. (1960), *Rev. Sci. Instrum.*, *31*, 576.

Walden, R. H., Krambeck, R. H., Strain, R. J., McKenna, J., Schryer, N. L., and Smith, G. E. (1972), *Bell Syst. Tech. J.*, *51*, 1635.

Weimer, P. K., Forgue, S. V., and Goodrich, R. R. (1950), *Electronics*, *23*, 70.

Weimer, P. K., Forgue, S. V., and Goodrich, R. R. (1951), *RCA Rev.*, *12*(1), 306.

Weimer, P. K., Pike, W. S., Sadasiv, G., Shallcross, F. V., and Meray-Horvath, L. (1974), "Multielement Self-Scanned Mosaic Sensors," in *Photoelectronic Imaging Devices*, Vol. 2, Biberman, L. M. and Nudelman, S. (Eds.), Plenum, New York.

Weimer, P. K., Sadasiv, G., Meyer, Jr., J. E., Meray-Horvath, L., and Pike, W. S. (1967), *Proc. IEEE*, *55*, 1591.

Wendland, P. H. (1967), *IEEE Trans. Electr. Devices*, *ED-14*(9), 285.

Wilcock, W. L., Emberson, D. L., and Weekly, B. (1960), *Nature*, *185*, 370.

Zworykin, V. K. and Morton, G. A. (1940), *Television—The Electronics of Image Transmission*, Wiley, New York.

CHAPTER

12

SYSTEMS APPROACH TO THE DESIGN OF INSTRUMENTS FOR MULTIELEMENT ANALYSIS

"The fox knows many things, the hedgehog knows one big thing."

The goal of this chapter is to provide a conceptualized framework of the philosophy of instrument development as it relates to the area of analytical science. The term *analytical science* as used here refers to the broad area consisting of the science of chemical measurements. The designation analytical science is used instead of the traditional, more common term *analytical chemistry* to express the notion that the discipline extends beyond the traditional boundaries of chemistry. At the same time, chemistry and a fundamental understanding of chemical principles remain a cornerstone of the discipline (i.e., most analytical scientists are, in fact, chemists by training). While the phenomenology remains inherently chemical, the means of achieving particular measurement objectives are essentially engineering in nature. The successful research analytical scientist today is therefore neither a pure chemist in the traditional sense nor a professional engineer but possesses manifestations or attributes of both. Engineering development of analytical systems without a thorough understanding of the underlying chemical phenomenology is pointless, while simple knowledge of phenomenology without a knowledge of the means of achieving particular measurement goals cannot lead to viable analytical systems.

Actually, today there are two broad categories of analytical scientists with quite different points of view. This chapter will be directed toward the research analytical scientist whose area of interest is instrumental development. In contrast to the research analytical scientist, whose interest is in the development of new analytical technology, the analytical applications specialist is concerned primarily with the application of analytical technology and must understand not only the sample under consideration but also the information that is actually desired in characterizing it as well as the various competing analytical techniques available that can be brought to bear on the problem. It is this second category that composes the bulk of the analytical community. Still a third party in this symbiotic relationship is the companies that build and market commercial analytical instrumentation. Figure 12.1 shows in abstract terms the relationship between these three constituencies. The instrument manufacturers supply instruments to the bulk of the analytical community concerned with applications. The applications chemist drives the triumvirate by supplying both the research analytical scientist and the instrument manufacturers with expanded measurement needs.

As long as the measurement needs of the applications community are satisfied by the current analytical methodology, the relationship remains static. As often happens, however, technological advances in other areas call for expanded measurement capabilities that cannot be achieved with current methodology. To satisfy the increased demands, the analytical research scientist attempts to develop new or improved analytical technology either by incorporating new technology such as the laser that has been developed in other areas outside of analytical science or by providing a wider phenomenological base through increased understanding and broader application of physiochemical effects. The research analytical scientist also has the responsibility of anticipating future measurement needs likely to be generated by technology as a whole. For example, interest in heterogeneous catalysis in chemical engineering may generate the need for improved surface analytical techniques. Concern over environmental issues may lead to the need for methods that provide information on the speciation of constituents dissolved in water. High technology involved with solid-state physics may demand

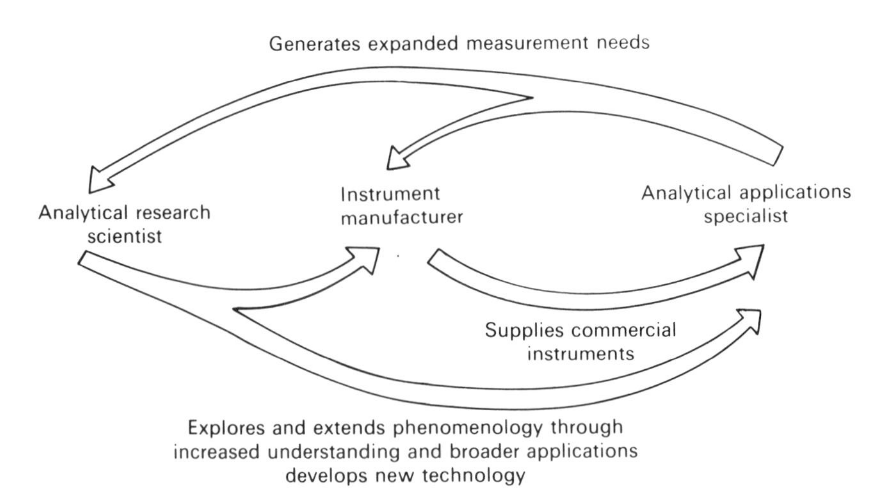

Figure 12.1. Schematic relationship between three constituencies of analytical community.

specialized microprobe methods to characterize very large scale integrated circuits. For the research analytical scientist with an interest in instrumental development who is aware of these needs, the opportunities for continued research and development are quite encouraging.

12.1. BASIC PHILOSOPHY OF INSTRUMENT DEVELOPMENT

Before becoming too heavily involved in the details of instrument development as it relates to multichannel spectrochemical analysis, it is worthwhile to consider some of the factors that are necessary for the development of any successful analytical system. The "embryology" of instrument development begins with the conception of a new idea for improving chemical measurement technology. The first stage in the development of the new concept is a feasibility evaluation in the laboratory. If laboratory studies support the feasibility of the concept, further development is warranted with the aim of demonstrating analytical applications. If the concept can be shown to fulfill real analytical needs, it stands a chance of being adopted by the analytical community as a whole as one of an arsenal of analytical techniques. As the technique (at this point it has matured from an embryonic concept to a fully developed technique) develops, it will eventually reach the stage of a fully mature technology. The lifetime of the technique before its ultimate death by obsolescence will depend on the advent of competing technology.

12.2. INTRODUCTORY SYSTEMS ENGINEERING

Since most multichannel spectroscopic instruments in use today are, in fact, a collection of more or less complex components or subsystems, it is worthwhile to approach the topic of spectrometer design from a systems engineering point of view (see Ellis and Ludwig, 1962). In the terminology of systems engineering, a *system* is anything that accomplishes an operational process. As shown in Figure 12.2, a system is connected to the real world by an input and an output. In the case of a multielement analytical system, for example, the

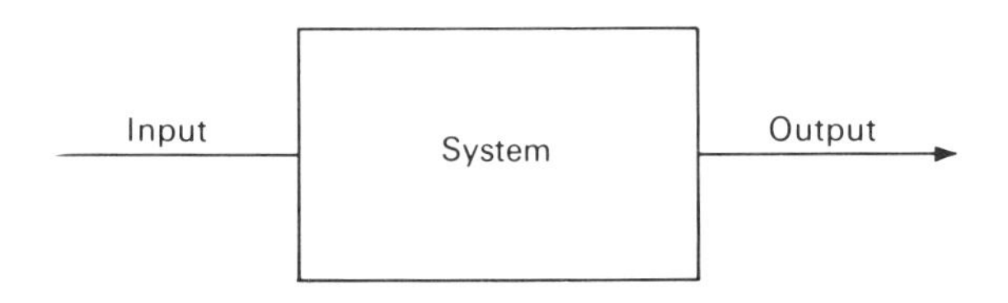

Figure 12.2. Block diagram of idealized system showing input and output.

input may be a sample and the output may be the array of numbers that express the concentrations of n elements. This definition of a system is important because it tends to emphasize the operational processes rather than a collection of pieces of hardware.

The operational processes that transform the input into the output are the goals and objectives of the *mission*. Use of the term mission may seem somewhat strange in the context of systems design; however, since most applications of systems engineering have been for military purposes, the terminology that has developed tends to reflect its origins. In the context of multielement spectrochemical analysis the mission in its broadest sense will be to determine the concentrations of n elements with a requisite precision and accuracy in the sample of interest.

12.2.1. Subsystems and Components

The physical complexes or entities that implement the operational process are systems. In the hierarchy of systems terminology, the lowest level of a system is known as a *component*. At the apex of the hierarchy is the complete system, which may be composed of several *subsystems*. In the context of multielement spectrochemical analysis the complete analytical system may consist of a source subsystem, a collection optics subsystem, a multichannel wavelength selection subsystem, a detector subsystem, and a data-processing subsystem, as shown in Figure 12.3. If a photomultiplier is used as a detector, a dc-regulated power supply may be a component of the detector subsystem. This terminology is, of course, relative to the given frame of reference. As a result, from the point of view of the supplier of the dc-regulated power supply, the power supply itself may be the system that is composed of transistors as components.

12.2.2. Factors Involved in Design Selection

The design of complex analytical systems is a multifaceted problem that, because of its nature, will require numerous trade-offs and compromises.

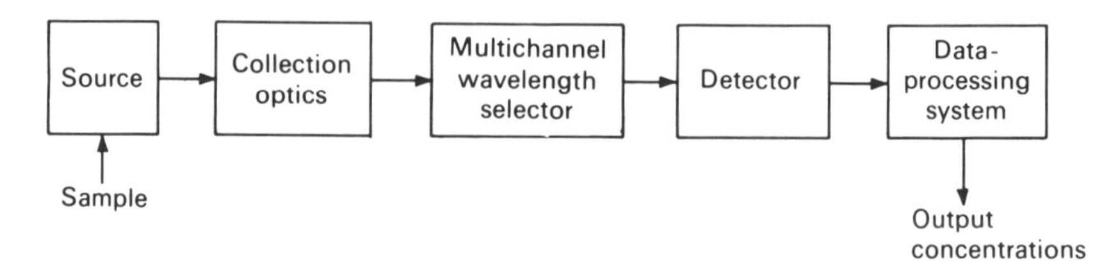

Figure 12.3. Block diagram of multielement spectroscopic analysis system.

Table 12.1. Possible Design Criteria for Selecting among Several Proposed Systems

(a) The configuration that gives the lowest detection limits.
(b) The configuration that can determine elements x, y, z with the highest accuracy and precision.
(c) The configuration that can monitor different combinations of elements conveniently.
(d) The configuration that can monitor some minimum number of elements.
(e) The configuration that has sufficient resolution to be used with the inductively coupled plasma (ICP).

Table 12.1 lists a series of design criteria that might be used in selecting from among several alternatively proposed systems for implementing a given mission. The criteria listed in Table 12.1 are all one-dimensional criteria that individually are too narrow to be used as a guide in establishing an overall design philosophy. What is needed for systems design is a multidimensional blend of criteria that is based on weighted value functions of the system parameters and constraints. The real purpose of systems engineering, therefore, is to examine all aspects of a design using balanced judgment to avoid unintentional overemphasis of certain criteria. In the analytical literature, detection limit often seems to be used as a one-dimensional criterion in selecting among various analytical configurations. This makes sense if a certain detection limit is needed to achieve a given mission. If detection limit is not an important factor in mission success, it does not make sense to weight this criterion too heavily. It is clear from the foregoing discussion that systems design is not an exact science since value judgments are inevitably involved.

Table 12.2 lists the factors that should be considered when selecting among several alternatively proposed systems. Items (c) in Table 12.2 can be illustrat-

Table 12.2. Factors to be Considered when Choosing between Alternatively Proposed Systems

(a) Determine what constitutes adequate performance for successful realization of the mission or objective.
(b) If two or more systems appear capable of achieving the mission, do additional performance specifications exist that will indicate superiority of one system over another?
(c) How sensitive is system performance to design details and/or operational factors?
(d) Do various penalty factors exist, and if so, what are they?
(e) Are economic factors such as cost constraints important?

ed in the context of multichannel spectroscopy by considering the problems associated with assembling a slew-scan mechanical system that can jump reproducibly from one atomic line to another. Since atomic lines are quite narrow, the mechanical tolerances required to build such a system might require considerable design effort. In addition, operational factors such as scan-drive wear may become important for purely mechanical systems. In using direct-reading spectrometers, temperature changes represent an operational factor that can alter system performance. The term *penalty factor* refers to undesirable characteristics such as increased size, weight, and complexity.

12.2.3. Optimal Design and Suboptimization

The term *optimum* was coined in 1710 by the German philosopher, statesman, and mathematician Baron Gottfried Wilhelm von Leibniz in his philosophical work entitled "Essays in Theodicy on the Goodness of God, the Liberty of Man, and the Origin of Evil." In a 1952 translation of this work by Huggard (1952), Leibniz states his now-famous philosophy on the best of all possible worlds:

> There is an infinitude of possible worlds, among which God must needs have chosen the best, since He does nothing without acting in accordance with supreme wisdom. Now this wisdom, united to a goodness that is no less infinite, cannot but have chosen the best . . . As in mathematics, when there is no maximum or minimum, everything is done equally or . . . nothing at all is done: so it may be said . . . that if there were not the best (optimum) among all possible worlds, God would not have produced any.

This Leibnizian universe was later satirized by the French author Francois Voltaire in his novel *Candide* (1759). From a mathematical standpoint, in the context of systems engineering, the concept of optimization refers to the maximization or minimization of selected criteria functions.

Suboptimization often results inadvertantly when one relevant factor is overweighted at the expense of other equally important factors. Strange as it may seem, however, as systems develop or evolve, suboptimization may occur. Probably the most famous example of suboptimal development familiar to most people today is the layout or arrangement of the typewriter keyboard. This topic is the subject of a delightful essay by the Harvard biologist and historian of science Stephen Jay Gould (1987).

From the standpoint of keyboard efficiency, in the English language 70% of the words use the letters *DHIATENSOR*, which should logically be located on (or close to) the home row of keys. Instead, the most common vowel in the English language (as known to fans of the television program "Wheel of

Fortune" in the United States) is *E*, which is located along with the other common vowels *I*, *O*, and *U* on the top row of letters. The letter *A*, while being located on the home row, is relegated to the little finger of the left hand.

In contemplating this obvious example of suboptimal design, we must believe that complex structures such as the typewriter keyboard layout arose for a reason other than complete randomness or the capricious nature of the original designer. In fact, rudimentary elements of a simple alphabetical arrangement can be seen in the home row of keys, where the sequence *DFGHJKL* is essentially in alphabetical order with the deletion of the vowels *E* and *I*. In fact, the typewriter and other technological innovations are products of history and, as such, are subject to the regulations governing the nature of temporal connections. The problem that arises in systems that evolve over time is that history may soon overtake the original purpose, and what was once a sensible solution becomes a liability.

This is in fact what happened in the development of the typewriter. The original typewriter was a mechanical arrangement of levers attached to keys. If the keys were struck too fast, it was not uncommon to have two keys jam so that the second and any subsequent keys would simply strike the jammed keys instead of striking the paper. This would result in lines of the same letter being produced. To compound the situation for the typist, in the original typewriter, invented by C. L. Sholes in the 1860s, the keys struck the paper from beneath so that you could not inspect what you were typing as you went along. In the original typewriter, jamming was a serious problem that became more prevalent as the typing speed increased. In this early configuration of the typewriter, too much speed was a definite liability. To minimize the possibility of jamming, the keyboard was deliberately arranged to reduce typing speed. In other words, the keyboard was arranged to optimize between the two conflicting factors of speed and jamming.

Clearly, jamming is no longer a problem with the modern word processor; nevertheless, the original suboptimal layout is still the only layout available for keyboards. To understand why suboptimal designs are not replaced once the factors that led to their origins are no longer in force, we must look in more detail at the early development and evolution of technological systems such as the typewriter. In the early days of typewriters, a number of competing systems were in existence. In the early development of a complex system from simple components, the influence of *contingency* is most strongly felt. The term contingency refers to the chance result of a long string of unpredictable antecedents. These tiny, seemingly trivial decisions (at the time) that are made during the early embryonic development of the technology begin to set the pathway for the future development of the system. As these choices from the distant past accumulate, they reinforce the stability of the pathway. The stabilization of the path of development is known as *incumbency*. Incumbency arises once the little quirks of early flexibility begin to force the sequence of further development into a particular firm channel.

In the development of the typewriter, the chance occurrence that set the channel for the future development of the typewriter was a widely publicized typing competition featuring two rival keyboard systems. Ironically, the two typists in the competition did not use the same typing style (Gould, 1987). Instead, one typed on one keyboard with four fingers, whereas the other, who had previously memorized the keyboard, had devised the system of touch-typing that we use today. When the touch-typist won the competition, the perception of the public was that the QWERTY keyboard (which we use today and is named in honor of the first six letters of the top row) was superior to the other keyboard arrangements. Since the QWERTY keyboard was never compared on an equal basis with other existing keyboard layouts at the time, we shall never know whether it was innately superior to alternative layouts or not.

Thus, the original typewriter was not deliberately designed suboptimally from a systems point of view. However, as the concept of the typewriter has evolved and the original need for slow typing speed has ceased to exist, the vested interests of the typing practitioners themselves have prevented subsequent reoptimization. This situation illustrates the important influence that practitioners themselves have on the direction and adoption of new technology.

With the advent of typing schools, the stage was set for the adoption of what we call today an "industry standard." Industry standards have both good and bad aspects. The impetus for industry standards arises as a result of vested interests. For example, in the case of touch-typing, those having the vested interest are the typists themselves, who already know how to type with the current keyboard and are not interested in learning to type on a new keyboard, even if it might be faster. Since stasis is the norm for complex systems, industry standards are frequently adopted to provide a dependable framework in what would otherwise be a chaotic situation. Once adopted, however, the overwhelming incumbency of industry standards makes even obvious improvements difficult to implement.

12.2.4. System Requirements, Imposed Constraints, and Design Objectives

The distinction between system requirements, imposed constraints, and design objectives is essentially one of degree. The *system requirements* are those primary characteristics (constraints) that are absolutely necessary if the system is to be able to fulfill its mission. On top of the basic system requirements there are often a series of additional constraints that may arise artificially as a result of commodity availability and other factors that are not directly related to mission requirements. These additional constraints are known as *imposed constraints*. The term *design objective* refers to design goals that are considered desirable but not essential to mission success. In the context of multielement analysis, acceptable accuracy and precision may be a system requirement, availability of certain optical components may be an imposed constraint, and compact size for the analytical system may be a design objective. As the system develops from prototype to advanced system, design objectives frequently become system requirements. Thus, the mission requirements for early automobiles called for reliable, functioning internal combustion engines. At that time, quiet operation might have been a design objective. Today, however, the system requirements for automobiles call not only for mufflers but for antipollution systems as well.

12.2.5. Instrument Development and Design Flow

Figure 12.4 illustrates some of the important aspects of design flow that are involved in instrument development. All forms of systems development must begin with the *mission objective*. Any well-developed mission objective for an analytical system must answer the basic question: What analytical need will be fulfilled by the instrument? Alternatively, one might ask: What is the primary purpose of the instrument? It is at this point that a fundamental understanding of analytical science is needed if the actual measurement needs of the analytical community are to be fulfilled.

From an analytical systems standpoint the mission objective includes considerations such as types of samples to be analyzed, number of elements to be determined, accuracy and precision needed, sensitivity requirements, sample size available, analysis time, and number of samples per unit time that can be analyzed. This is by no means a complete list of all the considerations that might be included in the mission objective; however, a clear formulation of the mission objective is essential in the development of viable analytical systems, and a great deal of effort should be spent on this aspect of design flow before investing time and money in the subsequent steps.

Once a clear formulation of the mission objective has been established, *requirements analysis* must be conducted to determine the specifications that will be required to accomplish the mission objectives. Thus requirements analysis will involve specifying the spectral resolution that might be needed with a given excitation source and sample type to avoid spectral interferences. What type of data-processing speeds will be necessary to achieve a given sample throughput per hour? What kinds of sample pretreatment will be necessary to avoid interferences and achieve the desired accuracy?

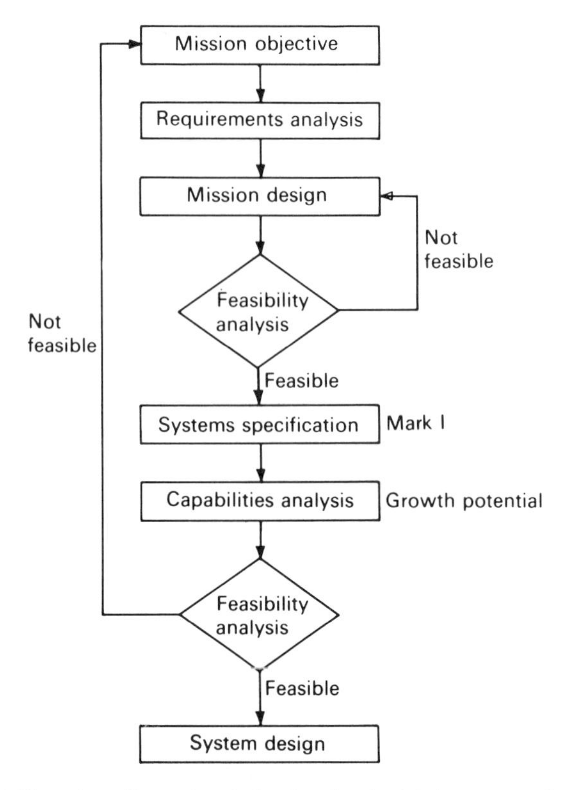

Figure 12.4. Flow chart illustrating design flow involved in instrument development.

Once the requirements analysis has been conducted, the next step in the development is *mission design*. In the mission design stage, the systems designer begins to consider the various subsystems that will be needed to achieve the mission objectives. In multichannel spectroscopy, the necessary subsystems required are relatively invariant. Nevertheless, at this stage of the development one can begin to consider the various choices that exist among the subsystem alternatives. For example, one could consider whether the wavelength selector subsystem should be a dispersive optical system or a transform optical system. If an array detector is selected, should the system use a photodiode array or a charge-coupled device array?

At this point, it may be difficult to decide between different possible alternative subsystems. These important choices must be made through the process of *feasibility analysis* in light of the mission objective and not on the basis of extraneous factors. The basic question that must be answered by feasibility analysis is whether or not the given overall system (or subsystem) can meet the requirements for the mission objective. If two or more alternative systems appear capable of meeting the mission objective, feasibility analysis may uncover additional performance specifications that will indicate the superiority of one system (or subsystem) over another. In the absence of available information in the literature on system performance, laboratory testing of various subsystems *under expected conditions of operation* may be necessary as part of feasibility analysis. If a particular subsystem cannot meet the mission requirements, mission design must continue until a suitable subsystem can be found. Of course, if no subsystem that can meet the mission requirements exists, it may be necessary to reconsider the mission objective to see if it is realistic with today's technology.

Once the feasibility analysis has been conducted, the various subsystems can be specified. The specification of the various subsystems that make up the overall analytical system represents the establishment of the prototype system (Mark I) from which more sophisticated configurations are expected to arise. At this point, it is important to look closely at any other secondary missions the system might conceivably accomplish. This process, known as *capability analysis*, attempts to predict whether the system can handle, or be modified conveniently to handle, expanded mission objectives as they arise. An important characteristic, from a capabilities analysis point of view, is the *growth potential* of the system. Table 12.3 illustrates the role of growth potential analysis in systems development. From a qualitative standpoint, the

Table 12.3. Role of Systems Analysis in Systems Design

Step 1. Operational Analysis Establish primary objectives

Step 2. Constraint Analysis Ascertain additional constraints that affect design.

Step 3. Feasibility Analysis Identify those classes of systems currently available that can satisfy requirements imposed by steps 1 and 2.

Step 4. Extension Analysis Determine characteristics of ultimate system meeting requirements imposed by step 1 and those of step 2 that are considered irrevocable.

Step 5. Growth Potential Analysis Identify path by which each class of feasible system is expected to evolve toward ultimate system.

(a) Requires technical breakthrough followed by total redesign.

(b) Evolution proceeds by discrete steps of technical advance, each followed by major redesign.

(c) Gradual modification based on advances in state-of-the-art without need for major redesign.

Systems in category (c) are said to possess growth potential growth potential is the ease with which the system can adapt to expanded mission objectives. In other words, can expanded mission objectives be accomplished by simple modification of the system or will it require that the total system be abandoned and completely redesigned from the beginning?

The final stage in the design flow is *system design*, which consists of optimizing the overall performance of the complete system. Since the various subsystems must be matched together to form the complete system, trade-offs and compromises will undoubtedly be required in producing the overall system. Whatever trade-offs are necessary, they should always be made with the overall system performance in mind so that the mission objective is not sacrificed.

12.3. THE EVOLUTION OF ANALYTICAL TECHNOLOGY

Up to this point, we have considered the development of a single analytical system as if it were the only choice available. Actually, there is a whole host of competing analytical technology from which the applications specialist can choose. In the commercial world, to sell a product in the presence of competition, it must be attractive to buyers. In the world of analytical technology, a technique, in order to survive, must have certain desirable features to be attractive to analytical applications specialists. Table 12.4 lists 10 criteria of importance in the selection of an analytical technique for a given measurement situation. Of the 10 criteria (which are not listed in order of importance), the first is clearly the most important. To be viable, any analytical technique must be able to fulfill *real* measurement needs. This is the primary basis upon which analytical technology is adopted. By analogy with biological systems, an analytical technique that can solve an important measurement problem that cannot be readily accomplished with competing technology can be said to occupy a *niche*. The broader the range of applications of a given analytical technology, the broader the niche and the greater the chances of survival that it has.

The various forms of analytical technology survive under the combined influence of many different types of measurement needs as dictated by a high-technology society. In spite of the great diversity of measurement needs, however, many analytical problems have certain features in common. As a result, an infinite number of analytical techniques is not necessary. Those techniques that satisfy certain key features become desirable to a wider audience of analytical specialists because they possess broad applications possibilities.

Table 12.4. Criteria for Selection of Analytical Technology

1. Does the technique fill a *real* measurement need?
2. Is the technique, in some way, simpler in terms of apparatus and difficulty of operation than competing technology?
3. What is the cost, both initial and upkeep (supplies and maintenance)?
4. How does the technique compare with competing technology in terms of speed (i.e., single element vs. multielement, samples per hour, etc.)?
5. Is the technique versatile? Can a variety of important measurement problems be solved with this technology?
6. Is the reliability of the analytical data in terms of accuracy and precision better than with competing technology?
7. Does the technique possess satisfactory sensitivity for the given application and for possible future applications?
8. Is the analytical system itself reliable or does it require significant amounts of downtime for servicing?
9. What are the space requirements for the analytical system?
10. What ancillary services are needed for the analytical system (cooling water, electricity, compressed gases, liquid helium, etc.)?

12.4. THE NEED FOR MULTIELEMENT ANALYSIS

Since measurement need is what drives analytical development, it is worthwhile at this point to examine briefly the need for simultaneous multielement analysis (SMA). The original major application of emission spectroscopy in chemical analysis was in the area of metallurgical analysis where it was used (and is still used today) to determine undesirable impurity elements and/or minor alloying constituents. The area of metallurgical analysis is customarily subdivided into ferrous materials and nonferrous materials. Nonferrous materials include aluminum, magnesium, zinc, tin, lead, and their alloys. This distinction is based on the need for somewhat higher dispersion and resolving power with the line-rich iron alloys compared with the simpler spectra emitted by the nonferrous materials. Another area where simultaneous multielement analysis has been used widely is in agriculture, where soil and plant analyses are conducted to determine trace-element deficiencies.

With the development of the semiconductor electronics industry and the rise of materials science as an important discipline, the need for survey analyses at trace and ultratrace levels spurred the development of various simultaneous multielement techniques. The term *survey analysis* refers to the qualitative and quantitative elemental characterization of a sample about

Table 12.5. Some Multielement Techniques

Technique	Number of Elements Simultaneously Determinable	Analysis Time	Sample Type
Spark source mass spectrometry	30–50	Photographic detection, slow; Electronic detection, fast	Solids
ICP-source mass spectrometry	30–50	Same as above	Liquids
Neutron activation analysis	30–45	Fast to slow depending on element	Solids or liquids
X-ray fluorescence	15–20	Fast	Solids or liquids
Emission spectroscopy	20–30	Fast	Solids or liquids

Source. Reprinted from Busch and Morrison (1973), with modifications.

which one has minimal previous information. Table 12.5 lists some of the more important competing multielement techniques. In the area of environmental science, the rising concern in the industrial nations of the impact of man on the environment has prompted numerous investigations on the possible biological hazard of the presence of potentially toxic elements in potable and waste waters. The development of various forms of high technology often calls for diverse and sometimes unexpected measurement needs. For example, the results obtained from the analysis of lubricating oils for wear metals have been used to aid in the maintenance of jet aircraft engines. All of these applications have pointed to the need for analytical systems capable of providing quantitative information on many elements in a sample simultaneously.

12.5. THE ULTIMATE SYSTEM

Having established that a need exists for analytical systems capable of simultaneous multielement analysis, it is instructive to speculate on the characteristics of the "ideal" spectroscopic system for multielement determinations. It should be clear from the foregoing discussion that no single system will be optimal for all applications, and in that sense, the ideal system does not exist. However, it may be possible to design a system that will be useful in a wide range of applications. The wider the range of applicability, the more desirable the instrument, and the closer it comes to being the ultimate system. Thus the characteristics of this hypothetical ultimate system provide a standard toward which all real systems should strive. As in any engineering design problem, it may be impossible to satisfy all the desired design criteria simultaneously in any real system. The eight essential characteristics (i.e., system requirements) of any ultimate multielement spectroscopic system are proper wavelength coverage, adequate resolution, wide dynamic range, efficiency, sensitivity, flexibility, speed, and reliability.

12.5.1. Wavelength Coverage

Any instrument intended for multielement analysis must be able to monitor atomic lines over the wavelength range from 200 to about 800 nm. Since a large number of important atomic lines fall in the short-wavelength ultraviolet region, good response in this region is essential for multielement work. The long-wavelength response limit of the system is less important because the density of spectral lines falls off rather quickly as one approaches 800 nm.

12.5.2. System Resolution

For simultaneous multielement analyses, line and adjacent background intensity measurements must be made for each element to be determined. In addition, the spectral lines of interest are not distributed uniformly throughout the spectrum. The nonuniform distribution means that information is required only over specific narrow wavelength intervals characteristic of each element of interest. Most of the complexity of atomic spectra arises from multiple line emission by each element or from radiation emitted by species in which the analyst has no interest. As a result of this spectral complexity, the wavelength selector must be capable of rejecting the unwanted portions of the spectrum from those desired. Even though large regions of the atomic spectrum may not contain any desired information (and could therfore be observed under conditions of low resolution), high resolution is often required to resolve the randomly situated atomic lines of interest from the unwanted spectral background.

Busch and Benton (1983) have discussed the effect of spectral density and distribution on multielement spectroscopic systems and have concluded that the ideal spectroscopic instrument for analytical atomic spectroscopy should have high resolution in the spectral regions of interest (i.e., in the vicinity of analytical lines) and low resolution in spectral regions that do not contain desired information. For a system with constant dispersion (like a conventional grating spectrometer used in a given order), the minimum resolution is set by the line–background combination, which is most difficult to resolve. This choice sets the dispersion and spectral resolution for the other elements even though they do not require it. High, uniform dispersion means that the spectrum from 200 to 800 nm, for example, will be quite long. The need for high dispersion by the wavelength selector subsystem has a profound effect on the successful application of currently available image detectors, which can only monitor a relatively small spectral region under these conditions.

In discussing the characteristics of the wavelength selector subsystem for multielement analysis, it should be realized that the analytical lines employed need not be dispersed across the focal plane of the instrument in order of increasing wavelength. Any convenient arrangement of lines is useful for multielement analytical spectroscopy as long as they are resolved from other potentially interfering radiation and the spectroscopist knows which line is associated with each element.

12.5.3. Dynamic Range

One of the major design problems in developing practical spectroscopic systems for simultaneous multielement anaysis is to match the linear dynamic range of the detection system to the dynamic range of the spectrum. The range of intensities that must be monitored simultaneously is quite large due to the presence of major, minor, and trace elements present in most samples. To be useful for simultaneous multielement analysis, a spectroscopic system must be capable of detecting weak trace-element lines in the presence of strong major-element lines that arise from the sample matrix and background from the spectral source. To accomplish this goal, two approaches can be taken. The first is to design a detection system with a linear dynamic range that matches that expected in the spectrum. Depending on the detection system envisioned, this may be difficult if not impossible. The second approach is to develop means for selective attenuation that will permit the attenuation of strong lines without simultaneously attenuating the weak lines.

12.5.4. Efficiency

A complete spectrum from 200 to 800 nm contains an enormous amount of information when examined under conditions of high resolution. Fortunately, not all of the information contained in a complete spectrum is required in analytical atomic spectroscopy. Since the linewidths associated with atomic transitions are quite narrow, and since most analyses do not require information about all the elements in the periodic table, intensity measurements are required only over specific narrow-wavelength intervals characteristic of each element of interest. For every element to be determined, the minimum information required for an analysis consists of the atomic line intensity and the adjacent background.

The efficiency of a spectroscopic system for multielement analysis can be expressed in terms of the amount of useful information collected by the system compared with the total number of resolution elements monitored. It can be defined, therefore, as the minimum number of intensities that need to be monitored divided by the total number of intensities monitored by the instrument. It should be realized that for purposes of routine multielement analysis where a fixed combination of elements is to be determined, a complete spectrum (in the sense of a continuous sequence of wavelengths) is not necessary or even desirable from a data-processing point of view. If, for example, only 12 specific elements are to be determined in a given sample, it is not necessary to use an instrument that will determine the intensities of all the resolution elements comprising the complete spectrum. For routine samples that are well characterized, a system with a higher efficiency will be faster and will require less time for data processing.

12.5.5. Sensitivity

Any viable spectroscopic system for multielement analysis must possess the requisite sensitivity to be useful for trace determinations. Many spectroscopic determinations are conducted to obtain information on trace constituents so the ultimate multichannel system must provide good detection limits for the elements of interest. Achieving good detection limits may depend on selecting the appropriate mode of observation of an atomic population, proper excitation source selection and optimization, and means for increasing spectrometer optical throughput. Since detection limits depend on the lmiting noise in the system, appropriate system design will call for an understanding of noise sources in the system and ways of reducing them.

12.5.6. Flexibility

Even for an instrument designed to determine a fixed combination of elements, a certain amount of flexibility is required in selecting the combination of analytical lines for a given analysis. That is to say that the analytical line for a given element is not necessarily fixed but is, in fact, sample dependent. The need for wavelength selection flexibility has an effect on the choice of wavelength selector–detector combination employed for a given application. It should be realized that the elements present in any complex sample can be classified into three categories on the basis of their concentration. Thus, a sample may be composed of major elements, minor elements, and trace elements on the basis of their relative abundance in the sample. It should be clear that the intensity from major-element lines will be much greater than for minor- and trace-element lines. Furthermore, because of self-absorption (see Mavrodineanu and Boiteux, 1965, for a discussion of self-absorption) and detector dynamic range limitations, an analytical line suitable for trace analysis (i.e., a sensitive resonance line) will not be satisfactory at higher concentrations. Spectrochemical procedures must be developed for specific sample types and the wavelengths selected on the basis of concentration and interference considerations from concomitants. Spectrochemical analyses are most useful, therefore, for routine analyses of particular sample types. When sample types are changed, new spectrochemical procedures must be developed that may call for different analytical lines even if the combination of elements monitored is unchanged.

12.5.7. Speed

For applications where a great number of analyses must be performed on a routine basis (such as in many metallurgical industries), speed may be one of the most important criteria for a multielement analytical system. Since procedures that involve chemical dissolution and digestion often require more time than the actual analysis itself, the ability to excite solid samples is highly desirable. In addition, minimum pretreatment reduces the chance of sample contamination by reagents and the loss of volatile elements during pretreatment. Aside from speed, all trace analyses should strive to reduce the amount of chemical manipulation required wherever possible to avoid contamination.

12.5.8. Reliability

In terms of reliability, a viable multielement analytical system must be capable of long-term operation with a minimum of downtime. In many industries that employ multielement analyses for quality control (such as metallurgical industries), the plant is operated 24 hours per day all year long. To be useful in this type of environment, the system must be as rugged and reliable as possible to withstand the rigors of daily usage. In this respect, a system with as few moving mechanical parts as possible is desirable.

12.5.9. Possible Configurations

Once an atomic population has been generated, its presence may be observed by its emission, absorption, or fluorescence as discussed in the next chapter. It is evident that the number of potential configurations for spectroscopic multielement systems is staggering when one considers the number of available options for each component combined with the three complementary modes of observation of the atomic vapor. The particular configuration selected must be developed from a systems point of view and must not only fill real analytical needs, but should also offer real advantages or increased capabilities to compete with alternative analytical technology. Those systems that can achieve the eight essential characteristics discussed previously are most likely to be useful in simultaneous multielement analysis. The following chapters will describe some of the approaches that have been taken to develop multielement spectroscopic systems and will discuss how close each comes to the ultimate system.

REFERENCES

Busch, K. W. and Benton, L. D. (1983), *Anal. Chem.*, *55*, 445A.

Busch, K. W. and Morrison, G. H. (1973), *Anal. Chem.*, *45*, 712A.

Ellis, D. O. and Ludwig, F. J. (1962), *Systems Philosophy*, Prentice-Hall, Englewood Cliffs, NJ.

Gould, S. J. (1987), *Natural History*, *96*(1), 14.

Huggard, E. M. (trans.) (1952), *Theodicy: Essays on the Goodness of God, the Freedom of Man and the Origin of Evil*, Yale University Press, New Haven.

Mavrodineanu, R. and Boiteux, H. (1965), *Flame Spectroscopy*, Wiley, New York.

CHAPTER

13

INTRODUCTION TO ANALYTICAL ATOMIC SPECTROSCOPY

All forms of atomic spectroscopy are predicated on the efficient generation of atomic populations. With the exception of volatile elements such as mercury and the rare gases, the generation of a free atomic population from a sample requires energy. Thus, some form of high-temperature atom reservoir is required. Once generated, however, the presence of the atomic population can be observed by three complementary modes of observation: atomic emission spectroscopy, atomic absorption spectroscopy, and atomic fluorescence spectroscopy. Because these modes of observation represent three ways of observing the same atomic population, they have many features in common. These common features are all related to factors that alter the efficiency of the atom reservoir to produce an atomic population. As a result, all three modes of observation suffer from ionization interferences and compound formation interferences (See Herrmann and Alkamade, 1963, and Mavrodineanu and Boiteux, 1965, for a discussion of interferences in flame spectroscopy). There are differences between the three modes of observation, however, and even though they appear to use similar components, the components are used for inherently different reasons.

13.1. ATOMIC EMISSION

Figure 13.1 shows the experimental arrangements used to implement the three modes of observation. Of the three modes, atomic emission is the simplest in terms of the number of subsystems required for implementation. It is also the mode of observation most amenable to simultaneous multielement analysis. However, it places the most stringent requirements on the atom reservoir subsystem and the wavelength selector subsystem.

The great popularity of atomic emission today is largely the result of the efforts of Professor Velmer Fassel of Iowa State University. During the heyday of the development of atomic absorption, Fassel continually fought the notion that was prevalent at the time, namely, that atomic absorption spectrometry must be clearly more sensitive than atomic emission spectros-

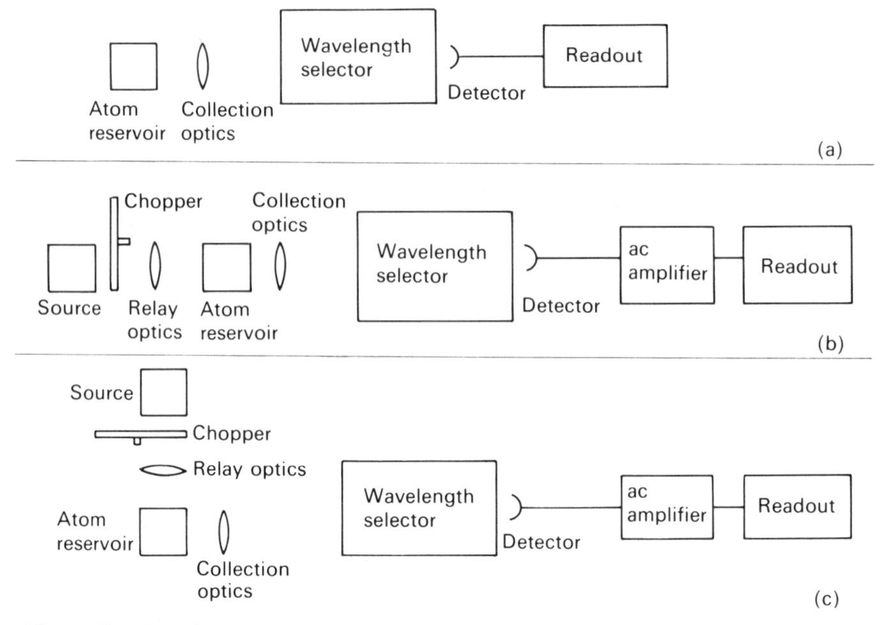

Figure 13.1. Experimental arrangements used to implement three modes of observation in atomic spectroscopy: (*a*) atomic emission configuration; (*b*) atomic absorption configuration; (*c*) atomic fluorescence configuration.

copy because with atomic absorption one observes the ground-state population, which is much larger than the tiny excited-state population observed with atomic emission.

The great fallacy of this argument, of course, was that it was irrelevant. In fact, the detection limit in atomic absorption with a photon-noise-limited detector might actually be expected to be worse rather than better compared with a similar measurement made in emission. The reason can be understood from the point of view of a simple statistical comparison of averages. As one approaches the detection limit in atomic absorption, one must determine the difference between two large signals, both of which are fluctuating because of photon shot noise (see Section 9.4). By comparison, the situation in atomic emission (and atomic fluorescence as well) at the detection limit is one of detecting the presence of a small signal over a small background. Since the intensities involved in this situation are less, the shot noise involved is also less, and it is easier to detect the presence of the small emission signal. Today, we know that flame atomic absorption and flame atomic fluorescence produce better detection limits compared with flame atomic emission for elements with resonance lines shorter than 350 nm (Alkemade, 1968).

The big increase in the popularity of atomic emission occurred with the development of the induction-coupled plasma (ICP) by Fassel and co-workers (Wendt and Fassel, 1965, 1966; Fassel and Kniseley, 1974a, b). The low background, superior detection limits, freedom from chemical interferences, and large linear dynamic range have made atomic emission–ICP spectroscopy the method of choice for trace-element determinations in solution.

13.1.1. The Function of the Atom Reservoir

To be suitable for atomic emission, the atom reservoir must be able to accomplish two primary functions simultaneously: (1) atom generation and (2) atom excitation. Both of these (but particularly, excitation, see Eq. 10.20) require high temperatures. High temperatures imply increased background radiation and complex spectra (many lines), both of which place increased demands on the dispersion and resolving power of the wavelength selector. Table 13.1 lists some common atom reservoirs for atomic emission.

In addition, for simultaneous multielement emission spectroscopy, the atom reservoir must be able to provide optimum excitation conditions for a wide variety of elements in as small a geometric region of the reservoir as possible so that this entire region can be viewed by the spectrometer at one time. Optimum excitation conditions vary from element to element and depend on the individual spectrochemical properties of the element. These spectrochemical properties include the excitation potential of the analytical line, the ionization potential of the element, and the free-atom population of the element as a function of the atom reservoir zone sampled by the spectrometer. Since these properties vary from element to element, individual optimization is not possible in simultaneous multielement analysis, and a compromise in operating conditions is necessary. The need for compromise operating conditions is an important characteristic of simultaneous multielement analysis that has a significant impact on system design (Busch and Morrison, 1973). Busch and co-workers have studied the need for compromise excitation conditions in simultaneous multielement analysis and have developed a mathematical procedure for determining the optimum compromise for

Table 13.1. Common Excitation Sources for Atomic Emission Spectroscopy

Source	Sample Type
Direct-current arc	Metallic self-electrodes and powders with graphite electrodes
Condensed high-voltage spark	Metallic self-electrodes and solution deposited on copper electrodes
Combustion flames	Trace constituents in aqueous solutions
Induction-coupled plasma	Trace constituents in aqueous solutions

n elements in terms of observation height in the atom reservoir and fuel-to-oxidant ratio of the flame (Brost, Malloy, and Busch, 1977).

If a flame is used as an atom reservoir, the fuel-to-oxidant ratio has an important effect in determining the chemical and physical environment in the particular observation zone (Busch, Howell, and Morrison, 1974). Boumans and DeBoer (1972) have studied the premixed nitrous oxide-acetylene flame as a potential excitation source for simultaneous multielement analysis. They evaluated this flame with a variety of burners and concluded that shielding with either nitrogen or argon is necessary for simultaneous multielement determinations. Shielding produces a large temperature gradient in the flame (Kirkbright and Vetter, 1971), and Boumans and DeBoer conclude that this provides the required wide range of excitation conditions over a small flame volume.

Today, however, the ICP is probably the most nearly optimum excitation source for liquid samples (Fassel and Kniseley, 1974a, b; Greenfield et al., 1975; Boumans and DeBoer, 1975a, b, 1976; Fassel et al., 1976; Winge et al., 1977; Olson, Haas, and Fassel, 1977; Boumans, 1987). Because of its high efficiency as an excitation source, the ICP produces complex spectra requiring high dispersion to avoid spectral interferences. Table 13.2 lists some of the advantages of the ICP. Because of the high temperature produced with the ICP, chemical interferences are reduced compared with those observed with flame sources. The freedom from chemical interferences has been achieved, however, at the expense of increased potential for ionization interferences and spectral interferences.

Table 13.2. Advantages of Induction–Coupled Plasma in Atomic Emission

1. One set of operating conditions nearly optimal for most elements, little compromise in operation required for multielement analysis
2. Working curves linear over 4–5 orders of magnitude in terms of concentration because of negligible self-absorption, permits determination of major, minor, and trace elements without need for multiple dilutions
3. High temperature of plasma reduces effect of chemical interferences common in analytical flame spectroscopy, permits determination of refractory elements such as Hf, Zr, U, B, W, and P
4. Simple plasma background spectrum consisting of argon lines and OH band emission, spectrum from 200 to 300 nm relatively clean
5. No contamination from electrodes
6. Superior detection limits, typically 0.1–10 ng/mL for most elements
7. Good stability over long-term operation, reduces need for frequent standardization of instrument

13.1.2. The Function of the Wavelength Selector

The role of the wavelength selector in atomic emission spectroscopy is to isolate the desired spectral line (i.e., the analytical line) from any source background continuum as well as all the other spectral features (atomic lines and molecular emissions) emitted by the source. As the spectral density emitted by the source increases, this function becomes increasingly difficult to satisfy. Because of the important role that this subsystem has on simultaneous multielement analysis, a large portion of this book has been devoted to this topic. The constraints imposed on the wavelength selector subsystem by the requirements necessary for simultaneous multielement analysis have been discussed in Section 12.5.2.

13.2. ATOMIC ABSORPTION

In examining the growth potential of atomic absorption for simultaneous multielement analysis, it is important to realize that atomic absorption developed as an inherently single-element technique. Its popularity can be attributed to the advantages that it possessed compared with its rival single-element technique at the time, absorption spectrophotometry. Prior to the development of atomic absorption, solution spectrophotometry was the preferred method of single-element analysis for metals in solution at trace levels (Sandell, 1959). Analyses were conducted by reacting the analyte metal ion with an appropriate complexing ligand to form an absorbing coordination complex that could be determined by absorption spectrophotometry. Although absorption spectrophotometry was quite satisfactory for simple samples, complex samples frequently required tedious, time-consuming prior separation of the element of interest from other species to prevent interferences. Interferences could arise from concomitant ions reacting with the complexing agent to form absorbing species. Since solution absorption spectra are typically quite broad, these extraneous absorbing species would often absorb strongly at the absorption maximum of the species of interest.

When atomic absorption was developed, it provided a competing analytical technique that could also perform single-element determinations on trace-level samples for a wide variety of metal ions in solution. Because of the specificity of atomic line spectra, analyses of many complex samples could be carried out routinely without the need for prior separation of concomitant ions. Since resolution requirements for the monochromator were modest when hollow-cathode lamps were used, table-top instruments not much larger than a conventional absorption spectrophotometer could be developed. Since separations were unnecessary in a large number of applications, analyses frequently could be performed faster by atomic absorption than by absorption spectrophotometry. Thus, while both techniques satisfied real measurement needs (i.e., single-element determinations of metals at trace levels in aqueous solution), atomic absorption provided additional advantages over solution spectrophotometry and was therefore adopted as the preferred method of analysis in many cases.

Although spectral interferences were greatly reduced compared with absorption spectrophotometry, they were not completely eliminated in atomic absorption either. In addition, as experience with the technique was gained, it became apparent that accurate results could be obtained by atomic absorption with complex samples only through a knowledge of various matrix effects that could alter the signal if present. To avoid these matrix effects, careful calibration of the instrument was required with standards that closely matched the sample. With ordinary technique, relative standard deviations of 2–5% could be achieved without difficulty. This precision is not as good as that obtainable with absorption spectrophotometry (0.2% relative standard deviation) but is satisfactory for a wide variety of applications. Finally, the need for a large selection of hollow-cathode lamps with limited shelf life represents a substantial investment with atomic absorption if a wide variety of elements is to be determined.

13.2.1. Source Requirements

Figure 13.1*b* shows the experimental arrangement for atomic absorption spectroscopy using the principle described by Walsh (1955). In conventional single-element atomic absorption, the absorption at the line center is determined by using a narrow-line source emitting the given resonance line of interest. For maximum sensitivity, the source emission line profile should be less than the absorption line profile of the analyte in the flame, as illustrated schematically in Figure 13.2. To achieve the linewidth requirements placed on the source shown in Figure 13.1*b*, low-pressure electrical discharges such as the hollow-cathode discharge tube are normally employed. By employing nonthermal excitation and low pressure, Doppler broadening and collision broadening (see Mavrodineanu and Boiteux, 1965, for a discussion of the sources of line broadening) are both reduced. Doppler broadening is reduced because the gas temperature within the lamp remains relatively low (the plasma is not in local thermodynamic equilibrium, excitation occurs primarily by electron impact). To be useful as a source in atomic absorption, the electronic excitation temperature in the source must be greater than that in the atom reservoir to satisfy Kirchhoff's law governing emission and absorption.

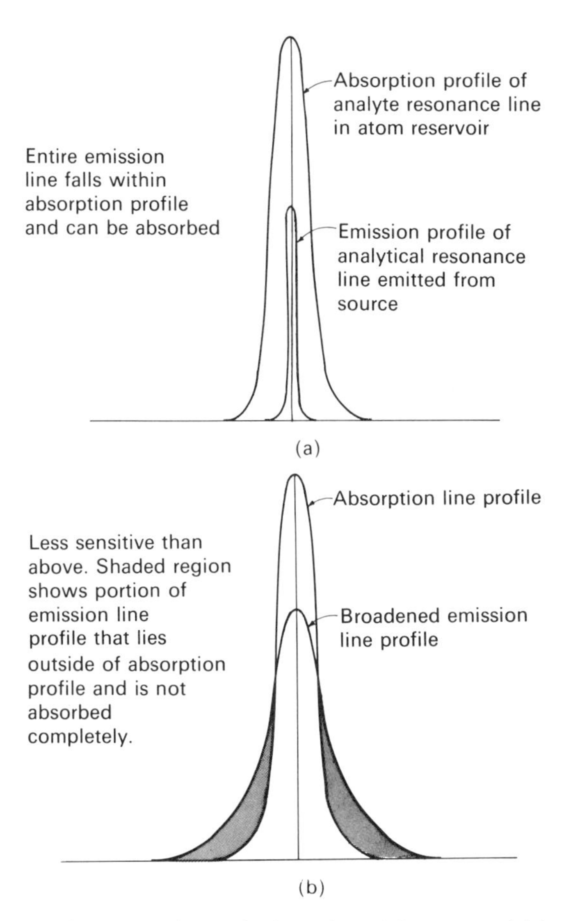

Figure 13.2. Effect of line profile in atomic absorption. (*a*) Greatest sensitivity occurs when entire emission line profile falls within absorption line profile and can be absorbed. (*b*) Less sensitivity occurs when portion of emission line profile lies outside of absorption profile and is not absorbed completely.

13.2.2. The Role of Chopping

Compared with the experimental arrangement for atomic emission (Figure 13.1*a*), atomic absorption requires additional components beyond the need for a suitable source. A chopper is employed along with an ac amplifier (often a lock-in amplifier or phase-sensitive detector is used to reduce noise) to modulate only the source emission and discriminate against the emission (within the monochromator bandpass at the resonance line wavelength) of the atom reservoir. Atom reservoir emission is not chopped and is therefore not amplified by the ac amplifier. The wavelength selector functions to isolate the desired resonance line from the other radiation emitted by the hollow-cathode discharge tube and also to reduce the amount of flame background incident on the detector. Reducing the amount of unmodulated flame background striking the detector reduces shot noise on the signal (which is proportional to the square root of the total intensity striking the detector). Since shot noise is white noise, a certain portion of it will appear within the frequency response bandwidth of the ac amplifier and be amplified. Compared with a simple ac amplifier, using phase-sensitive detection reduces the amount of shot noise transmitted by the system to that fraction of the noise within the amplifier frequency response bandwidth which has a fixed phase relationship with the reference signal generated by the chopper (see Horowitz and Hill, 1980, for a discussion of lock-in amplification). Since noise has no fixed phase relationship, lock-in amplification can reduce additive noise from unmodulated background dramatically.

13.2.3. The Function of the Atom Reservoir and Wavelength Selector

Although superficially similar at first glance, the functions of the atom reservoir and the wavelength selector in atomic absorption are different from those in atomic emission. In atomic absorption, the atom reservoir's primary purpose is to generate an atom population from the sample in as efficient a manner as possible. Since excitation of the sample in the atom reservoir is not necessary (or desirable, see following discussion), both flames and electrothermal atomization furnaces such as the carbon rod atomizer are suitable for use.

Compared with atomic emission spectroscopy, the requirements placed on the wavelength selector in conventional single-element atomic absorption are generally less. Since the spectrum emitted from the hollow-cathode lamp is often quite simple, short-focal-length monochromators can be used. The main requirement for the monochromator in this mode of observation is adequate resolution to reject any nearby source emission lines in the vicinity of the desired resonance line. When the resonance line is not close to any nearby extraneous source emission lines, wide slit widths can be employed to increase source radiation throughput within the monochromator. Except from a noise standpoint, it is not necessary for the monochromator to resolve spectral features from the atom reservoir emission because this radiation is unchopped and therefore not amplified by the ac amplifier.

Intense resonance line emission from the atom reservoir is undesirable, however, because it is passed by the monochromator and shows up as shot noise on the signal, adversely affecting the detection limit. The amount of this shot noise transmitted by the system can be reduced by phase-sensitive detection, as already mentioned. Since the atom reservoir excitation becomes more intense as the excitation potential of the resonance line decreases (i.e., the wavelength gets longer), atomic absorption generally produces better detection limits compared with atomic emission for elements with resonance lines shorter than 350 nm (Alkemade, 1968).

13.2.4. Sources for Simultaneous Multielement Analysis

In considering the feasibility of atomic absorption under the expanded mission requirements of multielement analysis, the need for a suitable multielement source is clearly evident. Although single-element hollow-cathode discharge tubes are available readily for conventional single-element atomic absorption, their use in multielement atomic absorption is limited by the optical requirements of the atomic absorption experiment itself. To employ a bank of hollow-cathode lamps for multielement analysis, some form of optical arrangement must be developed to combine the separate radiation beams produced by each lamp into a single beam prior to passage through the flame. Several approaches to this problem have been devised with limited success.

One approach to this problem, for example, was described by Mavrodineanu and Hughes (1968), who used the principle of reverse optics to produce a single beam from several hollow-cathode lamps. As shown in Figure 13.3, the lamps were arranged along the focal plane of a grating spectrograph at positions corresponding to the wavelength of the desired resonance lines. Resonance radiation from each lamp was recombined after striking the grating into a single beam that emerged from the entrance slit and was subsequently passed through the flame.

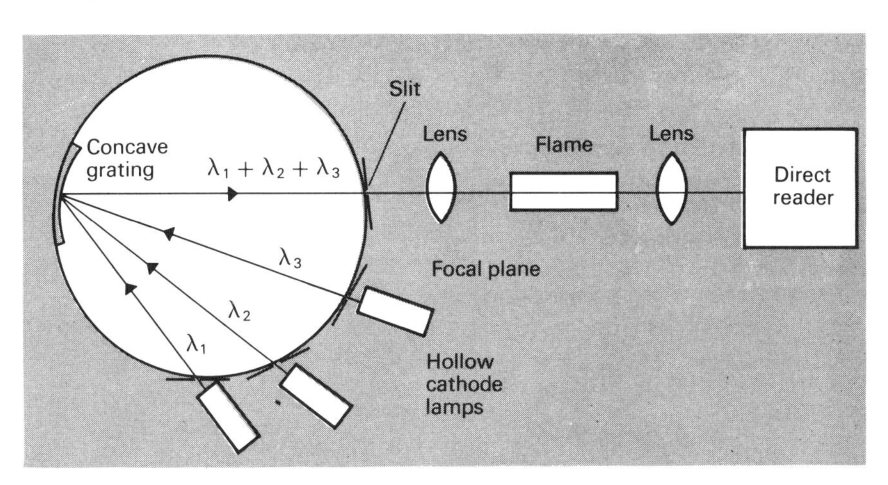

Figure 13.3. Optical arrangement for synthesis of multicomponent radiation beam from single-element hollow-cathode lamps. [Reprinted with permission from K. W. Busch and G. H. Morrison (1973), *Anal. Chem.*, *45*, 712A. Copyright 1974 American Chemical Society.]

Several workers have investigated the feasibility of producing multielement light sources. Massman (1963) studied the factors involved in the production of multielement hollow-cathode lamps. By employing concentric rings of different metals retained in a copper or steel sheath, lamps with up to four elements have been produced. Since sputtering rates and melting points of the individual elements making up the composite lamp are factors affecting the production of multielement hollow-cathode lamps, the number of multielement combinations likely to be produced in a single envelope seems limited. Since different combinations of elements will be needed for different samples, the development of multielement lamps for simultaneous multielement atomic absorption does not seem very promising.

Strasheim and Butler (1962) and Butler and Strasheim (1965) have described the development of specially designed hollow-cathode lamps that allow tandem mounting, as shown in Figure 13.4. This approach was found to be limited to three lamps because of light losses. By using the tandem method, a composite spectrum of 12 elements conceivably could be produced, assuming each lamp was a four-element multielement lamp.

While other approaches such as the use of a time-resolved spark (Strasheim and Human, 1968) have been tried, the most promising approach to multielement atomic absorption to date has been the use of a high-intensity continuum source (Fassel et al., 1966; Gibson et al., 1962; DeGalan, McGee, and Winefordner, 1967; McGee and Winefordner, 1967). These workers have demonstrated that a continuum source can be used for atomic absorption if a monochromator with sufficient resolving power is available. When a continuum source is employed, the function of the wavelength selector in Figure 13.1*b* is radically different from that described previously when hollow-cathode lamps were considered.

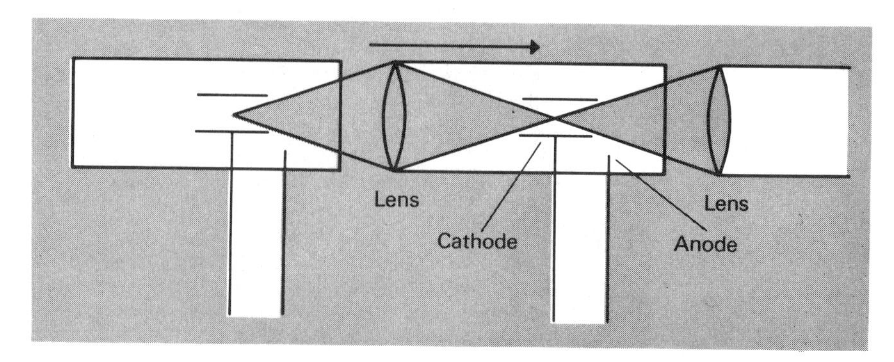

Figure 13.4. Tandem hollow-cathode lamps for production of multicomponent radiation beam. [Reprinted with permission from K. W. Busch and G. H. Morrison (1973), *Anal. Chem.*, *45*, 712A. Copyright 1973 American Chemical Society.]

As discussed, when line sources are used, the monochromator serves only to isolate the desired resonance line. When a continuum source is employed, however, much higher spectral resolution is required for adequate sensitivity, as shown in Figure 13.5. For adequate sensitivity in this situation, a monochromator with a spectral bandpass equal to or less than the absorption profile in the atom reservoir is desired.

Early research on the feasibility of continuum atomic absorption using a 150-W xenon arc gave detection limits that were generally poorer compared with those obtained with a hollow-cathode lamp. These poor detection limits were due, in part, to the low spectral radiance produced by the xenon arc below 250 nm. This spectral region is important because it contains many important elemental resonance lines. Low spectral radiance adversely affects the detection limits obtained with the continuum source because narrow slit widths are required to achieve the desired spectral bandpass. The combination of low spectral radiance and narrow slits results in low optical throughput to the detector, making it difficult to detect signals.

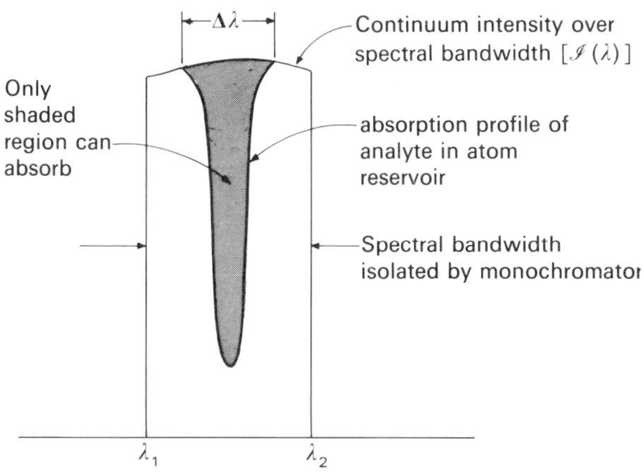

$$I_0 = \int_{\lambda_1}^{\lambda_2} \mathcal{I}(\lambda)\,d\lambda$$

$$I = \int_{\lambda_1}^{\lambda_2} \mathcal{I}(\lambda)\,d\lambda - \text{shaded area}$$

$$A = \log I_0/I = \log \frac{\int_{\lambda_1}^{\lambda_2} \mathcal{I}(\lambda)\,d\lambda}{\int_{\lambda_1}^{\lambda_2} \mathcal{I}(\lambda)\,d\lambda - \text{shaded area}}$$

A increases as denominator decreases, i.e., as the spectral bandwidth approaches the linewidth of absorption profile, $\lambda_2 - \lambda_1 \rightarrow \Delta\lambda$.

Figure 13.5. Effect of spectral bandwidth on absorbance in continuum atomic absorption spectrometry.

Recently, the use of a high-power Eimac xenon arc has been evaluated as a continuum source for multielement atomic absorption (for a review of continuum source atomic absorption, see O'Haver and Messman, 1986). As discussed by O'Haver and Messman (1986), these lamps are currently marketed under the trade name Cermax in the United States and provide improved performance because of the presence of an integral parabolic reflector as well as the use of a sapphire window. The use of the collimating reflector greatly improves the radiation output collection efficiency. Cochran and Hieftje (1977) found the 300-W Cermax lamp to produce a spectral radiant power that was 37 times more intense than that obtained with a conventional 150-W Hanovia xenon arc in the visible region of the spectrum.

O'Haver and co-workers (Zander, O'Haver, and Keliher, 1976; O'Haver, Harnly, and Zander, 1977; Zander, O'Haver, and Keliher, 1977; Harnly and O'Haver, 1977, 1981; Messman et al., 1983; Harnly et al., 1979, 1983) and others (Guthrie, Wolf, and Veillon, 1978; Veillon, Wolf, and Guthrie, 1979) have developed a multielement atomic absorption system that uses the high-intensity Cermax lamp in conjunction with an echelle dispersion system (see Section 4.3) to achieve the high resolution required. Simultaneous multi-element analyses for up to 16 elements have been demonstrated with this system. With present technology using continuum source atomic absorption, detection limits are typically a factor of 2 poorer compared with conventional narrow-line-source measurements except for elements like zinc and cadmium with short-wavelength resonance lines. In the case of elements with short-wavelength resonance lines, the detection limits obtained by continuum atomic absorption are an order of magnitude worse.

One advantage of multielement atomic absorption spectra of the type observed by O'Haver and co-workers compared with similar spectra in emission is simplicity. Absorption spectra are generally simpler than emission spectra because absorption is confined primarily to resonance transitions, whereas in emission transitions between upper states are observed in addition to resonance transitions. On the other hand, emission measurements are usually considered to have a greater dynamic range than absorption measurements, although O'Haver and Messman (1986) claim that with continuum source atomic absorption dynamic ranges of 10^5 are possible. In addition, because emission measurements are not restricted solely to resonance lines, greater flexibility is available in line selection compared with absorption measurements.

13.3. ATOMIC FLUORESCENCE

Atomic fluorescence spectrometry developed during the 1960s as a complementary technique to atomic absorption (Winefordner and Vickers, 1964).

In many respects, atomic fluorescence is a perfect example of an analytical technique that was developed to a high degree of refinement by analytical research scientists but was not widely adopted by the analytical applications specialists. While there is, no doubt, a variety of reasons for this state of affairs, it seems clear that the overwhelming incumbency (see Section 12.2.3) of atomic absorption had a lot to do with the failure of atomic fluorescence spectroscopy to be widely adopted as an analytical technique. As discussed in Chapter 12, in order for a technique to be adopted, it must fill a measurement need that cannot be met otherwise (i.e., occupy a niche) or have increased capabilities not found in competing techniques. Since the capabilities of atomic fluorescence overlap those of atomic absorption to a large extent, the priority of atomic absorption as an analytical technique has had a negative impact on the development of similar technology.

As shown in Figure 13.1*c*, the experimental arrangement for atomic fluorescence is similar to that used for atomic absorption except for the source and relay optics, which are arranged at right angles to the optical axis of the monochromator. Ullman (1980) has reviewed the area of multielement atomic fluorescence spectroscopy and has concluded that atomic fluorescence is ideally suited for multielement determinations because (1) beam combiners are not needed (as with atomic absorption) since the exciting radiation can be directed toward the atom reservoir from many angles; (2) the collection optics needed are simple since the fluorescence radiation is emitted in all directions; and (3) the wavelength selector subsystem can be simple since atomic fluorescence spectra are simpler than either atomic absorption spectra or atomic emission spectra.

13.3.1. Source Requirements

Compared with atomic absorption using a line source, the narrowness of the emission line profile is less important (except from the point of view of scatter) in atomic fluorescence because the detector does not view the source directly. Although source radiation outside the absorption line profile of the analyte in the atom reservoir cannot produce fluorescence, it is not monitored by the system and does not directly reduce the sensitivity in the same fashion as it would in atomic absorption (Figure 13.2). As is common for all fluorescence measurements, the fluorescent intensity produced by a given analyte concentration is directly proportional to the excitation source radiance. For high sensitivity, therefore, sources with high spectral radiance over the analyte absorption profile in the atom reservoir are needed. For atomic fluorescence, the important features of a suitable source are that it have a high radiance over the absorption linewidth and be unreversed.

Self-reversal is a form of self-absorption and is frequently seen for resonance transitions in sources with a hot inner core surrounded by a cooler layer. If the cooler layer contains absorbing atoms, these atoms will have a narrower absorption profile (less Doppler broadening) than the hotter atoms in the interior of the source. Radiation emitted by the atoms in the hotter interior of the source will be absorbed by the atoms in the cooler surrounding layer. Because the line profiles of the two populations of atoms are different, the cooler atoms will tend to absorb the line center from the emission profile of the hotter interior atoms. The effect of self-reversal is shown schematically in Figure 13.6. Since it is the central portion of the emission profile that is effective in exciting fluorescence of atoms in the atom reservoir (which needs to have a lower effective temperature than the source in order for absorption to occur), reversed sources, although they may seem quite intense, are actually poor sources for atomic fluorescence.

Compared with atomic absorption sources, sources for atomic fluorescence possessing high radiance are more difficult to produce. In the single-element mode, the most common light source for atomic fluorescence in the past has been the electrodeless discharge tube, which is nothing more than a quartz bulb containing argon at about 10 Torr which is excited by means of a microwave power supply at 2450 MHz. A small amount of a volatile halide salt is introduced into the bulb before it is sealed to provide the atoms of interest. Although electrodeless discharge tubes possess the necessary characteristics for a good atomic fluorescence source, the emitted intensity tends to drift with time, and they are less easily prepared for the less volatile elements. Some work has been done on the development of multielement electrodeless discharge tubes (Marshall and West, 1970; Fulton, Thompson, and West, 1970; Cresser and West, 1970; Patel, Brower, and Wineforder, 1972), but the number of elements successfully incorporated to date in a single lamp has

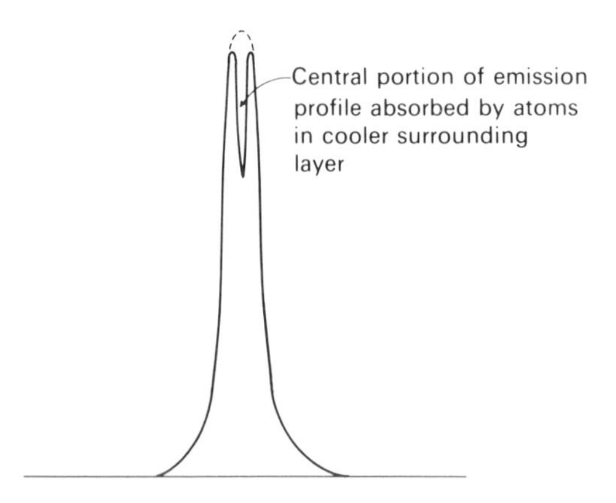

Figure 13.6. Idealized line profile produced by self-reversal with emission source having hot inner core surrounded by cooler layer.

been limited. Using a bank of single-element electrodeless discharge tubes would be prohibitively expensive unless one microwave power supply could be used to run them all by means of a microwave power divider.

A promising prospect as a practical source for atomic fluorescence is the hollow-cathode lamp. Although these lamps, when operated in their conventional mode, do not possess a sufficient radiance for satisfactory atomic fluorescence, previous work (Mitchell and Johansson, 1970, 1971; Malmstadt and Cordos, 1972; Cordos and Malmstadt, 1973) has shown that the radiance of these lamps can be increased to usable levels by operation in the pulsed mode. Using an intermittent pulsed mode of operation, Cordos and Malmstadt (1973) observed output intensities that were 20–300 times greater compared with a dc mode of operation. Since, as discussed, true multielement hollow-cathode lamps are not currently available, banks of lamps would be required.

In spite of the need for banks of lamps, the optical arrangement with the atomic fluorescence configuration is inherently more suited to multielement analysis than that needed for atomic absorption. Specifically, banks of lamps may be arranged around the atom reservoir without the necessity for complicated optical arrangements or multielement lamps. In the case of hollow-cathode lamps, power supplies capable of simultaneously operating banks of these lamps could easily be designed.

The use of a high-intensity Eimac (now Cermax) xenon arc continuum source for multielement atomic fluorescence has also been investigated by Winefordner and co-workers (Johnson, Plankey, and Winefordner, 1975; Ullman et al., 1979). For a complete review of this work and the area of multielement atomic fluorescence in general, the reader should consult the review by Ullman (1980).

13.3.2. The Atom Reservoir and Wavelength Selector

The roles of the atom reservoir, the wavelength selector subsystem, and the chopper–amplifier combination are all similar to those in atomic absorption. As with atomic absorption, the atom reservoir is not used as an excitation source, and good atomization efficiency for a wide variety of elements is its primary function. In addition, for atomic fluorescence work, the atom reservoir should produce a low concentration of quenching species and result in minimal scattering of the exciting radiation. Scattering is a serious interference in atomic fluorescence, which can occur as a result of both molecular (Rayleigh) scattering as well as scattering from unvaporized salt particles (Mie scattering) during analyte desolvation. Since scattering increases the amount of chopped resonance line intensity detected, it is often difficult to distinguish between true atomic fluorescence and scattering.

Since the wavelength selector does not view the source spectrum directly, its role is somewhat different from that in atomic absorption. As discussed previously, in atomic absorption the monochromator functions to isolate the desired resonance line from the other source emission. In atomic fluorescence, by contrast, the detector does not view the source directly, and so the only source emission seen by the detector is the resonance line (or lines) of the analyte. Since the fluorescence spectrum is generally quite simple, the monochromator typically used in atomic absorption measurements may be replaced with a simple bandpass filter. By using a solar-blind photomultiplier that does not respond to visible radiation (typical spectral response from 160 to 320 nm), the short-wavelength resonance lines can be observed under conditions of high optical throughput without using any filters at all. A chlorine filter is sometimes employed in these systems to block the OH-band radiation from 305 to 330 nm. These experimental arrangements are known collectively as *nondispersive atomic fluorescence*. A low-background atom reservoir is desirable in conventional atomic fluorescence, but it is especially important in nondispersive atomic fluorescence where the spectral bandwidth of the system is apt to be quite large.

A low-background flame commonly used for single-element determinations by atomic fluorescence is the turbulent hydrogen–argon–entrained air flame (Veillon et al., 1966). This flame is actually a hydrogen–air flame where the argon serves as an aspirating gas that has a low quenching cross section. Although this flame has been used in single-element atomic fluorescence determinations of volatile elements, the low flame temperature and non-reducing environment are likely to lead to severe interferences (Smith, Stafford, and Winefordner, 1968) in multielement determinations.

The use of shielded flames for atomic fluorescence has been studied by Slevin, Muscat, and Vickers (1972) for an air–acetylene flame sustained on a Meker burner. With this atom reservoir, an outer shielding flame surrounds an inner flame into which the analyte is introduced. The shielded flame was considered superior to the unshielded counterpart sustained on a standard Meker burner because, in the shielded case, the outer flame causes a more uniform temperature gradient across the inner flame. This is claimed to produce more uniform atom distributions compared with the standard Meker burner.

Separated flames such as the nitrous oxide–acetylene flame described by Kirkbright and West (1968) may provide a means for reduced background in atomic fluorescence measurements. Flame separation refers to the vertical displacement of the secondary reaction zone from the primary reaction zone by means of a Smithells separator (i.e., a silica tube, see Gaydon and Wolfhard, 1960) or inert-gas shielding. Flame separation allows observation of the reducing interconal zone of the flame without interference from the radiation emitted by the secondary reaction zone. With flame separation, the observed background radiation for the hottest flame region (which lies just above the primary reaction zone) is lower in certain wavelength regions than if a conventional unseparated flame is used.

13.4. THE ROLE OF NOISE IN SPECTROSCOPIC SYSTEMS

All measurements are subject, to a greater or lesser degree, to various forms of uncertainty. The nature and relative magnitude of the uncertainties that accompany spectroscopic measurements are highly system dependent. As a result, the noise sources of any proposed analytical system must be examined in detail to determine the limiting source of noise. In spectroscopy, there are two primary limiting situations that can be identified depending on whether the performance is limited by the instrument itself (generally the detector or preamplifier) or by the radiation source under investigation. Figure 13.7 shows the points in the system where various noise sources are injected.

If the radiation source itself limits the uncertainty of the measurement, the situation is referred to as *source-limited* performance. If a spectroscopic measurement is determined to be source limited, system performance improvements will depend on the nature of the source limitation. Source-limited performance arises from two primary causes: unintentional modulation or source flicker and photon shot noise. In addition to source flicker noise and photon shot noise, a third type of noise encountered in flames and plasmas is *whistle noise* (Epstein and Winefordner, 1984), which shows up as a high-frequency spike in the noise power spectrum of the flame or plasma (see Figure 9.13 for an example of a noise power spectrum).

For a comprehensive discussion of the role of noise in spectroscopy, the papers by Alkemade et al. (1978, 1980), Boutilier et al. (1978), and Epstein and Winefordner (1984) should be consulted.

13.4.1. Flicker Noise

Most spectral sources used in analytical atomic spectroscopy are plagued by a certain amount of source flicker noise, which arises from a combination of factors including small fluctuations in source temperature, fluctuations in atom number density in the source region sampled by the spectrometer, fluctuations in sample introduction, turbulence in the source, and positional

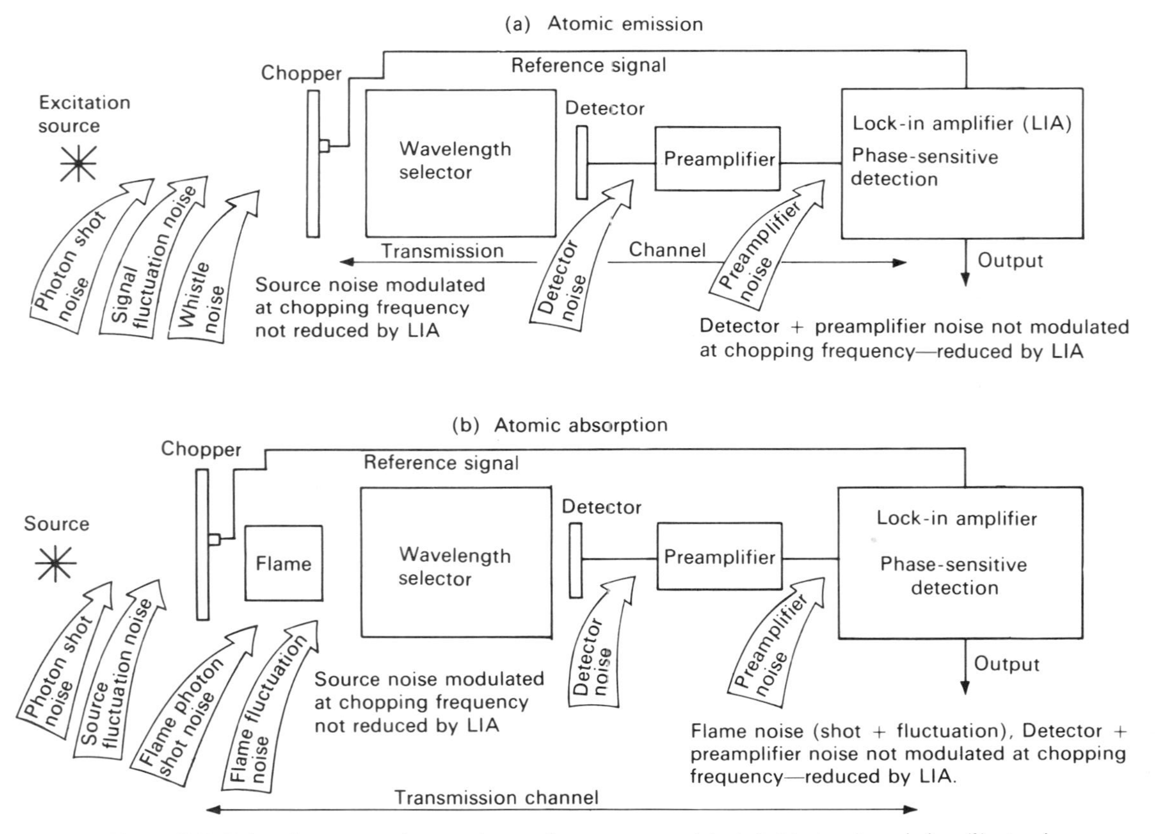

Figure 13.7. Points in system where various noise sources are injected: (*a*) atomic emission (*b*) atomic absorption.

instability. These source fluctuations result in a low-frequency noise at the detector that is known variously as fluctuation noise, flicker noise, excess low-frequency noise, $1/f^a$ noise, or pink noise. This form of noise is often called proportional noise because the root-mean-square (rms) value of the noise is proportional to the signal. When fluctuation noise is the limiting noise, signal averaging cannot be used to improve the signal-to-noise ratio because the noise accumulates at a rate directly proportional to the rate at which the signal accumulates.

If a measurement situation is flicker noise limited, there are two courses of action that can be implemented. The first and most obvious course of action is to improve the source itself by reducing the causes of the small fluctuations. This may be accomplished by shielding the source to reduce temperature fluctuations and improve positional stability. Improvements in sample nebulization stability can also be undertaken. A second approach to reducing unwanted source flicker can be referred to as *source compensation*. Source compensation is often used in astronomy to reduce the flicker caused by atmospheric modulation of stellar objects (or atmospheric scintillation). Atmospheric modulation can arise from atmospheric patchiness or variable cloud clutter or from refractive index variations arising from atmospheric density gradients. When source flicker noise is a dominant noise source, a *total intensity reference detector* (Decker, 1977) can be used to remove the majority of this noise source. To perform source compensation, the instantaneous signal is ratioed with the total instantaneous signal measured by the reference detector to give a ratio that is compensated with respect to source fluctuations. For multielement work over the normal spectral range, a silicon photodiode viewing the source makes a convenient reference detector. This procedure will work as long as the flicker component is not wavelength dependent in an unpredictable fashion.

In atomic emission spectroscopy, source compensation is often implemented by using an *internal standard*. The idea of using an internal standard was developed by Gerlach and Schweitzer (1931) as a means of compensating for the large degree of source fluctuation observed with a dc arc. The internal standard is an element present in all the samples and calibration standards in the same amount (i.e., it may be a matrix element). To be useful as a good internal standard, an element must have similar spectrochemical properties to those of the analyte (see Boumans, 1966). For multielement work, no single internal standard will be ideal for all the elements in the sample, and either several must be used or a single compromise internal standard must be selected. Internal standardization to reduce source fluctuation is accomplished by monitoring the ratio of the instantaneous analyte line intensity to the instantaneous internal standard line intensity, as described previously.

13.4.2. Photon Shot Noise

The other major category of source noise is known as *photon shot noise* (or photon noise or shot noise). Photon shot noise occurs in detectors that respond to the rate of arrival of photons on a photosensitive surface, and it arises as a result of the randomness in the arrival of photons at the surface of the detector. The fluctuation in arrival rate of photons is similar to the random arrival of raindrops in a certain area during a rain storm. Since, to a first approximation, the variation in the arrival of photons follows Poisson statistics, the standard deviation in the arrival of photons at the detector (i.e., the noise) is proportional to the square root of the average number of photons arriving in a certain time period. Since the average arrival of photons is proportional to the average intensity, photon shot noise increases as the square root of the average intensity incident on the detector. Figure 13.8 shows the influence of shot noise on a signal as the intensity increases. Using the signals and rms noise shown in Figure 13.8, a log–log plot of signal-to-noise ratio versus signal results in a linear relationship with a slope of 0.44. This slope is quite close to the theoretical value of 0.5 expected for square-root behavior under shot-noise-limited conditions. When photon shot noise is the limiting noise, signal averaging will improve the signal-to-noise ratio because

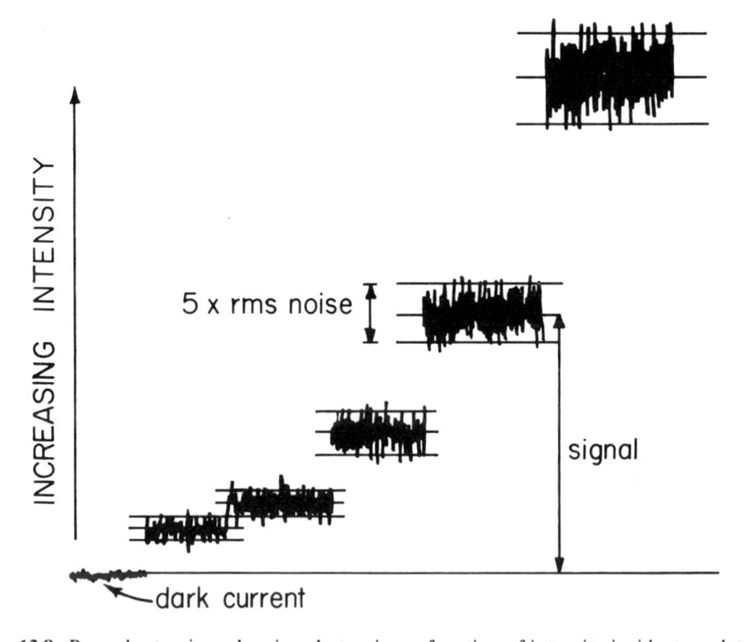

Figure 13.8. Recorder tracings showing shot noise as function of intensity incident on detector.

the signal accumulates at a rate proportional to the intensity whereas the noise accumulates at a rate proportional to the square root of the intensity. Under these conditions, the signal-to-noise ratio obtained with signal averaging increases as the square root of the number of observations in the average.

Photon shot noise is an inherent, fundamental source of noise in all spectroscopic observations. It cannot be eliminated because it is a characteristic of the light itself (except by light squeezing, see Slusher and Yurke, 1988). Its effect on a given measurement situation can be reduced only by statistical procedures that improve the photon-counting statistics. Such measures include signal averaging and bandwidth reduction (filtering). Both of these noise reduction procedures require time and have a potential impact on the ultimate success of various spectroscopic systems in terms of their potential success in multielement analysis. Bandwidth reduction means that the response time of the system will be longer while signal averaging implies that the signal and adjacent background must be measured repeatedly. The effectiveness of signal averaging in improving the signal-to-noise ratio will depend on the relative intensity of the spectral line compared with the background. Thus, a weak line with a poor signal-to-noise ratio will require more observations in the average to achieve a desired signal-to-noise improvement than an intense line with a large signal-to-noise ratio.

13.4.3. Detector Noise

The ultimate goal of instrument development in spectroscopy is to design spectroscopic systems that are photon noise limited. If the system performance is truly limited by photon shot noise, further development of the detection system must include either increases in the optical throughput of the system or more sophisticated signal averaging or bandwidth reduction if increased performance is to be achieved. If the system is not performance limited by the source, it must needs be limited by some aspect of the instrument itself. The two common components that limit the performance of a spectroscopic system are the detector and whatever amplifier is required to increase the signal to a usable level. If the performance of the system is limited by the noise originating in the detector, the system is said to be *detector noise limited*. For many classes of detectors, particularly those with long-wavelength response, detector noise arising from the thermal generation of charge carriers is much greater than the photon shot noise produced by the random arrival of photons.

For the UV–visible region of the spectrum where the photomultiplier is the detector of choice, detector noise is often negligible. Dark current in photomultipliers originates from ohmic leakage, thermionic emission, and regenerative effects (Engstrom, 1980). Assuming a typical photocathode dark emission of 10^{-15} A at room temperature and an electrical bandwidth of 1 Hz, the photocathode rms dark current noise will be given by

$$I_{\mathrm{rms}} = (2ei_{\mathrm{d}}B)^{1/2} = 1.8 \times 10^{-17}\ \mathrm{A} \tag{13.1}$$

where I_{rms} is the photocathode dark current noise, e is the charge on the electron, i_{d} is the magnitude of the average dark current, and B is the system bandwidth. In the presence of this dark current, the signal-to-noise ratio produced by a photocathode signal current i_{s} will be given by (Engstrom, 1980),

$$\mathrm{SNR} = i_{\mathrm{s}}/[2e(i_{\mathrm{d}} + i_{\mathrm{s}})B]^{1/2} \tag{13.2}$$

where SNR is the signal-to-noise ratio. From Eq. 13.2, it can be seen that for dark current to limit the signal-to-noise ratio, the photocathode signal must be less than 10^{-15} A. If a photocathode responsivity of 70×10^{-3} A/W of radiant energy at 420 nm is taken as the photocathode responsivity, photocurrents of 10^{-15} A would correspond to a light flux of 1.4×10^{-14} W incident on the photocathode. At 420 nm, this light flux corresponds to a photon flux of 3×10^{4} photons per second. This magnitude of photon flux is actually quite small, and most measurements made with a properly selected photomultiplier (i.e., one with an appropriate spectral response for the given wavelength range) will be photon noise limited (either by signal or by background) rather than detector noise limited. If the detector noise limit is approached, cooling the photomultiplier can be an effective means of reducing the dark current (Engstrom, 1980).

For comparison, let us consider a photodiode connected to a transimpedance amplifier (i.e., a current-to-voltage converter) with a load resistance of 2 MΩ, as shown in Figure 13.9. Neglecting detector noise itself, the

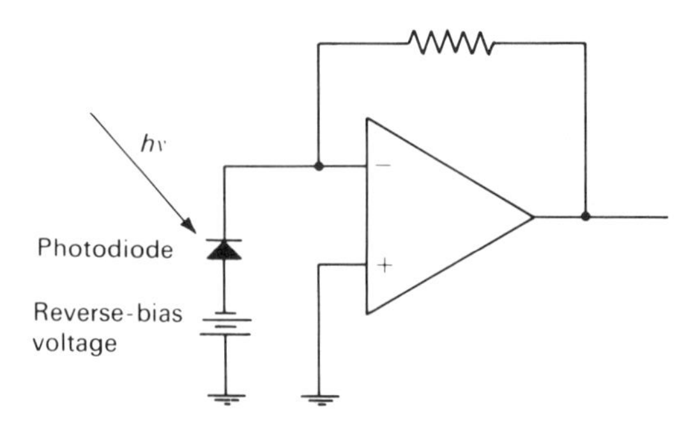

Figure 13.9. Transimpedance amplifier for photodiode.

Johnson noise of the load resistance for this experimental arrangement will be given by

$$I_{\rm rms} = (4kTB/R)^{1/2} \tag{13.3}$$

where k is Boltzmann's constant, T is the absolute temperature, B is the system bandwidth, and R is the resistance. For a temperature of 300 K and a bandwidth of 1 Hz, the thermal noise current generated by the load resistance is 9.1×10^{-14} A. The responsivity of a photodiode can be determined from a knowledge of its quantum efficiency (Keyes, 1980) by

$$R = q\eta\lambda G/hc \tag{13.4}$$

where R is the responsivity in amperes per watt of radiant energy, q is the charge on the electron, η is the quantum efficiency, λ is the wavelength, G is the gain of the detector ($G = 1$ for ordinary silicon photodiodes), h is Planck's constant, and c is the speed of light. Assuming a quantum efficiency of 0.8 for a silicon diode at 420 nm, one obtains a responsivity of 270 mA/W of radiant energy. The light flux that will produce a signal equal to the rms Johnson noise from the load resistor will be 3.4×10^{-13} W. Below an incident flux of 3.4×10^{-13} W, the photodiodie–amplifier combination will be limited by the Johnson noise in the load resistance. The photomultiplier used in the example, by contrast, is not detector limited until the incident flux drops to 1.4×10^{-14} W.

When evaluating the performance of different spectroscopic alternatives, it is important to perform the type of analysis shown in the preceding discussion to get some idea of the major noise sources. In comparing two systems, one using a photomultiplier with the characteristics described previously and the other using a silicon photodiode with a quantum efficiency of 0.8 at 420 nm, the photodiode system will be detector/amplifier noise limited for intensity levels less than 3.4×10^{-13} W, whereas the photomultiplier will approach photon-noise-limited behavior for light fluxes above 1.4×10^{-14} W. According to Eq. 13.2, there will be a flux range for the photomultiplier where photon shot noise and detector noise are of comparable magnitude, and in actual practice, true photon-noise-limited behavior may not occur until light levels of $10^{-13} - 10^{-12}$ W are attained. Similarly, for the photodiode system, true photon noise limiting performance may not be achieved until light levels of $10^{-12} - 10^{-11}$ W are reached. If the photomultiplier is actually photon–shot noise limited, however, attempts to improve its performance by detector cooling will not work. Under these conditions, only optical throughput improvements and bandwidth reduction or signal averaging will be effective. On the other hand, the photodiode system, if operated below a light flux of

3.4×10^{-13} W, can benefit from both source modulation and optical throughput improvements. One way to accomplish both of these goals is by multiplexing (see Chapter 14 for a discussion of multiplexing in the UV–visible spectral region).

13.5. THE ROLE OF MODULATION IN NOISE REDUCTION

Modulation is a process whereby the information to be transmitted by a system is shifted to a higher frequency regime for increased transmission efficiency or to avoid low-frequency noise in the transmission channel. Although a variety of forms of modulation have been developed (Schwartz, 1959; Goldman, 1948), we confine our discussion to amplitude modulation (AM). In the process of AM, the information signal to be transmitted is used to vary the amplitude of a much higher frequency carrier wave. In spectroscopy, the information signal may be a slowly varying light intensity that is chopped (i.e., modulated) by a rapidly rotating sectored disk. This produces an ac signal from what was previously a slowly varying dc signal, as shown in Figure 13.10. The ac signal produced by chopping or modulation can now be

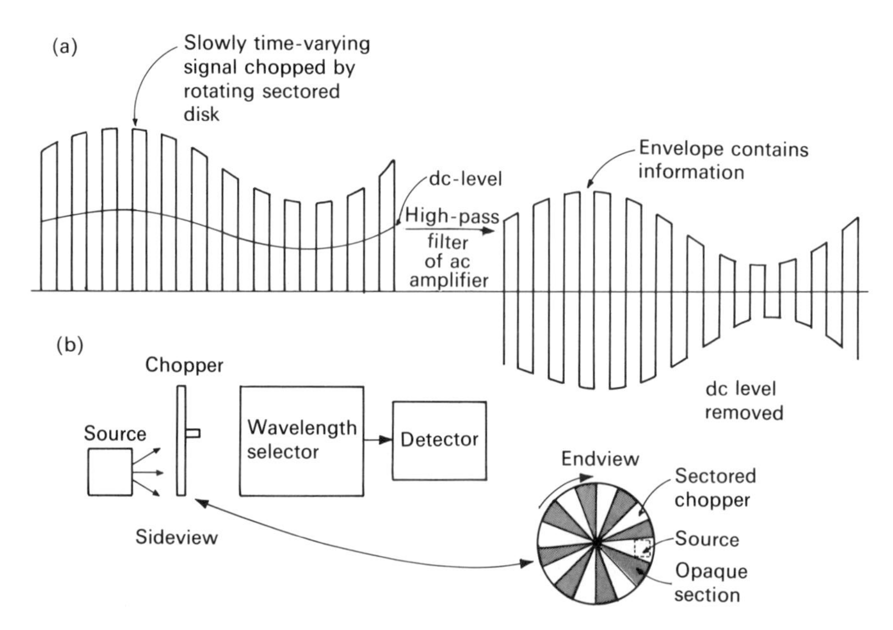

Figure 13.10. Modulation of source with slowly varying intensity by means of rotating sectored disk: (*a*) signal waveform produced by chopping, before and after passing through high-pass filter of amplifier; (*b*) block diagram of experimental arrangement used to produce modulation.

amplified by an ac amplifier with a high-pass filter input. Figure 13.10 shows the effect of passing the modulated signal through the input of the high-pass filter. Since the capacitor of the filter cannot transmit dc levels, the dc component of the chopped signal is removed upon passing through the filter, giving rise to the modulated carrier waveform shown on the right. This modulated waveform can now be amplified by the ac amplifier prior to demodulation. From Figure 13.10, it can be seen that the information to be transmitted is now contained in the envelope of the modulated carrier wave. The envelope of the modulated carrier wave can be recovered after amplification by passing the amplified signal through a diode (a half-wave rectifier) followed by a low-pass filter, as shown in Figure 13.11.

To understand how this process can reduce noise present on the transmission channel, we need to examine the frequency spectrum of the signal and the carrier. For simplicity, let us assume that the signal is a slowly varying sinusoidal wave that can be described by

$$f(t) = \cos \omega_m t \tag{13.5}$$

where w_m is the angular frequency of the information signal. Assume that the waveform in Eq. 13.5 is used to modulate the amplitude of a carrier wave whose time variation can be described by

$$v(t) = A \cos \omega_c t \tag{13.6}$$

where A is the amplitude of the carrier and the angular frequency of the

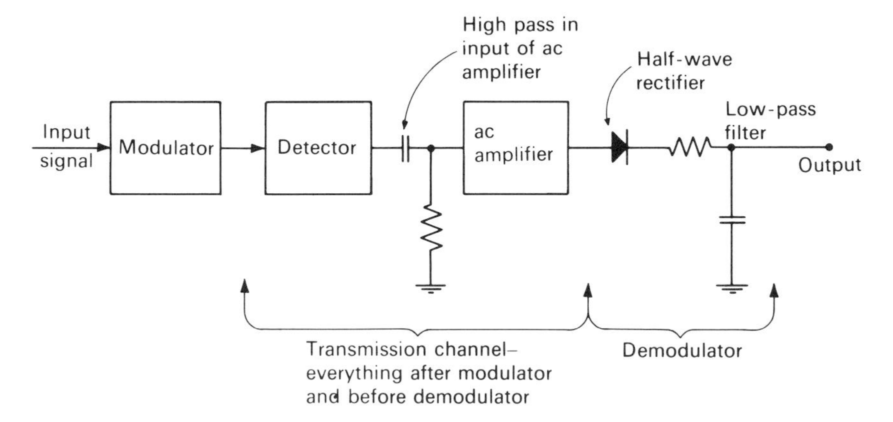

Figure 13.11. Information transmission pathway through system. Transmission channel consists of everything after modulator and before demodulator. Demodulator consists of half-wave diode rectifier followed by low-pass RC filter.

carrier, ω_c, is much greater than ω_m. The process of modulation is a nonlinear process equivalent mathematically to multiplication. Therefore, if we modulate the carrier wave given by Eq. 13.6 with the information given by Eq. 13.5, we produce the following results;

$$S(\mathrm{am}) = (1 + m\cos\omega_m t)[A\cos\omega_c t] \tag{13.7}$$

where $S(\mathrm{am})$ is the modulated waveform, m is the depth of modulation (which can be from 0 to 1), and the factor 1 has been added to bias the modulating signal up enough so that the envelop produced by modulation never goes through zero. If the dc bias on the information is not large enough, overmodulation will occur, and envelope detection will not retrieve the original information, as shown in Figure 13.12. Mathematically, Eq 13.7 can be replaced with its trigonometric identity to give

$$A\cos\omega_c t + Am(\cos\omega_m t)(\cos\omega_c t) = A\cos\omega_c t + (\tfrac{1}{2}Am) \times [\cos(\omega_c - \omega_m)t + \cos(\omega_c + \omega_m)t] \tag{13.8}$$

Equation 13.8 shows that the process of modulation has produced two new frequencies symmetrically located about the carrier frequency and known as

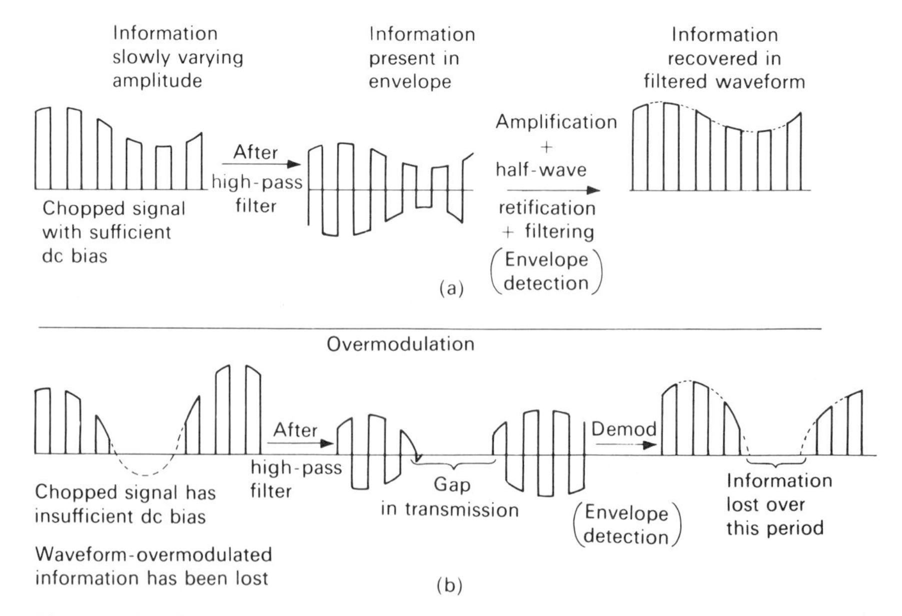

Figure 13.12. Effect of overmodulation on signal waveform showing loss of information: (*a*) chopped signal has sufficient dc bias; (*b*) chopped signal has insufficient dc bias resulting in loss of information.

sidebands. Thus, the process of modulation has shifted the frequency range of the information from a low-frequency to a higher frequency band, as shown in Figure 13.13. If the transmission channel over which the information must be transmitted has a noise distribution similar to that shown in Figure 13.14, modulation can improve the signal-to-noise ratio by shifting the information to higher frequencies where the noise may be less. It should also be evident that if the noise spectrum of the transmission channel has a flat noise distribution or one that for some reason peaks at higher frequencies (e.g., because of whistle noise), modulation will not be an effective form of noise reduction and may even degrade the signal-to-noise ratio.

13.5.1. Demodulation

Demodulation, or detection, is the process of recovering the desired information from the modulated carrier wave and is essentially the reverse of the modulation process. Thus, if the signal represented by Eq. 13.8 is multiplied again by the carrier, the result is

$$\begin{aligned} S(\text{am})[A\cos\omega_c t] &= A^2\cos^2\omega_c t + (\tfrac{1}{2}A^2 m)\{\cos\omega_c t\cos(\omega_c-\omega_m)t \\ &\quad + \cos\omega_c t\cos(\omega_c+\omega_m)t\} \\ &= A^2\cos^2\omega_c t + (\tfrac{1}{2}A^2 m)\{1/2[\cos(2\omega_c-\omega_m)t+\cos\omega_m t] \\ &\quad + 1/2[\cos(2\omega_c+\omega_m)t+\cos\omega_m t]\} \end{aligned} \tag{13.9}$$

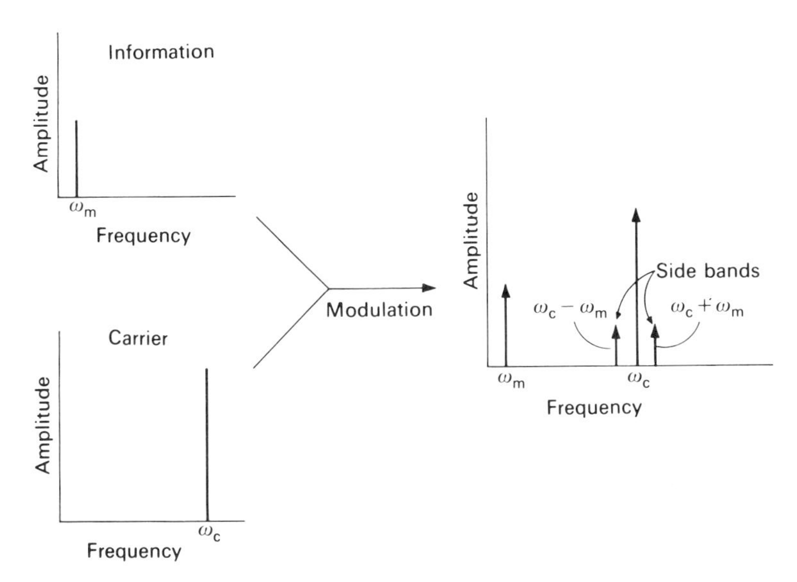

Figure 13.13. Amplitude-versus-frequency plots showing formation of sidebands as result of modulation.

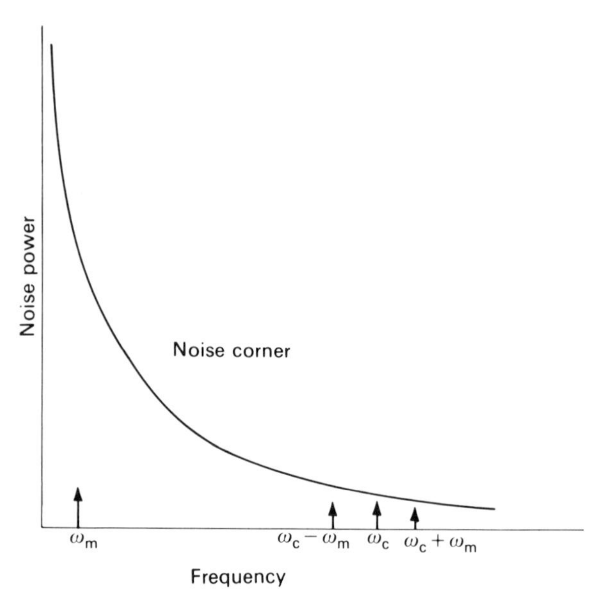

Figure 13.14. Noise power spectrum of transmission channel. Under favorable conditions, modulation can result in noise reduction if it shifts information from high–noise region of transmission channel (low frequency) to low-noise region (high frequency).

If the signal represented by Eq. 13.9 is filtered with a low-pass filter, the demodulated signal recovered is simply

$$D(\text{am}) = (A^2 m/2)\cos \omega_m t \tag{13.10}$$

which is proportional to the original information. Although not specifically indicated by Eq. 13.10, the signal recovered after demodulation is larger than the input signal because of the effect of the ac amplifier.

13.5.2. Benefits of Modulation

Modulation is an effective means of noise reduction only for noise sources that enter the system in the transmission channel. The transmission channel consists of that portion of the system *after the modulator and before the demodulator*. In atomic emission spectroscopy, where the source radiation is modulated with a chopper, the transmission channel is limited to the detector subsystem and the ac amplifier. In atomic absorption and atomic fluorescence, the transmission channel includes the atom reservoir in addition to the detector subsystem and ac amplifier. Thus, for the case of atomic emission, modulation cannot be used as a means of reducing source noise because the source is being modulated and the noise produced by it is being transmitted over the transmission channel along with the desired signal. In atomic absorption and atomic fluorescence, however, flame noise can be reduced by modulation because in these modes of observation, the flame is part of the transmission channel.

In atomic emission, therefore, modulation can be used as a means of noise reduction only when detector noise and/or amplifier noise are significant noise sources compared with source noise. It should also be remembered that even in the case where detector noise and/or amplifier noise are limiting, modulation can only result in noise reduction if the noise power spectrum of the transmission channel shows a $1/f^a$ power spectrum as shown in Figure 13.14. Figure 13.15 shows how low-frequency detector drift is reduced by source modulation. This discrimination against low-frequency transmission channel noise, which is added to the modulated signal, will be effective as long as the noise frequencies are much less than the half-power frequency of the high-pass input filter (see Appendix 9A), as shown in the Bode diagram of Figure 13.15. High-frequency noise that falls within the ac amplifier's bandpass can be reduced by increasing the time constant on the low-pass filter in the demodulator shown in Figure 13.11. This, of course, increases the response time of the system, making it difficult to track signals that vary rapidly with time.

These conclusions have a significant impact on the potential success of multiplex methods in atomic spectroscopy. All multiplex methods are actually

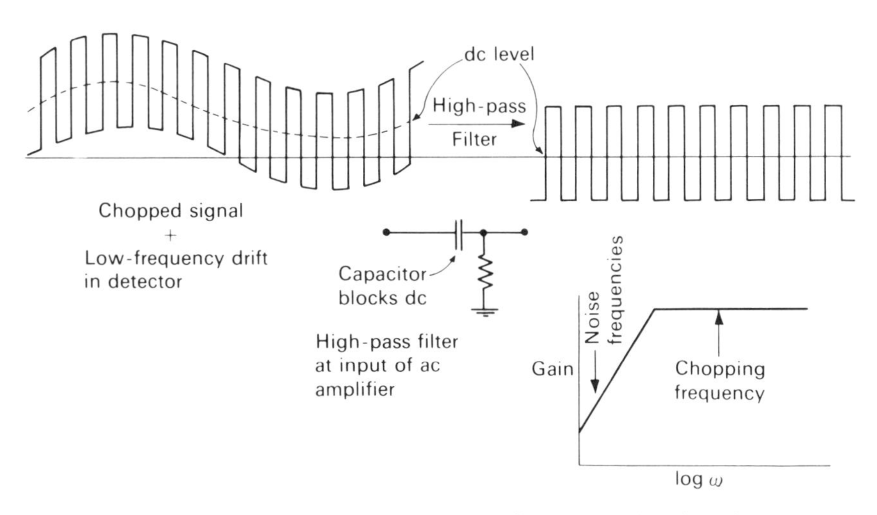

Figure 13.15. Effect of modulation on low-frequency drift in transmission channel components.

ways of performing spectromodulation (Harwit and Decker, 1974). As a result, they will be most effective in situations where detector noise and/or amplifier noise are significant noise sources (i.e., in detector-noise-limited situations). They cannot improve performance in photon-shot-noise limited situations because the noise is modulated along with the signal and also because photon shot noise is white noise so moving to a higher frequency will not reduce its magnitude. Multiplexing can be expected to result in some improvement in the UV–visible region of the spectrum for systems that employ a photodiode–amplifier combination. This result, however, should not be unexpected in light of the example discussed in this chapter because for this system, amplifier noise is an important source of noise. Thus, there is nothing unusual about spectroscopy in the UV–visible region compared with that in the infrared region (where multiplexing is effective). The difference in performance is not due to any inherent difference in the nature of photon behavior between the two spectral regions but instead is due to the different performance characteristics of the *detectors* used.

REFERENCES

Alkemade, C. T. J. (1968), *Appl. Opt.*, *7*, 1261.

Alkemade, C. T. J., Snellman, W., Boutillier, G. D., Pollard, B. D., Winefordner, J. D., Chester, T. L., and Omenetto, N. (1978), *Spectrochim. Acta*, *33B*, 383.

Alkemade, C. T. J., Snellman, W. Boutillier, G. D., and Winefordner, J. D. (1980), *Spectrochim. Acta*, *35B*, 261.

Boumans, P. W. J. M. (1966), *Theory of Spectrochemical Excitation*, Plenum, New York.

Boumans, P. W. J. M., Ed. (1987), *Inductively Coupled Plasma Emission Spectroscopy*, Wiley, New York.

Boumans, P. W. J. M. and DeBoer, F. J. (1972), *Spectrochim. Acta*, *27B*, 391.

Boumans, P. W. J. M. and DeBoer, F. J. (1975a), *Proc. Anal. Div. Chem. Soc.*, *12*, 140.

Boumans, P. W. J. M. and DeBoer, F. J. (1975b), *Spectrochim. Acta*, *30B*, 309.

Boumans, P. W. J. M. and DeBoer, F. J. (1976), *Spectrochim. Acta*, *31B*, 355.

Boutillier, G. D., Pollard, B. D., Winefordner, J. D., Chester, T. L., and Omenetto, N. (1978), *Spectrochim. Acta*, *33B*, 401.

Brost, D. F., Malloy, B., and Busch, K. W. (1977), *Anal. Chem.*, *49*, 2280.

Busch, K. W., Howell, N. G., and Morrison, G. H. (1974), *Anal. Chem.*, *46*, 575.

Busch, K. W. and Morrison, G. H. (1973), *Anal. Chem.*, *45*, 712A.

Butler, L. R. P. and Strasheim, A. (1965), *Spectrochim. Acta*, *21*, 1207.

Cochran, R. L. and Hieftje, G. M. (1977), *Anal. Chem.*, *49*, 2040.

Cordos, E. and Malmstadt, H. V. (1973), *Anal. Chem.*, *45*, 27.

Cresser, M. S. and West, T. S. (1970), *Anal. Chim. Acta*, *51*, 530.

Decker, Jr., J. A. (1977), "Hadamard-Transform Spectroscopy," in *Spectrometric Techniques*, Vol. 1, Vanasse, G. A. (Ed.), Academic, New York, pp. 189–227.

DeGalan, L., McGee, W. W., and Winefordner, J. D. (1967), *Anal. Chim. Acta*, *37*, 436.

Engstrom, R. W. (1980), *Photomultiplier Handbook*, RCA Solid State Division, Lancaster, PA.

Epstein, M. S. and Winefordner, J. D. (1984), *Prog. Analyt. Atom. Spectrosc.*, *7*, 67.

Fassel, V. A. and Kniseley, R. N. (1974a), *Anal. Chem.*, *46*, 1110A.

Fassel, V. A. and Kniseley, R. N. (1974b), *Anal. Chem.*, *46*, 1155A.

Fassel, V. A., Mossotti, V. G., Grossman, W. E. L., and Kniseley, R. N. (1966), *Spectrochim. Acta*, *22*, 347.

Fassel, V. A., Peterson, C. A., Abercrombie, F. N., and Kniseley, R. N. (1976), *Anal. Chem.*, *48*, 517.

Fulton, A., Thompson, K. C., and West, T. S. (1970), *Anal. Chim. Acta*, *51*, 373.

Gaydon, A. G. and Wolfhard, H. G. (1960), *Flames—Their Structure, Radiation, and Temperature*, 2nd ed., Chapman & Hall, London.

Gerlach, W. and Schweitzer, E. (1931), *Foundations and Methods of Chemical Analysis by the Emission Spectrum*, Adam Hilger, London, original German edition published in 1930.

Gibson, J. H., Grossman, W. E. L., and Cooke, W. D. (1962), *Appl. Spectrosc.*, *16*, 47.

Goldman, S. (1948), *Frequency Analysis, Modulation, and Noise*, McGraw–Hill, New York.

Greenfield, S., Jones, I. L., McGeachin, H. M., and Smith, P. B. (1975), *Anal. Chim. Acta*, *74*, 225.

Guthrie, B. E., Wolf, W. R. and Veillon, C. (1978), *Anal. Chem.*, *50*, 1900.

Harnly, J. M. and O'Haver, T. C. (1977), *Anal. Chem.*, *49*, 2187.

Harnly, J. M. and O'Haver, T. C. (1981), *Anal. Chem.*, *53*, 1291.

Harnly, J. M., O'Haver, T. C., Golden, B., and Wolf, W. R. (1979), *Anal. Chem.*, *51*, 2007.

Harnly, J. M., Patterson, K. Y., Veillon, C., Wolf, W. R., Marshall, J., Littlejohn, D., Ottaway, J., Miller–Ihli, N. J., and O'Haver, T. C. (1983), *Anal. Chem.*, *55*, 1417.

Harwit, M. and Decker, Jr., J. A. (1974), "Modulation Techniques in Spectrometry," in *Progress in Optics*, Vol. 12, Wolf, E. (Ed.), North-Holland, Amsterdam, pp. 101–162.

Herrmann, R. and Alkemade, C. T. J. (1963), *Chemical Analysis by Flame Photometry*, Interscience, New York.

Horowitz, P. and Hill, W. (1980), *The Art of Electronics*, Cambridge University Press, Cambridge.

Johnson, D. J., Plankey, F. W., and Winefordner, J. D. (1975), *Anal. Chem.*, *47*, 1739.

Keyes, R. J., Ed. (1980), *Optical and Infrared Detectors*, Springer-Verlag, Berlin.

Kirkbright, G. F. and Vetter, S. (1971), *Spectrochim. Acta*, *26B*, 505.

Kirkbright, G. F. and West, T. S. (1968), *Appl. Opt.*, *7*, 1305.

Malmstadt, H. V. and Cordos, E. (1972), *Am. Lab.*, *4* (8), 35.

Marshall, G. B. and West, T. S. (1970), *Anal. Chim. Acta*, *51*, 179.

Massman, H. (1963), *Z. Instrum.*, *71*, 225.

Mavrodineanu, R. and Boiteux, H. (1965), *Flame Spectroscopy*, Wiley, New York.

Mavrodineanu, R. and Hughes, R. C. (1968), *Appl. Opt.*, *7*, 128.

McGee, W. W. and Winefordner, J. D. (1967), *Anal. Chim. Acta*, *37*, 429.

Messman, J. D., Epstein, M. S., Rains, T. C., and O'Haver, T. C. (1983), *Anal. Chem.*, *55*, 1055.

Mitchell, D. G. and Johansson, A. (1970), *Spectrochim. Acta*, *25B*, 175.

Mitchell, D. G. and Johansson, A. (1971), *Spectrochim. Acta*, *26B*, 677.

O'Haver, T. C., Harnly, J. M., and Zander, A. T. (1977), *Anal. Chem.*, *49*, 665.

O'Haver, T. C. and Messman, J. D. (1986), *Prog. Analyt. Spectrosc.*, *9*, 483.

Olson, K. W., Haas. Jr., W. J., and Fassel, V. A. (1977), *Anal. Chem.*, *49*, 632.

Patel, B. M., Browner, R. F., and Winefordner, J. D. (1972), *Anal. Chem.*, *44*, 2272.

Sandell, E. B. (1959), *Colorimetric Determination of Traces of Metals*, 3rd ed., Interscience, New York.

Schwartz, M. (1959), *Information Transmission, Modulation, and Noise*, McGraw–Hill, New York.

Slevin, P. J., Muscat, V. I., and Vickers, T. J. (1972), *Appl. Spectrosc.*, *26*, 296.

Slusher, R. E. and Yurke, B. (1988), *Sci. Am.*, *258*(5), 50.

Smith, R., Stafford, C. M., and Winefordner, J. D. (1968), *Anal. Chim. Acta*, *42*, 523.

Strasheim, A. and Butler, L. R. P. (1962), *Appl. Spectrosc.*, *16*, 109.

Strasheim, A. and Human, H. G. C. (1968), *Spectrochim. Acta*, *23B*, 265.

Ullman, A. H., (1980), *Prog. Analyt. Atom. Spectrosc.*, *3*, 87.

Ullman, A. H., Pollard, B. D., Boutillier, G. D., Bateh, R. P., Hanley, P., and Winefordner, J. D. (1979), *Anal. Chem.*, *51*, 2382.

Veillon, C., Mansfield, J. M., Parsons, M. L., and Winefordner, J. D. (1966), *Anal. Chem.*, *38*, 204.

Veillon, C., Wolf, W. R., and Guthrie, B. E. (1979), *J. Agric. Food Chem.*, *27*, 490.

Walsh, A. (1955), *Spectrochim. Acta*, *7*, 108.

Wendt, R. H. and Fassel, V. A. (1965), *Anal. Chem.*, *37*, 920.

Wendt, R. H. and Fassel, V. A. (1966), *Anal. Chem.*, *38*, 337.

Winefordner, J. D. and Vickers, T. J. (1964), *Anal. Chem.*, *36*, 161.

Winge, R. K., Fassel, V. A., Kniseley, R. N., DeKalb, E., and Haas, Jr., W. J. (1977), *Spectrochim. Acta*, *32B*, 327.

Zander, A. T., O'Haver, T. C., and Keliher, P. N. (1976), *Anal. Chem.*, *48*, 1166.

Zander, A. T., O'Haver, T. C., and Keliher, P. N. (1977), *Anal. Chem.*, *49*, 838.

CHAPTER

14

SURVEY OF TRANSFORM SPECTROMETRIC SYSTEMS

All transform spectrometers obtain spectral information by means of an optical modulation process (Mertz, 1965). This is true for nondispersive, interferometric spectrometers such as the Fourier transform (FT) Michelson (Section 7.1) and Fabry–Perot (Section 6.8.6) systems, as well as for transform spectrometers that are dispersive, that is, the Mertz mock interferometer (Section 5.2.7; Fourier transformation), Girard's grill spectrometer (Section 5.2.6; Fresnel transformation), Golay's multislit spectrometers (Sections 5.2.4 and 5.2.5; white-noise transformation), and the Hadamard spectrometers (Section 5.5; Hadamard transformation).

Most transform spectrometers have higher throughout than scanning monochromators, and some are also capable of multiplexing. (See Mertz, 1965, and Fellgett, 1967, for further discussion.) Those transform spectrometers that can multiplex (i.e., monitor more than one resolution element at a time) would appear to offer considerable potential for simultaneous multi-element analysis. However, while the performance of FT spectrometers employing the Michelson interferometer and, to a much lesser extent, Hadamard transform (HT) spectrometers has been studied in the UV–visible portion of the spectrum, the value of multiplexing spectrometers for UV–visible emission spectrometry remains in question.

14.1. CHOICE OF SPECTROMETRIC SYSTEM

Transform spectrometers are generally more expensive to build and to operate than conventional scanning systems. Therefore, transform techniques are justified only if sufficient advantage results from their use, that is, the ability to determine more than one element at a time, an increase in signal-to-noise ratio (SNR), a greater spectral range, an increase in resolution, and so on. Several articles (Gebbie, 1969; Harwit and Decker, 1974; Decker, 1977; Horlick, Hall, and Yuen, 1982; Griffiths, Sloane, and Hannah, 1977; and Griffiths and deHaseth, 1986a,b) have discussed the factors to be considered when comparing Fourier and Hadamard transform spectrometers with each other and with conventional scanning monochromators. While some of these discussions (Harwit and Decker, 1974; Decker, 1977) were written with infrared applications in mind, many of the conclusions also apply to UV–visible emission spectrometry.

14.1.1. Signal-to-Noise Ratio (SNR)

As discussed in Sections 5.6 and 13.4, any advantage that can be achieved in SNR is fundamentally limited by the type of noise present in the system. If we can classify noise into three broad categories (Table 5.5)—detector noise, photon shot noise, and fluctuation ($1/f$, flicker, or proportional) noise—then we can state in a fairly general way that multiplexing is expected to improve the SNR only in the detector-noise-limited case, while a gain in throughput is expected to improve SNR in either the detector- or photon shot-noise-limited case. For other situations, high throughput and/or multiplexing either produce no SNR benefit at all or result in an SNR disadvantage.

In order to compare the potential SNR advantages gained by using Fourier and Hadamard transform methods, it should be noted that the entries in Table 5.5 under "Multiplexing, *S*-Matrix," apply to both Hadamard spectrometers as well as to FT spectrometers that employ a Michelson interferometer (Marshall and Comisarow, 1978). The factor of $\frac{1}{2}$ in Hadamard multiplexing results because each resolution element is detected for only half of the total time needed to complete the measurement. The factor of $\frac{1}{2}$ in FT multiplexing results because half of the spectral intensity is lost at the half-silvered mirror of the Michelson interferometer. (See Figure 7.1.)

It should also be recognized that not all authors agree on the exact value of the multiplex advantage, and for the detector-noise-limited case the increase in SNR may be given by $(\frac{1}{2}M)^{1/2}$ or even $(\frac{1}{8}M)^{1/2}$ rather than $(\frac{1}{4}M)^{1/2}$. Here, as in Table 5.5, M is the number of resolution elements being multiplexed. (See Khan, 1959; Filler, 1973; Chester, Fitzgerald, and Winefordner, 1976; Tai and Harwit, 1976; Winefordner et al., 1976; Hirschfeld, 1976a; Harwit and Tai, 1977; Treffers, 1977; Knacke, 1978; Luc and Gerstenkorn, 1978.) Whatever the value obtained for the multiplex advantage, most authors do agree that multiplexing will benefit the SNR of the entire spectrum only in a detector-noise-limited situation.

These mathematical generalizations regarding the benefits of multiplexing (Table 5.5) have been verified by computer simulation of spectra (Larson, Crosmun, and Talmi, 1974) for all three types of noise. In this computerized treatment, noise was simulated by choosing a random number from a normal distribution with a mean of zero and a standard deviation that behaved according to the type of noise under consideration, that is, independent of signal (detector noise), proportional to the square root of the signal (photon shot noise), or directly proportional to the signal (fluctuation noise). Comparison of computer-generated spectra for the multiplexed (Hadamard) and nonmultiplexed (scanning monochromator) cases demonstrated a multiplex advantage for the detector noise case (Figure 14.1) but a multiplex disadvantage for the photon shot noise (Figure 14.2) and fluctuation noise cases. Throughput advantage was not considered.

As a result of these mathematical predictions, transform methods have been utilized extensively for measurements where the system is dominated by detector noise (infrared, NMR, etc.) but not in the UV–visible region where detectors are typically more sensitive and generally dominated by photon shot noise. Ultraviolet–visible emission sources [flames, arcs, sparks, and induction-coupled plasmas (ICPs)] also tend to produce a considerable background signal. If this background signal is not modulated, then noise due to background will be additive, and total noise will still be proportional to the square root of source intensity. However, if this background noise is modulated (time or frequency dependent), then noise due to background will be multiplicative. In such cases, the system becomes dominated by fluctuation noise, which varies directly with the intensity of the source (Eq. 5.54).

Whenever fluctuation noise dominates, Table 5.5 predicts that neither Fellgett's advantage nor Jacquinot's advantage can be realized using transform techniques. Fortunately, fluctuation noise can often be reduced by changing the type of emission source employed in the experiment. For example, in flame emission, cylindrical burner heads (Aldous et al., 1970; Winefordner and Haraguchi, 1977) have been found to produce less fluctuation noise than normal slot burners (Horlick et al., 1982), and radiation emitted from an ICP may be even more stable (Griffiths and deHaseth, 1986a).

In favorable cases, fluctuation noise can also be reduced through the use of a "total intensity" reference detector (Hiltner and Code, 1950: Mertz, 1965; Decker, 1977). In this procedure, the ratio of the instantaneous detector readings to the total instantaneous intensity measurement by the reference detector is used as the primary spectral data. (The reference detector may also be used in a subtractive mode.) This procedure is said to work well unless the fluctuation noise varies with wavelength in an unpredictable way.

If fluctuation noise can be reduced to the point that photon shot noise dominates, then according to Table 5.5, only transform spectrometers with high throughput (Fourier transform and some doubly encoded HT spectrometers) should offer much SNR advantage over conventional scanning monochromators. For a high-luminosity spectrometer that does not multiplex, the SNR advantage will be $M^{1/2}$. If the spectrometer is also multiplexing, the net effect is found by multiplying the individual values for the throughput and multiplex advantages together (Luc and Gerstenkorn, 1978), that is, about $(\frac{1}{2}M)^{1/2}$ for the photon shot-noise-limited case.

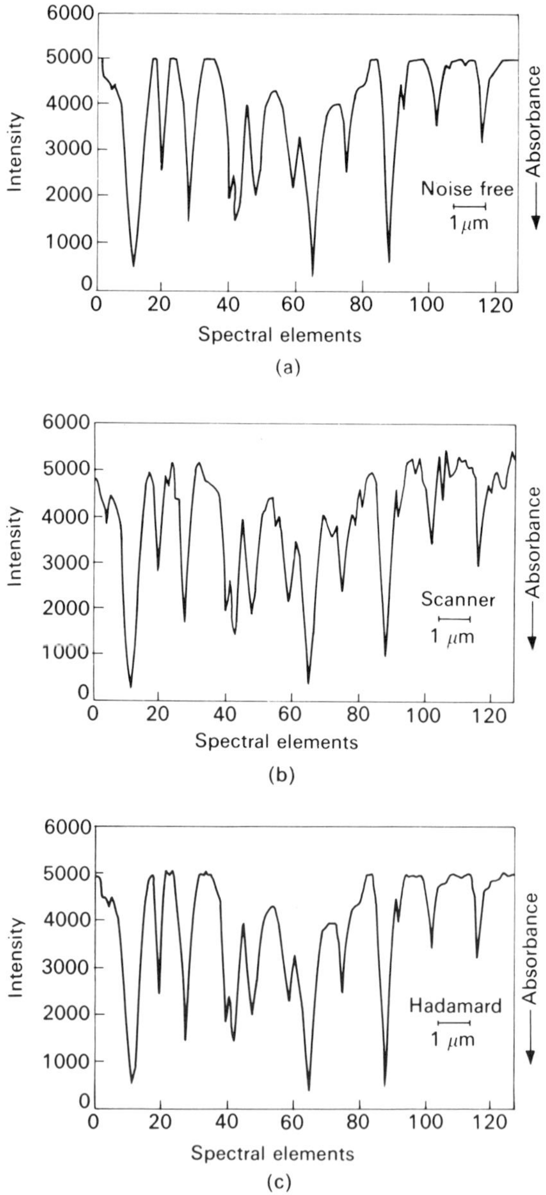

Figure 14.1. Computer-simulated infrared absorption spectra showing effect of multiplexing in detector-noise-limited case. (*a*) Noise-free spectrum. (*b*) Simulated spectrum obtained using scanning monochromator. (*c*) Simulated Hadamard transform spectrum for $M = 127$. [Reprinted with permission from N. M. Larson, R. Crosmun, and Y. Talmi (1974), *Appl. Opt.*, *13*, 2662. Copyright 1974 Optical Society of America.]

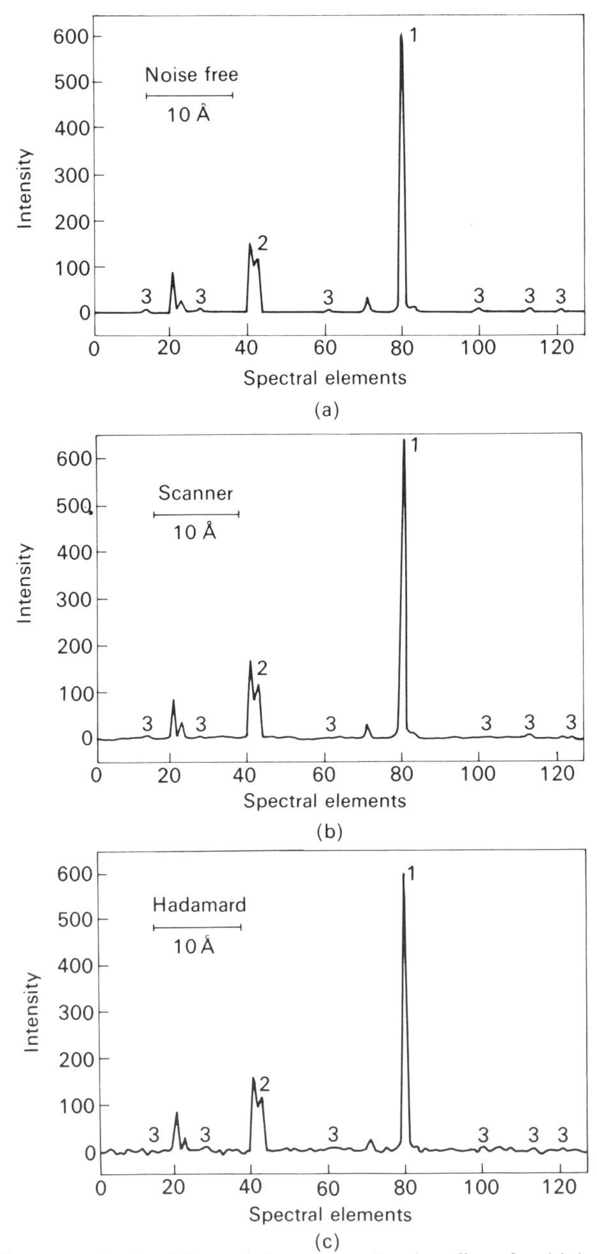

Figure 14.2. Computer-simulated line emission spectra showing effect of multiplexing in photon shot-noise-limited case. (*a*) Noise-free spectrum. (*b*) Simulated spectrum obtained using scanning monochromator. (*c*) Simulated Hadamard transform spectrum for $M = 127$. Numbers mark emission lines of (1) strong, (2) moderate, and (3) weak intensities. [Reprinted with permission from N. M. Larson, R. Crosmun, and Y. Talmi (1974), *Appl. Opt.*, *13*, 2662. Copyright 1974 Optical Society of America.]

It should be noted that the values in Table 5.5 predict the net effect of high throughput and/or multiplexing for the *entire spectrum*. A number of authors (Chester et al., 1976; Oliver, 1976; Oliver and Pike, 1974; Larson et al., 1974; Hirschfeld, 1976a; Luc and Gerstenkorn, 1978) have pointed out that in photon shot-noise-limited cases, there can still be a "distributive" (or "reduced") multiplex advantage for spectra that are optically dilute—those that contain a small number of emission lines superimposed on a weak background. This distributive effect originates from the multiplex property of averaging photon shot noise over the entire spectrum.

Following Hirschfeld's derivation (1976a, 1979), let $f(\nu)$ be defined as the spectral density function for the spectral region under study. Then

$$\mathrm{SNR}_{sc}(\nu)=f(\nu)^{1/2}E_{sc}^{-1/2}\,\Delta\nu\,T^{1/2}(\nu_2-\nu_1)^{-1/2} \tag{14.1}$$

where $\mathrm{SNR}_{sc}(\nu)$ is the SNR of a scanning monochromator, $\Delta\nu$ is the resolution, ν_1 and ν_2 are the limits of the scanning range, E_{sc} is the spectrometer efficiency, and T is the total measurement time. (Note that the term $(\nu_2-\nu_1)/\Delta\nu$ can also be written as M, the number of resolution elements being monitored.)

Assuming equal transmission and detectors for both instruments,

$$\mathrm{SNR}_{mx}(\nu)=f(\nu)E_{mx}^{-1/2}\Delta\nu\,T^{1/2}\left(\int_{\nu_1}^{\nu_2}f(\nu)d\nu\right)^{-\frac{1}{2}} \tag{14.2}$$

where $\mathrm{SNR}_{mx}(\nu)$ is the SNR of the multiplexed system and E_{mx} is its modulation efficiency. From Eqs. 14.1 and 14.2,

$$\frac{\mathrm{SNR}_{mx}(\nu)}{\mathrm{SNR}_{sc}(\nu)}=\left(\frac{f(\nu)E_{sc}(\nu_2-\nu_1)}{E_{mx}\int_{\nu_1}^{\nu_2}f(\nu)\,d\nu}\right)^{1/2}$$

$$=\left(\frac{f(\nu)E_{sc}}{f_{av}E_{mx}}\right)^{1/2} \tag{14.3}$$

From Eq. 14.3, it can be seen that the distributive multiplex advantage is proportional to the square root of the ratio between the intensity of the line under study, $f(\nu)$, and the mean intensity f_{av} per spectral element averaged over the entire range of spectral elements reaching the detector. (See also Nordstrom, 1982.)

The distributive multiplex effect is absent for detector-limited noise because noise is not affected by the strength of the signal. For the case of fluctuation noise (Hirschfeld, 1976a, b),

$$\frac{\mathrm{SNR}_{\mathrm{mx}}(\nu)}{\mathrm{SNR}_{\mathrm{sc}}(\nu)}=\frac{f(\nu)E_{\mathrm{sc}}}{f_{\mathrm{av}}E_{\mathrm{mx}}}\left(\frac{\Delta\nu}{\nu_2-\nu_1}\right)^{1/2} \tag{14.4}$$

Equation 14.4 is usually much less than one (Hirschfeld, 1979), meaning that there is no multiplex advantage of any kind for fluctuation-noise-limited situations.

Similar conclusions have been reached by Larson et al. (1974) using more complex mathematics. These authors predict that the multiplex advantage will be $(\frac{1}{2}M)^{1/2}$ for the detector-noise limited case (M is the number of slots in the mask) but $(x/2x_{\mathrm{av}})^{1/2}$ in the photon shot noise case. Here, x is the intensity of the line sought and x_{av} is the average intensity over the entire spectrum.

According to the results of Larson et al. (1974) and Hirschfeld (1976a), the distributive multiplex effect can be greater than 1 (an advantage) for a particular emission line provided the spectrum does not contain a large number of much more powerful lines. Whenever spectra are optically dense (i.e., contain a large number of very intense as well as very weak lines), the SNR of any strong spectral features will be increased at the expense of decreasing the SNR of the weak spectral features. This amounts to adding noise to an initially almost noise-free background.

Unfortunately, the majority of atomic emission spectra not only tend to be photon shot noise (or fluctuation noise) limited but also contain a large number of closely spaced spectral features superimposed on a strong background continuum. Such optically dense spectra are particularly likely to be encountered for sources that produce high temperatures and samples that contain a large number of elements.

The ICP is currently considered the most effective source for atomic emission spectrochemical analysis (especially multielement analysis) because it produces higher temperatures than flames, arcs, or sparks and is therefore capable of exciting more lines for a greater number of elements. (Other advantages of ICPs have been listed in Table 13.2. See also Dahlquist and Knoll, 1978; Winge, Fassel, and Kniseley, 1977; and Greenfield, McGeachin, and Smith, 1976.) Consequently, the best source for producing atomic emission may be the worst source, from an optical density standpoint, if transform methods are to be employed.

Figure 14.3 shows an optically dense emission spectrum (Ng and Horlick, 1985a) obtained for a solution of cobalt (2500 ppm), iron (100 ppm), nickel (1000 ppm), and vanadium (100 ppm). Emission was produced using an ICP source, and all spectral elements were detected simultaneously using a Michelson interferometer–photomultiplier tube (PMT) detector combination. Figure 14.4 shows another example of an ICP emission spectrum (Ng and Horlick, 1985a), this time of the five alkali metals (Na, 1 ppm; K, 2 ppm; Rb, 10 ppm; Cs, 5 ppm; and Li, 0.1 ppm). Again, all spectral elements were detected simultaneously using a Michelson interferometer–silicon photo-

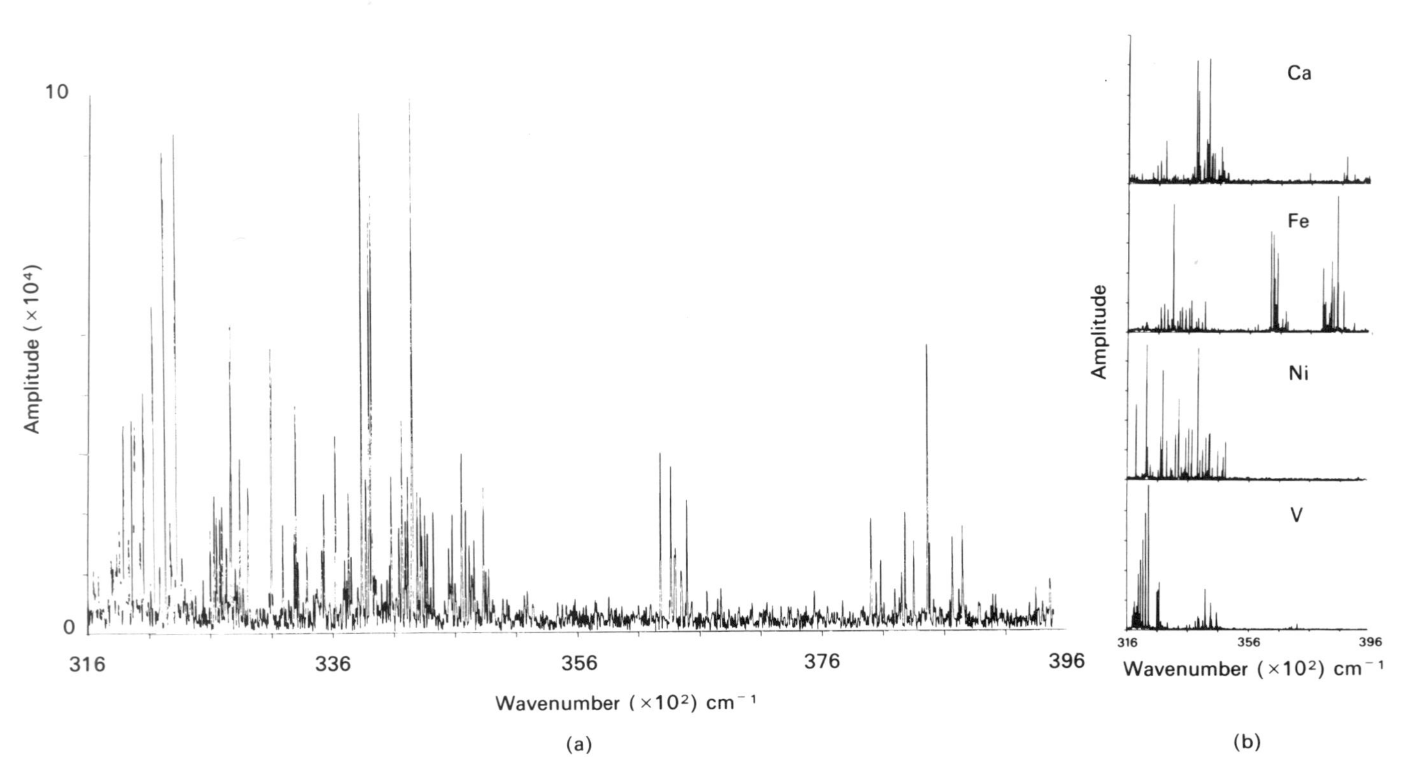

Figure 14.3. (*a*) An ICP emission spectrum resulting from Fourier transformation of interferogram obtained for multielement solution of Co, Fe, Ni, and V. (*b*) Spectra of individual elements. Detection system consists of 1P28 PMT used in combination with solar-blind optical filter. [Reprinted from R. C. L. Ng and G. Horlick (1985), *Appl. Spectrosc.*, *39*, 834, by permission from the Society for Applied Spectroscopy.]

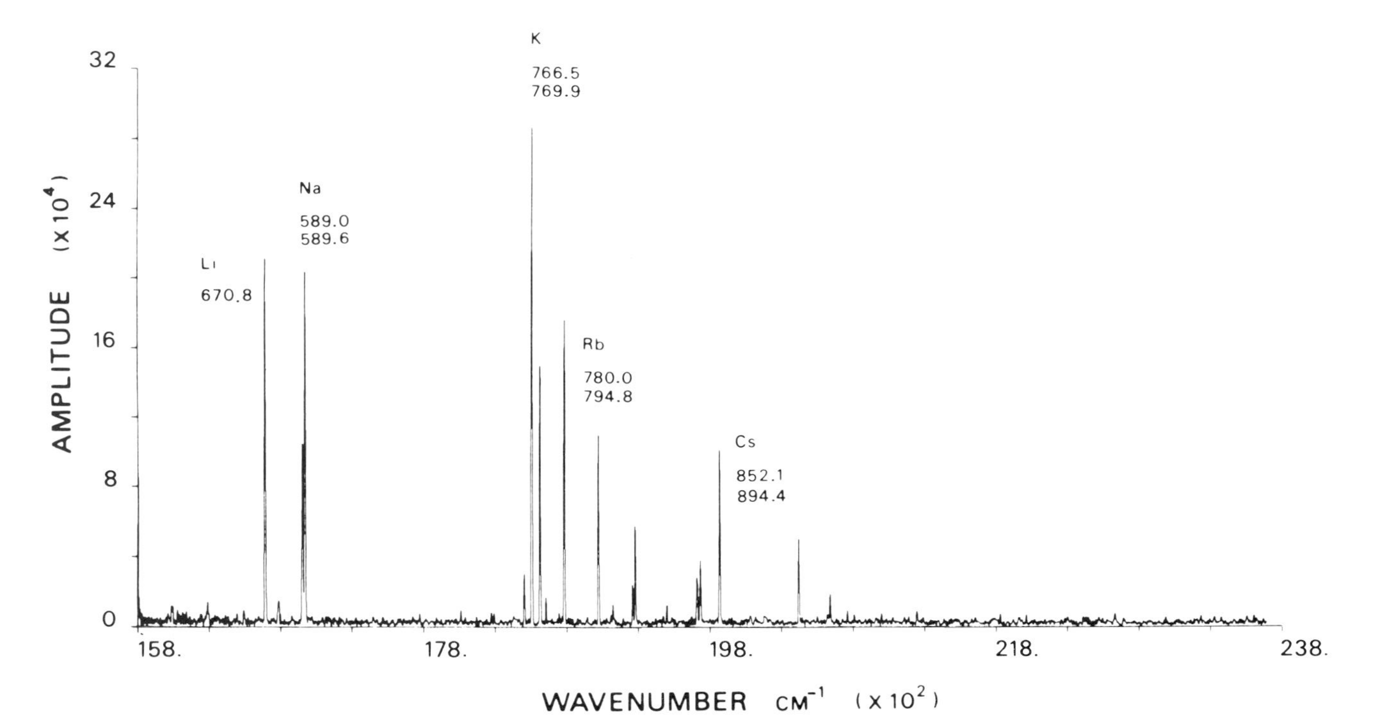

Figure 14.4. An ICP emission spectrum resulting from Fourier transformation of interferogram obtained for multielement solution of alkali metals using silicon photodiode detector. [Reprinted from R. C. L. Ng and G. Horlick (1985), *Appl. Spectrosc.*, *39*, 834, by permission from the Society for Applied Spectroscopy.]

diode detector combination. While the spectrum shown in Figure 14.4 is reasonably optically dilute, the range of spectral intensities is still rather large. If the spectrometer is photon shot noise limited, Eq. 14.3 predicts that improvement in SNR for the most intense lines can occur only at the expense of the SNRs of the weaker spectral lines. Consequently, the analyte lines must be chosen carefully to take maximum advantage of any SNR gain.

It should be noted that the validity of the assumption on which Eqs. 14.3 and 14.4 are based, that is, that multiplexing causes photon shot noise to be uniformly distributed across the entire spectral range, has been questioned by several authors for the case of atomic emission spectroscopy. (See Horlick et al., 1982, and Griffiths and deHaseth, 1986b). The behavior of noise and its effect on multiplexed spectra is of such importance that the experimental results of a number of noise studies in the UV–visible spectra region will be examined in more detail in Section 14.4.

14.1.2. Resolution

Another factor to be considered when choosing between transform and scanning spectrometers is spectral resolution. The spectral resolution of any spectrometer depends on the total path difference between resolution elements at the extremes of the spectrum. In a dispersive instrument (HT or scanning spectrometer), resolution is proportional to the size of the diffraction grating. In a Michelson interferometric spectrometer, resolution is proportional to the maximum retardation, that is, the maximum distance traveled by the moving mirror. For a resolution of 0.002 cm^{-1}, the diffraction grating must measure approximately 2 m on a side, or (for infrared) the mirror must travel approximately 2 m. For atomic emission work, however, phase correction of the interferograms generally requires that a double-sided interferogram be measured (Horlick and Yuen, 1975; Yuen and Horlick, 1977). If equal retardation on both sides of the zero-order centerburst must be scanned, the effective length traveled by the mirror, and therefore the maximum spectral resolution possible for the instrument, is reduced by one-half.

Given current technology, it is easier to construct an interferometer with a very long-travel mirror than to manufacture a very large diffraction grating. Consequently, interferometric FT spectrometers are capable of achieving much higher spectral resolution than dispersive instruments, although it must be noted that very long-travel interferometers are expensive and generally custom-made. Also, compared to infrared radiation, atomic emission is of much shorter wavelength. Ultraviolet–visible radiation places greater restrictions on both interferometric and dispersive spectrometer designs, generally requiring special retroreflectors to compensate for jitter in the moving mirror of the interferometer, narrower slots in the Hadamard mask, and closer ruling in the diffraction grating.

14.1.3. Spectral Range

Due to the overlap of higher orders of the diffraction grating, the maximum wavelength difference that can be simultaneously observed by a dispersive spectrometer is approximately one-half the longest wavelength. While it is true that the FT spectrometer does not use a grating, the beamsplitter (required to separate the two beams that will be brought into interference) only functions efficiently over a limited range of wavelengths, generally a factor of 2 or 3. To illustrate, Figure 14.5 shows the beamsplitter efficiency of an FT spectrometer that has been used to measure atomic emission spectra (Stubley and Horlick, 1985a). In the case of dispersive instruments (HT and scanning monochromators), spectral range may be extended by the use of filters, while the FT instrument requires the presence of more than one kind of beamsplitter.

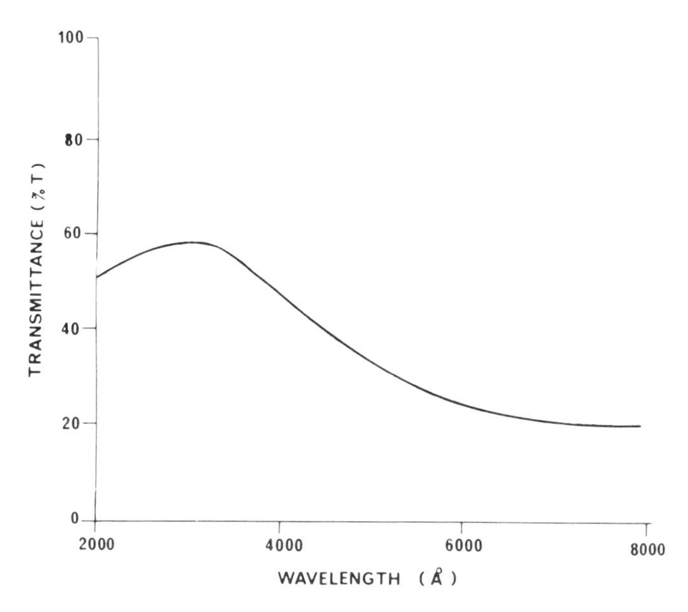

Figure 14.5. Efficiency of ultraviolet beamsplitter as measured by its transmission characteristics. [Reprinted from E. A. Stubley and G. Horlick (1985), *Appl. Spectrosc.*, *39*, 800, by permission from the Society for Applied Spectroscopy.]

14.1.4. Multiplex Number

Multiplex number refers to the number of spectral elements that can be monitored simultaneously. The FT spectrometer is readily capable of multiplexing many more spectral elements than the HT spectrometer, which is limited both by its diffraction grating and the width of the multislit array used to encode the radiation. The scanning monochromator, of course, has no multiplexing capability. Multiplexing, however, is not predicted to be of much advantage in the photon shot-noise-limited case unless the emission spectrum is optically dilute.

14.1.5. Throughput

Because resolution in FT interferometers does not depend on slit width, round apertures may be used instead of slits, resulting in a significant gain in available radiant energy, or throughput, through the spectrometer. For instruments of the same resolution and optical component size, the FT spectrometer can, in theory, have up to 200 times the throughput of a scanning dispersive monochromator (Bell, 1972). (It should be noted, however, that many commercially available FT spectrometers do not achieve anything close to this theoretical maximum. See Decker, 1977.) Singly encoded HT spectrometers have no increase in throughput over the scanning dispersive monochromator (Section 5.5). Doubly encoded HT spectrometers can be designed to operate with increased throughput (Section 5.5.5), but throughput is limited by the number and size of pairs of slits that are simultaneously open in the entrance and exit focal planes of the spectrometer.

14.1.6. Computational Requirements

Unlike scanning dispersive monochromators, both Fourier and Hadamard transform spectrometers require post measurement computation to convert the measured signal into its component frequencies and associated intensities. Due to an inherently more complex mathematics, fast Fourier transform runs approximately an order of magnitude slower than the equivalent fast Hadamard transform, per transform dimension (Pratt, Kane, and Andrews, 1969). In addition, double-sided interferograms are generally required for atomic emission studies, and due to the shorter wavelengths involved, the interferogram must be sampled at smaller intervals. Both these factors greatly increase the amount of memory required to store Fourier transform data. Thus, FT spectrometers require much more in terms of computer power and computer time than do HT spectrometers.

14.1.7. Construction

Fourier transform interferometers require fraction-of-wavelength-of-light tolerances in the motion of the mirror and are highly subject to vibrations. Hadamard transform spectrometers require optical tolerances similar to conventional scanning monochromators. Early forms of Hadamard spectrometers, however, used either stepped linear or rotating masks to attain spectral encodement. These moving masks were subject to a number of problems, including misalignment and instability during operation as well as irregularities in shape and size of the slits due to errors in construction (Section 6.1). More recent Hadamard spectrometers have employed stationary or "solid-state" masks fabricated from electro-optic or thermodiachromatic material (Hammaker et al., 1986). These stationary masks have eliminated or reduced many of the mechanical problems associated with the earlier moving-mask instruments (Section 14.3).

14.1.8. Spectral Manipulation

If the output from a spectrometer can be stored in computer memory as an array of numbers, then a variety of postmeasurement data manipulations are possible. For example, repeated spectral scans can be averaged to reduce SNR (detector and photon shot noise cases), spectra can be ratioed to reduce fluctuation noise, and a stored reference spectrum can be subtracted from the sample spectrum to produce a corrected sample spectrum. Figure 14.6 demonstrates the technique of background correction for an ICP/FT spectrometer in which the hydroxyl band centered at 326.1 nm is subtracted from the spectrum of an aqueous copper solution to yield only the copper emission at 327.4 nm. All three spectra in this figure have the same intensity scale (Stubley and Horlick, 1985b).

A necessary prerequisite for such data manipulation is instrumental stability, that is, precise tracking of wavelength scale and/or minimum fluctuation of the mechanical components of the spectrometer. Because of the number of moving parts present in the instrument, scanning monochromators may not reproduce spectra with sufficient frequency precision to make postmeasurement data manipulation worthwhile. In FT spectrometers, however, the use of an internal reference laser results in a high degree of wavelength reproducibility from scan to scan (the Connes or frequency precision advantage), and the calibration needed to ensure equally high wavelength accuracy is very simple. The ability to manipulate the resultant spectral data has been a major reason for the growth of FT techniques among spectroscopists (Griffiths, 1974; Hirschfeld, 1976b).

Hadamard transform spectrometers also have sufficient wavelength reproducibility to make data manipulation advantageous (Decker, 1977). More

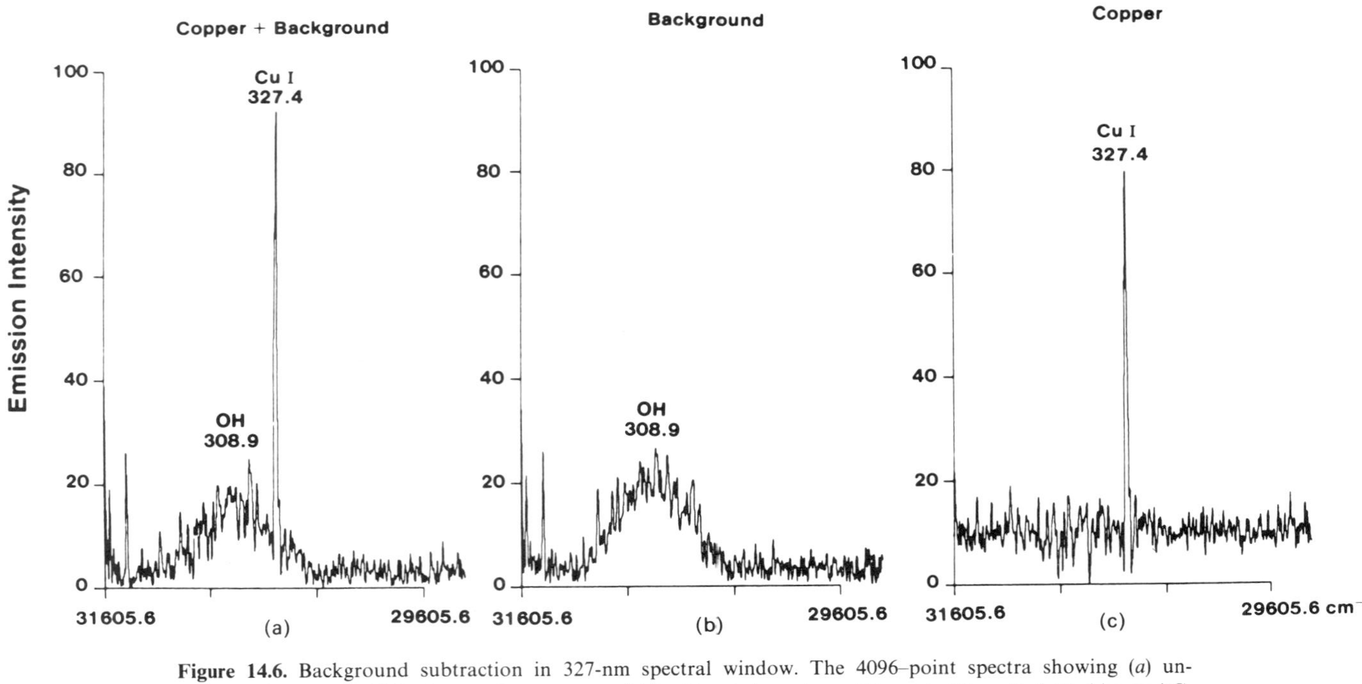

Figure 14.6. Background subtraction in 327-nm spectral window. The 4096–point spectra showing (*a*) unsubtracted spectrum, (*b*) background spectrum and (*c*) subtracted spectrum. [Reprinted from E. A. Stubley and G. Horlick (1985), *Appl. Spectrosc.*, *39*, 811, by permission from the Society for Applied Spectroscopy.]

recent Hadamard instruments employing stationary masks show even more promise in this regard due to the no-moving-parts nature of the spectrometer (Tilotta, Freeman, and Fateley, 1987).

14.2. FOURIER TRANSFORM SPECTROMETERS

As discussed in Section 14.1, the Michelson interferometric Fourier transform (FT) spectrometer has a number of attractive features that make it potentially useful for atomic emission spectrometry. True, improvements in SNR due to multiplexing and/or increased throughput may be reduced or even lost altogether because the system is photon shot noise or fluctuation noise limited. Nevertheless, FT instruments still offer high-wavelength precision, high-wavelength accuracy (coupled with simple calibration procedures), high resolution, the possibility of postmeasurement data manipulation (coaddition of spectra to increase SNR, background correction, etc.), and a large spectral range. This last features makes it possible for the FT spectrometer to monitor many elements simultaneously using the near-ideal light intensity measurement characteristics of the PMT. The advantages of the PMT include good linearity, wide dynamic range, high sensitivity, and direct electronic readout (Budde, 1983).

The two instruments now in common use for multielement spectrochemical analysis are the spectrograph (Chapter 4) and the direct reader (Chapter 15). While the spectrograph can permanently record thousands of spectral lines over a very wide spectral range, its photographic plate detection system suffers from nonlinear response to signal intensity, a limited dynamic range, and time-consuming, tedious development and reading (Chapter 8). Imaging devices such as the vidicon and silicon photodiode array (Chapters 10 and 11) can be substituted for the photographic plate. However, these detectors provide multichannel capability over a relatively narrow spectral range, and none has the sensitivity or dynamic range of the PMT. Direct readers are most useful for the routine detection of a more limited number of wavelengths (generally no more than 30; see Slavin, 1971) and require a separate exit slit–PMT combination for each wavelength. This has the disadvantage of limiting the resulting spectral information to those wavelengths that have been preselected and preset on the focal curve of the instrument.

14.2.1. Early FT Applications in UV–Visible Spectroscopy

Although FT methodology had its inception in the UV–visible spectral region (eg., Michelson, 1891, 1892), FT applications soon shifted to the infrared where experiments confirmed the theoretical predictions of a much improved

SNR. In 1965, Jacquinot suggested that FT spectroscopy should be reapplied to the UV–visible region, and a number of theoretical papers dealing with the subject soon followed (e.g., Filler, 1973). In 1972, Luc and Gerstenkorn equipped an old FT–IR spectrometer (Pinard, 1967) with a photomultiplier and successfully measured the magnesium lines in the visible region with a high degree of resolution. Following this initial experiment, a new Fourier spectrometer was built that was especially adapted for the near-infrared, visible, and near-UV (Luc and Gerstenkorn, 1978). Using this instrument, high-quality atomic emission and molecular absorption spectra were obtained. Up to 60,000 lines in the spectrum of iodine were identified with the strongest peaks having about $\pm 0.001\ \mathrm{cm}^{-1}$ uncertainty in peak position.

By the middle 1970s, analytical applications of UV–visible interferometry were also beginning to appear. Nitis Svoboda, and Winefordner (1972) employed an oscillating Fabry–Perot interferometer (Section 6.8.6) to modulate radiation from a continuum source for use in atomic absorption flame spectrometry. In this system, a near axial beam of radiation from a continuum source was passed through the oscillating interferometer. The transmitted beam, consisting of a series of interference fringes that were a function of the wavelength of the light and the rate of oscillation of the interferometer, was then passed through the absorbing cell (an air–acetylene flame). If no line absorption occurred, the energy contained in the spectral range of the interferometer was seen as a dc signal by the PMT detector. However, absorption of the analyte line caused the intensity at this wavelength to vary with a frequency equal to twice the oscillating frequency of the Fabry–Perot interferometer. The resulting signal was ac, which could be amplified, rectified by a lock-in amplifier, and recorded. The method had the potential advantages associated with a continuum source and increased throughput. However, Table 14.1 shows that the detection limits obtained for six elements using this technique were all considerably poorer than those obtained using a more conventional atomic absorption instrument with a mechanical chopper (Winefordner et al, 1970).

Pruiksma, Ziemer, and Yeung (1976) also employed a scanning Fabry–Perot interferometer to perform the simultaneous multielement analysis of a number of elements by atomic emission using an oxygen–hydrogen flame. (Scanning was accomplished by changing the distance between plates in the interferometer.) However, the Fabry–Perot interferometer can only monitor a fairly narrow spectral range (Section 6.8.6), and the method could be applied only to groups of elements that contained relatively few total emission lines in a rather small wavelength window. Table 14.2 shows the detection limits obtained for eight elements using this technique. While these data show that the technique did permit detection at semitrace-to-trace levels, the detection limits were quite poor compared to those obtained using a conventional scanning monochromator.

Table 14.1. Detection Limits (ppm) for Several Atomic Absorption Spectrometers

Element	Wavelength (nm)	AAFP[a]	AAMC[b]	AAMC[c]
Ag	328.1	2×10^{-1}	1×10^{-2}	5×10^{-4}
Ca	422.7	7×10^{-2}	3×10^{-2}	2×10^{-3}
Cd	228.8	2×10^{-1}	2×10^{-2}	1×10^{-2}
Cr	357.9	4×10^{-1}	2×10^{-1}	5×10^{-3}
Cu	324.7	1×10^{-1}	5×10^{-2}	5×10^{-3}
Mg	285.2	3×10^{-2}	1×10^{-2}	3×10^{-4}

[a] Atomic absorption flame spectrometer with continuum source and oscillating Fabry–Perot interferometer (Nitis et al., 1972).
[b] Atomic absorption flame spectrometer with continuum source and mechanical chopper (Fassel et al., 1966).
[c] Atomic absorption flame spectrometer with line source and mechanical chopper (Winefordner et al., 1970).
[Reprinted with permission from G. J. Nitis et al. (1972), "An Oscillating Interferometer for Wavelength Modulation in Atomic Absorption Flame Spectrometer," *Spectrochimica Acta, 27B*. Copyright 1972, Pergamon Press, Elmsford, NY.]

Table 14.2. Detection Limits for a Scanning Fabry-Perot Atomic Emission Spectrometer[a]

Element	Wavelength (nm)	Detection Limit[b] (ppm)
Na	589.0	0.02
Ca	422.7	2
Ba	553.6	8000
Mn	403.1	4000
Sr	460.7	20
Cr	425.4	120
Rb	420.2	200
Li	670.8	400

[a] From Pruiksma et al. (1976).
[b] Defined for SNR = 2.

The idea of performing multielement analyses in the UV–visible region with a Michelson interferometer was discussed by Horlick and Yuen in 1975. Since that time, a limited number of research groups have attempted to determine the potential of the method for atomic spectroscopy, and at least three reviews (Horlick et al., 1982; Nordstrom, 1982; Griffiths and deHaseth, 1986a) have appeared on the subject. While it may still be too early to predict the future role of the Michelson interferometer in multielement spectrochemical analysis, it is certainly safe to say that the results reported by these investigators have yet to promote a general acceptance of the method by the analytical community.

14.2.2. FT Applications in the UV–Visible: Horlick and Co-workers

Using a Michelson interferometer system based originally on a design developed for space flights (Chaney, Loh, and Such, 1969; Hanel and Conrath, 1969), Yuen and Horlick (1977) demonstrated that emission spectra from a flame and from a hollow-cathode lamp (HCL) could be measured with a high degree of wavelength precision and accuracy. The group then applied the method to multielement analysis using an ICP source (Horlick et al., 1982; Stubley and Horlick, 1984). While the instrumentation used in these studies has gradually evolved over a period of years, the basic block diagram of the ICP/FT system is shown in Figure 14.7.

Details of the Michelson interferometer are shown in Figure 14.8. The beamsplitter consists of a Suprasil 1 substrate coated with a 100-Å layer of aluminum and a Suprasil 1 compensator plate. The fixed mirror is attached to a ball-in-socket mount and can be aligned with the use of differential micrometers. The moving mirror is driven electromechanically, and the entire mirror drive system is supported by an air bearing. The detector assembly consists of a focusing lens (10 cm), lens holder, and detector mount. Several types of detectors have been employed: a silicon photodiode (600–1000 nm, near-infrared), a 1P28 photomultiplier tube (300–600 nm, visible), and an R166 solar-blind photomultiplier tube (UV).

The Michelson interferometer has three optical inputs (Figure 14.9): a He–Ne laser, a white-light (tungsten bulb) source, and the spectral signal of interest. Mirror alignment and alignment maintenance are considerably facilitated by this design because all three optical signals share the same beamsplitter and mirrors. Control signals derived from the white-light interferogram and the laser fringes allow multiple interferograms of the same signal to be time averaged exactly.

In FT-IR spectroscopy, single-sided interferograms can be employed, and the zero-order centerburst is used to mark the interferogram for signal averaging. Using the instrument shown in Figure 14.9, data acquisition may be started at some imprecise point before the position of zero path length difference. The centerburst is then used to trigger a counter that determines the precise stop pulse at which data acquisition ceases. Once a precise stop pulse has been determined, repetitive interferograms can be averaged backward, that is, by sequencing addition from the last data point. This technique

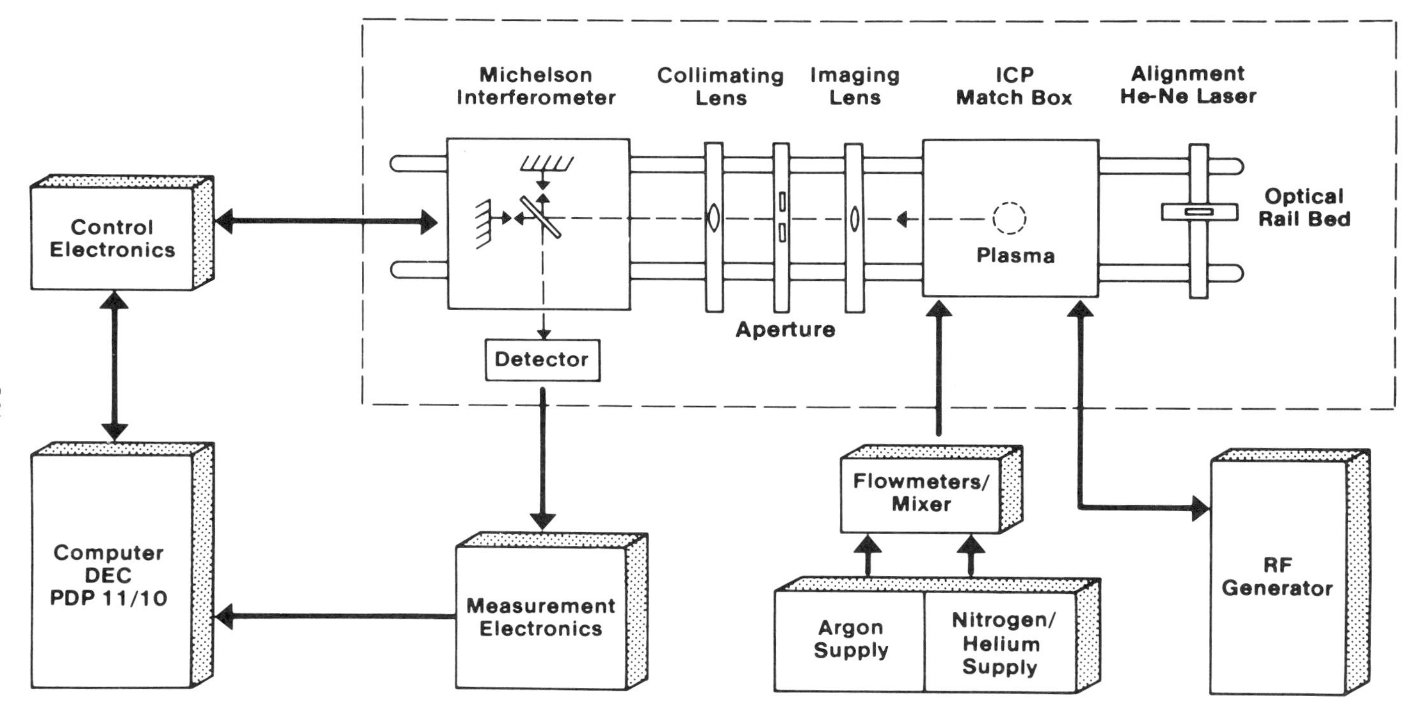

Figure 14.7. Block diagram of inductively coupled plasma/Fourier transform spectrometer system. [Reprinted from R. C. L. Ng and G. Horlick (1985), *Appl. Spectrosc.*, *39*, 834, by permission from the Society for Applied Spectroscopy.]

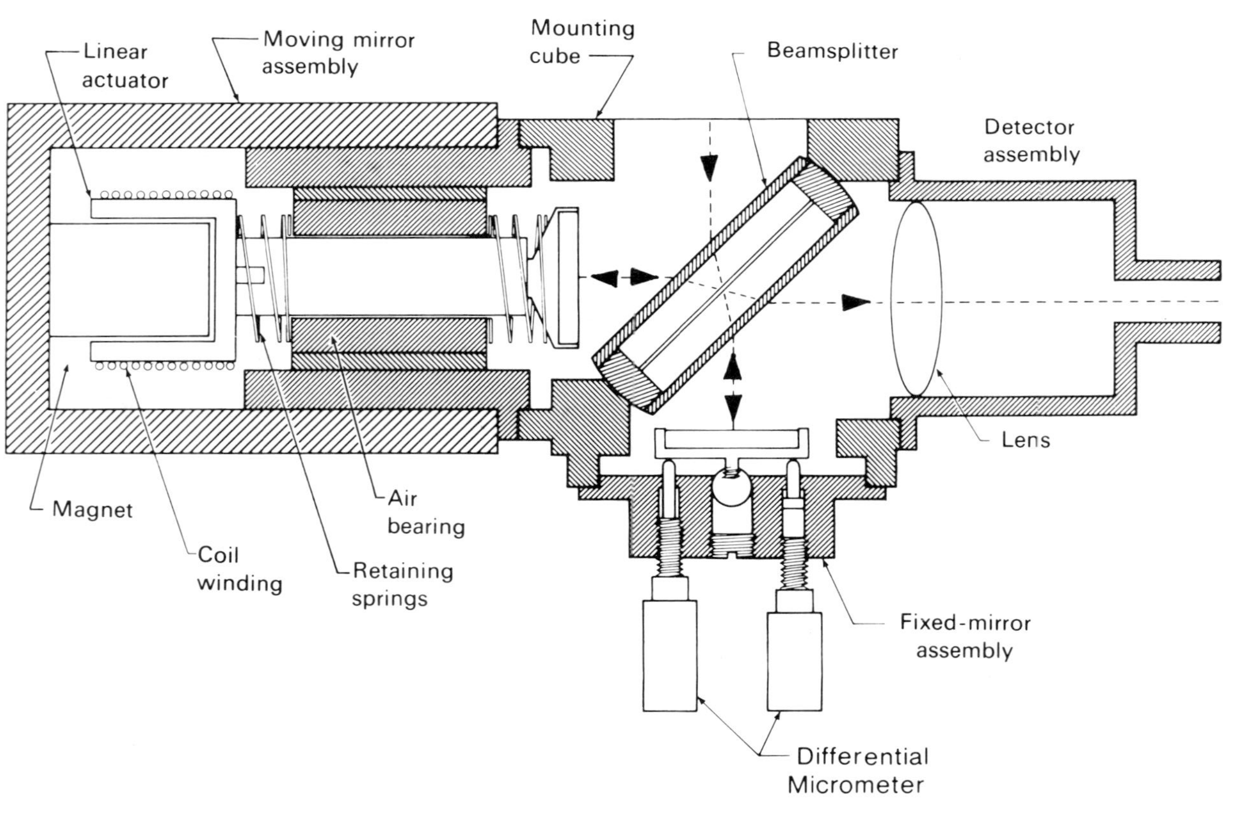

Figure 14.8. Schematic diagram of Michelson interferometer used by Stubley and Horlick (1985a). [Reprinted from E. A. Stubley and G. Horlick (1985), *Appl. Spectrosc.*, *39*, 800, by permission from the Society for Applied Spectroscopy.]

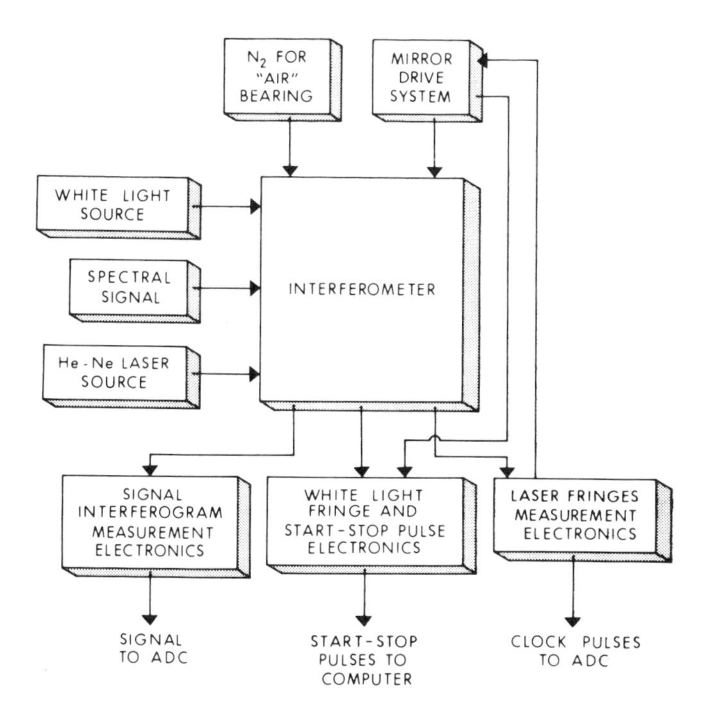

Figure 14.9. Block diagram showing origins of input and output signals. [Reprinted from G. Horlick and W. K. Yuen (1978), *Appl. Spectrosc.*, *32*, 38, with permission from the Society for Applied Spectroscopy.]

also produces double-sided interferograms that avoid the phase correction problems of one-sided interferograms (Horlick and Yuen, 1975; Yuen and Horlick, 1977). Counting and sampling of the interferogram are controlled through the fringe pattern produced by the He–Ne reference laser. Further details concerning data acquisition, the interferometer, and the control electronics can be found in the literature (Horlick and Yuen, 1978; Horlick et al., 1982; Stubley and Horlick, 1985a).

The standard He–Ne reference laser does not provide a basic sampling interval sufficient to avoid aliasing in the UV–visible spectral range. Aliasing results in fold-over of the undersampled modulation frequencies into other parts of the spectrum (Section 7.6.2) and should normally be avoided. Horlick et al. (1982) have shown how aliasing may be used to advantage by predictably folding lines in a sparce spectrum into close proximity with other analyte lines, thereby reducing data acquisition rate and memory requirements. (The effect is somewhat similar to using higher orders in a grating spectrometer.) Unfortunately, if noise is uniformly distributed across the spectrum, the SNR of the spectrum will be degraded by a factor of $2^{1/2}$ each time the spectrum is folded (Griffiths and deHaseth, 1985a). The SNR can be restored by signal averaging. However, measurement time must be doubled for each time the spectrum is folded to restore the original SNR of the unfolded spectrum.

Earlier work by the Horlick group (reviewed by Horlick et al., 1982) demonstrated that while suffering from some limitations in intensity, dynamic range, and detection limits, the ICP/FT spectrometer could be used to simultaneously measure most elements at sub-ppm ranges. Linear calibration curves were also obtained.

A more rigorous test of the analytical capabilities of this instrument (Stubley and Horlick, 1985c) employed a multielement solution containing 16 commonly analyzed elements, each at the 10-ppm concentration level. Emission spectra for this solution in the UV and visible regions are shown in Figures 14.10 and 14.11, respectively. Each spectrum is the result of 32 signal-averaged interferograms.

As can be seen from these figures, 5 of the 16 elements (K, Ni, Si, Na, and V) are not observed, and several prominent emission lines for the remaining elements are also absent. (For example, of the three Zn lines at 213.8, 206.1, and 202.6 nm, only the strongest line at 213.8 nm is apparent.) These missing

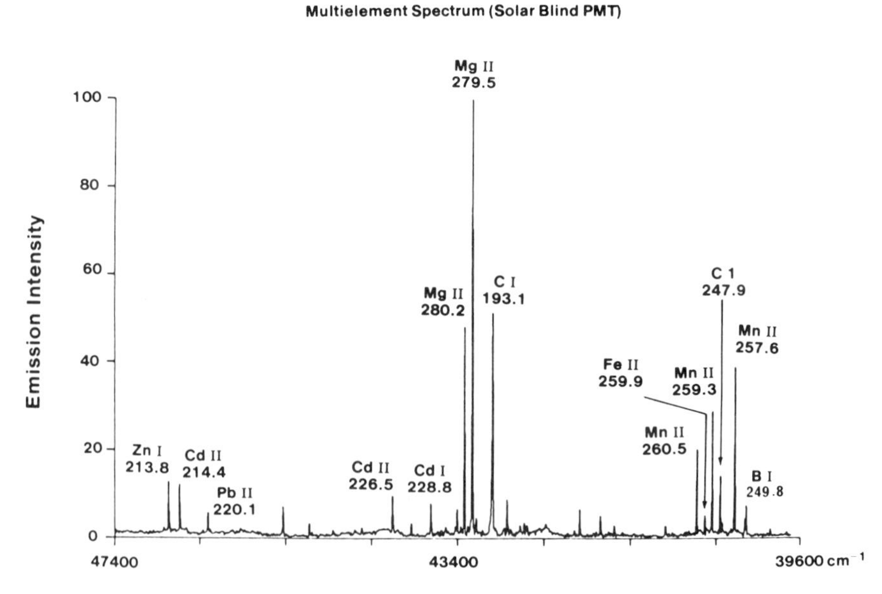

Figure 14.10. Emission spectrum of multielement solution as measured using solar-blind PMT. [Reprinted from E. A. Stubley and G. Horlick (1985), *Appl. Spectrosc.*, *39*, 805, with permission from the Society for Applied Spectroscopy.]

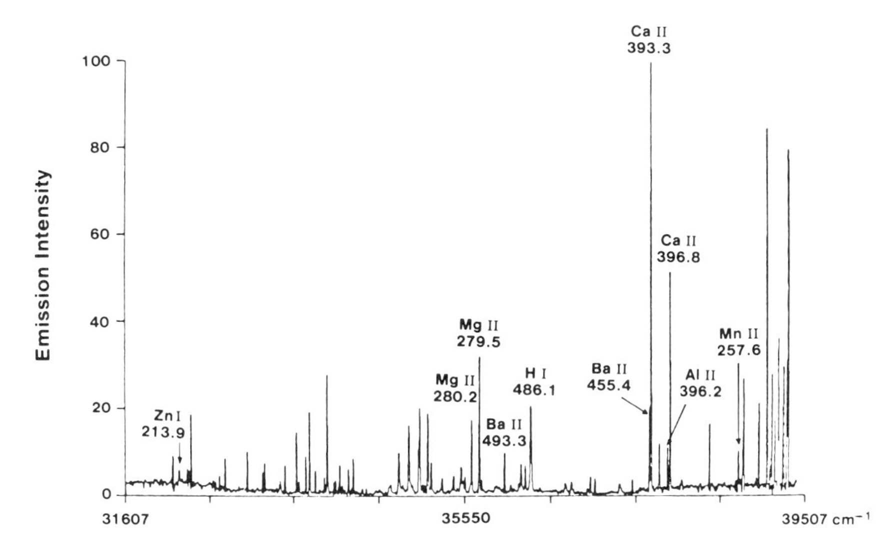

Figure 14.11. Emission spectrum of multielement solution as measured using 1P28 PMT. [Reprinted from E. A. Stubley and G. Horlick (1985), *Appl. Spectrosc.*, *39*, 805, with permission from the Society for Applied Spectroscopy.]

spectral features may be the result of spectral interferences, either direct or due to aliasing overlap, and deterioration of portions of the spectrum due to the distributive multiplex disadvantage. (The presence of intense emission lines imposes limitations on the observation of weaker emission lines in the same spectral region. See Section 14.1.1.) The effect of the intense Mg signal on the detection of the weaker Zn signal at 213.8 nm was studied, and the results confirmed the presence of a multiplex disadvantage for the weaker spectral line (Stubley and Horlick, 1985c). These results will be examined more closely in Section 14.4.

Table 14.3 shows the detection limits obtained by the ICP/FT spectrometer in the UV (R166 PMT) and visible (1P28 PMT) regions using 32 signal-averaged interferograms. In the six cases where comparisons can be made (B, Ba, Cd, Mn, Pb, and Zn), these detection limits are at least one order of magnitude poorer than those obtained by a commercially available direct-reader-based ICP instrument (ARL 34000 in Table 14.3).

Since this is a relatively simple, aqueous sample, the spectra shown in Figures 14.10 and 14.11 are still fairly optically dilute. For more complex samples, however, the multiplex disadvantage should become more apparent, and the ICP/FT detection limits are likely to be even worse than those listed

Table 14.3. Detection Limits Measured for a Multielement Solution[a]

Element	Wavelength (nm)	Detection Limits (ppb) 1P28 (Visible)	R166 (UV)	ARL 34000
B	249.7	—	62	3
Ba	455.4	77	—	0.6
Ca	393.3	16	—	—
Cd	226.4	—	48	2
Mg	279.1	58	5	—
Mn	257.6	145	17	0.7
Pb	220.7	—	130	37
Zn	213.8	—	37	4

[a] Values under 1P28 and R166 obtained from FT spectra in visible and ultraviolet, respectively. Values under ARL 34000 obtained from single-element spectra (nonmultiplexed) using commercially available direct-reader, ICP-based instrument. From Stubley and Horlick (1985c).

in Table 14.3. To improve the quantitative performance of the ICP/FT measurement, it is necessary to reduce the complexity of the spectrum being monitored. This can be accomplished in a number of ways, including optical filtering and/or reducing the size of the spectral window being monitored.

A block diagram of a windowed slew-scanning (WSS) ICP/FT spectrometer (WSS-FTS, Stubley and Horlick, 1985b) is shown in Figure 14.12. This instrument consists of a slew-scanning monochromator coupled to a Michelson interferometer (Figure 14.8). Throughput and spectral range are controlled by the monochromator, while resolution, wavelength accuracy, and wavelength precision are controlled by the interferometer. In this study, the entrance and exit slits of the monochromator were opened up to permit simultaneous observation of a narrow window having a total spectral range of about 8–10 nm. The position of this window could be preselected and easily altered by changing the monochromator setting. Slew scanning between spectral windows allowed multielement analysis to be performed at a reasonably rapid rate.

The same 16-element solution (Table 14.3) was reanalyzed using the WSS–FTS instrument over a spectral region that ranged from the vacuum UV (180 nm) to the near IR (925 nm). Figure 14.13 shows just the 259-nm window portion of the UV emission spectrum. (The complete UV spectrum is shown in Figure 14.10.) Since carbon now lies outside of the wavelength

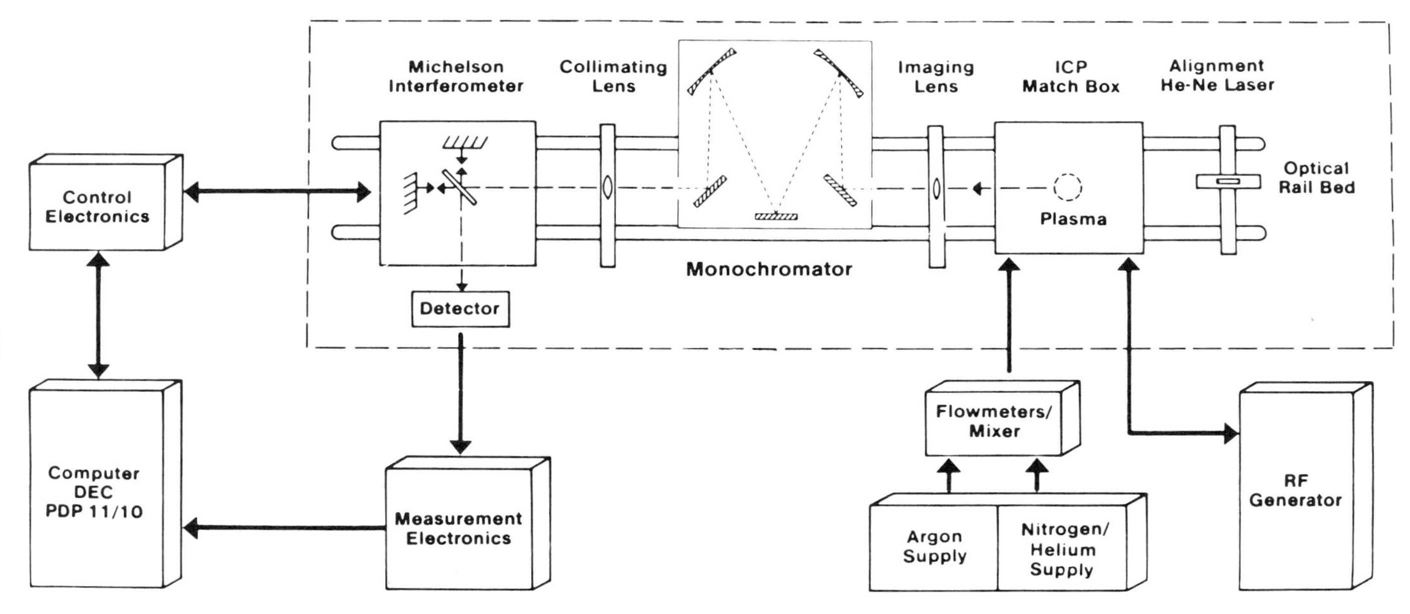

Figure 14.12. Block diagram of ICP–WSS–FTS measurement system. [Reprinted from E. A. Stubley and G. Horlick (1985), *Appl. Spectrosc.*, *39*, 811, with permission from the Society for Applied Spectroscopy.]

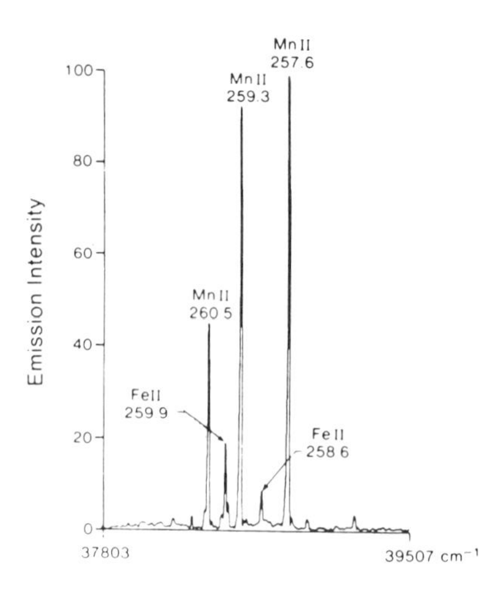

Figure 14.13. The 259-nm window emission spectrum as measured by ICP–WSS–FTS. [Reprinted from E. A. Stubley and G. Horlick (1985), *Appl. Spectrosc.*, *39*, 811, with permission from the Society for Applied Spectroscopy.]

window selected by the monochromator, the 247.9-nm C line has been eliminated, and the weak Fe line at 258.6 nm is clearly defined.

Table 14.4 compares the detection limits obtained for 15 out of the 16 elements using the WSS–FTS and FT systems in the UV (R166 PMT) and visible (1P28 PMT) with those obtained by a commercial direct-reader, ICP-based instrument. (Detection limits were calculated as the concentration needed to give a signal intensity of twice the standard deviation of a distilled water background for the appropriate spectral window.) As before, each spectrum represents an average of 32 interferograms. With the exception of Cu, Mn, and Na, the detection limits have been improved by the WSS–FTS system to the point that they are now within a factor of 2 or 3 of those measured by the commercial direct reader (ARL 34000). Data acquisition time has, however, been significantly increased. For example, the entire multielement spectrum can be acquired by the conventional ICP/FT system (Figure 14.7) in the same time required by the WSS–FTS instrument to acquire three to four elements or for the direct reader to provide quantitative information on every element it has been preset to monitor.

14.2.3. FT Applications in the UV–Visible: Faires and Co-workers

Faires et al. (1983) have utilized the interferometer developed for astronomical measurements in the visible and near infrared at Kitt Peak National Observatory (KPNO) for temperature determination (Faires et al., 1984) and

Table 14.4. A Comparison of Detection Limits[a]

	Detection Limits (ppb)				
	1P28 (Visible)		R166 (UV)		
Element	WSS–FTS	FTS	WSS–FTS	FTS	ARL 34000
Al	82	—	—	—	—
B	—	—	10	62	3
Ba	0.9	77	—	—	0.7
Ca	0.4	16	—	—	—
Cd	—	—	8	48	2
Cu	19	—	—	—	3
Fe	11	—	6	—	2
Mg	7	58	4	5	—
Mn	12	145	7	17	0.7
Na	135	0	—	—	14
Ni	—	—	26	—	—
Pb	—	—	18	130	—
Si	32	—	—	—	23
V	5	0	—	—	—
Zn	—	—	7	37	4

[a] Detection limits in each row obtained using same analytical wavelength. Whenever more than one emission line could be used for analysis of given element, value in table represents best FT detection limit obtained. From Stubley and Horlick (1985b).
[Reprinted from E. A. Stubley and G. Horlick (1985), *Appl. Spectrosc.*, *39*, 811, with permission from the Society for Applied Spectroscopy.]

linewidth and line shape analyses (Faires et al., 1985a) in the ICP. In addition to obtaining basic spectrophysical information about the ICP, these studies also explored the potentials and problems of the KPNO ICP/FT spectrometer combination for analytical applications.

The KPNO ICP/FT spectrometer is capable of obtaining unaliased spectra over the wavelength region 250 nm to 16.5 μm with a wavenumber accuracy on the order of 10^{-5} cm^{-1}, an intensity accuracy on the order of 0.1%, and a resolving power of up to 500,000 (Brault, 1976; Faires, 1984). The instrument is based on a folded Michelson interferometer design with an optical path difference of 1 m and a total internal path of about 12 m from input to output. (See Breckinridge and Schindler, 1981, for a discussion of folded interferometer designs.) The entire internal path is housed in a vacuum chamber and operated at 10^{-1}–10^{-2} Torr to reduce atmospheric absorption.

The relatively complicated optical path of the spectrometer is shown in Figure 14.14. Radiation is collimated at point 2 and divided by the beam-splitter at point 3. Points 4–6 describe the "cats-eye" retroreflectors, which are

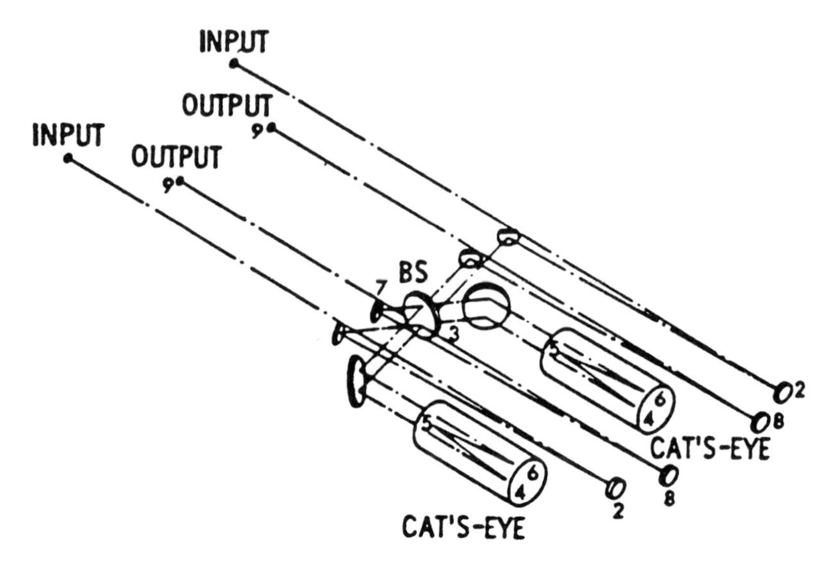

Figure 14.14. Optical path layout of 1-m interferometer at McMath solar telescope of the Kitt Peak National Observatory showing folded configuration required to fit it into vacuum tank and dual input–output capability. [Reprinted with permission from J. B. Breckinridge and R. A. Schindler (1981), "First Order Optical Design for Fourier Spectrometers," in *Spectrometric Techniques*, G. A. Vanasse (Ed.), Vol. 2, Academic, New York, Chapter 2, pp. 63–125.]

used to reduce errors arising from tilt or wobble in the mirror system. The beam is recombined at plane 7 and directed onto an optical element at point 8, which relays the ray onto the detector at point 9.

A cat's-eye retroreflector is simply a paraboloidal primary mirror with a small secondary mirror at the principal focus (Figure 14.15). It has the property of reversing the direction of wavefront propagation by 180°. Therefore, a light beam entering the cat's eye and deviating by angle θ relative to the direction it should have taken had mirror tilt misalignment not occurred will leave the cat's eye tilted back toward the direction of the original tilt by the same angle θ. (See Davis et al., 1980, for a more detailed discussion.)

The analytical applications of the KPNO FT spectrometer in conjunction with an ICP source and a silicon photodiode (blue-enhanced) detector was assessed between 280 and 560 nm. This spectral region contains many strong atomic argon lines that can act as spectral interferences. Over 109 ICP background emission lines with SNRs greater than 4 in this spectral range were identified, and the relative intensities, linewidths, and line positions (± 0.001 cm^{-1}) of the argon lines were also determined (Faires et al., 1985b). Three hundred prominent lines of 22 elements and 500 lines of Fe in this same wavelength range were studied in the same fashion (Faires, 1983; 1985a).

Linearity of the analytical working curves were verified for the 1–100 μg/mL concentration range, and detection limits in the nanogram-per-milliliter range were determined for 32 analytical lines.

Detection limits were found to be highly dependent on total sample composition and the relative intensity of the line selected for analysis. However, some "best case" examples in this spectral region are probably represented by the data given in Table 14.5, which gives *single*-element detection limits for the strongest observed lines of Al, Ni, Fe, and Ca. (Single-element values are for aqueous samples containing only one element.)

While the single-element Ca and Al detection limits are comparable to single-element values reported for commercial ICP instruments, the values for Ni and Fe are based on less analytically sensitive atomic lines and are poorer by about one order of magnitude than those reported for commercial ICP

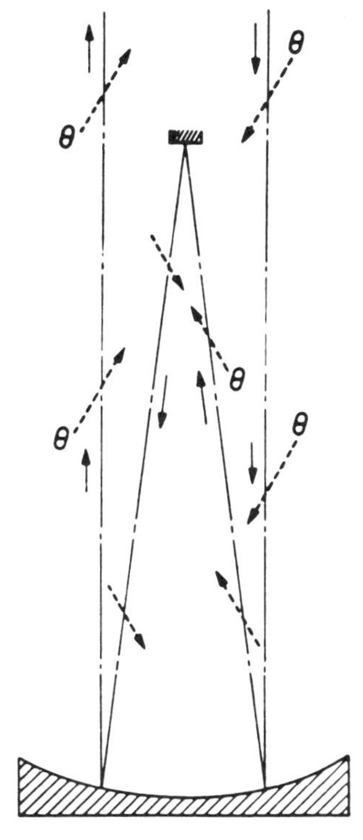

Figure 14.15. Cat's-eye retroreflector. Ray misaligned by θ strikes tilt-compensated cat's-eye. Misaligned ray leaves cat's-eye exactly opposite from its entrance. [Reprinted with permission from J. B. Breckinridge and R. A. Schindler (1981), "First Order Optical Design for Fourier Spectrometers," in *Spectrometric Techniques*, G. A. Vanasse (Ed.), Vol. 2, Academic, New York, Chapter 2, pp. 63–125.]

Table 14.5. Detection Limits for KPNO ICP/FT Spectrometer[a]

Element	Wavelength (nm)	Detection Limit (ppb)
Ca	393.37	0.11
Al	396.15	21
Ni	352.45	37
Fe	371.99	95

[a] From Faires (1985).
[Reprinted with permission from L. M. Faires, (1985), "Effects of Sample Matrix on Detection Limits in Analytical Inductively Coupled Plasma-Fourier Transform Spectrometry," *Spectrochimica Acta*, *40B*. Copyright 1985, Pergamon Press, Elmsford, NY.]

instruments using the more sensitive ion lines at shorter wavelengths. In some "worst case" examples, detection limits of the KPNO ICP/FT instrument were degraded by several orders of magnitude due to the presence of other strongly emitting elements in the sample matrix. These and other examples illustrating the multiplex disadvantage will be examined more closely in Section 14.4.

14.2.4. Other FT Applications in the UV–Visible

A limited number of other interferometric spectrometers have been used in the UV–visible spectral region. Most are hybrid instruments like the WSS–FTS system that combine interferometers with some kind of dispersive or imaging component. While these instruments generally offer some advantages in mechanical and/or computational simplicity, they also tend to suffer from all of the worst disadvantages associated with each of their hybrid parts.

Dohi and Sizuki (1971) and Winefordner and co-workers (Fitzgerald, Chester, and Winefordner, 1975; Chester and Winefordner, 1977) have described a selectively modulated interferometric dispersive spectrometer (SEMIDS) whose primary advantage is simplicity in transforming the interferogram. The instrument consists of a Michelson interferometer in which the stationary mirror has been replaced by a rotating grating (Figure 14.16). [SEMIDS somewhat resembles the SISAM spectrometer described by Connes (1958, 1959, 1961) except that in SISAM both mirrors of the Michelson interferometer are replaced by gratings. See Section 5.2.7.]

In conventional FT spectroscopy, all frequencies are modulated simultaneously, and the interferogram consists of the sum of a very large number of

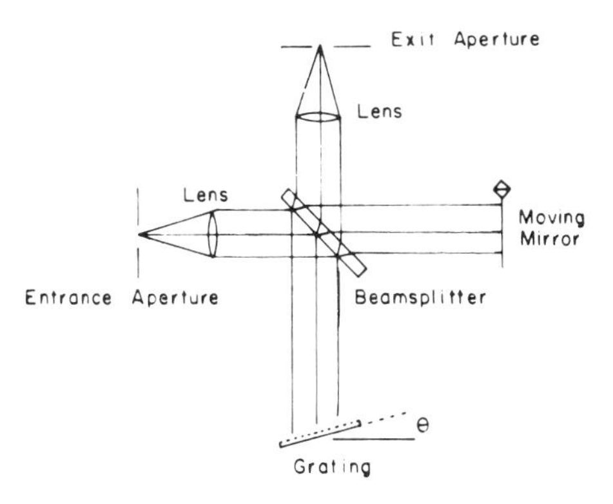

Figure 14.16. Schematic diagram of selectively modulated interferometric dispersive spectrometer (SEMIDS). [Reprinted with permission from T. L. Chester and J. D. Winefordner (1977), *Anal. Chem.*, *49*, 113. Copyright 1977 American Chemical Society.]

cosine waves that must be transformed by a computer using digital techniques. In SEMIDS, an oscillating mirror is used to modulate only those wavelengths that are diffracted by the grating in a direction perpendicular to the mirror–detector optical axis. This optical arrangement limits the number of wavelengths that can interfere with each other at any given time, and as a result the output contains the cosine functions of only a few waves. Simple modulation–demodulation techniques are sufficient to determine the radiation intensity in any spectral region, and the entire spectrum can be obtained by purely analog techniques.

The analytical capabilities of this instrument were investigated (Chester and Winefordner, 1977) using an air–acetylene flame as the source and a PMT as the detector. While linear calibration curves were obtained for a number of elements, the system was limited by fluctuation noise from the flame, and detection limits proved to be 10^2–10^3 times poorer than values obtained on a conventional single-channel atomic emission flame spectrometer.

In 1965, Stroke and Funkhouser proposed a FT spectrometer in which the interferogram was produced as a spatial pattern (equivalent to Young's interference pattern, Section 6.6), not as a signal in the time domain. This spatial pattern was recorded on a photographic plate, and the spectrum was reconstructed from the developed plate through Fourier transformation using coherent optics (Steward, 1987).

Japanese researchers (Yoshihara and Kitada, 1967; Kamiya, Yoshihara, and Okada, 1968; Yoshihara et al., 1976; Okamoto, Kawata, and Minami, 1984) have modernized this approach by using a stationary dual-source interferometer and replacing the photographic plate with a photodiode array. These types of instruments, called holographic spectrometers (Caulfield, 1976), have no moving parts and consequently avoid many of the mechanical

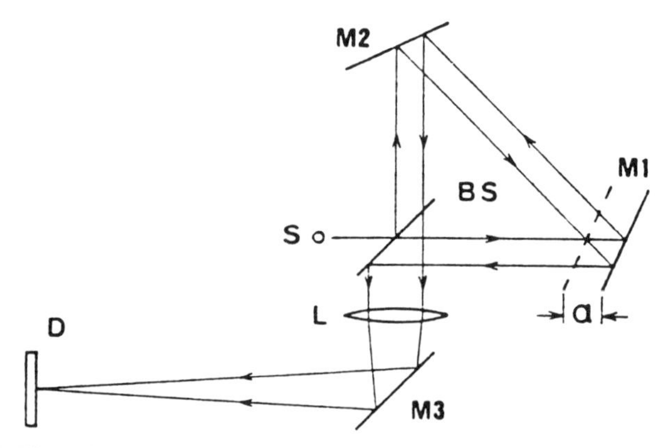

Figure 14.17. Block diagram of optics of Fourier transform spectrometer used by Okamoto et al. (1984). Symbols: S, light source; BS, beamsplitter; M_1, M_2, M_3, plane mirrors; L, lens; D, self-scanning photodiode array. [Reprinted with permission from T. Okamoto, S. Kawata, and S. Minami (1984), *Appl. Opt.*, *23*, 269. Copyright 1984 Optical Society of America.]

problems associated with the more traditional moving-mirror interferometers.

Figure 14.17 shows a block diagram of the optics of the dual source system. The beam from radiation source S is divided by the beamsplitter BS. Each of the resulting two beams travels the same path, but in opposite directions. After reflections at mirrors M_1 and M_2, the two beams are collimated by lens L, reflected by mirror M_3, and form an interference pattern on the photodiode array D. The optical equivalent of this arrangement is the double-source interferometer shown in Figure 14.18, where S_1 and S_2 are two extended virtual sources divided by BS. The distance d between the two virtual sources is $2^{1/2}a$, and all corresponding points in S_1 and S_2 have the same value of d.

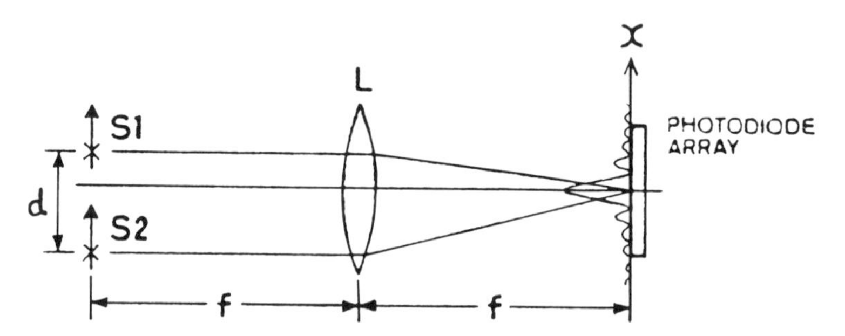

Figure 14.18. Schematic diagram of interferometer optics equivalent to Figure 14.17. Symbols: S_1, S_2, virtual sources; L, lens; d, distance between two corresponding points of virtual sources; f, focal length. (Compare Figure 6.11.) [Reprinted with permission from T. Okamoto, S. Kawata, and S. Minami (1984), *Appl. Opt.*, *23*, 269. Copyright 1984 Optical Society of America.]

Recently, Aryamanya-Mugisha and Williams (1985) have applied this same concept to the study of the UV, visible, and near-infrared regions. The instrument (Figure 14.19) is based on a Michelson interferometer but employs a tilted rather than a moving mirror. The instrument produces a static interference pattern whose bright and dark areas are determined by the optical path difference introduced by the mirror tilt. For example, ray B is divided at the beamsplitter into reflected, B_r, and transmitted, B_t, portions. These travel equal distances and hence produce a bright image of the source at the center of the array. The reflected and transmitted portions of rays A and C, on the other hand, travel different distances and recombine to yield a dimmer image of the source.

The interference pattern produced by this instrument is exactly the same as the interferogram generated by a standard Michelson interferometer, the only difference being that in this case the interferogram must be sampled as intensity versus distance (diode number) rather than intensity versus time. Since the spacing of the photodiodes in the array is fixed, the array always samples the interferogram at the same points. This eliminates the need for a referencing laser. Moreover, the problem of aliasing is completely avoided.

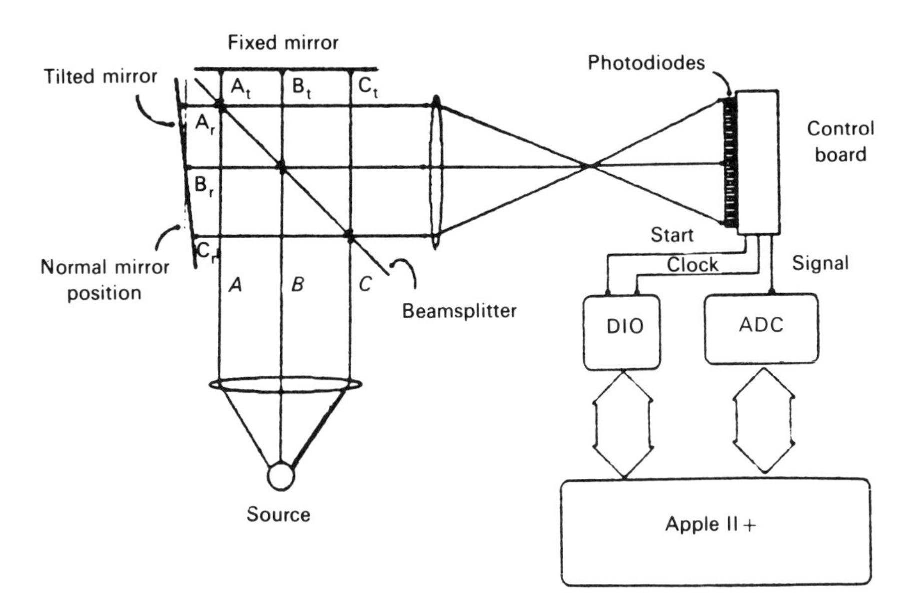

Figure 14.19. Block diagram of interferometric diode array spectrometer used by Aryamanya-Mugisha and Williams (1985). Mirror tilt is greatly exaggerated. [Reprinted from H. Aryamanya-Mugisha and R. R. Williams (1985), *Appl. Spectrosc.*, *39*, 693, with permission from the Society for Applied Spectroscopy.]

Nonuniform diode response, however, is a potentially serious problem, and a diode response function must be determined and used to correct the interferogram prior to transformation.

Molecular and broadband spectra (Aryamanya-Mugisha and Williams, 1985) as well as spectra of line sources (Okamoto et al., 1984) in the UV–visible region were recorded and found to compare favorably with results obtained using conventional scanning monochromators. While these instruments were not evaluated for multielement analysis, the potential for a compact, sturdy instrument capable of performing such a function is certainly evident. Low resolution, however, is a possible problem (on the order of 10 nm for the Aryamanya-Mugisha and Williams system and on the order of 56 cm^{-1} for the Okamoto instrument). While resolution in the tilted-mirror instrument can be improved by increasing the magnitude of the mirror tilt, there is a concomitant loss of SNR due to a decrease of the modulation efficiency of the interferometer. Noise was found to be distributed evenly throughout the transformed spectrum and was not signal dependent (Aryamanya-Mugisha and Williams, 1985). Since these optical systems do not need slits or apertures, the throughput advantage can be even larger than that of a conventional Michelson interferometer (Okamoto et al., 1984).

14.3. HADAMARD TRANSFORM SPECTROMETERS

Hadamard transform spectrometers employ the optical components of dispersive spectrometers. Radiation is dispersed by a stationary grating or prism and passed through a Hadamard encoding mask before being dedispersed and directed onto a single detector. The earliest Hadamard encodement was obtained using movable masks, which were subject to a number of errors that have already been discussed (Section 6.1). Hadamard instruments have had limited application, primarily in the infrared, where the noise is typically detector limited and it is possible to take advantage of the spectrometer's multiplexing ability. Most of these earlier instruments (Decker and Harwit, 1969; DeGraauw and Veltman, 1970; Decker, 1971; Phillips and Harwit, 1971; Hansen and Strong, 1972; Decker, 1972; Decker, 1973; Harwit, 1973; Swift et al., 1976), designed for applications in the infrared, have been reviewed elsewhere (Harwit and Decker, 1974; Harwit and Sloane, 1979) and will not be discussed here. More recent applications include photoacoustic imaging (Coufal, Moller, and Schneider, 1982; Hammaker et al., 1986), photothermal deflection imaging (Fotiou and Morris, 1986), FT–IR imaging (Kraenz and Kunath, 1982), and others (Sugimoto, 1986; Treado and Morris, 1989).

14.3.1. Early Hadamard Spectrometers for UV–Visible Spectroscopy

Following the successful application of Hadamard spectrometers in the infrared spectral region, several groups (Horlick and Codding, 1973; Santini, Milano, and Pardue, 1973) suggested that HT techniques might also be useful for multielement analysis by atomic spectroscopy. However, studies in the UV–visible region using photon shot-noise-limited detectors did not prove promising (Larson et al., 1974; Plankey et al., 1974; Keir, Dawson, and Ellis, 1977).

Plankey et al. (1974) evaluated an HT spectrometer using a multielement (Fe, Co, Ni) electrodeless discharge tube as well as a continuum source (Xe lamp). The atomic fluorescence measurement of cadmium and zinc in an air–acetylene flame was also investigated. The spectrometer was constructed from a conventional Czerny–Turner scanning monochromator modified only by removal of the folding mirror at the exit slit and installation of the Hadamard mask at the exit focal plane. The mask consisted of a bimetallic strip encoded with a 255 cyclic *S*-matrix (509 transparent and opaque slots). Radiation was detected by a photomultiplier, and the output was processed by a laboratory computer using the fast Hadamard transform.

This study indicated that HT spectroscopy was useful for spectral averaging, that the speed of the analysis (1 nm/sec) was relatively fast, and that useful spectra could still be obtained even when the mask was misaligned up to four or five slot widths. However, because of the multiplex nature of the measurement, the quality of the spectrum was reduced. Noise components from strong signals were distributed throughout the spectrum, causing signals with intensities $\sim 3\%$ of the most intense signal to be lost in the noise. Similar difficulties were observed in the analysis of lead and arsenic by atomic absorption using the HT technique (Keir et al., 1977). Compared to conventional scanning methods, the HT measurements had similar sensitivities, but detection limits were worse.

14.3.2. Stationary Hadamard Encoding Masks

Recently, a new generation of HT spectrometers has emerged that avoids the problems associated with the construction and operation of movable masks through the use of an array of stationary electro-optic switches (Tilotta and Fateley, 1988). The feasibility of using electrodiachromatic or thermodiachromatic masks for HT spectroscopy was first proposed for photoacoustic imaging in the infrared (Hammaker et al., 1986). The mask would be formed by depositing suitable material as a thin film onto a transparent substrate. This film would be divided into slots by a permanently opaque material that was insulated to prevent cross-talk between the slots (Figure

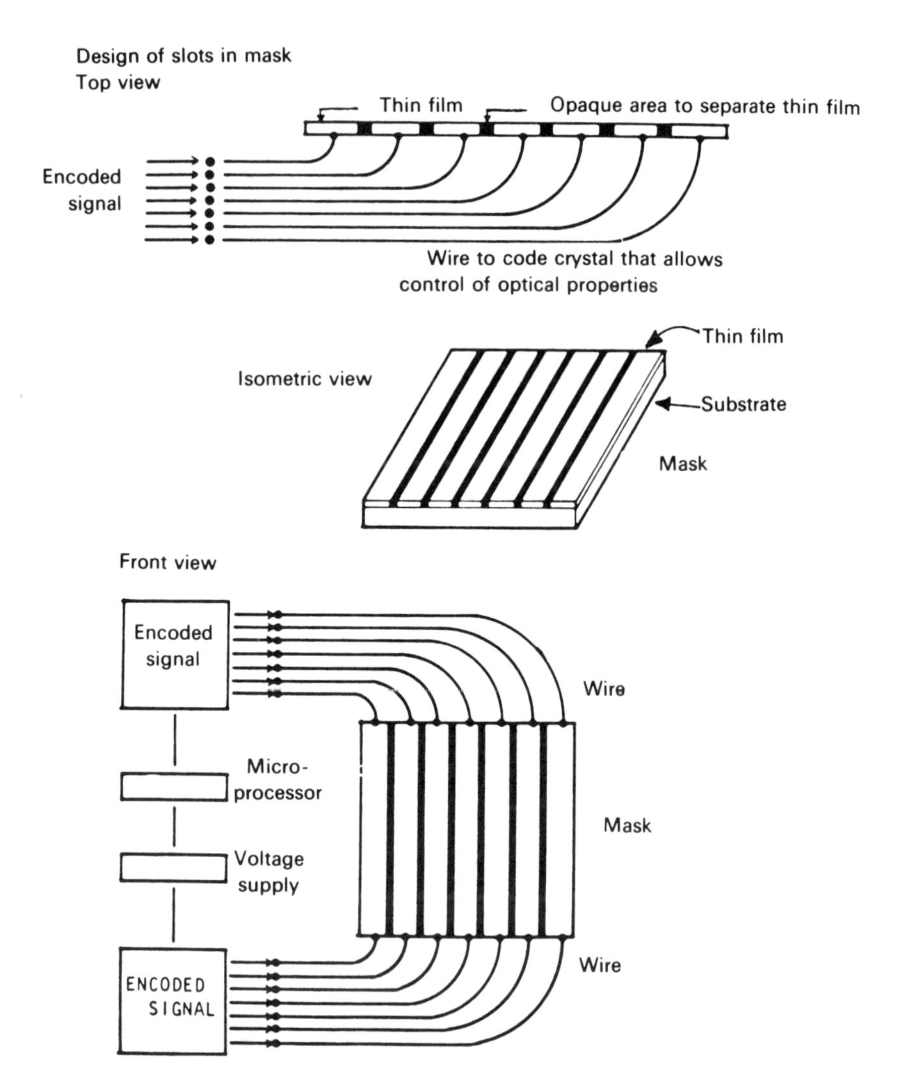

Figure 14.20. Stationary Hadamard encoding mask proposed by Hammaker et al. (1986). [Reprinted with permission from R. M. Hammaker, J. A. Graham, D. C. Tilotta, and W. G. Fateley (1986), "What is Hadamard Transform Spectroscopy?" in *Vibrational Spectra and Structure*, J. R. Durig (Ed.), Vol. 15, Elsevier Science, Amsterdam, Chapter 7, pp. 401–485.]

14.20). The Hadamard encodement would be controlled by a microprocessor, which directed a voltage supply to switch the individual slots on and off as desired.

The first practical application of this technique (Sugimoto, 1986) involved the measurement of atmospheric trace species in the 1.0–1.8-μm region by long-path differential optical absorption (Platt, Perner, and Patz, 1979). The spectrometer employed an array of 16 light-emitting diodes (LEDs) as the light source. The LEDs were modulated directly by a Hadamard matrix code, and the signal was detected by a 32-element Ge diode array. This instrumental configuration is equivalent to that of an HT spectrometer in which the encoding appears at the entrance focal plane and the resulting image of the mask is viewed by multiple detectors. (See, also, the discussion on multiple-entrance slit imaging spectrometers, Section 15.10.2). Because the mask was stationary, the system had no moving parts and was free from mask defects.

Extension of the concept to visible–near IR Raman spectroscopy was soon proposed (Tilotta, Hammaker, and Fateley 1987a) and demonstrated shortly thereafter (Tilotta et al., 1987). In this application, the instrument had a more traditional HT design with a single entrance slit and an encoding mask at the exit focal plane of the spectrometer. The mask consisted of an array of liquid crystals that were turned on and off by a microcomputer to generate a code that followed a left-circulant S-matrix.

Liquid crystal arrays had already been used in telecommunications, and their optical switching properties were being investigated for use in optical computers (Khan and Nejib, 1987). Figure 14.21 demonstrates the principle behind their method of operation.

Randomly polarized radiation E passing through a polarizer P_1 becomes linearly polarized with its light vector parallel to the polarizer transmission

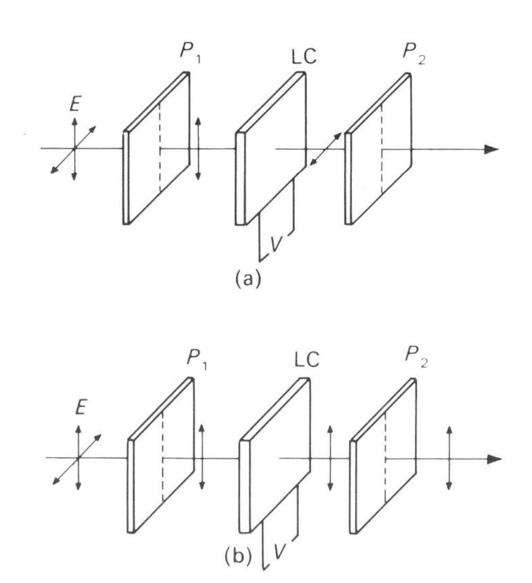

Figure 14.21. Liquid crystal electro-optic switch in (*a*) off and (*b*) on modes. [Reprinted from *Spectroscopy*, *3* (1), 20 (1988), with permission, Aster Publishing, Eugene, OR.]

axis. If this linearly polarized light enters a nonenergized ($V=0$) liquid crystal cell, it undergoes a 90° rotation upon exit (Figure 14.21*a*). Since the polarized component is now perpendicular to the transmission axis of polarizer P_2, the radiation will be absorbed by the second polarizer, causing the liquid crystal to function in the "off" mode. However, if the linearly polarized light enters an energized ($V \neq 0$) liquid crystal cell, it will not be rotated upon exit (Figure 14.21*b*), and the cell functions in the "on" mode.

While free from many of the errors associated with moving multislit masks, the solid-state encoding masks are subject to a number of other problems. For example, the *S*-matrices used in Hadamard encodement consist of an array of 0's and 1's. However, most materials that can be used as stationary encoding masks deviate from the ideal condition of providing 100% transmission in the on state and 0% transmission in the off state. This situation is analogous to a moving encoding mask with partially open and partially closed slits, or, mathematically, to an *S*-matrix in which the 1's have been replaced by a fraction less than 1 and the 0's have been replaced by a fraction greater than 0.

The effect of a uniform defect of this type has been treated mathematically (Tilotta, Hammaker, and Fateley, 1987b) by replacing the 1's and 0's in the *S*-matrix with T_t and T_0, the actual transmittance of the transparent (on) and opaque (off) states, respectively. The conclusions reached from this treatment are that such defects do not impair the multiplex capability of the instrument provided that M, the number of resolution elements being multiplexed, is large and the difference ΔT between the transmittance of the on and off states is not too small (Table 14.6). For large M, the improvement in SNR is approximately given by

$$\mathrm{SNR} = \Delta T(M+1)/(2M^{1/2}) \tag{14.5}$$

from which it can be seen that the improvement in SNR as a result of multiplexing is proportional to ΔT as well as to $\frac{1}{2}M^{1/2}$ for the detector-noise-limited case. (Compare to the ideal case where $T_t = 1$ and $T_0 = 0$ as given in Table 5.5.)

A further problem of the "leaky" optical state involves a baseline offset that varies with the total incident light level, that is, with changes in the concentration of emitting species in the sample. This baseline offset appears in the final encodegram and is carried through the Hadamard transformation into the spectrum.

One method of treating this baseline offset is to correct the encodegram before transformation (Tilotta, Fry, and Fateley, in press). The offset of an encodegram is determined by electronically disconnecting the modulator circuit and acquiring one additional encodegram under the same conditions as the actual spectrum is acquired (i.e., with the sample present). This offset is

Table 14.6. SNR Improvement[a] for Selected Values of N, T_t, and T_0 [b]

T_t	T_0	$N=15$	$N=255$	$N=4095$
1.000	1.000	0.000	0.000	0.000
1.000	0.9000	0.2070	0.8016	3.200
1.000	0.8000	0.4140	1.603	6.401
1.000	0.7000	0.6210	2.405	9.601
1.000	0.6000	0.8280	3.206	12.80
1.000	0.5000	1.035	4.008	16.00
1.000	0.4000	1.242	4.809	19.20
1.000	0.3000	1.448	5.611	22.40
1.000	0.2000	1.654	6.413	25.60
1.000	0.1000	1.860	7.214	28.80
1.000[c]	0.000	2.066	8.016	32.00

[a] Compared to a scanning, dispersive spectrometer.
[b] From Tilotta, Hammaker, and Fateley (1987b).
[c] Ideal case.

then subtracted directly from the encodegram of the spectrum, and the corrected encodegram can then be transformed to yield the corrected spectrum with no baseline offset. The subtraction procedure also has the effect of producing a constant mathematical attenuation of the overall system gain—a situation similar to placing a neutral density filter in the optical path of the radiation. Since both samples and calibration standards are treated in the same manner, spectral comparisons can still be made on the basis of relative intensity. Nearly linear calibration curves for K and Rb have been obtained up to 10.00 ppm using this correction procedure, with detection limits of 10 and 20 ppb for potassium and rubidium, respectively (Tilotta, Fry, and Fateley, in press).

Other problems involving liquid crystal arrays include hysteresis and time-delayed stabilization effects in the liquid crystal cells. This can be minimized by adopting an exposure period consisting of a "wait" delay followed immediately by a data accumulation sequence. The delay allows the mask pattern to stabilize after each liquid crystal energization (Tilotta, Fry, and Fateley, in press).

14.3.3. A Visible–Near Infrared HT Spectrometer With a Stationary Liquid Crystal Mask

Tilotta, Hammaker, and Fateley (1987c) have installed a liquid crystal HT encodement mask in a conventional dispersive instrument and obtained several emission spectra in the visible and near-infrared region to demonstrate the usefulness of the HT technique. The optical design of the instrument (Figure 14.22) is based on a double Czerny–Turner scanning spectrometer modified for reception of the encoding mask, M_a, by removal of the intermediate slit and replacement of two 1.5 × 4-cm 90° reflecting mirrors, M_3 and M_4, with 5 × 10-cm mirrors to minimize vignetting. As shown in Figure 14.22, radiation enters the single entrance slit, S_1, is dispersed in the upper monochromator, passes through the encoding mask, is dedispersed in the lower monochromator, and exits at the single slit S_2. The detector used for this study was an extended range (190–1150-nm) silicon photodiode.

The encoding mask was constructed from a modified two-dimensional, twisted nematic liquid crystal display module consisting of 320 × 64 cells, each

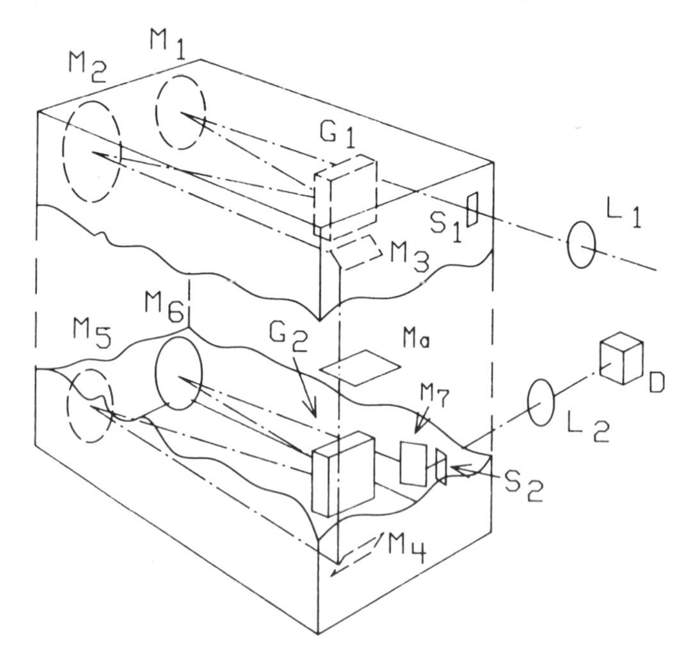

Figure 14.22. Optical design of Hadamard transform spectrometer employing liquid crystal spatial light modulator array. Radiation follows dot–dash line beginning with lens L_1, which focuses radiation onto entrance slit S_1. Concave collimating mirror M_1 collimates radiation onto plane grating G_1. Dispersed radiation is collected and focused onto encoding mask Ma by concave camera mirror M_2. Planar mirror M_3 deflects radiation onto mask. Radiation transmitted through encoding mask is deflected by planar mirror M_4 onto concave collimating mirror M_5. Mirror M_5 directs dispersed radiation onto plane grating G_2 for dedispersion. After dedispersion, pseudo–white light is collected by mirror M_6, which directs radiation towards mirror M_7. After exiting spectrometer at slit S_2, signal is focused onto detector D by lens L_2 and detected. [Reprinted from D. Tilotta, R. M. Hammaker, and W. G. Fateley (1987), *Appl. Spectrosc.*, *41*, 727, with permission from the Society for Applied Spectroscopy.]

of dimension $600 \times 720\ \mu m$ with $100\ \mu m$ spacing between the cells. The spectrometer operates in the spectral region of 300–2500 nm, but the liquid crystal efficiency is not constant over this range. For the wavelength region from 350 to 800 nm, the transmittance of each encoding element in the on state averaged 23%, while that of the off state averaged 9%. These values are increased to 58% and 49%, respectively, in the 800–928-nm spectral range.

The Hadamard encodement for the encoding mask was derived from a 127×127 two-dimensional, left circulant S-matrix. Mask control as well as data acquisition and transformation by the fast Hadamard transform algorithm was controlled by a laboratory microcomputer.

The spectra in Figure 14.23 illustrate the effect of operating the spectrometer in the HT mode. The spectrum 14.23*a* results when the instrument is operated like a scanning monochromator, that is, each of the 127 liquid crystal cells is opened sequentially. Spectrum 14.23*b* is produced when the encoding mask is used to multiplex the spectral information. Data acquisition time and total number of measurements are the same for both spectra, but Hadamard transformation has improved the SNR of the measurement by a factor of 4.

Another advantage of the HT spectrometer is its ability to selectively remove certain portions of the spectrum from the encodegram without the use of optical filters. This can be accomplished in two ways: positioning the dispersed radiation so that the undesirable radiation lies outside the "window" of the encoding mask (Tilotta et al., 1987) and turning off a preselected number of the encoding slots (liquid crystals) so that undesirable radiation striking these positions will not be included in the final encodegram (Tilotta, Fry, and Fateley, in press). Figure 14.24*a* illustrates the HT spectrum of a sample consisting of 1 ppm K, 0.1 ppm Rb, 500 ppm Li, and 3000 ppm Cs. Figure 14.24*b* shows the same spectrum in which appropriate slots in the liquid crystal array have been closed to eliminate the intense potassium signal. These two spectra illustrate the distributive nature of noise in a multiplexed spectrum and demonstrate that an SNR advantage (in this case a factor of 3) may be obtained by eliminating strong lines not needed in the analysis. This procedure has been of particular benefit in obtaining improved Raman spectra by elimination of the intense Rayleigh line (Tilotta et al., 1987).

14.4. EFFECT OF MULTIPLEXING ON NOISE: SOME EXPERIMENTAL RESULTS

Noise ultimately limits the detection of any element in atomic spectroscopy. Detection limits are, of course, not the only figure of merit to be considered when choosing a spectrometric system. However, it is probably safe to say

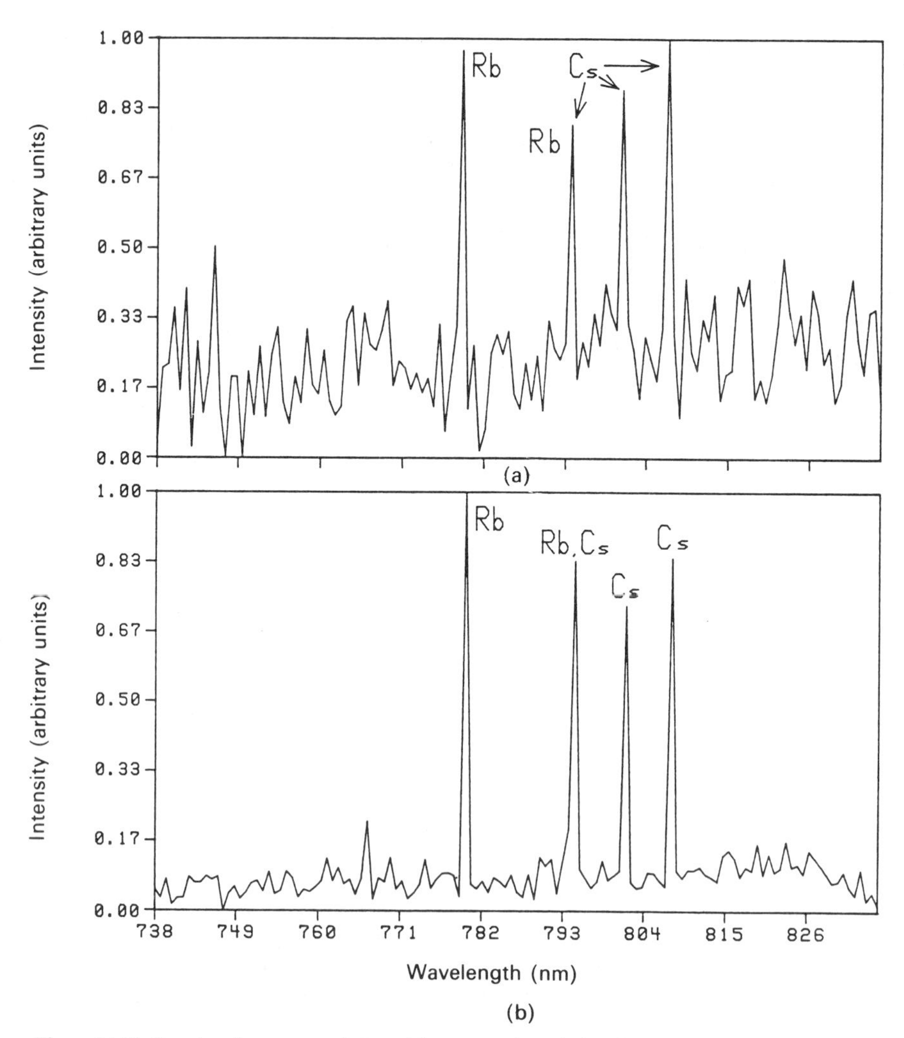

Figure 14.23. Spectra of some near-infrared flame atomic emission lines of rubidium and cesium. (*a*) Single-slit mask scan, monochromator simulation mode. (*b*) Hadamard multiplex scan taken in same time period. [Reprinted with permission from D. Tilotta, R. C. Fry, and W. G. Fateley (1990), "Selective Multiplex Advantage with an Electro-optic Hadamard Transform Spectrometer for Multielement Atomic Emission," *Talanta*, in press. Copyright 1990 Pergamon Press, Elmsford, NY.]

that if multiplexing spectrometers could provide the same SNR enhancement in the UV–visible as they have in the infrared, acceptance of FT and Hadamard spectrometers would be virtually assured in atomic spectroscopy.

The mathematical treatments of Hirschfeld (1976a,b), discussed in Section 14.1.1, have predicted that in atomic spectroscopy, multiplexing can result in an SNR enhancement only in photon shot-noise-limited cases and then only for strong spectral features in optically dilute spectra. But are these predictions confirmed experimentally? This section will examine the reported effects of multiplexing on noise in atomic spectroscopy and attempt to determine the validity of Eq. 14.3.

In one of the very earliest studies of this type, Horlick et al. (1982) examined some of the noise properties of his ICP/FT spectrometer (Section 14.2.2) using a solution containing 100 ppm of magnesium and 0.1 ppm of calcium. Under these conditions, the magnesium lines are about 10 times more intense than the calcium lines.

Spectra of the calcium ion emission lines at 393.4 and 396.8 nm with and without the magnesium emission detected are shown in Figures 14.25 and 14.26, respectively. (To obtain Figure 14.26, the magnesium emission was filtered out optically.) The variances of the intensity of the calcium 393.4 line were measured for both spectra (Table 14.7) and under a variance ratio F-test at $\alpha = 0.05$ were found to be statistically equivalent. This result was taken as direct evidence that in contradistinction to Eq. 14.3, noise is concentrated on the emission lines rather than uniformly distributed across the spectrum.

Although no information on baseline noise or SNRs is provided, the SNRs for the two calcium ion lines and the unidentified line ($\star$) at about 396 nm can easily be estimated from the figures. (Baseline noise is taken as approximately one-fifth the peak-to-peak noise envelope.) These estimates are given in Table 14.8 and clearly show that there is an SNR improvement for all emission lines when the Mg emission is not detected. The reduction in rms baseline noise may also be sufficient to reveal the presence of yet a fourth emission line (#) at about 394 nm (SNR = 5–6)—at the correct wavelength and of the right intensity relative to the unidentified line at 396 nm to suggest the presence of trace amounts of aluminum in the sample. Therefore, according to Table 14.8, there does appear to be a distributive multiplex effect since the quality (SNR) of the calcium spectrum has been improved by filtering out the more intense magnesium emission.

In a similar study (Stubley and Horlick, 1985c), the effect of the intense magnesium emission on the detection of the weaker 213.8-nm emission line of zinc was determined. (The entire emission spectrum is shown in Figure 14.27.) Calibration curves for zinc in the presence of 0, 100, and 500 ppm of magnesium were linear, but the detection limits for zinc (Table 14.9) decreased as the concentration of the magnesium increased. Background noise (Table

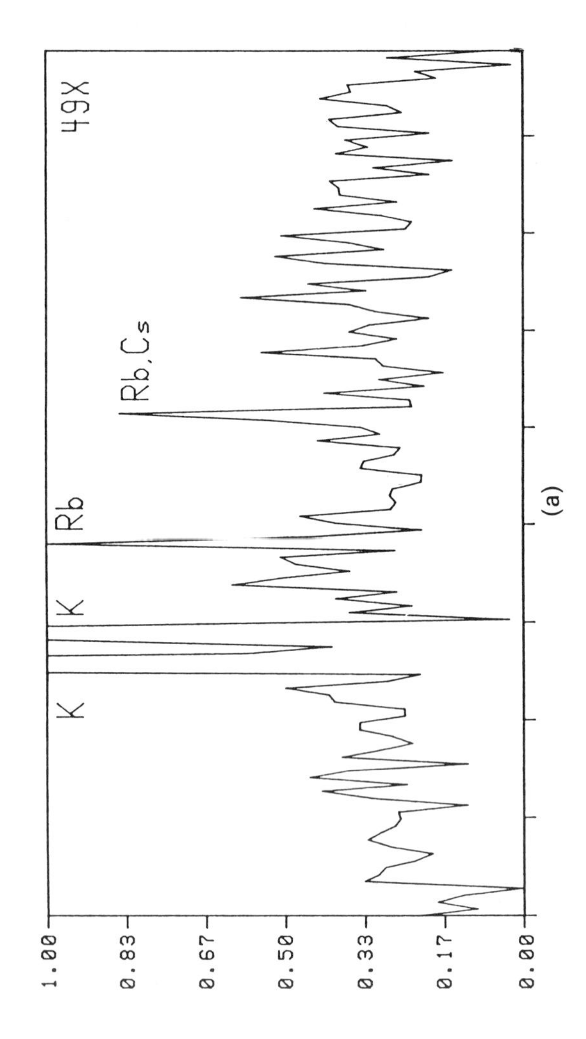
K
K
Rb
Rb,Cs
49X
1.00
0.83
0.67
0.50
0.33
0.17
0.00
Intensity (arbitrary units)

(a)

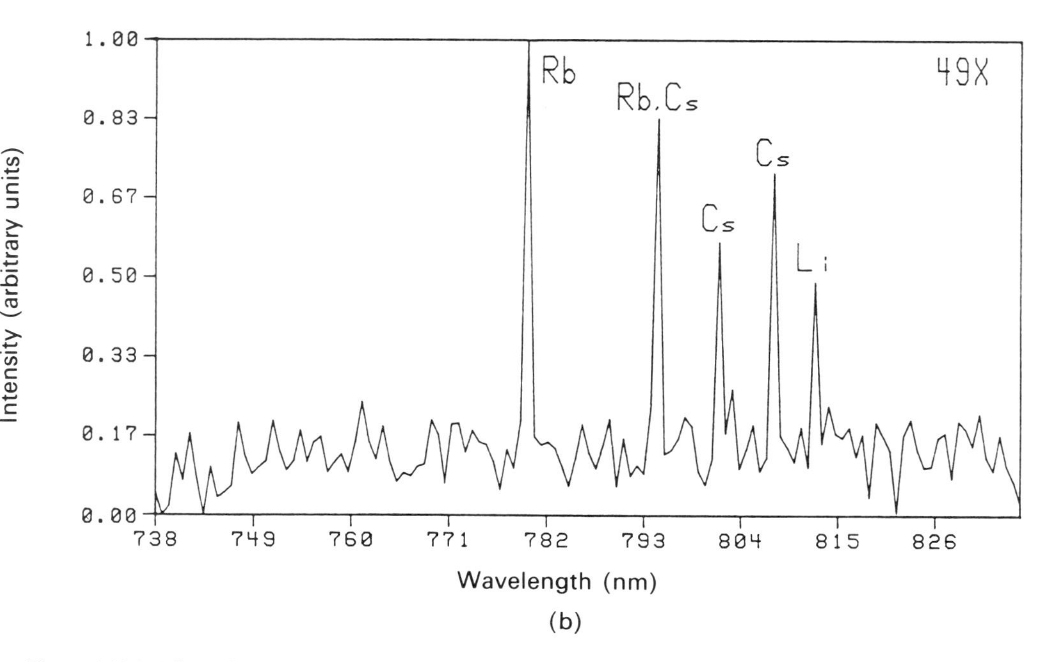

Figure 14.24. Effect of removing intense lines from spectrum. (*a*) Hadamard multiplex can of solution: 0.1 ppm K, 0.1 ppm Rb, 500 ppm Li, and 3000 ppm Cs. (*b*) Hadamard multiplex scan of same solution with K lines removed. [Reprinted with permission from D. Tilotta, R. C. Fry, and W. G. Fateley, "Selective Multiplex Advantage with an Electro-optic Hadamard Transform Spectrometer for Multielement Atomic Emission," *Talanta*, in press. Copyright 1990 Pergamon Press, Elmsford, NY.]

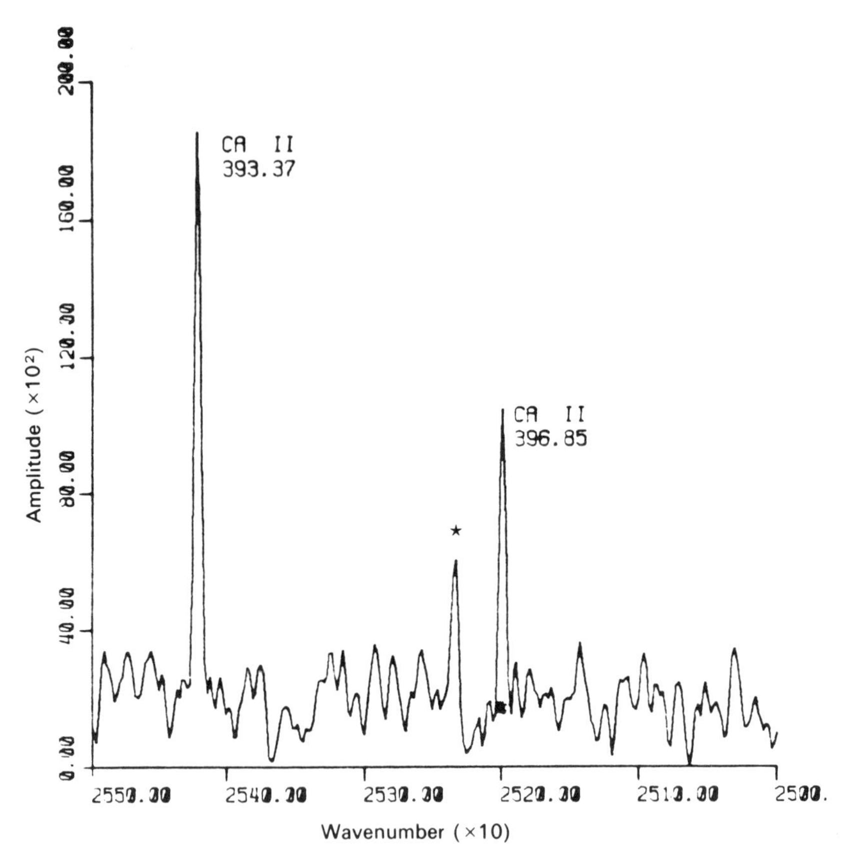

Figure 14.25. Fourier transform emission spectrum of Ca (0.1 ppm) ion lines in presence of unfiltered 100-ppm Mg. [Reprinted with permission from G. Horlick, R. H. Hall, and W. K. Yuen (1982), "Atomic Emission Spectrochemical Measurements with a Fourier Transform Spectrometer," in *Fourier Transform Infrared Spectroscopy*, J. R. Ferraro and L. J. Basile (Eds.), Vol. 3, Academic, New York, Chapter 2, pp. 37–81.]

14.9) also increased as the magnesium signal became more intense, another symptom of the distributive multiplex disadvantage. When the spectral window was reduced to only a few nanometers in the vicinity of the zinc emission line (Stubley and Horlick, 1985b) and magnesium could no longer be detected, background noise became virtually independent of magnesium concentration and similar in magnitude to that determined for the 100 ppm zinc–0 ppm magnesium solution (Figure 14.28).

The data in Table 14.9 represent 10 replicates of 32 signal-averaged interferograms and clearly show that noise in the ICP/FT spectrum is

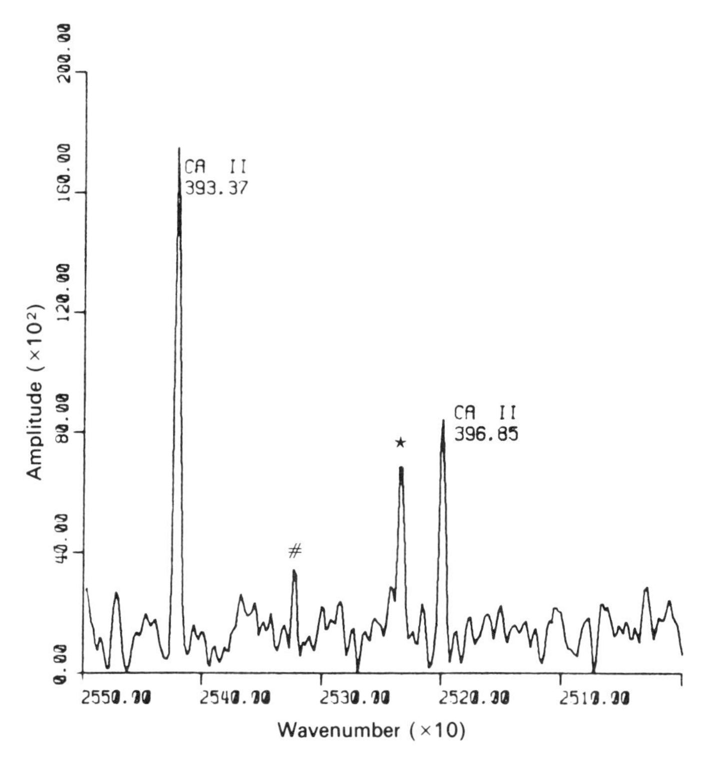

Figure 14.26. Fourier transform emission spectrum of Ca (0.1 ppm) ion lines in presence of filtered 100-ppm Mg. [Reprinted with permission from G. Horlick, R. H. Hall, and W. K. Yuen (1982), "Atomic Emission Spectrochemical Measurements with a Fourier Transform Spectrometer," in *Fourier Transform Infrared Spectroscopy*, J. R. Ferraro and L. J. Basile (Eds.), Vol. 3, Academic, New York, Chapter 2, pp. 37–81.]

Table 14.7. Variance Data for Ca–Mg Study[a]

Ca 393.3-nm Emission Peak	Ca + Mg (No Filter)	Ca + Mg (With Filter)
Mean	1.97765×10^4	1.86731×10^4
Variance	1.0240×10^7	4.3482×10^6
Relative standard deviation (%)	16.19	11.16

[a] Calcium = 0.1 ppm, Mg = 100 ppm. From Horlick et al. (1982).
[Reprinted with permission from G. Horlick, R. H. Hall, and W. K. Yuen (1982), "Atomic Emission Spectrochemical Measurements with a Fourier Transform Spectrometer," in *Fourier Transform Infrared Spectroscopy*, J. R. Ferraro and L. J. Basile (Eds.), Vol. 3, Academic, New York, Chapter 2, pp. 37–81.]

Table 14.8. SNR Estimates for Emission Lines in Figures 14.25 and 14.26

	Ca 393.37-nm Emission Line	Unknown 396-nm Emission Line	Ca 393.85-nm Emission Line
Unfiltered Mg[a]	32	8	17
Filtered Mg[b]	47	16	20

[a] Figure 14.25.
[b] Figure 14.26.

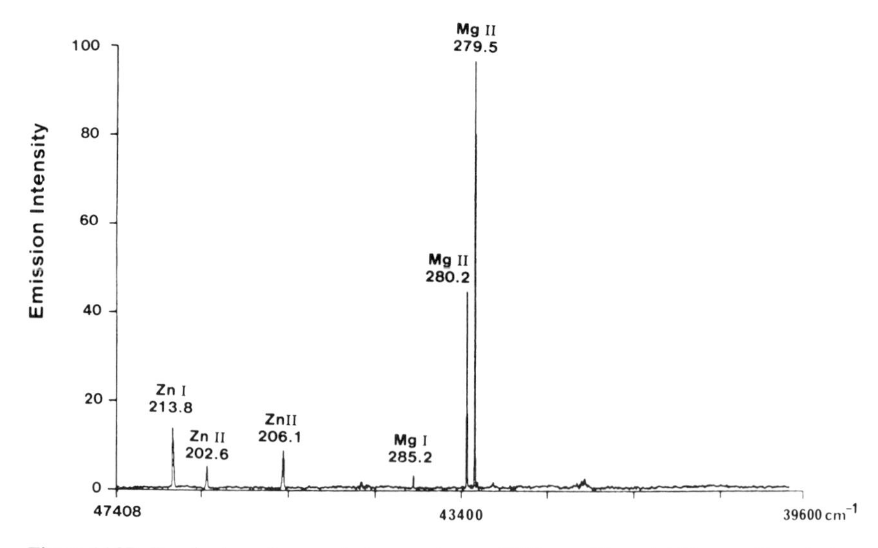

Figure 14.27. Fourier transform emission spectrum of 100 ppm Zn and 100 ppm Mg solution. [Reprinted from E. A. Stubley and G. Horlick (1985), *Appl. Spectrosc.*, *39*, 805, with permission from the Society for Applied Spectroscopy.]

distributed more or less according to Eq. 14.3. There does appear to be some frequency-dependent noise structure in the spectrum, with background noise somewhat higher in the vicinity of the magnesium and zinc emission lines (Table 14.9). However, variations in background noise must be interpreted with care if the spectra are aliased. Aliasing can produce ringing patterns in the vicinity of the emission peaks as well as other aberrations in the spectrum due to foldover (Faires, 1983; Griffiths and deHaseth, 1986b). Noise can also be affected by a great many other instrumental and/or computational factors including choice of apodization function, changes in mirror velocity, dynamic

Table 14.9. Detection Limits for Zinc at 213.8 nm and Standard Deviation of Background as a Function of Magnesium Concentration[a]

Mg Concentration (ppm)	Detection Limit Zn (ppm)	Standard Deviation[b]		
		Background	Near Zn	Near Mg
0	0.57	452	446	392
100	2.3	701	585	970
500	25.3	1042	1142	2364

[a] From Stubley and Horlick (1985c).
[b] Zinc concentration constant, 100 ppm.

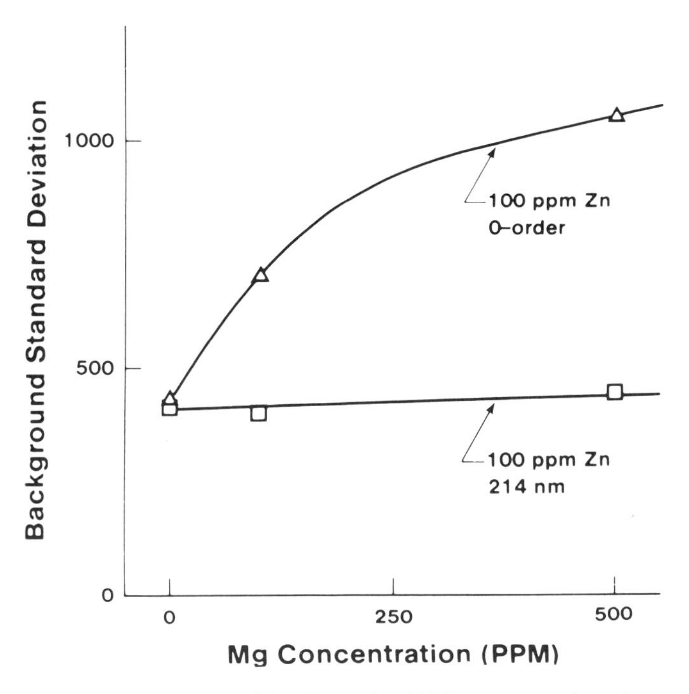

Figure 14.28. Standard deviations of baseline point 116 in presence of varying magnesium concentrations with normal FTS (zero-order) and with ICP–WSS–FTS (214-nm window). [Reprinted from E. A. Stubley and G. Horlick (1985), *Appl. Spectrosc.*, *39*, 811, with permission from the Society for Applied Spectroscopy.]

range of the A/D converter, irregularities in sampling the interferogram, and vibration present during a scan (Hirschfeld, 1979; Griffiths and deHaseth, 1986a).

Faires (1983, 1985) has also confirmed the presence of a distributive multiplex disadvantage in FT spectra that are photon shot-noise-limited. (The spectrometer used for these studies has been described in Section 14.2.3.) Table 14.10 shows the detection limits obtained for nickel (352.45-nm emission line) in the presence of aluminum, in the presence of calcium, and in the presence of both aluminum and calcium. (Analogous results are observed for the detection limit of the 396.15-nm aluminum emission line in the presence of iron, nickel, and calcium. See Faires, 1985.)

The detection limit for nickel in Table 14.10 does not appear to be affected by the presence of aluminum until a concentration of about 1000-ppm Al has been reached. This is readily explained by noting that there are a number of argon background emission lines in this region of the spectrum, and the aluminum emission intensity does not contribute significantly to this background until the Al concentration becomes quite high. Calcium, on the other hand, is almost totally ionized in the plasma, and its emission lines are quite intense. As shown in Table 14.10, the Ca emission begins to dominate over the argon background emission almost immediately, and the nickel detection limits are affected even at very low calcium concentrations. The effect of calcium is further illustrated by Figures 14.29–14.31, which show the same 10-wavenumber segment of the spectrum for three different experimental conditions. (The nickel 352.45-nm line lies near the center of the segment, and the intensity scale is identical for each plot.)

Table 14.10. Detection Limits for Nickel (352.45 nm) in Different Matrices[a]

Al (ppm)	Ni Detection Limit (ppb)	Ca (ppm)	Ni Detection Limit (ppb)
0	37	0	37
1	44	2.5	122
10	44	10	407
100	40	100	2667[b]
1000	187	1000	4506

[a] Reprinted with permission from L. M. Faires, (1985), "Effects of Sample Matrix on Detection Limits in Analytical Inductively Coupled Plasma-Fourier Transform Spectrometry," *Spectrochimica Acta, 40B.* Copyright 1985, Pergamon Press, Elmsford, NY.

[b] 100 ppm Ca plus 100 ppm Al: 2874 ppb nickel detection limit.

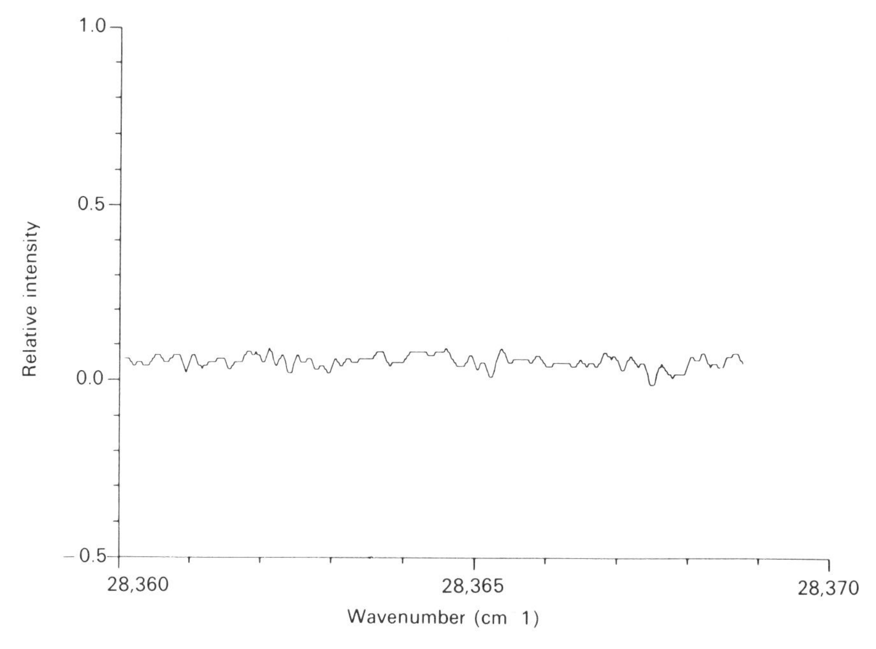

Figure 14.29. A 10-wavenumber segment of FTS spectrum of blank solution in vicinity of Ni I 352.45-nm (28364.39-cm^{-1}) line. Single-element detection limit for Ni line is calculated from noise level in this spectrum. [Reprinted with permission from L. M. Faires (1985), "Effects of Sample Matrix on Detection Limits in Analytical Inductively Coupled Plasma–Fourier Transform Spectrometry," *Spectrochimica Acta, 40B*. Copyright 1985 Pergamon Press, Elmsford, NY.]

Figure 14.30 shows the single-element spectrum of a 1-ppm solution of nickel in which the baseline noise is not significantly different from that of the blank solution (Figure 14.29). Figure 14.31 shows the spectrum of a 1-ppm nickel solution in the presence of 10-ppm calcium. Nickel emission is still present at the same intensity as in the single-element spectrum, but the presence of calcium has increased the baseline noise by a factor greater than 10.

Some frequency-dependent baseline noise structure was also observed in the KPNO ICP/FT spectra, which are unaliased (Faires, 1985). This is illustrated in Figure 14.32 for the aluminum 396.15-nm emission line. At concentrations above 10 ppm Al, baseline noise adjacent to the aluminum line was higher than the baseline noise measured far from the aluminum line. Whatever the cause of this nonuniformity of noise distribution, the results imply that detection limits may be further degraded when the analytical line is in close proximity to an intense interferent line.

Marra and Horlick (1986) have investigated frequency-dependent noise by calculating standard deviations and SNRs across the entire FT spectrum.

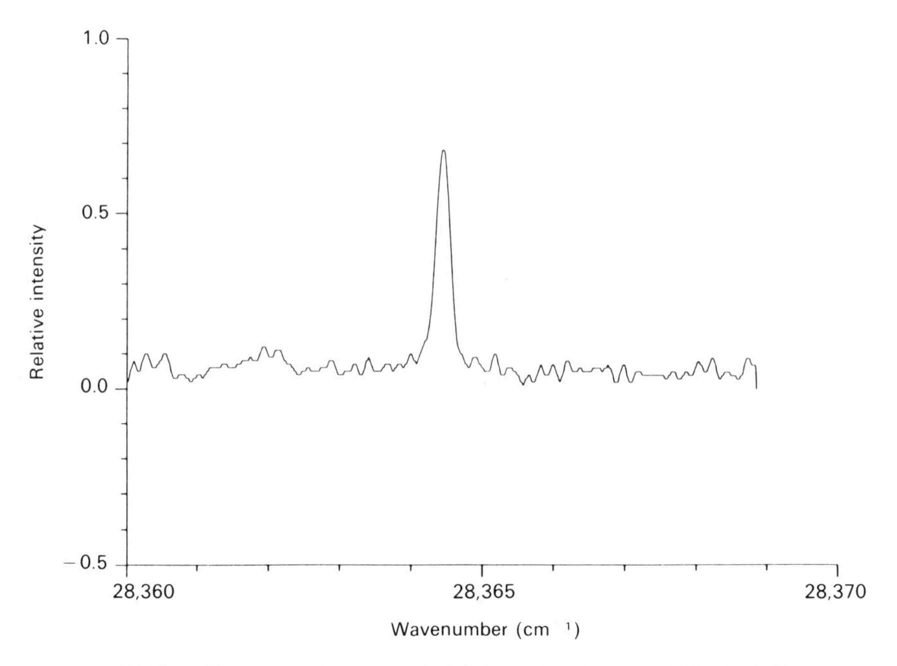

Figure 14.30. Same 10-wavenumber segment of FTS spectrum (Figure 14.29) of solution 1 ppm Ni. The Ni I 352.45-nm line is seen near center of segment. Noise level is not significantly different than that of blank. [Reprinted with permission from L. M. Faires (1985), "Effects of Sample Matrix on Detection Limits in Analytical Inductively Coupled Plasma–Fourier Transform Spectrometry," *Spectrochimica Acta, 40B*. Copyright 1985 Pergamon Press, Elmsford, NY.]

Data was obtained using the FT spectrometer discussed in Section 14.2.2. Sixteen repetitively scanned 4096-point interferograms were averaged to obtain one individual measurement, and this measurement was repeated eight times to obtain the standard deviation of each data point. The standard deviation of each data point was then plotted as a function of wavenumber to give a standard deviation spectrum. An SNR spectrum was obtained by dividing the standard deviation spectrum into the average signal spectrum. This method of presenting spectral data makes the noise distribution pattern in the spectrum readily apparent.

The emission spectrum, standard deviation spectrum, and SNR spectrum of a magnesium HCL are shown in Figures 14.33*a*, 14.33*b*, and 14.33*c*, respectively. Scale expansion of Figures 14.33*b*, *c* reveals a $1/f$ spectral noise distribution in the form of two satellite–like peaks on either side of the emission peaks. Since HCLs are quite stable, the nonuniform distribution of noise was attributed to a periodic error in sampling interval resulting from a 120-Hz pick-up in the mirror drive system. Larger satellite peaks occurred for emission lines at shorter wavelengths since these are more susceptible to

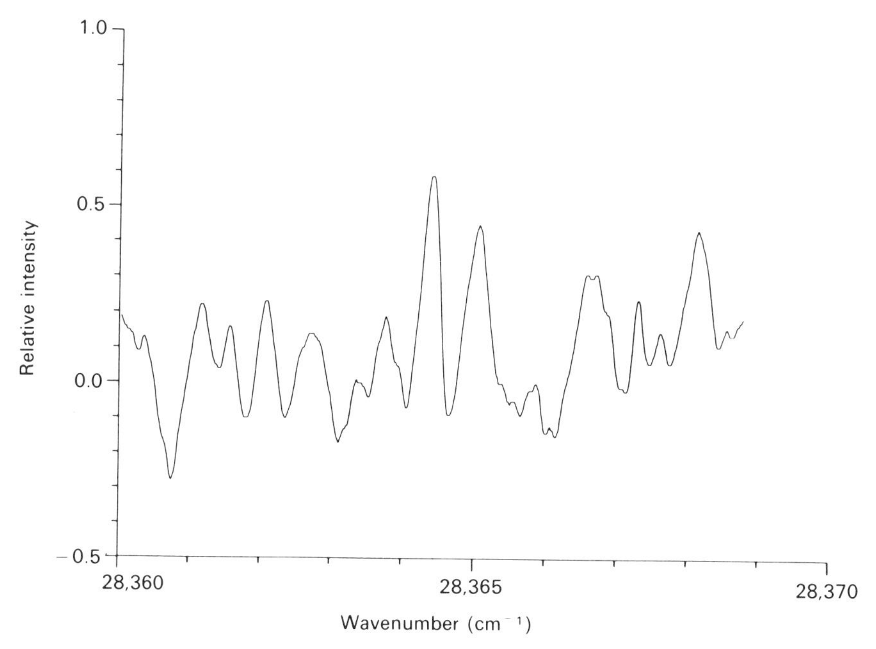

Figure 14.31. Same 10-wavenumber segment of FTS spectrum (Figures 14.29 and 14.30) of solution of 1 ppm Ni in 10 ppm Ca. The Ni I 352.45-nm line is still present at same intensity but noise level in spectrum has increased by factor of 11 due to calcium matrix. Plotted on same intensity scale as Figures 14.29 and 14.30. [Reprinted with permission from L. M. Faires (1985), "Effects of Sample Matrix on Detection Limits in Analytical Inductively Coupled Plasma–Fourier Transform Spectrometry," *Spectrochimica Acta, 40B*. Copyright 1985 Pergamon Press, Elmsford, NY.]

mirror drive irregularities. Not surprisingly, the stability of the interferometer itself significantly affects the observed noise characteristics of the spectrum.

The ICP emission spectrum, standard deviation spectrum, and SNR spectrum for a solution of 10 ppm Mg are shown in Figures 14.34*a*, 14.34*b*, and 14.34*c*, respectively. The basic characteristics of these ICP spectra are similar to their HCL counterparts (Figure 14.33), except that *two* pairs of satellite peaks are now observed. The closest pair, in the same location as the pair observed in the HCL spectra, was also attributed to instrumental instability. However, the second pair of satellites was attributed to modulation of the interferogram by rotation of the plasma, a noise component resulting from source instability. (Noise from plasma rotation has been previously reported by Walden et al., 1980, and Salin and Horlick, 1980.)

The effect of adding a high concentration of a second element (1600 ppm Fe, 400 ppm Mn, and 400 ppm Zn) to the 10-ppm solution of Mg was also studied. The presence of iron, an element having a very complex spectrum,

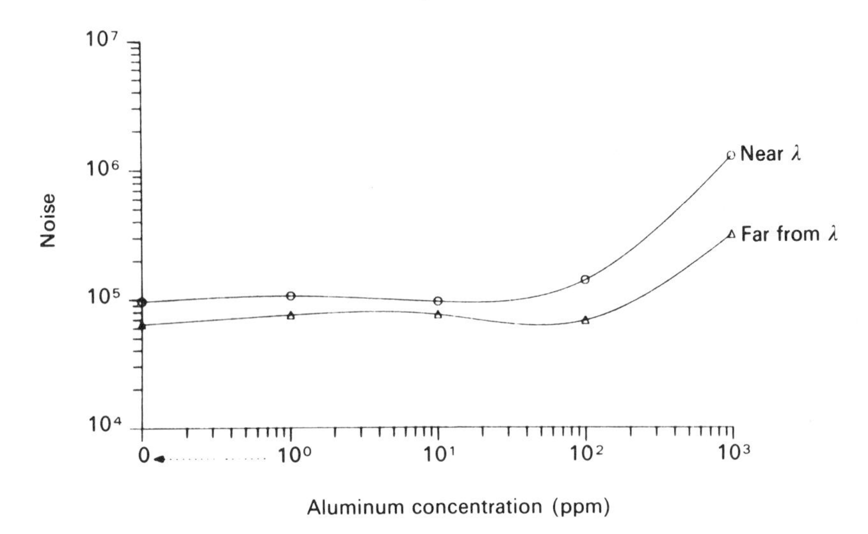

Figure 14.32. Noise as function of analyte concentration for Al I 396.15-nm line measured far from analytical wavelength and measured near analytical wavelength. At higher analyte concentrations, noise is seen to be higher near analytical line. Measured noise level in blank solution is indicated on far left as extrapolation to zero concentration. [Reprinted with permission from L. M. Faires (1985), "Effects of Sample Matrix on Detection Limits in Analytical Inductively Coupled Plasma–Fourier Transform Spectrometry," *Spectrochimica Acta, 40B*. Copyright 1985 Pergamon Press, Elmsford, NY.]

caused considerable degradation in the SNR of the Mg emission. The presence of Zn and Mn, elements having relatively simple spectra, caused only a small effect. These observations indicate that a strong emission line can result in noise distribution throughout the baseline, a distributive multiplex disadvantage.

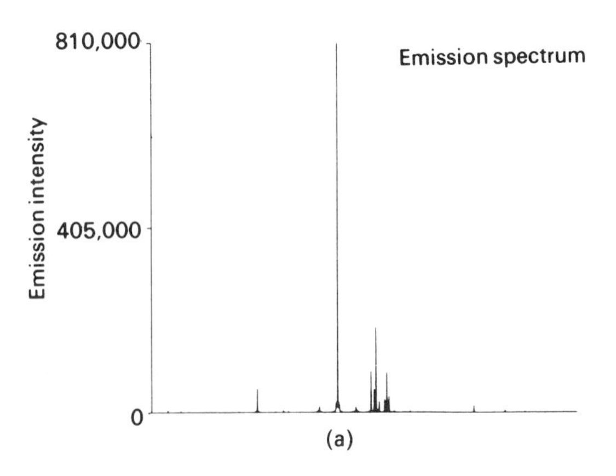

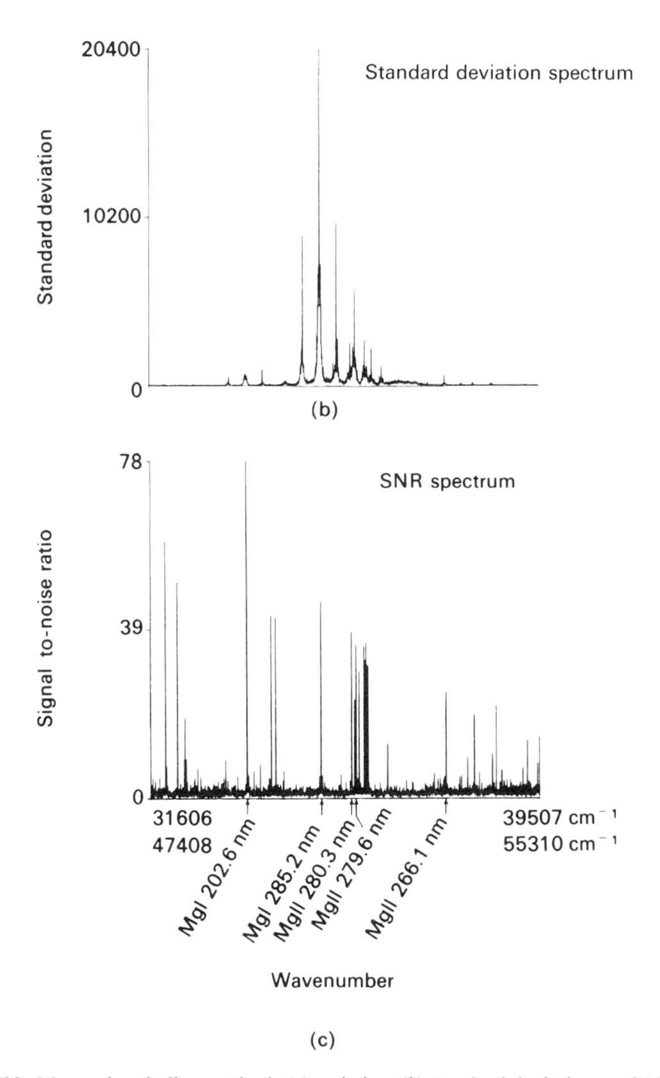

Figure 14.33. Magnesium hollow-cathode (*a*) emission, (*b*) standard deviation, and (*c*) signal-to-noise ratio spectra. [Reprinted from S. Marra and G. Horlick (1986), *Appl. Spectrosc.*, *40*, 804, with permission from the Society from Applied Spectroscopy.]

Recently, Voigtman and Winefordner (1987a) have analyzed the performance of an FT spectrometer mathematically and shown that source fluctuation ($1/f$) noise can remain localized about the generating spectral region(s). Therefore, a weak analyte line will not necessarily be buried in the noise from the flickering interferent line. The factors determining the magnitude of the interference include (1) the proximity of the analyte and interferent lines; (2) the strengths, density, and distribution of the interferent lines; (3) the type of interferent flicker noise present; and (4) the modulation frequency of the interferometer, that is, the rate of mirror movement. Generally speaking, however, the best interferometers employ sufficiently high modulation speeds and large mirror displacements to spread the flicker noise out over the baseline. Therefore, only in rather unusual cases can the multiplex disadvantage be avoided. Several methods of experimentally testing these conclusions have been proposed (Voigtman and Winefordner, 1987b).

14.5. CONCLUSIONS

The final word on noise and its distribution in multiplexed spectra has probably yet to be written. Nevertheless, it does seem clear that in photon shot-noise-limited cases, noise will be more or less uniformly distributed throughout the spectrum, resulting in a distributive multiplex disadvantage. Some frequency-dependent noise structure may be observed (Marra and Horlick, 1986) but in most cases can be traced to fluctuation ($1/f$) noise originating from the spectrometer and/or source. However, when the instrument is operated at high modulation frequencies, even $1/f$ noise tends to be spread throughout the baseline (Voigtman and Winefordner, 1987a).

Since the noise characteristics of the spectral measurement depend on many factors, the magnitude of the multiplex disadvantage will, to some degree, be determined by experimental conditions. Marra and Horlick (1986), for example, have shown that using a PMT detector, the dominant noise characteristics of their ICP/FT spectrometer (Section 14.2.2) changed with analyte concentration. At low concentrations (0.2–2-ppm Mg) the measure-

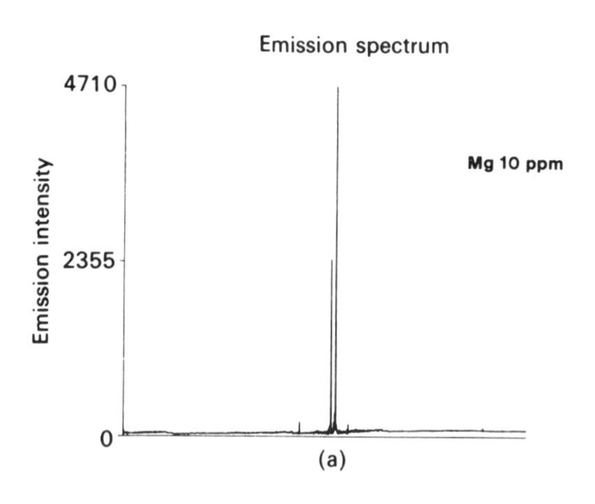

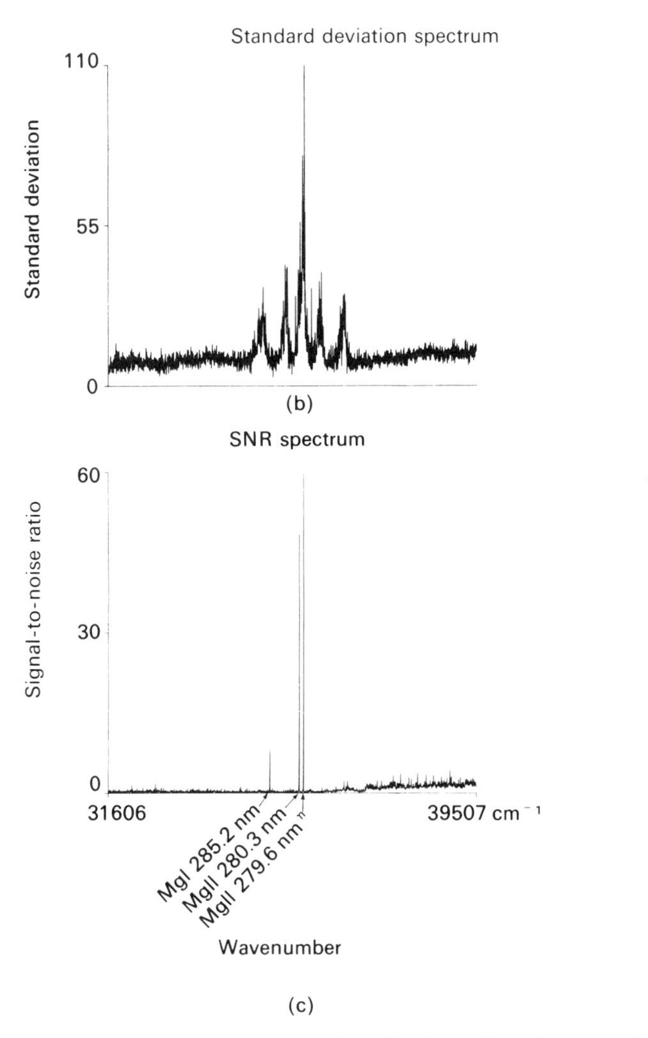

Figure 14.34. (*a*) Emission, (*b*) standard deviation, and (*c*) signal-to-noise ratio spectra for ICP emission of 10 ppm Mg. [Reprinted from S. Marra and G. Horlick (1986), *Appl. Spectrosc.*, *40*, 804, with permission from the Society for Applied Spectroscopy.]

ment appeared to be limited by a noise that was independent of signal level. Above 2-ppm Mg, the noise became directly proportional to signal intensity, that is, source flicker limited. Similar effects were observed when an air–acetylene flame served as the source and a silicon photodiode was substituted for the PMT detector (Horlick et al., 1982). At low analyte concentrations (0.03 ppm Li), the signal was detector noise limited (independent of signal intensity), while at higher concentrations the signal became fluctuation noise limited. Whenever the noise is independent of signal intensity, multiplexing has the potential to improve the SNR of the entire spectrum.

Silicon photodiodes (near infrared and visible) and the PMT (visible and UV) represent the two basic types of detectors that have been used to evaluate transform spectrometers for atomic emission. Studies by Moffatt, Buijs, and Murphy (1986) have shown that the silicon photodiode produces a significant background signal, resulting in a large detector noise component. Therefore, depending on specific conditions, spectra obtained using a silicon photodiode detector may evidence less of a distributive multiplex effect.

The distributive multiplex disadvantage has serious consequences for the analytical spectroscopist who must determine trace quantities of an element using a weak spectral signal in an optically dense spectrum. In a multielement analysis, the distributive multiplex disadvantage means that every analyte line will be affected differently, with the SNR gain or loss determined by the intensity of the line and the type and structure of the spectrum under consideration.

The problems associated with multiplexing in the UV–visible region force one to conclude that transform techniques will probably never be as generally useful for spectrochemical analysis as had been initially hoped. Applications still exist, but justification must involve factors in addition to or instead of the potential for SNR improvement.

One application in the visible–near infrared region, where transform methods are particularly beneficial, is Raman spectroscopy. Chase (1986, 1987) and Hirschfeld and Chase (1986) have shown that the use of a low-photon-energy exciting laser avoids many of the sample decomposition and sample fluorescence problems common to high-energy laser Raman spectroscopy. However, the intensity of the Raman-scattered radiation is also decreased according to the fourth-power dependence of Raman scatter intensity on exciting frequency. The worst of the multiplex disadvantage can be easily avoided by eliminating the Rayleigh line through optical filtering (FT–Raman; Chase, 1986) or by the use of framing masks (HT–Raman; Tilotta et al., 1987). The resulting spectrum is not only optically dilute but is also of relatively weak intensity. In this case, the ability to enhance even the strongest lines in the spectrum makes transform methods quite attractive.

Another potential application for multiplexing in the UV–visible region is on-line elemental analysis of peaks eluting from a gas chromatograph. Here, the rapid scanning capability of a Michelson interferometer can be used to great advantage. Schleisman, Fateley, and Fry (1984) have reported the determination of nonmetals such as carbon, hydrogen, and sulfur using an ICP/FT spectrometer and suggested that C–H ratios could be determined with good accuracy. Ng and Horlick (1985a, b) have also suggested using a simple cross-correlation technique for processing ICP/FT interferograms in the UV–visible for automatic qualitative analysis of up to 70 elements. In this case, the complexity of the ICP spectrum offers the advantage of greater reliability in determining the presence or absence of a particular element in the sample.

REFERENCES

Aldous, K. M., Browner, R. F., Dagnal, R. M., and West, T. S. (1970), *Anal. Chem.*, *42*, 939.

Aryamanya-Mugisha, H. and Williams, R. R. (1985), *Appl. Spectrosc*, *39*, 693.

Bell, R. J. (1972), *Introductory Fourier Transform Spectroscopy*, Academic, New York.

Brault, J. W. (1976), *J. Opt. Soc. Am.*, *66*, 1081.

Breckinridge, J. B. and Schindler, R. A. (1981), "First Order Optical Design for Fourier Spectrometers," in *Spectrometric Techniques*, Vol. 2, Vanasse, G. A. (Ed.), Academic, New York, Chapter 2, pp. 63–125.

Budde, W. (1983), "Physical Detectors of Optical Radiation," in *Optical Radiation Measurements*, Vol. 4, Grum, F. and Bartelson, J. (Eds.), Academic, New York, Chapter 5, pp. 143–214.

Caulfield, M. J. (1976), "Holographic Spectroscopy," in *Advances in Holography*, Vol. 2, Farhat, N. (Ed.), Marcel Dekker, New York.

Chaney, L. W., Loh, L. T., and Such, M. T. (1967), "A Fourier Transform Spectrometer for the Measurement of Atmospheric Thermal Radiation," Technical Report for ORA Project 05863, University of Michigan, Ann Arbor.

Chase, D. B. (1986), *J. Am. Chem. Soc.*, *108*, 7485.

Chase, D. B. (1987), *Anal. Chem.*, *59*, 881A.

Chester, T. L., Fitzgerald, J. J., and Winefordner, J. D. (1976), *Anal. Chem.*, *48*, 779.

Chester, T. L. and Winefordner, J. D. (1977), *Anal. Chem.*, *49*, 113.

Connes, P. (1958), *J. Phys. Radium*, *19*, 262.

Connes, P. (1959), *Rev. Opt.*, *38*, 157, 416.

Connes, P. (1961), *Rev. Opt.*, *39*, 402.

Coufal, H., Moller, U., and Schneider, S. (1982), *Appl. Opt.*, *21*, 116.

Dahlquist, R. L. and Knoll, J. W. (1978), *Appl. Spectrosc.*, *32*, 1.

Davis, D. S., Larson, H. P., Williams, M., Michel, G., and Connes, P. (1980), *Appl. Opt.*, *19*, 4138.

Decker, Jr., J. A. (1971), *Appl. Opt.*, *10*, 24.

Decker, Jr., J. A. (1972), "Hadamard Transform Analytical Spectrometer," in *Analysis Instrumentation*, Vol. 10, Chapman, R. L., McNeill, G. A., and Bartz, A. M. (Eds.), Instrument Society of America, Pittsburgh, PA, pp. 49–54.

Decker, Jr., J. A. (1973), *Appl. Opt.*, *12*, 1108.

Decker, Jr., J. A. (1977), "Hadamard-Transform Spectroscopy," in *Spectrometric Techniques*, Vol. 1, Vanasse, G. A. (Ed.), Academic, New York, Chapter 5, pp. 189–227.

Decker, Jr., J. A. and Harwit, M. (1969), *Appl. Opt.*, *8*, 2552.

DeGraauw, T. and Veltman, B. P. T. (1970), *Appl. Opt.*, *9*, 2658.

Dohi, T. and Suzuki, T. (1971), *Appl. Opt.*, *10*, 1359.

Faires, L. M. (1983), "Fourier Transform and Polychromator Studies of the Inductively Coupled Plasma," Ph.D. Thesis, University of New Mexico, Albuquerque, NM.

Faires, L. M. (1984), *ICP Inf. Newslett.*, *10*, 449.

Faires, L. M. (1985), *Spectrochim. Acta*, *40B*, 1473.

Faires, L. M., Palmer, B. A., and Brault, J. W. (1985a), *Spectrochim. Acta*, *40B*, 135.

Faires, L. M., Palmer, B. A., Engleman Jr., R., and Niemczyk, T. M. (1983), *Proceedings of the Los Alamos Conference on Optics '83* (April 11–15, 1983) Los Alamos and Sante Fe, New Mexico, SPIE Vol. 380, pp. 396–401, Society of Photo-Optical Instrumentation Engineers, Bellingham, WA.

Faires, L. M., Palmer, B. A., Engleman Jr., R., and Niemczyk, T. M. (1984), *Spectrochim. Acta*, *39B*, 819.

Faires, L. M., Palmer, B. A., Engleman Jr., R., and Niemezyk, T. M. (1985b), *Spectrochim. Acta*, *40B*, 545.

Fassel, V. A., Mossotti, V. G., Grossman, W. E. L., and Kniseley, R. N. (1966), *Spectrochim. Acta*, *22B*, 346.

Fellgett, P. (1967), *J. Phys.*, *28*, C2-165.

Filler, A. S. (1973), *J. Opt. Soc. Am.*, *63*, 589.

Fitzgerald, J. J., Chester, T. L., and Winefordner, J. D. (1975), *Anal. Chem.*, *47*, 2330.

Fotiou, F. K. and Morris, M. S. (1986), *Appl. Spectrosc.*, *40*, 704.

Gebbie, H. A. (1969), *Appl. Opt.*, *8*, 501.

Greenfield, S., McGeachin, H. McD., and Smith, P. B. (1976), *Talanta*, *23*, 1.

Griffiths, P. R. (1974), *Anal. Chem.*, *46*, 645A.

Griffiths, P. R. and deHaseth, J. A. (1986a), *Fourier Transform Infrared Spectrometry*, Wiley, New York, Chapter 16, pp. 520–535.

Griffiths, P. R. and deHaseth, J. A. (1986b), *Fourier Transform Infrared Spectrometry*, Wiley, New York, Chapter 7, pp. 248–283.

Griffiths, P. R., Sloane, H. J., and Hannah, R. W. (1977), *Appl. Spectrosc.*, *31*, 485.

Hammaker, R. M., Graham, J. A., Tilotta, D. C., and Fateley, W. G. (1986), "What is Hadamard Transform Spectroscopy?" in *Vibrational Spectra and Structure*, Vol. 15, Durig, J. R. (Ed.), Elsevier, Amsterdam, Chapter 7, pp. 401–485.

Hanel, R. A. and Conrath, B. J. (1969), *Science*, *165*, 1258.

Hansen, P. and Strong, J. (1972), *Appl. Opt.*, *11*, 502.

Harwit, M. (1973), *Appl. Opt.*, *12*, 285.

Harwit, M. and Decker, Jr., J. A. (1974), "Modulation Techniques in Spectrometry," in *Progress in Optics*, Vol. 12, Wolf, E. (Ed.), North-Holland, Amsterdam, pp. 101–162.

Harwit, M. and Sloane, N. J. A. (1979), *Hadamard Transform Optics*, Academic, New York.

Harwit, M. and Tai, M. H. (1977), *Appl. Opt.*, *16*, 3071.

Hiltner, W. A. and Code, A. D. (1950), *J. Opt. Soc. Am.*, *40*, 149.

Hirschfeld, T. (1976a), *Appl. Opt.*, *30*, 68.

Hirschfeld, T. (1976b), *Anal. Chem.*, *48*, 721.

Hirschfeld, T. (1979), "Quantitative FT–IR: A Detailed Look at the Problems Involved," in *Fourier Transform Infrared Spectroscopy*, Vol. 2, Ferraro, J. R. and Basile, L. J. (Eds.), Academic, Chapter 6, pp. 193–242.

Hirschfeld, T. and Chase, D. B. (1986), *Appl. Spectrosc.*, *40*, 133.

Horlick, G. and Codding, E. G. (1973), *Anal. Chem.*, *45*, 1490.

Horlick, G., Hall, R. H., and Yuen, W. K. (1982), "Atomic Emission Spectrochemical Measurements with a Fourier Transform Spectrometer," in *Fourier Transform Infrared Spectroscopy*, Vol. 3, Ferraro, J. R. and Basile, L. J. (Eds.), Academic, New York, Chapter 2, pp. 37–81.

Horlick, G. and Yuen, W. K. (1975), *Anal. Chem.*, *47*, 775A.

Horlick, G. and Yuen, W. K. (1978), *Appl. Spectrosc.*, *32*, 38.

Jacquinot, P. (1965), *Jpn. J. Appl. Phys.*, *4*, 402.

Kamiya, K., Yoshihara, K., and Okada, K. (1968), *Jpn. J. Appl. Phys.*, *7*, 1129.

Keir, J., Dawson, J. B., and Ellis, D. J. (1977), *Appl. Spectrosc.*, *32*, 59.

Khan, A. H. and Nejib, U. R. (1987), *Appl. Opt.*, *26*, 270.

Khan, F. D. (1959), *Astrophys. J.*, *129*, 518.

Knacke, R. F. (1978), *Appl. Opt.*, *17*, 6.

Kraenz, M. E. and Kunath, D. (1982), *J. Mol. Struct.*, *79*, 47.

Larson, N. M., Crosmun, R., and Talmi, Y. (1974), *Appl. Opt.*, *13*, 2662.

Luc, P. and Gerstenkorn, S. (1972), *Astron. Astrophys.*, *18*, 209.

Luc, P. and Gerstenkorn, S. (1978), *Appl. Opt.*, *17*, 1327.

Marra, S. and Horlick, G. (1986), *Appl. Spectrosc.*, *40*, 804.

Marshall, A. G. and Comisarow, M. B. (1978), "Multichannel Methods in Spectroscopy," in *Transform Techniques in Chemistry*, Griffiths, P. R. (Ed.), Plenum, New York, Chapter 3, pp. 39–68.

Mertz, L. (1965), *Transformations in Optics*, Wiley, New York.

Michelson, A. A. (1891), *Phil. Mag.*, *31*, 338.

Michelson, A. A. (1892), *Phil. Mag.*, *34*, 280.

Moffatt, D. J., Buijs, H., and Murphy, W. F. (1986), *Appl. Spectrosc.*, *40*, 1079.

Nitis, G. J., Svoboda, V., and Winefordner, J. D. (1972), *Spectrochim. Acta*, *27B*, 345.

Ng, R. C. L. and Horlick, G. (1985a), *Appl. Spectrosc.*, *39*, 834.

Ng, R. C. L. and Horlick, G. (1985b), *Appl. Spectrosc.*, *39*, 841.

Nordstrom, R. J. (1982), "Aspects of Fourier Transform Visible/UV Spectroscopy," in *Fourier, Hadamard, and Hilbert Transformations in Chemistry*, Marshall, A. G. (Ed.), Plenum, New York, Chapter 14, pp. 421–452.

Okamoto, T., Kawata, S., and Minami, S. (1984), *Appl. Opt.*, *23*, 269.

Oliver, C. J. (1976), *Appl. Opt.*, *15*, 93.

Oliver, C. J. and Pike, E. R. (1974), *Appl. Opt.*, *13*, 158.

Phillips, P. G. and Harwit, M. (1971), *Appl. Opt.*, *10*, 2780.

Pinard, J. (1967), *J. Phys.*, *C2*, 142.

Plankey, F. W., Glenn, T. H., Hart, L. P., and Winefordner, J. D. (1974), *Anal. Chem.*, *46*, 1000.

Platt, U., Perner, D., and Patz, H. W. (1979), *J. Geophys. Res.*, *84*, 6329.

Pruiksma, R., Ziemer, J., and Yeung, E. S. (1976), *Anal. Chem.*, *48*, 667.

Pratt, W. K., Kane, J., and Andrews, H. C. (1969), *Proc. IEEE*, *57*, 58.

Salin, E. D. and Horlick, G. (1980), *Anal. Chem.*, *52*, 1578.

Santini, R. E., Milano, M. J., and Pardue, H. L. (1973), *Anal. Chem.*, *45*, 915A.

Schleisman, A. J. J., Fateley, W. G., and Fry, R. C. (1984), *J. Phys. Chem.*, *88*, 398.

Slavin, M. (1971), *Emission Spectrochemical Analysis*, Wiley, New York, P. 56.

Steward, E. G. (1987), *Fourier Optics: An Introduction*, 2nd ed., Ellis Horwood, Chichester, Chapter 5, pp. 99–136.

Stroke, G. W. and Funkhouser, A. T. (1965), *Phys. Lett.*, *16*, 272.

Stubley, E. A. and Horlick, G. (1984), *Appl. Spectrosc.*, *38*, 162.

Stubley, E. A. and Horlick, G. (1985a), *Appl. Spectrosc.*, *39*, 800.

Stubley, E. A. and Horlick, G. (1985b), *Appl. Spectrosc.*, *39*, 811.

Stubley, E. A. and Horlick, G. (1985c), *Appl. Spectrosc.*, *39*, 805.

Sugimoto, N. (1986), *Appl. Spectrosc.*, *25*, 863.

Swift, R. D., Wattson, R. B., Decker, Jr., J. A., Paganetti, R., and Harwit, M. (1976), *Appl. Opt.*, *15*, 1595.

Tai, M. H. and Harwit, M. (1976), *Appl. Opt.*, *15*, 2664.

Tilotta, D. and Fateley, W. G. (1988), *Spectroscopy*, *3*, 14.

Tilotta, D., Freeman, R. D., and Fateley, W. G. (1987), *Appl. Spectrosc.*, *41*, 1280.

Tilotta, D., Fry, R. C., and Fateley, W. G. (1990), *Talanta*, in press.

Tilotta, D., Hammaker, R. M., and Fateley, W. G. (1987a), *Spectrochim. Acta*, *42A*, 1493.

Tilotta, D., Hammaker, R. M., and Fateley, W. G. (1987b), *Appl. Opt.*, *26*, 4285.

Tilotta, D., Hammaker, R. M., and Fateley, W. G. (1987c), *Appl. Spectros.*, *41*, 727.

Treado, P. J. and Morris, M. D. (1989), *Anal. Chem. 61*, 723A.

Treffers, R. R. (1977), *Appl. Opt.*, *16*, 3103.

Voigtman, E. and Winefordner, J. D. (1987a), *Appl. Spectrosc.*, *41*, 1182.

Voigtman, E. and Winefordner, J. D. (1987b), *Anal. Chem.*, *59*, 1364.

Walden, G. L., Bower, J. N., Nikel, S., Bolten, P. L., and Winefordner, J. D. (1980), *Spectrochim. Acta, 35B*, 535.

Winefordner, J. D., Avni, R., Chester, T. L., Fitzgerald, J. J., Hart, L. P., Johnson, D. J., and Plankey, F. W. (1976), *Spectrochim. Acta, 31B*, 1.

Winefordner, J. D. and Haraguchi, H. (1977), *Appl. Spectrosc., 31*, 195.

Winefordner, J. D., Svoboda, V., and Cline, L. J. (1970), *CRC Crit. Rev. Anal. Chem., 1*, 233.

Winge, R. K., Fassel, V. A., and Kniseley, R. N. (1977), *Spectrochim. Acta, 32B*, 327.

Yoshihara, K. and Kitade, A. (1967), *J. Appl. Phys., 6*, 116.

Yoshihara, K., Nakashima, K., and Higuchi, M. (1976), *Jpn. J. Appl. Phys., 15*, 1169.

Yuen, W. K. and Horlick, G. (1977), *Anal. Chem., 49*, 1446.

CHAPTER

15

NONTRANSFORM DETECTION SYSTEMS

Over the past 20 years, the development of multielement detection systems for spectrochemical analysis has been an active area of analytical research. During the course of this research, numerous spectroscopic systems have been developed and studied. Much of this work has been reviewed by Busch and Morrison (1973), Busch and Malloy (1979b), Busch and Benton (1983), and Winefordner, Fitzgerald, and Omenetto (1975). In organizing this monograph, we have found it expedient to divide the topic of multielement detection systems into transform-based systems and nontransform systems. The previous chapter dealt with transform systems and showed that these multiplex systems were most useful in dectector-noise-limited situations. This chapter will deal with the nontransform systems that have been studied and/or actually used for multielement spectrochemical analysis.

The majority of systems to be discussed in this chapter are not available commercially. In most cases, the instruments described represent one-of-a-kind research instruments developed to test a particular concept. As a result, the literature citations that describe the various systems actually represent a form of feasibility analysis (see Chapter 12) rather than discussions of extensive analytical applications (i.e., capability analysis). All three modes of spectroscopic observation (see Chapter 13) have been employed to monitor the atomic vapor produced in the atom reservoir, and examples of systems that employ atomic emission, atomic fluorescence, and atomic absorption, as well as combinations, are readily available. Because of the extensive amount of work that has been conducted on multielement detection systems, space does not permit an encyclopedic discussion of every system that has been developed. Instead, examples of the different approaches that have been investigated will be discussed in terms of the requirements imposed by multielement spectrochemical analysis (see Chapter 12).

Nontransform spectroscopic systems fall into two major categories (as shown in Table 15.1) based on the manner in which the spectral information has been encoded. The two major categories (or phyla, by analogy with biological classification) can be distinguished on the basis of the detector

Table 15.1. Nontransform Detection Systems for Multielement Analysis

1. Single-Detector Systems (essential characteristics: single detector, time domain encoding, sequential data acquisition)
   a. Nondispersive systems
   b. Dispersive systems
      Mechanical scan
         Linear scan
         Programmed scan
      Electronic Scan
         One-Dimensional Dispersion
         Two-Dimensional Dispersion
2. Multiple–Detector Systems (essential characteristics: multiple detectors, spatial encoding, simultaneous data acquisition)
   a. Direct-reading spectrometers
   b. Image devices
         One-Dimensional Dispersion
         Two-Dimensional Dispersion

subsystem employed. The *single-detector phylum* includes all nontransform systems that employ a single detector to monitor the radiant intensity and encode the spectral information in the time domain. As a result, single-detector systems are all temporal systems that form two classes on the basis of the wavelength selector subsystem: dispersive and nondispersive. The single-detector dispersive class can be divided into two orders depending on whether the particular family uses mechanical scanning or electronic scanning.

The other phylum of nontransform spectroscopic systems can be distinguished by its use of multiple detectors. These *parallel multichannel systems* encode the spectral information spatially and are, therefore, all dispersive systems. This phylum can be subdivided into two classes, those that employ image devices and those that use multiple photomultipliers (i.e., direct readers). Compared with the single-detector phylum, all of which gather spectral information in a sequential fashion, the multiple-detector phylum is composed of systems that are truly simultaneous in the manner in which they collect spectral data.

15.1. TIME DIVISION MULTIPLEXING

Hadamard and Fourier transform spectroscopies are examples of *frequency division multiplexing* where different optical components of the spectrum are modulated at different frequencies before striking the detector. In frequency division multiplexing, all the individually modulated spectral components strike the detector simultaneously, giving rise to a multiplex advantage in detector-noise-limited situations. By modulating each wavelength of interest at a different frequency, wavelength discrimination can be accomplished after detection either by transform procedures or, in the case of only two wavelengths, by using separately tuned amplifiers to respond to individual chopping frequencies (i.e., individual optical wavelengths). If electronic separation using tuned amplifiers is used, modulating frequencies must be sufficiently different to avoid cross-talk between transmission channels. This tends to restrict the number of channels because each channel requires a minimum bandwidth for signal transmission. Regardless of the means employed to achieve frequency division multiplexing, simultaneous sampling of several resolution elements generally leads to a multiplex disadvantage in the UV–visible region of the spectrum when a photomultiplier is used as a detector.

Chester and Winefordner (1976) evaluated the feasibility of using frequency-modulated sources in nondispersive atomic fluorescence. Their study demonstrated that a multiplex disadvantage exists for nondispersive systems similar to that observed with transform-based spectrometers (see Chapter 14). Their study also indicated that the limiting noise in nondispersive atomic fluorescence is flame background noise.

Another form of multiplexing known as *time division multiplexing* has been employed extensively in communications networks to transmit data from several sources over the same transmission channel (Schwartz, 1959). In time division multiplexing, the signals from the various sources are staggered in time prior to transmission over the transmission channel. This form of information transmission system is called a *sampled-data system* because the information from a given channel is sampled periodically. Because a given channel is sampled periodically, information from other sources (i.e., other channels) can be transmitted in the vacant intervals. Since information from each source is staggered in time, each source has full use of the transmission channel bandwidth.

15.2. INFORMATION TRANSMISSION

Since all analytical systems are, in reality, analogous to communication systems (they both transmit information from a source to a receiver), their performance can be analyzed in terms of elementary information theory. Intuitively, we know that the rate of information transmission over any transmission channel is limited in some fashion by the nature of the transmission channel itself. To understand the factors that limit information transmission, we must develop some idea about the information content of a signal.

The information content of any source is related to its rate of change and to its unpredictability. It is clear from this statement that an unvarying signal does not transmit any information at all! Only when the signal changes with time is information being transmitted. The more uncertain the signal level in the next instant, the more information content the signal possesses. To transmit large amounts of information, therefore, a transmission system must be able to change rapidly from one signal level to another. Unfortunately, all transmission systems have energy storage devices (i.e., capacitors and inductors) present, and changing the signal level implies changing the energy content of these components. The presence of these energy storage components limits the rate of change possible by the system. In addition, every system is limited at some point by small fluctuations (i.e., noise). The presence of noise limits the minimum signal amplitude change that can be detected.

To develop the notion of information content more fully, consider a hypothetical transmission system that can distinguish n different signal levels and requires t seconds to change from one level to another. How much information can be transmitted by a system such as this over a time interval of T? The amount of information that can be transmitted is related to the different and distinguishable signal amplitude combinations that can be transmitted over the time interval T. Since the number of signal changes that can be accommodated by the system in a time period T is T/t and the number of different levels that can be distinguished is n, it follows that the number of different possible combinations of signal levels that can be transmitted over this time period is

$$n^{T/t} \tag{15.1}$$

To make Eq. 15.1 consistent with our notion that the information content of a signal should be proportional to the length of the transmission time T, Eq. 15.1 can be modified by taking the logarithm of the number of different possible signal-level combinations. This gives

$$\text{Information} = (T/t)\log_2 n \tag{15.2}$$

where base 2 logarithms have been used to express the information content in *bits* (for *b*inary dig*its*).

Using Eq. 15.2, we can now define the *system capacity* as the maximum rate of transmitting information. From Eq. 15.2, the system capacity will be given by

$$C = \text{information}/T = (1/t)\log_2 n \tag{15.3}$$

To put Eq. 15.3 in more familiar terms, as far as analytical instruments are concerned, we can replace the term $1/t$ with its equivalent expression in terms of filter bandwidth (see Chapter 9):

$$\Delta f = B = (1/2t) \tag{15.4}$$

Similarly, the number of distinguishable signal levels can be replaced by

$$n = S_{max}/N \tag{15.5}$$

where S_{max} is the maximum signal level that can be transmitted and N is the noise. Substituting Eqs. 15.4 and 15.5 into Eq. 15.3 gives

$$C = 2B \log_2(S/N) \tag{15.6}$$

where S/N is the maximum possible signal-to-noise ratio of the transmission.

Equation 15.6 is known as *Hartley's law* in honor of R. V. L. Hartley of Bell Laboratories, who developed it in 1928.† It states that the channel capacity (in bits per second) is equal to 2 times the channel bandwidth times the logarithm to the base 2 of the maximum possible signal-to-noise ratio. For a transmission channel with a fixed capacity C, the wider the bandwidth of the transmission channel, the poorer the maximum possible signal-to-noise ratio that can be transmitted. Conversely, a large signal-to-noise ratio requires a narrow channel bandwidth. According to Eq. 15.4, a narrow channel bandwidth means a long sampling interval t. A long sampling time means that fewer signal amplitude changes can be sensed over a transmission period T.

It is interesting to note that Hartley's law and the sampling theorem of Nyquist (see Chapters 7, 11, and 14) are essentially the same. Nyquist's sampling theorem says that to completely specify a given waveform, the waveform must be sampled at a rate equal to twice the highest frequency component in the waveform. This is true because the amount of information present is proportional to $2BT$.

15.3. TIME DIVISION MULTIPLEXING IN ANALYTICAL SPECTROSCOPY

When different optical wavelengths are arranged sequentially to strike the detector, time multiplexing is occurring because time is being used as a means

† Hartley's law is sometimes written without the 2. This arises if Δf is taken as $1/t$ instead of $1/2t$ as in Eq. 15.4.

of encoding the spectral information prior to transmission over a single electronic channel (i.e., the single-detector output). With this form of signal transmission, a particular wavelength is always transmitted at a given time after the initiation of a spectral sweep. Since time is being used as a means of encoding the signal, a stable time base is required to avoid decoding errors that result in wavelength registration problems. Because time multiplexing is actually a form of sampled-data transmission, each wavelength in the spectrum is monitored for only a fraction of the total scan time. The sampling time required will therefore depend on the photon shot noise and source fluctuation noise (flicker) present in the spectral source being observed (McGeorge and Salin, 1985).

The presence of photon shot noise in the radiation field under observation imposes a fundamental limitation on the speed with which time-multiplexed spectrometers can gather data. To gather spectral information at high speed implies that each spectral resolution element must be sampled for as short a sampling time t as possible. This in turn imposes a wide system bandwidth according to Eq. 15.4 in order to keep up with the rate of data transmission. Since photon shot noise is white noise, the wider the system bandwidth, the more shot noise transmitted by the system. Thus a compromise exists between immunity from photon shot noise on the one hand and the speed of data acquisition on the other.

The signal-to-noise ratio obtained with a given spectral source and a time-multiplexed spectrometer can be improved in two ways (which actually amount to the same thing): time averaging (i.e., integration) and ensemble averaging (i.e., signal averaging). If the system behaves ergodically (see Chapter 9), both forms of averaging give the same result. By using a smaller frequency response bandwidth (which implies longer sampling times t and therefore a slower scan rate) or signal averaging, a longer total observation time T is required to obtain the spectral information. With an ideal instrument (where source flicker noise is absent), the ultimate rate of data transmission is limited by photon shot noise in the source. It should be recalled that neither bandwidth reduction nor signal averaging is an effective means of reducing source flicker noise (see Chapter 9).

15.4. NONDISPERSIVE TIME-MULTIPLEXED SYSTEMS

As a rule, nondispersive time-multiplexed analytical systems have all used atomic fluorescence as the preferred mode of observation of the atomic vapor because of the simplicity of the fluorescence spectra produced (see Chapter 13). Because of the inherent simplicity of atomic fluorescence spectra, it has been possible to develop instruments using optical filters as the means of wavelength isolation. One of the earliest of these nondispersive filter instruments to be developed was the one developed by Mitchell and Johansson (1970, 1971) and later patented by Mitchell (1971). This concept was extended and developed as a commercial instrument, known as the AFS-6† by Technicon Instruments Corporation of Tarrytown, New York, but was not successful in the marketplace and was later discontinued (Ullman, 1980).

Figure 15.1 shows a schematic diagram of the original optical and mechanical system used in the instrument conceived by Mitchell and Johansson (1970). With this analytical system, a bank of four hollow-cathode lamps was used to excite atomic fluorescence radiation sequentially in a hydrogen–air flame. The lamps were pulse modulated at 1000 Hz and were turned on sequentially in synchronism with a rotating filter wheel having optical interference filters located around the circumference at 90° intervals. The filter wheel was rotated at 1 Hz, and fluorescence signals were amplified with a phase-sensitive detector. Calibration curves for Ag, Cu, Fe, and Mg were obtained over a concentration range of 0.1–1000 ppm.

The original instrument was later improved (Mitchell and Johansson, 1971) by the use of expanded logic circuitry and independent gain controls for each element. An unmodulated emission mode was also added and found to function satisfactorily with synthetic "signals" but was not used in an actual analysis. An acetylene–air flame was used as the atom reservoir.

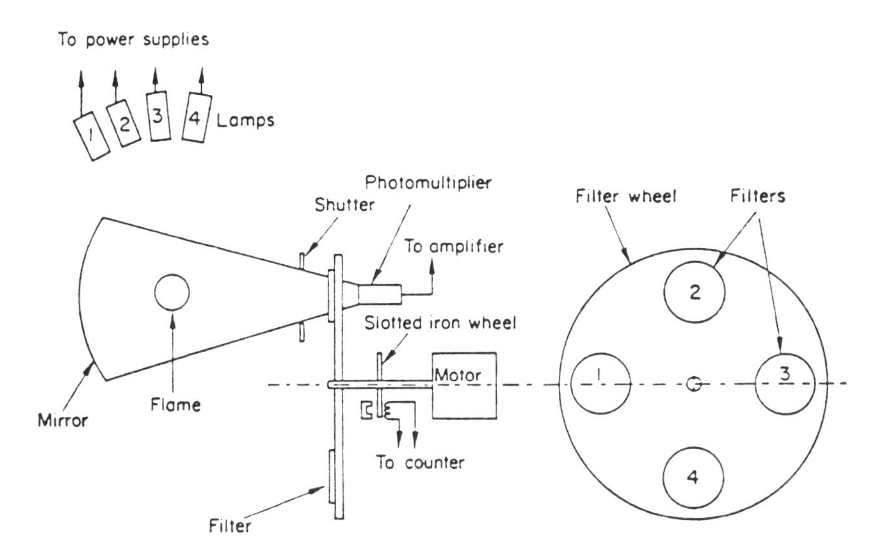

Figure 15.1. Schematic diagram of optical and mechanical system of instrument conceived by Mitchell and Johansson. [Reprinted with permission from D. G. Mitchell and A. Johansson (1970), "Simultaneous Multielement Analysis Using Sequentially Excited Atomic Fluorescence Radiation," *Spectrochimica Acta, 25B*, Copyright 1970 Pergamon Press, Elmsford, NY.]

† AFS-6 is currently a trademark of Bran & Luebbe Analyzing Technologies, Elmsford, NY.

The AFS-6™ developed by Technicon extended the original system developed by Mitchell and Johansson (1970, 1971) to permit the sequential determination of six elements. In the Technicon configuration, hollow-cathode excitation sources were arranged around the flame in a vertical plane at 60° to one another. Toroidal mirrors were used to focus the source radiation on the flame (which could be a hydrogen–air flame, an acetylene–air flame, or an acetylene–nitrous oxide flame). An inverse Cassegrain mirror system (known as an *Inca*, see Figure 15.2), which collected radiation over a solid angle of 0.82 sr, was used to collect the fluorescence radiation.

The AFS-6™ instrument gave linear dynamic ranges in excess of three orders of magnitude (Ullman, 1980) with detection limits comparable to other techniques. West and co-workers (Dagnall et al., 1971) used the instrument for the determination of Ca, Cu, Mg, Mn, K, and Zn in soil extracts. The six elements were determined in each of three dilutions of the extract medium (concentrated extract, 10- and 100-fold dilutions) in order to ensure that each element was measured on the linear portion of its calibration curve. Use of a single dilution for the analysis was not possible because of the widely differing concentrations of the elements in the soil extracts.

The instrument was also evaluated by West and co-workers (Dagnall et al., 1972) for the analysis of aluminum alloys.

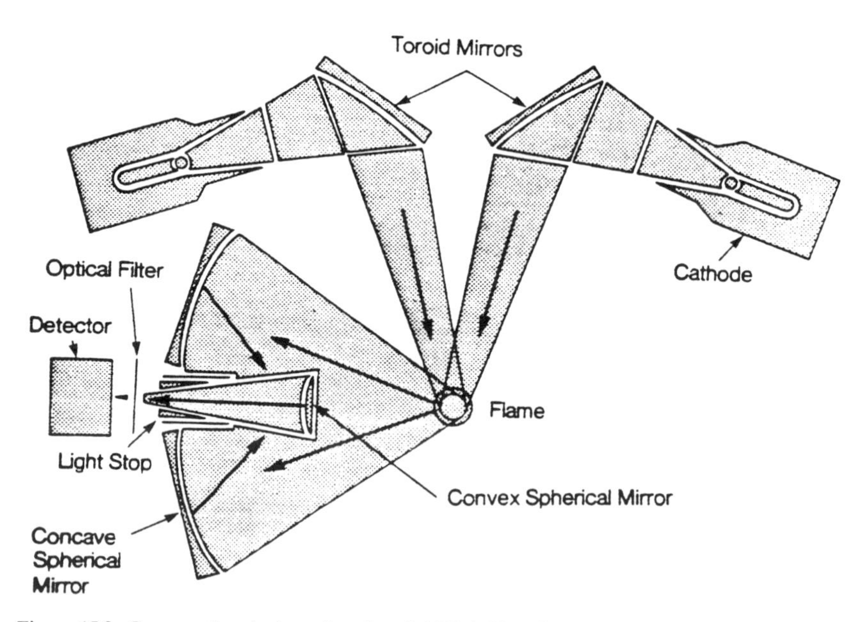

Figure 15.2. Cross-sectional view of optics of AFS-6. [Reprinted from A. Ullman (1980), *Prog. Anal. Atom. Spectrosc.*, *3*, 87, with permission from Pergamon Press, Elmsford, NY.]

Palermo, Montaser, and Crouch (1974) described a nondispersive atomic fluorescence system that employed computer-controlled hollow-cathode lamps as excitation sources. Time division multiplexing was achieved by pulsing several hollow-cathode lamps out of phase in a low-duty-cycle mode and detecting the fluorescence signal with a computer-controlled synchronous integrator. Using this approach, the fluorescence radiation from each element was transmitted over the same optical path but was separated from other elements in time. A solar-blind photomultiplier was used as the detector, and an argon-sheathed hydrogen–air flame was used as the atom reservoir. Sheathing was found to be necessary to reduce flame background. An observation height of 5–10 cm above the burner was used as another means of reducing flame background. Although good detection limits were obtained for Cd, Hg, Zn, and Pb, the system was not evaluated with real samples for an actual analytical problem. Its use is expected to be limited to easily volatilizable elements because of the relatively cool flame used. This may also lead to severe volatilization interferences with real samples.

Perhaps the most promising nondispersive atomic fluorescence spectrometer to be developed so far is the instrument commercially available from the Baird Corporation of Bedford, Massachusetts (see Demers, Busch, and Allemand, 1982; Demers, 1987). Known as the Baird Plasma/AFS-2000 Multielement Analyzer, the unit uses an induction-coupled plasma (ICP) as the atom reservoir. An array of up to 12 easily interchangeable element-specific modules (shown in Figure 15.3) encircles the ICP atom reservoir. Each element-specific module consists of a hollow-cathode lamp, an optical filter, and a photomultiplier. Each module's lamp (pulsed at 2 kHz) is turned on in sequence so that only one element is excited and measured at any one time. Sequential determination of 12 elements using this approach is said to require 15 sec following sample introduction. Spectral resolution (on the order of 0.001 nm) is achieved with the emission profile of the hollow-cathode lamp (see Chapter 13), while the optical filter in the module serves only to isolate the desired fluorescence radiation from extraneous background (Demers and Allemand, 1981).

Conceptually, the AFS-2000 offers a number of important analytical advantages. By combining the high-temperature ICP atom reservoir (with its large linear dynamic range and freedom from matrix effects) with the simplicity of atomic fluorescence spectra, a compact integrated instrument is possible. Because of the high resolution inherent in atomic fluorescence measurements, the manufacturer claims that the AFS-2000 can handle spectrally complex matrices such as tungsten- and chromium-based materials without difficulty. The stability of the system depends on the stability of the element-specific module rather than on complex mechanical drives or delicate optical components. Although the instrument is limited to 12 modules at a

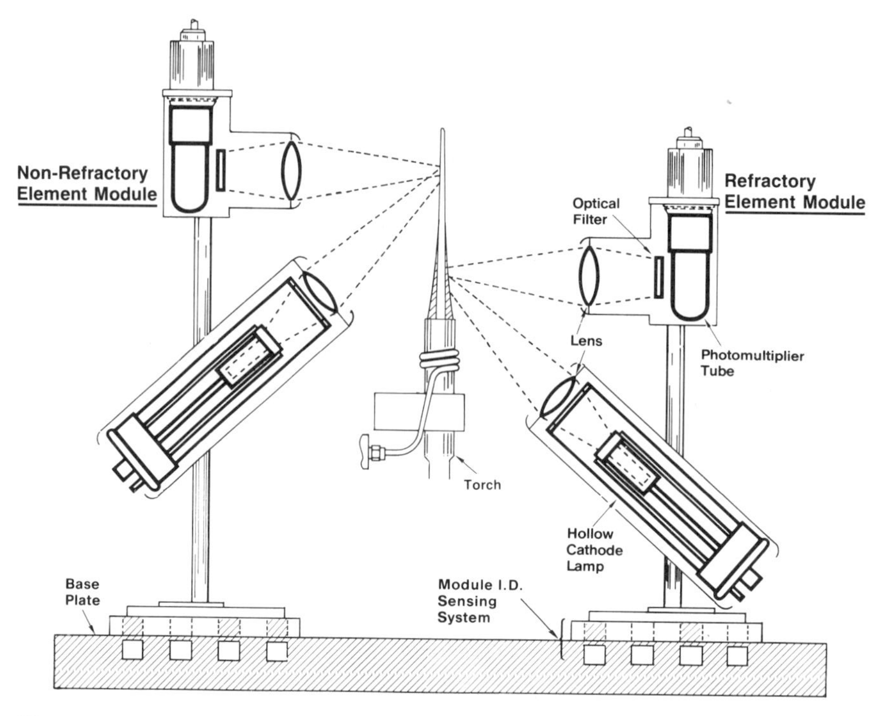

Figure 15.3. Arrangement of element-specific modules used in AFS-2000. (Courtesy of Baird Corporation.)

time (actually this is quite adequate for a large number of multielement analytical situations), it is quite flexible with regard to the particular combination of elements determined, and changing the combination requires only a simple exchange of modules. Modules are available for 65 elements according to the manufacturer.

When it was first introduced (Demers et al., 1982), the detection limits obtained with the prototype instrument for many elements (such as As, Hg, Sb, Se, B, Si, Ti, and V) were inferior to those obtained by ICP atomic emission measurements (Demers, 1987). By using more intense hollow-cathode lamp excitation sources known as *boosted-output hollow-cathode lamps* (see Lowe, 1971) and more efficient pneumatic nebulization, the detection capability of the AFS-2000 was improved greatly compared with the original instrument.

The improvement in detection capability with boosted-output hollow-cathode lamps is due to the more intense resonance line emission produced by these sources. Figure 15.4 shows a comparison of a conventional hollow-

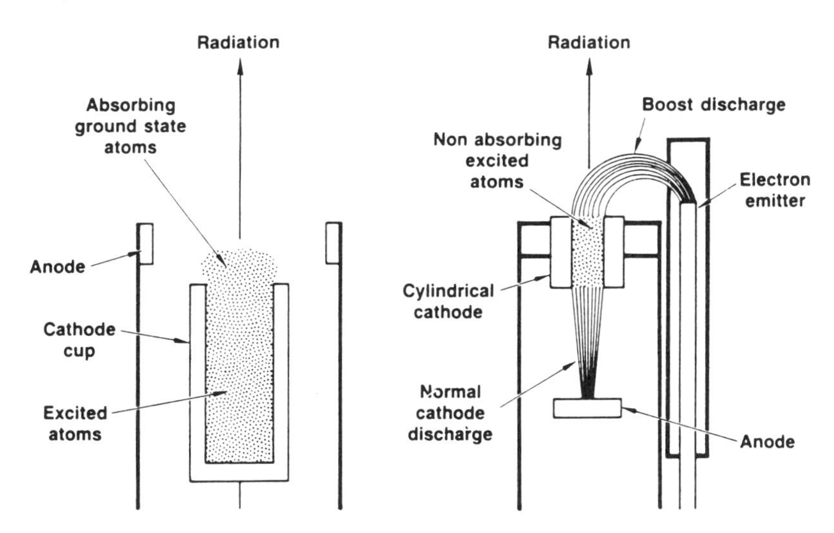

Figure 15.4. Schematic of conventional (left) and boosted-output (right) hollow-cathode lamp. (Reprinted from *American Laboratory*, Vol. 19, No. 8, pp. 30–41, 1987. Copyright 1987 International Scientific Communications, Inc.)

cathode lamp and its boosted counterpart. With a conventional hollow-cathode lamp, the cup-shaped cathode, which is formed from the desired element of interest, is surrounded by a metallic anode ring located near the open end of the cathode, as shown in Figure 15.4. The lamp is filled with a low pressure of an inert fill gas such as neon. When a sufficient voltage is applied across the tube, a gas discharge is established, and the positive fill-gas ions that are produced are accelerated toward the cathode where they collide. The positive fill-gas ions strike the cathode with energies of several hundred electron volts, dislodging atoms from the cathode surface in a process known as *sputtering*. Some of the sputtered atoms are excited by collisions with energetic species within the discharge region, which is confined to the cathode cup. Those atoms that diffuse to the mouth of the cup (i.e., outside the discharge region) tend to limit the emission intensity by self-absorption.

Boosted hollow-cathode lamps are more intense because the excitation process of the sputtered atoms is more efficient and self-absorption is reduced. Compared with an ordinary hollow-cathode lamp, the boosted version has a cathode in the form of a hollow cylinder with a flat circular metal anode located behind the cathode, as shown in Figure 15.4. In operation, a normal discharge, which sputters atoms of interest from the cathode into the cathode region, is maintained between the cylindrical cathode and the disk-shaped anode. A second, *booster discharge*, which is independent of the first discharge, is maintained between the anode and a filament cathode that has been coated specially to favor the copious emission of electrons. The electrons emitted by the filament cathode flow through the hollow cylindrical cathode where they increase the excitation of the ground-state sputtered atoms, thereby increasing the resonance emission line intensity. With booster discharge currents on the order of 300 mA, intensity gains of several-fold compared with ordinary hollow-cathode lamps are possible.

Although the AFS-2000 appears to offer definite analytical advantages, it is too early to tell whether it will be a commercial success in terms of being able to compete with existing technology such as the commercial emission ICP instruments.

15.5. DISPERSIVE TIME-MULTIPLEXED SYSTEMS

Dispersive time-multiplexed systems use a dispersive optical system to separate the radiation into individual resolution elements that can then be arranged to strike the detector sequentially, giving rise to a sequence of individual resolution element signals as a function of time. Two basic approaches can be employed to produce time-multiplexed signals in spectroscopy: (1) the detector may be scanned across a fixed spectrum or (2) the spectrum may be scanned across a fixed detector. In addition, time-multiplexed systems may be further differentiated on the basis of the manner in which the spectrum is scanned. With *linear-scan* systems, all resolution elements making up the complete spectrum are scanned at a constant, fixed rate. *Programmed-scan* systems, on the other hand, have been developed that do not sample all the resolution elements present in the complete spectrum but instead have the capability of stopping momentarily at wavelengths of analytical interest, while spectral regions of little interest are skipped entirely or rapidly scanned.

Scanning the entire spectrum one resolution element at a time is the approach used with *rapid-scanning spectrometers*. The system described by Dawson, Ellis, and Millner (1968) is representative of a typical linear-scanned system. With this system, a unique feedback loop was used to track the position of the oscillating diffraction grating. The wavelength setting was determined electro-optically be means of a mirror attached to the reverse side of the grating, as shown in Figure 15.5. The mirror projected an image of a finely ruled graticule onto a slit. A photomultiplier placed directly behind the slit produced a reference pulse train as the image of the graticule moved past the slit (i.e., as the grating rotated). The pulse train was used to gate the respective analytical signals into the appropriate integrator. The scan rate obtained with the system was 1800 nm/sec.

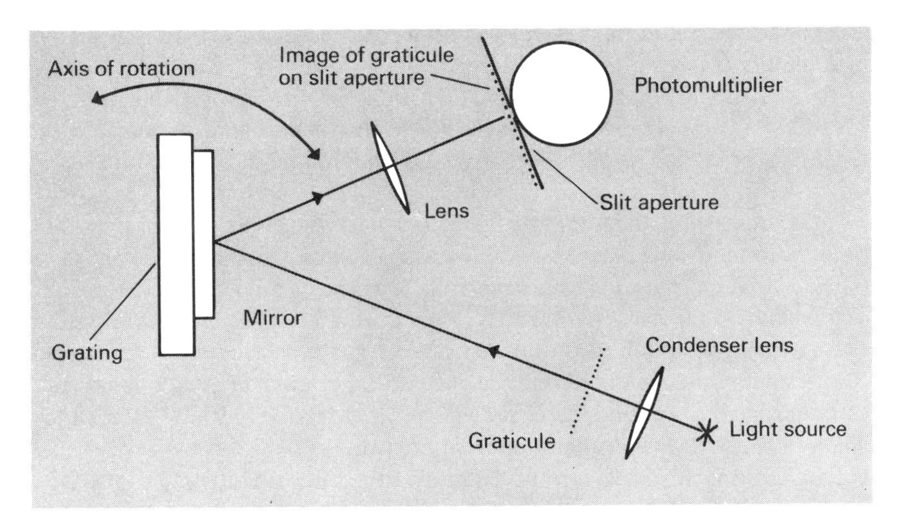

Figure 15.5. Electro-optical system for wavelength tracking for scanning monochromators. [Reprinted with permission from K. W. Busch and G. H. Morrison (1973), *Anal. Chem.*, *45*, 712A. Copyright 1973 American Chemical Society.]

Using an acetylene–air flame as the excitation source/atom reservoir, the system was used to determine Na, K, Ca, and Mg in clinical samples with a precision of 3%. Magnesium was determined by atomic absorption while the other three elements were determined by atomic emission.

While a great deal of effort has been expended in developing rapid-scanning spectrometers (see *Applied Optics*, 7(11), 1968, for a series of papers on rapid-scanning spectrometry), their role in multielement spectrochemical analysis is not expected to be significant because of their low information transfer efficiency (see following discussion). Although quite suited for the rapid acquisition of spectral information in molecular (i.e., solution) spectrophotometry (in the UV–visible region where absorption spectra consist of broad bands, i.e., low-information-content spectra), rapid-scanning spectrometers for atomic spectroscopy need short response times (i.e., wide bandwidths) to handle the large amount of information present in a complex line spectrum.

To get some idea of the amount of information possible in a complex line spectrum, consider the spectral region from 200 to 800 nm. Suppose that for purposes of atomic spectroscopy, the spectrum is divided into resolution elements of 0.01 nm. Suppose, further, that the maximum possible signal-to-noise ratio (limited by the maximum signal) that the spectrometer is capable of transmitting is 400. Using Eq. 15.2, for a message 60,000 resolution elements long with a maximum possible signal-to-noise ratio of 400, the maximum amount of information that could be present in the spectrum is 518,640 bits, or 64,830 bytes (8 bits per byte).

This calculation overestimates the amount of information that may be present in an actual spectrum because it assumes that a signal is equally likely to be present in each resolution element, which is not the case. Even for atomic line spectra, there is a certain amount of *redundancy* present in the spectrum (i.e., the presence of an aluminum line in one part of the spectrum implies that other aluminum lines may also be present at predictable locations in other parts of the spectrum). This *conditional probability* reduces the unpredictability of the message and therefore reduces the amount of information present in any actual spectrum. Nevertheless, the amount of information present in any complex line spectrum is quite large, and it does not make sense to monitor all resolution elements when only a relatively small number are actually required for any multielement analysis.

By comparison, a solution spectrum over the same wavelength range does not need to be sampled as frequently (say, every 5 nm) and contains a maximum of only 1037 bits of information if the spectrometer provides a maximum possible signal-to-noise ratio of 400. The actual information content of a solution spectrum may be much smaller than 10^3 bits because of the large redundancy in the spectrum (i.e., it changes only gradually from one resolution element to the next so it is highly predictable where the signal in the next resolution element will be). For these reasons, rapid-scanning spectrometry is quite suited for molecular spectroscopy in solution but ill-adapted to multielement spectrochemical analysis.

Although a number of reports on the application of linear-scan systems to simultaneous multielement analysis can be found in the literature (Cresser and West, 1970; Norris and West, 1971, 1972, 1973; Marshall and West, 1970; Fulton, Thompson, and West, 1970; Dawson et al., 1968; Rose et al., 1976), the primary criticism of these systems is that they are basically inefficient in terms of information transfer. Large fractions of time are spent scanning spectral regions of little interest (Johnson, Plankey, and Winefordner, 1975). In contrast, resolution elements of actual interest are sampled only for a very brief time, thereby degrading the precision of the measurement. The range of analytically useful scan speeds is bounded on the low end by excessive sample consumption and on the high end by the response time of the electronics. Repetitive scanning in combination with signal-averaging techniques can be used to improve the signal-to-noise ratio obtained with these systems (Rose et al., 1976).

Even with signal-averaging techniques, however, the important factor in evaluating the performance of linear-scan time-multiplexed systems is the total time required to collect the information on all the desured chemical elements in the sample. Since the entire spectrum is monitored, the minimum sampling time will be set by the time required for the weakest analyte signal of interest (in the presence of photon shot noise in the source) to achieve an adequate signal-to-noise ratio.

Consider a photon shot-noise-limited situation where a single-element instrument with signal-averaging capability is used to monitor a weak analytical signal. If N samples (i.e., observations) must be averaged to achieve an adequate signal-to-noise ratio, the total sampling time T required to obtain the N samples will be

$$T = N/r_s \tag{15.7}$$

where r_s is the sampling rate in samples per second that is set by the electronic bandwidth of the system. If a rapid-scanning spectrometer with the same electronic bandwidth as the single-element instrument is used to measure this same weak analytical signal, the scan time required to obtain a complete spectrum will be

$$T_{scan} = R/r_s \tag{15.8}$$

where R is the number of resolution elements present in the spectrum and r_s is the scan rate set by the sampling rate (i.e., you cannot scan any faster than you can sample). To accumulate N spectra will require

$$T_{total} = N T_{scan} \tag{15.9}$$

where T_{total} is the total time spent accumulating N spectral sweeps. The actual time spent observing the resolution element of interest, however, is only

$$T_{actual} = T_{total} f \tag{15.10}$$

where f is the fraction of a spectral sweep spent observing a given spectral element. Since $f = 1/R$, $T_{actual} = T$. In other words, with the time-multiplexed system, the same minimum observation time must be spent observing the resolution element of interest as was spent with the single-element instrument to achieve equivalent signal-to-noise performance. Compared with the single-element instrument, however, the scanning system requires R times as long (i.e., T_{total}/T).

The information transfer efficiency of a spectroscopic system can be defined as the time spent observing resolution elements of interest divided by the total time spent accumulating spectral information. The information transfer efficiency should not be confused with the resolution–luminosity product discussed in Chapter 1. For a multielement sample with n chemical elements of interest, the information transfer efficiency for a rapid-scanning spectrometer will be

$$E = 3nT/T_{\text{total}} = 3n/R \tag{15.11}$$

where E is the information transfer efficiency. The factor 3 has been included because it is necessary to measure adjacent blank resolution elements on either side of the analyte resolution element. With a linear-scan system, the information transfer efficiency for a 20-element determination extending over the wavelength range from 200 nm to 800 nm with a 0.01-nm resolution can be calculated from Eq. 15.11 to be 0.1%. Clearly, linear-scan systems are not highly efficient when it comes to measuring radiation of interest in a multielement determination. The information transfer efficiency does increase, however, with linear-scan systems as the number of chemical elements determined in the sample increases.

15.6. PROGRAMMED-SCAN DISPERSIVE SYSTEMS

Programmed-scan systems are expected to be more efficient in terms of the total observation time required to collect information on all the desired chemical elements in a sample because the fraction of the total scan time spent observing resolution elements of actual interest is higher. Naturally, the total observation time required with programmed-scan systems increases directly with the number of elements under consideration. In contrast with linear-scan systems, however, the length of time spent observing each resolution element need not be constant with programmed-scan systems. If the observation time for each resolution element can be adjusted independently, the weakest signal need not set the minimum required observation time for the stronger signals, as is true with linear-scan systems.

Because measurements with time-multiplexed spectrometers are made in a sequential manner, the time spent in *slewing* (i.e., rapid scanning) between resolution elements of interest can be used effectively for optimization of various instrumental parameters such as excitation/atomization source conditions, observation height, slit width, and photomultiplier voltage. The possibility of individual optimization for different elements (as opposed to compromise optimization; see Brost, Malloy, and Busch, 1977) is an important analytical advantage in multielement analysis. Because factors such as slit width, observation height, photomultiplier voltage, and so on, can all be adjusted under computer control from one chemical element to another, this capability relieves some of the problems associated with the determination of major, minor, and trace elements in a single sample dilution.

15.6.1. Computer-controlled Monochromators

One of the most sophisticated programmed-scan spectrometers for sequential multielement analysis by flame spectroscopy was the system developed by Malmstadt and co-workers (Malmstadt and Cordos, 1972; Cordos and Malmstadt, 1973; Spillman and Malmstadt, 1976a,b). In its final stage of development (Spillman and Malmstadt, 1976a,b), this instrument had a wavelength accuracy of 0.017 nm and could monitor any number of analytical lines between 213 and 1000 nm. By making measurements in both an atomic emission mode and an atomic fluorescence mode, a total of 33 elements could be monitored (8 by atomic fluorescence and 25 by atomic emission).

Figure 15.6 shows a block diagram of the automated multielement atomic emission–atomic fluorescence flame spectrometer developed by Malmstadt and co-workers. A bank of hollow-cathode lamps surrounding the flame and operated in an intermittent-pulsed mode was used for excitation of the atomic fluorescence. A programmable monochromator with a slew rate of 20 nm/sec and a wavelength accuracy of 0.02 nm was used as the wavelength selector subsystem. Slit width control enabled atomic fluorescence measurements to be made with wide slits (2 mm), while atomic emission measurements were

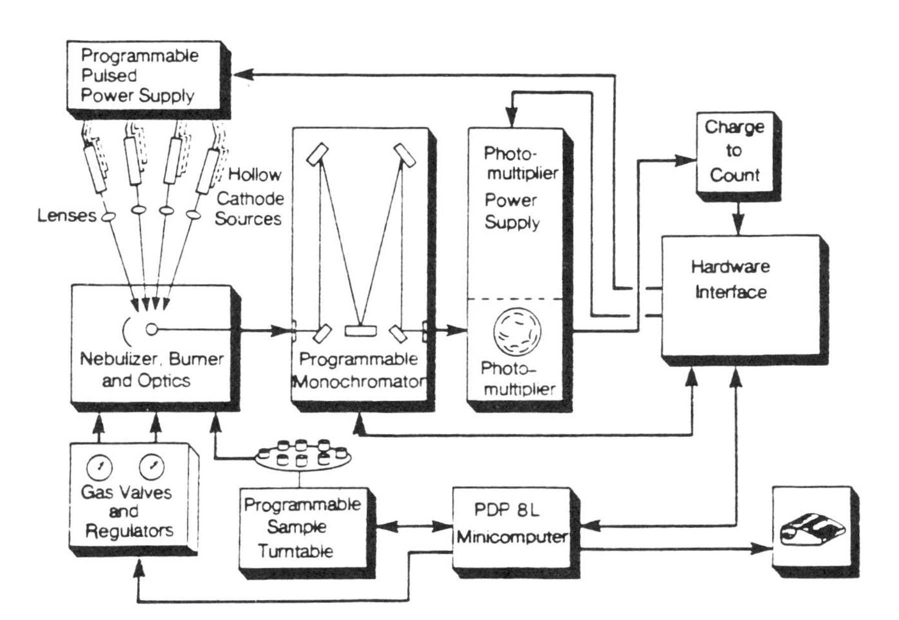

Figure 15.6. Block diagram of automated multielement AE/AF flame spectrometer used by Malmstadt and co-workers. (Reprinted from *American Laboratory*, Vol. 8, No. 3, pp. 89–97, 1976. Copyright 1976 International Scientific Communications, Inc.)

performed with narrow slits (80 μm). Wavelength accuracy was maintained using digital-shaft encoding from the monochromator.

One critical requirement of any programmed-scan spectrometer is wavelength accuracy and precision. For high-precision measurements, it is essential that the monochromator be able to locate a desired resolution element and stop reproducibly at the correct measurement wavelength. In analytical atomic spectroscopy, wavelength registration is a central concern. With most dispersive systems, wavelength selection is accomplished by moving a particular optical component in the wavelength dispersion system (i.e., either a mirror or the dispersing element itself). To design a dispersion system so that it can jump rapidly and reproducibly from one wavelength to another under the high-resolution conditions mandated by atomic spectroscopy so that selected spectral lines are imaged reproducibly on the detector calls for very high tolerances in the mechanical movement of the given optical element.

The mechanical tolerance required to position a spectral line reproducibly within the spectral bandpass of a monochromator can be eased somewhat by using unequal entrance and exit slits. When equal entrance and exit slit widths are used in an ideal monochromator, a triangular slit profile is produced, as shown in Figure 4.6, as the image of the entrance slit is scanned past the exit slit. If the dispersion drive train does not stop at the precise position where the image of the entrance slit coincides exactly with the exit slit, the relative signal that results will be less than the peak signal. This can lead to poor precision in the measurement of spectral line intensities with a programmed-scan system. If, on the other hand, the entrance slit is deliberately made smaller than the exit slit, a *trapezoidal slit function* similar to that shown in Figure 15.7 will result. If the scan drive train can be made to stop somewhere within the flat-topped region of the trapezoid, the precision of spectral line intensity measurements can be increased substantially. This approach can be used to advantage as long as resolution requirements are not too stringent.

Another approach for improving the precision of spectral line intensity measurements with a programmed-scan system is to perform *wavelength modulation.* Wavelength modulation actually accomplishes two desirable goals at once: improvement in the precision of spectral line intensity measurements and background correction. Wavelength modulation (Svoboda, 1968; Snelleman et al., 1970; Snelleman, 1968) is accomplished by placing a quartz refractor plate either after the entrance slit or before the exit slit. Figure 2.7 shows the effect of a rectangular cross section (i.e., parallel boundary) plate on the direction of an incident ray. If the plate has a rectangular cross section, the direction of the incident ray is not changed but the ray is displaced laterally by an amount that depends on the angle of incidence the ray makes with the refractor plate.

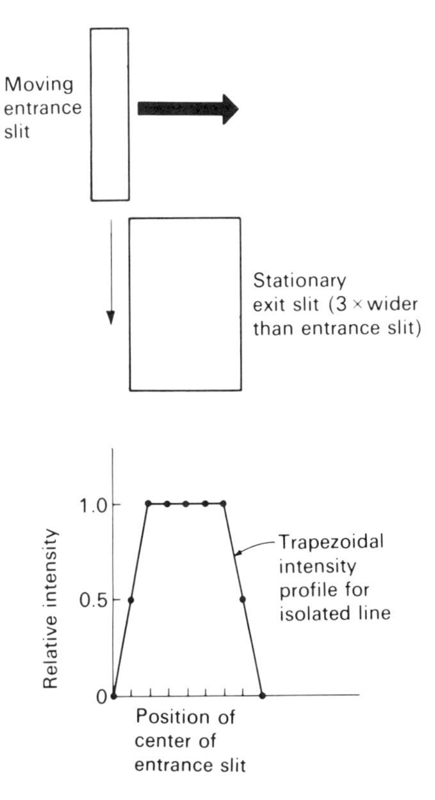

Figure 15.7. Generation of trapezoidal slit function by convolution of narrow entrance slit with wider exit slit.

If a refractor plate is mounted before the exit slit of a monochromator so that it can be rotated about a vertical axis parallel to the exit slit, the spectrum focused on the exit slit can be displaced laterally by varying the angle of rotation of the refractor plate. If the refractor plate can be made to oscillate at a low frequency (say, 10 Hz) about a rest position orthogonal to the incident beam, a spectral line focused on the exit slit can be made to move harmonically past the exit slit. If the background immediately adjacent to the spectral line is reasonably constant, this harmonic lateral shifting of the spectrum will not change the detector output signal. By contrast, the presence of the spectral line oscillating back and forth past the exit slit will create an ac detector signal that can be amplified selectively by means of an ac amplifier. Since a pulse will be produced at the detector by the presence of the spectral line for every zero crossing of the refractor plate modulating signal, the spectral line will be modulated at twice the refractor plate modulating signal.

Since a constant background intensity adjacent to the spectral line does not produce a time-varying signal, it is not amplified by the ac amplifier. As a result, the net effect of wavelength modulation is to subtract the adjacent background from the spectral line intensity. Furthermore, since the presence of the refractor plate causes the spectral line to oscillate over a small distance in the focal plane, wavelength registration is less critical because the line position is not fixed but oscillates over a small range that can be more easily located with mechanical scanning.

Using wavelength modulation, Spillman and Malmstadt (1976a,b) were able to improve the wavelength registration of the original instrument from 0.02 to 0.017 nm. By using square-wave modulation of the refractor plate, time averaging of the data was possible.

Winefordner and co-workers (Johnson et al., 1975; Ullman et al., 1979) have studied a programmed-scan dispersive system for atomic fluorescence that is similar to the Malmstadt system described. Instead of using a bank of hollow-cathode lamps as in the Malmstadt system, the Winefordner system used a high-intensity xenon arc continuum source to excite the atomic fluorescence from all the elements present in a flame. In the original system by Johnson et al. (1975), a slew-scan monochromator was used to skip from one analytical wavelength to another. An optical chopper was employed in conjunction with a synchronous photon counter to detect the signal.

An improved version of the instrument has been described by Ullman et al. (1979). In this configuration of the instrument, the optical chopper was replaced with a wavelength modulation system using a quartz refractor plate mounted behind the entrance slit of the monochromator and driven by a torque motor. By driving the torque motor with a stepped square wave, the wavelength monitored by the dispersion system was shifted to wavelengths above and below the analytical wavelength. The use of wavelength modulation in this system not only permitted continuous background correction as described in the preceding discussion but also corrected for broadband Mie scattering from undesolvated salt particles irradiated by the continuum source. Since the scattered radiation produced by salt particles irradiated by a continuum source is broadband and does not vary greatly in intensity over the optical bandwidth scanned by the refractor plate, its presence results in an essentially constant (dc) detector output similar to that produced by the background. As a result, scattering as well as background can be rejected selectively by the system.

Finally, since both the fluorescence radiation produced by the continuum source and the thermal emission from the flame were modulated by the refractor plate, the instrument responded to the *sum* of the atomic emission intensity and the atomic fluorescence intensity. As mentioned previously (Alkemade, 1968; Winefordner and Elser, 1971), for elements with resonance lines below 350 nm, the composite signal will be primarily from atomic fluorescence, whereas for elements with resonance lines above 500 nm the composite signal will be due primarily to atomic emission. For elements like strontium with a resonance line at 460.7 nm, monitoring the sum of the atomic emission signal plus the atomic fluorescence signal was observed to provide an enhancement in the signal-to-noise ratio compared with similar measurements made by either atomic emission alone or atomic fluorescence alone. By combining measurement modes in this fashion, it was not necessary to shift from the atomic emission mode to the atomic fluorescence mode during a scan.

The analytical potential of the Winefordner instrument was evaluated for wear metals analysis involving jet engine oil samples under the Spectrographic Oil Analysis Program of the U.S. Air Force. Standard reference materials obtained from the National Bureau of Standards were also analyzed to evaluate the instrument's capability.

A similar programmed-scan atomic fluorescence system using a separated flame was developed and described by Brinkman, Whisman, and Goetzinger (1979). Compared with the Malmstadt and Winefordner systems described previously, this system was less sophisticated in design. Although using a high-intensity xenon arc similar to that used by Winefordner, the Brinkman system did not use any form of modulation and used a dc detection system. As a result, even though the Brinkman system could monitor the sum of the atomic emission signal and the atomic fluorescence signal, it was not able to discriminate against flame background or source scatter the way the Winefordner instrument could. To avoid the effects of drift (i.e., $1/f$ noise) inherent in the transmission channel of any dc instrument, a blank, the sample, and two spiked samples had to be measured on each element before proceeding to the next element. The instrument was evaluated for the analysis of oil samples using standard addition as a means of sample calibration.

15.6.2. Multislit Systems

To avoid the wavelength registration problems associated with computer-controlled monochromators, a number of dispersive multislit systems with stationary dispersion optics have been developed.

One system that falls in this category and makes use of atomic absorption as the mode of observation is the instrument described by Lundberg and Johansson (1976). This system used a 0.35-m Czerny–Turner polychromator with three exit slits placed in the focal plane as the wavelength selector subsystem. A rotating chopper with three concentric slits was arranged so that light from only one fixed exit slit at a time reached the single photomultiplier tube. By arranging the position of the three exit slits properly so as to pass desired resonance radiation from a hollow cathode, rapid sequential determinations by atomic absorption were possible. With this configuration, it was not possible to access resonance lines whose wavelength separation was less than 2 nm because of the nature of the optical arrangement employed. Although the system was developed with three exit slits, it was anticipated that it could be extended to a maximum of five elements (i.e., five exit slits). Because of the practical upper limit of five elements and the limited ability to monitor various combinations of elements, this instrument appears most suited for specialized routine determinations.

Another system along the same line is the instrument described by Johansson and Nilsson (1976). This instrument used a composite grating assembled from five smaller gratings with different ruling characteristics and placed one above the other, as shown in Figure 15.8*a*. Each component grating making up the composite grating was selected to diffract a specific wavelength of radiation at a designated angle of incidence with respect to the grating. Since the angle of incidence was the same for all the component gratings that made up the composite grating, the entrance slit image produced at the exit focal plane was a line composed of five segments of differing wavelengths, as shown in Figure 15.8*c*. At one particular angle of diffraction, the entrance slit image produced in the exit focal plane consisted of the five desired analytical wavelengths arranged one above the other as segments of a line.

A slit-shaped bundle of fiber optics composed of five individual fiber-optic light guides was positioned in the exit focal plane to collect the analytical radiation from the five segments (i.e., the five analytical wavelengths). Each analytical wavelength (i.e., each segment of the slit image) was transmitted by one of the individual fiber-optic light guides. By arranging the distal ends of the five individual fiber-optic light guides around the periphery of a circle, a rotating disk with holes could be used to transmit radiation from one fiber-optic light guide at a time to a single photomultiplier tube. Because of the two-dimensional nature of the dispersion obtained with this system, it is similar to

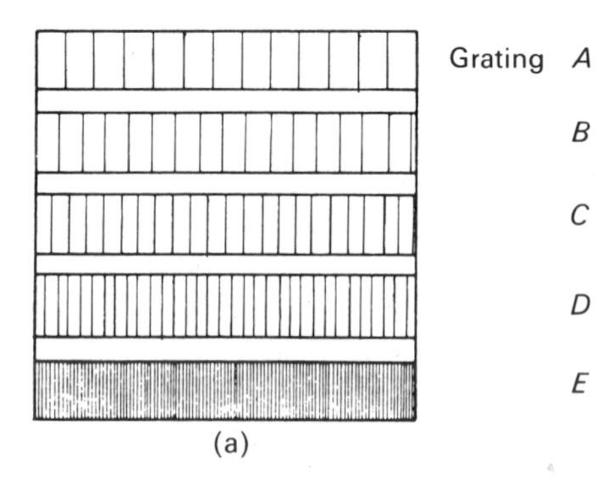

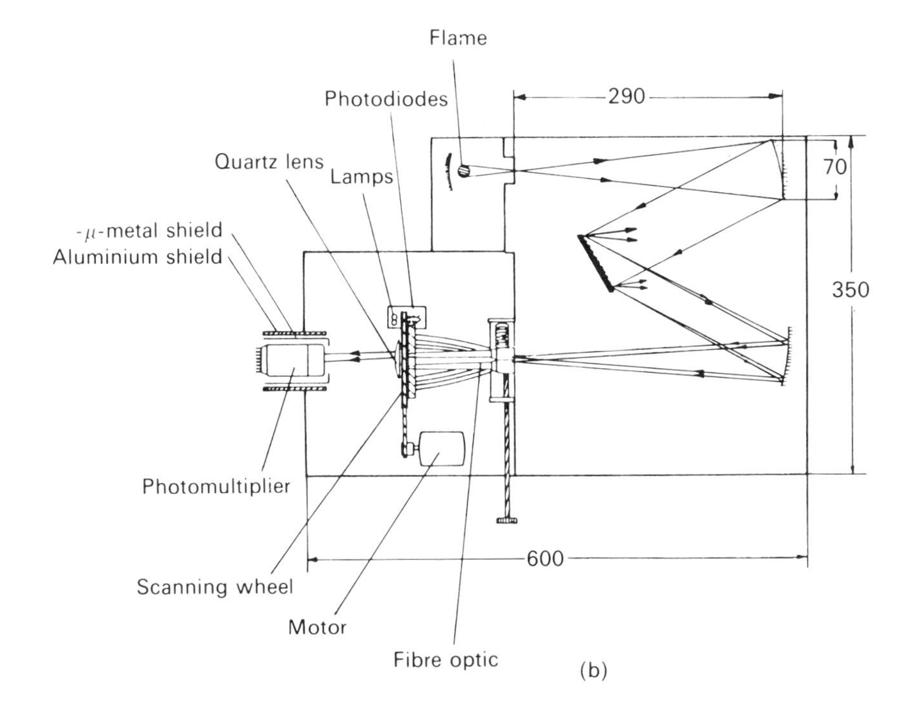

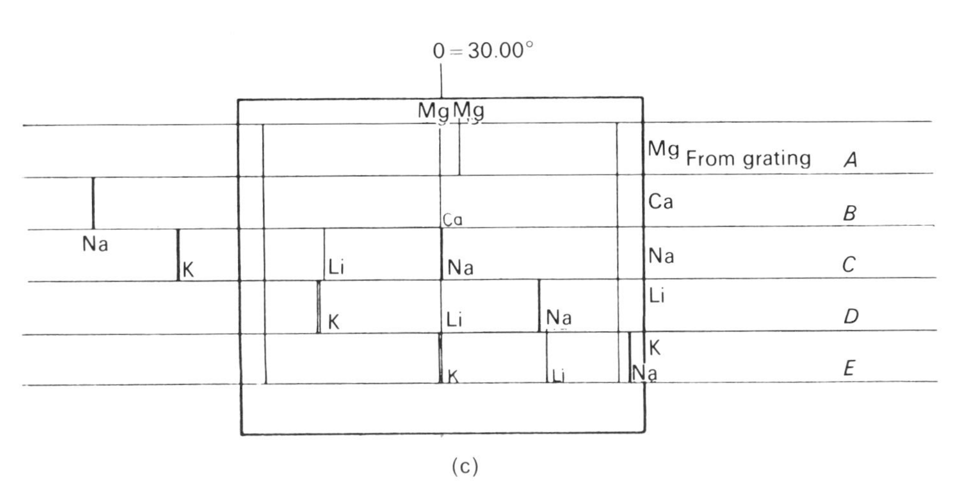

Figure 15.8. Instrument described by Johansson and Nilsson: (*a*) schematic diagram of composite grating; (*b*) schematic diagram of spectrometer; (*c*) spectra obtained with composite grating. [Reprinted with permission from A. Johansson and L. E. Nilsson (1976), "A Multiple Grating Flame Photometer for the Simultaneous Determination of Five Elements", *Spectrochimica Acta, 31B*. Copyright 1976 Pergamon Press, Elmsford, NY.]

the echelle spectrograph, although it does not produce the same high resolving power as obtained with an actual echelle grating. The instrument was designed for the routine determination of Na, K, Ca, and Mg in blood serum using Li as an internal standard. Because gratings with a high groove density were required to produce the composite grating, conventionally ruled gratings were not adequate, and holographic gratings were employed.

Another example of a time-multiplexed, multiple-exit-slit dispersive spectrometer is the system shown in Figure 15.9, developed by Salin and Ingle (1978a,b, 1979). This system has been used in the atomic absorption mode (Salin and Ingle, 1978a,b) as well as the atomic fluorescence mode (Salin and Ingle, 1979). In the atomic fluorescence configuration, a bank of four hollow-cathode lamps, which surrounded the hydrogen–air flame, was turned on and off sequentially by means of logic signals from a crystal-clock control circuit connected to the lamp power supply.

A multiple-exit-slit mask with four exit slits was placed at the exit focal plane of the Czerny–Turner dispersion system. The exit slit array was made from a quartz microscope slide by a photoreduction process of a 20-times-enlarged mask. The microscope slide was coated with aluminum by vacuum deposition prior to coating with a positive photoresist.

A *photon funnel* composed of two rectangular and two trapezoidal front-surface mirrors, shown in Figure 15.9, was used to direct the radiation passing through any of the four exit slits in the exit slit mask to a photomultiplier tube.

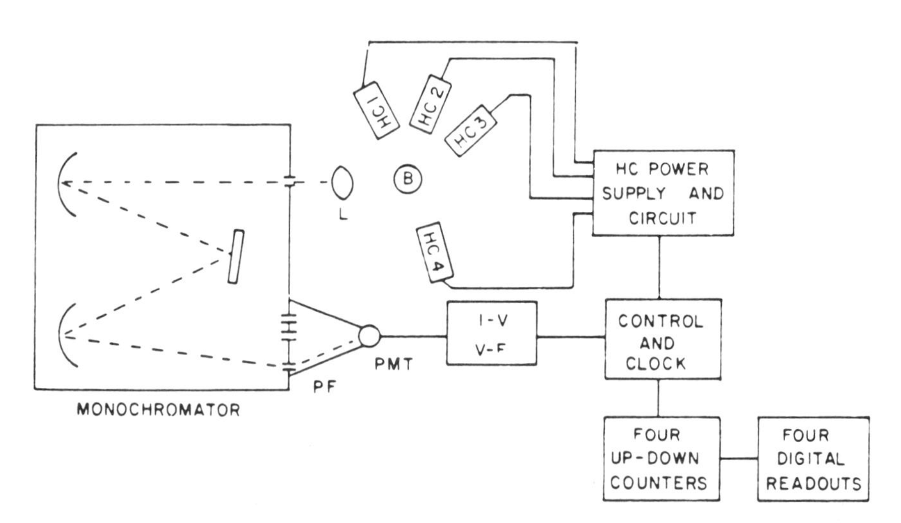

Figure 15.9. Block diagram of instrument developed by Salin and Ingle. Symbols: PMT, photomultiplier tube; *I–V*, current-to-voltage converter; *V–F*, voltage-to-frequency converter; HC, hollow-cathode tube; *L*, focusing lens; *B*, burner; PF, photon funnel. Lenses not shown. [Reprinted from E. Salin and J. D. Ingle, Jr. (1979), *Anal. Chim. Acta*, *104*, 267.]

Signals from the photomultiplier were converted from current to voltage and from voltage to frequency before being fed to four up–down counters and finally to four digital readouts. The crystal-clock control circuit that controlled the lamp sequence was also used to direct the voltage-to-frequency pulses from the signal-processing system to the appropriate up–down counter.

Another completely different multislit approach based on the principle of reverse optics and pioneered by Busch and co-workers (Busch and Benton, 1981, 1985; Benton, 1985; Busch, Busch and Benton, 1989a,b) employed a *multiple-entrance-slit spectrometer*. This instrument used a conventional 0.5-m Czerny–Turner dispersion system with a single exit slit and photomultiplier detector. A one-dimensional array of entrance slits was placed in the entrance focal plane of the spectrometer. Each entrance slit was located along the entrance focal plane in accordance with the grating equation so as to transmit an analytical line of interest through the exit slit to the photomultiplier. In operation, only one entrance slit was illuminated with source radiation at any given time by means of a specially designed predispersion optical system known as an *optical multiplexer*, or *optiplexer*. The complete system consisted of four subsystems as shown in Figure 15.10: (1) the source, (2) the predispersion optics, (3) a multiple-entrance-slit spectrometer, and (4) the data collection subsystem.

Using this approach, undispersed radiation from the excitation source (in this case, an acetylene–air flame supported on a Varian AA-6 burner/

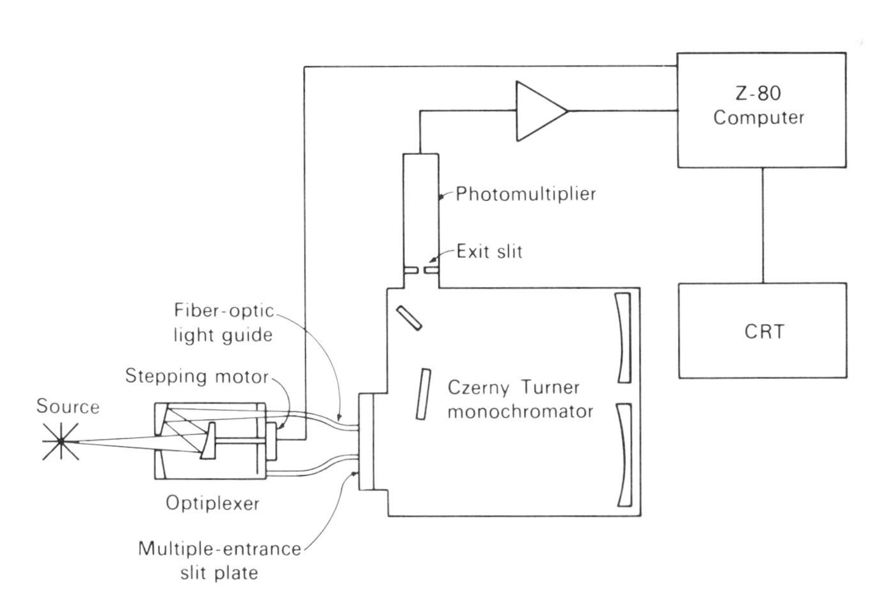

Figure 15.10. Block diagram of multiple-entrance-slit spectrometer developed by Busch and Benton.

nebulizer with a Meker burner head) was directed, by means of the predispersion mirror system (i.e., the optical multiplexer, see Figure 15.11), around an array of fiber-optic light guides arranged in a circle. Each individual fiber-optic light guide conveyed light from the source to a selected entrance slit, appropriately located in the entrance focal plane, so that light of a desired analytical wavelength was transmitted by the single exit slit to the photomultiplier. Because the entire dispersion system was stationary (i.e., only components of the predispersion optical system moved), wavelength registration problems associated with moving dispersive optical components were avoided.

Using the optical multiplexer, undispersed radiation from the source was made to illuminate the end of one particular fiber-optic light guide and stop for a given sampling time interval before moving on to the next fiber-optic light guide. During this sampling time interval, the photomultiplier voltage could be adjusted automatically under computer control as desired depending on the intensity of the spectral line being sampled. In addition, each entrance slit was independently adjustable, further extending the range of intensities that could be monitored sequentially. Since time division multiplexing was accomplished by directing the undispersed source radiation from one stationary entrance slit to another (i.e., one fiber-optic light guide to

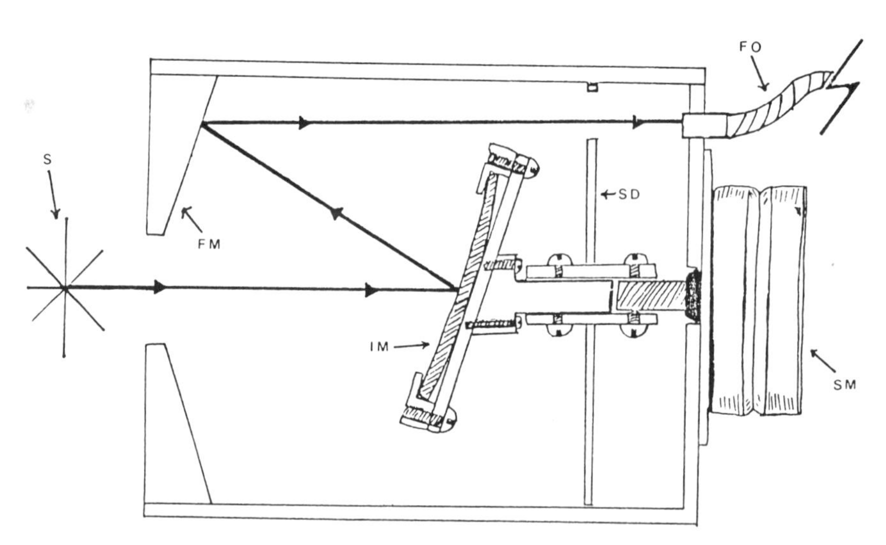

Figure 15.11. Schematic diagram of predispersion mirror system or optical multiplexer used in multiple-entrance-slit spectrometer. Symbols: *S*, source; IM, imaging mirror; FM, folding mirror; FO, fiber-optic light guide; SM, stepping motor; SD, shutter disk. [Reprinted from L. D. Benton (1985), M. S. Thesis, Baylor University.]

another), the time required to go between different wavelengths of interest was independent of the wavelength separation. The system was highly efficient in terms of information transfer because only wavelengths actually of interest were monitored.

Eight noncoherent fiber-optic light guides were arranged at 45° intervals around the periphery of the back plate of the optiplexer and plugged into the multiple entrance slit plate of the spectrometer. The multiple-entrance-slit plate was configured for Na (589.6 nm), Li (670.7 nm), Ca (422.7 nm), CaOH (547.5 nm), Ba (553.5 nm), Cs (455.5 nm), and Sr (460.7 nm). The remaining fiber-optic light guide monitored the MgO band at 507.4 nm.

Computer control of the system was implemented with a Z-80-based microcomputer running at 4 MHz. The analog signal from the photomultiplier was amplified using an RCA SK9173 operational amplifier prior to analog-to-digital conversion. The rotating mirror in the optical multiplexer was driven by a computer-controlled stepping motor. Home position for the rotating mirror was determined by means of a microswitch coupled to the shaft of the stepper motor. Mirror position for other settings of the optiplexer was determined from a knowledge of the number of steps from home position.

Although advanced only to the prototype stage by Busch and Benton, the multiple-entrance-slit spectrometer could be developed into a viable multielement spectroscopic system. With the proper multiple-entrance-slit plate, various combinations of elements could be determined as desired by plugging the fiber-optic light guides into the appropriate locations. Changing to a different combination of elements would simply require moving several fiber-optics from one location to another, similar to an old-fashioned telephone switchboard system. The number of fiber-optic light guides could easily be extended to 12 or higher. Although not implemented in the original instrument, background correction adjacent to a spectral line could be accomplished by wavelength modulation as described previously.

15.7. ELECTRONICALLY SCANNED TIME-MULTIPLEXED SYSTEMS

One way to avoid wavelength registration problems associated with computer-controlled monochromators is to use an electronically scanned detector such as the image dissector (see Chapter 10). Although the image dissector is a form of image device, it differs from other image detectors because it does not have a storage target. As a result, it is not technically a multichannel detector but is, in fact, a scanning photomultiplier. For this reason, it falls in the time-multiplexed category rather than the multichannel category along with the other image detectors.

Partial integration capability has been achieved with the so-called *smoothing dissector* consisting of an image dissector with a phosphor screen (see Talmi, 1975a,b). Although the finite decay time of the phosphor acts like a storage target, the phosphor tends to reduce the signal-to-noise performance of the tube as well as the dynamic range.

Image dissectors have been used with both one- and two-dimensional dispersive systems. For example, Harber and Sonnek (1966) described a simple electronic scanning spectrometer consisting of an image dissector mounted in the focal plane of a 12.7-cm Czerny–Turner dispersion system with a reciprocal linear dispersion of 13 nm/mm. With this arrangement, a spectral range of 250 nm could be scanned electronically at a scan rate of either 100 or 1000 scans per second.

The main disadvantage of the one-dimensional dispersion system for multielement analysis is the compromise required between wavelength coverage and resolution. Because the electronically scanned region is limited spatially by the dimensions of the tube itself, the image dissector mounted in the focal plane of a one-dimensional dispersion system monitors a *spectral window* whose extent depends on the reciprocal linear dispersion of the wavelength selector subsystem. To increase the wavelength coverage included in the spectral window, the dispersion of the optical system must be reduced. Reducing the dispersion adversely affects the performance of the system in terms of multielement analysis because it also reduces the resolution. To be most useful for multielement analysis, a detection system must be able to cover as wide a spectral window as possible at the highest possible resolution so that spectral lines of as many elements as possible can be monitored under conditions where spectral interferences are minimal. Since with a one-dimensional dispersion system, it is not possible to achieve wide wavelength coverage and high resolution simultaneously, a compromise is required.

The compromise between wavelength coverage and resolution can be avoided if a two-dimensional dispersion system such as the echelle dispersion system (see Chapter 4; Keliher and Wohlers, 1976) is used. This approach with the image dissector has been studied by a number of workers (Felkel and Pardue, 1978; Danielsson and Lindblom, 1972, 1976; Danielsson, Lindblom, and Söderman, 1974; Felkel and Pardue, 1979).

The system described by Danielsson and co-workers (1972, 1974, 1976) provides wide wavelength coverage at high resolution. This system covered the whole spectral range from 200 to 800 nm in 91 segments (i.e., orders) with a resolution of 0.001 nm. A stigmatic, coma-compensated echelle dispersion system was used to focus the spectral information on the image dissector. The optics of the dispersion system were specially designed to eliminate astigmatism at the long wavelengths, where the orders are closer together physically due to the nonuniform cross dispersion produced by the prism (see Chapter 4).

Astigmatism results in undesirable image elongation, which can result in partial overlap of lines from adjacent orders as they become closer together. The effect of astigmatism in the system described by Danielsson and co-workers was compounded by the slit-shaped aperture in the aperture plate of the image dissector tube.

Since scanning takes place electronically, accurate and precise control of the scanning electron beam by computer is necessary for reliable wavelength calibration. For computer control of the sweep circuitry, a high-resolution digital-to-analog converter is necessary. Much of the development work involved in designing an image dissector spectrometer has been in the area of software development for wavelength calibration. Since the spectrum obtained with an echelle dispersion system consists of a two-dimensional array of spectral "points" rather than spectral lines (see Figure 15.12), small errors in beam alignment can result in serious analytical errors. Since the image dissector is a real-time scanning device with no internal storage capability, computerized data acquisition is required for data storage. These factors preclude the use of a microcomputer as the main processor for an image dissector spectrometer, and most systems developed to date have employed minicomputers for data acquisition and control.

The dynamic range and sensitivity of the image dissector is comparable to a conventional photomultiplier with a similar photocathode, aperture, and dynode chain. For example, the image dissector used in the spectrometer by Felkel and Pardue (1979) had an S-20 photocathode, an electron multiplier with a gain of 10^6, and a linear dynamic range of 10^5–10^6. Compared with image detectors having charge storage targets that need to be erased after each scan, the image dissector is free from lag and blooming (see Chapter 10). Although not subject to blooming like other charge storage tubes, the image dissector does respond to back-scattered flux from internal tube parts. This effect is most serious when low light levels are being measured in the presence of bright areas on the photocathode and can reduce the dynamic range to two to three orders of magnitude.

In their system, Felkel and Pardue (1979) used a 0.75-m Spectraspan echelle grating spectrometer from Spectrametrics of Andover, Massachusetts, equipped with a 79-groove/mm echelle grating blazed at 63°26′. Auxiliary optics in the form of a Cassegrainian mirror system were needed to reduce the size of the focal plane produced by the standard instrument from 50×75 mm to a size that could be accommodated by the image dissector (13×20 mm). Image reduction had the side benefit of reducing the effective *f*-number of the spectrometer from *f*/10 to *f*/2.8, thereby increasing the flux per unit area striking the image dissector by a factor of about 13.

The image dissector camera system employed by Felkel and Pardue (1978, 1979) was a Model 658A system from EMR Photoelectric of Princeton, New

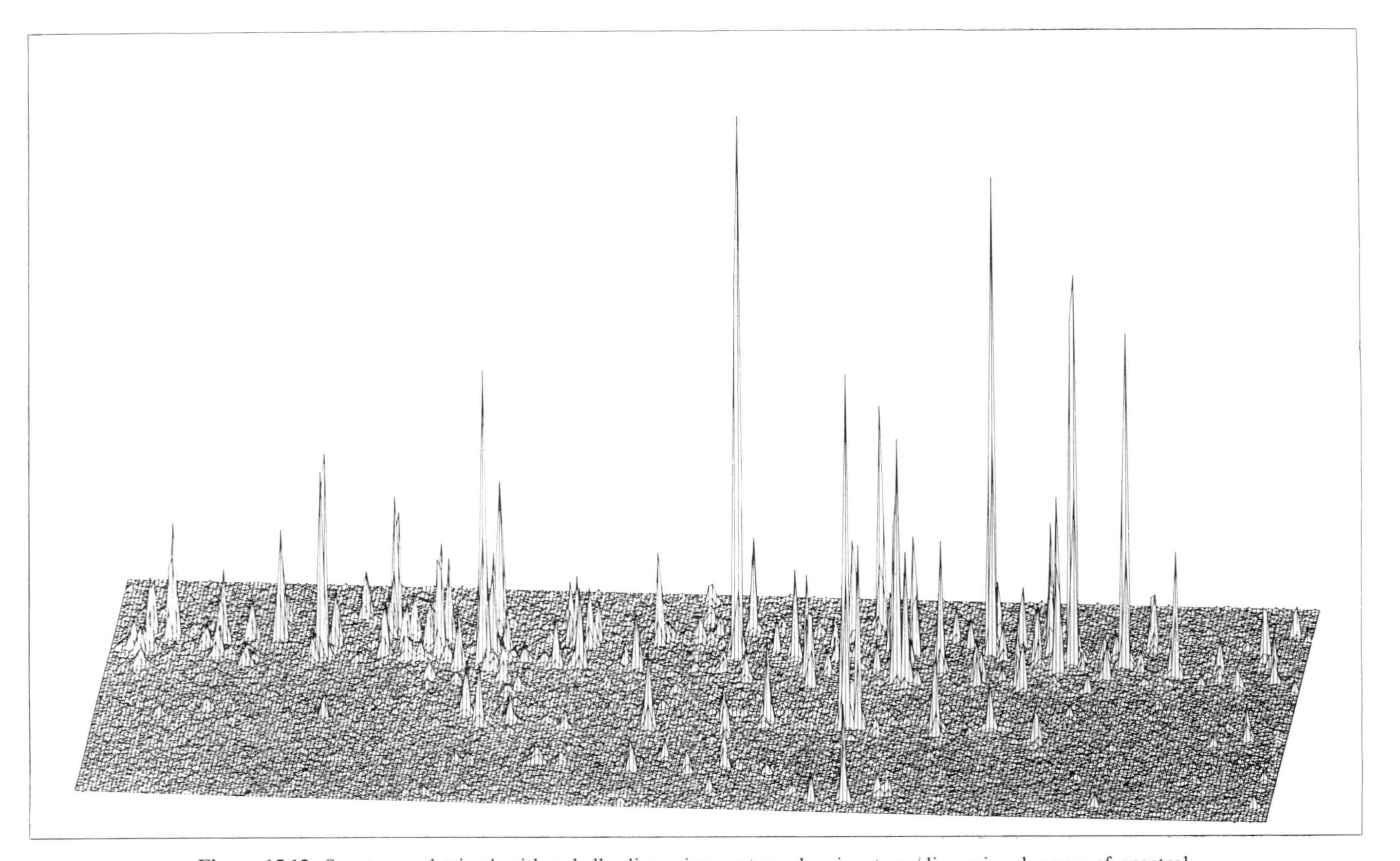

Figure 15.12. Spectrum obtained with echelle dispersion system showing two-/dimensional array of spectral information. (Courtesy of M. B. Denton, R. B. Bilhorn, P. M. Epperson, R. D. Jalkian, R. S. Pomeroy, G. R. Sims, and J. V. Sweedler, Department of Chemistry, University of Arizona.)

Jersey. The EMR 575E image dissector tube employed in this camera system had a sapphire window with a 43-mm-diameter photocathode. The aperture plate, on which the photoelectron image was focused magnetically, had a circular aperture 38 μm in diameter.

In their evaluation of the spectrometer, Felkel and Pardue (1978) compared the performance of the system in emission with a dc plasma using both an image dissector and a silicon vidicon. Their results showed that for Li, Na, K, Mg, Ca, Sr, and Ba the detection limits obtained with the image dissector were lower than those obtained with a silicon vidicon by a factor of 24.

Gustavsson and Ingman (1979) have described an image dissector echelle spectrometer that was used in the atomic fluorescence mode. The system, which was controlled by a laboratory computer, had a linear dynamic range of 2.5–4 orders of magnitude and used an ordinary xenon arc continuum source without an integral reflector as an excitation source. A flame atom reservoir was used, and the results obtained for eight elements were reported.

Although the image dissector is one of the most sensitive image tubes available today (because of the high gain of the electron multiplier), its widespread adoption as a detector will be limited ultimately by its high cost. Not only is the tube itself expensive, but the ancillary power supply and data acquisition/control electronics are sophisticated and, therefore, relatively costly.

15.8. NONDISPERSIVE MULTICHANNEL ATOMIC ABSORPTION

All of the systems discussed so far in this chapter have belonged to the single-detector, time-multiplexed category. At this point, we leave this domain and begin consideration of various examples of multiple-detector systems. Compared with single-detector systems that require sequential data acquisition, multiple-detector systems have the advantage of simultaneous data acquisition.

One such multiple-detector system for simultaneous multielement atomic absorption that avoids the problems associated with combining separate radiation beams into a single beam prior to passage through the flame (see Chapter 13) is the *resonance monochromator* approach described by Sullivan and Walsh (1968). The resonance monochromator shown in Figure 15.13 consists of a sealed tube containing an atomic vapor of the element of interest produced by either cathodic sputtering (as in a hollow-cathode lamp) or indirect electrical heating of the metal. When the resonance monochromator is irradiated by resonance radiation from the element of interest, the atomic vapor within the resonance monochromator absorbs some of the energy from the beam. The absorbed energy is subsequently reemitted as fluorescence radiation that is detected at right angles to the incident beam by means of a photomultiplier.

By arranging a series of hollow-cathode lamp–resonance detector combinations around a flame, simultaneous multielement determinations by atomic absorption are possible. Emission from each hollow-cathode lamp passes through the flame in the conventional fashion, and absorption by the sample takes place. Each resonance monochromator, which is located 180° around the flame from its corresponding hollow-cathode lamp, detects the decrease in intensity of its particular resonance line(s) transmitted through the flame.

The resolution of the resonance monochromator is quite high because the emission profile from the hollow-cathode lamp must overlap the absorption profile of the atoms in the resonance monochromator in order for a fluorescence signal to be produced. To date, the approach has been limited by variable performance and the limited life of the resonance detectors themselves. For many elements, it will be difficult to produce the required stable atomic vapor within the resonance monochromator and, for these elements, the maximum sensitivity is expected to be considerably less than with conventional atomic absorption. Nevertheless, the idea is a clever one and deserves further development because of its potential for providing a low-cost system for multielement atomic absorption.

15.9. DIRECT-READING SPECTROMETERS

It was not long after the development of multiplier phototubes in the 1940s that spectroscopists began to investigate the role that these new detectors

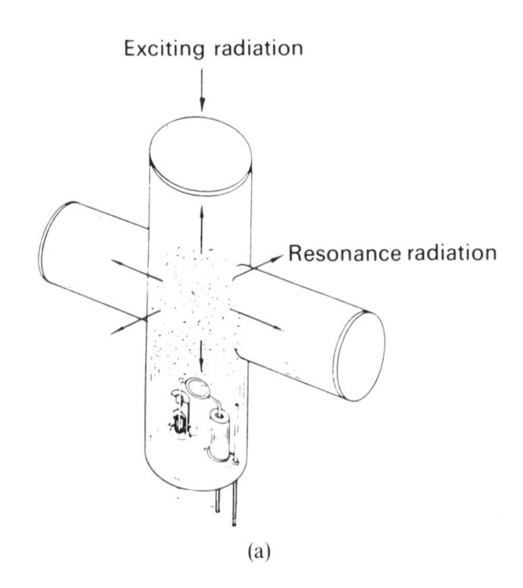

(a)

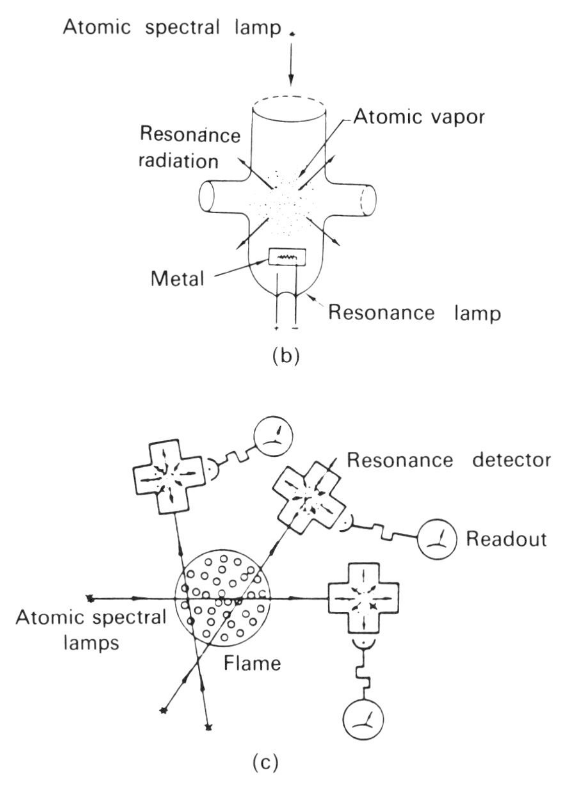

Figure 15.13. Resonance monochromator. (*a*) Schematic diagram illustrating resonance detector in which metal atomic vapor is produced by cathodic sputtering. (*b*) Schematic diagram illustrating resonance detector in which metal atomic vapor is produced by electrical heating. (*c*) Schematic diagram of multichannel atomic absorption spectrophotometer permitting variation of absorption path length through flame. [Reprinted from J. V. Sullivan and A. Walsh (1968), *Appl. Opt.*, *7*, 1271, by permission of the Optical Society of America.]

might play in spectrochemical analysis (Nahstoll and Bryan, 1945; Boettner and Brewington, 1944; Dieke and Crosswhite, 1945). In 1944, Hasler and Dietert were the first to describe a direct-reading instrument for simultaneous multielement spectrochemical analysis. In essence, a direct-reading spectrometer is similar to a spectrograph except that the photographic plate is replaced by a series of detector modules each consisting of an exit slit and a photomultiplier. By placing individual modules at various locations along the focal plane of the instrument, analytical lines of elements of interest can be monitored simultaneously.

Table 15.2 compares multielement photographic detection with the time required using a direct-reading approach. From the table, it was apparent to Hasler and Dietert (1944) that the direct-reading approach was about 4 times

Table 15.2. Time Requirements in Spectrochemical Analysis

Photographic Detection		Direct-Reading System	
Steps	Time[a] (min)	Steps	Time (min)
Preparation of sample	1.5	Preparation of sample	1.5
Discharge to sample and photography	0.75	Discharge to sample	0.75
Develop, wash, and dry plate	5	Read instrument for each element and record percentage	0.05n
Place plate in densitometer, read % T for internal standard line and each element	0.7 + 0.2n		
Convert % T values to percentage of composition	0.2n		

Total Time Required (min)

Number of elements	1	2	3	4	5	6	7	8	9	10
Photographic detection	8.3	8.7	9.1	9.5	9.9	10.3	10.7	11.1	11.5	11.9
Direct reader	2.3	2.35	2.4	2.45	2.5	2.55	2.6	2.65	2.7	2.7

[a] Number of elements being determined, n. From Hasler and Dietert (1944).

as rapid as photographic detection even if sample preparation time was included. Such time saving is important for "speed analysis" often associated with metallurgical applications of spectrochemical analysis. These analyses are used to answer the question arising from the foundry floor of whether to pour or not to pour a crucible of alloy. Under these circumstances, every minute saved is a minute less of furnace time required. By cutting a multielement determination from about 10 min to about 2.5 min, real savings in energy costs (i.e., fuel) can be realized by the foundry.

Using the arrangement described by Hasler and Dietert, detectors were placed on rails above and below the spectrum and each individual spectral line intensity, which was transmitted through a carefully placed exit slit associated with each detector, was relayed to that detector by means of a small mirror. With this layout, a total of 12 detectors could be mounted in the focal plane of the spectrograph at any wavelength between 200 and 600 nm. To reduce alignment problems associated with temperature changes in the spectrometer, exit slits were made three times wider than the primary entrance slit width to give a trapezoidal slit function that could accommodate a certain degree of entrance–exit slit mismatch before affecting the detector reading.

Today, direct-reading spectrometers are quite common in analytical laboratories that perform multielement spectrochemical analyses. In addition to their speed and the large volume of analyses these instruments can handle, direct readers provide a precision of 1–2%, which is not possible with routine photographic work (5%) (see Slavin, 1971). For routine samples involving the same combination of elements, the performance of direct-reading spectrometers is hard to beat. For this reason, most ICP spectrometers sold today use some form of direct-reading dispersion system. They are generally purchased for a particular analysis (i.e., sample type). In the area of metallurgical applications, they are programmed by the manufacturer for the elements to be determined and the anticipated sample matrix (i.e., the particular alloy). Once adjusted by the manufacturer, the instrument is not generally altered by the user because of the tolerances involved in spectral alignment.

The primary problems associated with the direct reader are (1) a lack of flexibility to monitor different combinations of elements; (2) large size owing to the high dispersion needed to allow detector placement; (3) sensitivity to temperature and other effects that result in spectral alignment problems; and (4) difficulties in monitoring adjacent spectral line background.

To overcome the problems associated with items 3 and 4, Fassel and coworkers (Golightly, Kniseley, and Fassel, 1970) have studied the use of image dissector tubes as detectors in place of ordinary photomultipliers for direct-reading spectrometers. By placing the image dissector in the focal plane of the direct-reading spectrometer, the need for an exit slit was eliminated. Because of the dimensions of the tube, thermal drift was avoided because the spectral line of interest always remained on the active faceplate of the tube. In addition, because of the scanning capability of the image dissector, spectral regions on either side of the analytical line could be monitored for background correction. Unfortunately, while this approach has certain desirable features as far as detecting a single chemical element is concerned, the cost of image dissector tubes today is such that it is not economically feasible to employ a bank of these detectors in a direct reader to achieve multielement capability.

To permit background correction, most direct readers employ a refractor plate placed directly after the entrance slit of the spectrometer. By rotating the refractor plate, the entire spectrum is displaced back and forth across each individual exit slit as described previously with wavelength modulation. Another method of spectral displacement that is sometimes used involves physically moving the entrance slit itself back and forth. Since moving the entrance slit is somewhat more difficult to accomplish, the refractor plate approach is more common.

To avoid some of the problems associated with direct readers, Applied Research Laboratories of Sunland, California, have developed an instrument that is a cross between a conventional direct reader and a time-multiplexed programmed spectrometer. The instrument, known as a Model 3520 fixed-grating sequential ICP spectrometer, has been described in detail by Routh and Paul (1985).

The spectrometer uses a 1.0-m Paschen–Runge mounting (see Chapter 4) with the grating maintained in a fixed position. An array of 255 exit slits equally spaced on 2-mm centers is arranged around the focal plane of the spectrometer, as shown in Figure 15.14. Any wavelength from 165 to 800 nm can be accessed by one of two photomultiplier tubes mounted on movable carriages. One photomultiplier tube is red sensitive and is used for long-wavelength lines, while the other is blue sensitive and is used to monitor short-wavelength lines.

With this arrangement the appropriate photomultiplier tube can be located on any desired spectral line by two independent movements. The first movement, which can be relatively coarse, involves locating the appropriate photomultiplier tube within 2 mm of any desired spectral line. Since the photocathode is relatively large compared with the exit slit, only moderate precision is required of the movable carriages, and high speed with little wear is possible. The remaining fine adjustment is done by moving the entrance slit ± 1 mm. Since this fine adjust takes place only over very short distances, it can be done easily with little wear and high speed to the requisite precision requirements. Since both movements are performed simultaneously, any accessible wavelength can be accessed within 1.5 sec. Using this system, Routh

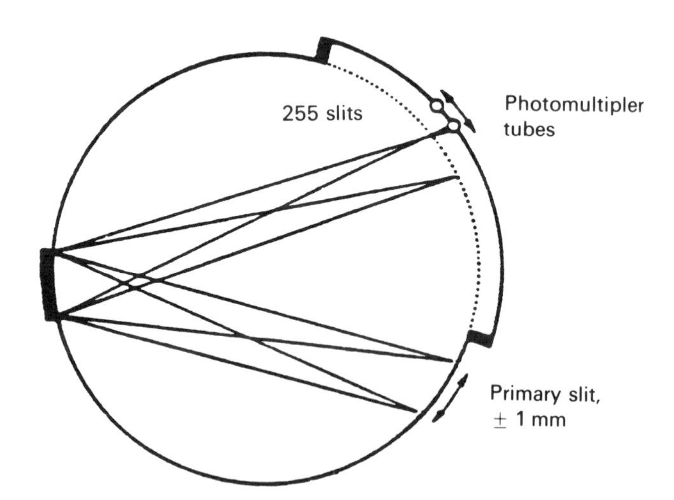

Figure 15.14. Paschen–Runge mounting used in ARL Model 3520 spectrometer. (Reprinted from *American Laboratory*, Vol. 17, No. 6, pp. 84–97, 1985. Copyright 1985 International Scientific Communications, Inc.)

and Paul (1985) report a wavelength reproducibility of 0.0005 nm over the wavelength range from 170 to 400 nm when using the second order.

Nygaard and Sotera (1988) have compared the performance of sequential and simultaneous plasma emission spectrometers using the Plasma 300 ICP sequential spectrometer (Thermo Jarrell Ash, Franklin, Massachusetts) and the ICAP 61 direct-reading polychromator-based system available from the same manufacturer. Their results show that while both systems perform satisfactorily with most samples, for very dilute samples where analyte concentrations are close to their detection limits, simultaneous spectrometers provide better analytical accuracy than sequential spectrometers.

The Plasma 300 spectrometer used in the study employed a 0.33-m double monochromator with a 1200-groove/mm grating to provide a spectral resolution of 0.02 nm for wavelengths between 190 and 365 nm in the second order. The ICAP 61 spectrometer used a concave grating with a 0.75-m focal curve and a 2400-groove/mm grating to give a spectral resolution of 0.03 nm in the first order. When these instruments were used to analyze a U.S. Environmental Protection Agency water sample (Water Purity 284, Sample 1) for nine elements with a 10-sec integration period, the simultaneous spectrometer was found to give an average relative error of 4% with five elements having positive errors and four having negative errors. By comparison, the sequential instrument had an average relative error of 21% with eight elements having negative errors and only one element showing a positive error.

The negative bias shown by the sequential instrument is due to the peak search protocol required prior to a measurement. Since the wavelength reproducibility of the sequential monochromator is not sufficient to locate the analyte peak maximum without additional help, a peak search is performed in the vicinity of the expected analytical line. As long as the *analytical line* is sufficiently intense to be located by the peak search protocol, wavelength reproducibility is excellent. If the analytical line is not intense enough to be identified as such during the peak search, however, there is a high probability that the line will not be located properly and that emission measurements will be made on the *side* of the analyte peak. As a result, when the analytes are close to their detection limits, peak searching typically results in erroneously low measurements.

Even when elements are present at levels much greater than the detection limit, where peak searching is expected to be reliable, problems can be encountered with sequential instruments. In the study by Nygaard and Sotera (1988), an anomalously low result was obtained with the sequential instrument for aluminum in the Environmental Protection Agency (EPA) water sample even though the aluminum concentration in the sample was well above the detection limit. This low result was produced by the peak search algorithm misidentifying a nearby hydroxyl peak at 308.21 for the aluminum line at 308.22. Because the hydroxyl peak is offset by 0.01 nm from the Al line, the Al measurements were made on the side of the aluminum line, and the concentrations were low as a result.

Problems of the type encountered with aluminum using sequential systems have been observed before (Nygaard Chase, and Leighty, 1983) and can be compensated for by using *side-line indexing* (Nygaard et al., 1984). Using side-line indexing, a nearby emission peak (such as the strong 308.14-nm hydroxyl line in the present case) is used as a wavelength reference, and emission measurements are made at a predefined offset from the reference line without benefit of any peak search protocol. While side-line indexing does improve the analytical accuracy of the sequential spectrometer, it makes the analytical method more complex.

15.10. IMAGE DETECTOR SPECTROMETERS

In 1970, Margoshes (1970a–c) proposed the possibility of using television camera tubes as detectors for multielement spectrochemical analysis. Although he did not construct an actual instrument, Margoshes suggested that a television detector would combine the best features of photographic and photoelectric detection. Thus, an ideal detector for multielement spectrochemical analysis would combine the inherent simultaneous integrating detection of the entire spectrum (that is characteristic of a photographic plate) with the inherent high sensitivity and direct conversion of light into an electric analog that is characteristic of photoelectric detection.

Actually, Margoshes's suggestion of using a television camera tube was not the first instance of anyone conceiving the use of such a detector in spectroscopy. As early as 1949, investigators studying transient spectra produced by combustion (Agnew et al., 1949; Benn, Foote, and Chase, 1949) employed an image orthicon (see Chapter 10) as a detector. These workers reported that for short exposure times the image orthicon was 50 times more sensitive than the photographic plate. Compared with photographic detection, the better time resolution of the image orthicon made it especially convenient for monitoring transient spectra.

In 1966, Anderson described a spectrometer designed to study aurorae and other night air glows that used an image orthicon as a detector. The detection system was able to monitor the spectrum from 380 to 820 nm simultaneously with a resolution of 3 nm by dividing it up into three segments stacked vertically in the focal plane. Although the system had a dynamic range of 200:1, only 20 different intensity levels could be distinguished within that range.

15.10.1. Imaging Spectrometers with Single Entrance Slits

Since this early work, numerous examples of the use of image detectors in spectroscopy can be found in the literature. Even in the limited area of multielement spectrochemical analysis, it is not possible to review in detail here all the studies that have been conducted with image detectors. For more comprehensive coverage of the topic, the reader should consult the reviews by Busch and Morrison (1973), Busch and Malloy (1979b), Busch and Benton (1983), Talmi (1975a,b), Talmi and Busch (1983), Talmi (1979, 1983), Winefordner et al. (1975), Jones (1985), McGeorge and Salin (1984), Mitchell, Jackson, and Aldous (1973), Bilhorn, Epperson et al. (1987), Billhorn, Sweedler et al. (1987), Epperson et al. (1988), and Sweedler et al. (1988).

In considering the development of image detector spectrometers four factors have been of paramount importance: (1) the type of image detector best suited for atomic spectroscopy (see Talmi and Busch, 1983, for a comparison of image detectors for low-light-level applications); (2) comparison of signal-to-noise performance of image detectors with conventional photoelectric detection; (3) various ways to alleviate the compromise between wavelength coverage and resolution; and (4) intrascenic dynamic range.

When the potential analytical utility of image devices in atomic spectroscopy is considered, it should be remembered that these devices were not developed or designed with spectroscopic applications in mind. As a result, it is not surprising that their performance as detectors in spectroscopy is often not ideal. In fact, Winefordner and co-workers have concluded on the basis of signal-to-noise considerations (Winefordner et al., 1975; Chester et al., 1976; Winefordner, et al. 1976; Cooney, Vo-Dinh, et al., 1977; Cooney, Boutillier, and Winefordner, 1977; Knapp et al., 1974) that present-day image devices are inadequate for simultaneous multielement analysis.

With regard to these theoretical signal-to-noise studies, it should be pointed out, however, that their validity depends heavily on how closely the actual system approaches the assumptions made in the derivation of the signal-to-noise expressions. In addition, the conclusions reached are not time invariant because image detector technology is producing improved multichannel array detectors with great regularity (see Epperson et al., 1988; Sweedler et al., 1988). Since the studies that conclude that image devices are inadequate for spectrochemical analysis invariably involve the comparison of image devices with the photomultiplier tube, they are strained inevitably by the fundamental differences that exist between these inherently different types of detectors (see Howell and Morrison, 1977). Since photomultiplier tubes respond to the rate of arrival of photons whereas most image detectors are integrating devices, comparison between the two forms of detection tends to suffer from the dilemma of what represents a "fair" comparison. Table 15.3 shows a

Table 15.3. Comparison of Photomultiplier (PMT) Detection Limits with Those Obtained Using Silicon Vidicon and SIT Vidicon

Element	Line (nm)	PMT (μg/mL)	UV Vidicon (μg/mL)	SIT Vidicon (μg/mL)	SIT Vidicon[c] (μg/mL)	Intensifier Potential (kV)
Ag	328.1	0.002	0.3	—	0.1	9.0
Al	396.1	0.003	0.05	0.01	0.008	4.5
Ba	553.5	0.001	0.02	0.001	0.001	9.0
Bi	306.8	20.0	5.0	—	1.0	7.0
Ca	422.7	0.0001	0.001	0.0001	0.0002	5.0
Co	345.4	0.03	0.6	—	0.5	9.0
Cr	425.4[a]	0.002	0.01	0.002	0.002	7.9
Cu	324.7	0.01	0.07	—	0.07	9.0
Fe	372.0	0.005	0.2	0.01	0.04	7.5
In	451.1	0.0004	0.02	0.006	0.003	7.5
K	766.5	0.00005	0.03	0.06	0.03	9.0
Li	670.8	0.00002	0.00002	0.00001	0.00001	9.0
Mg	285.2	0.005	0.2	—	0.07	9.0
Mn	403.1	0.001	0.02	0.003	0.002	9.0
Mo	390.2	0.1	0.1	0.07	0.04	4.8
Na	589.0	0.0005	0.0005	0.0002	0.00007	5.6
Ni	352.5	0.02	0.1	—	0.2	9.0
Pb	405.8	0.1	0.8	0.2	0.1	8.0
Rb	780.0	0.008	0.1	0.1	0.08	9.0
Sr	460.7	0.0002	0.002	0.0002	0.0001	7.3
Ti	399.8[b]	0.03	0.2	0.06	0.06	5.0
V	437.9	0.007	0.5	0.02	0.02	5.5
W	400.9	0.7	3.0	1.0	0.9	4.7

[a] PMT value obtained at 359.4 nm.
[b] PMT value obtained at 365.4 nm.
[c] SIT with scintillator

Source: Reprinted from Howell and Morrison (1977).

comparison of the detection limits obtained with a photomultiplier, a silicon vidicon, and an SIT vidicon with and without a scintillator.

Aside from questions regarding the detectivity of image devices in multi-element spectrochemical analysis, the two most serious problems that have to be solved involve the appropriate type of optical system to use with an image detector and matching the dynamic range of the detector to that inherent in the spectrum. The development of image detector spectrometers for multi-element spectrochemical analysis has proceeded along two basic lines of development based on the mode of dispersion. With one-dimensional systems, the spectral information is dispersed across the tube target as a single horizontal band of information. Because of the limited size of image detector targets, one-dimensional systems are based on a window concept (Busch, Howell, and Morrison, 1974a) that requires a compromise between wavelength coverage and resolution.

Several approaches have been taken to overcome the limited wavelength region that can be sensed with an image detector mounted in the focal plane of a conventional one-dimensional dispersion system. The most obvious way to access a wide variety of analytical atomic lines under resolution conditions adequate for atomic spectroscopy is to jump from one spectral window to the next in a slew-scan approach similar to that used with time-multiplexed spectrometers discussed earlier. This is the approach taken by Chester et al. (1976), Horlick (1976), Furuta et al. (1979), and Furuta (1980). Furuta (1980) discusses the results obtained with an instrument consisting of either an ICP source or nitrous oxide–acetylene flame combined with a programmable monochromator and an SIT vidicon (see Chapter 11) with an optical multichannel analyzer readout. A 1-m Czerny–Turner spectrograph equipped with a 2400-groove/mm grating produced a dispersion system with a reciprocal linear dispersion of 0.4 nm/mm in the first order. Unfortunately, the slew-scan window approach tends to defeat the goal of using a multichannel detector (i.e., the ability to monitor *all* the desired spectral information simultaneously).

Under certain favorable circumstances (Busch, Howell, and Morrison, 1974b; Howell, Ganjei, and Morrison, 1976; Ganjei et al., 1976) widely spaced spectral lines may be monitored with reasonable resolution without increasing either the extent of the wavelength window or slew scanning by monitoring spectral lines in overlapping orders (similar to aliasing with interferometers). Bystroff, Hirschfeld, and Pesek (1980) have termed the process of using overlapping orders *spectral folding*. Since the use of overlapping orders may not be possible in every situation, the limited wavelength coverage obtained by mounting an image detector in the focal plane of a conventional spectrograph remains the most serious problem associated with the development of viable image detector spectrometers for simultaneous multielement analysis.

To avoid the limited wavelength coverage associated with one-dimensional systems, various ways of modifying the spectral format focused on the image detector have been tried. One obvious way of improving the efficiency with which spectral information is imaged on an image detector is to use the two-dimensional format inherent with echelle dispersion systems that employ a cross-dispersing prism. This is the approach taken by Felkel and Pardue (1979), Sims and Denton (1983), and Denton, Lewis, and Sims (1983). Figure 15.12 shows a three-dimensional representation of a spectrum taken with an echelle spectrometer and a charge-coupled device (CCD) as a two-dimensional array detector.

Echelle dispersion systems, while providing high resolution, are not without disadvantages, however. When a prism is used as a cross-dispersion element, the various orders of spectral information are not evenly spaced owing to the nonuniform dispersion of the prism (see Figure 4.20). The nonuniform spectral format that results makes programming the scanning pattern of the image detector more complex. In addition, with the echelle dispersion system, spectral information is imaged as points of light rather than as slit-shaped lines (see Figure 15.12). Imaging an analytical wavelength as a point on an image detector tends to concentrate the radiation over a small number of pixels, which can lead to blooming problems that limit the overall dynamic range of the instrument when certain detectors are used.

Random-access target interrogation (Hirschfeld, 1973) can be used to improve the apparent dynamic range of image detectors compared with systems that use a raster scan pattern (see Chapter 10). In a random-access mode, regions of high illumination on the target are scanned more frequently than those that are weakly illuminated. This allows the charge pattern in weakly illuminated areas to develop more fully prior to readout while, at the same time, avoids saturation and blooming in the highly illuminated areas by frequent erasing. Random-access target interrogation is not without its own set of problems, however, as pointed out by Nieman and Enke (1976). These include (1) increased complexity of the software needed to control the scanning beam; (2) the need to scan the entire target at intervals appropriate to the signal level in order to avoid saturation and blooming; and (3) the need to measure and correct for the unequal integration times of the accessed areas. Felkel and Pardue (1977) point out, in addition, that random-access target interrogation results in a trade-off because the *entire target* must be primed (during which no information is accumulated) even though only a limited number of resolution elements are of interest.

Another approach to modifying the spectral format imaged on an image detector has been to modify the collimating/focusing system of a single-entrance-slit spectrometer to produce multiple angles of incidence on the diffraction grating. This may be accomplished by segmenting the collimating mirror or the grating itself. In the system described by Hoffman and Pardue (1979) and shown in Figure 15.15, the beam of radiation from the entrance slit is collimated with a mirror in the usual manner. The collimated diffracted beams from the grating that comprise the spectrum then strike an array of plane mirrors (M_2, M_3, etc.) set at slightly different angles as shown in Figure 15.15. As a result of this optical arrangement, the spectrum focused by camera mirror M_B consists of a series of horizontal strips displaced vertically with one horizontal strip for each plane mirror in the array. Because each mirror (M_2, M_3, etc.) is set at a different angle, the wavelength window focused by camera mirror M_B in the focal plane is displaced from one horizontal strip to

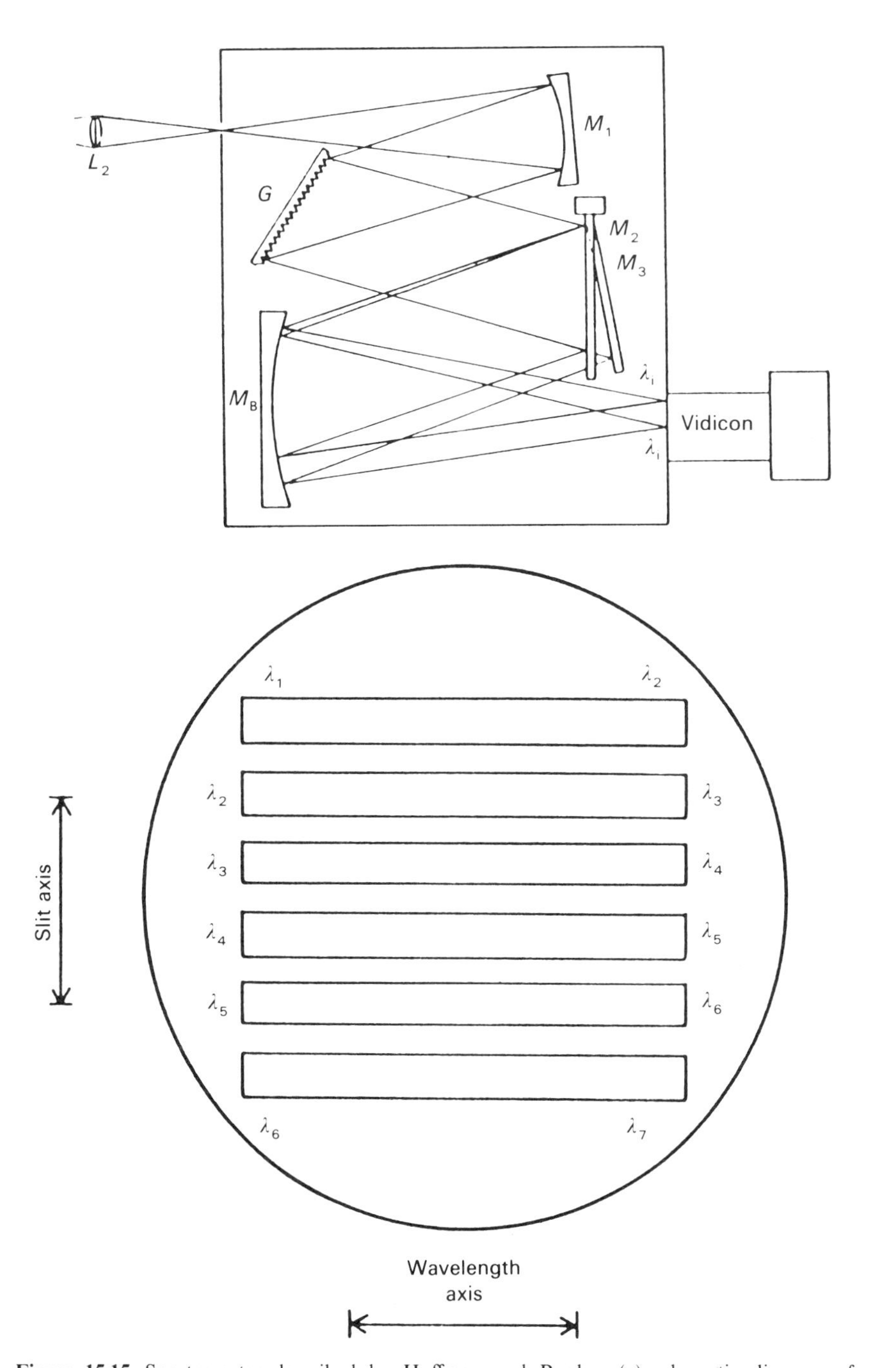

Figure 15.15. Spectrometer described by Hoffman and Pardue: (*a*) schematic diagram of dispersion systems (*b*) arrangement of spectral information on target of image detector. [Reprinted with permission from R. M. Hoffman and H. L. Pardue (1979), *Anal. Chem.*, *51*, 1267. Copyright 1979 American Chemical Society.]

another. The net result of this system is to focus different spectral regions on the target of the image detector as a series of vertically displaced horizontal strips. Using this approach, spectral regions that are not of interest (such as regions with a high source background but no desired analytical information) need not be imaged on the detector.

All single-entrance-slit systems, however, suffer from dynamic range limitations that are characteristic of the image detector itself. To avoid blooming when monitoring a source with both intense as well as weak lines, the slit width and/or the observation time must be reduced to accommodate the presence of the intense lines. In either case (reduction in the slit width or observation time to accommodate the presence of the intense lines on the target), the procedure results in a concomitant reduction in the weak lines, making them even more difficult to detect.

15.10.2. Multiple-Entrance-Slit Imaging Spectrometers

Two basic problems must be solved in order to use image detectors effectively for multielement spectrochemical analysis. The first is to devise an efficient imaging system to permit a large number of analytical lines to be imaged simultaneously at a spectral resolution acceptable for atomic spectroscopy. Since the image detector needs to be able to monitor weak lines in the presence of intense lines or background, the second problem that must be overcome is to match the dynamic range of the spectrum to the linear dynamic range of the detector. Both of these problems can be solved with multiple-entrance-slit systems.

Figure 15.16 shows the basic optical arrangement for the multiple-entrance-slit image detector spectrometer developed by Busch and co-workers (Busch and Malloy, 1979a,b; Busch, Malloy, and Talmi, 1979; Busch, 1983, 1985; Malloy, 1979; Busch, Busch and Malloy, 1989). A 0.5-m Czerny–Turner spectrograph with a camera mirror that was larger than the collimating mirror was modified by removing the entrance slit and mounting the image detector in what was formerly the entrance focal plane. A multiple-entrance-slit assembly consisting of a horizontal row of 29 holes (1.6 mm diameter and spaced on 4-mm centers) was mounted in what was formerly the exit port of the spectrograph. In the prototype instrument, each hole was masked from behind with two pieces of black tape to form an entrance slit 1.6 mm high.

To facilitate illumination of various combinations of entrance slits across the 12-cm horizontal row as desired, fiber-optic light guides were employed to convey light from the light source to the individual slits. The diameter of the entrance holes in the multiple-entrance-slit plate was selected so that each fiber-optic light guide could be plugged into the hole corresponding to any one of the 29 possible entrance slit positions. Those holes corresponding to

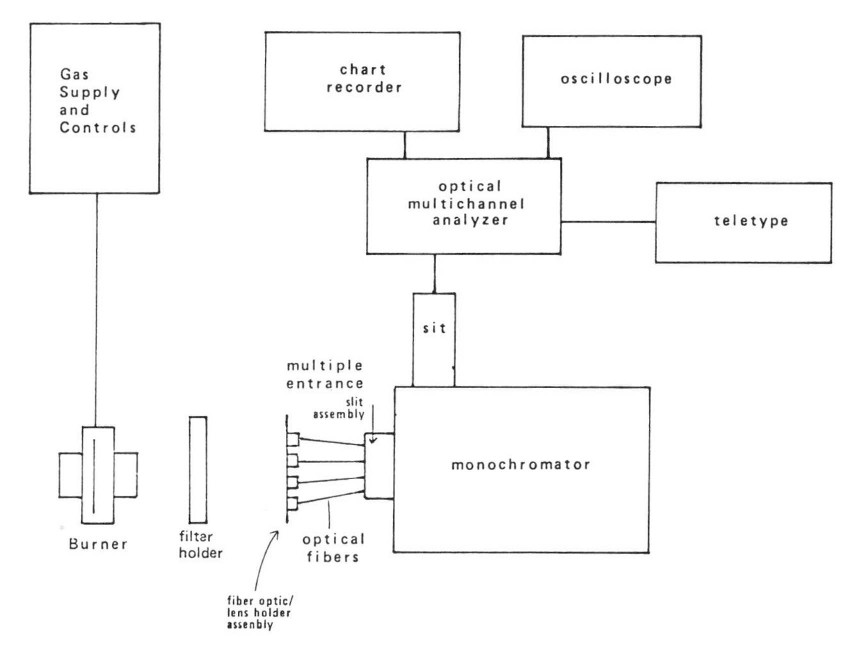

Figure 15.16. Block diagram of multiple-entrance-slit image detector spectrometer developed by Busch and Malloy. [Reprinted from B. Malloy (1979), Ph.D. Dissertation, Baylor University.]

entrance slit positions not in use were blocked by means of opaque plugs.

The input ends of each fiber-optic light guide were arranged around the source so that each could be individually adjusted vertically above the excitation source, as shown in Figure 15.17. For excitation sources like flames where different elements have different optimum sampling zones, the multiple-entrance-slit approach permits individual observation height adjustment of each fiber-optic light guide, thereby reducing the severity of the compromise analytical conditions required for multielement analysis by eliminating observation height as a factor. Input lenses were attached to each fiber-optic light guide to improve its light-gathering ability.

The easiest way to appreciate how a multiple-entrance-slit spectrometer works is to consider a direct-reading spectrometer operated in reverse. Thus, in contrast to the direct reader, which uses a single entrance slit and multiple exit slits each with its own detector, a multiple-entrance-slit spectrometer (or MESS) employs selected multiple entrance slits and a single multichannel detector. Since all the selected entrance slits are active simultaneously, the composite spectrum produced is integrated simultaneously, allowing the system to be used even with time-varying sources (such as might be encountered with atomic fluorescence measurements using an electrothermal

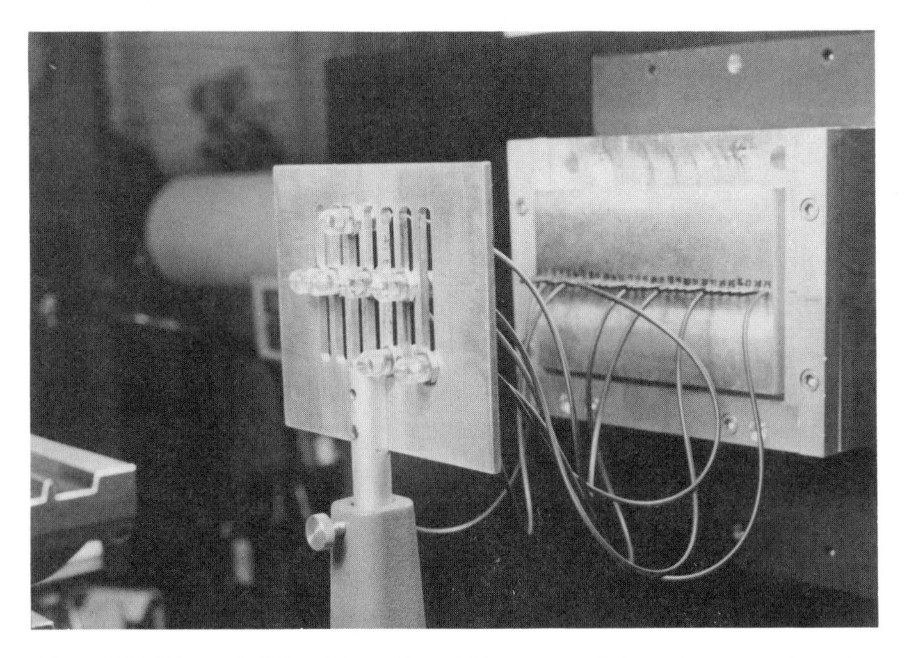

Figure 15.17. Fiber-optic light guides used in multiple-entrance-slit image detector spectrometer. [Reprinted with permission from K. W. Busch, B. Malloy, and Y. Talmi (1979), *Anal. Chem.*, *51*, 670. Copyright 1979 American Chemical Society.]

atomizer). [*Note*: The operation of the multiple-entrance-slit image detector spectrometer described here (where selected combinations of entrance slits are active simultaneously) should not be confused with the multiple-entrance-slit spectrometer described earlier, which used a photomultiplier detector. The latter is a time-multiplexed system where only one entrance slit at a time is accessed by the optical multiplexer.]

There are several important differences between a conventional direct reader and a MESS that should be emphasized, however. First, the direct-reading spectrometer typically requires a long focal length to achieve sufficient dispersion to permit multiple photodetector placement. Since a multichannel image detector with closely spaced channels (typically 25 μm apart) is used with the MESS, long-focal-length optics are not necessary.

The second difference between a direct reader and a MESS involves the tolerance required in slit location. With a direct reader, the spectral resolution element monitored is determined by the precise location of the multiple exit slits in the exit focal plane. By contrast, access of a particular spectral line with the MESS is *not* determined by the precise location of a given entrance slit in the multiple-entrance-slit array. In fact, since coarse entrance slit position merely controls the spectral region (i.e., wavelength window) focused on the image detector, slit placement is not as critical with the MESS as with the direct reader. Spectral resolution with the MESS depends on the entrance slit width, the reciprocal linear dispersion of the optical system, and pixel-to-pixel spacing of the image detector.

Using a spectrograph with a reciprocal linear dispersion of 3.2 nm/mm, Busch and Malloy (1979a) observed that each entrance slit focused a 40-nm spectral window on an image detector target of 13 mm diameter. Since the row of entrance slits was 12 cm long, any wavelength within a range of 384 nm could be imaged on the target of the detector with the MESS system. By employing several entrance slits simultaneously, widely separated spectral lines could be imaged on the detector with adequate resolution.

Figure 15.18 shows the system performance obtained with Mn, Cr, Sr, Ba, and Li introduced into an acetylene–nitrous oxide flame. To monitor this combination of elements simultaneously with a conventional one-dimensional dispersive system would require a spectral window of 267 nm. Such a window width is too large to provide adequate resolution for atomic spectroscopy. Figures 15.18*a–e* show the spectrum obtained with the MESS for each element when only one appropriate entrance slit was illuminated with a fiber-optic light guide. Figure 15.18*f* shows the composite spectrum obtained when all five fiber-optic light guides were plugged into their respective entrance slit holes simultaneously.

By employing a regularly spaced row of entrance slits illuminated by means of fiber-optic light guides, a "switchboard" system results that makes it convenient to monitor different composite spectra. Use of a composite spectrum is possible because, as mentioned earlier, the spectral information is not uniformly distributed throughout the spectrum and there are often broad areas with no spectral information at all. With the prototype system described by Busch and Malloy (1979a), a line of any given wavelength may be imaged onto the detector from any of three entrance slit positions. As a result, spectral lines can be moved across the target to avoid potential spectral interferences from other lines in the composite spectrum or to avoid regions of intense background such as shown in Figure 15.19 that would saturate the detector.

The MESS switchboard system allows the spectroscopist the freedom to alter the spectrum in a way that permits various combinations of elements to be imaged in a more efficient manner on a detector target of limited size. With a linear array, such as a linear photodiode array, folding multielement atomic spectra increases the useful information density that can be imaged onto the detector. With systems having a one-dimensional dispersion, the aim, in using the MESS successfully, is to plug various combinations of fiber-optic light guides into the entrance slit plate as needed to build up a composite spectrum in such a way that analytical wavelengths of interest are imaged simultaneously on the detector without interference from emissions by other sample constituents.

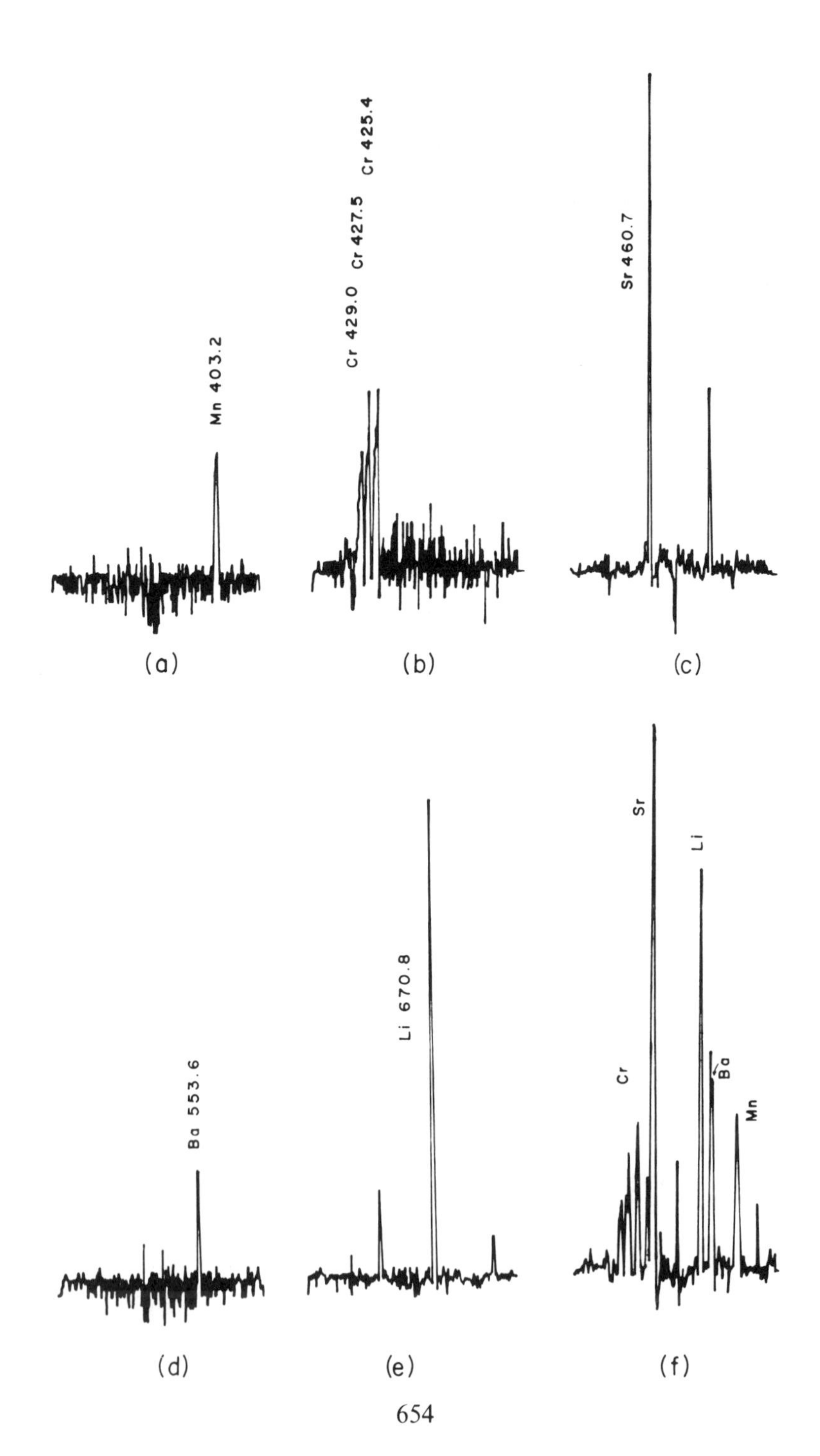
Mn 403.2
(a)
Cr 429.0
Cr 427.5
Cr 425.4
(b)
Sr 460.7
(c)
Ba 553.6
(d)
Li 670.8
(e)
Cr
Sr
Li
Ba
Mn
(f)

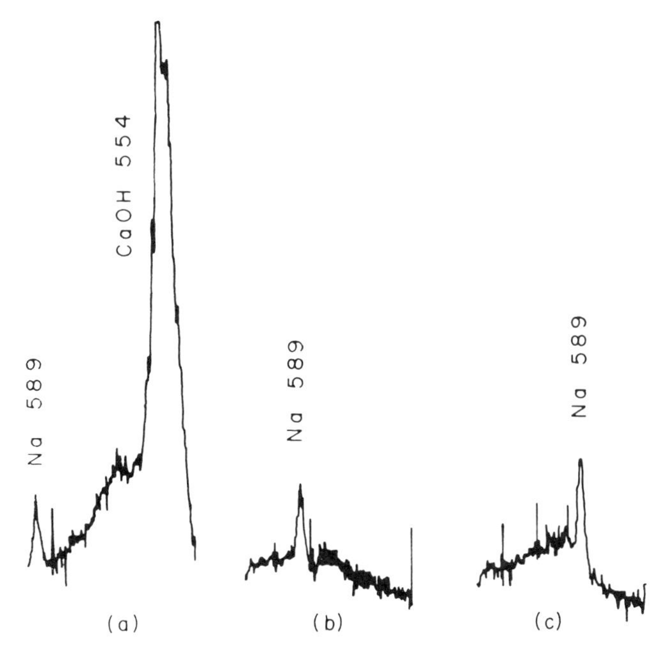

Figure 15.19. Spectrum of tap water obtained with multiple–entrance-slit image detector spectrometer (N_2O, 9.5 L/min; C_2H_2, 6.0 L/min; 500 accumulation cycles): (*a*) entrance slit 15; (*b*) entrance slit 16; (*c*) entrance slit 17. [Reprinted with permission from K. W. Busch, B. Malloy, and Y. Talmi (1979), *Anal. Chem.*, *51*, 670. Copyright 1979 American Chemical Society.]

It should be realized that in building up the composite spectrum by overlapping different spectral regions, the scrambled spectrum that results will not be arranged in order of increasing wavelength. For analytical purposes, however, it is not necessary that the spectrum be arranged in order of increasing wavelength. All that is necessary is that certain array channels be associated with particular, known chemical elements.

By using the array detector in this fashion, it is not being used to monitor a complete spectrum in the true sense. The complete spectrum contains far

Figure 15.18. Spectra obtained with multiple-entrance-slit image detector spectrometer: (*a*) Mn, 3 ppm, entrance slit 3 (402.4–442.2 nm); (*b*) Cr, 3 ppm, entrance slit 3 (402.4–442.4 nm); (*c*) Sr, 1 ppm, entrance slit 6 (440.8–480.8 nm); (*d*) Ba, 3 ppm, entrance slit 14 (543.2–583.2 nm); (*e*) Li, 1 ppm, entrance slit 23 (658.4–698.4 nm); (*f*) composite spectrum obtained by simultaneously using entrance slits 3, 6, 14, and 23. Multielement mixture contained 5 ppm Mn, Cr, and Ba and 1 ppm Li and Sr. [Reprinted with permission from K. W. Busch, B. Malloy, and Y. Talmi (1979), *Anal. Chem.*, *51*, 670. Copyright 1979 American Chemical Society.]

more information than is required for the typical multielement spectrochemical analysis. Instead, the image detector is being used more like a programmable direct reader where the information sought is not the complete spectrum but a group of analytical lines with adjacent backgrounds. When viewed in this fashion, the actual arrangement of the analytical lines across the target is of no real importance as long as the location of the analytical line for each element is known.

Although the goal of assembling an interference-free composite spectrum using the one-dimensional MESS approach (i.e., overlapping different spectral regions) may seem quite difficult, Bystroff et al. (1980) have developed a statistical model based on Poisson statistics that predicts that even for large numbers of chemical elements (>25), the chances are good that all can be determined in a single measurement with a 1000-element linear array detector. Using this model, the overall success probability for m chemical elements with $\langle n \rangle$ lines per element folded f times onto a 1000-resolution element array is given by

$$P = [1 - (1 - e^{(1 - m\langle n \rangle f)/1000})^{\langle n \rangle f}]^m \tag{15.12}$$

where P is the probability that a successful composite spectrum can be obtained by spectral folding. Using the National Bureau of Standards (NBS) tabulated lines (I and II) for 31 elements between 200 and 450 nm, Bystroff et al. (1980) found a predicted success probability of 93% in coming up with 10 lines for each chemical element that were overlap free.

Another important advantage of the multiple-entrance-slit switchboard system is the ability to modify the intensities of various spectral features to avoid target saturation and blooming. By using the switchboard system, each chemical element (or at most two or three chemical elements) is monitored by an individual fiber-optic light guide. Individual adjustment of the intensities transmitted by each fiber-optic light guide can be achieved in several ways. A variable neutral density filter may be used to attenuate the line or lines transmitted by a particular fiber-optic light guide without simultaneously attenuating the radiation transmitted by other light guides. Alternatively, smaller entrance slits may be employed with particular light guides to reduce the light transmitted without affecting other light guides. Finally, since the system uses a switchboard approach, a less sensitive (i.e., less intense) analytical line can be selected and the more intense emission can be moved off the detector target.

For detectors that permit a two-dimensional scan pattern, the multiple entrance slits can be arranged in a two-dimensional array in the entrance focal plane, as shown in Figure 15.20. If a stigmatic dispersion system is employed (such as a Czerny–Turner, Ebert, or Roland mounting), each row in the array

Figure 15.20. Multiple-entrance-slit image detector spectrometer with two-dimensional array of entrance slits. [Reprinted with permission from K. W. Busch and L. D. Benton (1983), *Anal. Chem.*, *55*, 445A. Copyright 1983 American Chemical Society.]

of entrance slits will image a particular spectral segment on the detector with one segment arranged above the other, as shown in Figure 15.21. Figure 15.21 shows a two-dimensional spectrum of a hollow-cathode lamp obtained with a prototype two-dimensional MESS developed to illustrate the concept (Busch, 1985; Busch, Busch and Malloy, 1989). The two-dimensional spectrum was obtained using three entrance slits, each entrance slit having a different column–row coordinate in the array. The vertical separation between spectral segments on the detector target was determined by the vertical separation between rows in the entrance slit array. The particular wavelength window sampled by a given row was determined by the horizontal position of the active slit in the particular row (i.e., the column coordinate.) Within a given row, several entrance slits may be employed simultaneously to produce a composite spectral segment in the same fashion as described for the one-dimensional configuration.

The two-dimensional multiple-entrance-slit configuration provides the maximum amount of flexibility in arranging spectral information on the

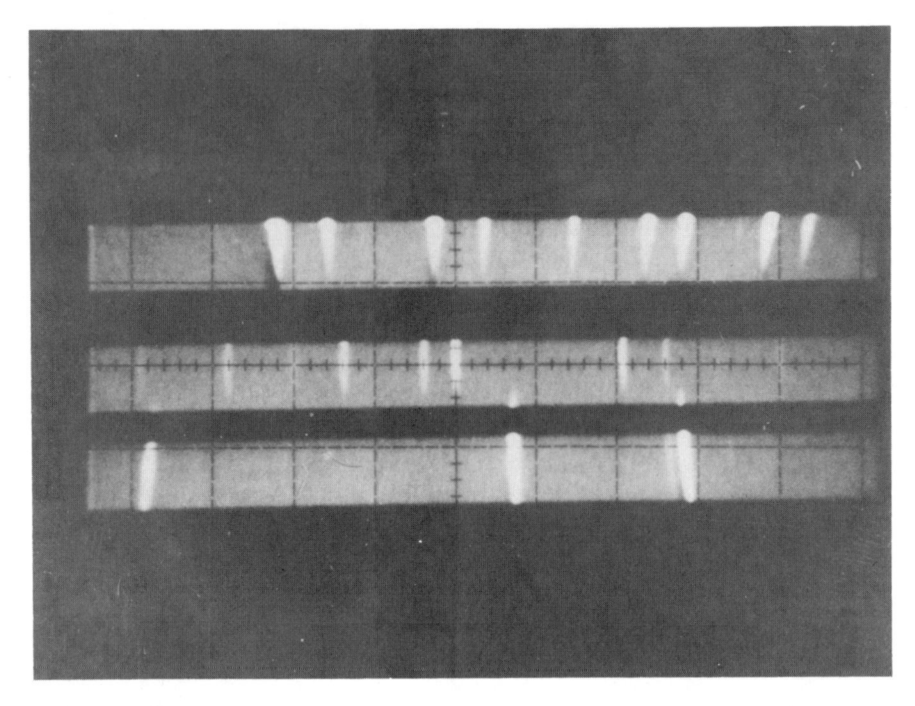

Figure 15.21. Spectrum of hollow-cathode lamp obtained with multiple-entrance-slit image detector spectrometer using two-dimensional array of entrance slits. [Reprinted with permission from K. W. Busch and L. D. Benton (1983), *Anal. Chem.*, *55*, 445A. Copyright 1983 American Chemical Society.]

image detector. Since there is less overlapping of spectral information with the two-dimensional configuration, there is less chance for spectral interference. In essence, the system behaves like a programmable echelle spectrometer, albeit without the high resolution inherent with an echelle grating.

Although the multiple-entrance-slit image detector spectrometer described by Busch and co-workers was never developed beyond the prototype stage, it demonstrated the type of optical system needed to make full use of the multichannel capabilities of image detectors. To be completely satisfactory for spectrochemical analysis, however, a present-day system would employ UV-transmitting fiber-optic light guides, a more sophisticated entrance slit array (i.e., a curved focal plane with regularly spaced adjustable entrance slits), a concave grating spectrometer, and a charge-coupled device detector.

The development of the MESS has led other workers to explore the possibilities for modified dispersion systems that make full use of the multichannel capabilities of image detectors. A novel spectroscopic system quite similar to the MESS developed by PRA International of London,

Ontario, Canada (and now marketed by Leco Instruments), has been evaluated by Karanassios and Horlick (1986). The system consists of a predispersion optical system, an echelle grating without cross-dispersion, and a linear photodiode array detector. Figure 15.22 shows an optical layout of the instrument, which is known as a Plasmarray spectrometer.

The predispersion optical system of the Plasmarray spectrometer consists of an entrance slit (S), a concave grating (G_1), and a spectral selection mask followed by a concave mirror (M_1) and plane grating (G_2), as shown in Figure 15.22. The preselection polychromator is based on a concave grating arranged around a 0.5-m Rowland circle to produce a low-resolution dispersion system with a reciprocal linear dispersion of 3.45 nm/mm. The spectral selection mask, shown in Figure 15.23, is located on the Rowland circle of G_1 and acts as a multiple spectral bandpass filter. Slots in the stainless steel mask allow small-wavelength regions around spectral lines of interest to pass through to the high-resolution echelle optical system. Slots in the mask are typically 100–150 μm wide to allow both desired analytical lines and adjacent background to pass through the mask to the subsequent portions of the optical system. The mask is mounted kinematically so that it may be replaced quickly and precisely to monitor different combinations of elements.

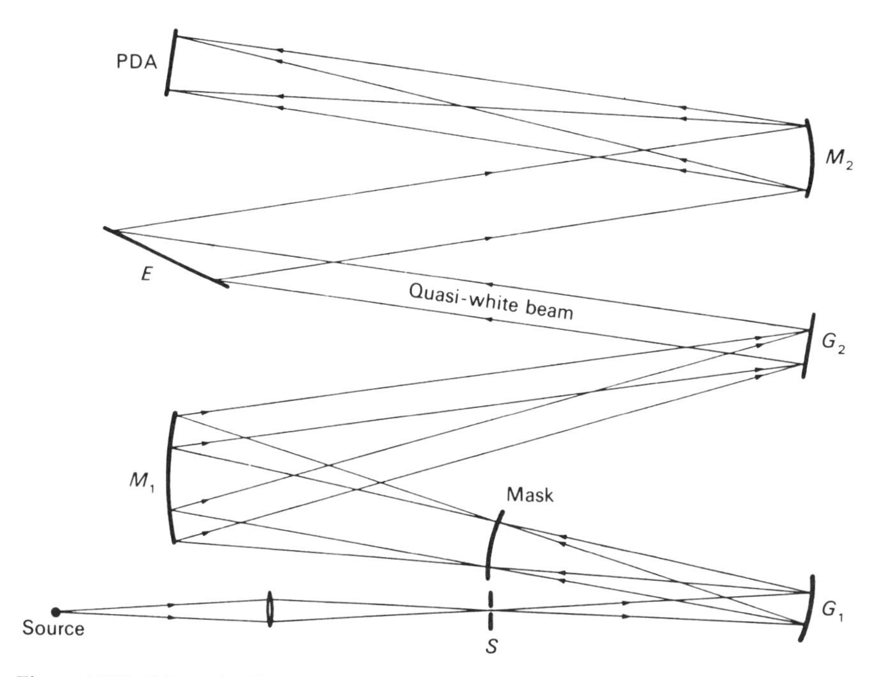

Figure 15.22. Schematic diagram of Plasmarray spectrometer by Leco Instruments, Ltd. [Courtesy of Leco Corporation.]

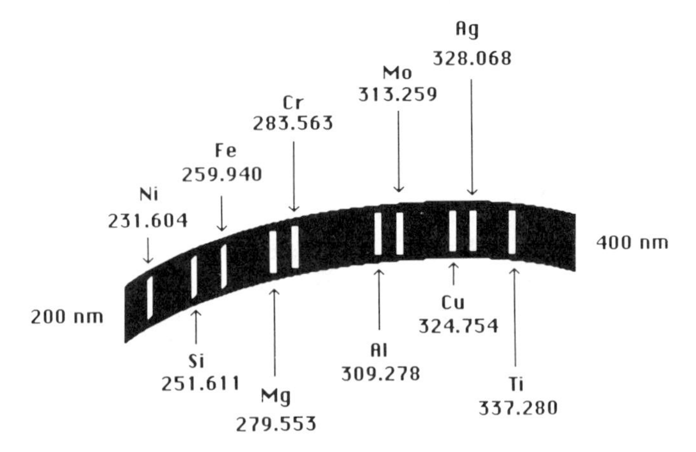

Figure 15.23. Schematic diagram of predisperser selection mask used in Plasmarray spectrometer. [Reprinted from V. Karanassios and G. Horlick (1986), *Appl. Spectrosc.*, *40*, 813, with permission from the Society for Applied Spectroscopy.]

Following the spectral selection mask, a recombination system consisting of the concave mirror (M_1) and plane grating (G_2) serves to recombine the desired spectral regions that have been selected by the mask into a collimated beam of polychromatic light where most of the radiation that is *not* of interest has been blocked by the mask. Following this collimation process, the modified polychromatic beam strikes the 31.6-groove/mm echelle grating (E) followed by a 1-m camera mirror (M_2) that focuses the radiation onto the photodiode array detector. A short-focal-length cylindrical lens mounted just prior to the detector adjusts the spectral line height to the 2.5-mm height of the photodiode array. Using this optical system, entrance slit heights of up to 12 mm can be used.

Compared with a conventional echelle spectrometer, the Plasmarray spectrometer does not employ a cross-dispersion prism to displace the various spectral orders produced by the echelle grating in the vertical direction. As a result, the spectral information presented to the photodiode array detector consists of a series of spectral *lines* (not points as with a conventional echelle system) from *overlapping* multiple orders. Figure 15.24 illustrates how the system works and compares it with the classical two-dimensional echelle dispersion system. In the figure, wavelength regions surrounding five spectral lines of interest are transmitted by the spectral selection mask.

In spite of the use of overlapping multiple orders (similar to the one-dimensional MESS), spectral interference is minimized with the Plasmarray spectrometer because most of the unwanted radiation is eliminated by the predispersion optical system (i.e., the spectral selection mask) prior to

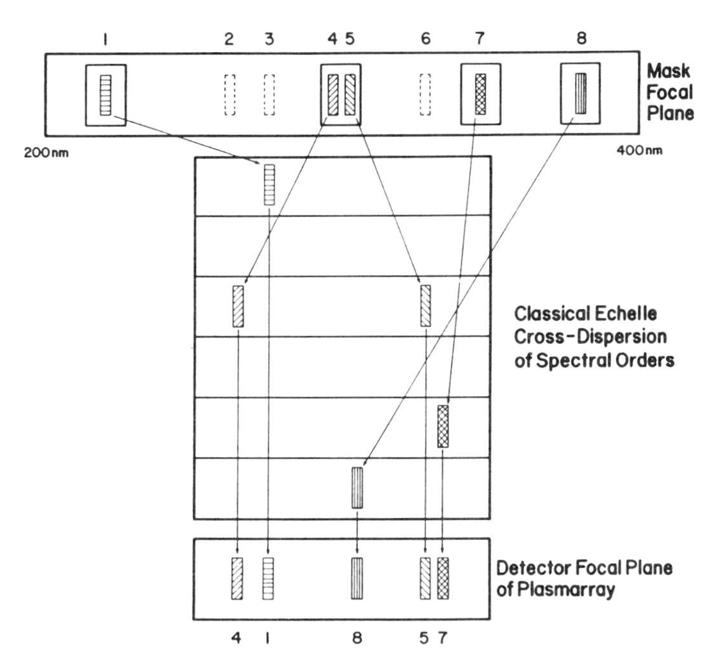

Figure 15.24. Schematic diagram illustrating principle behind dispersion system of Plasmarray spectrometer for five spectral lines of interest. Lines 2, 3, and 6 are not of interest and are not imaged at detector focal plane. [Courtesy of Leco Corporation.]

encountering the echelle grating. Because of the high dispersion produced by the echelle grating, spectral interference within any given order is minimal. Figure 15.25 shows a composite spectrum obtained with the Plasmarray spectrometer for a multielement sample containing Fe, Al, Si, Cu, Mg, Ni, Cr, Mo, and Ti.

15.11. QUO VADIMUS?

This book has attempted to present the various approaches that have been taken to permit multielement detection in spectrochemical analysis. Clearly, various approaches are viable depending on the particular analytical application under consideration. The "ultimate" analytical system for multielement spectrochemical analysis, if it can be said to exist at all, remains to be developed. Although one cannot predict the occurrence of major breakthroughs in achieving the goal of an ultimate multielement analytical system, improvements in detection systems continue to be made on a regular basis, and the future of spectrochemical analysis looks bright.

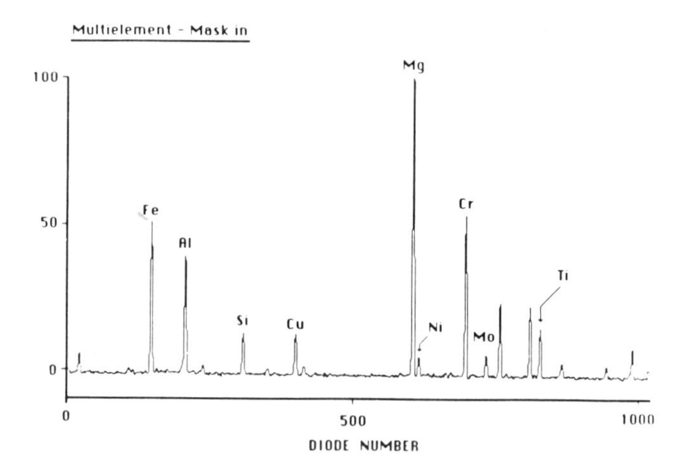

Figure 15.25. Composite spectrum obtained with Plasmarray spectrometer for multielement sample containing Fe, Al, Si, Cu, Mg, Ni, Cr, Mo, and Ti. [Reprinted from V. Karanassios and G. Horlick (1986), *Appl. Spectrosc.*, *40,* 813, with permission from the Society for Applied Spectroscopy.]

For the present, the direct-reading spectrometer and, to a lesser extent, the programmed monochromator combined with an ICP will remain the workhorse of the spectrochemical laboratory. Modular atomic fluorescence instruments with an ICP source look promising, particularly for nonspecialist use, but it is too early to tell whether this class of spectrometer can compete with existing technology. For survey analyses (i.e., both qualitative and quantitative determinations on nonroutine samples), the time-honored spectrograph with photographic detection will continue to be of service. For single-element trace determinations by nonspecialists, atomic absorption spectrophotometry will remain the method of choice both in terms of cost and ease of operation.

Finally, viable image detector spectrometers will be developed eventually as improvements in image detector technology occur. Most promising among the various image detectors today is the charge-coupled device. Clearly, some form of modified dispersion system (echelle or some other arrangement) will be required to make full use of the multichannel capability of such detectors. The development of the appropriate dispersion system for use with image detectors will continue to be an active area of spectrochemical research and must be conducted with full knowledge and understanding of the entire multielement spectrochemical process. To be successful at spectrochemical research, tomorrow's research worker will need to be an analytical scientist in the true sense of the term. This will require not only an understanding of spectrochemical analysis and the nature of the atomic spectrum itself but also an understanding of optics, electronics, spectroscopy, chemistry, and engineering.

REFERENCES

Agnew, J. T., Franklin, R. G. , Benn, R. E., and Bazarian, A. (1949), *J. Opt. Soc. Am.*, *39*, 409.

Alkemade, C. T. J. (1968), *Appl. Opt.*, *7*, 1261.

Anderson, J. E. (1966), *Rev. Sci. Inst.*, *37*, 1214.

Benn, R. E., Foote, W. S., and Chase, C. T. (1949), *J. Opt. Soc. Am.*, *39*, 529.

Benton, L. D. (1985), "A Study and Characterization of a New Single Detector Time-Multiplexed Spectrometer for Atomic Flame Emission Spectroscopy," M.S. Thesis, Baylor University, Waco, TX.

Bilhorn, R. B., Epperson, P. M., Sweedler, J. V., and Denton, M. B. (1987), *Appl. Spectrosc.*, *41*, 1125.

Bilhorn, R. B., Sweedler, J. V., Epperson, P. M., and Denton, M. B. (1987), *Appl. Spectrosc.*, *41*, 1114.

Boettner, E. A. and Brewington, G. P. (1944), *J. Opt. Soc. Am.*, *34*, 6.

Brinkman, D. W., Whisman, M. L., and Goetzinger, J. W. (1979), *Appl. Spectrosc.*, *33*, 245.

Brost, D. F., Malloy, B., and Busch, K. W. (1977), *Anal. Chem.*, *49*, 2280.

Busch, K. W. (1983), U.S. Patent 4,375,919.

Busch, K. W. (1985), U.S. Patent 4,494,872.

Busch, K. W. and Benton, L. D. (1981), 37th Southwest Regional Meeting of ACS, San Antonio, TX, Paper 7.

Busch, K. W. and Benton, L. D. (1983), *Anal. Chem.*, *55*, 445A.

Busch, K. W. and Benton, L. D. (1985), FACSS meeting, Philadelphia, PA, Paper 73.

Busch, K. W., Busch, M. A., and Benton, L. D. (1990a), *Talanta*, in press.

Busch, K. W., Busch, M. A., and Benton, L. D. (1990b), *Talanta*, in press.

Busch, M. A., Busch, K. W., and Malloy, B. B. (1990), *Talanta*, in press.

Busch, K. W., Howell, N. G., and Morrison, G. H. (1974a), *Anal. Chem.*, *46*, 575.

Busch, K. W., Howell, N. G., and Morrison, G. H. (1974b), *Anal. Chem.*, *46*, 1231.

Busch, K. W. and Malloy, B. (1979a), Joint ACS/Japanese Chemical Society Meeting, Honolulu, Hawaii, Paper 153.

Busch, K. W. and Malloy, B. (1979b), "The Role of Image Devices in Simultaneous Multielement Analysis," in *Multichannel Image Detectors*, Talmi, Y. (Ed.), American Chemical Society, Washington, DC, Chaper 2.

Busch, K. W., Malloy, B., and Talmi, Y. (1979), *Anal. Chem.*, *51*, 670.

Busch, K. W. and Morrison, G. H. (1973), *Anal. Chem.*, *45*, 712A.

Bystroff, R., Hirschfeld, T., and Pesek, J. (1980), Pittsburgh Conference on Analytical Chemistry and Applied Spectroscopy, Paper 338.

Chester, T. L., Haraguchi, H., Knapp, D. O., Messman, J. D., and Winefordner, J. D. (1976), *Appl. Spectrosc.*, *30*, 410.

Chester, T. L. and Winefordner, J. D. (1976), *Spectrochim. Acta, Part B*, *31B*, 21.

Cooney, R. P., Boutillier, G. D., and Winefordner, J. D. (1977), *Anal. Chem.*, *49*, 1048.

Cooney, R. P., Vo-Dinh, T., Walden, G., and Winefordner, J. D. (1977), *Anal. Chem.*, *49*, 939.

Cordos, E. and Malmstadt, H. V. (1973), *Anal. Chem.*, *45*, 425.

Cresser, M. S. and West, T. S. (1970), *Anal. Chim. Acta*, *51*, 530.

Dagnall, R. M., Kirkbright, G. F., West, T. S., and Wood, R. (1971), *Anal. Chem.*, *43*, 1765.

Dagnall, R. M., Kirkbright, G. F., West, T. S., and Wood, R. (1972), *Analyst*, *97*, 245.

Danielsson, A. and Lindblom, P. (1972), *Phys. Scripta*, *5*, 227.

Danielsson, A. and Lindblom, P. (1976), *Appl. Spectrosc.*, *30*, 151.

Danielsson, A., Lindblom, P., and Söderman, E. (1974), *Chem. Scripta*, *6*, 5.

Dawson, J. B., Ellis, D. J., and Millner, R. (1968), *Spectrochim. Acta, Part B*, *23B*, 695.

Demers, D. R. (1987), *Am. Lab.*, *19*(8), 30.

Demers, D. R. and Allemand, C. D. (1981), *Anal. Chem.*, *53*, 1915.

Demers, D. R., Busch, D. A., and Allemand, C. D. (1982), *Am. Lab.*, *14*(3), 167.

Denton, M. B., Lewis, H. A., and Sims, G. R. (1983), "Charge-Injection and Charge-Coupled Devices in Practical Chemical Analysis: Operation Characteristics and Considerations," in *Multichannel Image Detectors*, Vol. 2, Talmi, Y. (Ed.), American Chemical Society, Washington, DC.

Dieke, G. H. and Crosswhite, H. M. (1945), *J. Opt. Soc. Am.*, *35*, 471.

Epperson, P. M., Sweedler, J. V., Bilhorn, R. B., Sims, G. R., and Denton, M. B. (1988), *Anal. Chem.*, *60*, 327A.

Felkel, Jr., H. L. and Pardue, H. L. (1977), *Anal. Chem.*, *49*, 1112.

Felkel, Jr., H. L. and Pardue, H. L. (1978), *Anal. Chem.*, *50*, 602.

Felkel, Jr., H. L. and Pardue, H. L. (1979), "Simultaneous Multielement Determinations by Atomic Absorption and Atomic Emission with a Computerized Echelle Spectrometer/Imaging Detector System," in *Multichannel Image Detectors*, Talmi, Y. (Ed.), American Chemical Society, Washington, DC.

Fulton, A., Thompson, K. C., and West, T. S. (1970), *Anal. Chim. Acta*, *51*, 373.

Furuta, N. (1980), "Multielement Analysis Studies by Flame and Inductively Coupled Plasma Spectroscopy Utilizing Computer-controlled Instrumentation," Research Report from the National Institute for Environmental Studies, No. 12, Yatabe-machi, Tsukuba, Ibaraki 305, Japan.

Furuta, N., McLeod, C. W., Haraguchi, H., and Fuwa, H. (1979), *Bull. Chem. Soc. Jpn.*, *52*, 2913.

Ganjei, J. D., Howell, N. G., Roth, J. R., and Morrison, G. H. (1976), *Anal. Chem.*, *48*, 505.

Golightly, D. W., Kniseley, R. N., and Fassel, V. A. (1970), *Spectrochim. Acta, Part B*, *25B*, 451.

Gustavsson, A. and Ingman, F. (1979), *Spectrochim. Acta, Part B*, *34B*, 31.

Harber, R. A. and Sonnek, G. E. (1966), *Appl. Opt.*, *5*, 1039.

Hasler, M. F. and Dietert, H. W. (1944), *J. Opt. Soc. Am.*, *34*, 751.

Hirschfeld, T. (1973), U.S. Patent 3,728,029.

Hoffman, R. M. and Pardue, H. L. (1979), *Anal. Chem.*, *51*, 1267.

Horlick, G. (1976), *Appl. Spectrosc.*, *30*, 113.

Howell, N. G., Ganjei, J. D., and Morrison, G. H. (1976), *Anal. Chem.*, *48*, 319.

Howell, N. G. and Morrison, G. H. (1977), *Anal. Chem.*, *49*, 106.

Johansson, A. and Nilsson, L. E. (1976), *Spectrochim. Acta, Part B*, *31B*, 419.

Johnson, D. J., Plankey, F. W., and Winefordner, J. D. (1975), *Anal. Chem.*, *47*, 1739.

Jones, D. G. (1985), *Anal. Chem.*, *57*, 1207A.

Karanassios, V. and Horlick, G. (1986), *Appl. Spectrosc.*, *40*, 813.

Keliher, P. N. and Wohlers, C. C. (1976), *Anal. Chem.*, *48*, 333A.

Knapp. D. O., Omenetto, N., Hart, L. P., Plankey, F. W., and Winefordner, J. D. (1974), *Anal. Chim. Acta*, *69*, 455.

Lowe, R. M. (1971), *Spectrochim. Acta, Part B*, *26B*, 201.

Lundberg, E. and Johansson, A. (1976), *Anal. Chem.*, *48*, 1922.

Malloy, B. (1979), "An Improved Dispersion System for Simultaneous Multielement Analysis by Flame Spectroscopy," Ph.D. Dissertation, Baylor University, Waco, TX.

Malmstadt, H. V. and Cordos, E. (1972), *Am. Lab.*, *4*(8), 35.

Margoshes, M. (1970a), Pittsburgh Conference on Analytical Chemistry and Applied Spectroscopy, Cleveland, OH, March, Paper 99.

Margoshes, M. (1970b), *Opt. Spectra*, *4*, 26.

Margoshes, M. (1970c), *Spectrochim. Acta, Part B*, *25B*, 113.

Marshall, G. B. and West, T. S. (1970), *Anal. Chim. Acta*, *51*, 179.

McGeorge, S. W. and Salin, E. D. (1984), *Prog. Analyt. Atom. Spectrosc.*, *7*, 387.

McGeorge, S. W. and Salin, E. D. (1985), *Spectrochim. Acta, Part B*, *40B*, 447.

Mitchell, D. G. (1971), U.S. Patent 3,619,061.

Mitchell, D. G., Jackson, K. W., and Aldous, K. M. (1973), *Anal. Chem.*, *45*, 1215A.

Mitchell, D. G. and Johansson, A. (1970), *Spectrochim. Acta, Part B*, *25B*, 175.

Mitchell, D. G. and Johansson, A. (1971), *Spectrochim. Acta, Part B*, *26B*, 677.

Nahstoll, G. A. and Bryan, F. R. (1945), *J. Opt. Soc. Am.*, *35*, 646.

Nieman, T. A. and Enke, C. G. (1976), *Anal. Chem.*, *48*, 619.

Norris, J. D. and West, T. S. (1971), *Anal. Chim. Acta*, *55*, 359.

Norris, J. D. and West, T. S. (1972), *Anal. Chim. Acta*, *59*, 474.

Norris, J. D. and West, T. S. (1973), *Anal. Chem.*, *45*, 226.

Nygaard, D. D., Chase, D. S., and Leighty, D. A. (1983), *Appl. Spectrosc.*, *37*, 432.

Nygaard, D. D., Chase, D. S., Leighty, D. A., and Smith, S. B. (1984), *Anal. Chem.*, *56*, 424.

Nygaard, D. D. and Sotera, J. J. (1988) *Spectroscopy*, *3*(4). 22.

Palermo, E. F., Montaser, A., and Crouch, S. R. (1974), *Anal. Chem.*, *46*, 2155.

Rose, Jr., O., Mincey, D. W., Yacynych, A. M., Heineman, W. R., and Caruso, J. A. (1976), *Analyst*, *101*, 753.

Routh, M. W. and Paul, K. J. (1985), *Am. Lab.*, June, 84.

Salin, E. and Ingle, Jr., J. D. (1978a), *Anal. Chem.*, *50*, 1737.

Salin, E. and Ingle, Jr., J. D. (1978b), *Appl. Spectrosc.*, *32*, 579.

Salin, E. and Ingle, Jr., J. D. (1979), *Anal. Chim. Acta*, *104*, 267.

Schwartz, M. (1959), *Information Transmission, Modulation, and Noise*, McGraw-Hill, New York.

Sims, G. R. and Denton, M. B. (1983), "Multielement Emission Spectrometry Using a Charge-Injection Device Detector," in *Multichannel Image Detectors*, Vol. 2, Talmi, Y. (Ed.), American Chemical Society, Washington, DC.

Slavin, M. (1971), *Emission Spectrochemical Analysis*, Wiley, New York.

Snelleman, W. (1968), *Spectrochim. Acta, Part B*, *23B*, 403.

Snelleman, W., Rains, T., Yee, K.W., Cooke, H. E., and Menis, O. (1970), *Anal. Chem.*, *42*, 394.

Spillman, R. W. and Malmstadt, H. V. (1976a), *Anal. Chem.*, *48*, 303.

Spillman, R. W. and Malmstadt, H. V. (1976b), *Am. Lab.*, *8*(3), 89.

Sullivan, J. V. and Walsh, A. (1968), *Appl. Opt.*, *7*, 1271.

Svoboda, V. (1968), *Anal. Chem.*, *40*, 1385.

Sweedler, J. V., Bilhorn, R. B., Epperson, P. M., Sims, G. R., and Denton, M. B. (1988), *Anal. Chem.*, *60*, 282A.

Talmi, Y. (1975a), *Anal. Chem.*, *47*, 658A.

Talmi, Y. (1975b), *Anal. Chem.*, *47*, 697A.

Talmi, Y., Ed. (1979), *Multichannel Image Detectors*, American Chemical Society, Washington, DC.

Talmi, Y., Ed. (1983), *Multichannel Image Detectors*, Vol. 2, American Chemical Society, Washington, DC.

Talmi, Y. and Busch, K. W. (1983), "Guidelines for the Selection of Four Optoelectronic Image Detectors for Low-Light Level Applications," in *Multichannel Image Detectors*, Vol. 2, Talmi, Y. (Ed.), American Chemical Society, Washington, DC, Chapter 1.

Ullman, A. H. (1980), *Prog. Analyt. Atom. Spectrosc.*, *3*, 87.

Ullman, A. H., Pollard, B. D., Boutillier, G. D., Bateh, R. P., Hanley, P., and Winefordner, J. D. (1979), *Anal. Chem.*, *51*, 2382.

Winefordner, J. D., Avni, R., Chester, T. L., Fitzgerald, J. J., Hart, L. P., Johnson, D. J., and Plankey, F. W. (1979), *Spectrochim. Acta, Part B*, *31B*, 1.

Winefordner, J. D. and Elser, R. C. (1971), *Anal. Chem.*, *43*(4), 25A.

Winefordner, J. D., Fitzgerald, J. J., and Omenetto, N. O. (1975), *Appl. Spectrosc.*, *29*, 369.

INDEX

(continued from front)